Construction Introduction and Safety

Agricultural Technology

FIRST EDITION

Pearson

NCCER
National Center for Construction Education and Research

NCCER

President and Chief Executive Officer: Boyd Worsham
Vice President of Innovation and Advancement: Jennifer Wilkerson
Chief Learning Officer: Lisa Strite
Senior Curriculum Leader: Chris Wilson
Senior Manager, Production: Graham Hack
Digital Content Manager: Kelly Beck
Project Manager: Douglas Hancock
Project Coordinators: Colleen Duffy, Emma Morse, Emmah Walker
Technical Writing Manager: Gary Ferguson
Technical Writers: Leah Murphy, Mitch Ryan
Art Manager: Bree Rodriguez
Technical Illustrators: Judd Ivines, Liza Wailes
Production Artist: Chris Kersten
Permissions Specialist: Sherry Davis
Managing Editor: Hannah Murray
Lead Editors: Zi Meng, Alexandria Willbond
Editors: Lauren Tygrest, Terry ONeil, Melina Redondo
Desktop Publishing Manager: Eric Caraballoso III
Desktop Publishing Coordinator: Daphney Milian
Desktop Publishing Specialist: Julie Watkins
XML Developer: James Singer III
Digital Content Coordinator and Translation Manager: Yesenia Tejas
Digital Content Coordinator: Briana Rosa

Pearson

Commercial Product Manager – Associations: Andrew Dunaway
Senior Digital Content Producer: Shannon Stanton
Associate Project Manager: Monica Perez-Kim
Senior Content Producers: Alexandrina B. Wolf, Alma Dabral
Content Producers: Ashwathy Chandrasekar, Arup Kumar Ghosh, Mary Roth
Executive Marketing Manager: Mark Marsden
Designer: Mary Siener
Rights and Permissions: Kaitlyn Coddington

Composition: NCCER
Content Technologies: Gnostyx
Printer/Binder: Lakeside Book Company
Cover Printer: Lakeside Book Company
Text Typefaces: Palatino LT Pro and Helvetica Neue LT Pro

Cover Image

Cover photo provided by: Morton Buildings

1 2024

Pearson

NCCER | National Center for Construction Education and Research

Paperback
ISBN-13: 978-0-13-544448-1

PREFACE

To the Trainee

Welcome to the world of construction! You are joining the eight million Americans who have chosen a career in this lucrative field. Construction is one of the nation's largest industries, offering excellent opportunities for high earnings, career advancement, and business ownership.

People with many different talents and educational backgrounds—skilled craftspersons, managers, supervisors, and superintendents—find job opportunities in construction and related fields. As you will learn throughout your training, many other industries depend upon the work you will do in construction. From houses and office buildings to factories, roads, and bridges—everything begins with construction.

New with *Agricultural Technology: Construction Introduction and Safety*

NCCER is proud to present the first edition of *Agricultural Technology: Construction Introduction and Safety* in partnership with the National FFA Organization and the National Council for Agriculture Education. This textbook includes 11 modules for building foundational skills in construction. NCCER has added QR codes to videos exploring the different types of jobs available in the industry.

This first edition of *Agricultural Technology: Construction Introduction and Safety* features a module called *Build Your Future in Construction*. This introductory module explores the various careers available in the construction industry and introduces trainees to what it might be like to work on a construction site. We are also excited to present training on construction drawings and construction math. This first edition includes some to-scale drawings in NCCER Module 93106, *Introduction to Construction Drawings*, as well as prints throughout to illustrate the parts and pieces of drawings with which trainees must familiarize themselves. Both NCCER Module 93104, *Introduction to Hand Tools*, and NCCER Module 93105, *Introduction to Power Tools*, cover a wide variety of tools, including many power tools that are battery operated. All 11 modules have up-to-date Objectives and Performance Tasks that align with the best practices in education. NCCER Module 93102, *Basic Safety*, is aligned with OSHA's 10-hour program. This means that instructors who are OSHA 500 certified are able to issue 10-hour OSHA cards to their students who successfully complete the program.

We wish you success as you progress through this training program. If you have any comments on how NCCER might improve upon this textbook, please complete the User Update form using the QR code on this page. NCCER appreciates and welcomes its customers' feedback. You may submit yours by emailing **support@nccer.org**. When doing so, please identify feedback on this title by listing *#AgTech* in the subject line.

Our website, **www.nccer.org**, has information on the latest product releases and training.

NCCER

SCAN ME

NCCER Standardized Curricula

NCCER is a not-for-profit 501(c)(3) education foundation established in 1996 by the world's largest and most progressive construction companies and national construction associations. It was founded to address the severe workforce shortage facing the industry and to develop a standardized training process and curricula. Today, NCCER is supported by hundreds of leading construction and maintenance companies, manufacturers, and national associations. The NCCER Standardized Curricula was developed by NCCER in partnership with Pearson, the world's largest educational publisher.

Some features of the NCCER Standardized Curricula are as follows:

- An industry-proven record of success
- Curricula developed by the industry, for the industry
- National standardization providing portability of learned job skills and educational credits
- Compliance with the Office of Apprenticeship requirements for related classroom training (CFR 29:29)
- Well-illustrated, up-to-date, and practical information

NCCER maintains a secure online database that provides certificates, digital badges, transcripts, and wallet cards to individuals who successfully complete programs under an NCCER-accredited organization or through one of NCCER's self-paced, online programs. This system also allows individuals and employers to track and verify industry-recognized credentials and certifications in real time.

For information on NCCER's credentials, contact NCCER Customer Service at 1-888-622-3720 or visit **www.nccer.org**.

Digital Credentials

Show off your industry-recognized credentials online with NCCER's digital credentials!

NCCER is now providing online credentials. Transform your knowledge, skills, and achievements into digital credentials that you can share across social media platforms, send to your network, and add to your resume. For more information, visit **www.nccer.org**.

Cover

Morton Buildings, headquartered in Morton, Illinois, is the industry leader in post-frame manufacturing and construction. At Morton, the difference is in the details. From cutting-edge innovations and best-in-class warranty to craftsmen in the field, Morton Buildings is dedicated to surpassing expectations. The Morton legacy of excellence in construction spans more than a century and more than a quarter-million satisfied customers. When customers choose Morton, they experience the Morton advantage at every step of the building process — starting even before the walls go up.

Morton Buildings is 100% employee-owned and proud to be ranked #68 on the National Center for Employee Ownership (NCEO) 2023 Employee Ownership 100 list, which includes the nation's largest companies that are at least 50% owned by an employee stock ownership plan. For more information, please visit **www.mortonbuildings.com**.

DESIGN FEATURES

Content is organized and presented in a functional structure that allows you to access the information where you need it.

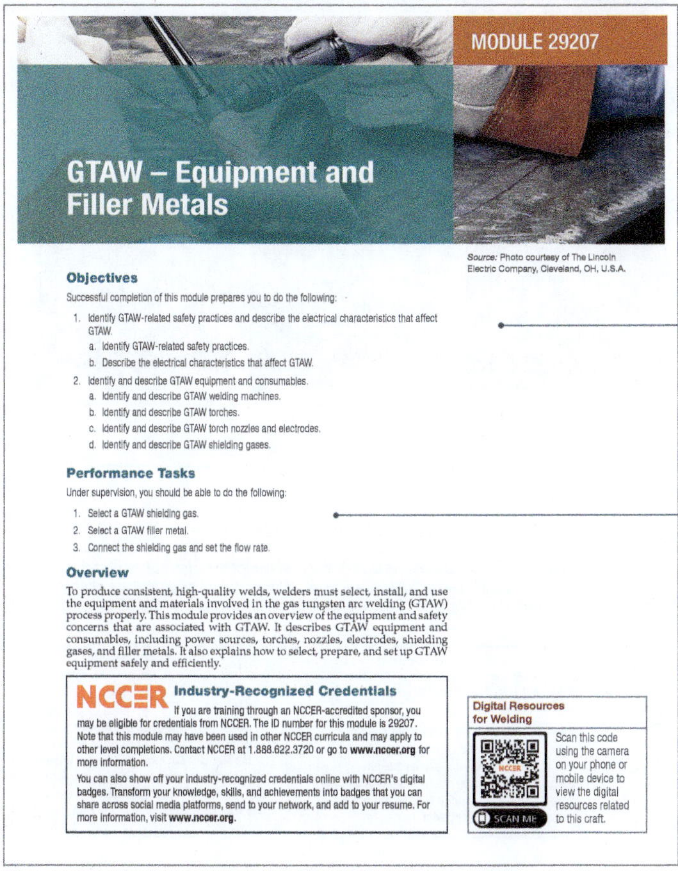

MODULE 29207

GTAW – Equipment and Filler Metals

Source: Photo courtesy of The Lincoln Electric Company, Cleveland, OH, U.S.A.

Objectives

Successful completion of this module prepares you to do the following:

1. Identify GTAW-related safety practices and describe the electrical characteristics that affect GTAW.
   a. Identify GTAW-related safety practices.
   b. Describe the electrical characteristics that affect GTAW.
2. Identify and describe GTAW equipment and consumables.
   a. Identify and describe GTAW welding machines.
   b. Identify and describe GTAW torches.
   c. Identify and describe GTAW torch nozzles and electrodes.
   d. Identify and describe GTAW shielding gases.

Performance Tasks

Under supervision, you should be able to do the following:

1. Select a GTAW shielding gas.
2. Select a GTAW filler metal.
3. Connect the shielding gas and set the flow rate.

Overview

To produce consistent, high-quality welds, welders must select, install, and use the equipment and materials involved in the gas tungsten arc welding (GTAW) process properly. This module provides an overview of the equipment and safety concerns that are associated with GTAW. It describes GTAW equipment and consumables, including power sources, torches, nozzles, electrodes, shielding gases, and filler metals. It also explains how to select, prepare, and set up GTAW equipment safely and efficiently.

NCCER **Industry-Recognized Credentials**

If you are training through an NCCER-accredited sponsor, you may be eligible for credentials from NCCER. The ID number for this module is 29207. Note that this module may have been used in other NCCER curricula and may apply to other level completions. Contact NCCER at 1.888.622.3720 or go to **www.nccer.org** for more information.

You can also show off your industry-recognized credentials online with NCCER's digital badges. Transform your knowledge, skills, and achievements into badges that you can share across social media platforms, send to your network, and add to your resume. For more information, visit **www.nccer.org**.

Digital Resources for Welding

Scan this code using the camera on your phone or mobile device to view the digital resources related to this craft.

You can navigate Agricultural Technology using color coded tabs on the upper right hand corner.

The Objectives list the skills and knowledge you need in order to complete the module successfully.

The Performance Tasks give you an opportunity to apply your knowledge to real-world tasks.

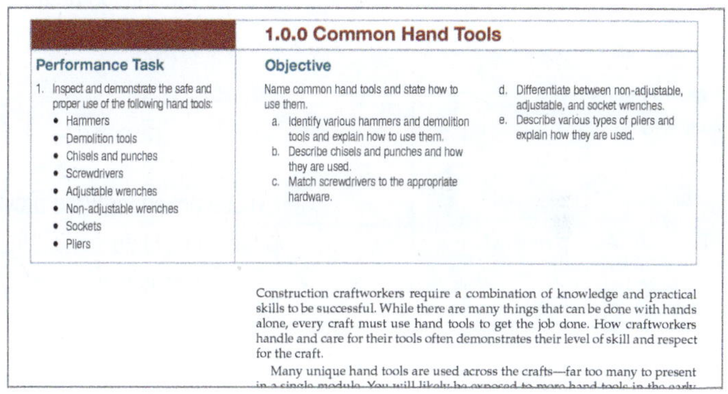

1.0.0 Common Hand Tools

Performance Task

1. Inspect and demonstrate the safe and proper use of the following hand tools:
   - Hammers
   - Demolition tools
   - Chisels and punches
   - Screwdrivers
   - Adjustable wrenches
   - Non-adjustable wrenches
   - Sockets
   - Pliers

Objective

Name common hand tools and state how to use them.

a. Identify various hammers and demolition tools and explain how to use them.
b. Describe chisels and punches and how they are used.
c. Match screwdrivers to the appropriate hardware.
d. Differentiate between non-adjustable, adjustable, and socket wrenches.
e. Describe various types of pliers and explain how they are used.

Construction craftworkers require a combination of knowledge and practical skills to be successful. While there are many things that can be done with hands alone, every craft must use hand tools to get the job done. How craftworkers handle and care for their tools often demonstrates their level of skill and respect for the craft.

Many unique hand tools are used across the crafts—far too many to present

Section Openers provide a visual organizational structure for the information. Objectives and Performance Tasks are broken out for each section.

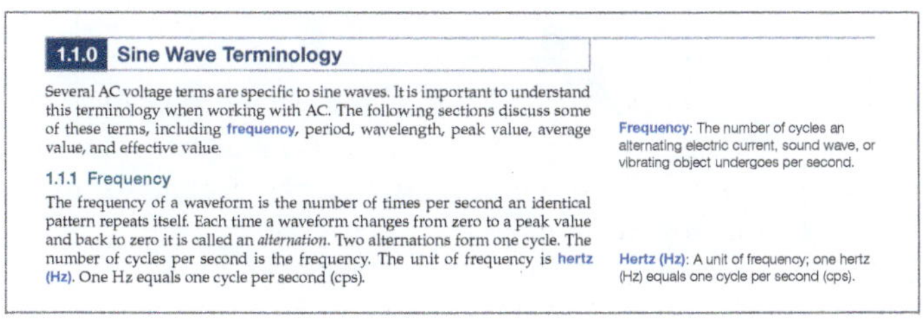

1.1.0 Sine Wave Terminology

Several AC voltage terms are specific to sine waves. It is important to understand this terminology when working with AC. The following sections discuss some of these terms, including **frequency**, period, wavelength, peak value, average value, and effective value.

1.1.1 Frequency

The frequency of a waveform is the number of times per second an identical pattern repeats itself. Each time a waveform changes from zero to a peak value and back to zero it is called an *alternation*. Two alternations form one cycle. The number of cycles per second is the frequency. The unit of frequency is **hertz (Hz)**. One Hz equals one cycle per second (cps).

Frequency: The number of cycles an alternating electric current, sound wave, or vibrating object undergoes per second.

Hertz (Hz): A unit of frequency; one hertz (Hz) equals one cycle per second (cps).

Trade Terms appear on the page adjacent to the text where they are first presented.

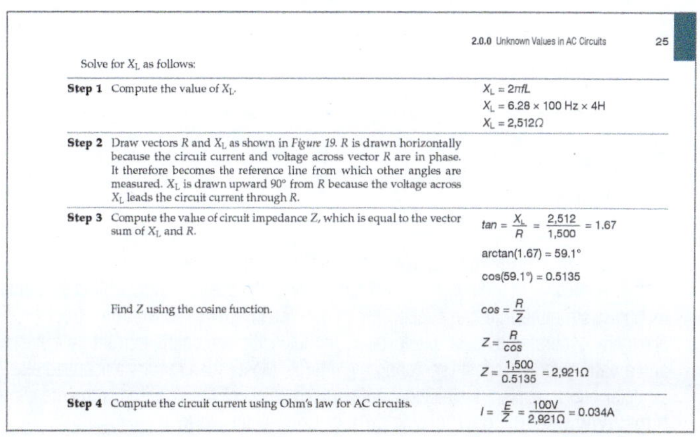

2.0.0 Unknown Values in AC Circuits 25

Solve for X_L as follows:

Step 1	Compute the value of X_L.	$X_L = 2\pi fL$ $X_L = 6.28 \times 100\text{ Hz} \times 4\text{H}$ $X_L = 2{,}512\Omega$
Step 2	Draw vectors R and X_L as shown in *Figure 19*. R is drawn horizontally because the circuit current and voltage across vector R are in phase. It therefore becomes the reference line from which other angles are measured. X_L is drawn upward 90° from R because the voltage across X_L leads the circuit current through R.	
Step 3	Compute the value of circuit impedance Z, which is equal to the vector sum of X_L and R.	$\tan = \frac{X_L}{R} = \frac{2{,}512}{1{,}500} = 1.67$ $\arctan(1.67) = 59.1°$ $\cos(59.1°) = 0.5135$
	Find Z using the cosine function.	$\cos = \frac{R}{Z}$ $Z = \frac{R}{\cos}$ $Z = \frac{1{,}500}{0.5135} = 2{,}921\Omega$
Step 4	Compute the circuit current using Ohm's law for AC circuits.	$I = \frac{E}{Z} = \frac{100\text{V}}{2{,}921\Omega} = 0.034\text{A}$

Step-by-step math equations help make the concepts clear and easy to grasp.

QR codes link directly to videos that highlight current content.

SUCCESS STORIES: *Credentials Matter: Start Building Your Career Today*

Scan this code using the camera on your phone or mobile device to view this video.

Important information is highlighted, illustrated, and presented to facilitate learning.

Placement of images near the text description and details such as callouts and labels help you absorb information.

Preparing Drills with Keyless Chucks

Most cordless drills use a keyless chuck. While the steps for preparing a cordless drill are similar, there are some small differences. Follow the steps below when preparing to use drills with keyless chucks:

Step 1 Disconnect the drill from its power source by removing the battery pack before loading a bit.

Step 2 As shown in (*Figure 7A*), open the chuck by turning it counterclockwise until the jaws are wide enough to insert the bit shank.

Step 3 Insert the bit shank into the chuck opening (*Figure 7B*). Keeping the bit centered in the opening, turn the chuck by hand until the jaws grip the bit shank.

Step 4 Tighten the chuck securely with your hand so that the bit does not move (*Figure 7C*). You are now ready to use the cordless drill.

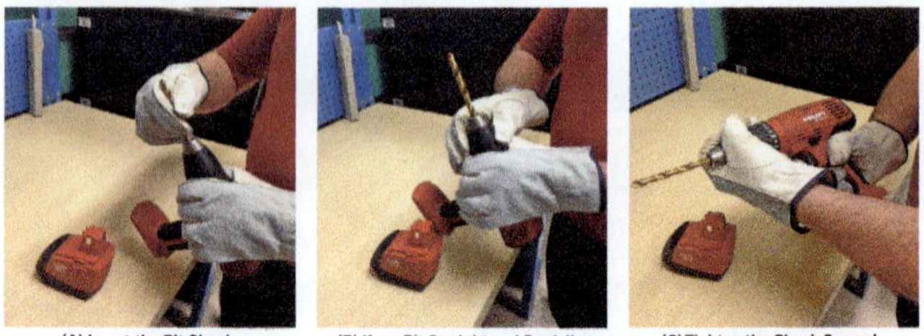

(A) Insert the Bit Shank. (B) Keep Bit Straight and Partially Tighten the Chuck. (C) Tighten the Chuck Securely.

Figure 7 Loading the bit on a keyless chuck.
Source: Cianbro Corporation

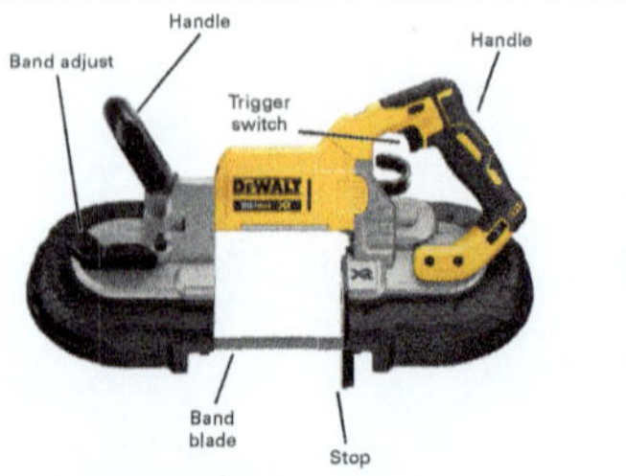

Figure 27 Portable handheld band saw.
Source: Image property of Stanley Black & Decker. Used with permission

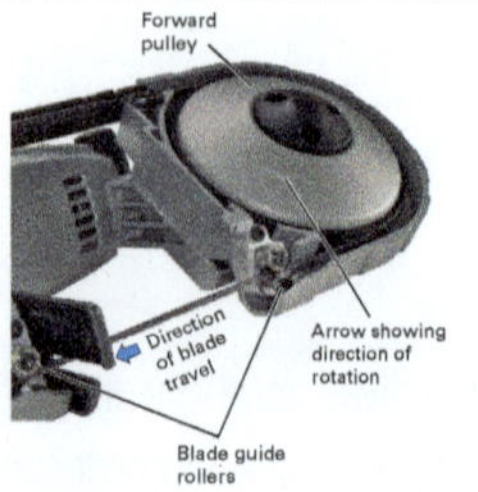

Figure 28 Band saw pulley and blade guide rollers.
Source: Courtesy of Milwaukee Electric Tool Corporation

blade length is determined by referencing the manufacturer's documentation. Like most blades, its coarseness is rated in teeth per centimeter or teeth per inch (TPI). As a general rule, higher TPIs produce smoother cuts. Thicker materials require coarser blades. However, if the blade is too coarse for the material, the individual teeth may begin to break off. *Figure 28* **shows how the blade is routed** around the pulleys and through the blade guides.

While some band saws have multiple speeds, most do not. The portable band saw generally cuts best at a low speed since higher speeds cause the blade's teeth to rub rather than cut. This can create heat through friction, which reduces the life of the blade.

New boxes highlight safety and other important information. Warning boxes stress potentially dangerous situations, while Caution boxes alert to dangers that may cause damage to equipment. Note boxes provide additional information on a topic.

WARNING!

A portable band saw always cuts in the direction of the user. For that reason, workers must be especially careful to avoid injury when using this type of saw. Always wear appropriate PPE and stay focused on the work.

CAUTION

Never assume anything. It never hurts to ask questions, but disaster can result if you don't ask. For example, do not assume that an electrical power source is turned off. First ask whether the power is turned off, then check it yourself to be completely safe.

NOTE

This training alone does not provide any level of certification in the use of fall arrest or fall restraint equipment. Trainees should not assume that the knowledge gained in this module is sufficient to certify them to use fall arrest equipment in the field.

These boxed features provide additional information that enhances the text.

Did You Know?

Lightning Rods

An interesting fact about grounding is that a lightning rod (air terminal) isn't meant to bring a bolt of lightning to ground. To do this, its conductors would have to be several feet (1 m or more) in diameter. The purpose of the rod is to dissipate the negative static charge that would cause the positive lightning charge to strike the house.

Orthographic Drawings

Orthographic drawings are used for elevation drawings. They show straight-on views of the different sides of an object with dimensions that are proportional to the actual physical dimensions. In orthographic drawings, the designer draws lines that are scaled-down representations of real dimensions. Every 12 inches, for example, may be represented by ¼ inch on the drawing. Similarly, in an example using metric measurements with a ratio of 1:2, every 30 millimeters may be represented by 15 millimeters on the drawing.

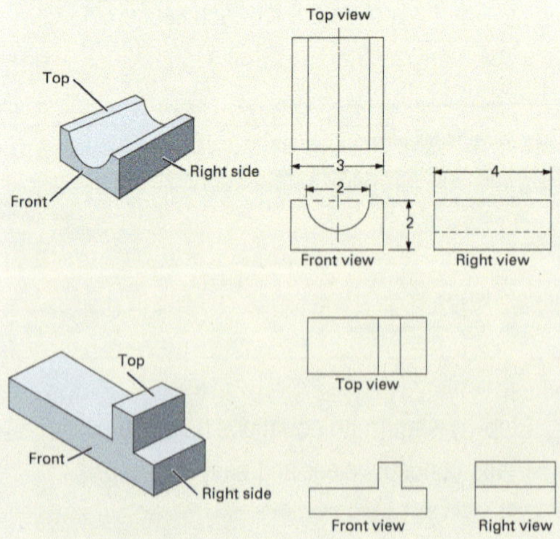

Around the World

Predicting Weather

Barometers measure atmospheric air pressure. Generally, an increased pressure indicates that the weather is pleasant. A reduced pressure indicates that the weather is cloudy and/or rainy.

Changes in barometric pressure are used to predict changes in weather conditions. For example, if it is raining outside, but the barometer is rising, it indicates that better weather is on the way. In the midst of a hurricane, the barometric pressure is an indicator of intensity. Extremely low pressures exist in powerful hurricanes. While traditional barometers work very well, sophisticated weather stations capture far more data and can transmit the information wirelessly.

Going Green

Residential Solar

As residential customers look for ways to reduce reliance on fossil fuels like coal and oil, they are turning to solar energy. Usually, they want to install photovoltaic (PV) panels on the roofs of their homes, or in their yards. A residential PV system usually includes the PV panels, a mounting structure that can support the panels, an inverter to convert the direct current (DC) electricity generated by solar photovoltaic modules into alternating current (AC) electricity, and a battery (or batteries) for storing the energy. Most states require a licensed electrician for solar energy system installation, but you should check the local requirements. (Source: **https://www.energy.gov/eere/solar/how-does-solar-work**)

Going Green looks at ways to preserve the environment, save energy, and make good choices regarding the health of the planet.

15302 Precision Measuring Tools 43

Cornerstone of Craftsmanship

How did you choose a career in the industry?

I was working at a gas station and wanted to make more money. So, I applied for a welder helper position, and then when they asked about a millwright helper, I asked, "What's that?" True story! I had no idea what it was. Alan Landry, the plant manager at the local chemical plant, got me into the millwright craft.

What types of training have you been through?

I've been through ABC Millwright training, Total Quality Management (TQM) programs, Rand McNally mechanical seals training, Lubritech lubrication system training, and laser alignment classes. There are many others that I've attended over the years to both broaden and sharpen my skills.

How important is education and training in construction?

It's very important! Nobody is born knowing how to do all this stuff. Knowledge can't be taken away from someone, so always take any training offered to you. The only other way to learn is hands-on, but classrooms and curriculums set the standard, and that's how you know you're learning the right way. Hands-on training and practice then become an extension of the classroom, instead of the only learning opportunity.

How important are NCCER credentials to your career?

My early training came before NCCER curricula was available, but I am an NCCER-certified instructor. Right now, many plants in my state are hiring rotating equipment people and many require the NCCER training to be eligible to be hired. I think that's a good indicator of how important the credentials have become.

How has training/construction impacted your life?

Without training, I certainly wouldn't be as successful as I have been. It is essential for career growth.

What kinds of work have you done in your career?

I've done just about everything from being a simple oiler to a project manager and served in many possible positions in between those two.

Tell us about your current job.

I'm currently a mechanical planner, and my primary task is to develop the scope for specific jobs. These are detailed, step-by-step plans of what we are going to do with a piece of equipment—take it down and/or put it back together—and I actually happen to have a scope on my desk right now. They outline what is required, what parts need to be ordered and acquired in advance, and the labor hours required.

Kim T. Panepinto
Mechanical Planner
Industrial Parts Specialties, LLC

Cornerstone of Craftsmanship sections feature career stories from people working in related fields.

Review questions at the end of each section and module allow you to measure your progress.

Section Review questions can be found at the end of each section to test your knowledge of the content.

Review Questions at the end of each module are provided to reinforce the knowledge you have gained.

2.0.0 Section Review

1. A straight pull contains two raceways. One of the raceways has a trade size of 3" and one has a trade size of 2". The length of the box *must* be _____.
   a. 16"
   b. 24"
   c. 26"
   d. 32"
2. Where angle or U pulls are made, to determine the distance between raceway entries enclosing the same conductor, you need to multiply the trade size of the largest raceway by _____.
   a. two
   b. four
   c. six
   d. eight
3. If possible, pull boxes should be installed _____.
   a. high on the wall for security
   b. at a height/location that makes it easy to pull conductors
   c. behind wall coverings for a neater look
   d. with extra knockouts left open for air flow

Module 26207-23 Review Questions

1. In which *NEC*® article is cable tray installation primarily addressed?
   a. *NEC Article 300*
   b. *NEC Article 392*
   c. *NEC Article 517*
   d. *NEC Article 550*
2. Which type of tray is *best* suited for corrosive areas and atmospheres?
   a. Stainless steel cable tray
   b. Coated aluminum cable tray
   c. Nonmetallic cable tray
   d. Basket tray
3. When a cable tray has a solid bottom, it is referred to as _____.
   a. basket tray
   b. ladder tray
   c. trough
   d. raceway
4. Cable tray is generally manufactured in _____.
   a. 2' and 4' lengths
   b. 6' and 8' lengths
   c. 6' and 12' lengths
   d. 12' and 24' lengths

Additional information can be found in Appendixes at the back of the book.

371

APPENDIX 29202A

Example of a Three-View Detail Drawing

Figure A01 is an example of an object that may be encountered in welding. *Figure A02* is an example of a three-view detail drawing representing that same object.

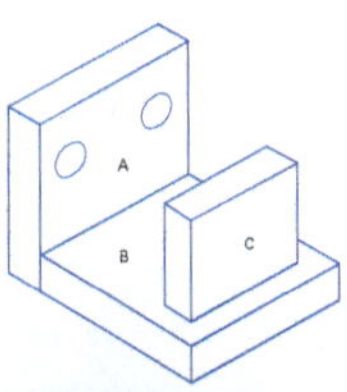

Details			
Part	Thickness	Length	Width
A	¾"	3½" (Tall)	3½"
B	¾"	4" (Long)	3½"
C	¾"	2½" (Tall)	2½"

Hole Dimensions and Location
- Hole diameter is ¾"
- Hole centers are 1¼" from the top edge of block "A"
- Hole centers are ¾" from the outside edges of block "A"

Figure A01

Some modules have corresponding Appendixes that provide supplementary information or activities to enhance your understanding of the material.

NCCERCONNECT

This interactive online course is a unique web-based supplement that provides a range of visual, auditory, and interactive elements to enhance training. Also included is a full eText.

Visit **www.nccerconnect.com** for more information!

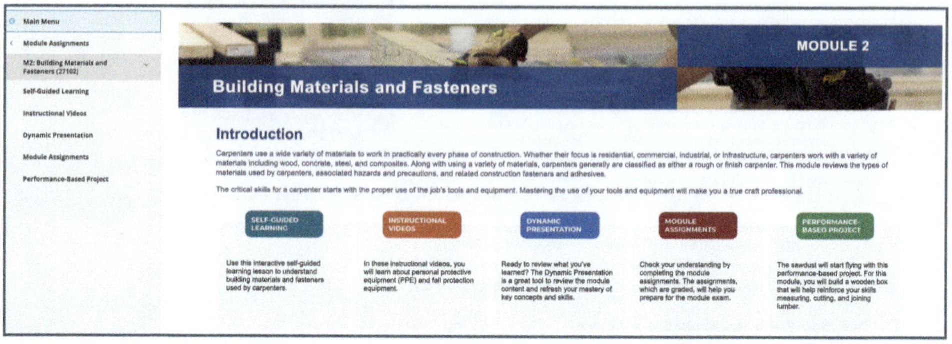

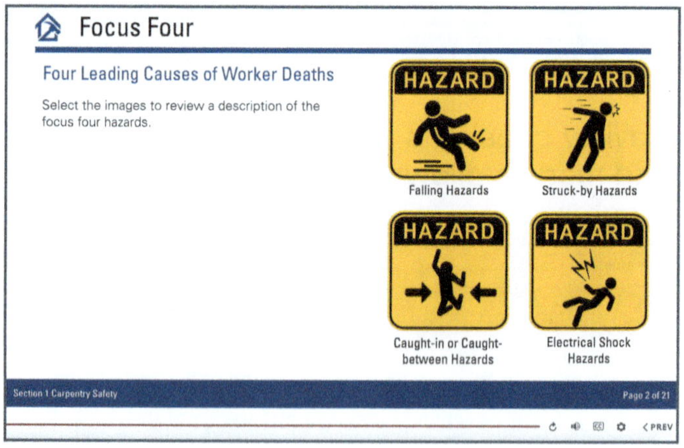

Use the interactive self-guided learning lesson to understand key concepts and terms needed for a career in construction.

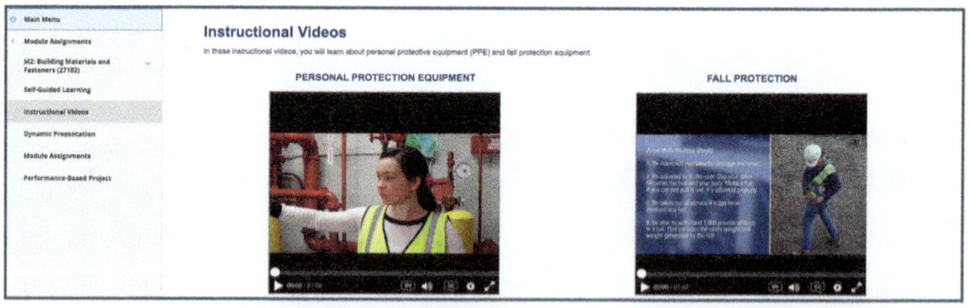

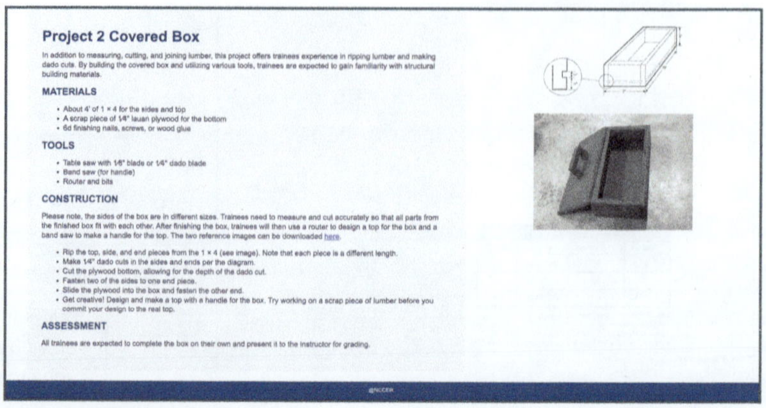

ACKNOWLEDGMENTS

This curriculum was revised as a result of the vision and leadership of the following sponsors:

ABC San Diego
Beaverhead County High School
Haskell
Hawkins High School
Holyoke High School
Hudson High School
Industrial Management & Training Institute Inc.
Knox School District
Louisiana Department of Education
MCTA
Montana Office of Public Instruction
The National Council for Agricultural Education
The National FFA Organization
Pittsburg State University
River Valley Technical Center
South Warren High School
Stanley Black & Decker/DeWalt
Sundt
TIC – The Industrial Company

This curriculum would not exist were it not for the dedication and unselfish energy of those volunteers who served on the Authoring Team. A sincere thanks is extended to the following:

Cody Berry
Mark Bonda
Ryan Camer
Jason Decker
Josh Garland
Harold Heintz
Erin Hunter
Caleb Igo
Jon Jones
Marcy Mahler
Chance Plaisance
Thomas Poole
Heidi Pratt
Melissa Rekeweg
Fernando Sanchez
HK Smith
Shauna Strecker
John Stronkowski
Scott Stump
Chad Sutton
Eric Tilleman
Tim Thompson
Darry Welker

A final note: This book is the result of a collaborative effort involving the production, editorial, and development staff at Pearson Education, Inc., and NCCER. Thanks to all of the dedicated people involved in the many stages of this project.

NCCER PARTNERS

To see a full list of NCCER Partners, please visit:

www.nccer.org/about-us/partners.

SCAN ME

You can also scan this code using the camera on your phone or mobile device to view these partnering organizations.

In Memory

To Brad Kibbe, whose unwavering passion for teaching the trades and dedication to agriculture laid the foundation for NCCER's *Agricultural Technology* curriculum. Brad's kindness, professionalism, collaborative spirit, and commitment to his colleagues left a lasting mark on all of us who were fortunate to work with him. He will be missed greatly.

CONTENTS

Module 93101 Build Your Future in Construction

Module 93102 Basic Safety

Module 93103 Introduction to Construction Math

Module 93104 Introduction to Hand Tools

Module 93105 Introduction to Power Tools

Module 93106 Introduction to Construction Drawings

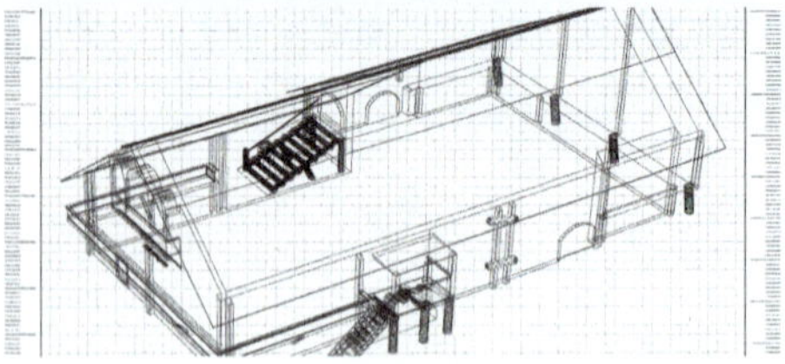

Module 93107 Introduction to Basic Rigging

Module 93108 Basic Communication Skills

Module 93109 Basic Employability Skills

Module 93111 Introduction to Leadership

Agricultural Technology Appendixes

Build Your Future in Construction

Source: iStock

Objectives

Successful completion of this module prepares you to do the following:

1. Describe the construction industry and how it relates to the agricultural industry.
   a. Define construction and summarize the current and future outlook for jobs.
   b. Identify some of construction's more prominent contributions in history.
2. Explain the benefits of a construction career.
   a. Recognize and describe how construction careers make a difference in the community.
   b. Describe the financial and professional benefits of pursuing a construction career.
3. Describe the typical career path for professionals in the construction and agricultural industries.
   a. Describe industry sectors and the progression path for construction careers.
   b. Identify different construction careers and the types of skills they require.
4. Identify ways to pursue a career in the construction industry.
   a. Explain the benefits of career and technical education programs.
   b. Describe the advantages of craft training programs and their relationship with apprenticeships.
   c. Summarize the path to a construction career through community colleges and universities.

Performance Tasks

This is a knowledge-based module. There are no Performance Tasks.

Overview

Construction is an exciting, well-paying industry that offers an abundance of career opportunities. Specifically, agricultural construction is the maintenance, renovation, or building of agricultural structures and operations. With a growing need for individuals who are ready to learn while getting paid, construction provides a great fit for people of all backgrounds, skills, and strengths. Carpenter, pipefitter, welder, electrician, and crane operator are just a few of the construction professions in high demand in both the agricultural industry and the broader community. This module will help you understand the state of the industry, the job opportunities that currently exist, and the training options that will lead you on a path to your new construction career.

Digital Resources for Agricultural Technology

NCCER

SCAN ME

Scan this code using the camera on your phone or mobile device to view the digital resources related to this craft.

NCCER Industry-Recognized Credentials

If you are training through an NCCER-accredited sponsor, you may be eligible for credentials from NCCER. The ID number for this module is 93101. Note that this module may have been used in other NCCER curricula and may apply to other level completions. Contact NCCER at 1.888.622.3720 or go to **www.nccer.org** for more information.

You can also show off your industry-recognized credentials online with NCCER's digital credentials. Transform your knowledge, skills, and achievements into credentials that you can share across social media platforms, send to your network, and add to your resume. For more information, visit **www.nccer.org**.

1.0.0 Overview of the Construction Industry

Performance Tasks

There are no Performance Tasks in this section.

Objective

Describe the construction industry and how it relates to the agricultural industry.

a. Define construction and summarize the current and future outlook for jobs.

b. Identify some of construction's more prominent contributions in history.

When you think of construction, what images come to mind? If you are like many people, you might imagine someone hammering a nail or driving a bulldozer. While these are important activities within a specific construction trade, the opportunities within the entire industry are much broader than you may imagine. The construction industry employs millions of skilled individuals working in trades that offer rewarding jobs, excellent benefits, and great opportunities to have a positive impact on society and the world. Whatever your skills or interests, construction has a place for everyone.

1.1.0 What Is Construction?

At its most basic level, the term *construction* means to build. As an industry, construction includes the creation of residential and commercial buildings, roads, bridges, dams, agricultural facilities, manufacturing plants, and more. It also includes lesser-known activities related to designing, managing, planning, repairing, maintaining, and even demolishing these structures.

Construction is both an art and a science and involves individuals skilled in a particular **craft/trade**. While **craft professionals** must rely on their technical knowledge, they often combine it with creativity and artistic skills. Creativity can mean many things, from using an artistic eye to looking at problems in an original way. It takes imagination to look at a jobsite or a piece of wood and know how to transform it into a finished product.

Craft/trade: A general term referring to one of the many construction specialties.

Craft professionals: Trained and skilled individuals who design and build things. Craft professionals conform to the technical and ethical standards of their trade.

Think about the work performed by carpenters, welders, and masons. In most cases, it is critical that the products they build are structurally sound. In many instances, they are also visually appealing. If you have ever admired a barn (*Figure 1*), a stained glass window, a bridge, or even someone's renovated kitchen, then you understand the value of construction and the different trades. The term *trade* is a general way of referring to one of the many construction specialties or crafts. From skyscrapers to small homes, highways to pipelines, and power lines to ocean liners, everything in society was built by individuals using a wide variety of materials, tools, and skills.

Figure 1 A renovated barn.
Source: iStock

Finding a way to follow one's passion in construction is easy. Restoring historic structures, like a renovated farmhouse or modernized greenhouse (*Figure 2*), is its own exciting specialization of the construction field. Even building a professional football stadium or basketball arena takes hundreds of trade professionals with a wide variety of expertise. Imagine the pride in creating lasting structures that benefit thousands of individuals and the opportunity to say you helped build it.

Credentials Matter:
Start Building Your Career Today

Scan the QR code using the camera on your phone or mobile device to view the accompanying resource.

1.1.1 State of the Industry

The construction industry is currently thriving. According to the Associated General Contractors of America (AGC), the industry builds over $1.3 trillion in structures per year and includes over 700,000 construction companies that employ more than 7 million total employees. In fact, construction contributed over $887 billion to the US economy in 2019 alone, which equates to about 4% of the total value of all goods and services produced in the United States.

Despite recent challenges in the overall US economy, the construction market is steadily growing. In addition to continued private spending, more construction growth is projected because of future plans to invest in the country's aging **infrastructure**. These are the agricultural, transportation, water, electrical, and telecommunications systems that allow countries to operate.

Infrastructure: The transportation, water, electrical, and telecommunications systems that allow communities and countries to operate.

Figure 2 A modernized greenhouse.
Source: iStock

1.1.2 Growing Demand for Skilled Craft Professionals

Opportunities abound for individuals entering the construction field (*Figure 3*). In fact, there simply are not enough skilled craft professionals to meet the industry's existing or anticipated demands in the coming years. Studies show that 80% of construction companies intend to hire more people over the next few years to meet the growing needs of the market. Many companies have encountered difficulties filling these positions, and employee shortages are now a top concern.

Since experienced professionals are leaving the industry and fewer new employees are replacing them, there is a significant workforce shortage and a high demand for talented craft professionals. In fact, construction companies are trying to hire and train individuals immediately for the projected future growth in construction. For those seeking to enter the construction industry, this means there are great jobs available, and they can earn a good income while being trained.

There are several reasons for this growing labor shortage. By 2030, an estimated 40% of the current construction workforce is expected to retire. The baby boomer generation, which includes people born between 1946 and 1964, currently makes up a large portion of the construction workforce. Many baby boomers are highly experienced craft professionals who have built their careers over decades and now hold leadership positions. As they retire, they will leave behind a large vacuum of labor and leadership in the trades industries.

Figure 3 Aerial view of apartment building under construction.
Source: iStock

At the same time, fewer young people are entering the construction industry. Instead, many individuals over the past few decades have opted to earn a degree at a four-year university. According to Statista, 34% of men and 35% of women in the United States received at least four years of college education in 2018. In 1968, these numbers were much lower—only 13% of men and 8% of women attended a four-year college. In just 50 years, college enrollment has nearly tripled. With a continued national emphasis on attending college, many young people are simply unaware of the abundant career opportunities offered by the construction and agricultural industries.

The agricultural industry offers a unique opportunity to those seeking to develop construction skills. Farms and ranches need construction craftworkers who are familiar with the agricultural industry and the infrastructure needed to sustain it. Statistics indicate that one-third of America's 3.4 million farmers are over the age of 65, and nearly a million more will reach retirement age within the next 10 years. These farmers will leave the workforce around the same time construction craftworkers retire. Many of the skills needed on a construction site are similar to the skills required to sustain an agricultural operation. Those developing construction skills in addition to their farm and ranch experience have a powerful skill set to benefit the agricultural industry and the greater community.

NOTE

Opportunities for women in construction are plentiful as well. There are 1.2 million women in the agricultural industry, representing 36% of the country's 3.4 million farmers. Advancing their skill set in construction further secures a profitable and functional career path.

1.1.3 Common Construction Industry Myths

Despite its essential role in keeping the world's infrastructure running smoothly, the current construction industry is widely misunderstood. These misconceptions may prevent some people from pursuing careers in construction. The following sections covering myths about construction explain why they are not true.

Myth: People in Construction Are Unskilled

Construction is often mistakenly thrown into the category of unskilled labor; that is, work that can be performed without any special training or knowledge. This misconception likely stems from the fact that a person does not need a college degree to get a construction job. In reality, craft professionals perform extremely detailed, specialized work that requires a high level of expertise gained through years of training, both in the classroom and on the job. Many professionals must also earn specialized licenses, credentials, and/or certifications that prove their expertise.

Myth: Construction Employees Make Minimum Wage

According to a 2018 survey by *U.S. News & World Report*, the lowest paid 25% of the construction workforce made almost double the minimum wage amount, at over $28,520 per year. The highest 25% of employees made almost double that amount. The average annual salary is over $40,000 and rising. Salaries do not include overtime or bonus pay, which means potential incomes are even higher.

Myth: Construction Skills Are Not Needed in the Agricultural Industry

The agricultural industry is unique in its immediate need for improvements to optimize farming operations, implement new technologies, and expand the agricultural enterprise. Skilled construction craftworkers are needed within the agricultural industry to ensure its longevity, maximize its resources, and keep up with a growing population's demand for a stable food supply.

Myth: Construction Is Not for Women

For most of history, men dominated the world of construction. Now, a greater focus on refined skills and knowledge of specific crafts means that the most successful and best paid construction professionals are those who can do the best work, regardless of gender.

According to the National Association of Women in Construction (NAWIC), 1.1 million women work in construction industry jobs that include field, office, and management positions. With women making up only 10% of the construction workforce (*Figure* 4), many construction companies see this underrepresented demographic as an opportunity to fill important job vacancies.

Figure 4 Construction opportunities abound for men and women.
Source: Getty Images

Success Story

Holley Thomas

Holley Thomas became interested in welding in 2007 while earning a degree in robotics from Central Alabama Community College. She discovered that not only did she have a great talent for welding, but she also loved the craft. She was recruited by KBR, a global engineering and construction firm, as part of their outreach and engagement with the college's technical programs.

Over the next 11 years, Thomas took advantage of KBR's training offerings and grew as an experienced craftsperson. She started as a welder, worked her way into supervision, and moved into a quality inspector role. She is the first woman welder to have taken top honors (gold) in welding at the Associated Builders and Contractors (ABC) National Craft Championship in 2010. In 2015, Thomas won the ABC's National Craft Professional of the Year award.

Thomas shares, "Being a skilled craft professional, and furthermore a female in the industry, helps me leave a great legacy. I've been able to show other women in the industry that you can start from the bottom and work your way up to where I am today."

In addition, women are paid more equally in construction than in many other industries. The U.S. Bureau of Labor Statistics (BLS) reports that women earn more than 99 cents for every dollar a man earns in construction jobs, compared to the US average of 81 cents.

Myth: Construction Does Not Use Technology

Construction has embraced technology and invested in it. According to **techcrunch.com**, funding in US-based construction technologies increased by about 320% in recent years and is expected to increase even more. Examples of technologies used in construction include:

- *Computers, mobile apps, and Internet of Things (IoT) devices* — Allow workers to share data, collaborate, and communicate more quickly (*Figure 5*). Examples of IoT devices include smartphones, tablets, residential and commercial sensors, and many other items.
- *Drones* — Survey jobsites and identify potential hazards in risky or hard-to-see areas.
- *Robots* — Lift and move heavy objects on jobsites.
- *Simulators* — Create realistic sensations of heights, stress, and hazards, allowing trainees and craft professionals to experience what it feels like to operate equipment and practice safety procedures before actually doing so on the job.
- *Software* — Create and update construction designs easily and seamlessly.

Figure 5 Many construction professionals use technology on the job.
Source: Getty Images

Going Green

Autonomous Flight Minimizes Use of Chemicals

Drones play a significant role in supporting modern agricultural construction and operations. Using aerial photography, farmers can use drones to plan new construction projects or evaluate irrigation systems. This helps them better evaluate the best use of land and resources, as well as avoid using areas prone to erosion. Drones can even be used to spray crops with fertilizer and pesticides in targeted areas. This reduces the amount of chemicals used in farming, which benefits the surrounding soils, air quality, and groundwater.

Source: iStock

Myth: Construction Is Dangerous

Jobs involving heavy machinery, power tools, and large components have inherent risks. However, the construction world has made great strides to create a safer and healthier work environment for craft professionals. Working in construction is safer than ever before due to many factors:

Occupational Safety and Health Administration (OSHA): An agency of the US Department of Labor with the mission to establish and enforce safe working conditions for the US workforce.

- *Higher government safety standards* — Since the formation of the **Occupational Safety and Health Administration (OSHA)** in 1971, construction injuries have dropped significantly. The mission of OSHA is to establish and enforce safe working conditions for the US workforce. Protective equipment requirements and dedicated safety managers on jobsites are two of the most influential reasons for safer work environments.
- *More emphasis on safety in training* — Many injuries can be avoided if craft professionals are well-trained and understand the proper procedures. Modern construction programs place a significant focus on teaching safe practices in the workplace.
- *Improved technology* — Digital technology has enabled construction companies to work more safely and efficiently. For example, drones allow construction crews to more easily spot potential hazards and monitor worksites remotely.
- *Safety matters in winning bids* — A construction company's safety records are a major factor in winning bids and being selected for projects.

Did You Know?

Establishing OSHA

President Nixon played an important role in the history of construction by signing the Occupational Safety and Health (OSH) Act in 1970, which established the government agency called OSHA.

Myth: Construction Is Bad for the Environment

Sustainability: The practice of minimizing negative impacts to Earth's climate and natural environment.

Green construction: Construction practices that involve the creation of structures that adhere to sustainability principles.

The construction industry is leading the charge on **sustainability**—the practice of minimizing negative impacts on Earth's climate and natural environment. Today, construction companies are attempting to reduce impacts on the environment by recycling materials, installing renewable energy sources, and using **green construction**, which involves creating structures that adhere to sustainability principles. In 1993, the US Green Building Council (USGBC) was created with a mission to promote sustainability-focused practices in the building industry. The council created the green building rating system known as Leadership in Energy and Environmental Design (LEED), which has become an international standard for certifying environmentally sound buildings.

Examples of construction practices that are good for the environment include:

- Preassembly of structures off-site, which allows for more material reuse
- Wind power technologies, which use wind to generate energy
- Solar panels, which harness sunlight to provide energy
- Water reclamation

An example of green construction is the use of steel buildings in agricultural structures and new discoveries in agrivoltaics, which consists of designing solar panels to complement farming by keeping crops cooler beneath the panels while providing a sustainable source of clean energy. Studies in agrivoltaics include the construction of a half-acre solar array built in Boulder, Colorado. The Nature Conservancy of Colorado partnered with two local teams to transform rocky terrain into an herb garden and solar farm, advancing research in this critical area of agricultural science (*Figure 6*).

Figure 6 Solar panels can be used in harmony with farming.
Source: iStock

1.2.0 History of Construction

Construction had its start thousands of years ago, establishing history through the built environment. Think about architectural wonders like the aqueducts of the Roman Empire (*Figure 7*), the Great Wall of China, or Machu Picchu in Peru. It's easy to forget that human beings created these amazing structures. Imagine the many aspects of work that were required to bring these structures to completion; the designing, planning, transporting of heavy materials, building, and so much more, were all done without the tools and technologies available today.

Figure 7 Construction of the first aqueduct occurred in Rome around 312 BC.
Source: iStock

Construction began by necessity, with humans building personal shelters using their hands or simple hand tools. Over time, people gained expertise in certain tasks and became specialists, such as carpenters, bricklayers, and blacksmiths. This allowed societies to build more complex structures and provide resources like water to communities for farming and daily use. As steam-powered machinery came about in the 19th century, more advanced structures and systems could be built, and they could be built much faster.

With the technological advancements that came with the Information Age, computer systems have revolutionized construction faster than in any other time period. Improvements in accuracy, consistency, speed, and complexity of construction exist today in ways never even imagined just 50 years ago. Construction has evolved into a high-tech business. Innovative individuals who tried new ideas, invented new products and processes, and manufactured easy-to-use materials have helped transform construction into a much more efficient and technology-driven business.

Did You Know?

A New Way to Water

In 1940, American farmer and inventor Frank Zybach invented the center-pivot irrigation sprinkling system. An improved version using a rolling pipe was developed shortly thereafter in 1950, as shown in the photo. The Stout-Wyss Irrigation System made its debut in Colorado and is the most popular method of irrigation on large modern farms today.

Source: iStock

1.2.1 Famous Construction Examples

All the famous and iconic man-made structures around the world would not be here today without the skilled people who built them. Consider these feats of construction that were groundbreaking at the time they were built:

- *Eiffel Tower in Paris, France* — Made of wrought iron in the late 1800s, it is still the second tallest building in France.
- *Empire State Building in New York, NY* — The world's tallest building upon its completion in 1931.
- *Golden Gate Bridge in San Francisco, CA* — The longest suspension bridge in the world upon its completion in 1937.
- *Hoover Dam on the Nevada/Arizona border* — A dam built with a workforce that peaked at more than 5,000 people in 1934.
- *Burj Khalifa in Dubai, United Arab Emirates* — A skyscraper completed in 2009 that is still the tallest artificial structure in the world at 2,722' (*Figure 8*).

Figure 8 The Burj Khalifa in Dubai is the tallest building in the world.
Source: iStock

1.2.2 Important Construction Innovations and Inventions

It is hard to imagine the world without inventions that make construction projects work more efficiently. The following are just a few examples of the construction industry's important innovations and inventions:

Personal protective equipment (PPE): Equipment or clothing designed to prevent or reduce injuries.

Computer-aided design (CAD): The use of computers to digitally sketch construction designs, allowing other people to see how aspects of a structure would look and work together when built.

Building information modeling (BIM): A process in which software reads computer files containing building data to create a virtual picture of a facility before it is built. BIM objects represent many dimensions of construction and allow builders to adjust the plans prior to beginning actual construction.

Virtual reality (VR): Provides a three-dimensional (3D) model of a construction site and allows users to directly immerse themselves into the virtual space. VR enables multiple people to envision the final product, feel what it is like to be in the space, and identify any issues before work begins.

- **Personal protective equipment (PPE)** — This equipment or clothing is designed to prevent or reduce injuries on a jobsite. Examples include hard hats, safety glasses, face shields, safety-toed footwear, gloves, and earplugs (*Figure 9*).
- **Computer-aided design (CAD)** — Before this invention in the 1960s, construction designs had to be drawn on paper by hand, using a tedious and time-consuming process. With CAD, users employ a more automated approach to sketching two-dimensional (2D) or three-dimensional (3D) designs, allowing other people to more clearly see how aspects of a structure would look and work together when built.
- **Building information modeling (BIM)** — In this process, software reads computer files containing information to get virtual pictures of the construction of a facility before it is built. BIM objects represent many dimensions of construction, such as time, cost, operation, and safety, to simulate the construction process and anticipate any issues that may arise. As a result, schedules, site layouts, and other details can be adjusted for the best outcome long before the physical building begins.
- **Virtual reality (VR)** — Like CAD and BIM, virtual reality takes what used to be paper drawings and information and allows users to visualize a structure before it is created. VR provides a 3D model of a construction site and allows users to directly immerse themselves into the virtual space. VR enables multiple people to envision the final product, feel what it is like to be in the space, and identify any issues before work begins.

Did You Know?

The First CAD

The first CAD system was developed in the 1960s by an American computer scientist named Ivan Sutherland. The system, called Sketchpad, was the first of its kind to store information and drawings and allow the user to alter designs digitally. This invention also laid the groundwork for BIM and VR.

Figure 9 A farmhand wears PPE to stay safe on the job.
Source: Image Provided as Courtesy of John Deere

1.0.0 Section Review

1. A major reason for the projected shortage of trained craft professionals is _____.
   a. construction salaries are too low
   b. technology has eliminated many construction careers
   c. demand for construction projects is declining
   d. the retirement of baby boomers

2. For every dollar made by men in construction, women make approximately _____.
   a. 50 cents
   b. 68 cents
   c. 87 cents
   d. 99 cents

3. During the 19th century, many structures could be built faster due to the invention of _____.
   a. hand tools
   b. steam-powered machinery
   c. self-healing concrete
   d. VR

2.0.0 Benefits of a Career in Construction

Objective

Explain the benefits of a construction career.

a. Recognize and describe how construction careers make a difference in the community.

b. Describe the financial and professional benefits of pursuing a construction career.

Performance Tasks

There are no Performance Tasks in this section.

When you choose a career in construction, you are choosing an industry that has a lasting impact on people and communities. At the same time, it provides an opportunity to combine knowledge, skills, and artistic expression. There are limitless opportunities for career expansion and advancement, plus the financial benefits that come from having a skill in high demand. With so many specialties from which to choose, the benefits of a career in construction are many.

2.1.0 Making a Difference

In many ways, the state of a society's infrastructure depends on the skills and expertise of craft professionals. The result of their work is what allows society to function and thrive. In short, craft professionals make a lasting difference in the world.

Craft professionals who work in construction take pride in being part of something greater than their individual contributions. They understand that it often takes the skills of many individuals working together to achieve great things.

2.1.1 Providing Essential Services

When you wake up and flip the light switch on, that switch is connected to wires that ultimately lead to a power plant miles away that craft professionals built and maintain. If you use pressurized irrigation, that water comes from a network of tubes or pipes connecting your farm to a water system that craft professionals built and maintain. In the west, this water might even travel for miles to reach your farm, using a system of canals built by craft professionals. The roads you drive on and the buildings in the schools you attend were all built by craft professionals. These services and structures are only possible because of skilled individuals who use their talents to install the components and systems that make activities like these a regular part of our daily lives.

During natural disasters or other catastrophes, craft professionals often act as first responders in their communities. Hurricanes, tornados, floods, and earthquakes can create destruction that interrupts many essential services and operations. Restoring essential services and rebuilding structures in a timely manner are critical to minimizing the impact of these emergencies. Examples of essential services include repairing barns for livestock shelter, restoring utilities such as water and electricity, and rebuilding residential structures that may have been destroyed in a disaster.

2.1.2 Working to Help the Environment

The current construction workforce has become increasingly focused on minimizing the environmental impacts of construction. As a result, the construction industry is changing, and the professionals who work in the industry are making positive differences for the environment. In addition to employing different building techniques, they are installing devices and systems that help businesses and homeowners conserve energy, reduce environmental impacts, and save money.

2.2.0 Construction Career Benefits and Opportunities

Construction careers can provide competitive salaries, a debt-free start, and financial security. With the sizeable shortage of construction professionals, the increasing demand for craft professionals has resulted in rising salaries and enticing benefits designed to attract talented individuals.

2.2.1 Competitive Salaries

Salaries for craft professionals are extremely competitive. Every other year, NCCER releases construction craft salary survey results based on the national average of base salaries for skilled craft professionals. The survey represents more than 350,000 industrial and commercial construction employees from across the United States. Results for some of the construction trades are shown in *Figure 10*. Construction salaries for all the crafts included in this survey range from $47,000 to $92,000. The salaries shown do not include overtime, bonuses, or other incentives. When these are combined, the overall benefits of construction jobs are even higher.

2.2.2 Starting Out Debt-Free

College tuition costs and student loans are the second highest source of debt in the United States. In fact, college tuition debt is second only to mortgage debt. *Forbes* reports that the total amount of student debt is over $1.5 trillion, which is equivalent to the value of about 500 National Football League franchises.

College tuition costs continue to rise at a fast rate. While some aspiring craft professionals may choose to obtain a bachelor's degree—particularly if they want to become civil engineers, project managers, or architects—the majority of construction careers do not require a degree. In fact, 7 out of every 10 jobs in the US economy require less than a four-year degree (*Figure 11*).

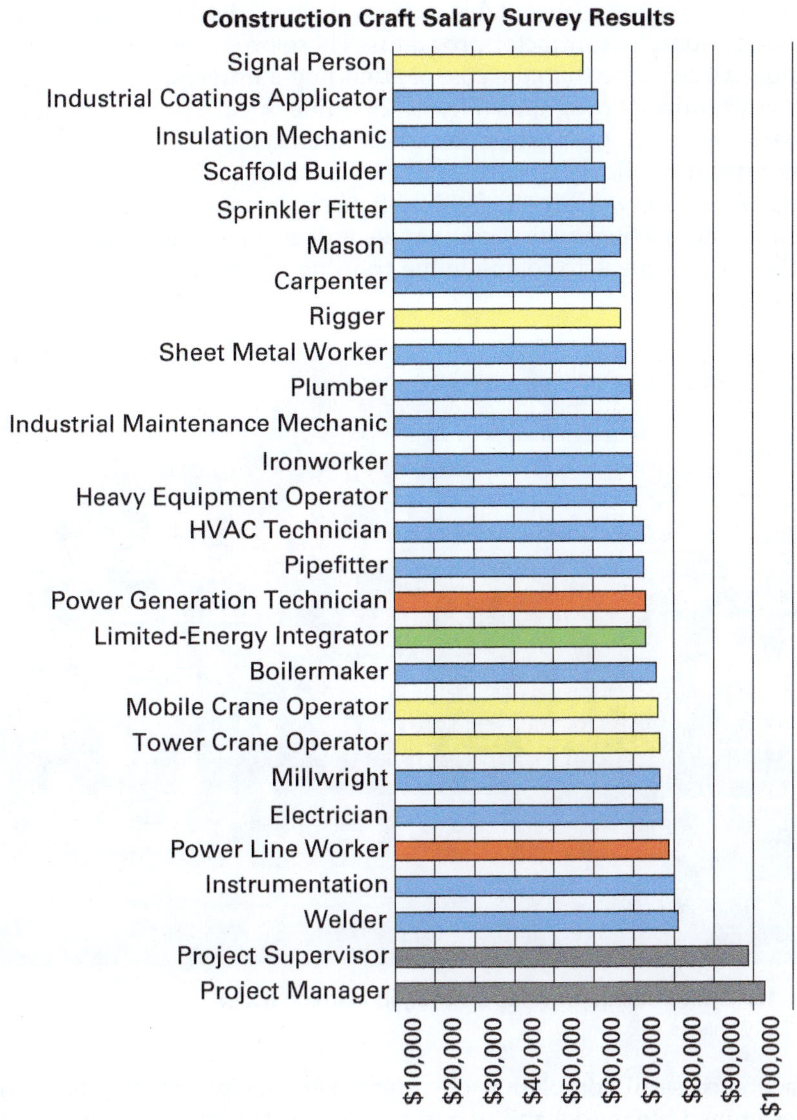

Figure 10 NCCER construction craft salary survey results.

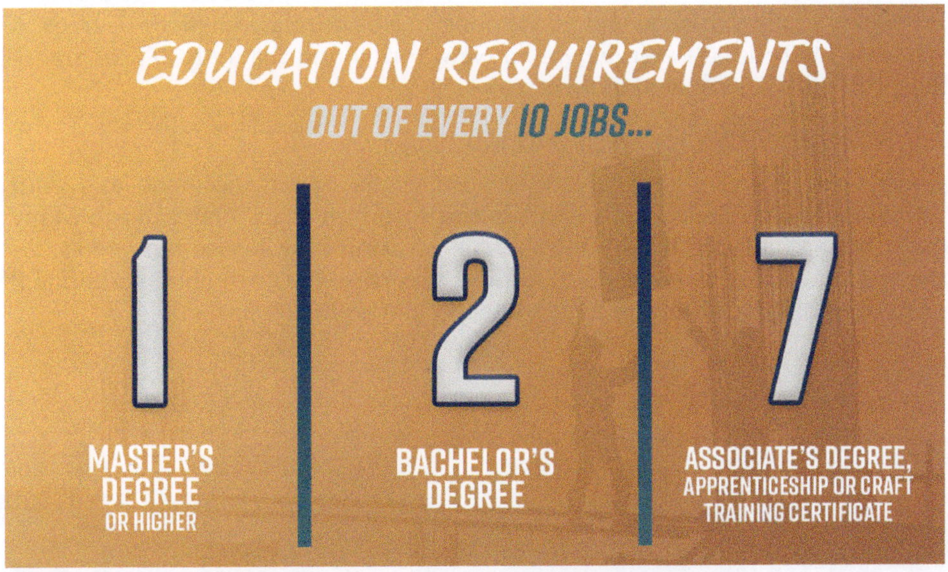

Figure 11 Education and construction career requirements.

Skilled craft training is available through technical schools, community colleges, associations, or contractor programs. These programs are shorter and far less expensive per year than the cost of attending a university.

Earn as you learn: An approach to education that pays trainees wages for work completed as part of an on-the-job learning (OJL) program.

With craft training programs (*Figure 12*), trainees learn both on the job and in the classroom under the instruction of experienced craft professionals. Craft training typically follows an **earn as you learn** model, which allows trainees to earn wages for their work in the field with pay increases as their skills advance. In addition, many construction companies will cover any additional education costs, allowing trainees to complete the training program debt-free and with a growing bank account.

Figure 12 Craft training provides on-the-job learning (OJL).
Source: Getty Images

Apprenticeship/craft training: A program in which you gain valuable skills on the job under the instruction of more experienced professionals, often making money as you obtain training.

In fact, successful completion of an **apprenticeship/craft training** program provides immediate marketability and acquisition of highly valuable skills—all while earning an income as you are being trained. This type of training makes it possible to avoid the high costs of attending a university. On the other hand, despite spending several years and considerable funds to earn a college diploma, university graduates often end up working in jobs unrelated to their degrees. A study by the Federal Reserve Bank of New York found that only 27% of graduates work in a field related to their major.

2.2.3 Career Advancement and Mobility

Construction careers typically offer clear paths to advancement, and craft professionals have excellent opportunities to progress in their careers. Many construction jobs begin with individuals learning how to use their tools. As they become proficient and develop their expertise, they can obtain additional licenses and certifications to increase their chances for promotion.

Success Story

Boyd Worsham, from Builder to President/CEO

As a boy growing up, Boyd Worsham always liked to build things. The carpentry program at his high school solidified his desire to enter the construction industry. He was inspired by his carpentry teacher, who taught not only skills in carpentry but also responsibility and how to be employable. His teacher introduced Boyd to a job opening as a carpentry helper, which he started in 1980, right out of high school.

As he developed his skills over the next 38 years, Worsham worked his way up, first as a journeyman carpenter, then carpenter foreman, assistant superintendent, superintendent, and, finally, to vice president of construction support for The Haskell Company. Haskell is a leader in design-build project delivery. At Haskell, he was involved in major projects throughout the United States. Haskell even paid for Worsham to obtain his Master of Business Administration from Jacksonville University, located in Jacksonville, Florida. Although he did not have a bachelor's degree, his 25 years of industry experience was counted in place of a degree. Ultimately, in September 2020, Worsham was appointed president and chief executive officer of NCCER, an international education foundation for the construction industry.

Source: Courtesy of Boyd Worsham

In addition to core craft skills, working in construction often requires leadership skills such as good communication, teamwork, and decision-making (*Figure 13*). These critical leadership skills allow advancement from trainee to craft professional to higher managerial and executive roles. Some craft professionals ultimately gain the knowledge and skills necessary for starting their own businesses and leading their own teams of employees.

During the development of their skills, these same individuals are often given the opportunity to travel or relocate to different areas because the skills they possess are in such high demand. Excellent construction jobs exist in cities and states across the United States and the world, meaning that skilled craft professionals are free to travel or move to locations as the demand allows. In some cases, companies even send their employees to construction sites in foreign countries.

Get Started in Construction:
Progress in Your Field

Scan the QR code using the camera on your phone or mobile device to view the accompanying resource.

Figure 13 Leadership skills help you advance in construction.
Source: iStock

2.0.0 Section Review

1. During disasters, craft professionals are considered first responders because they perform work to restore essential services and structures.
   a. True
   b. False

2. A major benefit of craft training is that _____.
   a. you have no financial responsibilities until you finish training
   b. you are able to earn as you learn
   c. you eventually make enough to pay off your training debt
   d. you have your apprenticeship after one year of training

3. One reason construction careers provide craft professionals with mobility is _____.
   a. subcontractors often give company automobiles to employees
   b. most craft professionals must travel due to the lack of work
   c. their skills are often in demand everywhere
   d. most good contractors demand that their employees travel long distances

3.0.0 Construction Craft Careers

Performance Tasks

There are no Performance Tasks in this section.

Objective

Describe the typical career path for professionals in the construction and agricultural industries.

a. Describe industry sectors and the progression path for construction careers.
b. Identify different construction careers and the types of skills they require.

Skilled construction professionals are part of a growing industry that offers a wide array of career opportunities. While some people still hold an outdated view that lumps all construction jobs together, today's reality is much different. The construction industry offers many career paths and specializations. In fact, NCCER offers specialized training programs for more than 45 construction crafts. This means individuals can choose from a large variety of construction jobs, many of which offer tremendous earning potential and great opportunities for advancement.

3.1.0 Construction Industry Sectors

Sectors: Categories of construction distinguished by specific types of work, materials, equipment, and skills. Sectors of construction include residential, commercial, agricultural, industrial, and heavy civil/infrastructure.

The construction industry comprises several **sectors**. Each sector includes specific types of construction, materials, equipment, and skills:

- *Residential* — This sector includes design, construction, and maintenance of single-family homes, multifamily homes such as apartment buildings, public housing developments, and even separate garages and sheds.
- *Commercial* — This sector includes design, construction, and maintenance of schools, government buildings, medical facilities, hotels, sports arenas and stadiums, shopping centers, and large office buildings.
- *Agricultural* — This sector includes the design, construction, and maintenance of structures used for the purposes of raising livestock, farming, food processing, logging, forestry, and ranching.
- *Industrial* — This sector includes construction of public works such as manufacturing plants, oil refineries, electrical generating plants, chemical processing plants, and large mills.

- *Heavy civil/infrastructure* — Often referred to as *horizontal construction,* this sector includes professionals who build bridges, roadways, airports, tunnels, and dams.

Each construction sector has its own approach to initiating and paying for projects, selecting the required equipment, and ensuring its workforce is adequately trained and prepared for the job.

Although each sector and construction craft have specific requirements for skills and training, the career path for construction careers is usually the same as the one shown in *Figure 14*.

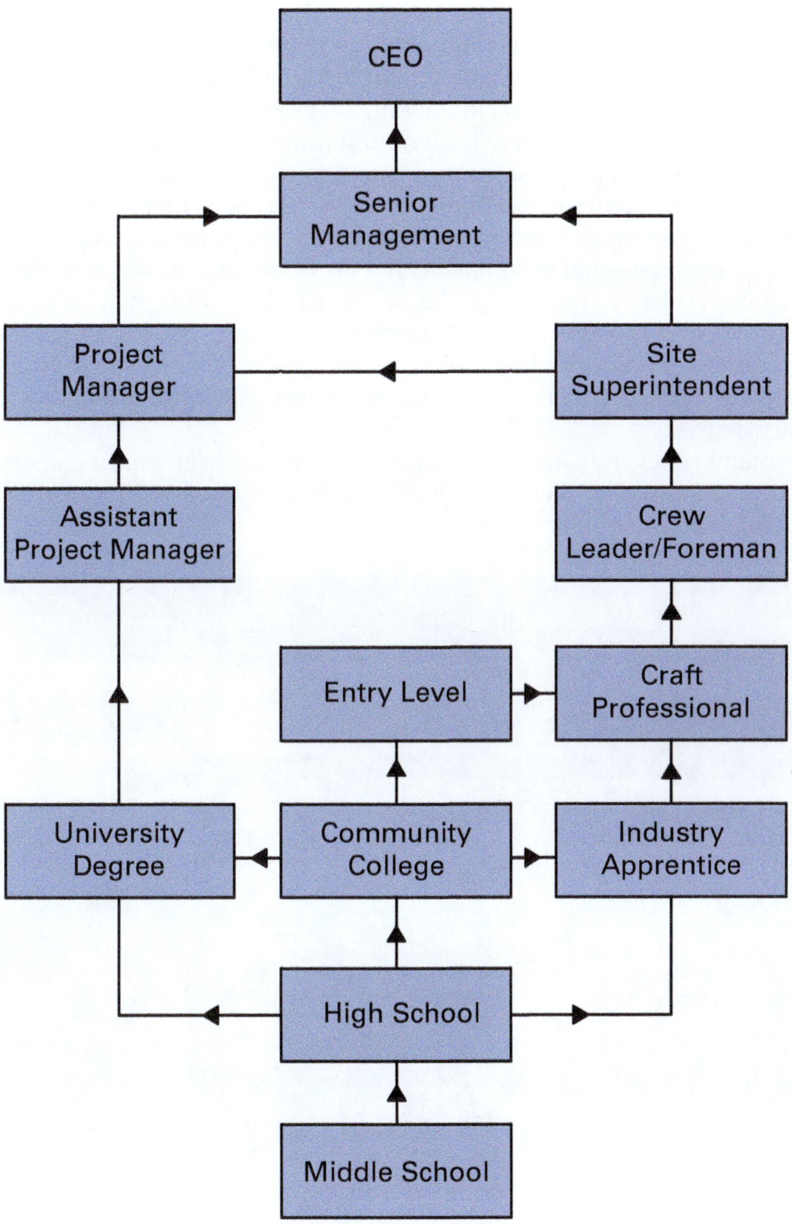

Figure 14 Career path for construction careers.

Contractor: Oversees many parts of a construction project, including resource management, budgeting, code adherence, quality assurance, and construction materials.

Project owner: The initiator of the project who usually finances the building endeavor.

Subcontractor/specialty contractor: Typically employs craft professionals who have specialized skills needed to complete a particular part of the construction project (e.g. HVAC, electrical, plumbing).

Crew leader: Also called the *foreman*, this supervisory role oversees a crew of craft professionals. It is the crew leader's job to make sure that work is completed correctly and on time. Crew leaders are responsible for the safety and work of those under them.

Site superintendent: Manages day-to-day activities on a construction site and supervises work performed by the subcontractors.

Project manager: Oversees the planning and delivery of construction projects on time and within budget.

A **contractor** oversees many parts of a construction project, including resource management, budgeting, code adherence, quality assurance, and construction materials. A contractor typically works directly with the construction **project owner** or the owner's representative. The project owner is the initiator of the project and usually finances the building endeavor. A contractor hires a **subcontractor/specialty contractor** to complete specific construction tasks. In most cases, subcontractors employ craft professionals with expertise and skills that can be used across multiple construction sectors. For example, while some craft professionals are responsible for tasks related to erecting a structure, others are involved in installing and maintaining electrical, communication, HVAC, and plumbing systems.

Many construction professionals start their careers working for a subcontractor and may continue there or move up to become general contractors. As they refine their skills and develop their leadership capabilities, they may be promoted to **crew leader**, superintendent, or even progress to higher senior management positions. The crew leader, sometimes called the *foreman*, manages a team or crew of craft professionals assigned to work on a construction project (*Figure 15*). A **site superintendent** manages day-to-day activities on a construction site and supervises work performed by the subcontractors. A superintendent can be a craft professional who has been a crew leader and has completed the necessary supervisory training, or he or she can be a person who graduated from a university construction management program. On the other hand, a **project manager** plans, schedules, and executes projects within a set budget but is not always on the construction site. A project manager typically holds a university degree.

As in many other industries, opportunities for advancement in construction are directly related to workers' skill levels, their continued willingness to grow and develop, and the quality of their work.

Figure 15 Construction project team.

3.2.0 Types of Construction Careers

One of the biggest benefits of pursuing a construction career is the variety of available jobs. Depending on your personal interests, professional goals, and desire to learn, construction offers great opportunities for practically everyone willing to learn and develop their skills.

Another benefit is the earning potential offered by many of the construction careers. Salary ranges differ based on specific construction career paths and an individual's skill level, quality of work, region, and desire to succeed. Other factors include ongoing classroom education and craft-specific training and certifications.

While performing the role of construction project manager or civil engineer requires a college education, most construction careers require only a high school diploma or equivalent to get started. Formal training programs are generally offered through technical schools, community colleges, and construction-oriented training organizations, and many use the NCCER curricula. The headquarters of NCCER is shown in *Figure 16*. Most NCCER programs are designed to provide multiple levels of training, allowing a trainee to learn and obtain craft credentials in stages.

Figure 16 NCCER.

The following sections describe several popular construction careers, providing information related to each career path's responsibilities, salary potential, needed skills, and job outlook.

3.2.1 Carpenter

Carpenters construct, repair, and install building frameworks and structures made from wood and other materials (*Figure 17*). Carpenters are involved in many kinds of construction from the restoration of historic buildings to the construction of homes, barns, commercial buildings, and more.

Examples of carpentry specializations include:

- *Rough carpenter* — This type of carpenter does the framing, formwork, and other structural work.
- *Finish carpenter* — This type of carpenter puts finishing touches on structures after they are almost fully built. Most of the visible wood inside of homes and buildings is placed there by finish carpenters. Typical jobs for a finish carpenter include window and door trim, crown molding, bookshelves, baseboards, and wooden stairways.

Figure 17 Carpenters build both large and small structures.
Source: Getty Images

Salary

The average salary for a carpenter is approximately $57,000 per year. The top 10% of carpenters earn more than $84,690.

Job Growth Estimates

The demand for carpenters in the United States is expected to remain the same, with about 90,000 new openings per year expected through the next decade.

Skills

To be successful, a carpenter should possess the following traits or skills:

- Physical fitness
- Sense of balance
- Good hand-eye coordination
- Attention to detail
- Ability to read blueprints and follow instructions for installing certain products
- Basic math skills (to calculate sizes and amounts of materials accurately)
- Good problem-solving skills

Specific Qualifications

In some cases, carpenters are expected to earn an apprenticeship, which requires a specific number of hours on the job and in the classroom. NCCER offers a four-year carpentry training course that can be used to fulfill the classroom instruction requirement of an apprenticeship.

Success Stories:
SkillsUSA 2018 Carpentry Competition Highlights

Scan the QR code using the camera on your phone or mobile device to view the accompanying resource.

3.2.2 Electrician

Electricians install and maintain the electrical and power systems for homes, agricultural operations (*Figure 18*), businesses, and factories. In large factories, electricians usually do maintenance work that is more complex. These kinds of electricians may repair motors, transformers, generators, and electronic controllers on machine tools and industrial robots. They also advise management regarding electrical hazards.

Figure 18 Electricians install and maintain lighting and power for farm operations.
Source: iStock

Another related electrical career is focused more on installation and maintenance of low-voltage electronic cables and devices. Some of the systems installed by these professionals include audio and video, security, fire protection, residential and commercial building networks, CCTV, and access control systems. These electronics professionals are known as limited-energy integrators or electronic systems technicians.

Salary

The average salary for an electrician is approximately $67,000 per year. The top 10% of electricians earn more than $96,580.

Job Growth Estimates

The demand for electricians in the United States is expected to grow by almost 10% during the next decade, which equates to almost 80,000 new openings per year over the same time period.

Skills

To be successful, an electrician should possess the following traits or skills:

- Physical fitness
- Good color vision (recognize electrical wires by color)
- Attention to detail
- Good math skills
- Good problem-solving skills
- Customer service skills
- Ability to work with hands

Discover Success:
Following in Footsteps

Scan the QR code using the camera on your phone or mobile device to view the accompanying resource.

Specific Qualifications

Credentials needed by electricians vary by state. In most states, they must be licensed. To earn a license, an electrician must pass a test with questions related to national, state, and local codes. These codes are in place to ensure safe installation of electrical wiring and equipment.

Electricians are generally required to take continuing education courses related to safety, code changes, and product training to maintain their licenses. To work in a specialty area, electricians are often required to obtain additional training.

3.2.3 Welder

Welders join metals together using a high-intensity electrical arc at temperatures between 6,000°F and 10,000°F (3,316°C and 5,538°C). By joining metal together, welders create a variety of structures, including buildings, ships, bridges, automobiles, and other smaller items. They also cut steel using oxyfuel, air carbon arc equipment, or plasma arc equipment. A strong focus on precision and safety is key to success in this craft. Examples of welding specializations include:

- *Pipe welder* — As the name implies, pipe welders construct and repair sections of pipe and related components (*Figure 19*).
- *Stainless steel welder* — This type of welder uses either tungsten inert gas (TIG) welding, spot welding, or metal inert gas (MIG) welding to join stainless steel metals. The method used typically depends on the qualities of the stainless steel being welded.
- *Weld inspector* — This type of welder assures the quality of existing welds by using high-tech testing methods.
- *Underwater welder* — While wearing full underwater diving gear, this type of welder performs different types of welding using a power source supplied through cables and hoses connected to welding equipment.
- *Robotic welder* — This type of welder programs robots and computer-controlled machines to perform welding.

Figure 19 Pipe welders join sections of pipe.
Source: iStock

Salary

The average salary of a pipe welder is approximately $69,000. Structural welders make an average salary of approximately $56,160. When other advanced specialty welding salaries are combined, the average salary of an experienced welder rises to $71,067.

Job Growth Estimates

The demand for welders in the United States is expected to grow by about 3% during the next decade. As the nation's infrastructure ages, the expertise of welders will be needed to rebuild roads, bridges, and buildings. Aspiring welders have more opportunities if they are open to traveling or relocating to work on what are often short-term projects.

Skills

To be successful, a welder should possess the following traits or skills:

- Attention to detail
- Hand-eye coordination
- Physical fitness
- Strength in reading 2D and 3D drawings
- Basic understanding of electricity

Specific Qualifications

Some welding positions require a general welding certification. The American Welding Society (AWS) offers the Certified Welder credential. To become a specialist, a welder must obtain a certificate focused on a particular area, such as pipe welding. Underwater welders must also have a diver's certification.

3.2.4 Pipefitter

Pipefitters install, assemble, fabricate, maintain, and repair mechanical piping systems. They work with many kinds of pipe in manufacturing, commercial, and industrial settings such as chemical plants, oil refineries, food processing plants, and paper mills. They work with piping systems that carry water (*Figure 20*), gases, liquid chemicals, solids, and fuels. Pipefitters need to have a good understanding of blueprints, basic math, welding, pipe bending, fittings, and valves.

Figure 20 Pipefitters connect a large farm field water system.
Source: iStock

Did You Know?

Welding Makes History

Two men of the 19th century are credited with making significant advancements in the welding field. French engineer Auguste de Méritens was the first to use heat from electricity to weld together lead plates. Around the same time, Russian inventor Nikolay Benardos discovered a similar method of welding using carbon rods.

Discover Success:
Earn More through Welding Jobs

Scan the QR code using the camera on your phone or mobile device to view the accompanying resource.

Salary

The average salary for a pipefitter is approximately $63,000 per year. The top 10% of pipefitters earn more than $97,170.

Job Growth Estimates

The demand for pipefitters in the United States is expected to grow approximately 4% in the next decade, which works out to approximately 50,000 new openings per year.

Skills

To be successful, a pipefitter should possess the following traits or skills:

- Mechanical abilities
- Physical fitness
- Problem-solving skills
- Mathematical skills, particularly in geometry
- Ability to measure, cut, thread, and assemble pipe
- Ability to sketch piping systems

Specific Qualifications

Pipefitters typically go through a four-year craft training or apprenticeship program to become a journey-level fitter. Several states require pipefitters to be licensed when working on gas lines.

Discover Success:
Family Means Everything

Scan the QR code using the camera on your phone or mobile device to view the accompanying resource.

3.2.5 Heavy Equipment Operator

Heavy equipment operators drive and control heavy construction equipment (*Figure 21*) to move construction materials, dirt, and other heavy objects at construction sites, large mills, mining operations, and distribution centers. On large construction sites, heavy equipment may be used to clear, grade, and lift equipment prior to and during the construction of roads, buildings, bridges, airports, and power generation facilities.

Figure 21 Heavy equipment operators drive large construction machinery.
Source: iStock

Examples of heavy equipment include backhoes, bulldozers, dump trucks, forklifts, and excavators. Almost all operators choose to specialize in one or more of these types of equipment.

Salary

The average salary for a heavy equipment operator is approximately $61,000 per year. The top 10% of heavy equipment operators earn more than $84,650. Professionals operating more complex equipment typically get paid higher salaries.

Job Growth Estimates

The demand for heavy equipment operators in the United States is expected to grow by about 4% in the next decade. Metropolitan areas usually offer the most job opportunities, and operators trained to work on multiple types of equipment typically have the best opportunities.

Skills

To be successful, a heavy equipment operator should possess the following traits or skills:

- Physical strength
- Good sense of balance
- Hand-eye-foot coordination
- Mechanical skills

Specific Qualifications

Several OSHA standards apply to operating different types of heavy equipment. *OSHA Standard 1910.178(l)(3)* requires operators of powered industrial trucks (forklifts) to be licensed or certified. When operators are tasked with hauling their own equipment between jobsites, they may also be required to obtain a commercial driver's license (CDL).

Success Stories:
James Snyder: I Built This

Scan the QR code using the camera on your phone or mobile device to view the accompanying resource.

3.2.6 Crane Operator

Crane operators use state-of-the-art machinery to hoist heavy construction materials above and around a jobsite. Joysticks, levers, and pedals allow operators to control the equipment. Professional crane operators are regularly tasked with using trigonometry and physics equations to calculate maximum load size based on the configuration of the crane and load.

There are many different types of cranes, including tower cranes that are seen high in the sky over tall buildings in cities around the world. Tower crane operators frequently control equipment that remains in place but grows taller as the building progresses upward. In some cases, cranes can rise many stories above their original starting point (*Figure 22*).

Salary

The average salary for a tower or mobile crane operator is approximately $66,000 per year. The top 10% of tower or mobile crane operators earn more than $91,840.

Job Growth Estimates

The demand for crane operators in the United States is expected to grow by about 8% in the next decade, which is above the average growth for all occupations. Metropolitan areas usually have the largest need for crane operators and pay the highest salaries. Because crane operators have unique skills, job prospects are extremely good for this profession.

Figure 22 Crane operators construct a modern grain silo.
Source: iStock

Skills

To be successful, a crane operator should possess the following traits or skills:

- Endurance
- Agility
- Physical coordination
- Good sense of balance
- Ability to judge distance
- Hand-eye-foot coordination
- Good math skills
- Comfortability with technology

Specific Qualifications

According to *OSHA Standard 1926.1400(a)*, crane operators must be certified or licensed, with the type of certification based on the type or capacity of the crane. Licensing usually consists of a combination of written exams and physical tests in which individuals must demonstrate safe operating practices.

3.2.7 Ironworker

Ironworkers place and install iron or steel beams, columns, and other construction materials to form and reinforce structures. Contrary to the name, ironworkers primarily work with steel and are often referred to as *structural ironworkers*. These professionals primarily build structural frameworks that support large construction (*Figure 23*).

Figure 23 Ironworkers build steel structural frameworks.
Source: iStock

Salary

The average salary for an ironworker is approximately $60,000 per year. The top 10% of ironworkers earn more than $89,790.

Job Growth Estimates

The demand for ironworkers in the United States is expected to grow by about 5% in the next decade. Because of the ongoing need to repair aging highways and bridges, an ironworker's skills should remain in high demand for years to come.

Skills

To be successful, an ironworker should possess the following traits or skills:

- Physical fitness
- Agility
- Good sense of balance
- Hand-eye coordination
- Ability to work at heights
- Mechanical skills
- Math skills
- Problem-solving skills

Specific Qualifications

Ironworkers with a welding certification often have more job prospects.

Discover Success:
Looking for Something Different

Scan the QR code using the camera on your phone or mobile device to view the accompanying resource.

3.2.8 HVAC Technician

HVAC technicians install, maintain, and repair heating, ventilation, and air conditioning systems. HVAC work includes both mechanical and electronic systems such as motors, pumps, fans, thermostats, and computerized switches that control systems in residential, commercial, and industrial structures (*Figure 24*). Indoor agricultural operations often operate HVAC systems and evaporative-cooling systems throughout the year. Technicians can specialize in specific equipment to support these operations, such as water-based heating systems, ventilation requirements, or livestock cooling.

Figure 24 HVAC technicians install and repair air conditioning systems.
Source: Getty Images

Salary

The average salary for an HVAC technician is approximately $62,000 per year. The top 10% of HVAC technicians earn more than $77,920.

Job Growth Estimates

Demand for HVAC technicians in the United States is expected to grow by about 4% over the next decade. Increased opportunities exist for technicians willing to move to higher-growth areas.

Skills

To be successful, an HVAC technician should possess the following traits or skills:

- Good hand-eye coordination
- Mechanical aptitude
- Basic math skills
- Good problem-solving skills

Specific Qualifications

Some states and localities require technicians to be licensed; others require technicians to be certified when working with different types of chemicals.

Success Stories:
No One Day Is the Same—Discover Success

Scan the QR code using the camera on your phone or mobile device to view the accompanying resource.

3.2.9 Project Manager

Project managers are essential to completing construction projects on time and on budget. Their duties include planning, coordinating, budgeting, and supervising projects from beginning to end. They typically prepare cost estimates, explain project contracts to other professionals, manage personnel, resolve project issues, and ensure that projects adhere to safety codes and regulations. Project managers usually spend most of their time on-site, where they monitor projects and ensure construction activities stay on schedule (*Figure 25*).

Figure 25 Project managers monitor project schedules and activities.
Source: Getty Images

Salary

The average salary for a construction project manager is approximately $92,000 per year. The top 10% of construction project managers earn more than $164,790. Project managers may also earn bonuses based on their performance.

Job Growth Estimates

The demand for construction project managers in the United States is expected to grow by about 8% over the next decade, which is well above the average growth for all occupations. As construction processes and mobile technology become more sophisticated, the need for project managers who stay current with technological changes is expected to increase.

Skills

To be successful, a project manager should possess the following traits or skills:

- Good oral and written communication skills
- Ability to make sound decisions
- Business acumen
- Ability to manage time and oversee multiple ongoing activities
- Strong customer service skills
- Good leadership skills

Specific Qualifications

Unlike most construction careers, project managers typically need a bachelor's degree from an accredited college or university. Large construction companies tend to seek candidates who have a bachelor's degree in a construction-related field and direct construction experience.

Some states require construction project managers to be licensed. Additionally, certifications in construction management and project management can increase the chances of landing a project management job.

3.2.10 More Construction Careers

NCCER's skilled trade certifications and curricula offer many advantages. These include *cross-functional skills* that can support a variety of industries, especially agricultural operations. Certification not only provides skills that can be used on your own farm or ranch—it also opens new opportunities to prepare you for a career path in the field of construction. These skills establish a firm foundation for employment or to start your own construction business supporting the agricultural community.

While some of the more popular construction careers have been highlighted, NCCER offers training in more than 45 crafts. You can discover more about them on NCCER's Build Your Future website, **www.byf.org**. This site offers information about salaries by craft, career paths, labor demands by state, and much more.

3.0.0 Section Review

1. Which construction sector includes schools, government buildings, hotels, and shopping centers?
   a. Residential
   b. Commercial
   c. Public
   d. Industrial

2. The construction role that typically reports to the project owner is the _____.
   a. superintendent
   b. foreman
   c. subcontractor
   d. contractor

3. To start your career path in most construction crafts, you generally need a _____.
   a. high school diploma or equivalent
   b. bachelor's degree
   c. master's degree
   d. state permit

4. The construction craft professional regularly tasked with using trigonometry and physics equations to calculate maximum load sizes is the _____.
   a. heavy equipment operator
   b. ironworker
   c. crane operator
   d. pipefitter

4.0.0 Starting Your Construction Career

Performance Tasks

There are no Performance Tasks in this section.

Objective

Identify ways to pursue a career in the construction industry.

a. Explain the benefits of career and technical education programs.
b. Describe the advantages of craft training programs and their relationship with apprenticeships.
c. Summarize the path to a construction career through community colleges and universities.

You have several training options to consider if you are interested in pursuing a construction career. If you are still in high school, you should find out if any local schools offer additional construction classes. You can also look for summer jobs with local contractors. The internet is a great resource for learning about construction crafts while you earn your diploma.

After earning your high school diploma or equivalent, choose which training path works best for you. Available training options include:

- Technical school
- Apprenticeship program
- Community college or university
- Industry training program

Most craft professionals are taught through a combination of technical schools, on-the-job learning (OJL), craft training programs, and/or apprenticeships.

4.1.0 Career and Technical Education

Career and technical education (CTE) classes are generally offered at high schools and technical education centers. Some schools start earlier by offering middle school classes that teach introductory concepts in electronics, engineering, carpentry, or other construction-related crafts. In either case, students attending these programs can explore career options available for each of the construction crafts. In some cases, students attending high school CTE classes can develop their hands-on skills through construction competition events like SkillsUSA. School guidance counselors can provide information about which CTE courses are offered and how students can enroll.

Career and technical education (CTE): Programs offered in high school and technical education centers that allow students to explore career paths in construction.

CTE programs place an emphasis on career readiness and hands-on learning (*Figure 26*). These programs emphasize the following approaches:

- *Real-life application* — Being able to practice skills learned has been shown to increase students' retention and understanding. For many students, being able to link traditional lessons directly to their future careers means they are more committed to succeeding in their education. Being able to do so in a low-pressure environment makes learning easier.
- *Teaching nontechnical skills* — Having a combination of craft-specific technical skills along with other important but nontechnical skills increases a student's employability. Some of these nontechnical skills include effective communication, time management, attention to detail, critical thinking, and a customer service attitude.
- *Career exploration* — By learning about careers early on and from teachers with industry experience (*Figure 27*), students can discover their own strengths and interests, find a career that fits their passions, and set a path to achieve their goals.

After completing CTE classes, students generally have the qualifications needed to progress to one of these next steps:

- Industry apprenticeship
- Community college
- University degree

Figure 26 CTE programs emphasize hands-on learning.

Figure 27 Students learn from industry experts in CTE programs.
Source: iStock

Success Story

Construction in The Villages

The Villages is a growing retirement community located in North Central Florida. In 2017, The Villages Charter School (VCS) added a construction management academy to introduce students to construction, which is one of the most prevalent career fields in their community.

Each year, VCS construction teacher Bruce Haberle has his senior class help construct a home for Habitat for Humanity. This project gives students real-life opportunities to apply the skills they are learning, like reading blueprints, working on a schedule, and keeping an efficient pace. The program allows students to see what a career in construction is really like. Most recently, students worked to construct a 1,200-square-foot, 3-bedroom, 2-bathroom home.

Haberle is still refining the program but says he has three primary goals for his students: prepare them, get them in the industry, and set them up for success in their careers.

4.2.0 Craft Training and Apprenticeships

The craft training and apprenticeship approach enables students to learn skills from experienced construction professionals. Regardless of students' previous exposure in high school, many construction companies provide training themselves, either through in-house programs or by paying for trainees to attend classes at a local construction association or community college. Ultimately, craft training programs and apprenticeships are excellent opportunities to be paid while working on the job.

A craft training or apprenticeship program (*Figure 28*) features a split between learning in the classroom (typically 20% of learning) and on the job (typically 80% of learning). It usually takes two to four years and requires at least 2,000 hours of OJL and 144 hours of classroom instruction during each year of the apprenticeship. According to the US Department of Labor, there are five components to an apprenticeship that are also included in craft training:

- Business involvement
- Structured OJL
- Related instruction
- Rewards for skill gains
- Nationally recognized credentials

Craft training programs and apprenticeships offer many benefits, including full-time, paid work while learning skills on the job. This model of education not only gives trainees needed experience, but it keeps them from incurring high education debt. It also allows them to make valuable connections in the industry through relationships built with mentors.

Figure 28 Apprenticeships and craft training offer paid work while learning.
Source: Courtesy of LeChase Construction Services, LLC

You can find craft training programs and apprenticeships in your area in the following ways:

- Search the national apprenticeship database at **www.apprenticeship.gov**.
- Contact local construction associations like Associated Builders and Contractors (ABC) as well as Associated General Contractors (AGC).
- Search online job boards or other social media job search sites.
- Reach out to local employers.
- Research information on websites for national construction organizations.

4.3.0 Community Colleges and Universities

Success Stories:
How Many Jobs Need a Four-Year Degree

Scan the QR code using the camera on your phone or mobile device to view the accompanying resource.

More and more colleges and universities offer construction-related degree programs. These institutions offer a variety of construction-oriented academic degrees. Community colleges usually provide two-year programs and offer associate degrees, while four-year universities are more often used for bachelor's degrees and beyond.

4.3.1 Community College

Community colleges often collaborate with contractors and construction companies by helping facilitate their training programs. Rather than teach employees in-house, these companies can send trainees to a community college. Doing so allows companies to fulfill the knowledge-based component of the program by taking advantage of the college's dedicated teaching staff and classroom facilities.

Earning an associate degree from a community college allows you to receive college credit as well as industry-recognized certifications and credentials upon graduation. Options for the next steps include:

- Industry apprenticeship
- Entry-level job
- University degree

4.3.2 Four-Year Universities

An increased number of four-year universities offer degree programs in construction-related areas such as design, construction management, estimating, human resources, architecture, engineering, and safety (*Figure 29*).

Did You Know?

Father of Skyscrapers

Louis Henry Sullivan, an American architect in the late 19th century, created a new style of architecture that resulted in buildings that were tall but still considered beautiful, a unique concept at the time. Called the "Father of Skyscrapers," he is most known for his design of the Wainwright Building in St. Louis.

Degrees signify formal academic achievement and offer in-depth learning. A bachelor's degree focuses on an area of study (called a *major*) and typically requires at least 120 hours of credit work. Typical degree programs will consist of general education courses (science, math, history, and English) along with specific construction courses.

A big advantage of a college degree is that it may allow you to move up the construction career ladder more quickly after gaining initial experience in your craft. You may also be more qualified for higher-paying roles that involve decision-making responsibilities.

However, there are financial challenges related to obtaining a four-year degree. The tuition can be very expensive, and it can be difficult to work a full-time job while attending classes. In addition, a college degree is not a guarantee that you will be offered a specific construction job, and unfortunately, unless you have on-the-job construction experience, you may still have to start your career in an entry-level position.

Figure 29 Many universities offer construction-related degrees.
Source: iStock

4.0.0 Section Review

1. CTE programs place an emphasis on preparing students for construction careers through _____.
   a. hands-on learning
   b. lectures and quizzes
   c. virtual lessons
   d. group projects

2. Depending on the particular craft, an apprenticeship or craft training program can generally be completed within _____.
   a. six months
   b. two to four years
   c. six to eight years
   d. ten years

3. Earning a construction degree guarantees you will be offered a construction job upon graduation.
   a. True
   b. False

Module 93101 Review Questions

1. What percentage of construction companies anticipate having to hire more people in the coming years to replace retiring craft professionals?
   a. 60%
   b. 70%
   c. 80%
   d. 90%

2. What percentage of construction field and office positions are currently filled by women?
   a. 10%
   b. 20%
   c. 30%
   d. 40%

3. The international standard used for certifying environmentally sound buildings is known as _____.
   a. LEED
   b. OSHA
   c. BIM
   d. NCCER

4. A developing technology that allows craft professionals to gather and share data and reports is _____.
   a. IoT devices
   b. robots
   c. wireless tools
   d. drones

5. The Empire State Building is important in construction history because, upon completion, it was the tallest building in _____.
   a. New York City
   b. the United States
   c. the Western Hemisphere
   d. the world

6. The fact that craft professionals restore essential services and structures following disasters is one reason they can be classified as _____.
   a. EMTs
   b. first responders
   c. public servants
   d. OSHA workers

7. Recent surveys indicate construction careers for all craft professionals typically pay salaries that fall between _____.
   a. $18,000 and $31,000
   b. $27,000 and $47,000
   c. $37,000 and $72,000
   d. $47,000 and $92,000

8. In addition to possessing specific craft skills, what other traits are essential for construction professionals to advance in their careers?
   a. Being friends with the contractor.
   b. Having good math and drafting skills.
   c. Good communication and decision-making skills.
   d. Willingness to travel and work off the clock.

9. The construction of single-family homes falls under the construction sector called _____.
   a. small structure
   b. infrastructure
   c. industrial
   d. residential

10. The construction role that is sometimes referred to as the *foreman* is also called the _____.
   a. site superintendent
   b. crew leader
   c. project owner
   d. contractor

11. The construction career that typically requires a four-year college education is a(n) _____.
   a. project manager
   b. finish carpenter
   c. underwater welder
   d. limited-energy integrator

12. One of the biggest advantages of pursuing a construction career is the _____.
   a. amount of vacation
   b. ability to make your own schedule
   c. large variety of jobs
   d. lack of needed skills for most crafts

13. A craft professional who creates structures by joining metals using high heat and pressure is called a(n) _____.
   a. heavy equipment operator
   b. ironworker
   c. crane operator
   d. welder

14. A craft profession that requires a good knowledge of blueprint reading and geometry is a(n) _____.
   a. electrician
   b. pipefitter
   c. HVAC technician
   d. welder

15. Apprenticeship programs are typically structured so that learning is _____.
   a. 10% in the classroom and 90% on the job
   b. 20% in the classroom and 80% on the job
   c. 30% in the classroom and 70% on the job
   d. 40% in the classroom and 60% on the job

Answers to odd-numbered questions are found in the Review Question Answer Keys at the back of this book.

Answers to Section Review Questions

Answer	Section Reference	Objective
Section 1.0.0		
1. d	1.1.2	1a
2. d	1.1.3	1a
3. b	1.2.0	1b
Section 2.0.0		
1. a	2.1.1	2a
2. b	2.2.2	2b
3. c	2.2.3	2b
Section 3.0.0		
1. b	3.1.0	3a
2. d	3.1.0	3a
3. a	3.2.0	3b
4. c	3.2.6	3b
Section 4.0.0		
1. a	4.1.0	4a
2. b	4.2.0	4b
3. b	4.3.2	4c

MODULE 93102

Basic Safety

Source: iStock

Objectives

Successful completion of this module prepares you to do the following:

1. Explain the benefits of safety, the cost of workplace incidents, and ways to reduce related hazards.
   a. Describe the types of workplace incidents along with their physical and monetary impacts.
   b. Summarize the causes and consequences of common incidents.
   c. Explain how to recognize, evaluate, and control workplace hazards.
2. Describe common fall hazards and methods to prevent them.
   a. Summarize the most common types of construction fall hazards.
   b. Describe components of effective fall arrest systems and how they prevent or halt falls.
   c. Explain how to use ladders and stairs safely.
   d. Identify key steps to ensuring scaffolds are assembled and used safely.
3. Recognize and avoid struck-by, caught-in, and caught-between hazards.
   a. Describe struck-by hazards and how to avoid them.
   b. Describe common caught-in/caught-between hazards and steps that can prevent them.
4. Identify common electrical hazards and how to avoid them.
   a. Summarize basic jobsite electrical safety guidelines.
   b. Explain the importance of disabling equipment as well as basic lockout/tagout (LOTO) procedures.
5. Associate personal protective equipment (PPE) with the hazards they reduce or eliminate.
   a. Explain how PPE is used to protect craftworkers from different types of injuries.
   b. Explain how respirators protect craftworkers from respiratory dangers.
6. Describe safety practices used with other common jobsite hazards.
   a. List other types of hazards craftworkers may encounter.
   b. Describe common environmental hazards and how craftworkers should respond to them.
   c. Summarize hazards associated with hot work.
   d. Identify fire hazards and describe basic firefighting procedures.
   e. Name different types of confined spaces and how to avoid related hazards.

NOTE

Trainees who successfully complete this module will be awarded a Construction Site Safety Orientation credential.

Digital Resources for Agricultural Technology

Scan this code using the camera on your phone or mobile device to view the digital resources related to this craft.

SCAN ME

Performance Tasks

Under supervision, you should be able to do the following:

1. Properly set up and climb/descend an extension ladder, demonstrating proper three-point contact.
2. Inspect the following PPE items and determine if they are safe to use:
   - Eye protection
   - Hearing protection
   - Hard hat
   - Gloves
   - Fall arrest harnesses
   - Lanyards
   - Connecting devices
   - Approved footwear
3. Properly don, fit, and remove the following PPE items:
   - Eye protection
   - Hearing protection
   - Hard hat
   - Gloves
   - Fall arrest harness
4. Inspect a typical power cord and ground fault circuit interrupter (GFCI) to ensure their serviceability.

Overview

Work at construction, agricultural, and industrial jobsites can be hazardous. Most jobsite incidents are caused by at-risk behavior, poor planning, lack of training, or a failure to recognize present hazards. To help prevent incidents, every company and operation must have a proactive safety program. Safety must be incorporated into all phases of the job and involve workers at every level, including management.

NCCER Industry-Recognized Credentials

If you are training through an NCCER-accredited sponsor, you may be eligible for credentials from NCCER. The ID number for this module is 93102. Note that this module may have been used in other NCCER curricula and may apply to other level completions. Contact NCCER at 1.888.622.3720 or go to **www.nccer.org** for more information.

You can also show off your industry-recognized credentials online with NCCER's digital credentials. Transform your knowledge, skills, and achievements into credentials that you can share across social media platforms, send to your network, and add to your resume. For more information, visit **www.nccer.org**.

1.0.0 Safety and Hazard Recognition

Objective

Explain the benefits of safety, the cost of workplace incidents, and ways to reduce related hazards.

a. Describe the types of workplace incidents along with their physical and monetary impacts.
b. Summarize the causes and consequences of common incidents.
c. Explain how to recognize, evaluate, and control workplace hazards.

Performance Tasks

There are no Performance Tasks in this section.

When you take a job, you have an obligation to your employer, coworkers, family, and yourself to work safely. You also have an obligation to make sure anyone you work with is working safely. Your employer is likewise obliged to maintain a safe workplace for all employees. The ultimate responsibility for on-the-job safety, however, rests with you; safety is part of everyone's job. In this module, you will learn to ensure your safety, and that of the people you work with, by obeying the following rules:

- Follow safe work practices and procedures, both regulatory and corporate.
- Inspect safety equipment before use.
- Use safety equipment properly.

To take full advantage of the wide variety of training, job, and career opportunities that the construction, agricultural, and mining industries offer, you must first understand the importance of safety. Successful completion of this module will be your first step toward achieving this goal. Later modules offer more detailed explanations of safety procedures, along with opportunities to practice them.

Whether you are on a construction jobsite or working on an agricultural site, there are often workers from many trades in one place. These workers are all performing different tasks. As a result, the jobsite is constantly changing and hazards are continually emerging. These hazards can jeopardize your safety. Your employer or supervisor should make every effort to plan safety into each job and to provide a safe and healthful jobsite. Ultimately, however, your safety is in your own hands.

Safety training is provided to make you aware that hazards exist all around you every day. The time you spend learning and practicing safety procedures can save your life and the lives of others.

Safety is a learned behavior and attitude. It is a way of working that must be incorporated into the company as a culture. A **safety culture** is created when all the workers within a jobsite or organization see the value of a safe work environment and support it through their actions (*Figure 1*). Creating and maintaining a safety culture is an ongoing process that includes a sound safety structure and attitude, and relates to organizations as well as individuals. From management to laborers, everyone in the company and at the jobsite must be responsible for safety every day they come to work.

There are many benefits to having a safety culture. Companies with strong safety cultures usually have the following characteristics:

- Fewer at-risk behaviors
- Lower **incident** and **accident** rates
- Less turnover
- Lower absenteeism
- Higher productivity

CAUTION

Every jobsite carries risk, even if you are located at a rural jobsite. The fatality rate for farm workers is seven times higher than that of workers in the private sector. Stay safe by assessing on-site risks before work begins, identifying jobsite hazards, and developing potential solutions to those hazards.

Safety culture: The culture created when the whole company sees the value of a safe work environment.

Incident: Per the US Occupational Safety and Health Administration (OSHA), an unplanned event that does not result in personal injury but may result in property damage or is worthy of recording.

Accident: According to the US Occupational Safety and Health Administration (OSHA), an unplanned event that results in personal injury and/or property damage.

Figure 1 A safety culture promotes safe practices.
Source: iStock

A strong safety culture can also improve an organization's safety record, which leads to winning more bids, keeping workers employed, and maintaining productivity. Construction contractors with poor safety records are sometimes excluded from bidding, so good safety performance is essential. Agricultural operations with poor safety records may face the devastation of injury and loss of life. Safety is personal for everyone but must be supported by the entire team. Factors that contribute to a strong safety culture include the following:

- Embracing safety as a core value
- Strong leadership
- Establishing and enforcing high standards of performance
- The commitment and involvement of all employees
- Effective communication and commonly understood and agreed-upon goals
- Using the workplace as a learning environment
- Encouraging workers to have a questioning attitude and empowering them to stop work when faced with potential hazards
- Good organizational learning and responsiveness to change
- Providing timely responses to safety issues and concerns
- Continually monitoring performance
- Positive reinforcement when proper safety practices are demonstrated by employees

Did You Know?

Safety First

Safety training is required for all activities. Never operate tools, machinery, or equipment without prior training. Always refer to the manufacturer's instructions.

1.1.0 Incidents and Accidents

Incidents and accidents can occur at any jobsite. Both at-risk behavior and poor working conditions can cause these undesirable events. You can help prevent such events by using safe work habits, understanding what causes them, and learning how to prevent them.

The terms *incident* and *accident* are often used interchangeably. However, according to the US **Occupational Safety and Health Administration (OSHA)**, an incident is an unplanned event that may or may not result in property damage. When an incident occurs, no personal injury has occurred. However, an incident is worthy of being documented so that steps can be taken to prevent it from recurring.

Occupational Safety and Health Administration (OSHA): An agency of the US Department of Labor. Also refers to the Occupational Safety and Health Act of 1970, a law that applies to more than 130 million workers and seven million jobsites in the United States.

Case History

A Farm Fatality Under Investigation

Safety training is imperative for agricultural heavy equipment operators. A 2003 fatality within the Oregon agricultural community highlights common mistakes that can lead to devastating losses on the farm. According to Oregon Health and Science University, an authority on agricultural safety, a lack of formal training in the operation of heavy equipment led to the death of a 32-year-old farmer. "He had considerable experience repairing, operating, and maintaining farm machinery and was aware of the potential hazards of working around large farm machinery," recalls the victim's father. That morning, the victim completed several combine setups to process grass seed in the field that day. Relying on knowledge passed down from other experienced farmers, sales representatives, and family, the victim performed the initial course adjustments on his combine and then entered the field. A nearby operator then confirmed, "The victim exited the combine as it slowly moved forward to assess seed loss at the discharge end of the machine." The operator then reported, "the victim fell under the left rear wheel." The operator and the victim's father surmise that the victim tripped while observing the seed loss. Unable to recover his balance due to the uneven ground, the combine rolled over him and crushed him. In fact, the practice of exiting the combine to observe its operation after initial adjustments is common to this farm and widespread in the agricultural community, despite the manufacturer's recommendation to shut down the machine to perform this task.

The Bottom Line: Proper training should advise farm operators to only assess for seed loss after the combine is shut down and the engine is turned off. Observers should always maintain a safe distance from all moving machinery.

Source: Oregon Fatality Assessment Evaluation, Fatality Investigation Report

An accident is defined as an unplanned event that results in personal injury and/or property damage. Therefore, an event that results in property damage alone could be considered an incident or an accident. If personal injury or a fatality has occurred, the event is an accident.

Although there are varying opinions on the use of these two terms, the most important thing to understand is that both incidents and accidents are undesirable events that have a negative effect on both projects and workers. Do not be surprised when you hear the terms used interchangeably by other workers. The definitions provided by OSHA are used here to provide context for these terms as they are used throughout this module.

The lessons you will learn in this module will help you work safely, whether you are working around traffic or operating heavy machinery in a rural community. You will be able to spot and avoid hazardous conditions on the jobsite. By following safety procedures, you will help keep your workplace free from incidents and accidents and protect yourself and others from injury or even death.

NOTE

The US National Safety Council defines an incident as an unplanned, undesired event that adversely affects the completion of a task. In this definition, there is no mention of injury or property damage.

1.1.1 Incident and Accident Categories

Incidents and accidents are often categorized by their severity and impact, as follows:

- *Near-miss* — An unplanned event in which no one was injured and no damage to property occurred, but during which either could have happened. Near-miss incidents are warnings that should always be reported rather than overlooked or taken lightly.
- *Property damage* — An unplanned event that results in damage to tools, materials, or equipment, but no personal injuries.
- *Minor injuries* — Personnel may have received minor cuts, bruises, or strains, but the injured workers returned to full duty on their next regularly scheduled work shift.
- *Serious or disabling injuries* — Personnel received injuries that resulted in temporary or permanent disability. Included in this category would be lost-time incidents, restricted-duty or restricted-motion cases, and those that resulted in partial or total disability.
- *Fatalities* — Deaths resulting from unplanned events.

Studies have shown that for every serious or disabling injury, there were 10 injuries of a less serious nature and 30 property damage incidents. Further study showed that 600 near-miss incidents occurred for every serious or disabling injury.

There are four leading causes of death in construction work. These are often referred to as the "big four," the "fatal four," or the "focus four." They include falls, struck-by hazards, caught-in or caught-between hazards, and electrical hazards (*Figure 2*). Deaths from falls far exceed all other causes.

The following are explanations of the leading hazard groups:

- Falls from elevation are incidents involving the failure to provide, failure to use, or failure of appropriate fall protection. Falls from farm equipment include sustaining injuries after falling from tractors, grain bins, or ladders.
- Struck-by accidents involve unsafe operation of equipment, machinery, and vehicles, as well as improper handling of materials, such as through unsafe rigging operations or being struck by falling objects.
- Caught-in or caught-between accidents involve unsafe operation of equipment, machinery, and vehicles, such as farm machine entanglements. This category also includes improper safety procedures at **trench** sites and in other **confined spaces**.
- Electrical shock accidents involve contact with overhead wires, use of defective tools, failure to disconnect power source before repairs, or improper **ground fault** protection.
- Other hazards include suffocation accidents, where a worker is exposed to a suffocation hazard in an enclosed space. In the agricultural industry, suffocation accidents occur in silos or grain bins, where oxygen levels are dangerously low, or the oxygen is contaminated with silo gas or manure gas. Agricultural operations use chemicals, such as herbicides, insecticides, and pesticides. These chemicals present inhalation and exposure risks to workers on agricultural jobsites as well.

Trench: A narrow excavation made below the surface of the ground that is generally deeper than it is wide, with a maximum width of 15' (4.6 m). Also see *excavation*.

Confined spaces: Work areas large enough for a person to work, but arranged in such a way that an employee must physically enter the space to perform work. Confined spaces have a limited or restricted means of entry and exit. They are not designed for continuous work. Tanks, vessels, silos, pits, vaults, and hoppers are examples of confined spaces. Also see *permit-required confined space*.

Ground fault: Incidental grounding of a conducting electrical wire.

1.1.2 Costs

Incidents and accidents cost billions of dollars each year and cause much needless suffering. The National Safety Council estimates that the organized safety movement has saved more than 4.2 million lives since it began in 1913. This section examines why incidents and accidents happen and how you can help prevent them.

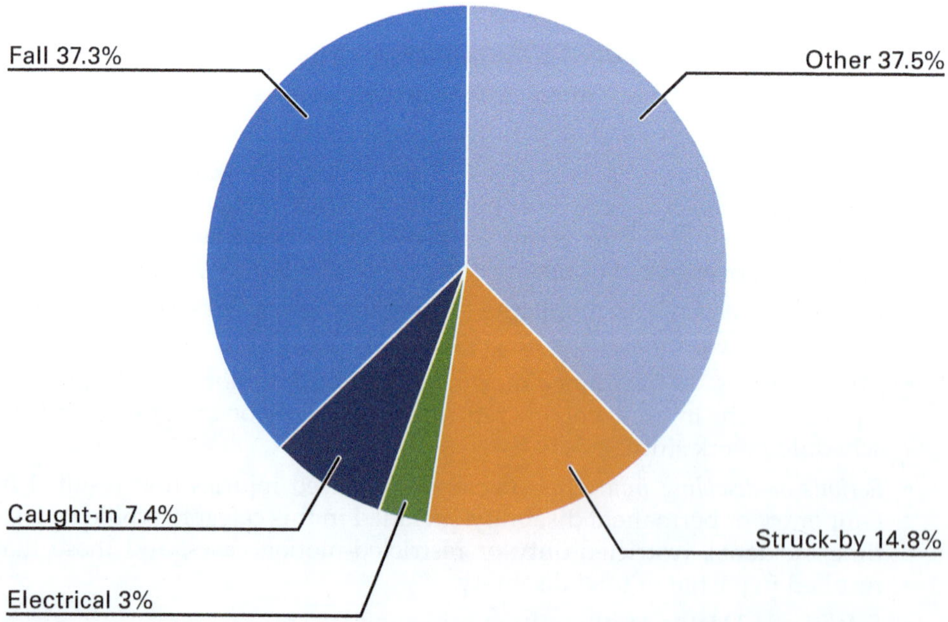

Figure 2 The four high-hazard areas.

Accidents that result in injury or death can have a lasting effect not only on the victims, but on their families, coworkers, and employers. An injured worker who is disabled by an accident faces potentially large medical bills. On top of those expenses, the worker's family faces a loss of income they rely on. It's convenient to think that insurance will take care of the costs, but it may not. Workers who are injured because they violated established safety rules may have their insurance claims denied and may be dismissed from the company because of the violation. A worker who is injured in a fall because he or she was not using fall protection could be refused compensation. How these issues are handled varies dramatically among companies, industries, and countries.

Employers and coworkers can be affected because many contract awards are based, in part, on a company's safety record. In addition, an employer faces lost productivity and a loss in available labor resources. In some agricultural operations, it may even be the loss of a family member. Incidents and accidents can also result in the loss of future jobs, which affects the company's financial position and the surrounding community. A poor financial position means layoffs, hiring freezes, or inability to purchase new equipment or tools. In this way, these events affect not only injured employees and their families, but everyone on the jobsite.

1.2.0 Incident and Accident Causes

When an incident or accident occurs on the jobsite, it can often be attributed to one of the following causes:

- Failure to communicate
- At-risk work habits
- Alcohol or drug abuse
- Lack of skill
- Intentional acts
- Unsafe acts
- Rationalizing risks
- Unsafe conditions
- Poor housekeeping
- **Management system** failure

Each of these causes is discussed further in the sections that follow.

Management system: The organization of a company's management, including reporting procedures, supervisory responsibility, and administration.

The Fatal Four

Out of 4,779 worker fatalities in US private industry during the 2018 calendar year, 21.1% of them (1,008 in total) were in construction. In other words, one out of five worker deaths were in construction. The leading causes of worker deaths on construction sites were falls, followed by struck-by-object, electrocution, and caught-in/caught-between. These fatal four were responsible for nearly three out of five (58.6%) construction worker deaths in 2018 reports, as reported by the US Bureau of Labor Statistics. Overall, a total of 591 workers died in 2018 as a result of the fatal four.

NOTE

The agricultural community is rooted in family and community. This gives the industry its strength, but family members and close associates tend to hand down knowledge and information informally. Agricultural craftworkers gain extensive knowledge by talking to experienced farmers, reviewing machinery with sales representatives, and passing knowledge down through their families.

CAUTION

Never assume anything. It never hurts to ask questions, and disaster can result if you don't ask. For example, do not assume that an electrical power source is turned off. First ask whether the power is turned off, then check it yourself to be completely safe.

1.2.1 Failure to Communicate

Many incidents happen because of a lack of communication. For example, you may learn how to do things one way on one job, but what happens when you go to a new jobsite? You need to communicate with the people at the new jobsite to find out whether they do things the way you have learned to do them. If you do not communicate clearly, incidents can happen. Remember that different people, companies, and jobsites do things in different ways.

Making assumptions about what other workers know and what they will do can cause incidents. Don't assume, for example, that all workers understand what you are saying; some workers have limited language skills, especially outside of their native language. Also, don't use terms or jargon that other people may not understand.

All worksites have specific markings and signs to identify hazards and provide emergency information (*Figure 3*). Learn to recognize these types of signs:

- Informational
- Safety
- Caution
- Danger
- Temporary warnings

Toolbox Talks

Toolbox talks are one way to effectively keep all workers aware and informed of safety issues and guidelines. Toolbox talks are meetings, 5 to 10 minutes in length, that review specific health and safety topics. These are very common at construction sites of all types.

Figure 3 Communication tags and signs.

Informational markings or signs provide general information. These signs are blue. The following are considered informational signs:

- No Admittance
- No Trespassing
- For Employees Only

Safety signs give general instructions and suggestions about safety measures. These signs have white backgrounds; most have a green panel with white letters. These signs tell you where to find or avoid important areas such as the following:

- First aid stations
- Emergency eye wash stations
- Evacuation routes
- **Safety Data Sheet (SDS)** stations
- Exits (usually have white letters on a red field)
- Limited or no entry areas (*Figure 4*)

Safety Data Sheet (SDS): A document that must accompany any hazardous substance. The SDS identifies the substance and gives the exposure limits, the physical and chemical characteristics, the kind of hazard it presents, precautions for safe handling and use, and specific control measures of each substance.

CAUTION

An SDS provides instructions on the hazards associated with handling chemicals, such as herbicides, pesticides, fertilizers, and fungicides. You may encounter these chemicals during construction activities on an agricultural jobsite. Always review the SDS for safe handling, proper storage, appropriate personal protective equipment (PPE), and first aid recommendations in case exposure occurs.

Figure 4 Sign warns workers and the public not to enter an area.
Source: iStock

Caution markings or signs tell you about potential hazards or warn against unsafe acts. When you see a caution sign, protect yourself against a possible hazard. Caution signs are yellow and have a black panel with yellow letters. They may give you the following information:

- Hearing and eye protection are required
- **Respirators** are required
- Smoking is not allowed

Respirators: Devices that provide clean, filtered air for breathing, no matter what is in the surrounding air.

Danger markings or signs tell you that an immediate hazard exists and that you must take certain precautions to avoid an incident. Danger signs are red, black, and white. They may indicate the presence of the following:

- Defective equipment
- **Flammable** liquids and compressed gases
- Safety barriers and barricades
- Emergency stop button
- High voltage (*Figure 5*)

Flammable: Capable of easily igniting and rapidly burning; used to describe a fuel with a flash point below 100°F (38°C).

Half Measures

Most workers who die from falls are wearing harnesses but failed to tie off properly. Always follow the manufacturer's instructions when wearing a harness. Know and follow your company's safety procedures and the safety features of the equipment you are using. Whenever you are working on roofs, ladders, and other elevated locations, make sure you always have an adequate anchor point.

Source: iStock

Figure 5 Sign warning of electrified fence.
Source: iStock

TABLE 1 Tags and Signs

Basic Stock (Background)	Safety Colors (Ink)	Message(s)
White	Red panel with white or gray letters	Do Not Operate Do Not Start
White	Black square with a red oval and white letters	Danger Unsafe Do Not Use
Yellow	Black square with yellow letters	Caution
White	Black square with white letters	Out of Order Do Not Use
Yellow	Red/magenta (purple) panel with black letters and a radiation symbol	Radiation Hazard
White	Fluorescent orange square with black letters and a biohazard symbol	Biological Hazard

Safety tags are temporary warnings of immediate and potential hazards. They are not designed to replace signs or to serve as permanent means of protection. Learn to recognize the standard incident and accident prevention signs and tags (*Table 1*).

1.2.2 At-Risk Work Habits

Some examples of at-risk work habits are procrastination, carelessness, and horseplay. Procrastination (putting things off) is a common cause of incidents. For example, delaying the repair, inspection, or cleaning of equipment and tools can cause incidents and even accidents. If you try to push machines and equipment beyond their operating capacities, you risk injuring yourself and those working around you.

Machines, power tools, and even a pair of pliers can hurt you if you don't use them safely. It is your responsibility to be careful; tools and machines don't know the difference between wood or steel, and flesh and bone.

Your safety is affected not only by how you do your work, but also by how you act on the jobsite. This is why most companies have strict policies for employee behavior. Horseplay and other inappropriate behavior are forbidden. Workers who engage in such behavior on the jobsite may be fired.

CAUTION

Work habits and work attitudes are closely related. If you resist taking orders, you may also resist listening to warnings. If you let yourself be easily distracted, you won't be able to concentrate. If you aren't concentrating, you could cause a significant problem.

Industry-Specific Signs

Some signs are made to warn of local or industry-specific hazards and limitations. These signs can be permanent or temporary, depending on the hazard.

Source: iStock

These strict policies are provided for the worker's protection. There are many hazards on construction sites. Each person's behavior—at work, on a break, or at lunch—must follow the principles of safety.

A person pulling a practical joke on a coworker could consider it just having fun, but in fact, it could cause the coworker serious, even fatal, injury. If you horse around on the job, play pranks, or don't concentrate on what you are doing, you are showing a poor work attitude that can lead to a serious incident.

1.2.3 Alcohol and Drug Abuse

Alcohol and drug abuse costs the construction industry millions of dollars a year in incidents, accidents, lost time, and lost productivity. The true cost of alcohol and drug abuse is much more than just money, of course. Substance abuse can cost lives. Just as drunk driving kills thousands of people on our highways every year, alcohol and drug abuse kills on the construction site.

Using alcohol or drugs creates a risk of injury for everyone on a jobsite. Many states have laws that prevent workers from collecting insurance benefits if they are injured while under the influence of alcohol or illegal drugs.

Would you trust your life to a combine operator who was under the influence of drugs? Would you bet your life on the responses of a coworker under the influence of alcohol or drugs? Alcohol and drug abuse have no place in the workplace. A person on a jobsite who is under the influence of alcohol or drugs is an incident or accident waiting to happen—possibly a fatal accident.

WARNING!

People who work while using alcohol or drugs are at risk of incident or injury; their coworkers are at risk as well. That's why most employers have a formal substance abuse policy that you should be aware of and follow. Avoid any substances that can affect your job performance; the life you save could be your own.

You do not have to be abusing drugs such as marijuana, cocaine, or heroin to create a job hazard. Many prescription and over-the-counter drugs, taken for legitimate reasons, can affect your ability to work safely. Amphetamines, barbiturates, and antihistamines are only a few of the prescription-controlled legal drugs that can affect your ability to work or operate machinery safely. The main thing is to understand and follow your company's substance abuse policy.

Did You Know?

Stress Effects

Stress creates a chemical change in your body. Although stress may heighten your hearing, vision, energy, and strength, long-term stress can harm your health. Not all stress is job-related; some stress develops from the pressures of dealing with family, friends, and daily living. In the end, your ability to handle and manage your stress determines whether stress hurts or helps you. Use common sense when you are dealing with stressful situations. For example, consider the following:

- Keep daily occurrences in perspective. Not everything is worth getting upset, angry, or anxious about.
- When you have a particularly difficult workday scheduled, get plenty of rest the night before.
- Manage your time. The feeling of always being behind creates a lot of stress. Waiting until the last minute to finish an important task adds unnecessary stress.
- Talk to your supervisor. Your supervisor may understand what is causing your stress and may be able to suggest ways to manage it better.

CAUTION

If your doctor prescribes any medication that you think might affect your job performance, ask about its effects. Your safety and the safety of your coworkers depend on everyone being alert on the job.

1.2.4 Lack of Skill

Every worker needs to learn and practice new skills under careful supervision. Never perform new tasks alone until you have been checked out by a supervisor or mentor. Lack of skill can cause incidents and accidents quickly. For example, suppose you are told to cut some boards with a circular saw, but you aren't skilled with that tool. A basic rule of circular saw operation is never to cut without a properly functioning guard. Because you haven't been trained, you don't know this. You find that the guard on the saw is slowing you down, so you jam the guard open. The result could be a serious accident. Proper training can prevent this from happening.

Never operate a power tool or machine until you have been trained to use it. Some power tools require that a worker be trained and certified in their use. You can greatly reduce the chances of incidents and accidents by learning the safety rules for each task you perform.

1.2.5 Intentional Acts

When someone purposely causes property damage or injury, it is called an intentional act. An angry or dissatisfied employee may purposely create a situation that leads to an incident or accident. If someone you are working with threatens to get even or pay back someone, let your supervisor know at once.

Unfortunately, terrorism must also be a consideration on the jobsite. The level of concern is typically dependent on the site itself and the nature of the structure. Pipelines, for example, may represent a significant target due to their size, importance, location (sections of pipeline are in remote areas), and potential for destruction. Controversial sites also tend to attract attention, perhaps from terrorist groups operating within a country that oppose the construction or what the structure represents.

As an individual worker, the most important thing you can do is to pay attention. Become familiar with people on the jobsite and look for strangers that do not seem to fit. A terrorist attack can come from any direction and be delivered in many different ways. It can begin with a single individual gaining access to the jobsite.

1.2.6 Unsafe Acts

An unsafe act is a change from an accepted, normal, or correct procedure that can cause an incident or accident. It can be any conduct that causes unnecessary exposure to a jobsite hazard or that makes an activity less safe than usual. Here are examples of unsafe acts:

Personal protective equipment (PPE): Equipment or clothing designed to prevent or reduce injuries.

- Failing to use required **personal protective equipment (PPE)**
- Failing to warn coworkers of hazards
- Lifting improperly
- Loading or placing equipment or supplies improperly
- Making safety devices (such as saw guards) inoperable
- Operating equipment at improper speeds
- Operating equipment without authority
- Servicing equipment in motion
- Taking an improper working position
- Using defective equipment
- Using equipment improperly

1.2.7 Rationalizing Risk

Everybody takes risks every day. When you get in your car to drive to work, you know there is a risk of being involved in an accident. Yet when you drive using all the safety practices you have learned, you know that there is a very good chance you will arrive at your destination safely. Driving is an acceptable risk because you have some control over your own safety and that of others.

Some risks are not acceptable. On the job, you must never take risks that endanger yourself or others just because you think you can justify doing so. This is called rationalizing risk, which means ignoring safety warnings and practices. For example, because you are late for work, you might decide to run a red light. Perhaps you feel the risk is worth the time saved. This is not only a dangerous driving habit, but this kind of thinking is unacceptable on the jobsite.

The following are common examples of rationalized risks on the job:

- Crossing barricades or boundaries because there is no apparent activity
- Not wearing gloves because it's easier to do the job with bare hands
- Removing your hard hat because you are hot and you cannot see anyone working overhead
- Not tying off your fall protection because you only have to lean over by about a foot

Think about the job before you do it. If you think that it is unsafe, then it is unsafe. Stop working until the job can be done safely. Bring your concerns to the attention of your supervisor. Your health and safety, and that of your coworkers, make it worth taking extra care.

Self-Care

Practicing good self-care not only benefits you personally, but also increases your chances of a long and successful construction career. Self-care strategies can be done in simple, cost-effective ways and include the following:

- *Quality sleep* — Sleep deprivation negatively impacts every system in your body and can result in significant errors in judgment, reduced ability to safely operate machinery, and increased risk of becoming distracted. These consequences are extremely dangerous on a jobsite. Quality sleep is enhanced by other kinds of self-care and limiting screen time before bed.
- *Healthy eating* — The high energy levels and physical strength required by construction work are supported by a healthy diet filled with lean proteins, vegetables, fruits, and whole grains. If your grocery store has limited fresh produce, look for plain frozen fruits and vegetables. Dozens of apps and blogs exist to assist with budgeting and recipes. Hydration, especially in hot weather, is also critical.
- *Exercise* — In addition to supporting the physical requirements of construction work, regular exercise aids in regulating sleep and reducing stress. Many workouts can be done at home using just your body weight.
- *Massage therapy* — Regular massage, particularly for the upper back, shoulder, and neck, keeps your muscles and connective tissue flexible, which helps prevent and heal injuries.
- *Chiropractic care* — Many people who suffer from chronic low back pain benefit from chiropractic care, which involves the manipulation of joints with a focus on the spine.
- *Acupuncture* — This ancient practice has become increasingly popular as a minimally invasive, drug-free way of addressing chronic pain and stress management.

Most schools offering massage therapy, acupuncture, or chiropractic programs have student clinics that provide these services at a heavily discounted rate. Insurance also covers many self-care strategies.

Make self-care a priority as you embark on your career path. Your long-term health and career longevity depend on it.

1.2.8 Unsafe Conditions

An unsafe condition is a physical state that is different from the acceptable, normal, or correct condition found on the jobsite. It usually results in an incident or accident. An unsafe condition can be anything that reduces the degree of safety normally present. The following are some examples of unsafe conditions:

- Congested workplace
- Defective tools, equipment, or supplies
- Excessive noise
- Fire and explosive hazards
- Hazardous atmospheric conditions (such as gases, dusts, fumes, and vapors)
- Inadequate supports or guards
- Inadequate warning systems
- Cluttered work area
- Poor lighting
- Poor ventilation
- Radiation exposure
- Unguarded moving parts such as pulleys, drive chains, and belts

All employees should be given the authority to stop work when an unsafe condition is observed.

1.2.9 Poor Housekeeping

Housekeeping means keeping your work area clean and free of scraps, clutter, or spills. It also means being orderly and organized. Store your materials and supplies safely and label them properly. Arranging your tools and equipment to permit safe, efficient work practices and easy cleaning is also important.

If the worksite is indoors, make sure it is well-lit and ventilated. Don't allow aisles and exits to be blocked by materials and equipment. Make sure that flammable liquids are stored in safety cans. Oily rags must be placed only in approved, self-closing metal containers (*Figure 6*).

Figure 6 Container for oily rags and waste.
Source: Courtesy of Justrite Safety Group

Remember that the major goal of housekeeping is to prevent incidents. Good housekeeping reduces the chances for slips, fires, explosions, and falling objects. Here are some good housekeeping rules:

- Remove all scrap material and lumber with nails protruding from work areas.
- Clean up spills to prevent falls.
- Remove all **combustible** scrap materials regularly.
- Make sure you have containers for the collection and separation of refuse. Containers for flammable or harmful refuse must have covers.
- Keep floors clean and clear of waste and combustible materials, like straw or hay.
- Dispose of waste often.
- Store all tools and equipment in the proper location and condition when you are finished with them.
- Keep all aisles and walkways clear of materials and tools.
- Provide adequate ventilation when working in livestock structures to prevent the buildup of toxic fumes and humidity.

Another term for good housekeeping is *pride of workmanship*. If you take pride in what you are doing, you won't let trash build up around you. The saying "a place for everything and everything in its place" is the right attitude on the jobsite.

Combustible: Capable of easily igniting and rapidly burning; used to describe a fuel with a flash point at or above 100°F (38°C).

Tool Blades

Dull blades cause more accidents than sharp ones. If you do not keep your cutting tools sharpened, they won't cut very easily. When you have a hard time cutting, you exert more force on the tool. When that happens, something is bound to slip. And when something slips, you can be injured.

1.2.10 Management System Failure

Sometimes the cause of an incident or accident is failure of the management system. The management system should be designed to prevent or correct the acts and conditions that can create safety hazards. If the safety management system is not functioning properly, problems are likely to occur.

What traits could mean the difference between a management system that fails and one that succeeds? A company or operation implementing a good management system will do the following:

- Put safety policies and procedures in writing.
- Distribute written safety policies and procedures to each employee.
- Review safety policies and procedures periodically.
- Enforce all safety policies and procedures fairly and consistently.
- Evaluate supplies, equipment, and services to see whether they are safe.
- Provide regular, periodic safety training for employees.
- Prioritize investments in health and safety.

1.3.0 Hazard Recognition, Evaluation, and Control

The process of hazard recognition, evaluation, and control is the foundation of an effective safety program. When hazards are identified and assessed, they can be addressed quickly, reducing the hazard potential. Simply put, the more aware you are of your surroundings and the dangers in them, the less likely you are to be involved in an incident.

There is a standard rule in the United States that affects every worker in most industries. It is referred to as the **Hazard Communication Standard (HCS)**. The US HCS also aligns with the Globally Harmonized System of Classification and Labeling of Chemicals (GHS). HCS may also be called the Right to Know requirement. It requires all US contractors to educate their employees about the hazardous chemicals they may be exposed to on the jobsite. Employees must be taught how to work safely around these materials. Other countries have similar programs.

Hazard Communication Standard (HCS): The Occupational Safety and Health Administration standard that requires contractors to educate employees about hazardous chemicals on the jobsite and how to work with them safely.

Many people think that there are very few hazardous chemicals on construction or agricultural jobsites. That is simply not true. In practice, the term *hazardous chemical* applies to solvents, paint, concrete, and even wood dust, along with more obvious substances such as acids.

As an employee, you have the following responsibilities under HCS:

- Know where SDSs are on your jobsite.
- Report any hazards you spot on the jobsite to your supervisor.
- Know the physical and health hazards of any hazardous materials on your jobsite, and know and practice the precautions needed to protect yourself from these hazards.
- Know what to do in an emergency.
- Know the location and content of your employer's written hazard communication program.

The final responsibility for your safety rests with you. Your employer must provide you with information about hazards, but you must know this information and follow safety rules.

1.3.1 Hazard Recognition

There are many potential hazard indicators. The best approach in determining if a situation or equipment is potentially hazardous is to ask these questions:

- How can this situation or equipment cause harm?
- What types of energy sources are present that can cause an incident?
- What is the magnitude of the energy?
- What could go wrong to release the energy?
- How can the energy be eliminated or controlled?
- Will I be exposed to any hazardous materials?
- Are there chemicals being used in the surrounding area?

Before you can fully answer these questions, you need to know the different types of incidents and accidents that can happen and the energy sources behind them. Some of the different types of events that can cause injuries include the following:

NOTE

You might recognize the first four high-hazard conditions as the fatal four previously discussed. Remember, these four types of accidents cause more than half of all construction fatalities.

- Falls on the same elevations or falls from elevations
- Being caught in, on, or between equipment
- Being struck by falling objects
- Contact with acid, electricity, heat, cold, radiation, pressurized liquid, gas, or toxic substances
- Being cut by tools or equipment
- Exposure to high noise levels
- Repetitive motion or excessive vibration
- Inhalation of toxic vapors, fumes, or sprays

When equipment is the cause of an incident or accident, it is usually because there was an uncontrolled release of energy. The different types of energy sources that can be released include the following:

Pneumatic: Powered by air pressure such as a pneumatic tool.

Hydraulic: Powered by fluid under pressure.

- Mechanical
- **Pneumatic**
- **Hydraulic**
- Electrical
- Chemical
- Thermal (heat or cold)
- Radioactive
- Gravitational
- Stored energy

There are several ways to recognize hazards and potential hazards on a jobsite. Some techniques are more complicated than others. To be effective, they all must answer this question: What could go wrong with this situation or operation? No matter what hazard recognition technique you use, answering that question in advance will save lives and prevent equipment damage.

1.3.2 Job Safety Analysis (JSA) and Task Safety Analysis (TSA)

Performing a job safety analysis (JSA), also known as *job hazard analysis (JHA)*, is one approach to hazard recognition. Another common technique is performing a task safety analysis (TSA), also called a *task hazard analysis (THA)*.

In a JSA, the task at hand is broken down into its individual parts or steps and then each step is analyzed for its potential hazards. Once a hazard is identified, certain actions or procedures are recommended that will correct that hazard. For example, during a JSA, it is determined that using a chain hoist to install a pump motor in a tight space would be safer than having a worker do it manually. By using the chain hoist, the chance that the worker's hand would get crushed during installation is reduced. Using the JSA process saved the worker from injury. *Figure 7* shows an example of a form used to conduct a JSA.

JSAs can also be used as pre-planning tools. This helps to ensure that safety is planned into the job. You may be asked to take part in a JSA during job planning. When JSAs are used as preplanning tools, they contain the following information:

- Tools, materials, and equipment needs
- Staffing or manpower requirements
- Duration of the job
- Quality concerns

Task safety analysis is similar to job safety analysis in that both require workers to identify potential hazards and needed safeguards associated with a job they are about to do. The difference is the form used to report the hazards. During a TSA, a fill-in-the-blank checklist or electronic form is often used to document any hazard found during analysis. Before work begins, the first-line supervisor or team leader should discuss the conclusions found during the TSA with the crew. Some companies require workers to sign the completed TSA forms or checklists before they start work (*Figure 8*). This helps companies document the hazards and ensure that workers have been told of the potential hazards and safety procedures.

Job Safety Analysis

Title of Job or Task

Task	Start	End	Hazards	Controls
1.				
2.				
3.				
4.				
5.				
6.				
7.				
8.				

Required Training: Required Personal Protective Equipment (PPE):

Job Name: ______

Job Number: ______

Supervisor: ______

Date: ______

Weekly Vehicle Checklist: ___ Tire Pressure ___ Transmission Fluid ___ Oil ___ Lights ___ Air Filter ___ Weekly Mileage

Name of Employees:

Print Name	Sign Name	Total Hours

Figure 7 Job safety analysis form.

Figure 8 A supervisor reviews the TSA for a combine harvester.
Source: iStock

1.3.3 Risk Assessment

Whether an action is considered safe is often a matter of evaluating risk. Risk is a measure of the probability, consequences, and exposure related to an event. Probability is the chance that a given event will occur. Consequences are the results of an action, condition, or event. Exposure is the amount of time and/or the degree to which someone or something is exposed to an unsafe condition, material, or environment.

A safe operation is one in which there is an acceptable level of risk. This means there is a low probability of an incident and that the consequences and exposure risk are all acceptable. For example, climbing a ladder has risk that is considered to be acceptable if the proper ladder is being used as intended, if it is set up correctly, and if it is in good condition. The probability of exposure to a hazard and its potential consequences are all low. If any one of these conditions were different, climbing the ladder would have an unacceptable level of risk.

1.3.4 Reporting Injuries, Incidents, and Near-Misses

All on-the-job incidents and accidents, no matter how minor, must be reported to your supervisor or jobsite manager. Some workers think they will get in trouble if they report minor injuries, but that is not the case. Small injuries, like cuts and scrapes, can later become big problems because of infection and other complications.

NOTE

When a farm has 10 or more nonfamily workers, they are subject to OSHA regulation, but only 9% of farms employ 10 or more workers. Regardless, OSHA recommends all farm owners, operators and managers follow the recommended safety practices provided for the health and safety of all workers.

US employers with more than 10 employees are required to maintain a log of significant work-related injuries and illnesses using specific OSHA forms and documents. Employee names can be kept confidential in certain circumstances. A summary of these injuries must be posted at certain intervals, although employers do not need to submit it to OSHA unless requested. Employers can calculate the total number of injuries and illnesses and compare the result with the average national rates for similar companies. By analyzing incidents and accidents, companies and OSHA can improve safety policies and procedures. By reporting an incident, you can help keep similar events from happening in the future. For details on the operation of OSHA and its important mission, refer to *Appendix 93102A*.

1.3.5 Safety Data Sheets

Safety Data Sheets (SDSs) are fact sheets prepared by the chemical manufacturer or importer. Each product used on a construction site must have an SDS. An SDS describes the substance, along with its hazards, safe handling, first aid, and emergency spill procedures. The HCS requires new SDSs (formerly known as material safety data sheets or MSDSs) to be in a standardized format. The sections of the form include the following:

- *Section 1* — Product identification, manufacturer contact information, recommended uses and restrictions
- *Section 2* — Hazard identification
- *Section 3* — Composition/information on ingredients
- *Section 4* — First aid measures
- *Section 5* — Firefighting measures
- *Section 6* — Accidental release measures
- *Section 7* — Handling and storage
- *Section 8* — Exposure controls/personal protection
- *Section 9* — Physical and chemical properties
- *Section 10* — Stability and reactivity
- *Section 11* — Toxicological information
- *Section 12* — Ecological information
- *Section 13* — Waste disposal considerations
- *Section 14* — Transport information
- *Section 15* — Regulatory information
- *Section 16* — Other information

Appendix 93102B shows a sample SDS for regular bleach. The most important things to look for on an SDS are the specific hazards, personal protection, handling procedures, and first aid information. Most SDSs have a 24-hour emergency response number.

Using *Appendix 93102B*, try to find the information you would need to use the bleach described on the sample SDS. First, locate the hazards; Section 2 of the SDS shows that the adhesive is flammable, and is an eye and skin irritant that can cause respiratory irritation, dizziness, and drowsiness.

Section 4 lists the first aid measures for eye contact, skin contact, or inhalation. Section 5 explains fire hazards and firefighting measures. Now you have the information you need in case of an emergency.

Next, find out how to minimize these hazards. Section 7 gives general handling and storage information. It indicates when ventilation is needed to reduce hazardous vapors, which can be as simple as a fan in an open window. If ventilation is not enough, respiratory protection is needed. Section 8 tells you how to protect your eyes and skin.

Remember to familiarize yourself with the safe handling procedures not only for all chemicals and materials being used for the construction activity, but also for chemicals and materials that are at the jobsite. Each industry uses chemicals that are unique to the operation and the agricultural industry is no different. Agricultural chemicals can present hazards at the jobsite. You might need to move them away from the construction site or dispose of containers that were previously contaminated with chemicals, such as pesticides, herbicides, or fungicides. If these chemicals must be moved, consult the SDS to ensure that proper handling, labeling, and storage requirements are followed. The SDS also provides clean up requirements, which might include rinsing containers with a soap and water solution or triple rinsing with a liquid capable of dissolving the residue left by chemicals previously used in the container.

WARNING!

Exposure to chemicals, such as herbicides, fertilizers, pesticides, and insecticides can cause acute reactions, such as stinging eyes, blisters, and dizziness. Long-term exposure can also lead to chronic adverse health effects. Always wear appropriate PPE, consult the SDS for proper handing, and follow first aid recommendations if exposed.

Competent person: A worker who can identify hazards and take the corrective action necessary to address them.

The SDSs must be kept in the work area and be readily accessible to all workers. The company's safety officer or **competent person** should review the SDS before a hazardous material is used. Ask your supervisor to point out where the SDSs are located if you are not sure. Have him or her point out the sections that relate to your job; the health and safety of you and your coworkers may depend on it.

1.0.0 Section Review

1. The person primarily responsible for your safety is _____.
   a. your foreman
   b. your instructor
   c. yourself
   d. your employer

2. The color commonly used for informational signs is _____.
   a. green
   b. red
   c. yellow
   d. blue

3. The SDS for any chemical used at a jobsite *must* be available _____.
   a. at the jobsite
   b. online
   c. at the contractor's office
   d. at the nearest hospital

2.0.0 Elevated Work and Fall Protection

Objective

Describe common fall hazards and methods to prevent them.

a. Summarize the most common types of construction fall hazards.
b. Describe components of effective fall arrest systems and how they prevent or halt falls.
c. Explain how to use ladders and stairs safely.
d. Identify key steps to ensuring scaffolds are assembled and used safely.

Performance Tasks

1. Properly set up and climb/descend an extension ladder, demonstrating proper three-point contact.
2. Inspect the following PPE items and determine if they are safe to use:
   - Fall arrest harnesses
   - Lanyards
   - Connecting devices
3. Properly don, fit, and remove the following PPE item:
   - Fall arrest harness

Falls from elevated areas are the leading cause of fatalities in the workplace. Falls from elevated heights account for about one-third of all deaths in the construction trade. Approximately 85% of injured workers lose time from work, approximately 33% require hospitalization, and some never return to the job.

While the risk of falls is high in construction, agricultural operations, and some other trades, there are many things that workers can do to safeguard themselves. Using the appropriate PPE, following proper safety procedures, practicing good housekeeping habits, and staying alert at all times will help you stay safe when working at an elevation. Employers are required to provide for both fall prevention and fall arrest, and to make sure workers are trained and certified in the use of fall protection equipment. Fall prevention consists of covered floor openings, climbing aids, barricades, and guardrails that are designed to protect against falls. Fall arrest consists of equipment such as body harnesses, **lanyards**, connection devices, lifelines, and safety nets that are intended to protect a worker in case a fall occurs. All these topics are covered in this section.

Lanyards: Short sections of rope or strap, one end of which is attached to a worker's safety harness and the other to a strong anchor point above the work area.

2.1.0 Fall Types and Hazards

Falls are classified into two groups: falls from an elevation and falls from the same level. Falls from an elevation can happen during work from **scaffolds**, work platforms, decking, concrete forms, ladders, stairs, and work near **excavations**. Falls from an elevation often result in death unless the fall is arrested. Falls on the same level are usually caused by tripping or slipping. Sharp edges and pointed objects, such as exposed concrete reinforcing bars (rebar), could cut and otherwise harm a worker. Other bodily injuries are also common results of tripping or slipping.

In the United States, fall protection is required for platforms or work surfaces with unprotected sides or edges that are 6' (roughly 2 m) above the ground (*Figure 9*). This is commonly referred to as the **six-foot rule**. However, some international regulations and company policies may require fall protection for heights less than 6'.

Scaffolds: Elevated platforms for workers and materials.

Excavations: Any man-made cuts, cavities, trenches, or depressions in an earth surface, formed by removing earth. They can be made for anything from basements to highways. Also see *trench*.

Six-foot rule: A rule stating that platforms or work surfaces with unprotected sides or edges that are 6' (roughly 2 m) higher than the ground or level below it require fall protection.

Figure 9 A worker wears fall protection while standing on a grain silo platform.
Source: iStock

2.1.1 Walking and Working Surfaces

Slips, trips, and falls on walking and working surfaces cause 15% of all incidental deaths in the construction industry. Some incidents occur due to environmental conditions, such as snow, ice, or wet surfaces. Others happen because of poor housekeeping and careless behavior, such as leaving tools, materials, and equipment out and unattended. You can avoid slips, trips, and falls by being aware of your surroundings and following the rules on your site. Remember these general walking and working surface guidelines to avoid incidents:

- Keep all walking and working areas clean and dry. If you see a spill or ice patch, clean it up, or barricade the area until it can be properly attended to.
- Keep all walking and working surfaces clear of clutter and debris.
- Run cables, extension cords, and hoses overhead or through crossover plates so that they will not become tripping hazards.
- Do not run on scaffolds, work platforms, decking, roofs, or other elevated work areas.

2.1.2 Unprotected Sides, Wall Openings, and Floor Holes

Any opening in a wall or floor is a safety hazard. There are two types of protection for these openings: **guarded** or covered. Cover any hole in the floor when possible. Hole covers must be clearly marked. When it is not practical to cover the hole, use barricades. If the bottom edge of a wall opening is less than 39" (1 m) above the floor and would allow someone to fall 6' (roughly 2 m) or more, then place guards around the opening.

Guarded: Enclosed, fenced, covered, or otherwise protected by barriers, rails, covers, or platforms to prevent dangerous contact.

The types of barriers and barricades used vary from one jobsite to another. There may also be different procedures for when and how barricades are put up. Learn and follow the policies at your jobsite. Several different types of guards are commonly used:

- Railings are used across wall openings or as a barrier around floor openings to prevent falls (*Figure 10* [A]).
- Warning barricades alert workers to hazards but provide no real protection (*Figure 10* [B]). Warning barricades can be made of plastic tape or rope strung from wire or between posts. The tape or rope is color coded as follows:
  - *Red means danger.* No one may enter an area with a red warning barricade. A red barricade is used when there is danger from falling objects or when a load is suspended over an area.
  - *Yellow means caution.* You may enter an area with a yellow barricade, but be sure you know what the hazard is, and be careful. Yellow barricades are used around wet areas or areas containing loose dust. Yellow with black lettering warns of physical hazards such as bumping into something, stumbling, or falling.
  - *Yellow and purple together mean radiation warning.* No one may pass a yellow and purple barricade without authorization, training, and the appropriate PPE. These barricades are often used where piping welds are being X-rayed.
- Protective barricades give both a visual warning and protection from injury (*Figure 10* [C]). They can be wooden posts and rails, posts and chain, or steel cable. People should not be able to get past protective barricades.
- Blinking lights are placed on barricades so they can be seen at night (*Figure 10* [D]).
- Hole covers are used to cover open holes in a floor or in the ground (*Figure 10* [E]).

WARNING!

Never remove a barricade unless you have been authorized to do so. Follow your employer's procedures for putting up and removing barricades.

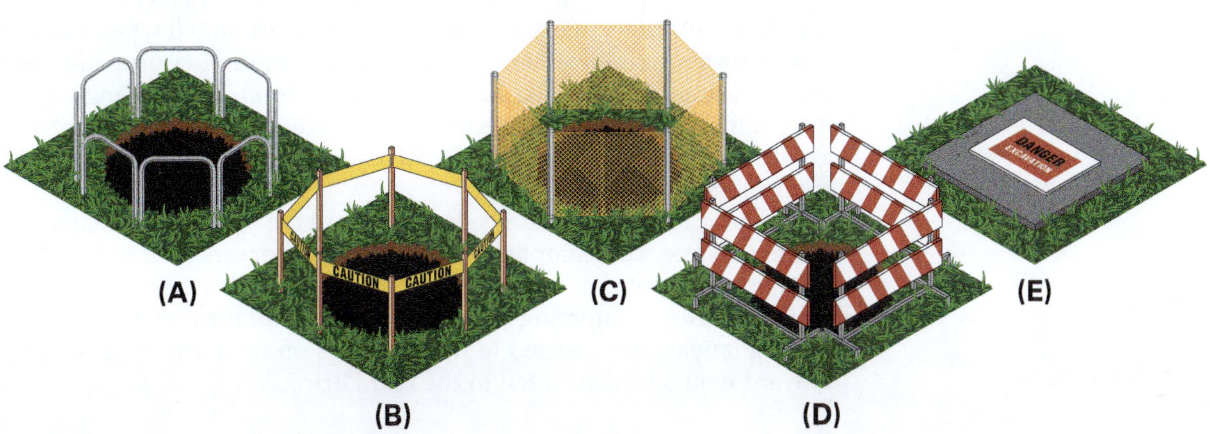

Figure 10 Common types of barriers and barricades.

Follow these guidelines when working near unprotected sides, floor holes, and wall openings:

- Hole covers must be cleated, wired, or otherwise secured to prevent them from slipping sideways or horizontally beyond the hole.
- Covers must extend adequately beyond the edge of the hole.
- Hole covers must be strong enough to support twice the weight of anything that may be placed on top of them. They must also be clearly marked. Use $^{3}/_{4}$" (2 cm) plywood as a hole cover, provided that one dimension of the opening is less than 18" (46 cm); otherwise, 2" (5 cm) lumber is required.
- Never store material or equipment on a hole cover.
- Guard all stairway floor openings, except for the entrance, with standard railing and **toeboards**.
- Guard all wall openings from which there is a drop of more than 6' (roughly 2 m) and for which the bottom of the opening is less than 39" (1 m) above the working surface.
- Guard all open-sided floors and platforms 6' (roughly 2 m) or more above adjacent floor or ground level using a standard railing or the equivalent.

Toeboards: Vertical barriers at floor level attached along exposed edges of a platform, runway, or ramp to prevent materials and people from falling.

2.2.0 Fall Arrest

NOTE

This training alone does not provide any level of certification in the use of fall arrest or fall restraint equipment. Trainees should not assume that the knowledge gained in this module is sufficient to certify them to use fall arrest equipment in the field.

The key to preventing serious injury or death should a fall occur is the personal fall arrest system (PFAS). A complete PFAS (*Figure 11*) consists of anchor points, a body harness, and connecting devices. Anchor points are related to the structure, and the type and availability of anchor points help determine what other equipment should be chosen. The body harness comprises the system of belts, rings, or hooks worn by the worker. Connecting devices and lanyards are used to maintain attachments between anchor points and the PFAS.

2.2.1 Anchor Points

There are both permanent anchor points and temporary, reusable anchor points. Anchor points must be rated at or equal to 5,000 lb (2,268 kg) breaking or tensile strength, or twice the intended load. Anchorage points should be permanently engineered and certified or specifically manufactured as temporary anchor points (wire rope slings, anchor straps) and attached to a strong, secure anchorage structure (*Figure 12*).

Ideally, the fall arrest anchor point will be located directly above the back D-ring (*Figure 13*) to minimize any swing zone hazards, as well as the free fall distance. The maximum free fall distance is 6' (roughly 2 m). To understand swing zones, picture a tied-off worker standing 3' (1 m) away from a point directly under the anchor point. If the worker falls, gravity will cause their body to swing toward the anchor point axis and momentum will cause the body to swing past that point. If there is a wall or other solid object close by, the worker may strike it and be injured. Swing zones are minimized when the anchor point is directly above the worker when he falls. Serious injury and damage can occur when a human body strikes an immovable object while swinging as a pendulum. Although the PFAS may do its job by preventing the worker from falling a great distance, serious injury or death can still occur by striking an object in the swing zone.

Anchor points are often needed to secure the position of the worker, leaving the hands free to accomplish a task. Ideally, workers will select anchor points to maintain a potential fall distance of no more than 2' (61 cm). For example, a worker placing reinforcing bars in a concrete wall form may use two short positioning lanyards connected to the hip rings on the harness. The fall protection lanyard would be connected to the rear D-ring.

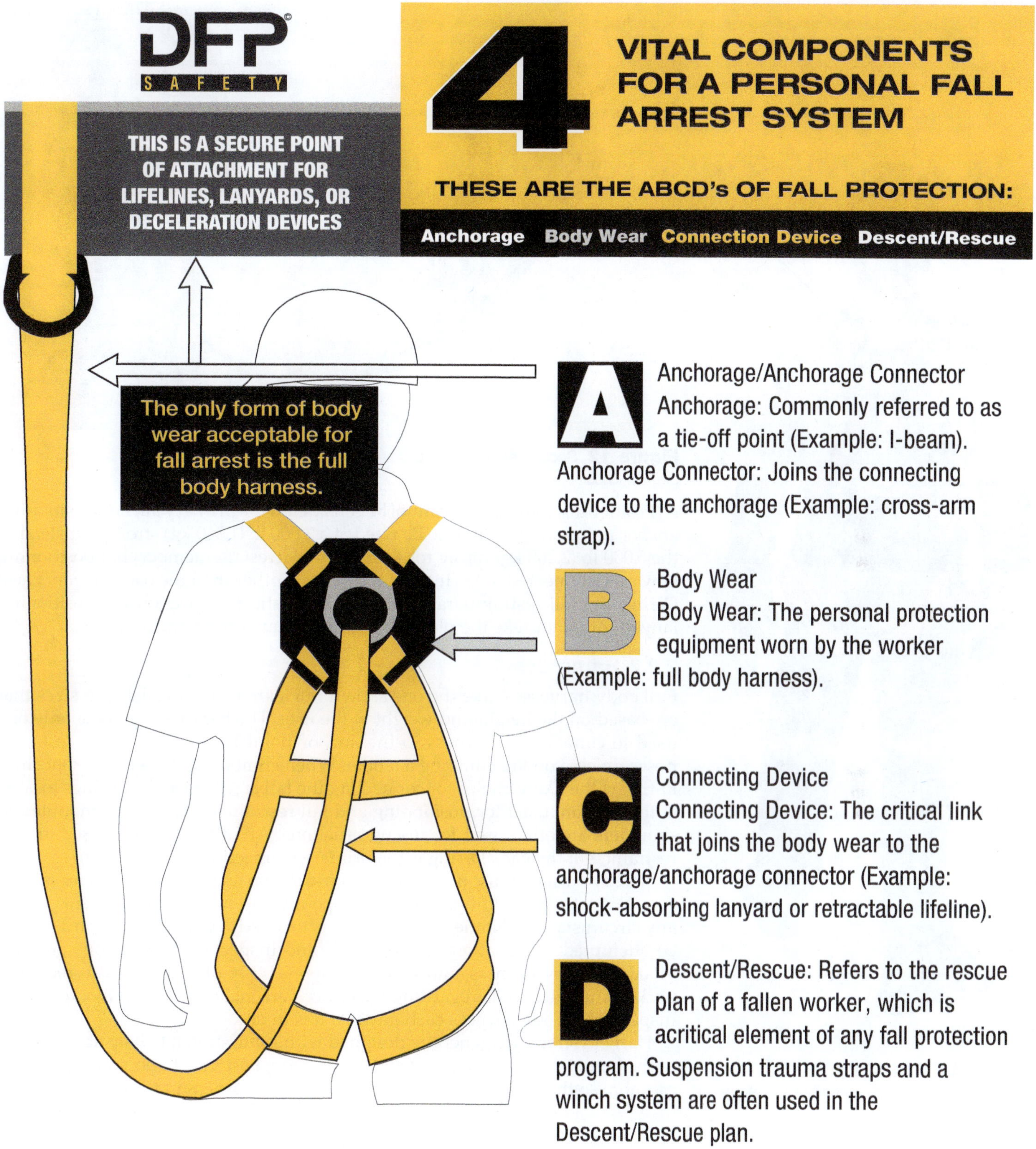

Figure 11 Personal fall arrest system.
Source: Image property of Stanley Black & Decker. Used with permission.

Figure 12 Secure anchor point.
Source: iStock

Figure 13 Back D-ring.
Source: Used with permission from MSA

Positioning connections cannot be considered the primary anchor. Positioning anchor points are required to be rated at a 3,000 lb (1,360 kg) strength instead of the 5,000 lb (2,268 kg) rating for primary fall arrest. Remember that positioning lanyards, connected to D-rings on the harness other than the back or front chest D-ring, are fall restraints rather than fall arresting connections. A positioning lanyard does not take the place of a fall arrest lanyard or anchor point.

2.2.2 Harnesses

Full body harnesses, like the one shown in *Figure 14*, are available in sizes that are based on the height and weight of the user. The back D-ring is the only one used to connect the harness to the anchor point for primary fall arrest purposes unless you are climbing a ladder. When climbing a ladder, the front chest D-ring is the likely choice for connecting the fall arrest harness. D-rings located at the hips are used for positioning and fall restraint only. D-rings mounted to shoulders are often used for rescue situations. All of them can be used for fall restraint, but the back D-ring is the primary connection for fall arrest.

A body harness must fit correctly to ensure that it will provide proper protection. Do not place additional holes or openings in harness components under any circumstances. No field modifications to a body harness or lanyard should be attempted. Installation, maintenance, and inspection instructions are provided for every harness, and it is the responsibility of the worker to read and understand the details regarding his or her personal equipment. Workers must inspect their body harness each day that it is in use.

Harness straps are generally designed with some stretch to help absorb some of the potential force of a fall. This means good, taut installation on the body is essential so that the worker cannot fall out of the harness in the event of a fall.

Figure 14 Full body harness with belt support.
Source: Image property of Stanley Black & Decker. Used with permission.

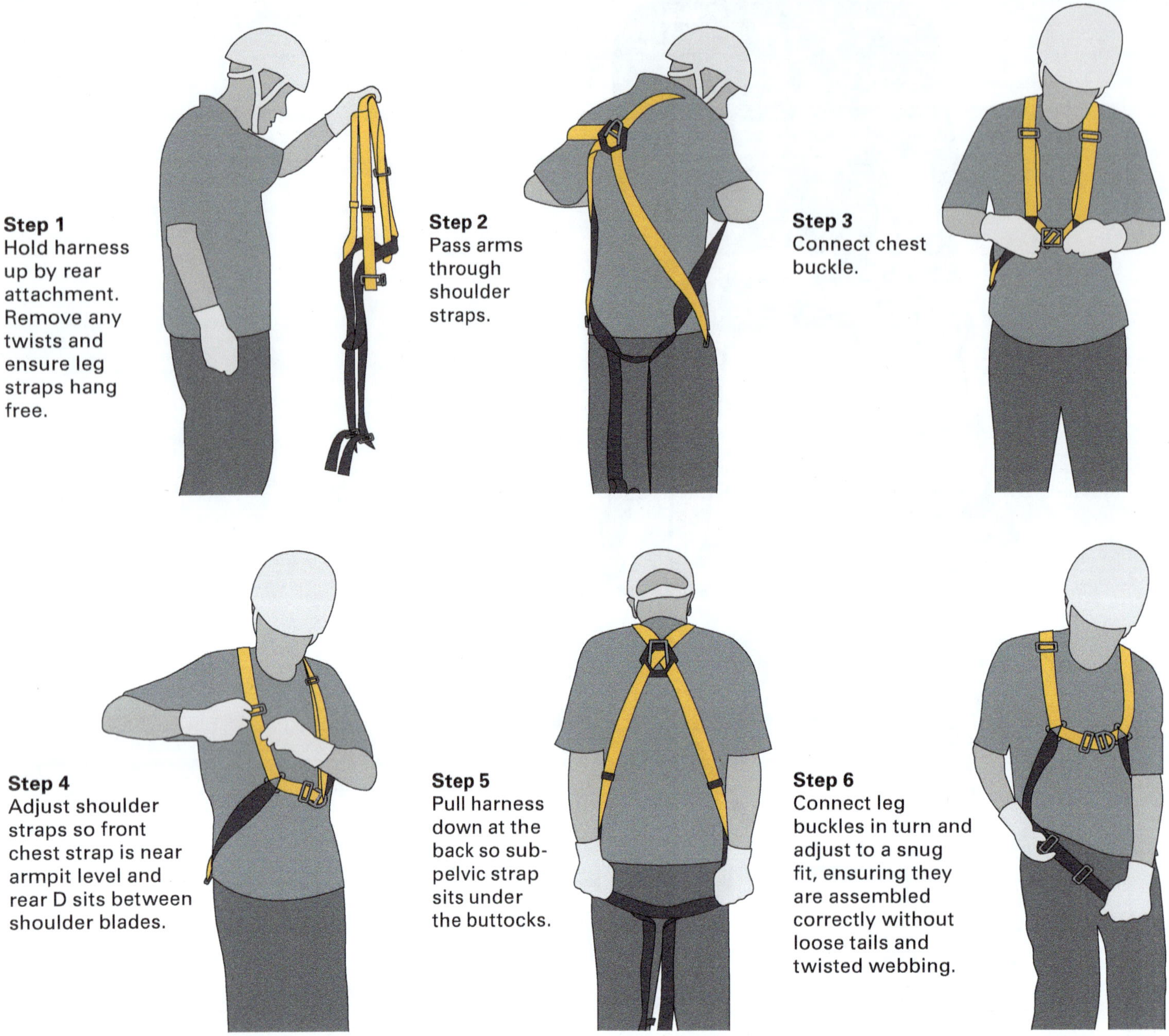

Figure 15 Installing the body harness.
Source: Provided by RIDGEGEAR Ltd.

NOTE

These guidelines are general in nature and the instructions provided for specific equipment by the manufacturer must always take precedence. It is the worker's responsibility to be intimately familiar with the duty of each ring and strap on a given harness.

Figure 15 shows the proper procedure for donning a common full body harness. The most important adjustments to be made include the chest straps, the groin straps, and the final position of the back D-ring. The related details that follow must be considered during the fitting and wearing of a full body harness:

- The back D-ring location is vital to proper fall arrest. Position this ring between the shoulder blades. If it is too low, it will tend to cause the body to hang in a more horizontal position during fall arrest, increasing pressure on the diaphragm and affecting breathing. If it is positioned too high or with too much slack, the D-ring may strike the worker's head at the base of the fall, and the shoulder straps may be pulled too tightly into the neck and restrict blood flow. The impact at the base of the fall arrest can be dramatic, so the force must be spread all around the body to prevent injury to any one portion.

- Chest straps generally form either an H-pattern or an X-pattern. Adjust H-pattern straps to land between the bottom of the sternum and the belly button. This helps ensure the horizontal portion of the H does not contact the throat during a fall, choking the worker. Some harness designs may not allow for this adjustment, with the final position of the H being based solely on a properly sized harness.
- The position of the chest straps is also crucial for X-pattern harnesses. Position the X at or just below the sternum.
- Leg straps are an integral and required part of a PFAS. Adjust the groin straps for a good, snug fit. Too much slack here will cause extreme discomfort in a fall when the impact snatches them up tight and you are left suspended this way.
- A suspension trauma strap (*Figure 16*) is recommended as part of the PFAS gear. The suspension trauma strap is stored in a pouch connected to the harness within easy reach. This is done by either one end of the strap being permanently sewn to the harness (by the manufacturer), one end of the strap attached to the harness with a carabiner (*Figure 17*), or by choking the pouch around a harness strap or hip D-ring. The strap can then be quickly removed and used without any possibility of the user dropping it. Once connected, the strap allows the worker to stand up in the harness, relieving suspended weight and pressure from the hips and groin. This helps open the path for blood flow from the legs back to the heart, preventing blood from pooling in the lower extremities.

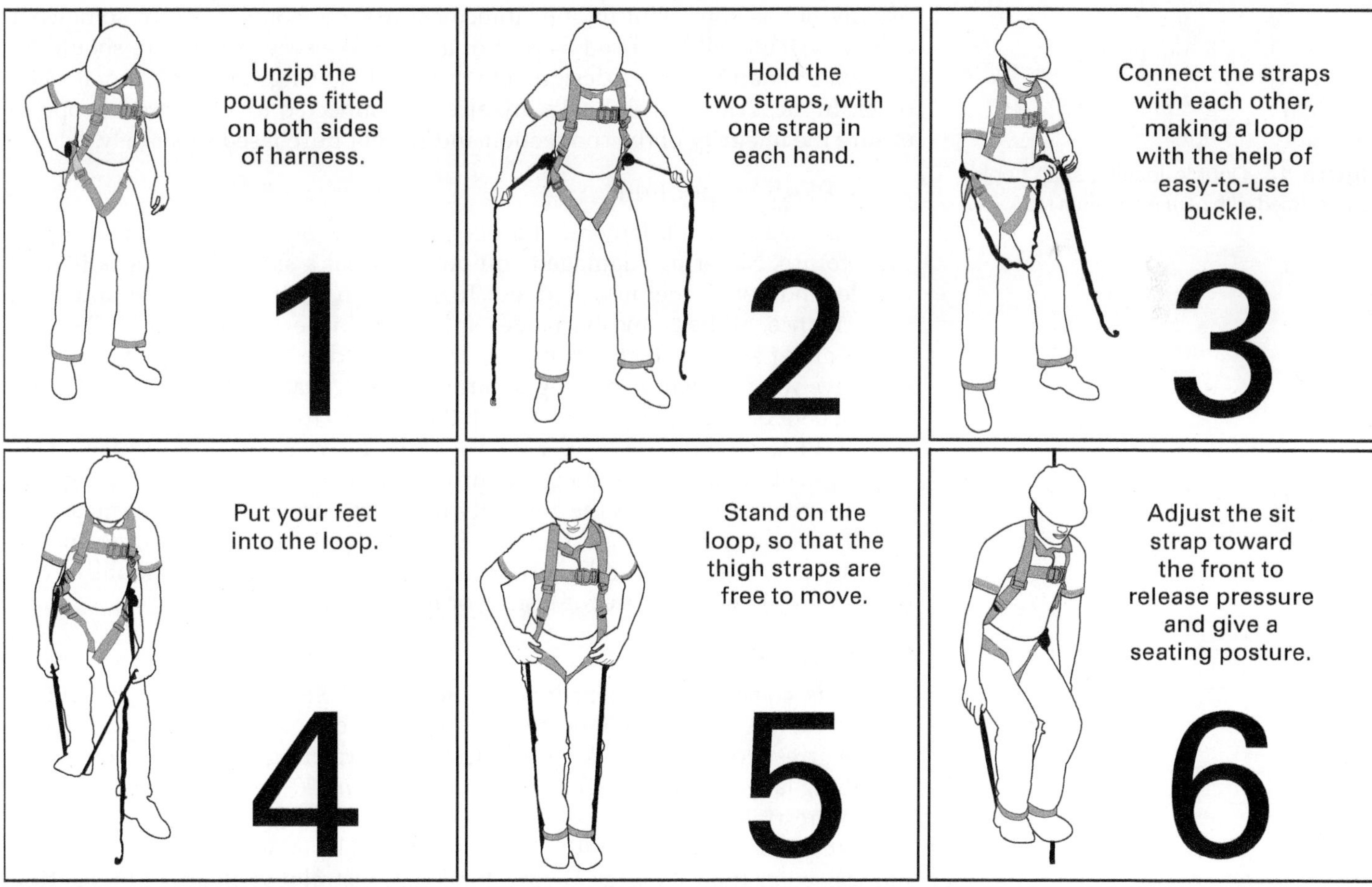

Figure 16 Suspension trauma strap use.
Source: KStrong Inc. (https://kstrong.com)

Figure 17 A carabiner.
Source: Used with permission from MSA

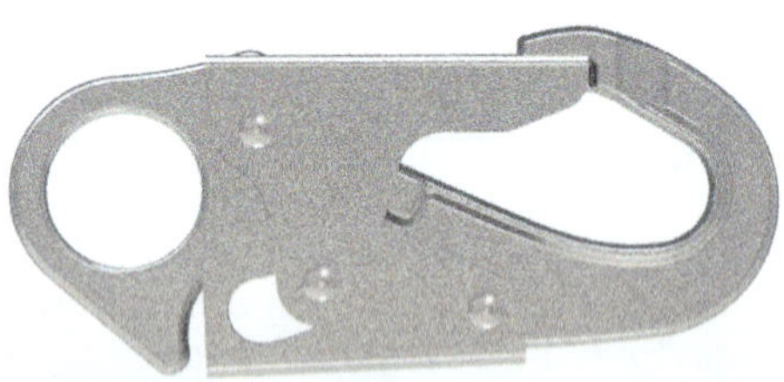

Figure 18 Double-locking snap hook.
Source: Used with permission from MSA

- A separate waist or tool belt, while not considered a necessary component of the fall arrest system, must be fitted properly. It is best used for body positioning with the D-rings precisely located at the hips, rather than in the front or rear. Do not adjust it in a way that could apply pressure to the kidneys or lower back. If the waist belt is not an integral part of the harness, it must not be worn on the outside of the harness—put it on first, then add the harness over it. This is also true of any added tool belts.
- Saddles, like waist belts, are also not considered an integral or required part of the PFAS. They are optional and often detachable. They are generally used by workers to allow a seated, suspended position when the task may require long periods in the same location.

It is important to note that not all hardware will qualify as a component of a PFAS. Some hardware is to be used only for attaching tools and equipment to the worker or to structures. Hardware used as connecting devices as part of the PFAS must be drop forged steel and have a finish resistant to corrosion and salt spray per ANSI standards. Any type of hook or carabiner must be equipped with safety gates or keepers to prevent the hooked object from being disconnected incidentally. In most cases, these safety gates will be required to be two-step, also called *double action*. Designs for these features vary, but those designed so that both movements required to open the gate can be done with a single hand are generally better. Using both hands to manipulate a single connector can be a hazard in itself.

Double-locking snap hooks (*Figure 18*) are usually curved and have an opening to allow connection to a line that can then be securely closed. They are usually not as consistent in appearance as carabiners and come in somewhat different shapes. When used as part of a PFAS, the security closure should be automatic. Snap hooks are designed to connect to D-rings primarily, not to each other. In most cases, snap hooks are already connected to a lanyard or rope to ensure the integrity of the connection and are not purchased separately.

2.2.3 PFAS Inspection

Always inspect your lanyard, harness, and any other fall protection gear prior to use. Never use damaged equipment. Treat a safety harness as if your life depends on it, because it does! To maintain their service life and high performance, all belts and harnesses should be inspected frequently. Damage to fall arrest systems includes burns, hardening due to chemical contact, and excessive wear. When inspecting a harness, check that the buckles and D-ring are not bent or deeply scratched. Check the harness for any cuts or rough spots.

The PFAS should be inspected monthly by a competent person. This requirement should be established through your company's safety program. The competent person has the authority to impose prompt corrective measures to eliminate any hazards. If there is any question about a defect, no matter how small it may seem to be, err on the side of caution. Take the fall arrest component(s) out of service for testing or replacement.

2.2.4 Lanyards

Lanyards consist of some primary material of construction (rope, webbing, aircraft cable, etc.) with a connecting device attached to the ends. Lanyards used for fall arrest are either shock absorbing or self-retracting (*Figure 19*). Non-shock absorbing lanyards, such as the one shown in *Figure 20*, are used for positioning and fall restraint. Lanyards used for positioning are not considered part of the fall arrest system; they are fall restraints and will be attached to D-rings on the harness other than the back D-ring. Since fall restraint is all about preventing a fall from happening, lanyards used for positioning should not allow a fall or movement greater than 2' (0.61 m). Two such lanyards are usually required to permit movement from one place to another.

(A) Shock-Absorbing (B) Retractable

Figure 19 Fall arrest lanyards.
Source: Used with permission from MSA

Figure 20 Non-shock absorbing lanyard.
Source: Used with permission from MSA

The retractable fall arrest lanyard shown in *Figure 19* (B) is rapidly becoming the preferred device. Since it automatically retracts or feeds lanyard as the worker moves about, there is never a great deal of slack that can pose a risk. In addition, the potential fall distance is significantly reduced.

Lanyards for fall restraint or arrest must never be field fabricated and must never be connected to increase their length. Depending on the use, padding may be added during fabrication or added in the field to protect against sharp edges. Never tie a knot in a lanyard, as knots can severely reduce the load limit. Never wrap a lanyard around a structure and then choke it back into itself unless it is designed for this use. A special large D-ring is usually attached to the lanyard for this purpose. It is important to remember that whenever you are wearing a PFAS, 100% tie-off is required. A PFAS is absolutely useless unless you are tied off.

The D-ring used on the PFAS for fall arrest may only be connected to one live connection at a time. This can be challenging when trying to move from one point to another, especially horizontally. The user must be able to reach back and disconnect a lanyard, while maintaining one lanyard connected at all times. A Y-configured lanyard (*Figure 21*) can be used for this purpose. A Y-configured lanyard has a single point of attachment at the D-ring that is used to accommodate two lanyards. They are also referred to as double-leg or tieback lanyards.

Before using a shock absorbing or self-retracting lanyard, the potential fall distance is determined. Then the proper equipment is selected to meet available fall clearance (*Figure 22*). Note in the figure that the fall distance with the self-retracting lanyard is significantly shorter than it is with a shock absorbing lanyard, hence its increasing popularity. Failure to select proper equipment and calculate fall distance may result in serious personal injury or death. These calculations must be done by experienced and qualified personnel on the jobsite. If a personal fall arrest system is actuated due to a fall, it must be inspected by a competent person before it can be used again. In most cases, replacement will be necessary.

Figure 21 Y-configured shock absorbing lanyard.
Source: Used with permission from MSA

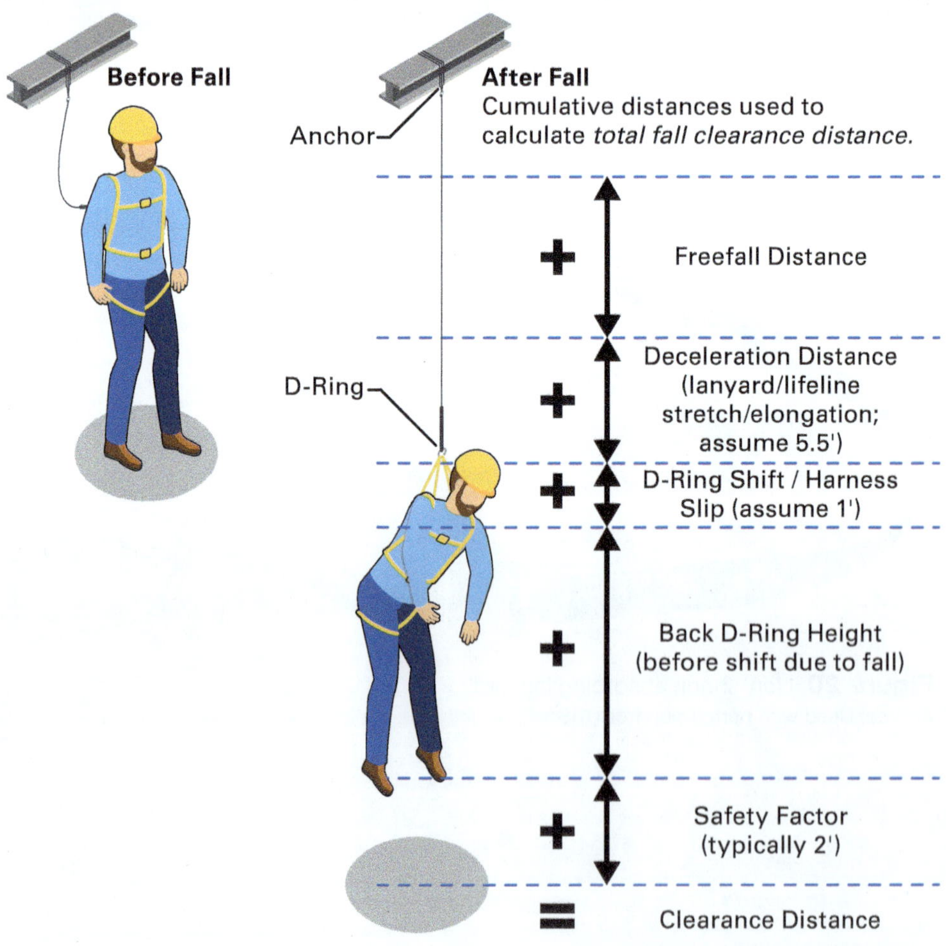

Figure 22 Potential fall distances for self-retracting and shock absorbing lanyards.

2.2.5 Lifelines

A lifeline is a flexible line such as a cable or rope connected vertically to an anchorage at one end (vertical lifeline), or horizontally to an anchorage at both ends (horizontal lifeline). It serves as a means for connecting other components of a personal fall arrest system to the available anchorage.

Vertical lifelines are suspended from a fixed anchorage. A fall arrest device such as a rope grab or self-retracting lifeline (*Figure 23*) is attached to the lifeline. A beam grab, or beamer (*Figure 24*), is sometimes used as an anchorage for a lifeline. Vertical lifelines must have a minimum breaking strength of 5,000 lb (2,268 kg). Workers must use separate vertical lifelines. A vertical lifeline must have a termination on the end unless it extends to the ground or to the next lower level of the structure.

Figure 23 Self-retracting lifeline.
Source: Used with permission from MSA

Figure 24 Beam grab.

Horizontal lifelines (*Figure 25*) are connected between two fixed anchorages. A lanyard is attached to the lifeline. Horizontal lifelines must be designed, installed, and used under the supervision of a competent person. The required breaking strength of any horizontal lifeline is determined by the number of workers that will be attached to it.

WARNING!

Keep in mind that horizontal lifelines are rated for a maximum number of connected workers. Check with your supervisor or the manufacturer before connecting to a lifeline that is being used by other workers.

Figure 25 Horizontal lifeline.
Source: Honeywell Safety Products

2.2.6 Guardrails

Guardrail systems (*Figure 26*) are a common type of fall prevention. They protect workers by providing a barrier between the work area and the ground or lower work areas. They may be made of wood, pipe, steel, or wire rope and must be able to support 200 lb (90 kg) of force applied in any direction to the top rail and 150 lb (68 kg) for the **midrail**. A guardrail must be 42" ± 3" (106 cm ± 8 cm) high to the top rail and have a toeboard that is a minimum of 4" (10 cm) high. This helps to prevent the inadvertent loss of tools or material through the bottom rail. The toeboard must be securely fastened with not more than ¼" (6 mm) clearance above the floor level.

Midrail: Mid-level, horizontal board required on all open sides of scaffolds and platforms that are more than 14" (35 cm) from the face of the structure and more than 10' (3 m) above the ground. It is placed halfway between the toeboard and the top rail.

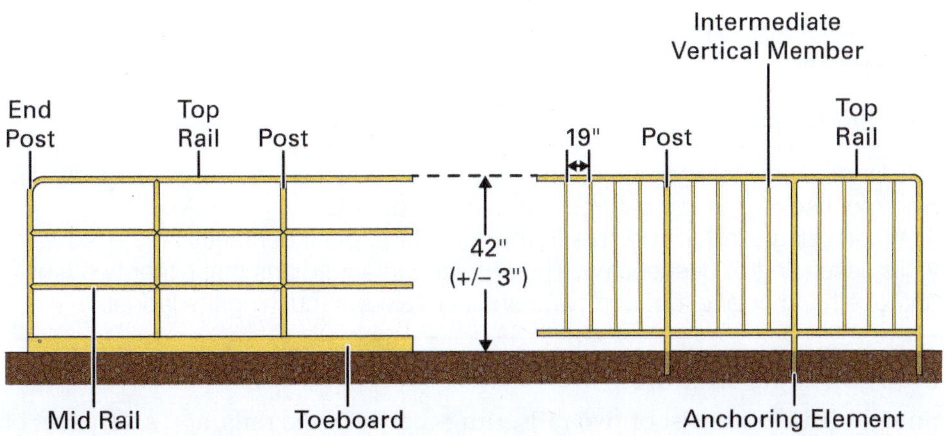

Figure 26 Typical components of a guardrail system.

2.2.7 Safety Nets

Safety nets are used for fall protection on bridges and similar projects. They must be installed not more than 30' (9 m) beneath the work area. There must be enough clearance under a safety net to prevent a worker who falls into it from hitting the surface below it. There must also be no obstruction between the work area and the net.

Depending on the actual vertical distance between the net and the work area, the net must extend 8' to 13' (2 m to 4 m) beyond the edge of the work area. Mesh openings in the net must be limited to 36 in^2 (232 cm^2) and 6" (15 cm) from the side. The border rope must have a 5,000 lb (2,268 kg) minimum breaking strength, and connections between net panels must be as strong as the nets themselves. Safety nets must be inspected at least once a week and after any event that might have damaged or weakened them. Worn or damaged nets must be removed from service.

2.3.0 Ladders and Stairs

Ladders are used daily to perform work in elevated locations. Any time work is performed above ground level, there is a risk of incidents. This risk can be reduced by carefully inspecting ladders before you use them and by using them properly.

Overloading means exceeding the **maximum intended load** of a ladder. Overloading can cause ladder failure, which means that the ladder could buckle, break, or topple. The maximum intended load is the total weight of all people, equipment, tools, materials, carried loads, and other loads that the ladder can hold at any one time. Check the manufacturer's specifications to determine the maximum intended load. Ladders are usually given a duty rating that indicates their load capacity as shown in *Table 2*. Note that ladders designed for the metric market are not usually direct equivalents to ladders built for the US market. Capacity ratings of 286 lb (130 kg) and 330 lb (150 kg) are the most common for the trades. Ladder heights will also be stated in metric units and may not be exactly the same size as those designed for the US market.

Maximum intended load: The total weight of all people, equipment, tools, materials, and loads that a ladder can hold at one time.

Figure 27 Portable straight ladder.

Drop Testing Safety Nets

Safety nets should be drop tested at the jobsite after the initial installation, whenever relocated, after a repair, and at least every six months if left in one place. The drop test consists of a 400 lb (181 kg) bag of sand of 29" to 31" (74 cm to 79 cm) in diameter that is dropped onto the net from at least 42" (107 cm) above the highest walking/working surface at which workers are exposed to fall hazards. If the net is still intact after the bag of sand is dropped, it passes the test.

TABLE 2 Ladder Duty Ratings and Load Capacities

Duty Rating	Load Capacities
Type IAA	375 lb, extra heavy-duty/professional use
Type IA	300 lb, extra heavy-duty/professional use
Type I	250 lb, heavy-duty/industrial use
Type II	225 lb, medium-duty/commercial use
Type III	200 lb, light-duty/household use

There are different types of ladders for different jobs. Selecting the right ladder for the job at hand is important to complete a job as safely and efficiently as possible. Ladder types include portable straight ladders, extension ladders, and stepladders.

WARNING!

Use ladders only for their intended purposes. Ladders are not interchangeable; incorrect use of a ladder can result in injury or damage.

When using a ladder, be sure to maintain three-point contact with the ladder while ascending or descending. Three-point contact means that either two feet and one hand or one foot and two hands are always touching the ladder.

2.3.1 Straight Ladders

Straight ladders consist of two rails, rungs between the rails, and safety feet on the bottom of the rails (*Figure 27*). Straight ladders are generally made of aluminum, wood, or fiberglass.

Metal ladders conduct electricity and should never be used around electrical equipment. Any portable metal ladder must have "Danger! Do Not Use Around Electrical Installations" stenciled on the rails in 2" red letters. Ladders made of dry wood or fiberglass, neither of which conducts electricity, should be used around electrical equipment. Check that any ladder, especially a wooden ladder, is completely dry before using it; even a small amount of water will conduct electricity.

Different types of ladders are intended for use in specific situations. Aluminum ladders are corrosion-resistant and can be used where they might be exposed to the elements. They are also lightweight and can be used where they must often be lifted and moved. Fiberglass ladders are very durable, so they are useful where some amount of rough treatment is unavoidable. Wooden ladders, which are heavier and sturdier than fiberglass or aluminum ladders, can be used where heavy loads must be moved up and down. However, wooden ladders are subject to more rapid deterioration as the wood swells and shrinks. Both fiberglass and aluminum are easier to clean than wood.

Figure 27 shows the safety feet attached to a straight ladder. Make sure the feet are securely attached and that they are not damaged or worn down. Do not use a ladder if its safety feet are not in good working order.

It is very important to place a straight ladder at the proper angle before using it. A ladder placed at an improper angle will be unstable and could cause a fall. *Figure 28* shows a properly positioned straight ladder.

The distance between the foot of a ladder and the base of the structure it is leaning against must be $\frac{1}{4}$ of the distance between the ground and the point where the ladder touches the structure. Stated another way, there should be a 4:1 ratio between the distances. For example, if the height of the wall shown in *Figure 28* is 16' (4.9 m), the base of the ladder should be 4' (1.2 m) from the base of the wall. If you are going to step off a ladder onto a platform or roof, the top of the ladder should extend at least 3' (0.9 m) above the point where the ladder touches the platform, roof, side rails, etc.

Ladders should be used only on stable and level surfaces unless they are secured at both the bottom and the top to prevent any incidental movement (*Figure 29*). Never try to move a ladder while you are on it. If a ladder must be placed in front of a door that opens toward the ladder, the door should be locked or blocked open. Otherwise, the door could be opened into the ladder.

Ladders are made for vertical use only. Never use a ladder as a work platform by placing it horizontally. Make sure the ladder you are about to climb or descend is properly secure before you do so. Check to make sure the ladder's feet are solidly positioned on firm, level ground. Also check to make sure the top of the ladder is firmly positioned and in no danger of shifting once you begin your climb. Remember that your own weight will affect the ladder's steadiness once you mount it. It is important to test the ladder first by putting some of your weight on it without beginning to climb. This way, you can be sure that the ladder will remain steady as you climb.

When climbing a straight ladder, keep both hands on the rails or rungs (*Figure 30*). Always maintain three points of contact, as shown in the figure. This can be two feet and one hand, or one foot and two hands. Always keep your body's weight in the center of the ladder between the rails. Face the ladder at all times. Never go up or down a ladder while facing away from it.

To carry a tool while you are on a ladder, use a **hand line** or tagline attached to the tool. Climb the ladder and then pull up the tool. Don't carry tools in your hands while you are climbing a ladder.

2.3.2 Extension Ladders

An extension ladder is two straight ladders connected so that the overlap between them can be altered to increase or decrease the length of the ladder (*Figure 31*).

Extension ladders are positioned and secured following the same rules as straight ladders. When adjusting the length of an extension ladder, always reposition the movable section from the bottom, not the top, to ensure that the rung locks (*Figure 32*) are properly engaged after you make the adjustment. Check to make sure the section locking mechanisms are fully hooked over the desired rung. Also check to make sure that all ropes used for raising and lowering the extension are clear and untangled.

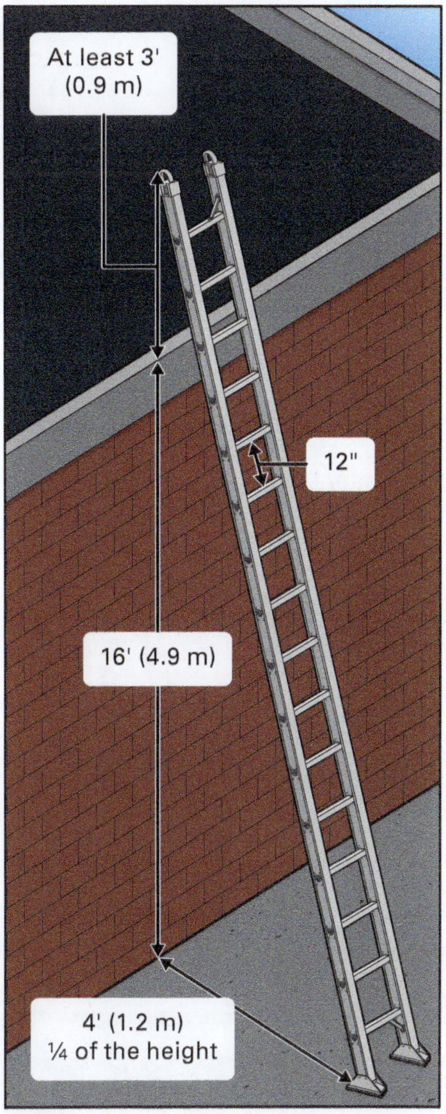

Figure 28 Proper positioning of a straight ladder.

Hand line: A line attached to a tool or object so a worker can pull it up after climbing a ladder or scaffold.

(A) Bottom Secured **(B) Top Secured**

Figure 29 Securing a ladder.

Figure 30 Moving up or down a ladder.

Figure 31 Examples of extension ladders.
Source: Courtesy of Werner

Figure 32 Rung locks.
Source: Courtesy of Werner

Extension ladders are positioned and secured following the same rules as straight ladders. There are, however, some safety rules that are unique to extension ladders:

WARNING!

Extension ladders have a built-in extension stop mechanism. Do not remove this mechanism. If the mechanism is removed, it could cause the ladder to collapse under a load.

Haul materials up on a line rather than carrying them up an extension ladder by hand.

- Make sure the extension ladder overlaps between the two sections (*Figure 33*). For ladders up to 36' (11 m) long, the overlap must be at least 3' (0.9 m). For ladders 36' to 48' (11 m to 14.6 m) long, the overlap must be at least 4' (1.2 m). For ladders 48' to 60' (14.6 m to 18.3 m) long, the overlap must be at least 5' (1.5 m).
- Never stand above the highest safe standing level on a ladder. On an extension ladder, this is the fourth rung from the top. If you stand higher, you may lose your balance and fall. Some ladders have colored rungs to show where you should not stand.
- Avoid carrying anything on a ladder by hand because it will affect your balance and may cause you to fall. Haul materials up on a line instead.
- Keep yourself centered on the ladder. Do not overreach, lean to one side, or try to move a ladder while standing on it.

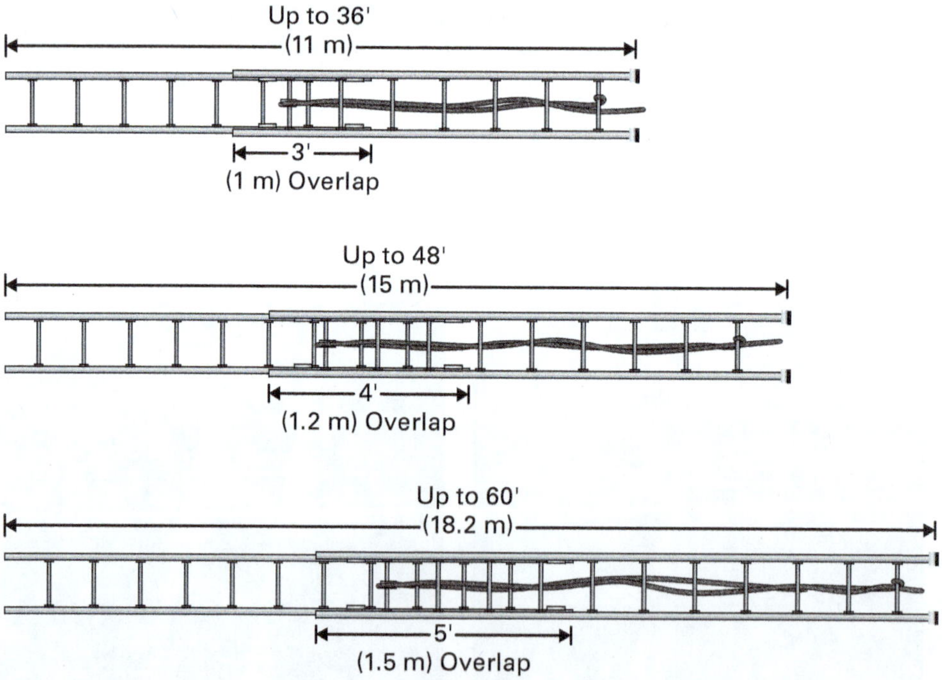

Figure 33 Overlap lengths for extension ladders.

2.3.3 Stepladders

Stepladders are self-supporting ladders made of two sections hinged at the top (*Figure 34*). The section of a stepladder used for climbing consists of rails and rungs like those on straight ladders. The other section consists of rails and braces. Spreaders are hinged arms between the sections that keep the ladder stable and stop it from folding while in use.

When positioning a stepladder, be sure that all four feet are on a hard, even surface. Otherwise, the ladder can rock from side to side or corner to corner when you climb it. With the ladder in position, be sure the spreaders are locked in the fully open position. The following safety precautions must be followed when using a stepladder:

- Never stand on the top step or on the top of a stepladder. Putting weight this high will make the ladder unstable. The top of the ladder is made to support the hinges, not to be used as a step.
- Although the rear braces may look like rungs, they are not designed to support weight. Never use the braces for climbing. However, there are specially designed two-person ladders available with steps on both sides.
- Check the load capacity of the ladder and do not exceed it.

WARNING!

Never stand on a step with your knees higher than the top of a stepladder. You need to be able to hold on to the ladder with your hand. Keep your body centered between the side rails.

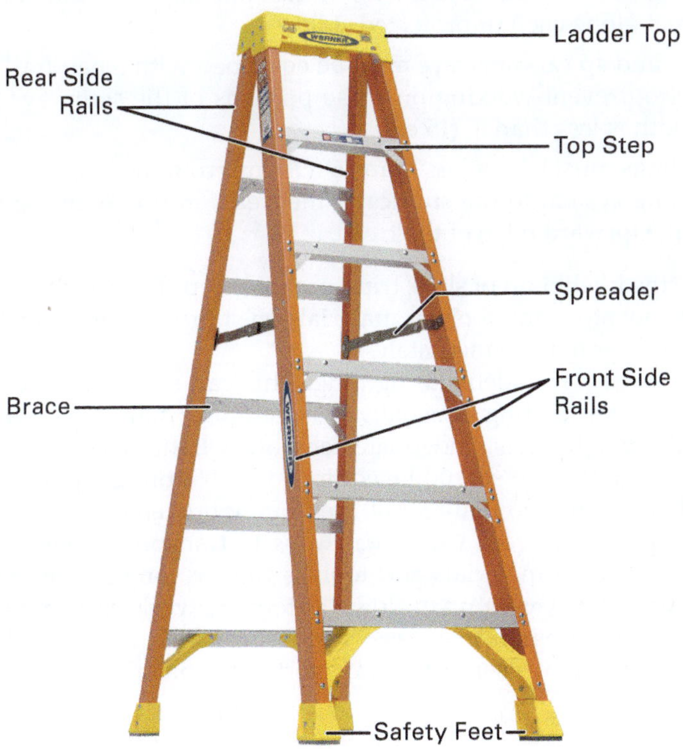

Figure 34 Typical fiberglass stepladder.
Source: Courtesy of Werner

CAUTION

Wooden ladders should never be painted. The paint could hide cracks in the rungs or rails. Clear varnish, shellac, or a preservative oil finish will protect the wood without hiding defects.

2.3.4 Inspecting Ladders

Always inspect a ladder before you use it. Check the rails and rungs for cracks or other damage. Also check for loose rungs. If you find any damage, do not use the ladder. Check the entire ladder for loose nails, screws, brackets, or other hardware. If you find any hardware problems, tighten the loose parts or have the ladder repaired before you use it. OSHA requires regular inspections of all ladders, including an inspection just before each use.

The same rules for inspecting straight ladders apply to extension ladders. In addition, the rope that is used to raise and lower the movable section of the ladder should be inspected. If the rope is frayed or has worn spots, it should be replaced before the ladder is used.

The rung locks support the entire weight of the movable section and the person climbing the ladder. Inspect them for damage before each use. If they are damaged, they should be repaired or replaced before the ladder is used.

Inspect stepladders the way you inspect straight and extension ladders. Pay special attention to the hinges and spreaders to be sure they are in good condition. Also, be sure the rungs are clean. The rungs of a stepladder are usually flat, so oil, grease, or dirt can build up and make them slippery.

2.3.5 Stairways

Stairways are also routinely used on construction sites where there is a break in elevation of 19" (48 cm) or more, and no ramp, runway, sloped embankment, or personnel hoist is provided. Observe the following regulations, based on OSHA standards, when using stairways on a jobsite:

- Stairways having four or more risers or rising more than 30" (76 cm), whichever is less, must be equipped with at least one handrail and one stair railing system along each unprotected side.
- Winding and spiral stairways must be equipped with a handrail offset sufficiently to prevent walking on those portions of the stairways where the tread width is less than 6" (15 cm).
- Stair railings must be not less than 36" (91 cm) from the upper surface of the stair railing system to the surface of the tread, in line with the face of the riser at the forward edge of the tread.

To reduce the likelihood of slips, trips, or falls, keep stairways clean and clear of debris. Do not store any tools or materials on stairways, and clean up liquid spills, rainwater, or mud immediately.

Stairways must have adequate lighting. This can sometimes be a problem because permanent lighting is usually installed after stairway construction is completed. If the lighting is inadequate, temporary lighting should be installed in the stairway. Each bulb should be equipped with a protective cover, and the string should be inspected daily for burned-out or broken bulbs.

Whenever possible, avoid using stairways to transport materials between floors. Carrying small materials and tools is fine, as long as the materials do not block your vision. Going up or down a stairway while carrying large items is physically demanding and increases the chance of injuries and falls. Use the building elevator or crane service to transport large materials from one floor to another.

2.4.0 Scaffolds

Scaffolds provide safe elevated work platforms for people and materials. They are designed and built to comply with high safety standards, but normal wear and tear or accidentally putting too much weight on them can weaken them and make them unsafe. That's why it is important to inspect every part of a scaffold before each use. Personnel who assemble scaffolds must be certified to do so.

2.4.1 Types of Scaffolds

Two basic types of scaffolds—self-supporting scaffolds and suspended scaffolds—are used in the construction industry. The rules for safe use apply to both. Self-supporting scaffolds can be manufactured units or can be assembled at the site.

Manufactured scaffolds (*Figure 35*) are made of painted steel, stainless steel, or aluminum. They are stronger and more fire-resistant than wooden scaffolds. They are supplied in ready-made, individual units which are assembled on-site. A rolling scaffold has wheels on its legs so that it can be easily moved. The scaffold wheels have brakes to inhibit scaffold movement while workers are standing on it.

CAUTION

Only a competent person has the authority to supervise the setup, move, and takedown of scaffolds. Only a competent person can approve the use of scaffolds on the jobsite after inspecting the scaffolds.

WARNING!

Never unlock the wheel brakes of a rolling scaffold while anyone is on it. People on a moving scaffold can lose their balance and fall.

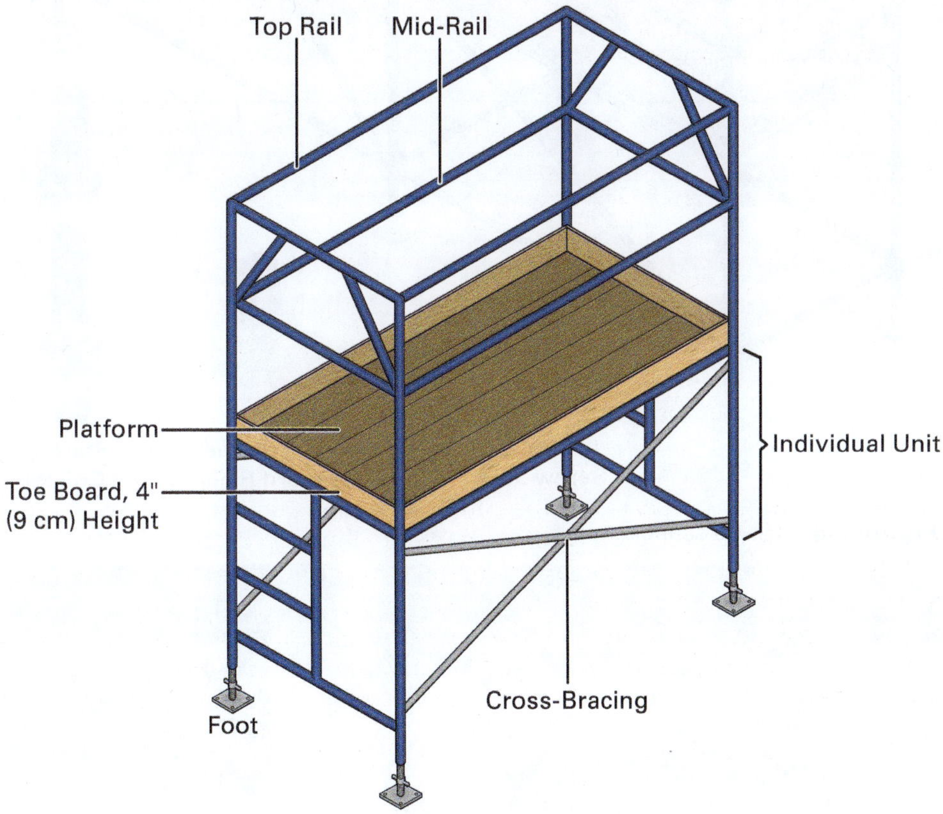

Figure 35 Typical manufactured scaffold.

Built-up scaffolds (*Figure 36*) are built from the ground up at a jobsite using steel framework sections and lumber. Swing (suspended) scaffolds (*Figure 37* [A]) are suspended by ropes or cables in a manner that allows them to be raised or lowered as needed. Another type of suspended scaffolding is a work cage (*Figure 37* [B]). A work cage is typically suspended with rigging devices that attach to I-beams with various sizes of clamps and rollers.

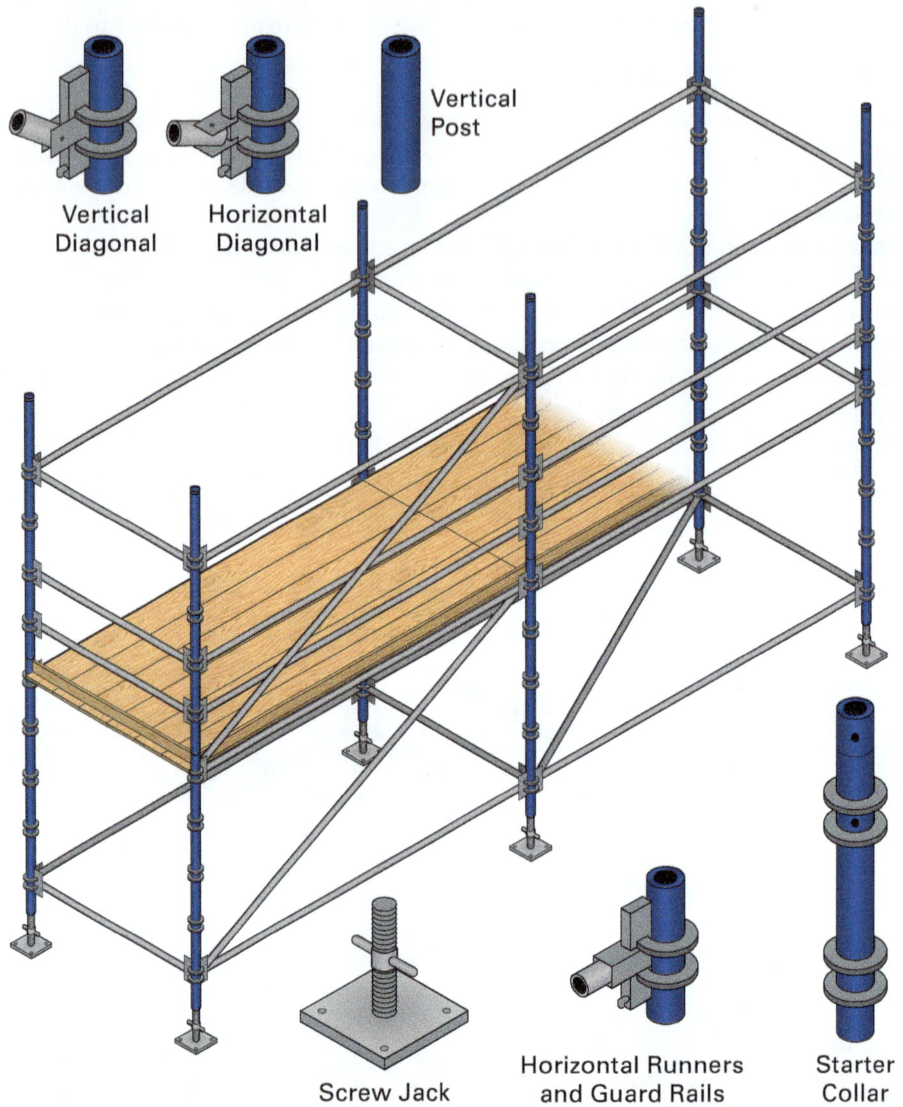

Figure 36 Built-up scaffold.

(A) Suspended Scaffold

(B) Work Cage

Figure 37 Suspended scaffold and work cage.
Source: iStock

2.4.2 Inspecting Scaffolds

Any scaffold that is assembled on the jobsite must be tagged to indicate whether the scaffold meets OSHA standards and is safe to use. Three colors of tags are used: green, yellow, and red (*Figure 38*).

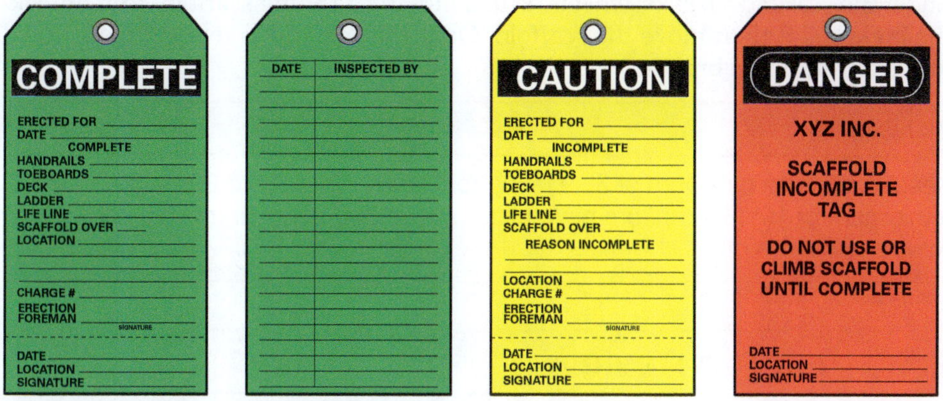

Figure 38 Typical scaffold tags.

The color coded tags indicate the following:

- A green tag means the scaffold meets all OSHA standards and is safe to use.
- A yellow tag means the scaffold does not meet all OSHA standards. An example is a scaffold on which a railing cannot be installed because of equipment interference. To use a yellow-tagged scaffold, you must wear a safety harness attached to a lanyard. You may have to take other safety measures as well.
- A red tag means a scaffold is being put up or taken down. Never use a red-tagged scaffold.

Don't rely on the tags alone; inspect all scaffolds before you use them. Check for bent, broken, or badly rusted tubes. Also check for loose joints where the tubes are connected. Any of these problems must be corrected before the scaffold is used.

Make sure you know the weight limit of any scaffold you will be using. Compare this weight limit to the total weight of the people, tools, equipment, and material you expect to put on the scaffold. Scaffold weight limits must never be exceeded.

If a scaffold is more than 10' (3 m) high, check to see that it is equipped with **top rails**, midrails, and toeboards; otherwise, use a PFAS. All connections must be pinned. That means they must have a piece of metal inserted through a hole to prevent connections from slipping. **Cross-bracing** must be used. A handrail is not the same as cross-bracing. The walking area must be completely **planked**.

If it is possible for people to walk under a scaffold, the space between the toeboard and the top rail must be screened. This prevents objects from falling off the work platform and injuring those below.

When you examine a rolling scaffold, check the condition of the wheels and brakes. Be sure the brakes are working properly and can stop the scaffold from moving while work is in progress. Be sure all brakes are locked before you use the scaffold.

Top rails: Top-level, horizontal boards required on all open sides of scaffolds and platforms that are more than 14" (36 cm) from the face of the structure and more than 10' (3 m) above the ground.

Cross-bracing: Braces (metal or wood) placed diagonally from the bottom of one rail to the top of another rail that add support to a structure.

Planked: Having pieces of material 2" (5 cm) thick or greater and 6" (15 cm) wide or greater used as flooring, decking, or scaffold decks.

2.4.3 Using Scaffolds

Be sure that a competent person inspects the scaffold before you use it. There should be firm footing under each leg of a scaffold before putting any weight on it. If you are working on loose or soft soil, you can put planks or matting under the scaffold's legs or wheels. When moving a rolling scaffold, first unlock the brakes and then move the scaffold. Once the scaffold is repositioned, don't forget to relock the brakes.

WARNING!

Keep scaffolds a minimum of 3' (91 cm) from power lines up to 300 volts (V) and a minimum of 10' (3.1 m) from power lines above 300 V. Never move a scaffold while someone is on it.

2.0.0 Section Review

1. A barrier with yellow and purple markings indicates a _____.
   a. fire hazard
   b. fall hazard
   c. radiation hazard
   d. confined space hazard

2. Positioning lanyards are used only as fall restraint devices and may *not* be used as fall arrest devices.
   a. True
   b. False

3. When positioning a straight ladder against a wall, how far from the wall should the base of the ladder be?
   a. 4' (1.2 m)
   b. $\frac{1}{4}$ the distance from the ground to the point where the ladder touches the wall
   c. The height of the wall minus 4' (1.2 m)
   d. $\frac{1}{2}$ the distance from the ground to the point where the ladder touches the wall

4. If a scaffold has a yellow tag, it means _____.
   a. the scaffold may only be used by one person at a time
   b. the scaffold is condemned and cannot be used
   c. a safety harness must be worn when using it
   d. the scaffold is under assembly and cannot be used

3.0.0 Struck-By, Caught-In, and Caught-Between Hazards

Objective

Recognize and avoid struck-by, caught-in, and caught-between hazards.

a. Describe struck-by hazards and how to avoid them.

b. Describe common caught-in/caught-between hazards and steps that can prevent them.

Performance Tasks

There are no Performance Tasks in this section.

There is an apparent overlap in struck-by and caught-in/caught-between hazards that needs to be clarified as they can involve the same kinds of equipment. *Struck-by* means being hit by a moving object, while *caught-between* means being trapped between a moving object and a solid surface. Assume a worker is working near a crane that is swinging a load and the load hits the worker. That is an example of a struck-by incident. However, if the worker is behind the crane and the crane backs up, pinning the worker against a wall, it becomes a caught-between incident.

3.1.0 Struck-By Hazards

On any jobsite, there is a risk of being struck by falling objects, such as tools dropped from above. Flying objects such as debris from grinding and chipping metal are another struck-by hazard. On any site where there is moving equipment or where workers are near roadways, there is also the danger of being struck by a vehicle.

3.1.1 Falling Objects

Workers are at risk from falling objects when they are beneath machinery and equipment such as cranes and scaffolds, where overhead work is being performed, or when working around stacked materials. To protect against struck-by injuries from falling objects, always wear an approved hard hat. Employers generally require workers to always wear hard hats on construction sites.

When working near machinery and equipment such as cranes, never stand or work beneath the load or in the fall zone. Barricade hazard areas where rigging equipment is in use, and post warning signs to inform other workers of falling object hazards. Inspect cranes and rigging components before use and do not exceed the rated load capacity.

When performing overhead work, be sure all tools, materials, and equipment are secured to prevent them from falling on people below. Use protective measures such as toeboards, debris nets, catch platforms, or canopies to catch or deflect falling objects. Use tool lanyards to prevent tools from falling.

Many workers are hurt or killed by falling stacks of material. Do not stack materials higher than 4:1 height-to-base ratio. Secure all loads by blocking and interlocking them. Interlocking means placing alternate layers at right angles to each other. Be aware of changing weather conditions, such as wind, that may lift and shift loads.

3.1.2 Flying Objects

There is a danger from flying objects when power tools or activities such as pushing, pulling, or prying cause objects to become airborne. Chipping, grinding, brushing, or hammering are all examples of job tasks that may cause flying objects. Tools that move at very high speeds, like pneumatic and powder-actuated tools, can be very dangerous. Injuries from flying objects can range from minor abrasions to concussions, blindness, or death. The workers shown in *Figure 39* are using a pneumatic chipping hammer and a pneumatic grinder, and are properly using protective face shields.

(A)

(B)

Figure 39 Protection from flying particles.

To protect against flying object hazards, follow these guidelines:

- Use eye protection such as safety glasses, goggles, or face shields where machines or tools may cause flying particles.
- Inspect tools and machines to ensure that protective guards are in place and in good condition.
- Make sure you are trained in the proper operation of pneumatic and powder-actuated tools.
- Use shielding devices such as a welding screen or similar equipment to block flying debris.

3.1.3 Vehicle and Roadway Hazards

A common cause of incidents for workers on large jobsites and highway projects are vehicle-related hazards such as being run over by vehicles or equipment (especially backing equipment) or by equipment tip over. The most common cause of death for equipment operators is equipment rollover.

WARNING!

Tractor overturns are the leading cause of fatal injuries on farms in the United States. One in ten operators will overturn their tractor at least once during their career. Tractor overturns lead to about 130 deaths each year.

Tractor rollovers are preventable. Tractors should be equipped with rollover protection structures to keep the operator protected when they are properly secured with seat belts. Operators should understand the hazards that contribute to rollovers, such as unstable or uneven ground, turning forces that push on the tractor, and the location of the center of gravity (CG) of a tractor. The CG is the point where all the weight of the tractor is in balance. This can be altered due to surface factors or uneven forces during a turn, resulting in unstable conditions that cause the tractor to roll over.

Operators can take several precautions to avoid side rollovers:

- Practice proper braking
- Avoid weak soils on banks and shoulders
- Lower buckets during transport
- Drive at appropriate speeds
- Use additional caution in sloped areas

Rear overturns, when the front end of the tractor flips backward, are just as dangerous and happen very quickly. Several additional precautions are necessary to avoid this situation:

- Use caution when operating on an incline (*Figure 40*)
- Back out a stuck tractor and never use boards in front of the rear wheels
- Add rear counterweights when transporting front loads
- Avoid hitching above the drawbar of the tractor

If vehicle safety practices are not observed at your site, you risk being struck by swinging backhoes, moving vehicles, or swinging crane loads. If you work near public roadways (*Figure 41*), you risk being struck by passenger or commercial vehicles. When working near moving vehicles and equipment, follow these guidelines:

- Always stay alert and keep a safe distance from vehicles and equipment.
- Maintain eye contact with vehicle or equipment operators to ensure that they see you.
- Never get into blind spots of equipment operators.

Figure 40 Learn to safely operate a tractor on a slope.
Source: iStock

Figure 41 A busy jobsite.

> **Did You Know?**
>
> **Struck-By Fatalities**
>
> Nearly one in four struck-by vehicle deaths involve construction workers—more than any other occupation.

- Keep off equipment unless authorized.
- Wear reflective or high-visibility vests or other suitable garments.
- Never stand between pieces of equipment unless they are secured.
- Never stand under loads handled by lifting or digging equipment, or near vehicles being loaded or unloaded.

Operators must also use caution when driving vehicles. The operator of any vehicle is responsible for the safety of passengers and the protection of the load. Follow these safety guidelines when you operate a vehicle on a jobsite:

- Always wear a seat belt.
- Be sure that each person in the vehicle has a firmly secured seat and seat belt.
- Obey all speed limits. Reduce speed in crowded areas.
- Look to the rear and sound the horn before backing up. If your rear vision is blocked, get a **signaler** to direct you.
- Every vehicle must have a backup alarm. Make sure the backup alarm works.
- Always turn off the engine when fueling.
- Turn off the engine and set the brakes before leaving the vehicle.
- Never stay on or in a truck that is being loaded by excavating equipment.
- Keep windshields, rearview mirrors, and lights clean and functional.
- Carry road flares, fire extinguishers, and other standard safety equipment at all times.
- Never use a cell phone while operating a motor vehicle.

Signaler: A person who is responsible for directing a vehicle when the driver's vision is blocked in any way.

WARNING!

When a vehicle is operated indoors, the carbon monoxide in its exhaust can kill or sicken anyone in the vicinity unless there is good ventilation. Carbon monoxide is especially dangerous because you cannot see, smell, or taste it. It is never a good idea to operate a vehicle, generator, or any other machine with an internal combustion engine indoors. Make sure the area is well ventilated before operating motorized vehicles or using equipment, such as a generator, indoors.

3.2.0 Caught-In and Caught-Between Hazards

Congested worksites, heavy equipment, and the presence of multiple trades can contribute to caught-in/caught-between hazards. The primary causes of caught-in/caught-between fatalities include trench/excavation collapse, rotating equipment, and unguarded parts.

One of the most disastrous caught-in hazards on a construction jobsite is being trapped by a cave-in of the walls of a trench or excavation. Other caught-in or caught-between hazards involve protective guards on power tools and machines, as well as the risk of being crushed by heavy equipment.

3.2.1 Trenches and Excavations

Trenches and excavations are common hazards, especially in construction and pipeline work. Anyone working in or around a trench or excavation must know and follow safety procedures aimed at protecting workers from cave-ins.

An excavation is any man-made cut, cavity, trench, or depression formed by removal of earth or soil (*Figure 42*). Sometimes the terms *excavation* and *trench* are used interchangeably, but there is a difference. A trench is an excavation that is deeper than it is wide, and usually not wider than 15' (4.6 m). Nearly all trenches are dangerous if not protected. Because trenches are narrow, workers can easily become trapped.

Tool Lanyards

When working at elevations, a dropped tool can become a serious hazard. If the tool has moving parts, like battery-powered drills, the fall will likely destroy it.

Tool lanyards specifically designed for work in the elevated environment have been introduced by Snap-on®. Tethering tools to the tool belt or wrist can prevent injuries, save tools, prevent component damage, and prevent lost time in retrieval.

Source: Image property of Stanley Black & Decker. Used with permission.

Hazards involved with trench and excavation work include the following:

- Cave-ins
- Water accumulation
- Falling objects
- Collapse of nearby structures
- Hazardous atmospheres produced by toxic gases in the soil

WARNING!

Just 2' to 3' (0.61 m to 0.91 m) of soil can put enough pressure on your lungs to prevent you from breathing. In as little as four to six minutes without oxygen, you can sustain considerable brain damage.

Cave-ins are the most common and deadly hazard in excavation work. When dirt is removed from an excavation, the surrounding soil becomes unstable and gravity can force it to collapse. Cave-ins occur when soil or rock falls, or slides, into an excavation. Most cave-ins occur in trenches 5' to 15' (1.53 m to 4.6 m) deep and happen suddenly with little or no warning. On average, about 1,000 trench collapses occur each year in the United States.

Did You Know?

Excavation Fatalities

The fatality rate for excavation work is 112% higher than the rate for general construction.

Figure 42 A trench for drainage pipes.
Source: iStock

TABLE 3 Soil Types

Name	Type/Characteristics
Solid Rock	Excavation walls stay vertical as long as the excavation is open.
Type A Soil	This soil is fine-grained and cohesive: clay, hardpan, and caliche. Particles are too small to see with the naked eye.
Type B Soil	This includes angular rock, silt, and similar soil.
Type C Soil	This soil is coarse-grained and granular: sand, gravel, and loamy sand. Particles are visible to the naked eye.

Soil conditions can change, so they must be constantly evaluated. There are certain factors that could change the surroundings of the site, making a cave-in more likely. These factors include the following:

- Changes in weather conditions, such as freezing, thawing, or sudden heavy rain
- An excavation dug in unstable or previously disturbed soil
- Excessive vibration around the excavation
- Water accumulation in an excavation

Sometimes there are visible warning signs around the excavation that can be spotted before a cave-in occurs. Being aware of warning signs increases your chances of getting out before a collapse occurs. Visible warning signs of a potential cave-in include the following:

- Ground settlement or narrow cracks in the sidewalls, slopes, or surface next to the excavation
- Flakes, pebbles, or clumps of soil separating and falling into the excavation
- Changes or bulges in the wall slope

If you notice any of these signs, get out of the excavation immediately and alert your coworkers.

Soil type is a key factor in determining the type of protective system needed to ensure that the trench will be safe. Solid rock is the most stable, while sandy soil is the least stable. The four types of soil are shown in *Table 3*.

To be safe, treat soil as if it is Type C soil, per *Table 3*, unless proven otherwise. It is better to overprepare for a stronger soil than to not prepare enough for a weaker one.

A competent person must inspect excavations daily and decide whether cave-ins or failures of protective systems could occur and whether there are any other hazardous conditions present. The competent person must conduct inspections before any work begins, as needed throughout the shift, and after every rainstorm or other hazard-increasing incident.

If the inspection reveals indications of protective system failure, hazardous atmospheres, or a possible cave-in, workers must be removed from the hazardous area and may not return until corrective action has been taken. Always ask the competent person on-site or your immediate supervisor if you have questions about proper safety practices.

Once visual and manual tests are performed and the soil type is determined, a protective system must be chosen. Protective systems are required in nearly all excavations. There are various types of trench protective systems to meet each type of soil condition. Selecting a protective system for an excavation depends on soil conditions, the depth of the trench, and the environmental conditions surrounding the site.

Regardless of what type of system is used, if the excavation is more than 20' (6.1 m) deep, the entire excavation protective system must be designed by a registered professional engineer. There are two basic systems of trench protection: sloping and benching systems and support systems. The method to be used is determined by the engineer and is based on the types of soil and the site conditions.

Sloping and benching are forms of trench protection that cut away and slant the excavation face. A sloping system is a method in which the sides of an excavation are cut back to a safe angle using relatively smooth inclines (*Figure 43* [A]). A benching system is similar to a sloping system, but instead of smooth inclines, the sides of the trench wall are cut back using a series of steps (*Figure 43* [B]). Benching systems cannot be used with Type C soil.

Did You Know?

Heavy Soil

One cubic yard of earth weighs approximately 2,700 lb (1,225 kg)—that's the weight of a small car. Since a meter is longer than a yard, a cubic meter of soil weighs more—about 3,527 lb (1,600 kg).

Did You Know?

Cave-In Fatalities

Each year in the United States, more than 100 people are killed and many more are seriously injured in cave-in incidents. The chances of a trapped worker being killed can be as high as 50%.

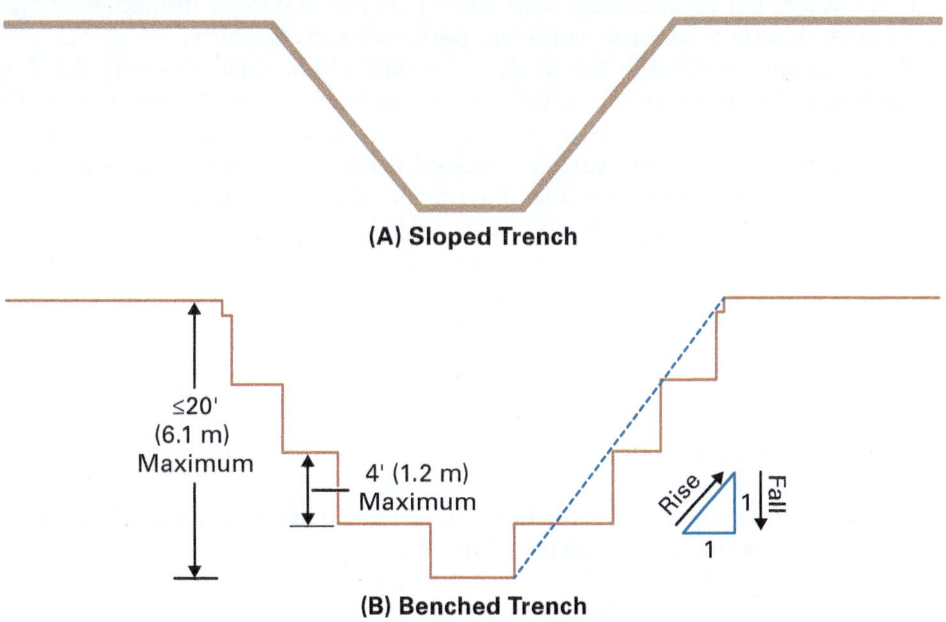

Figure 43 Sloped and benched trenches for Type B soil.

Trenches are often located in narrow places, so sloping and benching are not options. In these situations, support systems like **shoring** or **shielding** must be used. Shoring structures are typically made of metal or wood and are used to support the sides of a trench and prevent soil from caving in. They consist of plating held firmly in place with expandable braces (*Figure 44*). There are many types of shoring systems. Some of them are easy to install, while others require experience and engineering.

Figure 44 Shoring structure.
Source: Kundel Industries

Shoring: A support system designed to prevent a trench or excavation cave-in.

Shielding: A structure used to protect workers in trenches.

Shielding structures, also known as trench boxes, are placed inside trenches or excavations, and are strong enough to protect workers in the event of a cave-in, so long as the workers are within the confines of the box (*Figure 45*). Trench shields are used only to provide a protected space for workers. Shoring not only protects workers, but also prevents the trench walls from collapsing.

Spoil: Material such as earth removed while digging a trench or excavation.

Figure 45 Shielding structure.
Source: Kundel Industries

Spoil piles, comprised of the material removed from an excavation and other materials, represent a hazard if not handled properly. Loose rock, soil, materials, and equipment on the face or near the excavation can fall or roll into the excavation, or overload and possibly collapse excavation walls. Keep all spoil, materials, and heavy equipment at least 2' (0.6 m) away from the edge of an excavation, or set up barricades to contain falling material. Scale the excavation face to remove loose material, and place spoil so that rainwater runs away from the excavation. Use a retaining device strong enough and high enough to resist expected loads. If the spoil cannot be safely stored on-site, remove it to a temporary site.

When working in a trench, there must be a safe means of entry and exit for workers, such as a stairway, ladder, or ramp. There must be an exit every 25' (7.6 m) for every trench over 4' (1.2 m) deep. Lifting equipment such as loader buckets and backhoe shovels are not safe means for entering or exiting a trench. Once you are in the trench and before you begin work, take a moment to look around and find the nearest ladder so that you can plan your exit, if necessary.

Your company is required to have an emergency action plan that must be communicated to every worker. If you are not sure what the emergency action plan is for your site, don't be afraid to ask questions. Your knowledge could help prevent serious injury or even death, and your supervisor wants you and everybody working with you to be as safe as possible.

Most importantly, try to prevent emergencies before they happen. When in doubt, get out! If you notice potentially dangerous conditions while working in an excavation, get yourself and your coworkers out of danger immediately and inform your supervisor or a competent person of your concerns.

Call Before You Dig: Dial 811 to Keep Yourself and Others Safe

Additional information on safe digging practices may be found at **www.call811.com**.

Source: Common Ground Alliance

Digging and excavation are likely to be part of all construction projects. It is essential that the site supervisor or project manager gets approval to dig before work begins; most states have laws that require this. Millions of lines of cable, piping, and utility fixtures are buried throughout the United States, and breaking into one of them could lead to power outages, electrocution, gas leaks, explosions, and the loss of life.

In addition to the physical dangers of digging, contractors are at risk for fines, penalties, and legal action if they start digging without having called 811 to get the okay. Local utility representatives need to come out and mark the ground to indicate where their lines may be buried. Contractors cannot assume that the risk for hitting a line is low because a worksite is in a rural location. Utility lines are everywhere. Calling 811 a week or two in advance of the project allows local utility operators to respond and offer guidance as to where a worker may safely dig.

All states have a One-Call Notification System center that makes the process easy. Sometimes this can be done online. Regardless of state availability, however, 811 works nationwide. This could save your life.

The Common Ground Alliance Initiative, who leads the nation's campaign to "Call Before You Dig," has five rules for contractors. These rules include:

1. Make a call to either a local center or the national 811 line and place a request to dig.
2. Wait to hear back from all known and potential utilities. This can take several days to a couple of weeks, so be sure to call a few weeks ahead.
3. Confirm that everything is marked on the property before getting started. Utilities will place flags or other markers to indicate where it is or is not safe to dig.

4. Respect the process. This includes maintaining the markings made by utility operators and keeping current with the approval to dig. Where questions arise, call 811.
5. Proceed with caution. When digging, be aware of the tolerance zone, which is 1.5' to 2' (0.46 m to 0.6 m) away from a marked line. If the dig must happen closer to the line than 1.5' to 2' (0.46 m to 0.6 m), it must be done carefully by hand or with a vacuum excavation tool.

3.2.2 Tool, Machine, and Equipment Guards

Almost all tools and machines used in construction and industrial work are equipped with guards that protect workers from rotating parts. *Figure 46* shows the guards on a grinding machine. All tools and machines that could harm workers must have a guard shielding the hazard. The following types of tools and machines must have guards:

- Grinding tools
- Shearing tools
- Presses
- Punches
- Cutting tools
- Rolling machines
- Tools or machines with pinch points
- Tools or machines with sharp edges

Figure 46 Bench grinder with guards.
Source: iStock

Machine guards should prevent moving parts of the machine from coming in contact with your arms, hands, or any other part of the body, while allowing you to use the machine comfortably and efficiently. Some workers find machine guards to be aggravating and try to remove them from machines. Guards should be secure and should not be easily removed. They should be maintained in good condition, made of durable material, and bolted or screwed to the machine so that tools are needed for their removal. *Figure 47* shows a coupling guard installed over the coupling that connects a motor shaft to a pump shaft. The guard is bolted in place and should only be removed when the motor is not running. The guard prevents someone from being caught in the coupling, which rotates at high speed. Its highly visible markings indicate that it is hazardous.

Figure 47 Coupling guard.

Follow these guidelines for using and caring for tool and machine guards:

- Do not remove a guard from a tool or machine except for cleaning purposes or to change a blade or perform other services. Make sure the machine is turned off and tagged out.
- When you are finished with cleaning or maintenance, replace the guard immediately.
- Do not use any material to wedge a guard open.
- Guards and attachments are designed for the specific tool or machine you are using. Only use attachments that are specifically designed for that tool or machine.

3.2.3 Cranes and Heavy Equipment

Motorized equipment is used in many different jobs, including construction, agriculture, mining, plant maintenance and operations, road maintenance, equipment transportation, and snow removal. Working with motorized equipment can be dangerous. Workers can be crushed by falling loads, fall from equipment, be electrocuted by power lines, or be struck or trapped by vehicles. The swing radius of equipment can also be a hazard if the job is not carefully planned and properly barricaded. Most heavy equipment has pinch points. *Figure 48* shows some of the pinch point hazards on an excavator.

Figure 48 Pinch points.

Dangers exist for both equipment operators and other workers on the site. In one example, a contractor was operating a backhoe when another employee attempted to walk between the swinging back end of the backhoe and a concrete wall. As the employee approached the backhoe from the operator's blind side, the back end hit the victim, crushing him against the wall. A similar problem can occur with a crane or excavator, as shown in *Figure 49*. As the cab swings, the rear end extends beyond the base of the machine. For that reason, barricades must be placed around the working perimeter of the machine.

Because working with or near motorized equipment is dangerous, you must understand that your first responsibility on a job is safety. This includes your own safety, the safety of others on the site, and the safe use of equipment on the site. You must know the hazards and safety procedures of every job you are on, regardless of the work you are doing.

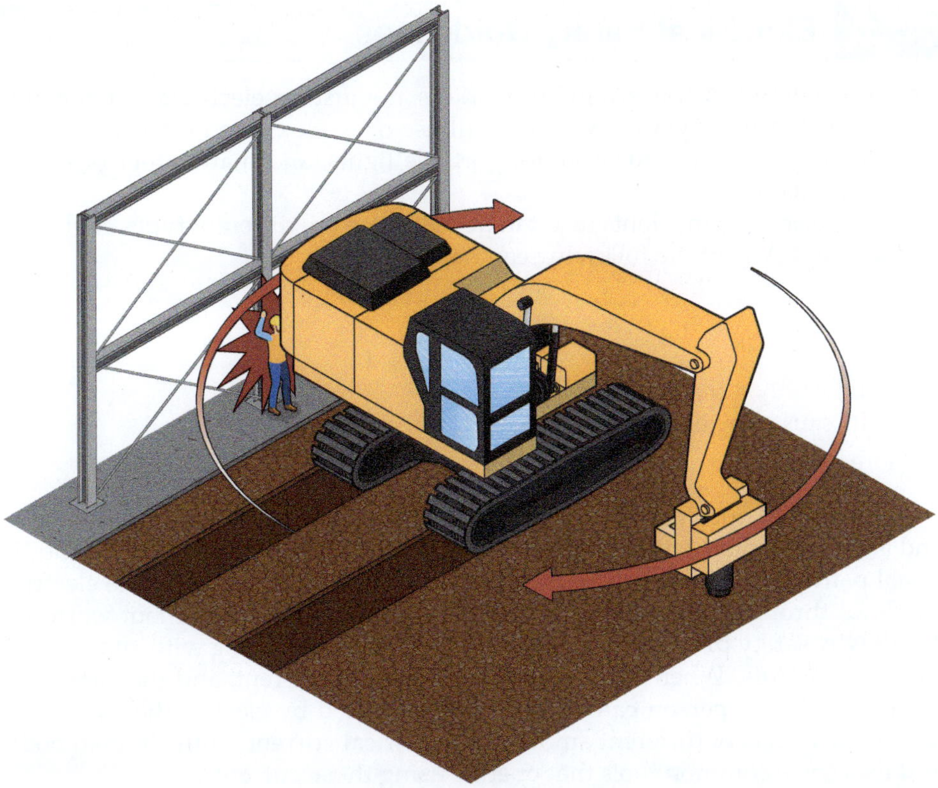

Figure 49 Example of a caught-between hazard.

3.0.0 Section Review

1. Which of these activities is *most* likely to produce flying objects?
   a. Using a chipping hammer
   b. Using a screwdriver
   c. Painting a wall
   d. Climbing a ladder

2. The type of soil *most* likely to result in a trench cave-in is _____.
   a. solid rock
   b. Type A
   c. Type B
   d. Type C

4.0.0 Energy Release Hazards

Objective

Identify common electrical hazards and how to avoid them.

a. Summarize basic jobsite electrical safety guidelines.

b. Explain the importance of disabling equipment as well as basic lockout/tagout (LOTO) procedures.

Performance Task

4. Inspect a typical power cord and ground fault circuit interrupter (GFCI) to ensure their serviceability.

Whenever possible, you will de-energize equipment before you begin working on it. Your employer will have written guidelines that you must follow to de-energize equipment. It is important to follow all guidelines, as some circuits receive power from multiple sources. To place equipment in a safe work condition, all energy sources must be removed. Once equipment is de-energized, **lockout/tagout (LOTO)** devices must be attached to all sources to prevent someone from unknowingly restoring the energy before the work has been completed.

Lockout/tagout (LOTO): A formal procedure for taking equipment out of service and ensuring that it cannot be operated until an authorized person has removed the lock and/or warning tag.

4.1.0 Electrical Safety Guidelines

Electrical safety is a concern for all workers, not just for electricians. On many jobs, no matter what your trade, you will use or work around electrical equipment. Extension cords, power tools, portable lights, and many other pieces of equipment use electricity.

Not all electrical incidents result in death. There are different types of electrical incidents. Any of the following can happen:

- Burns
- Electric shock
- Explosions
- Falls caused by electric shock
- Fires

CAUTION

The National Ag Safety Database's 2023 report identified 62 farm worker deaths by electrocution in the United States.

Ground: The conducting connection between electrical equipment or an electrical circuit and the earth.

If the human body comes in contact with an electrically energized conductor and is also in contact with a **ground** at the same time, the body becomes an additional path of resistance for the electrical current to flow through. The electricity flows through the body in less than the blink of an eye without warning. That's why safety precautions are so important when working with and around electrical circuits. When a body conducts electrical current, and the current is high enough, the person can be electrocuted (killed by electric shock). *Table 4* shows the effects of different amounts of electrical current on the human body and lists some common tools that operate using those currents.

TABLE 4 Effects of Electrical Current on the Human Body

Current in Amperes (A)	Common Item/Tool	Reaction to Current
0.001 A	Watch battery	A faint tingle.
0.005 A	9 V battery	A slight shock.
0.006 A to 0.025 A (women) 0.009 A to 0.030 A (men)	Christmas tree bulb	A painful shock. Muscular control is lost.
0.050 A to 0.9 A	Small electric radio	Extreme pain. Breathing stops; severe muscular contractions occur. Death may result.
1.0 A to 9.9 A	Jigsaw (4 A), Sawsall® or Port-a-Band® saw (6 A), portable drill (3 A to 8 A)	Ventricular fibrillation and nerve damage occur. Death may result.
10 A and above	ShopVac® (15-gallon), circular saw	Heart stops beating; severe burns occur. Death may result.

Here's an example: A craftworker is operating a portable power drill while standing on damp ground. The power cord inside the drill has become frayed, and the electric wire inside the cord touches the metal drill frame. 3 A of current pass from the wire through the frame, then through the worker's body and into the ground. *Table 4* shows that this worker will probably die.

There are specific policies and procedures to keep the workplace safe from electrical hazards. You can do many things to reduce the chance of an electrical incident. If you ever have any questions about electrical safety on the jobsite, ask your supervisor.

Did You Know?

Electrocution

Electric shocks or burns are a major cause of incidents in the construction industry. According to the Bureau of Labor Statistics, electrocution is the fourth leading cause of death among construction workers.

WARNING!

Less than 1 A of electrical current can kill. Always take precautions when working around electricity.

4.1.1 Grounding

Grounding is a method of protecting humans from electric shock; however, it is normally a secondary protective measure. The term *ground* refers to a conductive body, usually the earth. A ground is a conductive connection, whether intentional or incidental, by which an electric circuit or equipment is connected to earth or to an engineered grounding system. By grounding a tool or electrical system, a low-resistance path to the earth is intentionally created. When properly done, this path offers low resistance and has enough current-carrying capacity to prevent the buildup of voltages that could create a personnel hazard. This does not guarantee that no one will receive a shock. It will, however, greatly reduce the possibility of such incidents, especially when used in combination with your company's safety program.

Use three-wire extension cords for portable power tools and make sure they are properly connected (*Figure 50* [A]). The three-wire system is one of the most common safety grounding systems used to protect you from incidental electrical shock. The third wire is connected to a ground system. If the insulation in a tool fails, the current will pass to ground through the third wire—not through your body. Double-insulated tools (*Figure 50* [B]) are also very effective in preventing shocks. In fact, it has become more common to use double-insulated tools because they are safer than relying on a three-wire cord alone. *Figure 50* (C) also shows the double-insulated symbol that can be found on double-insulated tools. Double-insulated tools use a two-wire power cord with no ground pin. One prong of the plug is larger than the other so it can only be connected to a polarized receptacle.

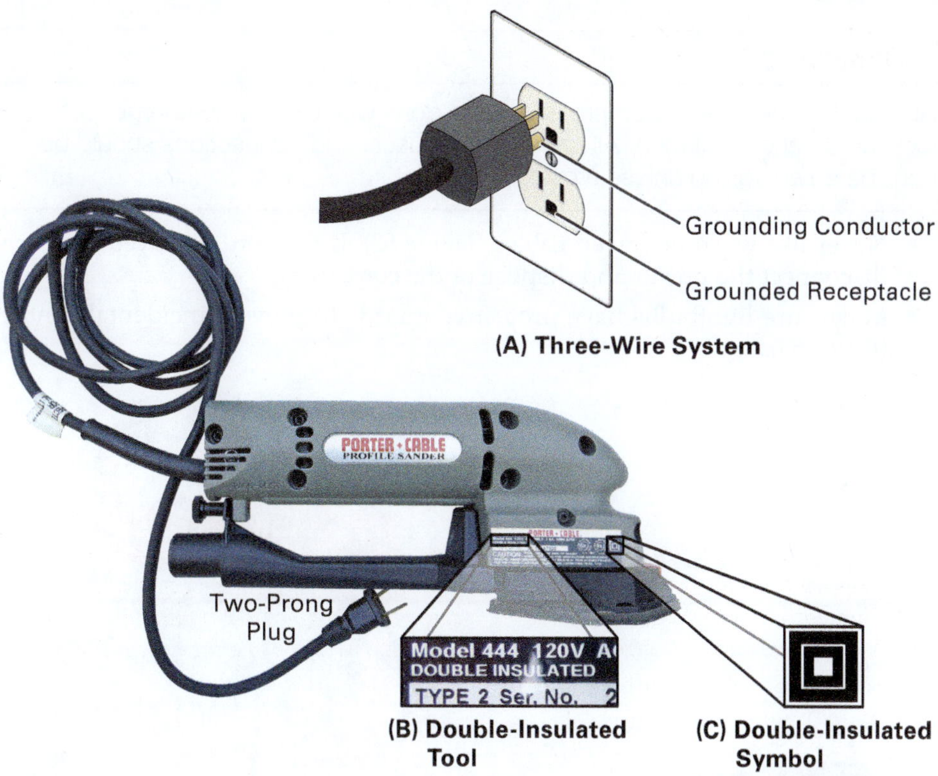

Figure 50 Three-wire system, double-insulated tool, and double-insulated symbol.

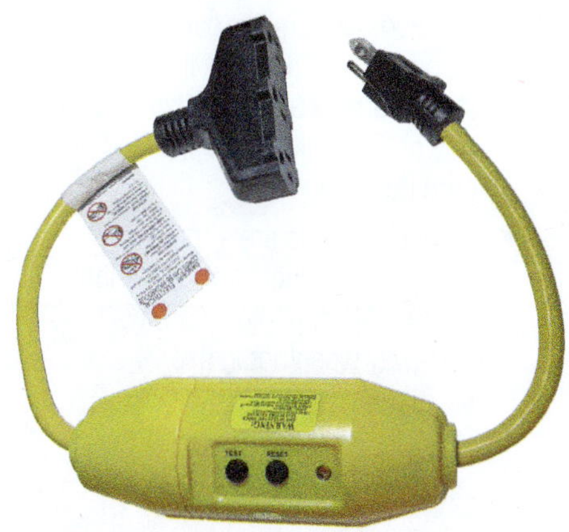

Figure 51 Extension cord with a GFCI.

4.1.2 Ground Fault Circuit Interrupters

A **ground fault circuit interrupter (GFCI)** is a fast-acting circuit breaker that senses small imbalances in the circuit caused by current leakage to ground. A GFCI continually matches the amount of current going to an electrical device against the amount of current returning from the device. Whenever the two values differ by more than 5 milliamps (mA), the GFCI interrupts the electric power within $\frac{1}{40}$ of a second. *Figure 51* shows an extension cord with a GFCI.

Ground fault circuit interrupter (GFCI): A device designed to protect people from electric shock and equipment from damage by interrupting the flow of electricity if a circuit fault occurs.

A GFCI must be used on all receptacles that are not part of the building's permanent wiring, such as temporary power and extension cords. A GFCI provides protection against a ground fault, which is the most common form of electrical shock. It also provides protection from fires, overheating, and wiring insulation deterioration.

Tripping of GFCIs—interruption of circuit flow—is sometimes caused by wet connectors and tools. Limit the amount of water that tools and connectors come into contact with by using watertight or sealable connectors. Tripping may also be caused by cumulative leakage from several tools or from extremely long circuits. GFCIs should be periodically inspected in accordance with the manufacturer's recommendations or site safety practices. GFCIs have a Test button that can be pressed to verify that the GFCI is working. The GFCI should be tested before any use.

Electrical Cord Safety

Electrical cords are frequently seen on construction sites, yet they are often overlooked. Use the following safety guidelines to ensure your safety and the safety of other workers.

- Every electrical cord should have a UL Solutions label attached to it. Check the UL Solutions label for specific wattage. Do not plug more than the specified number of watts into an electrical cord.
- A cord set not marked for outdoor use is to be used indoors only. Check the UL Solutions label on the cord for an outdoor marking.
- Do not remove, bend, or modify any metal prongs or pins of an electrical cord.
- Do not run a cord through doorways or through holes in ceilings, walls, and floors that might pinch the cord. Also, check to see that there are no sharp corners along the cord's path. Any of these situations will lead to cord damage.
- Extension cords are a tripping hazard. They should never be left unattended and should always be put away when not in use.

4.1.3 Summary of Electrical Safety Guidelines

The following are some basic jobsite electrical safety guidelines:

- All power tools used in construction should be ground fault protected.
- Make sure that panels, switches, outlets, and plugs are grounded.
- Never use bare electrical wire.
- Never use metal ladders near any source of electricity.
- Inspect electrical power tools before you use them.
- Never operate any piece of electrical equipment that has a danger tag or lockout device attached to it.

WARNING!

All work on electrical equipment should be done with circuits de-energized, locked out, and confirmed. All conductors, buses, and connections should be considered energized unless proven otherwise.

- Never use worn or frayed cables (*Figure 52*). If the cord is frayed or worn, disconnect the power and dispose of the cord.
- Make sure lightbulbs have protective guards to prevent incidental contact (*Figure 53*).

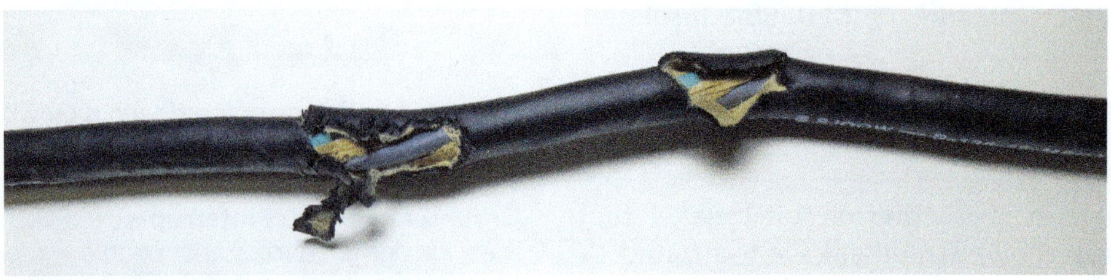

Figure 52 Never use damaged cords.

Figure 53 Work light with protective guard.

4.1.4 Working Near Energized Electrical Equipment

No matter what your trade, your job may include working near exposed electrical equipment or conductors. This is one example of **proximity work**. Often, electrical distribution panels, switch enclosures, and other equipment must be left open during construction. This leaves the wires and components in them exposed. Some or all the wires and components may be energized. Working near exposed electrical equipment can be safe, but only if you keep a safe working distance.

Proximity work: Work done near a hazard but not actually in contact with it.

Regulations and company policies tell you the minimum safe working distances from exposed conductors. The safe working distance can be very small or span a significant distance, depending on the voltage. The higher the voltage, the greater the required safe working distance.

You must learn the safe working distance for each situation. Make sure you never get any part of your body or any tool you are using closer to exposed conductors than that distance. You can get information on safe working distances from your instructor, your supervisor, company safety policies, and regulatory documents. This subject is also covered in greater detail in other craft curricula.

One of the common causes of electrical shock, especially in the agricultural industry, is contacting overhead wires. These risks are present when using metal ladders, cranes, grain spouts on combines, tilling chisels, or excavating equipment. A distance of at least 10' (3 m) must be maintained from any conductor carrying 50,000 V or less. Greater distances are required for higher voltages.

4.1.5 Electrical Shock

If you are there when someone gets an electrical shock, you can save a life by taking immediate action. The best thing to do is to immediately shut off the power to the circuit. Do not touch the victim while they remain in contact with the power source. Once the circuit is disconnected, call an ambulance or the emergency number at your jobsite. Render first aid only if you are trained to do so.

WARNING!

Do not touch the victim or the electrical source with your hand, foot, or any other part of your body or any object or material. You could become part of the circuit—and another victim.

4.2.0 LOTO Requirements

Failure to disable machinery before working on it is a major cause of injury and death on jobsites. LOTO procedures safeguard workers against unexpected releases from various energy sources. An energy source can be electrical, mechanical, hydraulic, pneumatic, chemical, or thermal. After the power has been turned off, energy can still be stored in a device or component. For example, even when a hydraulic system has been turned off, hydraulic pressure may remain in all or part of the system. This high-pressure energy can be released if the system is opened during service or repairs, creating an extremely hazardous situation. Additional dangers exist if the device or system contains chemicals, flammable liquids, high-temperature liquids, or gases.

When anyone is working on or around active systems, all equipment that could release energy must be shut down, drained, de-energized, or otherwise rendered harmless whenever possible. Switches, circuit breakers, valves and other components are switched off or closed, and then locks or tags (*Figure 54* [A]) are applied so they cannot be re-energized while work is ongoing.

(A) Electrical Lockout

(B) Pneumatic Lockout

Figure 54 LOTO devices.
Source: Brady Corporation (54B)

Figure 55 Typical safety tags.

Generally, each lock has its own key, and the individual who applies the lock keeps the key. A variety of tags are used, depending on the circumstances and the organization. Tags (*Figure 55*) typically have the word *DANGER* on them, along with other printed information, writing space for comments, and a line for the installer's signature and date. The authorized person who applied the lock must be the one to remove it. In an emergency, a supervisor or other person authorized by the employer may remove a LOTO.

In a lockout, an energy-isolating device such as a disconnect switch or circuit breaker is placed in the Off position and a lock is applied. Padlocks are popular and versatile, but other styles (*Figure 56*) are often specifically suited for the equipment being serviced. Multiple lockout devices (*Figure 57*) are used when more than one person is accessing the equipment. When a worker needs to service a system and sees that another person's lock has been applied, it is tempting to accept this as a safe situation. However, the first worker may remove the lock and re-energize the circuit while the second worker is still vulnerable. In these cases, the multiple lockout devices allow several workers to apply locks, and each is ensured that the system cannot be restored until theirs is removed.

(A) Electrical Panel Lockout

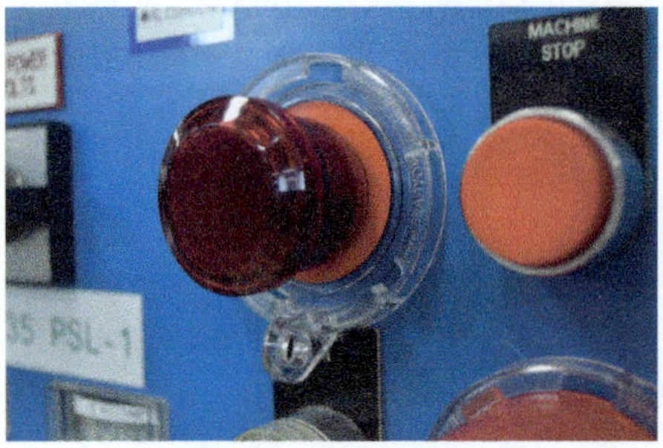

(B) Push Button Lockout

(C) Electrical Plug Lockout

Figure 56 Lockout devices.
Source: Brady Corporation

In a tagout, components that control power to equipment and machinery are set to a safe position and a written warning is attached. This method may be more appropriate for valves and other energy-controlling devices that are difficult or impossible to lock, since tags alone cannot prevent a control device or switch from being repositioned.

It is important to emphasize that LOTO procedures are not solely related to electrical energy sources. Technicians cannot forget that other forms of energy may need to be rendered safe as well.

The exact procedures for LOTO will vary by organization and site. Check with your instructor or site supervisor regarding the detailed LOTO procedure at your site and be sure you are familiar with the proper steps.

Figure 57 Multiple LOTO devices.

4.2.1 Pressurized or High-Temperature Systems

Many jobs require that workers be close to tanks, piping systems, and pumps that contain pressurized or high-temperature fluids. Be aware that touching a container of high-temperature fluid can cause burns. Many industrial processes involve fluids that are as hot as several thousand degrees. Also, if a container holding pressurized fluids is damaged, it may leak and spray dangerous fluids. Any work around pressurized or high-temperature systems is considered proximity work. Barricades, a monitor, or both may be needed to ensure the safety of those working nearby.

4.0.0 Section Review

1. The purpose of double insulating an electrically powered tool is to _____.
   a. prevent the tool from overheating
   b. protect users from electrical shock
   c. keep the tool from getting too cold
   d. allow it to use a three-prong receptacle

2. When a lockout is performed, the key to the lock is _____.
   a. left in the lock
   b. kept by the person who applied the tag
   c. given to the foreman for safekeeping
   d. hidden in a convenient place

5.0.0 Personal Protective Equipment

Objective

Associate personal protective equipment (PPE) with the hazards they reduce or eliminate.

a. Explain how PPE is used to protect craftworkers from different types of injuries.
b. Explain how respirators protect craftworkers from respiratory dangers.

Performance Tasks

2. Inspect the following PPE items and determine if they are safe to use:
   - Eye protection
   - Hearing protection
   - Hard hat
   - Gloves
   - Approved footwear
3. Properly don, fit, and remove the following PPE items:
   - Eye protection
   - Hearing protection
   - Hard hat
   - Gloves

Every worker is responsible for wearing appropriate PPE on the job. When worn correctly, PPE is designed to protect you from injury. You must keep it in good condition and use it when you need to. When workers are injured on the job, it is often because they are not using required PPE.

You will not see all the potentially dangerous conditions just by looking around a jobsite. It is important to stop and consider what type of incidents could happen on any job that you are about to do. Knowing how to use PPE will greatly reduce your chance of getting hurt. Keep in mind that PPE requirements are usually established by the jobsite. Some jobsites will have different requirements than others, so always make sure you check. For example, some sites require high-top safety boots instead of the more common standard 6" tops.

5.1.0 PPE Items

Remember, while PPE is the last line of defense against personal injury, using it properly and taking care of it are the first steps toward protecting yourself on the job.

The first line of defense is represented by the engineering and administrative controls used by employers to eliminate or reduce the need for some PPE. Engineering controls are implemented by making physical changes, such as reducing noise by the use of mufflers on equipment engines, installing guards on moving equipment, and making sure equipment is properly maintained. An example of an administrative control is operating noisy equipment on a shift when fewer workers are present.

The best protective equipment is of no use unless you follow these rules:

- Regularly inspect it.
- Properly care for it.
- Use it properly when it is needed.
- Never alter or modify it in any way.

The following sections describe protective equipment commonly used on construction, agricultural, and mining sites, and tell how to use and care for each piece of equipment. Be sure to wear the equipment according to the manufacturer's specifications.

WARNING!

Your clothing must comply with good general work and safety practices. Do not wear clothing or jewelry that could get caught in machinery or otherwise cause an incident, such as loose clothing, baggy shirts, or dragging pants. You must always wear a shirt; some tasks will require long-sleeve shirts. Your shirt should always be tucked in unless you are welding.

Protective equipment commonly worn by craftworkers on any jobsite usually includes the following:

- Hard hats
- Eye and face protection
- Gloves
- Foot and leg protection
- Hearing protection
- Respiratory protection
- High-visibility clothing

Sharing PPE

It is not a good idea for workers to share PPE. If one person has an infection, it can be passed to another through the shared equipment.

5.1.1 Hard Hats

Figure 58 shows a typical hard hat. The outer shell of the hat can protect your head from a hard blow. The webbing inside the hat keeps space between the shell and your head. Adjust the headband so that the webbing fits your head and there is at least one inch of space between your head and the shell.

Figure 58 Typical hard hat.
Source: Photo courtesy of Bullard

Do not alter your hard hat in any way. Inspect your hard hat every time you use it. If there are any cracks or dents in the shell, or if the webbing straps are worn or torn, get a new hard hat. Wash the webbing and headband with soapy water as often as needed to keep them clean. Wear the hard hat only as the manufacturer recommends. Never wear anything but approved clothing under your hard hat.

WARNING!

No articles should be worn under the hard hat that could interfere with fit and visibility. That includes ball caps or hoodies that obscure peripheral vision. Only employer-approved gear is to be worn under the hard hat.

Did You Know?

Hard Hats

Hard hats were once made of metal. However, metal conducts electricity, so most hard hats are now made of reinforced plastic or fiberglass.

5.1.2 Eye and Face Protection

Wear eye protection (*Figure 59*) whenever there is even the slightest chance of an eye injury. In general, eye protection is required any time you are on a jobsite. Face shields (*Figure 60*) are added over safety glasses for certain tasks, such as grinding. Appropriate eye or face protection is required when there is a possibility of exposure to hazards from flying particles, molten metal, liquid chemicals, acids or caustic liquids, chemical gases or vapors, or potentially injurious light radiation. The following are examples of tasks requiring eye protection:

- Grinding and chipping metal
- Using power saws and other tools/equipment that can throw out solid material
- Working with molten lead, tar pots, and other molten materials
- Working with chemicals, acids, and corrosive liquids
- **Arc welding**

Arc welding: The joining of metal parts by fusion, in which the necessary heat is produced by means of an electric arc.

(A) Safety Glasses

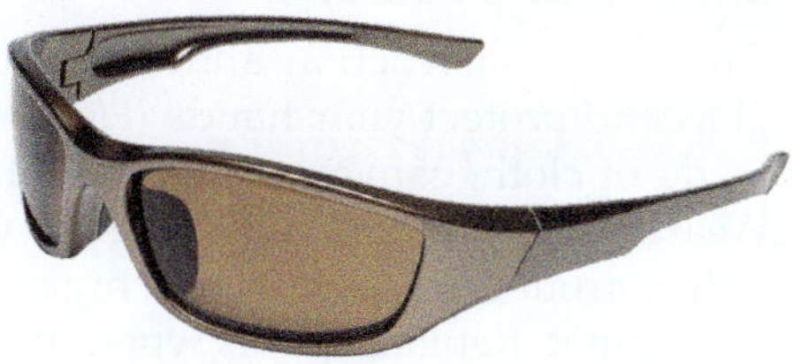

(B) Tinted Safety Glasses

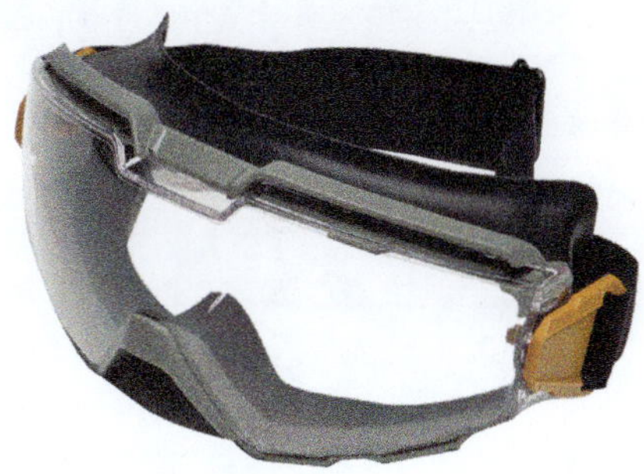

(C) Safety Goggles

Figure 59 Typical safety glasses and goggles.
Source: Used with permission from MSA

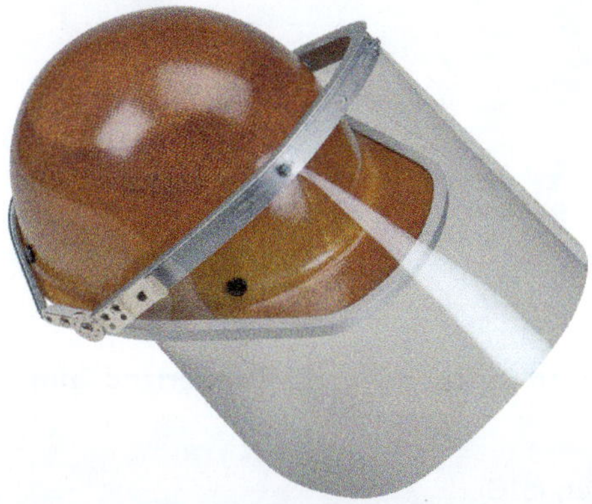

(A) Clear Face Shield

(B) Tinted Face Shield

Figure 60 Full face shields.
Source: Used with permission from MSA

CAUTION

Standard eyeglasses are not adequate protection in high-hazard work zones. Shatterproof lenses and side shields are required for eye protection in those areas.

On the jobsite, potential eye hazard areas are usually identified, but always be on the lookout for possible hazards.

Regular safety glasses will protect you from falling objects or from objects flying toward your face. Side shields provide added protection and are required whenever there is risk of flying debris. Safety goggles provide the best protection from all directions. A face shield is required, in addition to safety glasses or goggles, when there is likely to be flying debris. Grinding and chipping activities are good examples of such activities.

Handle your safety glasses and goggles with care. If they get scratched, replace them; the scratches will interfere with your vision. Clean the lenses regularly with lens tissues or a soft cloth.

Welders must use tinted goggles or welding hoods. The tinted lenses protect the eyes from the bright welding arc or flame. Welders must use filter lenses as specified by jobsite requirements. Welder's helpers and all employees working in the vicinity of arc welding should not look directly at the welding process and are also required to use eye protection with the prescribed level of shading. Oxyacetylene welding and burning also require the use of a filter lens.

Follow these general precautions for eye care:

- Always report all eye injuries and suspected foreign material in your eye to your supervisor immediately. Do not try to remove foreign matter yourself.
- Keep your hands away from your eyes.
- Keep materials out of your eyes by regularly clearing debris from your hard hat brim, the top of your goggles, and your face shield. When removing a cap, hard hat, or face shield, be careful to remove it in a way that prevents accumulated dirt and debris from falling into your eyes.
- Flood your eyes with water if you feel something in them. Never rub them, as this can make the problem worse.
- Know the location of eyewash stations and how to use them.

5.1.3 Hand Protection

On many construction and agricultural jobsites, you must wear heavy-duty gloves to protect your hands (*Figure 61*). Construction work gloves are usually made of cloth, canvas, or leather. Make sure your work gloves are a good fit. Wearing gloves that are too big for your hands can lead to injury. Never wear gloves around rotating or moving equipment. They can easily get caught in the equipment. Replace gloves when they become worn, torn, or soaked with oil or chemicals.

(A) Fitted Work Gloves **(B) Gloves with Cuffs** **(C) Rubberized Gloves**

Figure 61 Work gloves.
Sources: MCR Safety (61B); Courtesy of LIFT Safety (61C)

Gloves help prevent cuts and scrapes when you handle sharp or rough materials. Heat-resistant gloves are needed for handling hot materials. Craftworkers should be familiar with standards governing hand protection, such as *ANSI/SEA Standard 105-2011, Hand Protection Selection Criteria* and *European Standard EN388, Protective Gloves Against Mechanical Risks*. These standards establish a rating system for gloves tested for their resistance to various hazards, including chemical burns, abrasions, cuts, punctures, heat and flame resistance, and cold resistance. Additional European standards govern protection from chemicals and microorganisms (EN374), thermal hazards (EN407 and EN511), and ionizing radiation and radioactive contamination (EN421).

Electricians use special rubber-insulated gloves when they work on or around live circuits. When working with solvents or other chemicals, it may be necessary to wear chemical-resistant gloves.

WARNING!

Only specially trained employees are allowed to use dielectrically tested rubber gloves to work on energized equipment. Never attempt this work without proper training and authorization.

5.1.4 Foot and Leg Protection

Approved safety footwear must be worn on all jobsites. The best shoes to wear on a construction site are shoes with a safety toe to protect toes from falling objects (*Figure 62*). This type of shoe is generally required on every construction-related jobsite.

Figure 62 Safety shoe.

Some specialized work will require different footwear or gear. For example, safety-toed rubber boots are needed on jobsites that are subject to chemically hazardous conditions or standing water. When climbing a ladder, your shoes or boots must have a well-defined heel to prevent your feet from slipping off the rungs. You will need foot guards when using jackhammers and similar equipment.

Never wear canvas shoes or sandals on a construction, agricultural, or mining site. They do not provide adequate protection. Always replace boots or shoes when the sole tread becomes worn or the shoes have holes, even if the holes are on top. Because of the risk of fire, do not wear oil-soaked shoes when you are welding or cutting metal.

Remember these general guidelines relating to leg protection:

- Never carry pointed tools, such as scissors or screwdrivers, in your pants pockets. Use a canvas or leather tool sheath with all sharp ends pointing down.
- When using certain special equipment, such as chainsaws and brush hooks, use shin guards.
- Consider stability when stepping into or onto locations where materials are stored. Materials may shift and pinch your legs and/or feet.

Case History

Safety Gear Matters

A training specialist traveled nearly three hours to reach a remote jobsite in order to observe and photograph the installation of high-voltage transmission lines. On arriving, he parked his car, put on his work boots, hard hat, reflective vest, and safety glasses, and walked over to the location where the work was being done. He was introduced to the project manager, who shocked him by saying: "Sir, I have to ask you to leave my jobsite because you are not wearing electrically insulated boots." Given no choice, the training specialist headed toward his car for the long trip home. Fortunately, the project manager needed to go back to the office for a meeting and offered the loan of his boots, thus preventing six hours of wasted time, as well as the loss of a rare opportunity.

The Bottom Line: Project managers and superintendents in construction, agricultural, and industrial work are serious about safety. If you are not fully prepared, you may find yourself sitting on the sidelines. A worker who shows up at a jobsite without the required PPE could lose a day's pay and possibly face disciplinary action.

5.1.5 Hearing Protection

Unlike damage to most parts of the body, ear damage does not always cause pain. Exposure to loud noise over a long period of time can cause hearing loss, even if the noise is not loud enough to cause pain. Vibrations caused by sound waves enter the ear and are received by the cochlea, which consists of tiny hair cells inside the ear that convert the vibrations into sound signals. Exposure to intense sound waves over time can damage the cochlea and reduce the ability to hear. Damage to the cochlea can also cause tinnitus, which is a permanent condition characterized by ringing or buzzing in the ears that never stops. Hearing loss reduces your quality of life and makes simple, daily tasks more complicated. Save your hearing by using hearing protection whenever you are working in a noisy environment.

Here is a good rule to follow: if the noise level is so great that you have to raise your voice to be heard by someone who is less than 2' (61 cm) away, you need to wear hearing protection.

Most construction companies follow rules from official safety standards in deciding when hearing protection must be used. A type of hearing protection includes specially designed earplugs that fit into your ears and filter out noise (*Figure 63* [A]). Another type of hearing protection is earmuffs, which are large, padded covers for the entire ear (*Figure 63* [B]). The headband on earmuffs must be adjusted for a snug fit. If the noise level is very high, you may need to wear both earplugs and earmuffs.

(A) Earplugs **(B) Earmuffs**

Figure 63 Hearing protection.
Sources: iStock (63A); Used with permission from MSA (63B)

Noise-induced hearing loss can be prevented by using noise control measures and personal protective devices. *Table 5* shows the recommended maximum length of exposure to sound levels rated 90 decibels (dB) and higher. When noise levels exceed those outlined in *Table 5*, an effective hearing conservation program is required. A company-appointed program administrator will oversee this program. If you have questions about the hearing conservation program on your site, see your supervisor or the program administrator.

TABLE 5 Maximum Noise Levels

Sound Level (dB)	Maximum Hours of Continuous Exposure per Day	Example
90	8	Power lawn mower
92	6	Belt sander
95	4	Tractor
97	3	Hand drill
100	2	Chain saw
102	1.5	Impact wrench
105	1	Spray painter
110	0.5	Power shovel
115	0.25 or less	Hammer drill

5.2.0 Respiratory Hazards and Protection

Wherever there is danger of an inhalation hazard, you must use a respirator. There are a number of jobsite conditions that require workers to wear respirators, including the following:

- Dust from metal grinding
- Drilling or cutting concrete
- Toxic vapors from chemical sprays, animal waste, or fuels
- Toxic fumes from welding or flame cutting of some metals
- Working with cleaning solvents
- Working in low-oxygen environments such as confined spaces
- Spray painting
- Sandblasting
- Working with chemicals such as chlorine or ammonia

Construction sites and materials present hazards on the jobsite, but you might also be exposed to inhalation hazards from operations around a jobsite. Agricultural operations present such a risk due to the use of herbicides, pesticides, insecticides, and other chemicals (*Figure 64*). You should always be aware of the surrounding activities whenever working on an agricultural jobsite. If airborne chemicals are being used to support the daily activities of a farm or ranch, protect yourself by using an air-purifying respirator (APR).

5.2.1 Silica Standard

Silica is a mineral found in concrete, masonry, and rock. During construction, silica may be found in a dust form. Silica dust is created when cutting, drilling, or performing similar tasks on materials containing crystalline silica. Prolonged exposure to silica dust can cause **silicosis**, which is an incurable, and sometimes fatal, lung disease. The time it takes for silicosis to develop varies based on the length of time and the amount of silica someone is exposed to. In addition to silicosis, the inhalation of silica dust increases the risk of lung cancer and **chronic obstructive pulmonary disease (COPD)**. Individuals working closely with materials containing silica, such as workers operating abrasive blasting (sandblasting) equipment, also risk the development of kidney disease.

WARNING!

When you are working in an area where silica dust is present, you must use the appropriate respiratory protection. Never work around silica dust without proper training, authorization, and PPE.

Silicosis: A serious lung disease resulting from the inhalation of crystalline silica particles.

Chronic obstructive pulmonary disease (COPD): Lung disease(s) that results in the obstruction of lung airflow and interferes with normal breathing. Chronic bronchitis and emphysema generally fall under the diagnosis of COPD.

Figure 64 Farmhand wearing APR while spraying.
Source: iStock

OSHA holds employers responsible for protecting workers from crystalline silica. *OSHA Standard 29 CFR 1926.1153*, known in the workplace as the silica standard, outlines the steps for controlling exposure when working with sand, stone, concrete, and similar materials. Several options are available for keeping workers safe, as well as methods to monitor exposure.

The basic requirements for employers include the establishment and implementation of a written silica exposure control plan that identifies tasks leading to exposure and the approaches to protecting workers. In addition, employers must do the following:

- Designate a competent person to ensure the exposure control plan is effectively implemented.
- Establish procedures to restrict access to areas where silica dust is actively generated.
- Train workers on tasks that create silica dust and ways to limit exposure.
- Limit common housekeeping practices that expose workers to silica dust when an alternative is available.
- Provide medical exams that include lung function tests and X-rays every three years for workers that wear a respirator for 30 days or more each year.
- Keep records of all the above.

The OSHA silica standard guides employers and workers in managing silica dust exposure. *Table 1* of the standard provides a great deal of the information needed by craftworkers. All construction craftworkers are encouraged to review the table and become familiar with the requirements. Note that the requirements are also subject to change, and this topic is one that is consistently monitored to ensure the guidelines remain effective.

NOTE

Table 1 of *OSHA Standard 29 CFR 1926.1153* is an extensive table and subject to change; therefore, its contents have not been included herein. All trainees are encouraged to access the standard at **www.osha.gov/laws-regs/regulations/standardnumber/1926/1926.1153** and become familiar with the requirements for various activities related to silica dust.

The primary means of management is to use saws and similar power tools that are equipped with a built-in water delivery system (*Figure 65*). Water prevents the dust from becoming airborne. Handheld and stand-mounted concrete drills must have a dust collection system connected to a vacuum (*Figure 66*). Vacuums used to capture silica dust must be equipped with **high-efficiency particulate arrestance (HEPA) filters** (*Figure 67*). These filters are extremely effective for capturing the tiniest particles and are also used in critical environments such as hospital surgical areas and **clean rooms**.

High-efficiency particulate arrestance (HEPA) filters: Special filters designed to capture very small particles. HEPA filters have a minimum efficiency of 99.97%, capturing particles that are 0.3 microns (0.0000117") or larger in size.

Clean rooms: A room or area that is used for scientific research or specialized processes, such as the manufacture of pharmaceuticals and semiconductors. The cleanliness level of a clean room is generally based on the number of particles measured per cubic meter of area.

Figure 65 Concrete saw with a water delivery system (wet sawing).
Source: Robert I. Carr, Ph. D., P.E.

Figure 66 Rotary hammer drill with OSHA Table 1-compliant onboard dust extractor.
Source: Image property of Stanley Black & Decker. Used with permission.

Figure 67 HEPA vacuum cleaner.
Source: Photo used with permission of Husqvarna Group

There are some conditions that require the use of breathing filtration. For silica dust applications, OSHA requires filtration to meet an assigned protection factor (APF) of 10. This is a level of performance that is typical of many respirators. However, not all disposable filters meet the criteria.

It is important for your long-term health and the health of others to be aware of the sources of silica dust and how they are managed. Epoxy grout, for example, also contains silica. Anytime concrete, brick, stone, grout, and similar materials are being ground or sawed, it is best to assume silica is present and act accordingly.

5.2.2 Other Respiratory Hazards

Asbestos is another hazardous material that can be harmful to your lungs. Prolonged exposure can cause lung cancer, asbestosis (scarring of the lung tissue), and a cancer called mesothelioma. It may take more than 20 years for these diseases to develop. Smoking significantly increases the risk of lung disease. If you smoke, tar from cigarettes will stick to the asbestos fibers in the lungs, making it more difficult for your body to get rid of the asbestos.

Sources of asbestos include older insulation and other building materials, such as floor tiles, pipe insulation, drywall compounds, and reinforced plaster, along with some roofing and siding materials. The United States banned production of most asbestos products in the 1970s, meaning that asbestos-containing materials (ACMs) are generally found in structures built before 1980. During renovation or maintenance operations, asbestos may be dislodged and the fibers become airborne. Asbestos fibers are particularly hazardous and must not be disturbed unless special techniques and procedures are used to remove it safely. Although you will not feel the effects immediately, this exposure can be a serious health hazard. Never handle asbestos without proper training, authorization, and PPE.

Only workers who have been trained and licensed are authorized to perform work related to asbestos including, but not limited to, the following tasks:

- Demolition
- Removal
- Alteration
- Repair
- Maintenance
- Installation
- Cleanup
- Transportation
- Disposal
- Storage

If asbestos is encountered or suspected on the jobsite, a supervisor must be notified and all work must stop until the asbestos is either sealed off or removed by licensed professionals.

Around the World

Asbestos Use

Asbestos was identified as a health risk many years ago. However, countries have reacted to the hazard in different ways. While some countries banned its use in products and construction decades ago, other countries, such as India, continue to use asbestos without regulation of any form. Indeed, asbestos use in India has risen over 80% in the last decade.

The World Health Organization (WHO) estimates that 125 million people worldwide are exposed to asbestos each year in the workplace. Even in the United States, asbestos may still be encountered in old buildings, flooring, and insulation components. Be aware of the regulations concerning the use of asbestos in your region and stay alert to sources that might do you harm if not avoided.

5.2.3 Types of Respirators

Federal law specifies which type of respirator to use for different types of hazards. There are four general types of respirators:

- Half mask or mouthpiece with a mechanical filter
- Full facepiece mask with a chemical canister (gas mask)
- Supplied-air mask or powered air-purifying respirator (PAPR)
- Self-contained breathing apparatus (SCBA)

A full facepiece mask with a chemical canister is used to protect against brief exposure to dangerous gases or fumes. A half mask or mouthpiece with a mechanical filter is used in areas where you might inhale dust or other solid particles (*Figure 68*).

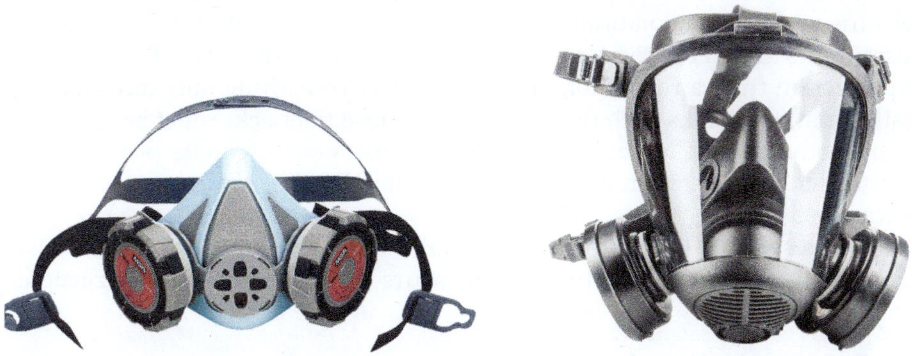

(A) Half Mask **(B) Full Facepiece Mask**

Figure 68 Examples of air-purifying respirators.
Sources: Used with permission from MSA (68A); iStock (68B)

A SCBA carries its own air supply in a compressed air tank. It is used where there is not enough oxygen or where there are dangerous gases or fumes in the air. A supplied-air mask uses a remote compressor or air tank to provide breathable air in oxygen-deficient atmospheres. Supplied-air masks can generally be used under the same conditions as SCBAs (*Figure 69*).

(A) PAPR **(B) SCBA**

Figure 69 Examples of air-supplied respirators.
Source: iStock

5.2.4 Wearing a Respirator

If you need to use a respirator, your employer must provide you with the appropriate training to select, test, wear, and maintain this equipment. Your employer is also responsible for having you medically evaluated to ensure you are fit enough to wear respiratory protection equipment without being harmed. You must also be tested to ensure a proper fit and a good seal.

Wearing a respirator generally places a burden on the employee. A self-contained breathing apparatus is heavy and may be difficult for some workers to carry; negative-pressure respirators restrict breathing, and other respirators may cause claustrophobia. For these and other reasons, workers must be medically evaluated by a physician or other licensed health care professional to determine under what conditions they may safely wear respirators.

Respirators are ineffective unless properly fit-tested to the user. To obtain the best protection from a respirator, perform positive (breathing out) and negative (breathing in) fit checks each time it is worn. These fit checks must be repeated until a good face seal is obtained. In order to provide adequate protection, a respirator must be clean, in good condition, and all of its components must be in place.

When a respirator is required, a personal monitoring device is usually also required. This device samples the air to measure the concentration of hazardous chemicals.

5.2.5 Selecting a Respirator

CAUTION

The National Institute for Occupational Safety and Health (NIOSH) issues guidelines and ratings for respirators and masks. Be sure to know the NIOSH rating (N, R, or P) of the mask you are using. The N stands for "**n**ot resistant to oil," the R is "somewhat **r**esistant to oil," and the P signifies that the mask is "oil **p**roof."

CAUTION

All respirator instructions, warnings, and use limitations contained on each package must be read and understood by the wearer before use.

NOTE

Often, adjuvants are added to pesticides, which enhance their performance and change their physical properties. These adjuvants may contain oil, which may change the type of mask or respirator needed for protection.

Follow company and government procedures when choosing the type of respirator for a particular job. Also be sure that it is safe for you to wear a respirator. Under current regulations, workers must fill out a questionnaire to identify potential problems in wearing a respirator. Depending on the answers, a medical exam may be required. When a respirator is not required, workers may voluntarily use a dust or particle mask for general protection. These masks do not require fit testing or a medical examination.

Always use the appropriate respiratory protective device for the hazardous material involved and the extent and nature of the work to be performed. Before using a respirator, you must determine the type and concentration level of the contaminant, and whether the respirator can be properly fitted on your face. If you wear contact lenses, practice wearing a respirator with your contact lenses to see whether you have any problems. That way, you will identify any problems before you use the respirator under hazardous conditions.

The use of air-purifying respirators (APRs) is common in the agricultural industry. These respirators use special filters and cartridges to remove specific gases, vapors, and particles from the air. The respirator cartridges contain charcoal, which absorbs certain toxic vapors and gases. The filters remove particles such as dust, mist, and metal fumes by trapping them within the filter material. When the wearer detects any taste or smell, it indicates the charcoal's absorption capacity has been reached and breathing may become difficult. This means the cartridge can no longer remove the contaminants and must be changed. APRs should be used for protection against only the types of contaminants listed on the filters and cartridges.

5.0.0 Section Review

1. Personnel working in the vicinity of arc welding work *must* wear eye protection with tinted lenses.
   a. True
   b. False

2. Drilling concrete requires use of a respirator because it produces _____.
   a. asbestos
   b. lead
   c. toxic fumes
   d. silica dust

6.0.0 Jobsite Hazards

Objective

Describe safety practices used with other common jobsite hazards.

a. List other types of hazards craftworkers may encounter.
b. Describe common environmental hazards and how craftworkers should respond to them.
c. Summarize hazards associated with hot work.
d. Identify fire hazards and describe basic firefighting procedures.
e. Name different types of confined spaces and how to avoid related hazards.

Performance Tasks

There are no Performance Tasks in this section.

It is impossible to list all the hazards that can exist on a construction, agricultural, or industrial jobsite. This section describes some of the more common hazards and explains how to deal with them. For your safety, you must know the specific hazards where you are working and how to prevent incidents and injuries. If you have questions specific to a jobsite, ask your supervisor.

6.1.0 Jobsite Exposure Hazards

The term *exposure* refers to contact with a chemical, biological, or physical hazard. Exposure can be chronic or acute. Chronic exposure is long-term, repeated, and may be mild or severe. Acute exposure is short-term and intense.

According to the HCS, your employer must inform you of any hazards to which you might be exposed. In order to better protect you, your employer may require preemployment and periodic medical examinations to ensure that your health is not being negatively affected by chemical exposure during work. Exams are usually conducted annually, and generally involve targeted organ testing. Because most medical tests cannot directly check for specific chemical exposure, blood tests are used to check the status of the organs (called *target organs*) known to be most affected by the chemicals to which a person has been exposed.

The results of these tests are compared to preemployment test results and any other annual or periodic exams you may have had. Employers also frequently request a final screening when a worker leaves the company.

Workers can be exposed to chemicals and other hazardous materials in a variety of ways. Routes of exposure include breathing (inhalation), eating or drinking (ingestion), and skin contact (absorption). Chemical exposure is subject to permissible exposure limits (PELs). In other words, some level of exposure to certain chemicals and fumes is considered acceptable and does not represent a hazard. A PEL is the maximum concentration of a substance that a worker can be exposed to in an eight-hour shift. When a worker approaches or reaches the PEL, he or she must be removed from that environment.

There are various types of hazardous materials you may come in contact with during various types of construction and agricultural work, including the following:

- Lead
- Bloodborne pathogens
- Chemicals

Taking the time to understand each of these hazards will help you stay healthy and safe. Working around these hazards requires special training and PPE. Never handle any of these materials without proper training, authorization, and PPE.

6.1.1 Lead

Lead occurs naturally in the Earth's crust and is spread through human activities such as mining and the burning of fossil fuels. As an element, lead is indestructible—it does not break down. Once it is released into the environment, it can move from one medium to another. For example, lead in dust can be carried long distances, dissolve in slightly acidic water, and find its way into soil where it can remain for years. Lead is difficult to detect because it has no distinctive taste or smell.

Lead has many useful properties. It is soft and easily shaped, durable, resistant to some chemicals, and fairly common. Because lead is so versatile, it is used in the production of piping, batteries, and casting metals. However, lead is a toxic metal that can cause serious health problems. You can be exposed to lead by breathing air, drinking water, eating food, and swallowing or touching dust or dirt that contains lead.

You may encounter lead-based paints during demolition or renovation of structures built before 1978, when lead-based paint was banned. Dust created from sanding lead-based paints is hazardous. Protective clothing and equipment must always be used when lead levels are above the PEL. If you are unsure whether lead is present, consult your supervisor.

All waste contaminated with lead is considered hazardous. Those who handle hazardous waste must have special training. Never handle hazardous materials or waste without proper training, authorization, and PPE.

6.1.2 Bloodborne Pathogens

Bloodborne pathogens are another health hazard you may encounter on the jobsite. Sometimes you will hear them referred to as bloodborne infectious diseases. They are diseases that can be transmitted by contact with an infectious person's blood or other bodily fluids, the most common being HIV (the virus that causes AIDS) and hepatitis B and C.

You could be exposed to another person's blood on the jobsite when administering first aid to an injured person or being involved in a multi-victim incident. There is also a risk of exposure to bloodborne pathogens if you work with or around livestock. To safeguard yourself, you must know the universal precautions to prevent exposure and follow them closely:

- Always use appropriate gloves, eye protection, and a mask when administering first aid.
- If you come in contact with a person's or animal's blood, immediately wash the affected areas with soap and water. Otherwise, use antiseptic hand cleanser or antiseptic towelettes until washing with soap and running water is possible.
- Notify your supervisor of the contact right away.

You may need to seek medical attention for precautionary screening, particularly if any bloodborne infectious diseases are known to be present. Your employer should have a written exposure control plan and procedure for such occurrences.

6.1.3 Chemical Splashes

Chemicals such as acids, solvents, pesticides, and fertilizers can pose physical hazards, including acid reactions or chemical burns. Acids can create toxic vapors or react violently when mixed with other chemicals. Other chemicals are flammable, combustible, or explosive. For example, solvents and compressed gases are often flammable. These materials can catch fire in their liquid or gaseous state. Some materials, such as blasting caps, are dangerous solids. These items should be kept away from open flames, sparks, intense heat, or other ignition sources to prevent fires or explosions. Know the physical hazards of the chemicals being used to prevent fires and chemical reactions.

Workers should always check the SDS for a chemical before working with or around it. That way, you will be prepared to take appropriate corrective action if you or another worker is exposed.

WARNING!

Chemical splashes can become medical emergencies. Before working with any chemicals, ensure that you know where the nearest shower and eyewash station are, and verify that they are functioning properly.

Remember to always wear prescribed clothing and PPE on the jobsite. This will help guard against your skin and eyes being splashed with chemicals. If you are not sure of the PPE needed on your jobsite or for a certain task, consult your supervisor before beginning work.

WARNING!

Never handle any chemicals without proper training and PPE. Review the SDS for each chemical, and strictly follow the SDS's first aid instructions if you suspect exposure.

Process Safety Management

One way to avoid the hazards associated with chemical spills is to evaluate those hazards and plan ahead for dealing with unwanted releases. Forward-thinking companies adopt a method known as process safety management for this purpose. Process safety management involves training employees and contractors, making information available, establishing formal procedures, analyzing potential hazards, and taking steps to mitigate the hazards.

6.1.4 Container Labeling

On construction and agricultural sites, any material in a container must have a label. Labels describe what is in a container and warn of chemical hazards. The HCS states that hazardous material containers must be labeled, tagged, or marked. The label must include the name of the material, the appropriate hazard warnings, and the name and address of the manufacturer. On December 1, 2015, OSHA required that all shipments be labeled as outlined by the GHS, which had been adopted as a worldwide standard to enhance awareness and improve the safety and health of workers exposed to these substances.

In today's global marketplace, chemicals and similar hazardous materials are in constant movement between countries. With many countries having their own hazardous material labeling system, products must often be labeled multiple times to support the systems of both the shipping and receiving countries. The GHS will solve this problem by providing some standardized labeling for various materials. There are nine pictograms that workers must become familiar with. Those that are related to flammable, combustible, and explosive materials are shown in *Figure 70*, and those that represent other hazards are shown in *Figure 71*.

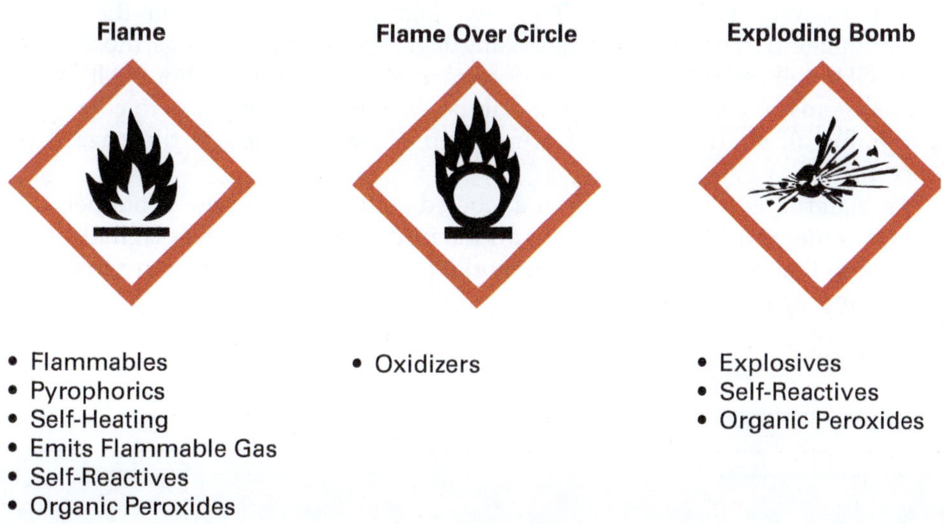

Figure 70 GHS labels for flammable, combustible, and explosive materials.

Figure 71 GHS labels for various materials.

Pictograms have been in use for several years to depict a variety of hazards or to communicate information. The advantage of pictograms is that they cross any language barrier. Regardless of language, pictograms such as those used in the GHS tell the user what type of hazard to be concerned about. Although other systems of container labeling will be encountered for years to come, the pictograms all share common features and visually represent the hazard of concern.

If a material is transferred from a labeled container to a new container, the new container must be labeled with all the information from the original label. Make sure that any materials being worked with are labeled. Be sure that you understand your company's labels.

WARNING!

Never use chemicals from an unlabeled container.

6.1.5 Radiation Hazards

Radioactive materials are used in construction during radiographic testing of welds in piping, vessels, and pumps. These hazards are also found at nuclear power plants. Radioactive materials must be properly labeled, and warning signs must be posted in the work area. Only trained workers are allowed in these areas. If you see signs containing symbols like those shown in *Figure 72*, stay away from that area unless properly trained.

Radioactive materials are a special type of physical hazard. Excessive radiation exposure can cause skin burns, nausea, vomiting, infertility, and cancer. Employees can minimize radiation exposure by limiting the amount of time they are exposed and/or increasing their distance from the source. Always use the proper shielding or PPE. Check with your supervisor to find out if there are any radioactive hazards and how to avoid them.

Figure 72 Radioactivity warning symbols.

6.1.6 Biological Hazards

Biological hazard signs are used to warn workers of the actual or potential presence of a biological hazard (*Figure 73*). Biological hazards, or biohazards, can be any infectious agents that create a real or potential health risk. Biological hazard signs are commonly used to identify contaminated equipment, containers, rooms, and materials.

Background colors on signs may vary but must contrast enough for the symbol to be easily identified. Wording is used on the sign to indicate the nature of the hazard or to identify the specific hazard.

Figure 73 Biological hazard symbol.

6.1.7 Evacuation

In many work environments, specific evacuation procedures are needed. These procedures go into effect when dangerous situations arise, such as fires, chemical spills, and gas leaks. In an emergency, you must know the evacuation procedures. You must also know the signal (usually a horn or siren) that tells workers to evacuate.

When the evacuation signal sounds, follow the evacuation procedures precisely. That usually means taking a certain route to a designated assembly area and telling the person in charge that you are there. If hazardous materials are released into the air, you may have to determine which way the wind is blowing. Some sites may have a **wind sock** that indicates the wind direction. Different evacuation routes are planned for different wind directions. Taking the right route will keep you from being exposed to the hazardous material.

Wind sock: A cloth cone open at both ends mounted in a high place to show which direction the wind is blowing.

6.2.0 Environmental Extremes

Workers on construction, agricultural, and industrial sites are often required to work outdoors in extreme heat and cold. It is important to know how to protect yourself in such environments and to recognize the symptoms of conditions that could be caused by environmental extremes.

6.2.1 Heat Stress

Heat stress occurs when abnormally hot air and/or high humidity, or extremely heavy exertion, prevents your body from cooling itself fast enough. When this happens, you may suffer heat cramps, heat exhaustion, or heatstroke. To prevent heat stress, take the following precautions:

- Drink plenty of water.
- Avoid alcoholic or caffeinated drinks.
- When possible, perform the most strenuous work during cooler parts of the day.
- Wear lightweight, light-colored clothing.
- Wear loose-fitting cotton clothing if it does not create a hazard.
- Keep your head covered and your face shaded.
- Take frequent, short breaks.
- Rest in the shade whenever possible.

Workers are encouraged to drink 4 cups, approximately 1 liter (L), of water per hour when the heat index is 103°F (39.5°C) or higher during work. Start by having one or two glasses of water before beginning work, and then drink 1 cup, approximately 237 milliliters (mL), every 15 to 20 minutes when working in hot conditions. It is possible to drink too much water as well. Water intake should not exceed 6 cups (1.4 L) per hour or 3 gallons (11.4 L) per day. Drinking water should be cool but not cold.

6.2.2 Heat Cramps

Heat cramps are muscular pains and spasms caused by heavy exertion. Any muscles can be affected, but most often it is the muscles that have been used the most. Loss of water and electrolytes from heavy sweating causes these cramps. Symptoms of heat cramps include the following:

- Painful muscle spasms and cramping
- Pale, sweaty skin
- Normal body temperature
- Abdominal pain
- Nausea

As noted above, body temperature is not normally elevated during the onset of heat cramps. An increase in body temperature usually indicates that the problem is escalating to a more serious stage. If you experience heat cramps, take the following steps: move to a cool area, drink some water, and gently stretch and massage cramped muscles.

6.2.3 Heat Exhaustion

Heat exhaustion typically occurs when people exercise heavily or work in a warm, humid place where body fluids are lost through heavy sweating. When it is humid, sweat does not evaporate fast enough to cool the body properly.

Symptoms of heat exhaustion include the following:

- Cool, pale, and moist skin
- Heavy sweating
- Headache, nausea, vomiting
- Dilated pupils
- Dizziness
- Possible fainting
- Fast, weak pulse
- Slight elevation in body temperature

If someone is showing the symptoms of heat exhaustion, get the victim to a shaded area and have him or her lie down. Try to cool the victim by applying cold wet cloths, fanning, removing heavy clothing, and giving small amounts of cool water if the victim is conscious. If the victim's condition does not improve within a few minutes, call emergency medical services.

6.2.4 Heat Stroke

Heat stroke is life-threatening. The body's temperature control system, which produces sweat to cool the body, stops working. Body temperature can rise so high that brain damage and death may result if the body is not cooled quickly.

WARNING!

If you suspect someone has heat stroke, call emergency medical services immediately.

The symptoms of heat stroke include the following:

- Hot, dry, or spotted skin
- Extremely high body temperature
- Very small pupils
- Mental confusion
- Headache
- Vision impairment
- Convulsions
- Loss of consciousness

While waiting for help to arrive, try to cool the victim using the methods described for heat exhaustion. Ice packs placed in the armpits and the groin should be used, if possible.

6.2.5 Cold Stress

When your body temperature drops even a few degrees below normal, which is about 98.6°F (37°C), you can begin to shiver uncontrollably and become weak, drowsy, disoriented, unconscious, or even fatally ill. This loss of body heat is known as cold stress, or hypothermia.

People who work outdoors during the winter need to learn how to protect against loss of body heat. The following guidelines can help you keep your body warm and avoid the dangerous consequences of hypothermia, frostbite, and overexposure to the cold.

WARNING!

Always seek immediate medical attention if you suspect hypothermia or frostbite.

Whether outdoors, indoors, or in mild or cold weather, it is a good idea to dress in layers. Layering your clothes allows you to adjust what you are wearing to suit changing temperature conditions. In cold weather, wear cotton, polypropylene, or lightweight wool next to your skin, and wool layers over your undergarments. For outdoor activities, choose clothing made of waterproof, wind-resistant fabrics such as nylon. Since a great deal of body heat is lost through the head, always wear a hat for added protection.

Water chills the body far more rapidly than air or wind. Always take along an extra set of clothing whenever working outdoors. If clothes become damp, change to dry clothes to prevent your body temperature from dropping in cold weather. Wear waterproof boots in damp or snowy weather and always pack rain gear even if the forecast calls for sunny skies.

6.2.6 Frostbite

Frostbite is a dangerous condition that can have lifelong effects on your body. It usually affects the hands, fingers, feet, toes, ears, and nose. Symptoms of frostbite include a pale, waxy-white skin color and hard, numb skin.

Remember the following when providing first aid for frostbite:

- Never rub the affected area. This can damage the skin and tissue.
- After the affected area has been warmed, it may become puffy and blister.
- The affected area(s) may have a burning feeling or numbness. When normal feeling, movement, and skin color have returned, the affected area(s) should be dried and wrapped to keep it warm.
- If there is a chance the affected area may get cold again, do not warm the skin. If the skin is warmed and then becomes cold again, it will cause severe tissue damage. Seek medical attention as soon as possible. First aid for frostbite includes moving the victim to a warm area, removing wet clothing, and applying warm water of about 105°F (41°C) to the affected area. Do not pour the warm water directly onto the area, however.

Around the World

Temperature Extremes

Some places in the world reach incredibly hot or cold temperatures. Although work may need to come to a halt in the worst of conditions, the job must continue on between the extremes. In Chita, Russia, for example, the average low temperature in January is roughly –22°F (–30°C). The average high during this period is around 1°F (–17°C). Since these temperatures are a simple reality of life in Chita, residents have no choice but to adapt and continue working. Many workers around the globe need to work in such extreme environments. Their resilience and commitment to their jobs is worthy of admiration.

6.2.7 Hypothermia

Hypothermia is a serious, potentially fatal condition caused by loss of body heat. You do not have to be in below-freezing temperatures to be at risk for hypothermia. Hypothermia can happen on land or in water.

The effects of hypothermia can be gradual and often go unnoticed until it is too late. If you know you'll be working outdoors for an extended period of time, work with a buddy. At the very least, let someone know where you'll be and what time you expect to return. Ask your buddy to check you for overexposure to the cold and do the same for your buddy. Check for shivering, slurred speech, mental confusion, drowsiness, and weakness. If anyone shows any of these signs, call for emergency medical assistance and see that the victim is moved indoors as soon as possible to warm up.

WARNING!

If a coworker exhibits uncontrolled shivering, slurred speech, clumsy movements, fatigue, and confused behavior, immediately call for medical assistance.

Symptoms of hypothermia include the following:

- A drop in body temperature
- Fatigue or drowsiness
- Uncontrollable shivering
- Slurred speech
- Clumsy movements
- Irritable, irrational, confused behavior

If a person is showing symptoms of hypothermia from exposure to cold air, immediately call for medical assistance. The victim should be moved to a warm, dry area. Any wet clothing should be removed and replaced with dry clothing or blankets. If possible, give the victim warm, sweet drinks; avoid drinks with caffeine or alcohol. Have the victim move their arms and legs to create muscle heat. If they are unable to do this, place warm bottles or hot packs around the armpits, groin, neck, and head. Do not rub the victim's body or place them in a warm bath. This could cause additional harm.

Body heat is lost up to 25 times faster in water than on land, and the methods used to help victims while waiting for medical help to arrive are different. If you are a victim, first of all, do not remove any clothing. Close up and tighten all clothing to create a layer of trapped water close to the body. This provides insulation that slows heat loss. Keep your head out of the water and put on a hat or hood if possible. Get out of the water as quickly as possible or climb onto a floating object. Do not attempt to swim unless you can reach a floating object or another person. Swimming or other physical activity uses the body's heat and reduces survival time by about 50%.

If it is not possible to get out of the water, wait quietly, and conserve your body heat by folding your arms across your chest, keeping your thighs together, bending your knees and crossing your ankles. If another person is in the water, huddle together with your chests pressed tightly together.

6.3.0 Hot Work Hazards

Welding and torch cutting of metals is done using heat that is high enough to melt steel. Most flame cutting and some welding processes are done with flammable gases. For those reasons, such processes are classified as hot work and are subject to special safety precautions. Grinding with a pneumatic or electric grinding machine (*Figure 74*) is also classified as hot work, as it gives off sparks that could cause a fire or explosion in an environment containing flammable gases or concentrated dust. Hot work always carries a risk of severe burns as well as fire hazards. Being aware of these risks is of great importance and will help keep you safe.

Figure 74 Grinding creates sparks.

6.3.1 Arc Welding Hazards

Arc welding is a process in which metals are joined using a high-intensity electric arc (*Figure 75*) at temperatures in the range of 1,500°F to 3,000°F (815°C to 1,648°C). The arc melts and fuses the base metals. In most cases, a filler metal is melted along with the base metal to strengthen the joint. Welders are required to wear specialized PPE that protects their eyes from damage and protects their bodies from burns caused by sparks and molten metal. Such eye protection includes tinted goggles as well as a full face shield with a tinted viewing window. Anyone working in the vicinity of arc welding must also be protected, especially from damage to their eyes caused by looking at the welding arc. Never look at an arc welding operation without wearing the proper tinted eye protection because the ultraviolet (UV) light from the arc will burn your eyes. Even a reflected arc can harm your eyes. It is extremely important to follow proper safety procedures at and around all welding operations. Serious eye injury or even blindness, as well as burns, can result from unsafe conditions.

Figure 75 Arc welding.
Source: Photo courtesy of The Lincoln Electric Company, Cleveland, OH, U.S.A.

When people are working near a welding operation, **welding curtains** (*Figure 76*) must be set up and everyone in the vicinity must wear tinted protective eyewear.

Welding curtains: Protective screens set up around a welding operation designed to safeguard workers not directly involved in that operation.

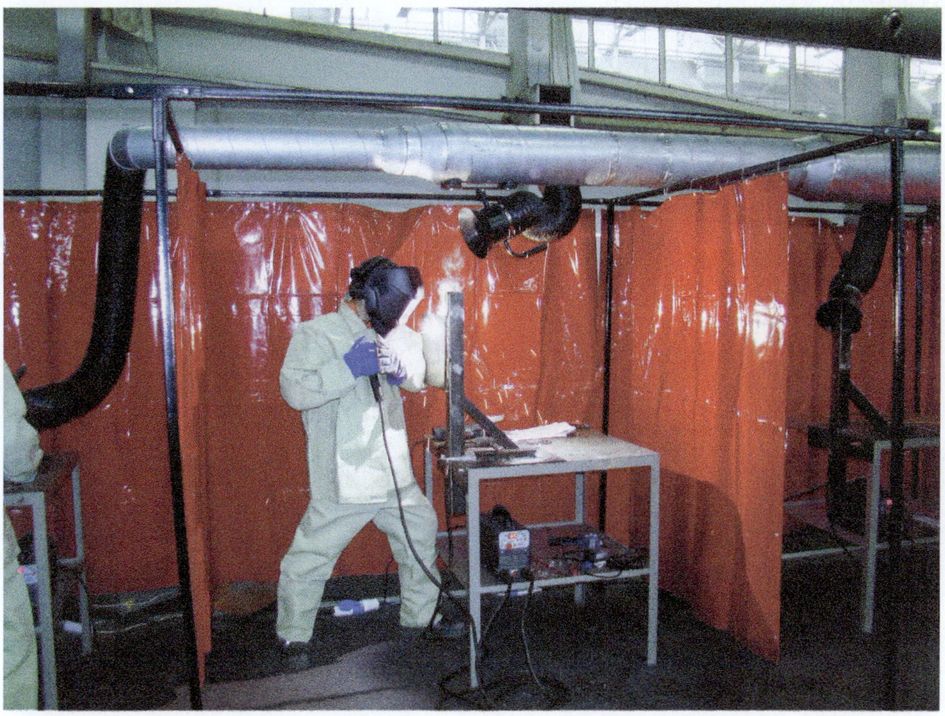

Figure 76 Welding curtain.
Source: Shutterstock

Even brief exposure to the UV light from arc welding can cause a **flash burn** and severely damage your eyes. You may not notice the symptoms until sometime after the exposure. Here are some symptoms of flash burns to the eye:

Flash burn: The damage that can be done to eyes after even brief exposure to UV light from arc welding. A flash burn requires medical attention.

- Headache
- Feeling of sand in your eyes
- Red or weeping eyes
- Trouble opening your eyes
- Impaired vision
- Swollen eyes

Welded material is dangerously hot. Mark it with a sign and stay clear for a while after the welding has been completed.

6.3.2 Oxyfuel Cutting, Welding, and Brazing

Brazing: A process using heat in excess of 800°F (427°C) to melt a filler metal that is drawn into a connection. Brazing is commonly used to join copper pipe.

Oxygen and acetylene (oxyfuel) gases are combined to produce a flame with a temperature high enough to melt steel. So-called oxyfuel is commonly used to cut and weld steel and to join copper pipe by **brazing**.

Figure 77 shows a worker using an oxyfuel torch to cut steel. Persons working with or around welding, brazing, or metal cutting equipment can be severely burned from the intense heat and flames generated in these processes. In addition, the gases used in these processes pose an additional danger because they are stored under pressure in metal cylinders (*Figure 78*).

Figure 77 Oxyfuel cutting.
Source: Photo courtesy of The Lincoln Electric Company, Cleveland, OH, U.S.A.

Many hazards are involved in the handling, storage, and use of compressed gases. It takes energy to compress and confine the gas. That energy is stored until purposely released to perform useful work or until incidentally released by container failure or other causes.

Some gases, such as acetylene, are highly flammable. Oxygen is an explosion hazard if it comes into contact with grease or oil. Flammable compressed gases have additional stored energy besides simple compression-released energy. In case of a fire, for example, escaping oxygen will make the fire more intense. Other compressed gases, such as nitrogen, can cause asphyxiation simply because they displace oxygen.

The cylinders containing oxygen and acetylene must be transported, stored, and handled very carefully. Always follow these safety guidelines:

- Keep the work area clean and free from potentially hazardous items such as combustible materials and petroleum products.

WARNING!

Keep oxygen away from sources of flame and combustible materials, especially substances containing oil, grease, or other petroleum products. Compressed oxygen mixed with oil or grease will explode. Never use petroleum-based products around fittings that serve compressed oxygen lines.

- Use great caution when you handle compressed-gas cylinders.
- Store cylinders in an upright position where they will not be struck, where they will be away from corrosives, and where they cannot tip over or fall. Cylinders must be stored in the upright position in a secure container or secured to a wall with chains. When in storage, oxygen and fuel cylinders must be separated by 20' (6.1 m) or by a 5' (1.5 m) high half-hour fire-rated barrier (*Figure 79*).

Figure 78 Compressed-gas cylinders used for oxyfuel cutting.
Source: Courtesy of Vestil Manufacturing

WARNING!

Do not remove the protective cap unless a cylinder is secured. If the cylinder falls over and the nozzle breaks off, the cylinder could shoot off like a rocket, injuring or killing anyone in its path.

6.3.3 Transporting and Securing Cylinders

Always handle cylinders with care. They are under high pressure and should never be dropped, knocked over, rolled, or exposed to heat in excess of 140°F (60°C). When moving cylinders, always be certain that the valve caps are in place. Cylinders must be transported to the workstation in the upright position on a hand truck or bottle cart such as the one shown in *Figure 78*. Oxygen and cutting gases should not be transported on the same cart unless the cart has a divider, as shown. Be aware that some jobsites will not permit oxygen and cutting gases to be transported on the same cart under any circumstances.

Never attempt to lift a cylinder using the holes in a safety cap; use an approved lifting cage (*Figure 80*). Make sure that the cylinder is secured in the cage. Cages of various sizes are available for high-pressure cylinders and cylinders containing liquids.

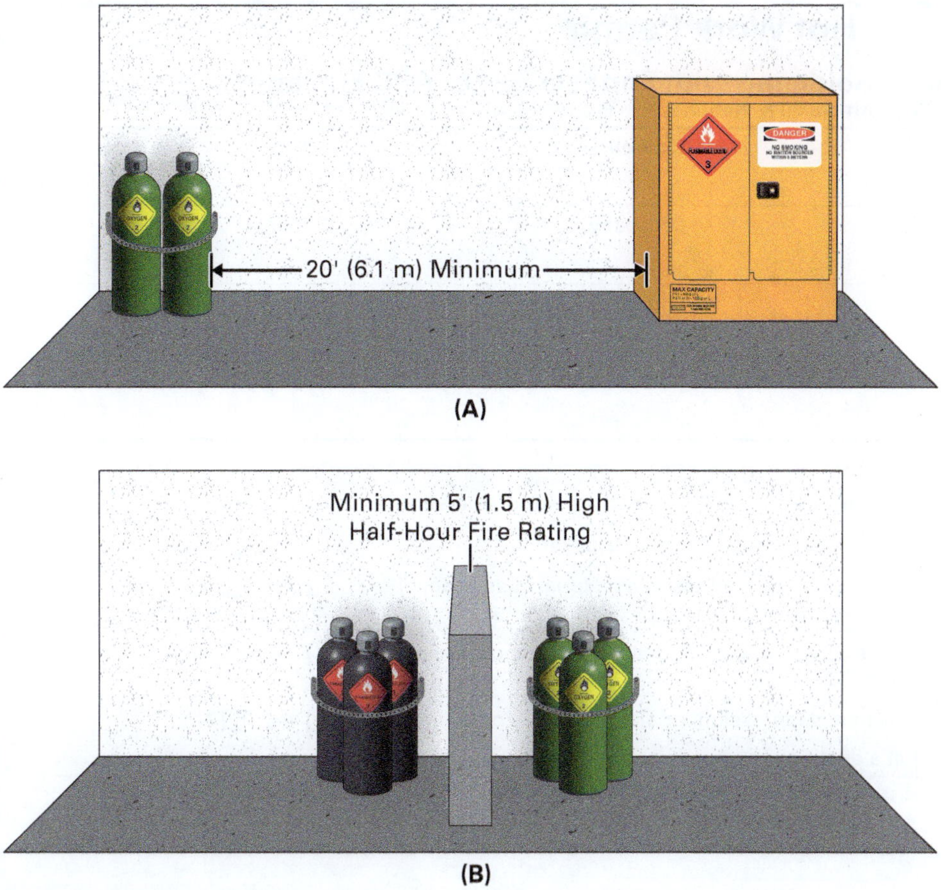

Figure 79 Cylinder storage.

Figure 80 Cylinder lifting cage.
Source: SafTCart

6.3.4 Hot Work Permits

A hot work permit (*Figure 81*) is an official authorization from the site manager to perform work that may pose a fire hazard. The permit includes information such as the time, location, and type of work being done. The hot work permit system promotes the development of standard fire safety guidelines. Permits also help managers keep records of who is working where and at what time. This information is essential in the event of an emergency in which personnel need to be evacuated.

Most sites require the use of hot work permits and fire watches. When these requirements are violated, severe penalties may be imposed because of the risk it represents. Before a hot work permit is issued, a competent person must inspect the work area to ensure that no flammable or combustible materials are present. This inspection includes the areas above and below where the work will be performed, as well as the surrounding area.

A fire watch is posted when welding or cutting work is being done in many areas. One person other than the welding or cutting operator must constantly scan the work area for fires. Fire watch personnel should have ready access to fire extinguishers and alarms and know how to use them. Welding and cutting operations should never be performed without a fire watch. The area where welding is done must be monitored until there is no longer a risk of fire. The person on fire watch must be dedicated to that activity and may not perform other work during the time when hot work is being performed.

6.4.0 Fire Hazards and Firefighting

Fire is always a hazard on construction jobsites. Many of the materials used in construction are flammable. In addition, welding, grinding, and many other construction activities create heat or sparks that can cause a fire. Fire safety involves two elements: fire prevention and firefighting.

Hot Work Permit

For Cutting, Welding, or Soldering with Portable Gas or ARC Equipment

(References: 2021 Uniform Fire Code Article 49 & National Fire Protection Association Standard NFPA 51B.)

Job Date	Start Time	Expiration	WO #
Name of Applicant		Company	Phone
Supervisor		Phone	
Location / Description of Work			

Is FireWatch Required?

1. Are combustible materials in building construction closer than 35 feet to the point of operation? ❑ Yes ❑ No
2. Are combustibles more than 35 feet away but would be easily ignited by sparks? ❑ Yes ❑ No
3. Are wall or floor openings within a 35 foot radius exposing combustible material in adjacent areas, including concealed spaces in floors or walls? ❑ Yes ❑ No
4. Are combustible materials adjacent to the other side of metal partitions, walls, ceilings, or roofs, which could be ignited by conduction or radiation? ❑ Yes ❑ No
5. Does the work necessitate disabling a fire detection, suppression, or alarm system component? ❑ Yes ❑ No

Answering "Yes" to any of the above indicates that a qualified fire watch is required.

FireWatcher Name(s)	Phone

Notifications

Notify the following groups at least 72 hours prior to work and 30 minutes after work is completed.

Write in names of persons contacted.

Notify in person OR by phone ONLY if question #5 above is answered "Yes":

- Facilities Management Fire Alarm Supervisor

Notify by phone or in person: (If by phone, write down name of person and send them a completed copy of this permit.)

- Facilities Management Fire Protection Group
- Environmental Health & Safety Industrial Hygiene Group

Signatures Required

University Project Manager	Date	Phone

I understand and will abide by the conditions described in this permit. I will implement the necessary precautions which are outlined on both sides of this permit form. Thirty minutes after each hot work session, I will reinspect work areas and adjacent areas to which spark and heat might have spread to verify that they are fire safe, and contact Facilities Management Alarm Technicians to have any disabled fire protection systems reactivated.

Permit Applicant	Company or Department	Date	Phone

01/17/2024

Figure 81 Hot work permit.

6.4.1 How Fires Start

For a fire to start, three things are needed in the same place at the same time: fuel, heat, and oxygen. This is known as the *fire triangle*; if one of these three is missing, a fire will not start.

Fuel is anything that will combine with oxygen and heat to burn. Oxygen is always present in the air. When pure oxygen is present, such as near a leaking oxygen hose or fitting, material that would not normally be considered fuel (including some metals) will burn.

Heat refers to a source of ignition. It may be as simple as a single spark. Heat is anything that will raise a fuel's temperature to the **flash point**. The flash point is the temperature at which a fuel is encouraged to burn. The flash points of some fuels are quite low—room temperature or less. When the burning gases raise the temperature of a fuel to the point at which it ignites, the fuel itself will burn—and keep burning—even if the original source of heat is removed.

Flash point: The temperature at which fuel gives off enough gases (vapors) to burn.

Research has added a fourth side to the fire triangle concept, resulting in the development of a new model called the *fire tetrahedron* (*Figure 82*). A tetrahedron is a triangular pyramid with four triangular faces. The fourth element involved in the combustion process is referred to as the chemical chain reaction. Specific chemical chain reactions between fuel and oxygen molecules are essential to sustaining a fire once it has begun. If any one element of the tetrahedron is missing, a fire cannot start. If a fire has started, removing any one element will put it out.

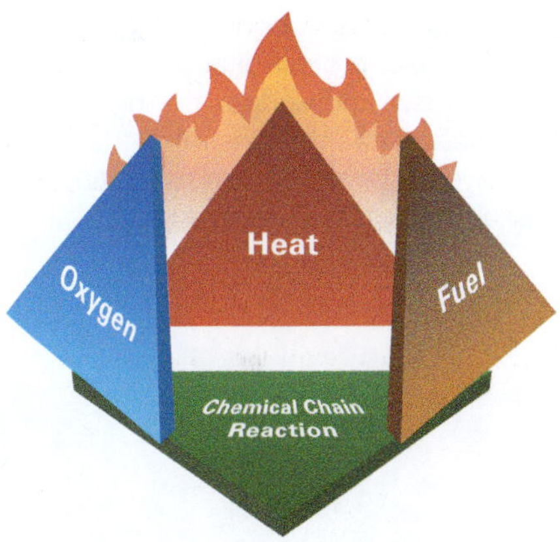

Figure 82 Fire tetrahedron.

6.4.2 Combustibles

Combustibles are categorized as liquid, gas, or ordinary combustibles. The term *ordinary combustibles* refers to paper, wood, dust from organic materials (grain, wood, flour, sugar), cloth, and similar fuels. Liquids can be flammable or combustible. Flammable liquids have a flash point below 100°F (38°C). Combustible liquids have a flash point at or above 100°F (38°C). Flammable gases used on construction sites include acetylene, hydrogen, ethane, and propane. To save space, these gases are compressed so that a large amount is stored in a small cylinder or bottle. As long as the gas is kept in the cylinder, oxygen cannot get to it and start a fire. The cylinders should be stored away from sources of heat. If oxygen is allowed to escape and mix with a flammable gas, the resulting mixture will explode under certain conditions.

The easiest way to prevent fire in ordinary combustibles is to keep a neat, clean work area. If there are no scraps of paper, cloth, or wood lying around, there will be no fuel to start a fire. Establish and maintain good housekeeping habits. Use approved storage cabinets and containers for all waste and other ordinary combustibles.

6.4.3 Fire Prevention

The best way to ensure fire safety is to prevent a fire from starting. Fire can be prevented by the following actions:

- *Removing the fuel* — Liquid does not usually burn. What generally burns are the gases (vapors) given off as the liquid evaporates. Keeping liquids in an approved, sealed container prevents evaporation. If there is no evaporation, there is no fuel to burn.
- *Removing the heat* — If the liquid is stored or used away from a heat source, it will not be able to ignite.
- *Removing the oxygen* — The vapor from a liquid will not burn if oxygen is not present. Keeping safety containers tightly sealed prevents oxygen from contacting fuel.

The best way to prevent a fire is to make sure that fuel, oxygen, and heat are never present in the same place at the same time.

The following are some basic safety guidelines for fire prevention:

- Always work in a well-ventilated area, especially when using flammable materials such as shellac, lacquer, paint stripper, or construction adhesives.
- Never smoke or light matches when working with or near flammable materials.
- Keep oily rags in approved, self-closing metal containers.
- Only store combustible materials in approved containers.

6.4.4 Basic Firefighting

Qualified person: A person who, by possession of a recognized degree, certificate, or professional standing, or by extensive knowledge, training, and experience, has demonstrated the ability to solve or prevent problems relating to a certain subject, work, or project.

You are not expected to be an expert firefighter, but you may have to deal with a fire to protect your safety and the safety of others. You need to know the locations of firefighting equipment on your jobsite as well as which equipment to use on different types of fires. However, only a **qualified person** is required to have firefighting skills.

Most companies tell new employees where fire extinguishers are kept. If you have not been told, be sure to ask. Also, ask how to report fires. The telephone number of the nearest fire department should be clearly posted in your work area. If your company has a fire brigade, learn how to contact them. Learn your company's fire safety procedures. Know what type of extinguisher to use for different kinds of fires and how to use them. Make sure all extinguishers are fully charged. Never remove the tag from an extinguisher—it shows the date the extinguisher was last serviced and inspected.

A fire watch is required to have a fire extinguisher while on duty. A portable extinguisher such as the one shown in *Figure 83* is commonly provided for that purpose.

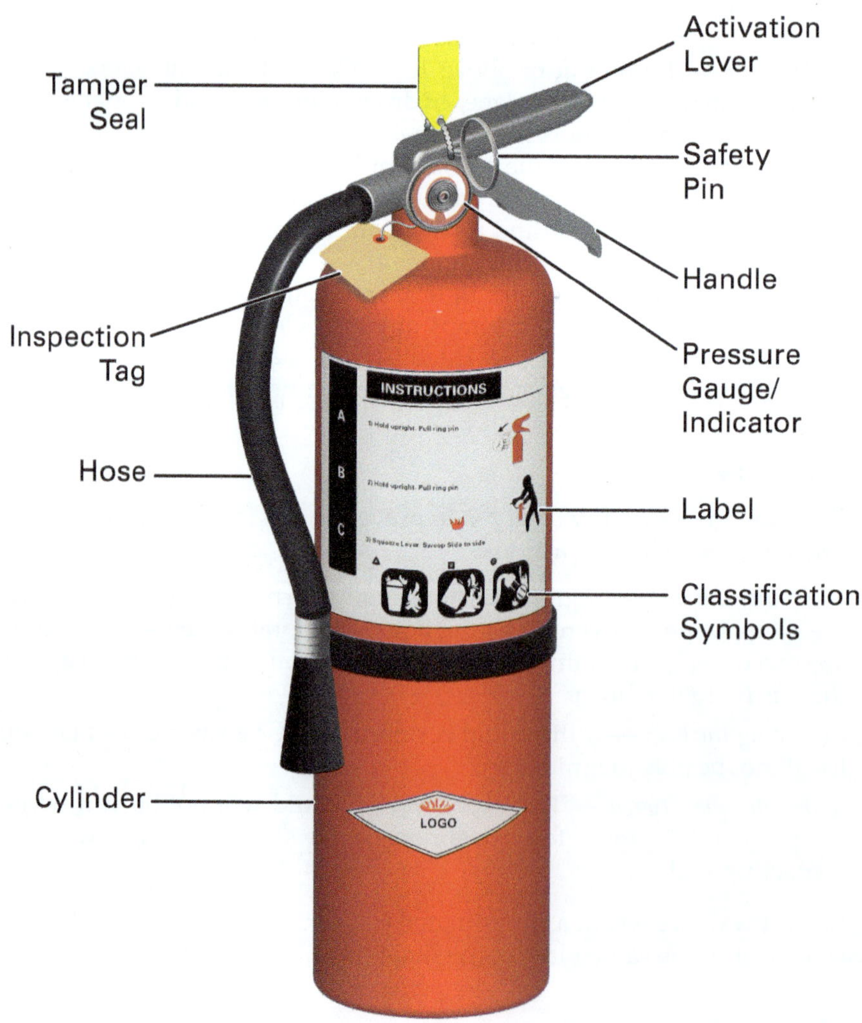

Figure 83 A portable fire extinguisher.

The function of a fire extinguisher is to remove one of the three elements (oxygen, heat, fuel) needed to sustain a fire. A fire extinguisher cannot remove fuel, so it is designed to remove either heat or oxygen. Heat is removed by using a coolant such as water; oxygen is removed by smothering the fire with dry chemicals or CO_2.

Fire extinguishers are rated for specific types of fires. Class A extinguishers, for example, are intended for fires involving ordinary combustibles such as paper, wood, and fabric. An extinguisher that is rated only for Class A fires might contain water, which could not be used on electrical or grease fires. Using water on a Class B fire (flammable liquids, grease, or gases) could spread the fire or splash burning fuel. Using water on a Class C (electrical) fire could cause you to be electrocuted because water conducts electricity. Class D fire extinguishers are required for fires involving reactive metals such as sodium, potassium, magnesium, titanium, zirconium, and the metal hydrides. A Class D fire extinguisher contains a powder designed to coat the metal to either smother the fire or keep oxygen from reaching the fire. If the powder coat fails to extinguish the fire, the best solution is to reapply the powder to keep the fire from spreading.

A pictogram label (*Figure 84*) on the extinguisher denotes the type(s) of fire for which the extinguisher is intended. The extinguisher shown is a dry chemical extinguisher, which can be used on a Class A, B, or C fire. This type is the most common because it can be used to fight most fires. Some CO_2 extinguishers can also be used on Class A, B, or C fires, but others are designed for use on only Class B and C fires. This can be seen in *Figure 85*.

CO_2 is heavier than oxygen, so it smothers the fire by displacing the available oxygen. It also cools as it expands, so it can serve as a coolant. Some companies only permit CO_2 extinguishers because the dry chemical in a dry chemical extinguisher is corrosive and can damage electrical systems. In addition, the fine powder emitted by the material in a dry chemical extinguisher can cause eye and respiratory irritation when it becomes airborne. One advantage of the dry chemical extinguisher is that its working range is 10' to 20' (3 m to 6.1 m), while the range of the CO_2 extinguisher is 3' to 8' (0.9 m to 2.4 m). A pressurized water extinguisher, which is only used for Class A fires, has a range of about 50' (15 m).

WARNING!

Do not use a CO_2 extinguisher on a Class D fire, as it can cause the fire to spread. Also note that CO_2 is heavier than air, so it will concentrate in low areas, displacing oxygen. For that reason, a CO_2 extinguisher must never be taken into, or used in, a confined space. Before it can be used in a confined space, all personnel in the space must first be evacuated. After the extinguisher has been used in the confined space, the space must be cleared by a competent person before it can be reentered.

Finally, CO_2 is discharged from the extinguishers at a temperature of –109°F (–78°C), so it can cause frostbite if it contacts the skin. Never discharge it while it is pointed at someone and do not hold it by the activation lever unless the activation lever is insulated.

WARNING!

When using a CO_2 extinguisher, do not hold the activation lever with your bare hand as it will become extremely cold.

Figure 84 Fire extinguisher pictograms.

Prevention and Preparation Are the Keys to Fire Safety

Any fire in the workplace can cause serious injury or property damage. When chemicals are involved, the risks are even greater. Prevention is the key to eliminating the hazards of fire in the workplace. Preparation is the key to controlling any fires that do start. Take the following precautions to ensure fire safety in the workplace:

- Keep work areas clean and clutter-free.
- Know how to handle and store chemicals.
- Know what you are expected to do in case of a fire emergency.
- Should a fire start, call for professional help immediately. Don't let a fire get out of control.
- Know what chemicals you work with. You might have to tell firefighters at a chemical fire what kinds of hazardous substances are involved.
- Make sure you are familiar with your company's emergency action plan for fires.
- Use caution when using power tools near flammable substances.

Extinguisher Type	Type of Fire	Extinguisher Label
	**Ordinary Combustibles** Fires in paper, cloth, wood, rubber, and many plastics require a water-type extinguisher labeled A.	
	**Flammable Liquids** Fires in oils, gasoline, some paints, lacquers, grease, solvents, and other flammable liquids require an extinguisher labeled B.	
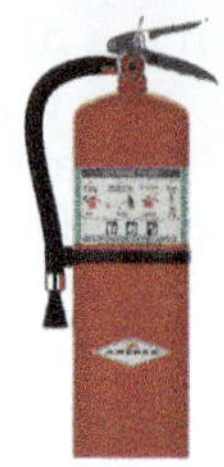	**Electrical Equipment** Fires in liquids, fuse boxes, energized electrical equipment, computers, and other electrical sources require an extinguisher labeled C.	
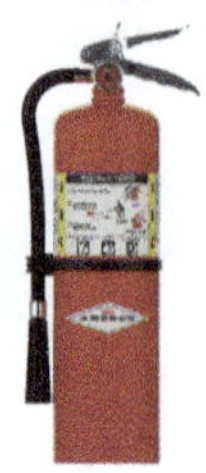	**Ordinary Combustibles, Flammable Liquids, or Electrical Equipment** Multipurpose dry chemical extinguishers are suitable for use on Class A, B, and C fires.	
	**Metals** Combustible metals such as magnesium and sodium require special extinguishers labeled D.	

Note: CO_2 or dry chemical fire extinguishers may be used on flammable liquids and electrical equipment.

Figure 85 Fire extinguisher applications.
Source: Courtesy of Amerex Corporation

Figure 86 Fire extinguisher pressure gauge.

Fire extinguishers must be periodically inspected to make sure they are properly charged. The needle on the extinguisher's pressure gauge must register in the green area (*Figure 86*). The inspection must be documented on the tag attached to the extinguisher. Whenever you check out an extinguisher, make sure its inspection tag (*Figure 87*) is up to date and that the pressure gauge registers the correct charge. Not all extinguishers have gauges, however. Extinguishers without gauges must be weighed to determine if they are fully charged.

6.4.5 Using a Fire Extinguisher

Use the PASS method (*Figure 88*) when attacking a fire:

- **P**ull the pin from the handle, breaking the tamper seal in the process (*Figure 89*).
- **A**im the nozzle at the base of the fire while 8' to 10' (2.4 m to 3 m) away.
- **S**queeze the discharge handle.
- **S**weep the nozzle back and forth at the base of the fire.

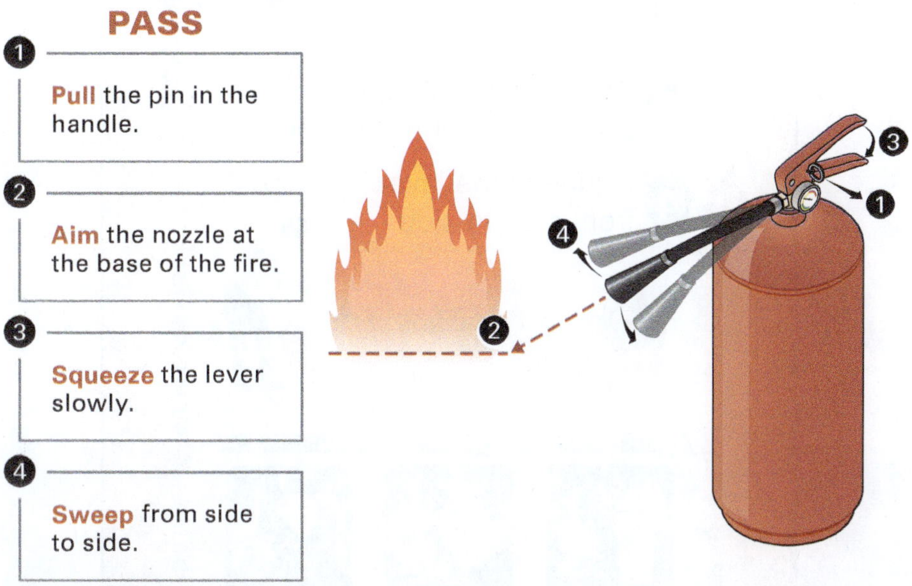

Figure 88 PASS method.

Figure 87 Fire extinguisher inspection tag.

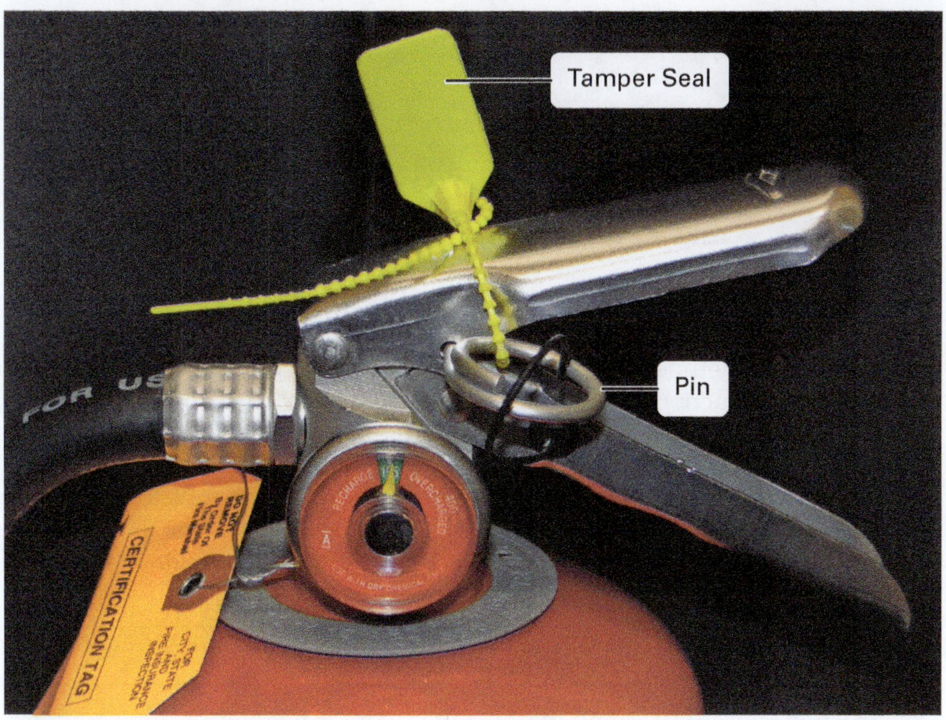

Figure 89 Pin and tamper seal.

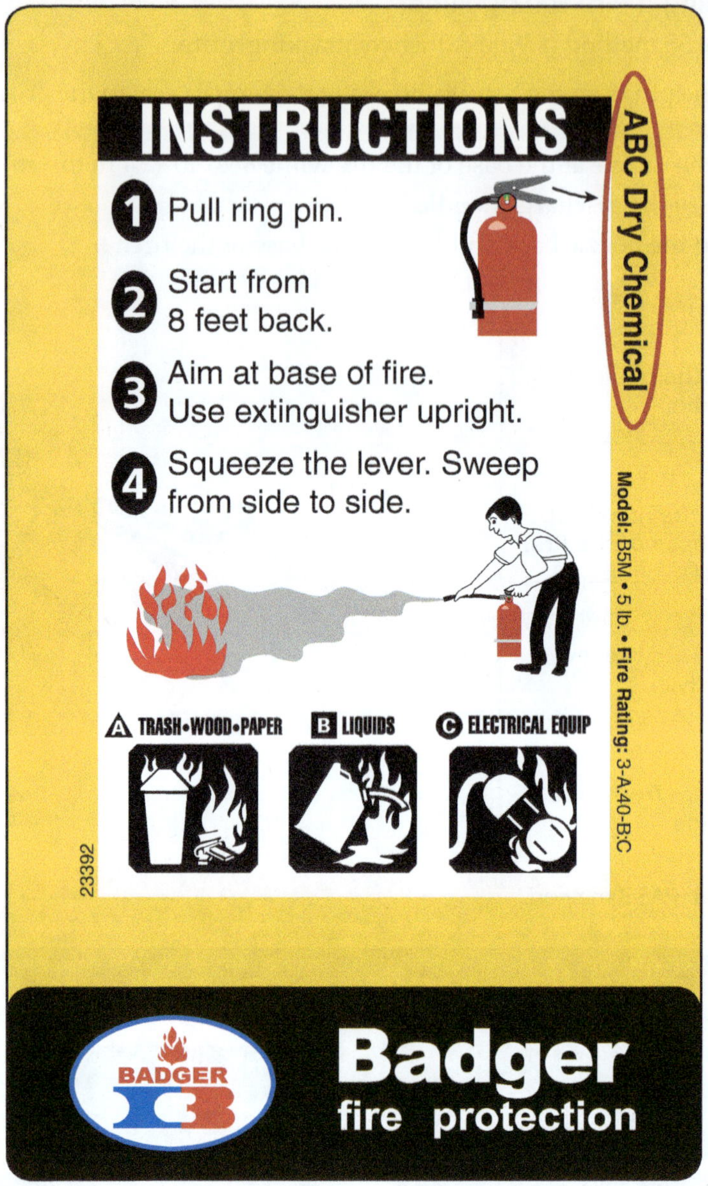

Figure 90 Example of fire extinguisher instructions.
Source: Badger Fire Protection

Note that the method of attacking the fire can vary from one extinguisher to another. The instructions for a specific fire extinguisher are printed on the label, as shown in *Figure 90*. Be sure you are familiar with the types of extinguishers that are commonly provided on your current jobsite.

As the fire is extinguished, move in closer until it is completely out. Stop shooting every three or four sweeps to check progress. Try not to use all the contents on the initial attack in case the fire flares up again. Once the extinguisher has been used, it must be recharged.

6.5.0 Confined Spaces

Construction, agricultural, and maintenance work is not always done outdoors; a lot of it is done in confined spaces. A confined space is a space that is large enough to work in but that has limited means of entry or exit. A confined space is not designed for human occupancy and it has limited ventilation. Examples of confined spaces are tanks, vessels, silos, storage bins, hoppers, vaults, pits, and certain compartments on ships and barges (*Figure 91*).

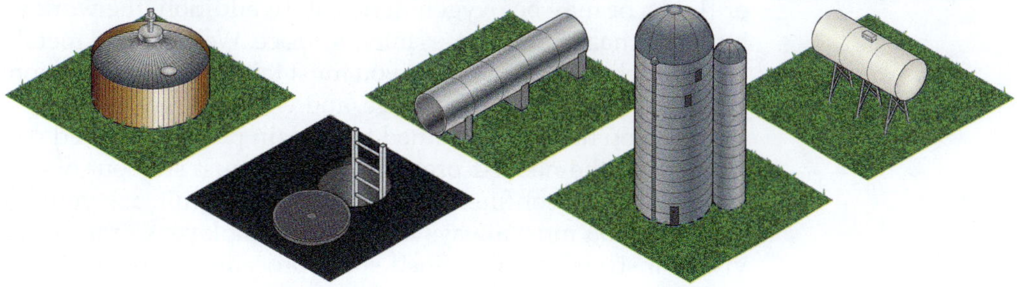

Figure 91 Examples of confined spaces.

OSHA defines a confined space as a space that has the following characteristics:

- Large enough and so configured that employees can bodily enter and perform their assigned work.
- Limited or restricted means of entry or exit.
- Not designed for continuous employee occupancy.

OSHA further defines a **permit-required confined space** as a space that has one or more of the following characteristics:

Permit-required confined space: A confined space that has been evaluated and found to have actual or potential hazards, such as a toxic atmosphere or other serious safety or health hazard. Workers need written authorization to enter a permit-required confined space. Also see *confined space*.

- Contains or has the potential to contain a hazardous atmosphere.
- Contains a material that has the potential to engulf an entrant.
- Has an internal configuration such that an entrant could be trapped or asphyxiated by inwardly converging walls or by a floor that slopes downward and tapers to a smaller cross section.
- Contains any other recognized serious safety or health hazard.

Atmospheric hazards are a concern in a confined space. The air in the space can contain flammable or explosive vapors or toxic gases. The space can also contain too much or too little oxygen. For safe working conditions, the oxygen level in a confined space must range between 19.5% and 23.5% by volume as measured with an oxygen analyzer, with 21% being considered the normal level. Oxygen concentrations below 19.5% by volume are considered deficient; those above 23.5% by volume are considered enriched. Special meters are available to test atmospheric hazards. Forced ventilation can be used to overcome these hazards, but that does not eliminate the need to satisfy other confined space safety requirements, including inspection and permitting.

WARNING!

If too much oxygen is introduced into a confined space, it can be absorbed by a worker's clothing and ignite. If too little oxygen is present, it can lead to death in minutes. For this reason, the following precautions apply:

- Make sure confined spaces are ventilated properly for cutting or welding purposes.
- Never use oxygen in confined spaces for ventilation purposes.
- Always remain aware of the work going on around you, as conditions can change.

A permit-required confined space is a type of confined space that has been evaluated by a qualified person and found to have actual or potential hazards. Written authorization, commonly known as an entry permit, is required to enter a permit-required confined space. Once the entry permit is issued, it must be posted at the entry to the confined space.

When equipment is operating, confined spaces may contain hazardous gases or fluids, or may be oxygen-deficient. In addition, the work you are doing may introduce hazardous fumes into the space. Welding and metal cutting are examples of such work. For safety, you must take special precautions both before you enter and leave a confined space, and while you work there.

Until you have been trained to work in permit-required confined spaces and have taken the needed precautions, you must stay out of them. If you are not sure whether a confined space requires a permit, ask your supervisor or farm manager. You must always follow your employer's procedures and your supervisor's instructions. Confined space procedures may include getting clearance from a safety representative before starting the work. You will be told what kinds of hazards are involved and what precautions you need to take. You will also be shown how to use the required PPE. Remember, it is better to be safe than sorry, so ask!

WARNING!

Without proper training, no employee is allowed to enter a permit-required or non-permit-required confined space. Employers are required to have programs to control entry to, and hazards in, both types of confined spaces.

Never work in a confined space without an attendant. An attendant must remain outside a permit-required confined space. The attendant monitors entry, work, and exit (*Figure 92*).

Figure 92 Permit-required confined space.

6.0.0 Section Review

1. The term *PEL* refers to a worker's _____.
   a. personal energy level
   b. maximum exposure limit
   c. blood lead level
   d. product efficiency level

2. A condition in which body temperature drops a few degrees below normal is known as _____.
   a. heat stress
   b. cold stroke
   c. cold stress
   d. heat exhaustion

3. Swollen eyes and a sandy feeling in the eyes are symptoms of _____.
   a. a flash burn
   b. heat stroke
   c. asbestos exposure
   d. frostbite

4. The three elements needed for a fire to start are fuel, combustible liquid, and oxygen.
   a. True
   b. False

5. A confined space is an enclosed space that has _____.
   a. only one door
   b. limited access
   c. hazardous gases
   d. locked doors

Module 93102 Review Questions

1. The four leading causes of death in the construction industry include electrical hazards, struck-by hazards, caught-in or caught-between hazards, and _____.
   a. vehicular incidents
   b. falls
   c. radiation exposure
   d. chemical burns

2. A sign that has a white background and a green panel with white lettering is a _____.
   a. general information sign
   b. safety instruction sign
   c. caution sign
   d. danger sign

3. To properly dispose of oily rags, they *must* be _____.
   a. stored in a container designed for the purpose
   b. washed thoroughly and returned to use
   c. taken outdoors and thrown into a dumpster
   d. burned at the end of the shift

4. Keeping your work area clean and free of scraps or spills is referred to as _____.
   a. managing
   b. organizing
   c. housekeeping
   d. stacking and storing

5. The HCS classifies all paint, concrete, and wood dust as _____.
   a. hazardous materials
   b. common materials
   c. inexpensive materials
   d. nonhazardous materials

6. Under the HCS, if you spot a hazard on your jobsite, you *must* _____.
   a. report it to your supervisor
   b. leave immediately
   c. notify your coworkers
   d. correct the problem

7. Which of the following *must* be reported to your supervisor?
   a. Only major injuries
   b. Only incidents and major injuries
   c. All injuries, incidents, and accidents
   d. Only incidents in which a death occurred

8. Metal ladders should *not* be used near _____.
   a. stairways
   b. scaffolds
   c. electrical equipment
   d. windows

9. If you lean a straight ladder against the top of a 16' (4.9 m) wall, the base of the ladder should be _____.
   a. 3' (0.9 m) from the base of the wall
   b. 4' (1.2 m) from the base of the wall
   c. 5' (1.5 m) from the base of the wall
   d. 6' (1.8 m) from the base of the wall

10. The two basic types of scaffolds are _____.
    a. self-supporting and suspended scaffolds
    b. fixed and portable scaffolds
    c. metal and wooden scaffolds
    d. assembled and deliverable scaffolds

11. Interlocking stacked material is done by _____.
    a. applying a chain and padlock to it
    b. driving stakes around the stack
    c. securing the objects with rope
    d. placing the objects at right angles

12. To reduce the risk of workers being hurt or killed by falling materials, the maximum height-to-base ratio of a stack of materials should be _____.
    a. 2:1
    b. 4:1
    c. 8:1
    d. 10:1

13. The *most* common cause of death for equipment operators is _____.
    a. hit and run
    b. equipment rollover
    c. brake malfunction
    d. head-on collisions

14. Most cave-ins happen suddenly with little or no warning and occur in trenches 5' (1.5 m) to _____.
    a. 10' (3.1 m) deep
    b. 15' (4.6 m) deep
    c. 20' (6.1 m) deep
    d. 25' (7.6 m) deep

15. In every trench over 4' (1.2 m) deep, there *must* be an exit every _____.
    a. 10' (3.1 m)
    b. 12' (3.7 m)
    c. 18' (5.5 m)
    d. 25' (7.6 m)

16. The *minimum* distance that a spoil pile must be located from the edge of an excavation is _____.
    a. 6" (15.2 cm)
    b. 2' (0.6 m)
    c. 5' (1.5 m)
    d. 100 yards (91.4 m)

17. The type of trench protection designed to prevent trench wall cave-ins is _____.
    a. a trench shield
    b. a trench box
    c. a spoil pile
    d. shoring

18. Protective guards are provided on power tools and machines to keep _____.
    a. dirt out of the tool or machine
    b. workers from being caught in rotating or moving parts
    c. the tool or machine from being struck by moving equipment
    d. unauthorized personnel from using the tool or machine

19. The *minimum* safe working distance from exposed electrical conductors _____.
    a. depends on the voltage
    b. is 6" (15 cm)
    c. is 1' (30 cm)
    d. is unlimited

20. Work that is performed near a hazard but not in direct contact with it is called _____.
    a. close-call work
    b. near-miss work
    c. proximity work
    d. barricade work

21. Circuit breakers and disconnect switches are examples of _____.
    a. energy-isolating devices
    b. energy removal devices
    c. LOTO devices
    d. multiple lockout devices

22. Which of the following provide the *best* eye protection from all directions?
    a. Welding hoods
    b. Face shields
    c. Safety goggles
    d. Strap-on glasses

23. The type of respirator that has its own clean air supply is the _____.
    a. half mask
    b. mouthpiece with mechanical filter
    c. self-contained breathing apparatus
    d. full facepiece mask

24. A worker can be exposed to hazardous materials by inhalation, ingestion, and _____.
    a. osmosis
    b. absorption
    c. radiation
    d. proximity

25. Hypothermia is a condition brought on by _____.
    a. excessive alcohol consumption
    b. prolonged exposure to cold
    c. excessive sweating
    d. prolonged exposure to heat

26. When in storage, oxygen and fuel cylinders used in oxyfuel cutting *must* be separated by _____.
    a. 20' (6.1 m)
    b. a metal barrier
    c. a 20' (6.1 m) wall
    d. 10' (3.05 m)

27. Grease and oil *must* be kept away from oxygen tanks because _____.
    a. grease or oil will contaminate the oxygen
    b. oxygen causes oil to freeze
    c. grease or oil can cause oxygen to explode
    d. oxygen will contaminate the oil or grease

28. Which of these *must* be present in the same place at the same time for a fire to occur?
    a. Oxygen, carbon dioxide, and heat
    b. Oxygen, heat, and fuel
    c. Hydrogen, oxygen, and wood
    d. Grease, liquid, and heat

29. A fire extinguisher labeled C would be used to fight a(n) _____.
    a. electrical fire
    b. magnesium fire
    c. paper fire
    d. gasoline fire

30. Which of the following characteristics is typical of a confined space?
    a. It has a limited amount of ventilation.
    b. There is no means of escape.
    c. It is too small to work in.
    d. It may be entered by untrained employees.

Answers to odd-numbered questions are found in the Review Question Answer Keys at the back of this book.

Answers to Section Review Questions

Answer	Section Reference	Objective
Section 1.0.0		
1. c	1.0.0	1
2. d	1.2.1	1b
3. a	1.3.5	1c
Section 2.0.0		
1. c	2.1.2	2a
2. a	2.2.1	2b
3. b	2.3.1	2c
4. c	2.4.2	2d
Section 3.0.0		
1. a	3.1.2	3a
2. d	3.2.1	3b
Section 4.0.0		
1. b	4.1.1	4a
2. b	4.2.0	4b
Section 5.0.0		
1. a	5.1.2	5a
2. d	5.2.1	5b
Section 6.0.0		
1. b	6.1.0	6a
2. c	6.2.5	6b
3. a	6.3.1	6c
4. b	6.4.1	6d
5. b	6.5.0	6e

User Update

SCAN ME

Did you find an error? Submit a correction by visiting **www.nccer.org/olf** or by scanning the QR code using your mobile device.

MODULE 93103

Introduction to Construction Math

Source: iStock

Objectives

Successful completion of this module prepares you to do the following:

1. Identify whole numbers and solve basic arithmetic problems with them.
   a. List the key qualities of whole numbers and summarize their place values.
   b. Add and subtract whole numbers.
   c. Multiply and divide whole numbers.
2. Name fraction types and calculate with fractions.
   a. Define equivalent fractions and calculate their lowest common denominators.
   b. Define improper fractions and convert them into mixed numbers.
   c. Add and subtract fractions.
   d. Multiply and divide fractions.
3. Identify decimal numbers and calculate with them.
   a. List the key qualities of decimal numbers and summarize their place values.
   b. Add, subtract, multiply, and divide decimal numbers.
   c. Convert between decimals, fractions, and percentages.
4. Name common length-measuring tools and use them to measure lengths accurately.
   a. Describe US standard and metric rulers, using them correctly to measure lengths.
   b. Describe US standard and metric measuring tapes, using them correctly to measure lengths.
5. Name common length, weight, volume, and temperature units in both US standard and metric systems and convert them into other comparable units.
   a. List and convert between common US standard and metric length units.
   b. List and convert between common US standard and metric weight units.
   c. List and convert between common US standard and metric volume units.
   d. List and convert between common US standard and metric temperature units.
6. Classify angles and geometric shapes, as well as calculating their areas or volumes.
   a. List each angle type.
   b. Name common geometric shapes and summarize their qualities.
   c. Calculate the area of two-dimensional shapes.
   d. Calculate the volume of three-dimensional shapes.

NOTE

Unless the text indicates otherwise, attempt all math problems without a calculator. However, you may use a calculator to check your answers.

Digital Resources for Agricultural Technology

SCAN ME

Scan this code using the camera on your phone or mobile device to view the digital resources related to this craft.

Performance Tasks

Under supervision, you should be able to do the following:

1. Using a measuring tape, measure lumber pieces in both US standard and metric units.
2. Using a measuring tape, measure a room-sized space in both US standard and metric units.
3. Using a measuring tape, determine a short inside measurement in both US standard and metric units.
4. Add US standard measurements that include fractions.

Overview

Craft professionals rely on math to do their jobs accurately and efficiently. Plumbers calculate pipe lengths, plan drain slopes, and interpret dimensioned plans. Carpenters frame walls and ceilings properly by using math to meet code requirements. HVAC professionals develop ductwork and calculate airflow with practical geometry. Whichever craft lies in your future, math will play a role in it. This module will review the math you need and sharpen the skills you will be using in future modules.

NCCER Industry-Recognized Credentials

If you are training through an NCCER-accredited sponsor, you may be eligible for credentials from NCCER. The ID number for this module is 93103. Note that this module may have been used in other NCCER curricula and may apply to other level completions. Contact NCCER at 1.888.622.3720 or go to **www.nccer.org** for more information.

You can also show off your industry-recognized credentials online with NCCER's digital credentials. Transform your knowledge, skills, and achievements into credentials that you can share across social media platforms, send to your network, and add to your resume. For more information, visit **www.nccer.org**.

1.0.0 Whole Numbers

Performance Tasks

There are no Performance Tasks in this section.

Objective

Identify whole numbers and solve basic arithmetic problems with them.

a. List the key qualities of whole numbers and summarize their place values.
b. Add and subtract whole numbers.
c. Multiply and divide whole numbers.

No matter which construction craft you enter, you can be sure that you will be working with **whole numbers**. Accurately reading, writing, and communicating whole numbers to others is extremely important on any jobsite. Carpenters often use whole numbers during the planning phase to estimate the square footage of drywall or the linear footage of baseboard. Sheet metal workers use whole numbers for almost every process they do.

Never forget that mathematical mistakes waste time, effort, and materials. Mistakes hurt the bottom line. Even more importantly, your skill with numbers directly reflects on you. Your reputation as a craft professional heavily depends upon your ability to develop a friendly relationship with numbers and math.

In this section, you will learn how to work with whole numbers. Whole numbers are complete number units without a **fraction** or **decimal**. This section examines whole numbers only. Later sections cover working with fractions and decimals.

Whole numbers: Complete number units without fractions or decimals.

Fraction: A portion of a whole number represented by two numbers. The upper number of a fraction is called the *numerator*, while the bottom number is called the *denominator*.

Decimal: The part of a number represented by digits to the right of a point, called a *decimal point*. For example, in the number 1.25, .25 is the decimal portion of the number. In this case, it represents the fraction $^{25}/_{100}$.

1.1.0 Place Values of Whole Numbers

Each **digit** making up a whole number has a specific **place value**. A digit is any of the numerical symbols from 0 to 9. *Figure 1* shows an example of a whole number with seven digits. To read this seven-digit whole number out loud, say "five million, three hundred sixteen thousand, two hundred forty-seven."

Each of this whole number's seven digits represents a specific place value. The digit's value depends on its place, or location, within the whole number. For example, in this whole number, the place value of the 5 is five million, while the place value of the 2 is two hundred.

Keep in mind the following facts about whole numbers:

- Numbers larger than zero are called **positive numbers** (such as 1, 2, 3, and so on). Except for zero, all numbers without a negative (–) sign in front of them are positive.
- Numbers less than zero are called **negative numbers** (such as –1, –2, –3, and so on). Negative numbers have a negative (–) sign in front of them.
- Zero (0) is neither positive nor negative.
- Whole numbers can be positive, negative, or zero.

Some whole numbers may contain the digit zero. For example, the whole number 7,093 has a zero in the hundreds place. When reading that number out loud, say "seven thousand ninety-three."

Digit: Any of the numerical symbols 0 to 9.

Place value: The value that a digit represents in a whole number, determined by its place within the whole number or by its position relative to the decimal point. In the number 124, the number 2 represents 20, since it is in the tens position.

Positive numbers: Numbers greater than zero. For example, 1, 2, and 3 are positive numbers. Except for zero, any number without a negative (–) sign in front of it is treated as a positive number.

Negative numbers: Numbers less than zero. For example, –1, –2, and –3 are negative numbers.

1.1.1 Practice Problems: Place Values of Whole Numbers

1. A number with 9 in the hundreds place, 4 in the ones place, 3 in the thousands place, and 6 in the tens place would be written as _____.
   a. 3,964
   b. 4,693
   c. 30,964
   d. 39,064

2. In the number 25,718, the numeral 5 is in the _____.
   a. ones place
   b. tens place
   c. hundreds place
   d. thousands place

3. An estimate for a commercial flooring job requires one thousand, six hundred ninety-three square meters of carpet to complete the first floor. Written as a whole number, this value is _____.
   a. 163
   b. 1,693
   c. 10,693
   d. 16,093

4. A rancher estimates that his livestock pen will require sixteen thousand, five hundred feet of copper piping to provide indoor water needs for his cattle. Written as a whole number, this value is _____.
   a. 1,650
   b. 16,500
   c. 160,500
   d. 16,000,500

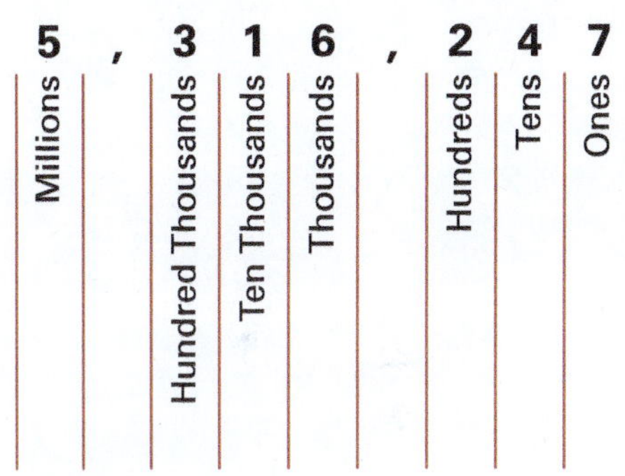

Figure 1 Place values.

Did You Know?

Whole Numbers

The following are whole numbers:
1 5 67 335 2,654
The following are not whole numbers:
$\frac{1}{2}$ $\frac{3}{4}$ $7\frac{1}{8}$ 0.45 4.25
Whole numbers never have a fractional or decimal part.

5. A poultry farmer estimates that the total cost to install an evaporative cooling system for her indoor chicken coop will be three hundred twenty-two thousand nine hundred seven dollars. Written as a whole number, this value is _____.
   a. 3,297
   b. 300,297
   c. 322,907
   d. 322,000,907

1.2.0 Adding and Subtracting Whole Numbers

Sum: The result of an addition problem. For example, in the problem 7 + 8 = 15, 15 is the sum.

To add means to combine the values of two or more numbers together into one **sum** or total. To add whole numbers, perform the following steps:

Step 1 Line up the digits in the top and bottom numbers by place value columns.

$$\begin{array}{r} 723 \\ +\ 84 \\ \hline \end{array}$$

Step 2 Beginning at the right side, add the numbers in the ones column (3 and 4) together first.

$$\begin{array}{r} 723 \\ +\ 84 \\ \hline 7 \end{array}$$

Step 3 Continue adding the digits in each column, moving from right to left, one column at a time. In this example, adding the 2 and 8 in the tens column gives 10. This requires carrying the 1 from the tens column over to the next column on the left. To do so, place the 0 in the tens column and carry the 1 over to the top of the hundreds column as shown.

$$\begin{array}{r} \overset{1}{7}23 \\ +\ 84 \\ \hline 07 \end{array}$$

Step 4 Add the 7 already in the hundreds column to the 1 carried over. The resulting sum is 807.

$$\begin{array}{r} \overset{1}{7}23 \\ +\ 84 \\ \hline 807 \end{array}$$

Some problems may require several carries. The following shows an example:

$$\begin{array}{r} \overset{1}{6}\overset{1}{6},\overset{1}{7}23 \\ +\ 5,784 \\ \hline 72,507 \end{array}$$

Difference: The result of subtracting one number from another. For example, in the problem 8 – 3 = 5, 5 is the difference between the two numbers.

To subtract means to take away the quantity of one number from the quantity of a second number. The result is the **difference**. To subtract whole numbers, perform the following steps:

Step 1 Line up the digits in the top and bottom numbers by place value columns. Generally, position the larger number on the top. If not, the result will be a negative number. Negative numbers are uncommon in jobsite math.

$$\begin{array}{r} 12,766 \\ -\ 1,483 \\ \hline \end{array}$$

Step 2 As with addition, start with the right column—the ones column. Subtract the 3 from the 6 to get 3 and record it under the ones column. Move to the tens column, noting that 8 cannot be subtracted from 6. Solve this problem by borrowing a 1 from the hundreds column. To borrow, cross out the 7 in the hundreds column, change it to a 6, and bring a 1 over to the 6 in the tens column. This action makes it 16 instead of 6. Subtract 8 from 16 to get 8 and record it in the tens column.

```
       6 16
   12,7 6 6
 -  1,4 8 3
 ----------
        8 3
```

Step 3 Continue working toward the left, column by column, completing each subtraction for the remaining place values.

```
       6 16
   12,7 6 6
 -  1,4 8 3
 ----------
   11,2 8 3
```

Some problems may require several borrowing operations. The following shows an example:

```
     3 12  10 15
   9 4 3 , 1 5 3
 - 4 3 6 , 3 7 2
 ---------------
   5 0 6 , 7 8 1
```

1.2.1 Practice Problems: Adding and Subtracting Whole Numbers

Use addition and subtraction to solve the following problems. Read each question carefully to determine the appropriate procedure. Be sure to show all your work.

1. In calculating a bid for a roof restoration, a contractor estimates $850 for lumber, $460 for roofing shingles, and $170 for hardware. What is the total cost for the materials portion of the bid?

2. A plumbing contractor allots $10,236 to complete an irrigation control system for modernizing a greenhouse. If piping will cost $2,477, labor will cost $2,263, and computer control equipment will cost $3,218, how much money remains to handle unexpected costs?

3. An HVAC contracting company sent out three work crews to complete three installations over the past week. If Crew One worked 10-, 9-, 11-, 12-, and 9-hour days, Crew Two worked 9-, 12-, 12-, 9-, and 9-hour days, and Crew Three worked 12-, 12-, 10-, 9-, and 11-hour days, how many total hours did the three crews work for the week?

4. A general contractor ordered three different sized windows to complete the restoration of an original farmhouse. She ordered a bow window that cost $874; one 36" × 36" double-hung window that cost $67; and one 36" × 54" double-hung window that cost $93. If she budgeted $1,250 for windows, how much will she have left after buying them?

5. Russia's currency is called the ruble and is identified with the ₽ symbol. An electrical contractor has 850,360 ₽ in the business's bank account. After depositing 119,980 ₽ in payments made by clients and paying out 79,205 ₽ in wages, how much remains in the account?

Did You Know?

Numeral Systems

People in ancient Egypt, Babylon, Greece, and Rome developed different ways of writing numbers. Some of these early systems were complex and difficult to use. A new system slowly emerged from India over several centuries. Arab traders discovered and then popularized it, leading to Europeans adopting it during the Middle Ages. The Hindu-Arabic system uses only ten symbols—1, 2, 3, 4, 5, 6, 7, 8, 9, 0—and continues as the standard today. Combining these ten symbols (also called *numerals* or *digits*) produces any desired number.

1.3.0 Multiplying and Dividing Whole Numbers

Multiplication is the repeated addition of a number to itself. For example, assume six different people gave their crew leader \$8 to pick up lunch. Adding \$8 together six times could give the total amount. However, multiplying 6 and \$8 to get a **product** of \$48 is easier and faster. With larger numbers, adding repeatedly takes a long time. Multiply whole numbers with the following steps:

Product: The result of a multiplication problem. For example, the product of 6 × 6 is 36.

Step 1 Line up the digits of the top and bottom numbers by place value columns. Position the larger number on top.

$$\begin{array}{r} 374 \\ \times\ 26 \\ \hline \end{array}$$

Step 2 Start with the ones column and multiply the 6 by the 4 (6 × 4 = 24). Write the 4 in the ones column and carry the 2 up to the tens column.

$$\begin{array}{r} 3\overset{2}{7}4 \\ \times\ 26 \\ \hline 4 \end{array}$$

Step 3 Multiply the 6 by the 7 next, and then add the carried-over 2 (6 × 7 = 42, 42 + 2 = 44). Write the 4 in the tens column and carry the 4 up to the hundreds column.

$$\begin{array}{r} \overset{4}{3}\overset{2}{7}4 \\ \times\ 26 \\ \hline 44 \end{array}$$

Step 4 Next, multiply the 6 by the 3, and then add the carried-over 4 (6 × 3 = 18, 18 + 4 = 22). Write a 2 in the hundreds column and place the remaining 2 in the thousands column (since there are no places left in the upper number).

$$\begin{array}{r} \overset{4}{3}\overset{2}{7}4 \\ \times\ 26 \\ \hline 2{,}244 \end{array}$$

Step 5 Now, multiply each digit in 374 by the 2 in 26. Multiply the 2 by the 4 in the ones column (2 × 4 = 8). Write the 8 in the tens column, directly under the 2,244. Do not write the 8 in the ones column, because the 2 in the number 26 is in the tens column. To help keep things aligned, write a zero (0) in the ones column.

$$\begin{array}{r} 374 \\ \times\ 26 \\ \hline 2{,}244 \\ 80 \end{array}$$

Step 6 Multiply the 2 by the 7 in the tens column (2 × 7 = 14). Write down a 4 in the hundreds column and carry the 1.

$$\begin{array}{r} \overset{1}{3}74 \\ \times\ \ 26 \\ \hline 2{,}244 \\ 480 \end{array}$$

Step 7 Multiply the 2 by the 3, and then add the carried-over 1 (2 × 3 = 6, 6 + 1 = 7). Write a 7 in the thousands column.

$$\begin{array}{r} \overset{1}{3}74 \\ \times\ 26 \\ \hline 2{,}244 \\ +7{,}480 \end{array}$$

Step 8 Complete the operation by adding the two products to get the final product of 9,724. There will be as many products to add together as there are digits in the bottom number. In this case, there are two digits in the bottom number (2 and 6). So, there are two products to add.

$$\begin{array}{rl} \overset{1}{3}74 & \\ \times\ 26 & \\ \hline 2{,}244 & \text{(product of } 6 \times 374\text{)} \\ +\ 7{,}480 & \text{(product of } 20 \times 374\text{)} \\ \hline 9{,}724 & \text{(final product)} \end{array}$$

Division is the opposite of multiplication. Multiplication adds a number to itself several times (5 + 5 + 5 = 15, or 5 × 3 = 15). Division subtracts one number from another several times. The number being divided is the **dividend**. The number being divided into the dividend is the **divisor**. The result is the **quotient**. For example, in 15 ÷ 3 = 5, 15 is the dividend, 3 is the divisor, and 5 is the quotient.

Adding, subtracting, or multiplying whole numbers always gives a whole number result. With division, however, the outcome may not be so neat. If the divisor does not divide perfectly into the dividend (such as 10 ÷ 3), the answer will be a whole number plus a leftover value. The leftover, called the **remainder**, completes the answer. For example, dividing 10 by 3 gives a quotient of 3 and a remainder of 1.

Long division requires a sequence of operations. Division, multiplication, and subtraction all contribute to the process.

Divide whole numbers with the following steps:

Step 1 Set up the problem using the division bar as shown. The ÷ symbol has the same meaning, but the division bar keeps things organized on paper. Position the divisor on the left side of the division bar and place the dividend on the right side. In this example, 24 is the divisor and 2,638 is the dividend.

$$24\overline{)2{,}638}$$

Step 2 Addition, subtraction, and multiplication operations begin in the ones column. By contrast, division starts on the dividend's left side. In this example, try to divide 24 into 2. This operation fails because 24 is larger than 2. Next, try using the two leftmost digits together (26). Divide 24 into the 26. This operation works because 24 divides into 26 once. Write a 1 above the 6 and write a 24 under the 26. The 24 equals 1 × 24. Subtract 24 from 26 to get 2. Using zeros (0) helps keep everything aligned.

$$\begin{array}{r} 01?? \\ 24\overline{)2{,}638} \\ -24\downarrow \\ \hline 02 \end{array}$$

Step 3 Bring down the next number (3) in 2,638 and place it next to the 2. Determine if 24 divides into 23. The answer is no, so place a 0 on the answer line above the 3. Write zeros in the appropriate place value columns since 24 × 0 = 0. Subtract the two numbers to get 23.

$$\begin{array}{r} 010? \\ 24\overline{)2{,}638} \\ -24\downarrow \\ \hline 023 \\ -000\downarrow \\ \hline 023 \end{array}$$

Dividend: The number being divided in a division problem. For example, in the problem 16 ÷ 8 = 2, 16 is the dividend.

Divisor: The number that is divided into another number in a division problem. For example, in the problem 16 ÷ 8 = 2, 8 is the divisor.

Quotient: The result of a division problem. For example, when dividing 6 by 2, the quotient is 3.

Remainder: The amount left over in a division problem. For example, in the problem 34 ÷ 8, 8 goes into 34 four times (8 × 4 = 32) with 2 left over, so the remainder is 2.

Construction Estimates

Creating accurate estimates for construction is an essential job skill. Many agricultural operations have limited cash flow and projects must remain within budget. For example, solar panels cannot be included in an update to the electrical system on a farm unless the farm's budget contains enough money. Once the design is ready, the estimated costs include materials, labor, and overhead. Overhead refers to costs of putting outside labor on the job and paying supporting staff, as well as administrative expenses. Farmers must adjust designs, material, and labor expenses to stay within budget.

Numerators and Denominators

This module introduces fractions. Think of a fraction as another way to write a division problem. The number on top of a fraction is called the *numerator*. It is identical to the dividend in a long division problem. The lower number, called the *denominator*, is the same as the divisor. The biggest difference between fractions and long division is that most fractions encountered on the jobsite result in a quotient less than one. In a typical long division problem, the quotient is usually greater than one. Overall, fractions simply represent another way to write a division problem.

Step 4 Bring down the next number (8) in 2,638 and place it next to the 3. Determine how many times 24 can go into 238. Knowing the multiplication tables helps here. Think about numbers that are easier to work with, like 10. Multiplying 10 by 24 is 240, which is close to 238 but too large. However, 24 fits into 238 nine times (24 × 9 = 216). Write a 9 on the answer line next to the 0. Write 216 below the 238 and subtract to get a remainder of 22. The final answer indicates that 24 goes into 2,638 a total of 109 times with a remainder of 22. Write the complete answer as *109 r22*.

```
     010?          0109 r22
 24)2,638      24)2,638
   -24             -24
    023             023
   -000            -000
    0238            0238
                   -0216
                      22
```

Always remember that the answer to each individual division step should be a number from 0 to 9. If the answer is 10 or more, something went wrong. Either a multiplication error has happened, or the divisor really could be divided into the digits at least one full time.

1.3.1 Order of Operations

Many math problems require several operations to solve them. Problems with multiple operations often appear on one or both sides of an **equation**. Regardless of the complexity of equations, they all require a specific order of operations.

Equation: A mathematical statement indicating that the values of two mathematical expressions, such as 2 × 2 and 1 × 4, are equal. An equation is written using the equal sign (=) in this manner: 2 × 2 = 1 × 4.

To get an accurate answer, operations must be done in the proper order. For simple equations, the correct order is multiplication, division, addition, and subtraction. Use a simple memory aid to remember this order: MDAS ("My Dear Aunt Sally").

The following equation is a good example:

$$6 + 3 \times 5 = A$$

Adding 6 to 3 and then multiplying the sum by 5 results in an answer of 45, which is wrong. However, multiplying 3 by 5 first and then adding 6 results in an answer of 21, which is correct. Following the proper order is essential.

Use a second memory aid—PEMDAS—for more complicated equations. Remember this memory aid with the phrase "Please Excuse My Dear Aunt Sally." The *P* represents parentheses, and the *E* represents exponents. If an equation has an operation in parentheses, such as (6 × 3), do that operation first.

Handle any exponents next. This module examines them only briefly, so do not worry if the concept is unfamiliar. An *exponent* is a superscript number that appears next to a normal-sized number—for example, 6^2. The exponent, 2, specifies that two 6s get multiplied together (6 × 6 = 36). If the exponent is 3 (6^3), three 6s get multiplied together (6 × 6 × 6 = 216).

Number of Courses

Masons often use the term *courses* when erecting block walls. This refers to the number of block rows stacked up to make the wall. A 15-course wall made from 8" × 8" × 16" concrete blocks is 10' high.

Block height = 8"
Number of courses = 15
12" = 1'
Therefore:
8" × 15 = 120"
120" ÷ 12" = 10'

1.3.2 Practice Problems: Multiplying and Dividing Whole Numbers

Use multiplication and division to solve the following practical problems. Read each question carefully to determine the appropriate procedure. Addition or subtraction may also be required. Be sure to show all your work.

1. A farmhand sends a trainee to the truck for 180 special fasteners. The fasteners come in bags of 15. How many bags will the trainee bring back?

2. If the following amounts of lumber need to be delivered to each of 2 different staging areas at 4 different jobsites, how many boards of each size are needed?
   a. 65 2' × 4' boards
   b. 45 2' × 8' boards
   c. 25 2' × 10' boards

3. One irrigation job requires 45 meters of PVC pipe, and a second job requires 30 meters. The pipe comes in 6 meter lengths only. How many sections should the plumber buy? How much pipe will be left over, assuming that there are no errors?

4. A crane rental company charges $800 per day, $3,400 per week (5 days), and $10,500 per month.
   a. How much would it cost to rent the crane for 3 days?
   b. How much would it cost to rent the crane for 12 days?
   c. How much would it cost for one month and 11 days?

1.0.0 Section Review

1. Given the number 92,475, which digit is in the ten-thousands place value column?
   a. 2
   b. 4
   c. 7
   d. 9

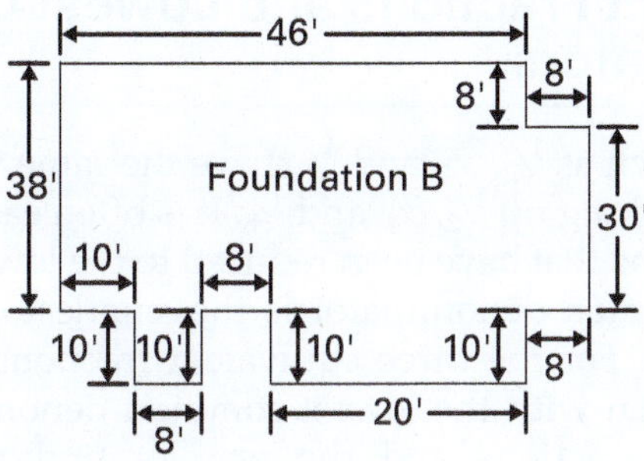

Figure SR01

2. What is the distance around the foundation shown in *Figure SR01*?
   a. 206'
   b. 214'
   c. 224'
   d. 234'

3. How many concrete blocks like the one shown in *Figure SR02* are required to erect a 2' high wall for the foundation shown in *Figure SR01*? (Hint: multiply the distance around the foundation by 12 to convert it from feet to inches first.)
   a. 336
   b. 504
   c. 514
   d. 528

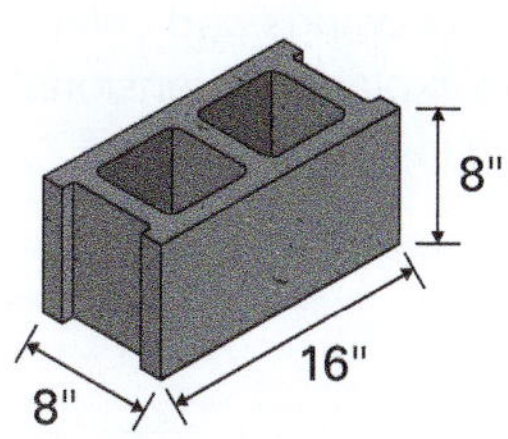

Figure SR02

2.0.0 Fractions

Performance Tasks

There are no Performance Tasks in this section.

Objective

Name fraction types and calculate with fractions.

a. Define equivalent fractions and calculate their lowest common denominators.
b. Define improper fractions and convert them into mixed numbers.
c. Add and subtract fractions.
d. Multiply and divide fractions.

A fraction divides whole numbers into parts. Fractions often appear as two numbers separated by a slash or a horizontal line, such as $\frac{1}{2}$.

The slash or horizontal line means the same thing as the ÷ sign. Think of a fraction as another way to write a division problem. The fraction $\frac{1}{2}$ means 1 divided by 2, or one divided into two equal parts. This fraction is spoken as "one-half."

Denominator: The part of a fraction below the dividing line. For example, the 2 in $\frac{1}{2}$ is the denominator. It is equivalent to the divisor in a long division problem.

Numerator: The part of a fraction above the dividing line. For example, the 1 in $\frac{1}{2}$ is the numerator. It is equivalent to the dividend in a long division problem.

The bottom number of the fraction, the **denominator**, indicates how many parts the upper number is being divided into. In a typical division problem, it is the divisor. The upper number, the **numerator**, is the whole number being divided into parts. It is the dividend of the division problem. In the fraction $\frac{1}{2}$, the 1 is the upper number, or numerator, and the 2 is the bottom number, or denominator. A fraction in which the numerator and denominator are identical (such as $\frac{2}{2}$, $\frac{8}{8}$, $\frac{16}{16}$) always equals 1.

2.1.0 Equivalent Fractions and Lowest Common Denominators

Equivalent fractions: Fractions having equal values with different numerators and denominators, such as fractions $\frac{1}{2}$ and $\frac{2}{4}$.

Equivalent fractions, such as $\frac{1}{2}$, $\frac{2}{4}$, and $\frac{4}{8}$, have the same value. *Figure 2* shows the individual units making up $\frac{1}{2}$, $\frac{1}{4}$, and $\frac{1}{8}$. It is often easier and more practical to work with fractions that have been reduced to the lowest common denominator. The lowest common denominator is the smallest denominator that all the fractions can share. For the three equivalent fractions, $\frac{1}{2}$, $\frac{2}{4}$, and $\frac{4}{8}$, the fraction $\frac{1}{2}$ is the version with the lowest common denominator. This section introduces equivalent fractions and the way to find the lowest common denominator.

2.1.1 Finding Equivalent Fractions

To someone who is used to working with fractions, $\frac{1}{2}$, $\frac{2}{4}$, and $\frac{4}{8}$ are obviously equivalent fractions. However, there is a mathematical way to confirm this. Determine each fraction's value by dividing the denominator by the numerator. So, 2 divided by 1 equals 2; 4 divided by 2 equals 2; and 8 divided by 4 also equals 2. Since the quotients (2) match, the fractions are equivalent. A physical example proves equivalence too. Cut pieces of wood $\frac{1}{2}$" long, $\frac{2}{4}$" long, and $\frac{4}{8}$" long. All three pieces will match.

NOTE

The dimensional units *foot* and *inch* often appear as the symbols ' and " in construction drawings and calculations. This module explains dimensional units and symbols in a later section.

When measuring objects, try to record all measurements using the same fractions—in sixteenths of an inch, for example. In other words, use the same denominator for each measurement. Doing this makes comparing, adding, and subtracting fractional measurements easier.

Sometimes it is necessary to convert a fraction to an equivalent one. Multiplying the numerator and denominator by the same number creates an equivalent fraction. For example, to find out how many sixteenths of an inch are equal to $\frac{1}{2}$ inch, work out what number times 2 gives 16. The answer is 8, so multiply both the numerator and the denominator by 8:

$$\frac{1 \times 8}{2 \times 8} = \frac{8}{16}$$

This calculation shows that $\frac{8}{16}$" equals $\frac{1}{2}$".

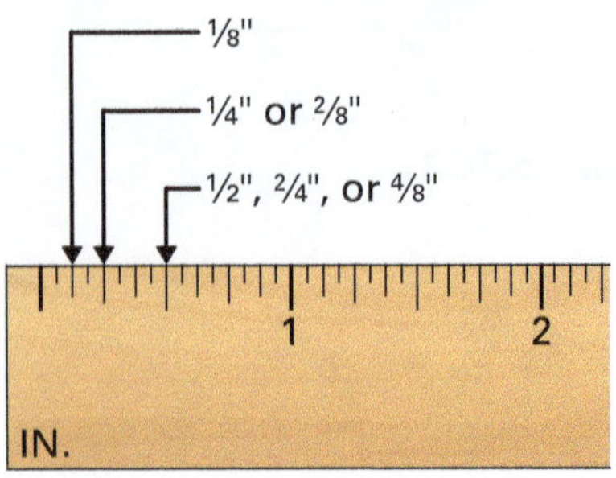

Figure 2 Equivalent fractions.

2.1.2 Reducing Fractions to Their Lowest Terms

If a measurement is $\frac{4}{16}$", it may be easier to work after reducing the number to its lowest terms. Calling out a measurement of $\frac{4}{16}$" on a construction jobsite could cause confusion. To find the lowest terms of $\frac{4}{16}$", use division as follows:

Step 1 Determine the largest number that divides perfectly into both the numerator and the denominator. With a little practice, this is an easy mental task. If there is no number (other than 1) that divides perfectly into both numbers, the fraction is already in its lowest term form. For this example, however, the largest number is 4.

Step 2 Divide both the numerator and the denominator by the number determined in *Step 1*. For this example, divide both the numerator and the denominator by 4.

$$\frac{4 \div 4}{16 \div 4} = \frac{1}{4}$$

This operation shows that the lowest term of $\frac{4}{16}$ is $\frac{1}{4}$. Always remember to divide the numerator and the denominator by the same number. Be sure that both numbers divide perfectly, with a whole number answer as the result.

2.1.3 Comparing Fractions and Finding Lowest Common Denominators

Which measurement is larger, $\frac{3}{4}$" or $\frac{5}{8}$"?

Tackle this question by asking a different question: which is longer, three sections of a board cut into four equal pieces or five sections of a board cut into eight equal pieces (*Figure 3*)?

Comparing fractions with different denominators is challenging. While the visual example makes it obvious that $\frac{3}{4}$ is more than $\frac{5}{8}$, this method is inappropriate for the jobsite. A mathematical approach makes the task more accurate. If the fractions have the same denominator, simply compare the numerators. Fractions with different denominators require another method.

To compare $\frac{3}{4}$ and $\frac{5}{8}$, find a common denominator. The common denominator is a number that both denominators can divide into exactly. The following method calculates a common denominator:

Step 1 Multiply the two denominators together (4 × 8 = 32). The result (32) is a common denominator because each denominator divides into it perfectly, giving a whole number result.

Step 2 Convert the two fractions so both have the same denominator (32):

$$\frac{3 \times 8}{4 \times 8} = \frac{24}{32}$$

$$\frac{5 \times 4}{8 \times 4} = \frac{20}{32}$$

Now, comparing the two fractions is simple: $\frac{3}{4}$ is longer than $\frac{5}{8}$.

Working with fractions like $\frac{24}{32}$ or $\frac{20}{32}$ can be challenging. To make the task easier, find the lowest common denominator possible between them, which means reducing the fractions to their lowest terms. Although 32 is a common denominator for the two fractions, it is not the lowest common denominator.

Around the World

Fractions on the Job

In the real world, most measurements are not whole numbers. Typically, craftworkers measure and cut pipe, lumber, and other materials to fractional lengths, such as $\frac{3}{8}$, $\frac{5}{16}$, or $\frac{3}{4}$ of an inch or a foot. All craftworkers must be comfortable with fractions, especially those who use the US standard measuring system. The metric system almost eliminates the need for fractions in most day-to-day measurements.

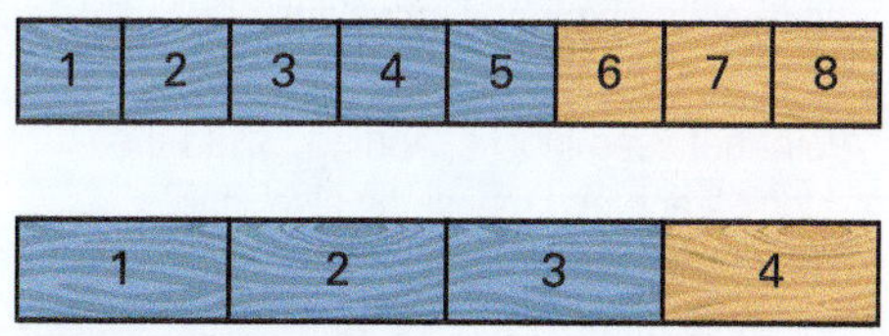

Figure 3 Lumber divided into equal sections.

To find the lowest common denominator, follow these steps:

Step 1 Reduce each fraction to its lowest terms.

Step 2 Find the lowest common multiple of the denominators. Sometimes one denominator is already a multiple of the other. If this is the case, find the equivalent fraction for the term with the smaller denominator.

Step 3 If no common multiples exist, multiply the denominators together to get a common denominator.

Consider the lumber example again. The fractions $3/4$ and $5/8$ are already in their lowest terms. Comparing the denominators shows that 8 is a multiple of 4. Calculate the equivalent fraction with a denominator of 8 for $3/4$:

$$\frac{3 \times 2}{4 \times 2} = \frac{6}{8}$$

Comparing $6/8$ to $5/8$ reveals that $6/8$ is the larger fraction. This method gives the same result as comparing $24/32$ to $20/32$.

The job determines the method craft professionals use. For example, carpenters often work in sixteenths, so they favor this fraction in calculations.

Did You Know?

Nominal Measurements

A so-called 8" × 8" × 16" concrete block actually measures $7\frac{5}{8}$" × $7\frac{5}{8}$" × $15\frac{5}{8}$". The masonry industry refers to it by its nominal dimensions of 8" × 8" × 16" to allow for a $3/8$" mortar joint between blocks. Adding $3/8$" to the block's actual dimensions also makes the block's measurements come out in whole number inches, rather than fractions. Other block types and sizes use this same practice.

2.1.4 Practice Problems: Finding Equivalent Fractions

Find the equivalents for the following fractions:

1. $1/4$ equals how many sixteenths?
   a. $2/16$
   b. $4/16$
   c. $6/16$
   d. $8/16$
2. $2/16$ equals how many thirty-seconds?
   a. $1/32$
   b. $2/32$
   c. $4/32$
   d. $8/32$
3. $3/4$ equals how many eighths?
   a. $2/8$
   b. $4/8$
   c. $5/8$
   d. $6/8$
4. $3/4$ equals how many sixty-fourths?
   a. $48/64$
   b. $50/64$
   c. $52/64$
   d. $54/64$
5. $3/16$ equals how many thirty-seconds?
   a. $2/32$
   b. $4/32$
   c. $6/32$
   d. $8/32$

Without using a calculator, find the lowest term for each of the following fractions:

6. $2/16$ = _____
7. $2/8$ = _____
8. $12/32$ = _____
9. $4/8$ = _____

10. $^4/_{64}$ = ____

Find the lowest common denominator for the following pairs of fractions:

11. $^2/_6$ and $^3/_4$
    a. 6
    b. 8
    c. 12
    d. 16

12. $^1/_4$ and $^3/_8$
    a. 4
    b. 8
    c. 12
    d. 18

13. $^1/_8$ and $^1/_2$
    a. 3
    b. 5
    c. 7
    d. 8

14. $^1/_4$ and $^3/_{16}$
    a. 8
    b. 16
    c. 18
    d. 20

15. $^4/_{32}$ and $^5/_8$
    a. 64
    b. 32
    c. 16
    d. 8

2.2.0 Improper Fractions and Mixed Numbers

An **improper fraction** has a larger numerator than its denominator. The fractions $^{43}/_6$, $^{19}/_{16}$, and $^{55}/_8$ are examples of improper fractions.

Converting an improper fraction to a **mixed number** makes calculations easier. A mixed number consists of a whole number and a fraction. The mixed numbers $7^1/_6$, $1^3/_{16}$, and $6^7/_8$ are equivalent to the improper fractions $^{43}/_6$, $^{19}/_{16}$, and $^{55}/_8$. Some improper fractions convert to a whole number without a fraction. For example, the improper fraction $^{42}/_6$ converts to 7.

Remember, fractions are just another way of writing a division problem. To convert an improper fraction to a mixed number, perform the division. The following steps demonstrate converting the improper fraction $^{67}/_{32}$ to a mixed number:

Improper fraction: A fraction whose numerator is larger than its denominator. For example, $^8/_4$ and $^6/_3$ are improper fractions.

Mixed number: A combination of a whole number with a fraction or decimal. Examples of mixed numbers are $3^7/_{16}$, 5.75, and $1^1/_4$.

Step 1 Divide the numerator (dividend) by the denominator (divisor). The denominator must be able to be divided into the numerator at least once, or the fraction is not an improper fraction. Set up the problem on paper if necessary.

$$32\overline{)67}$$

Step 2 Perform the long division.

$$\begin{array}{r} 02 \text{ r3} \\ 32\overline{)67} \\ -64 \\ \hline 3 \end{array}$$

Step 3 The quotient (2) of the problem is the whole number part of the mixed number. Any remainder (3 in this example) becomes the fractional part's numerator. Retain the original denominator.

$$2\frac{3}{32}$$

2.2.1 Practice Problems: Changing Improper Fractions to Mixed Numbers

Change these improper fractions to mixed numbers:

1. $^{35}/_{8}$ = _____
2. $^{97}/_{16}$ = _____
3. $^{13}/_{4}$ = _____
4. $^{14}/_{5}$ = _____
5. $^{49}/_{16}$ = _____

2.3.0 Adding and Subtracting Fractions

To add fractions, first find a common denominator and convert them to equivalent fractions. Once the denominators are the same, add the numerators. The following steps show the process for adding $^{3}/_{4}$ and $^{5}/_{8}$:

Step 1 Find the fractions' common denominator. One common denominator is 32. The lowest common denominator is 8. Both choices work fine and give equally valid results.

Step 2 Convert the fractions to equivalent fractions with the same denominator.

$$\frac{3 \times 8}{4 \times 8} = \frac{24}{32}$$

$$\frac{5 \times 4}{8 \times 4} = \frac{20}{32}$$

Step 3 Add the numerators. Place this sum over the common denominator (32).

$$\frac{24}{32} + \frac{20}{32} = \frac{44}{32}$$

Step 4 Reduce the fraction to its lowest terms. For this example, both the numerator and denominator divide by 4. Doing this reduces the fraction to $^{11}/_{8}$, an improper fraction. Finally, change the improper fraction $^{11}/_{8}$ to the mixed number $1^{3}/_{8}$.

Any common denominator will work, but the lowest common denominator gives a reduced result. Of course, the result may still be an improper fraction that must be changed to a mixed number.

Subtracting fractions is similar to adding them. Begin by finding a common denominator. For example, a painter has $^{7}/_{8}$ of a liter of paint. Another worker asks for $^{1}/_{4}$ of a liter. How much will be left?

Step 1 Find a common denominator. In this case, 8 is a common denominator and is also the lowest common denominator.

$$\frac{7}{8} - \frac{1}{4}$$

Step 2 Since one fraction is already in eighths, only convert $\frac{1}{4}$ to eighths. Multiply the numerator and denominator by 2 to get a fraction with a denominator of 8.

$$\frac{1 \times 2}{4 \times 2} = \frac{2}{8}$$

Step 3 Subtract the numerators. Place the difference over the common denominator (8). The result is $\frac{5}{8}$.

$$\frac{7}{8} - \frac{2}{8} = \frac{5}{8}$$

Sometimes a project requires subtracting a fraction from a whole number. To do this, change the whole number into a fraction. Then, find a common denominator and subtract the numerators.

Changing a whole number into a fraction is easy. Make the whole number the numerator and use 1 for the denominator. For example, the whole number 5 becomes the fraction $\frac{5}{1}$. To subtract $\frac{2}{7}$ from 5, find the lowest common denominator 7. Converting $\frac{5}{1}$ into sevenths gives the fraction $\frac{35}{7}$. Subtract $\frac{2}{7}$ from $\frac{35}{7}$, giving a difference of $\frac{33}{7}$. Change the improper fraction to the mixed number $4\frac{5}{7}$.

Around the World

The Roots of Fractions

The word *fraction* comes from a Latin word that means "to break." Other English words, such as *fracture*, also come from the same source. Ancient cultures, including the Egyptians, the Babylonians, and the Indians devised their own ways of writing and using fractions. The Arabs added the line between the numerator and denominator.

2.3.1 Practice Problems: Adding and Subtracting Fractions

Add the following fractions. Reduce the results to the lowest terms and change any improper fractions to mixed numbers.

1. $\frac{1}{8} + \frac{4}{16}$ = _____
2. $\frac{4}{8} + \frac{6}{16}$ = _____
3. $\frac{2}{4} + \frac{3}{4}$ = _____
4. $\frac{3}{4} + \frac{2}{8}$ = _____
5. $\frac{14}{16} + \frac{3}{8}$ = _____

Subtract the following fractions. Reduce the results to the lowest terms.

6. $\frac{3}{8} - \frac{5}{16}$ = _____
7. $\frac{11}{16} - \frac{5}{8}$ = _____
8. $\frac{3}{4} - \frac{2}{6}$ = _____
9. $\frac{11}{12} - \frac{4}{8}$ = _____
10. $\frac{11}{16} - \frac{1}{2}$ = _____

Subtract the following fractions and whole numbers. Reduce the results to the lowest terms and change any improper fractions to mixed numbers.

11. $8 - \frac{3}{4}$ = _____
12. $12 - \frac{5}{8}$ = _____
13. Two punches are made from $9\frac{7}{16}$" long steel bar stock. If one punch is $4\frac{1}{64}$" long and the other is $4\frac{3}{32}$" long, how many inches of stock are wasted?
    a. $\frac{11}{16}$"
    b. $\frac{15}{16}$"
    c. $1\frac{15}{64}$"
    d. $1\frac{21}{64}$"

14. After sawing $12\frac{1}{16}$" off a board that is $20\frac{3}{4}$" long, the length of the remaining board will be _____.
    a. $8\frac{1}{4}$"
    b. $8\frac{11}{16}$"
    c. $11\frac{1}{4}$"
    d. $17\frac{3}{8}$"

15. A rough opening for a window measures $36\frac{3}{8}$". The window itself measures $35\frac{15}{16}$". The total clearance that will exist between the window and the rough opening is _____.
    a. $\frac{7}{16}$"
    b. 1"
    c. $1\frac{7}{16}$"
    d. $1\frac{3}{4}$"

2.4.0 Multiplying and Dividing Fractions

Multiplying and dividing fractions differs from adding and subtracting fractions. A common denominator is not required. Follow these steps to multiply $\frac{4}{8}$ and $\frac{5}{6}$:

Step 1 Multiply the numerators together to get a new numerator. Multiply the denominators together to get a new denominator.

$$\frac{4}{8} \times \frac{5}{6} = \frac{20}{48}$$

Step 2 If possible, reduce the result to its lowest terms. In this example, $\frac{20}{48}$ reduces to $\frac{5}{12}$, since both the numerator and denominator are divisible by 4.

Fractions do not need to be reduced before multiplying, but what may seem like an extra step can save time in the long run. Reducing them first makes the multiplication easier, since the numbers are smaller. It makes reducing the product to the lowest terms easier too. Regardless of the method, however, the final answer will be the same.

Dividing fractions is similar to multiplying fractions but with one difference. Before multiplying, you need to **invert**, or *flip*, the divisor fraction. Follow these steps to divide $\frac{1}{2}$ and $\frac{3}{4}$:

Invert: To reverse the order or position of numbers. In fractions, inverting means reversing the positions of the numerator and denominator, so $\frac{3}{4}$ becomes $\frac{4}{3}$. When dividing by fractions, one fraction is inverted.

Step 1 Invert the divisor fraction ($\frac{3}{4}$).

$$\frac{3}{4} \text{ becomes } \frac{4}{3}$$

Step 2 Change the division sign (÷) to a multiplication sign (×).

$$\frac{1}{2} \times \frac{4}{3}$$

Step 3 Multiply the numerators and denominators in the normal way.

$$\frac{1}{2} \times \frac{4}{3} = \frac{4}{6}$$

Step 4 Reduce to the lowest terms when possible.

$$\frac{4}{6} \text{ reduces to } \frac{2}{3}$$

Check a division result for correctness by using multiplication. Multiplying $\frac{3}{4}$ with $\frac{2}{3}$ gives a product of $\frac{1}{2}$, which is the problem's original dividend.

To divide a fraction by a whole number, convert the whole number to a fraction by putting it over 1. Remember that $^4/_1$ is the same as 4. Perform the following steps to divide $^1/_2$ by 4:

Step 1 Change the whole number (4) into a fraction.

$$4 = \frac{4}{1}$$

Step 2 Invert the fraction.

$$\frac{4}{1} \text{ becomes } \frac{1}{4}$$

Step 3 Multiply the fractions.

$$\frac{1}{2} \times \frac{1}{4} = \frac{1}{8}$$

Convert a mixed number, such as $2^1/_3$, into an improper fraction before inverting it. Follow these steps to convert $2^1/_3$ into an improper fraction:

Step 1 Convert the whole number portion to a fraction.

$$2 = \frac{2}{1}$$

Step 2 Multiply it by $^3/_3$ to match the denominators.

$$\frac{2 \times 3}{1 \times 3} = \frac{6}{3}$$

Step 3 Add the converted whole number portion and the fraction portion together.

$$\frac{6}{3} + \frac{1}{3} = \frac{7}{3}$$

2.4.1 Practice Problems: Multiplying and Dividing Fractions

Without using a calculator, find the answers to the following multiplication problems. Reduce results to their lowest terms and change any improper fractions to mixed numbers.

1. $^4/_{16} \times {}^5/_8$ = _____
2. $^3/_4 \times {}^7/_8$ = _____
3. $^2/_8 \times 15$ = _____
4. $^3/_7 \times 49$ = _____
5. $^8/_{16} \times {}^{32}/_{64}$ = _____

Without using a calculator, find the answers to the following division problems. Reduce results to their lowest terms and change improper fractions to mixed numbers.

6. $^3/_8 \div 3$ = _____
7. $^5/_8 \div {}^1/_2$ = _____
8. $^3/_4 \div {}^3/_8$ = _____

9. On construction drawings, smaller dimensions often represent larger ones. This technique allows large objects or structures to fit on the paper. If $\frac{1}{4}$" represents 1', then a line on the drawing measuring $8\frac{1}{2}$" represents _____.
   a. 34'
   b. 36'
   c. 38'
   d. 40'

10. How many $\frac{7}{8}$" long strips can be cut from a single 7" long strip of material?
    a. 5
    b. 6
    c. 7
    d. 8

2.0.0 Section Review

1. Which of the following fraction pairs are equivalent fractions?
   a. $\frac{1}{2}$ and $\frac{17}{32}$
   b. $\frac{1}{2}$ and $\frac{32}{64}$
   c. $\frac{1}{2}$ and $\frac{3}{4}$
   d. $\frac{1}{2}$ and $\frac{63}{128}$

2. The improper fraction $\frac{76}{64}$ converts to the mixed number $1\frac{9}{32}$.
   a. True
   b. False

3. $\frac{11}{16} + \frac{3}{8} =$ _____.
   a. $\frac{9}{16}$
   b. $\frac{7}{8}$
   c. $1\frac{1}{16}$
   d. $1\frac{1}{8}$

4. $1\frac{1}{4} - \frac{7}{16} =$ _____.
   a. $\frac{5}{16}$
   b. $\frac{5}{8}$
   c. $\frac{3}{4}$
   d. $\frac{13}{16}$

5. $\frac{1}{8} \times \frac{1}{10} =$ _____.
   a. $\frac{1}{80}$
   b. $\frac{1}{40}$
   c. $\frac{4}{5}$
   d. $1\frac{1}{4}$

6. $\frac{7}{16} \div \frac{7}{8} =$ _____.
   a. $\frac{49}{128}$
   b. $\frac{3}{8}$
   c. $\frac{1}{2}$
   d. 2

3.0.0 The Decimal System

Objective

Identify decimal numbers and calculate with them.

a. List the key qualities of decimal numbers and summarize their place values.
b. Add, subtract, multiply, and divide decimal numbers.
c. Convert between decimals, fractions, and percentages.

Performance Tasks

There are no Performance Tasks in this section.

The number 10 lies at the heart of the decimal system. Learning the place values of digits earlier in this module established the groundwork for the decimal system. Decimal numbers offer numerous advantages, including simpler calculations and easier fractions.

3.1.0 Decimals

Decimal values include a dot (or period), called a *decimal point,* within the number. Digits to the left of the decimal point represent a whole number. Digits to the right of the decimal point represent a fraction of the number 1. In fact, decimals are sometimes called *decimal fractions,* since they represent part of a whole number. Even if a decimal point is not visible, assume one to the right of the ones place. For example, with the number 5,316,247, the decimal point is on the right side of the number 7.

Figure 4 shows the place values of digits to the left and right of the decimal point. For applications requiring extreme precision, extending a number five places to the right of the decimal point usually provides the required precision. For more common applications, extending the number two places to the right of the decimal point is usually adequate. The application determines the requirements, however. In some cases, a whole number alone may be sufficient.

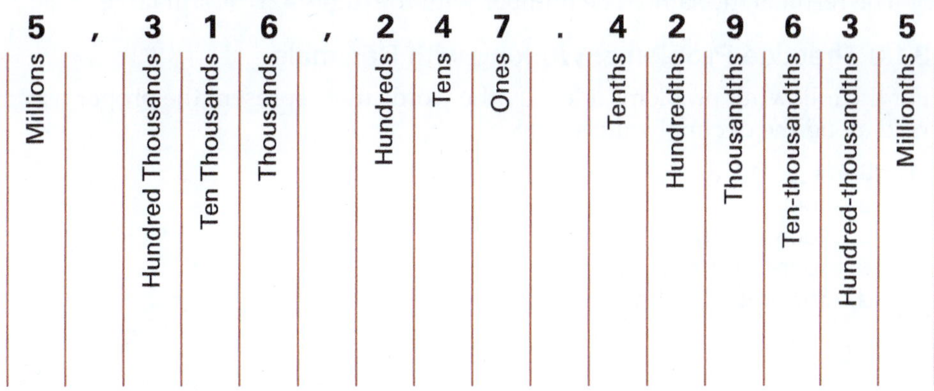

Figure 4 Place values on both sides of the decimal point.

To read a decimal fraction aloud, say the number as it is written. Then say the name of the place value for the rightmost digit. For example, 0.56 is spoken as "fifty-six hundredths." The rightmost digit (6) is in the hundredths place.

Note that the number 0.56 begins with a zero to the left of the decimal point. It adds clarity when the decimal represents a value less than one but usually is not spoken.

Mixed numbers also appear in decimal form. The number 15.7 is an example. Speak it aloud as "fifteen and seven-tenths." Notice that the word "and" separates the whole number from the decimal fraction.

3.1.1 Rounding Decimals

Sometimes an answer can be too precise. For example, assume tubing costs €3.76 (the € symbol represents the *euro*, the currency of the European Union) per meter. If the budget specifies €800 for the tubing, how much tubing can be purchased?

The answer, approximately 212.7659574 meters, goes on for many decimal places. However, a supplier will not sell 0.7659574 meters of tubing. An answer to the nearest tenth suffices for this application. Therefore, round the number to a single digit to the decimal point's right. The following steps show the process:

Step 1 Underline the rounding place column.

212.7659574

Step 2 Look at the digit one place to its right (6).

212.7659574

Step 3 If the digit to the right is 5 or more, round up by adding 1 to the underlined digit. If the digit is 4 or less, leave the underlined digit alone. In this example, the digit to the right is 6, which is more than 5, so round up by adding 1 to the underlined digit. Finally, drop the remaining digits to the right of it.

212.8

NOTE

In this case, following the rounding rules causes a slight budget overrun (€800.13). Budgets sometimes follow their own rounding rules to avoid financial problems. Unless instructed otherwise, however, always follow standard rounding rules.

3.1.2 Comparing Decimals with Decimals

Deciding whether one decimal value is larger or smaller than another is simple. The process resembles comparing whole numbers. Examine the following numbers:

4,214 and 4,217

Obviously 4,217 is larger. Adding a decimal point does not change anything:

0.4214 and 0.4217

The result is the same. The number with the digits 4217 is still larger.

3.1.3 Practice Problems: Working with Decimals

For the following problems, identify the words that represent the proper way to read aloud the decimal values shown.

1. 0.4 = _____.
   a. four
   b. four-tenths
   c. four-hundredths
   d. four-thousandths

2. 0.05 = _____.
   a. five
   b. five-tenths
   c. five-hundredths
   d. five-thousandths

3. 2.5 = _____.
   a. two and five-tenths
   b. two and five-hundredths
   c. two and five-thousandths
   d. twenty-five-hundredths

For the following problems, identify the written numerical value of the spoken number.

4. Eighteen-hundredths = _____.
   a. 0.0018
   b. 0.018
   c. 0.18
   d. 1.8

5. Five and eight-tenths = _____.
   a. 5.0
   b. 5.008
   c. 5.08
   d. 5.8

For the following problems, select the answer that arranges the decimals from *smallest* to *largest*.

6. 0.400, 0.004, 0.044, and 0.404
   a. 0.400, 0.004, 0.044, 0.404
   b. 0.004, 0.044, 0.404, 0.400
   c. 0.004, 0.044, 0.400, 0.404
   d. 0.404, 0.044, 0.400, 0.004

7. 0.567, 0.059, 0.56, and 0.508
   a. 0.508, 0.56, 0.567, 0.059
   b. 0.059, 0.56, 0.508, 0.567
   c. 0.567, 0.059, 0.56, 0.508
   d. 0.059, 0.508, 0.56, 0.567

8. 0.320, 0.032, 0.302, and 0.003
   a. 0.003, 0.032, 0.302, 0.320
   b. 0.320, 0.302, 0.032, 0.003
   c. 0.302, 0.320, 0.003, 0.032
   d. 0.003, 0.032, 0.320, 0.302

9. 0.867, 0.086, 0.008, and 0.870
   a. 0.870, 0.867, 0.086, 0.008
   b. 0.008, 0.086, 0.867, 0.870
   c. 0.086, 0.008, 0.867, 0.870
   d. 0.008, 0.870, 0.867, 0.086

10. 0.626, 0.630, 0.616, and 0.641
    a. 0.616, 0.641, 0.630, 0.626
    b. 0.616, 0.626, 0.630, 0.641
    c. 0.641, 0.616, 0.626, 0.630
    d. 0.630, 0.616, 0.626, 0.641

3.2.0 Adding, Subtracting, Multiplying, and Dividing Decimals

Calculating with decimals requires the same skills as working with whole numbers. However, a correct answer depends on placing the decimal point correctly.

3.2.1 Adding and Subtracting Decimals

When adding or subtracting decimals, always keep the decimal points aligned.

To add 4.76 and 0.834, line up the problem as shown. Adding a 0 keeps the numbers aligned correctly.

$$\begin{array}{r} \overset{1}{4}.760 \\ +\ 0.834 \\ \hline 5.594 \end{array}$$

> ### Did You Know?
> **Dots and Commas**
>
> In the United States, the dot (or period) acts as the decimal point. Commas separate every three digits of a whole number (for example, 14,203,897). But many European, African, and South American countries do the opposite. The comma acts as the decimal point, and the dot separates every three whole number digits. So, the expressions *845,302.125* (United States) and *845.302,125* (Europe) are the same. This difference can be very confusing! Be alert for it, especially when working with products and documentation developed outside the United States.

Remember, all the decimal points, including the answer's, must align vertically. The same rule applies when subtracting decimals. To subtract 2.724 from 5.6, line up the decimal points vertically. Again, add zeros to keep the numbers aligned and to make borrowing easier.

$$\begin{array}{r} \overset{4}{\cancel{5}}.\overset{15}{\cancel{6}}\overset{9}{\cancel{0}}\overset{10}{0} \\ -\ 2.724 \\ \hline 2.876 \end{array}$$

3.2.2 Multiplying Decimals

Multiplying decimal numbers closely resembles whole number multiplications. For example, an office cubicle partition panel is 4.5' wide. How wide a wall will seven side-by-side panels make?

Step 1 Set up the problem just like a whole number multiplication.

$$\begin{array}{r} 4.5 \\ \times\ 7 \\ \hline \end{array}$$

Step 2 Multiply the two numbers.

$$\begin{array}{r} \overset{3}{4}.5 \\ \times\ 7 \\ \hline 315 \end{array}$$

Step 3 To place the decimal point, count the number of digits to the right of the decimal points of both numbers being multiplied. In this example, only 4.5 has a decimal point. There is just one digit to its right.

Step 4 Place the answer's decimal point by counting over the same number of digits from the right of the product.

$$\begin{array}{r} 4.5 \\ \times\ 7 \\ \hline 31.5 \end{array}$$

If there are more digits to the right of the decimal points than there are in the answer, add extra zeros to the left of the digits.

$$\begin{array}{r} 0.507 \\ \times\ 0.022 \\ \hline 1014 \\ 10140 \\ 000000 \\ +\ 0000000 \\ \hline 11{,}154 = 0.011154 \end{array}$$

Total up the digits (6) to the right of the decimal points in the two multiplied numbers. Place the answer's decimal point six digits from the right. In this case, add an extra zero since there are only five digits in the answer.

3.2.3 Dividing with Decimals

Dividing decimal numbers can take one of three forms:

Type 1: Those with a decimal point in the number being divided (the dividend):

$$22\overline{)44.5}$$

Type 2: Those with a decimal point in the number being divided by (the divisor):

$$0.22\overline{)4{,}450}$$

Type 3: Those with decimal points in both numbers (the dividend and the divisor):

```
0.22)44.5
```

The following examples demonstrate each type.

Divide Type 1 numbers using the following steps (dividend = 44.5 and divisor = 22):

Step 1 Place the answer's decimal point directly above the decimal in the dividend.

```
     .
22)44.5
```

Step 2 Divide as usual.

```
   02.0
22)44.5
  − 44
   00.5r
```

The answer is 2, with a remainder of 0.5. Note that the quotient's decimal point aligns with the dividend's decimal point. This alignment is critical!

Divide Type 2 numbers using the following steps (dividend = 4,450 and divisor = 0.22):

Step 1 Move the divisor's decimal point all the way to the right. Count the number of places (2) that it moved.

0.22 becomes 22

Step 2 Next, move the dividend's decimal point to the right by the same number of places (2). Since the dividend is a whole number, add extra zeros. Place the answer's decimal point directly above the dividend's.

4,450 becomes 445,000

Step 3 Divide as usual.

```
     20,227.2
22)445,000.0
  − 44
    0050
  − 0044
    00060
  − 00044
    000160
  − 000154
    0000060
  − 0000044
    000001.6r
```

The answer is 20,227.2 with a remainder of 1.6.

Divide Type 3 numbers using the following steps (dividend = 44.5 and divisor = 0.22):

Step 1 Move divisor's decimal point all the way to the right. Count the number of places (2) it moved.

0.22 becomes 22

Step 2 Move the dividend's decimal point the same number of places (2) to the right. Add extra zeros if required.

44.5 becomes 4,450

Did You Know?

Scientific Notation

Scientific notation is a system that manages very large or very small decimal numbers. Numbers in this form have two parts: the base and the exponent. The base is a regular whole number or decimal number. The exponent represents the number of times to multiply 10 by itself ($10^2 = 10 \times 10 = 100$). A negative exponent represents the number of times to multiply $\frac{1}{10}$ by itself ($10^{-2} = \frac{1}{10} \times \frac{1}{10} = \frac{1}{100}$). Scientists, engineers, and mathematicians often use scientific notation. The following are some examples:

Decimal Notation	Scientific Notation
1	1×10^0
−50	-5×10^1
7,530,000,000	7.53×10^9
0.0000000082	8.2×10^{-9}

Step 3 Place the answer's decimal point directly above the decimal point in the dividend.

```
      .
22)4,450.
```

Step 4 Divide as usual.

```
    202
22)4,450
  - 44
    005
  - 000
    0050
  - 0044
    0006r
```

The answer is 202 with a remainder of 6.

Double-check any division answer by multiplying the quotient by the divisor and adding the remainder. The result should be the original dividend.

3.2.4 Using the Calculator to Add, Subtract, Multiply, and Divide Decimals

Doing decimal math with a calculator resembles calculating with whole numbers. Follow these steps using the problem 45.6 + 5.7 as an example.

Step 1 Turn the calculator on.

Step 2 Press 4, 5, . (decimal point), and 6. The number 45.6 appears.

NOTE

For the next step, press whichever operation key the problem specifies: + to add, – to subtract, × to multiply, ÷ to divide.

Step 3 Press the + key. The 45.6 should stay on the display.

Step 4 Press 5, . (decimal point), and 7. The number 5.7 appears.

Step 5 Press the = key. The answer should appear.

Step 6 Press the On or C key to clear the calculator fully. The CE key clears the most recent entry only. A 0 should appear.

Step 7 Repeat the previous steps, but select a different operation (–, ×, or ÷). Continue repeating the sequence to perform all four operations.

$$45.6 + 5.7 = 51.3$$

$$45.6 - 5.7 = 39.9$$

$$45.6 \times 5.7 = 259.92$$

$$45.6 \div 5.7 = 8$$

Did You Know?

Other Counting Systems

Ancient Babylonians developed one of the earliest place value systems—called *base 60*. It used special symbols combined in 59 different ways to represent numbers. Interestingly, the system had no zero symbol. Without the zero, calculations were harder and required remembering the location of empty place values. While nobody uses base 60 today, it survives in one important way—we still measure time in units of 60 (60 seconds per minute and 60 minutes per hour).

Computer programmers sometimes use a special system called *base 16*, or *hexadecimal*. It uses 16 numerals instead of 10: 0–9, plus A, B, C, D, E, and F. In the hexadecimal system, each place value is 16 times greater than the one to the right. For example, the hexadecimal number 1E is the same as the decimal number 30 (1 = 16, E = 14, so 16 + 14 = 30).

3.2.5 Practice Problems: Decimals

Without using a calculator, find the answers to these addition and subtraction problems. Be sure to keep the decimal points aligned.

1. 2.50 + 4.20 + 5.00 = _____

2. 1.82 + 3.41 + 5.25 = _____

3. 6.43 + 86.4 = _____

4. The combined thickness of a piece of sheet metal 0.078 cm thick and a piece of band iron 0.25 cm thick is _____.
   a. 0.308 cm
   b. 0.328 cm
   c. 3.08 cm
   d. 32.8 cm

5. Yesterday, a hay field contained 6.7 tons of baled hay. Since then, 2.3 tons were removed. How many tons of baled hay remain?
   a. 3.4 tons
   b. 4.4 tons
   c. 5.4 tons
   d. 6.4 tons

Use the following information to answer Questions 6 and 7:
A machinist mills a part for a pasture gate on a ranch. The starting thickness is 6.18". The machinist makes three cuts, each $^3/_{10}$ of an inch thick.

6. How many inches of material have been removed at this point?
   a. 0.6"
   b. 0.8"
   c. 0.9"
   d. 1.09"

7. The remaining part's thickness is _____.
   a. 5.28"
   b. 6.08"
   c. 6.10"
   d. 6.15"

8. Ceramic tile weighs 4.75 pounds per square foot. Therefore, 128 square feet of ceramic tile weighs _____.
   a. 598 pounds
   b. 608 pounds
   c. 908 pounds
   d. 1,108 pounds

Without using a calculator, find the answers to the following division problems. Round answers to the nearest hundredth.

9. 45.36 ÷ 18 = _____

10. 4.536 ÷ 18 = _____

11. 0.4536 ÷ 18 = _____

12. $25\overline{)10.20}$

13. $6\overline{)31.2}$

Without using a calculator, find the answers to the following division problems. Round answers to the nearest hundredth.

14. 282 ÷ 14.1 = _____

15. 694 ÷ 3.2 = _____

16. 99 ÷ 0.45 = _____

17. $2.5\overline{)102}$

18. $0.6\overline{)312}$

Without using a calculator, find the answers to the following division problems. Round answers to the nearest hundredth.

19. 20.82 ÷ 4.24 = _____

20. 38.9 ÷ 3.7 = _____

21. 9.9 ÷ 0.45 = _____

22. $0.25\overline{)10.20}$

23. $0.6\overline{)31.2}$

Use a calculator to find the answers to the following problems. Round answers to the nearest hundredth.

24. $\begin{array}{r} 45.89 \\ +\ 7.85 \\ \hline \end{array}$

25. $\begin{array}{r} 7.6 \\ \times\ 0.12 \\ \hline \end{array}$

26. $\begin{array}{r} 685.79 \\ -\ 562.66 \\ \hline \end{array}$

27. 6.45 ÷ 3.25 = _____

28. $\begin{array}{r} 34.76 \\ +\ 3.64 \\ \hline \end{array}$

Solve these problems to practice rounding decimals. Round answers based on each problem's requirement.

29. A 90.5" pipe must be cut into 3.75" pieces. How many complete 3.75" pieces are possible?
    a. 14
    b. 24
    c. 34
    d. 44

30. If a tractor ran for 1,001 kilometers on 151.8 liters of gas, to the nearest tenth of a kilometer, how many kilometers per liter did it achieve?
    a. 0.1
    b. 0.2
    c. 6.5
    d. 6.6

31. A supplier sells wire for $4.30 per pound. If the total sale is $120.95, to the nearest hundredth of a pound, how many pounds did the customer buy?
    a. 0.28
    b. 2.8
    c. 28.1
    d. 28.13

32. XYZ Supply Company sells a particular vent pipe at a special price of $0.37 per linear foot. To the nearest tenth of a foot, how many linear feet of pipe can be purchased with $115.38?
    a. 308.1
    b. 310.8
    c. 311.8
    d. 311.9

33. The same vent pipe's regular price is $0.48 per linear foot. To the nearest tenth of a foot, how many linear feet can be purchased for the same price ($115.38)?
    a. 240.38
    b. 240.4
    c. 241
    d. 241.4

3.3.0 Converting Decimals, Fractions, and Percentages

Sometimes craft professionals must convert one number form into another to make work easier. Some numbers may appear as decimals, some as percentages, and others as fractions. Decimals, percentages, and fractions are all different ways of expressing the same thing. The decimal 0.25, the percentage 25%, and the fraction $^1/_4$ all represent the same quantity. Each form can change into one of the others.

3.3.1 Converting Decimals to Percentages and Percentages to Decimals

To understand percentages, think of a whole number divided into 100 parts. A $1.00 bill is a good example, since it equals 100 equal parts—pennies. Each penny represents 1% of a dollar. Any part of a whole can be expressed as a percentage. *Figure 5* shows a good craft example. The tank holds 100 gallons. At present, it contains 50 gallons. To what percentage is the tank filled?

The correct answer is 50%. Percentage reflects a quantity based on there being 100 parts of something. How many gallons out of 100 does the tank contain? It contains 50 out of 100, or 50%. Percentages are an easy way to express parts of a whole.

Decimals and fractions also express parts of a whole. The tank in *Figure 5* is 50% full. Expressed as a fraction, it is $^1/_2$ full. The decimal equivalent is 0.50 full.

Percentages can include decimals too. Adding a half gallon of liquid to the tank in *Figure 5* increases it to 50.5% full.

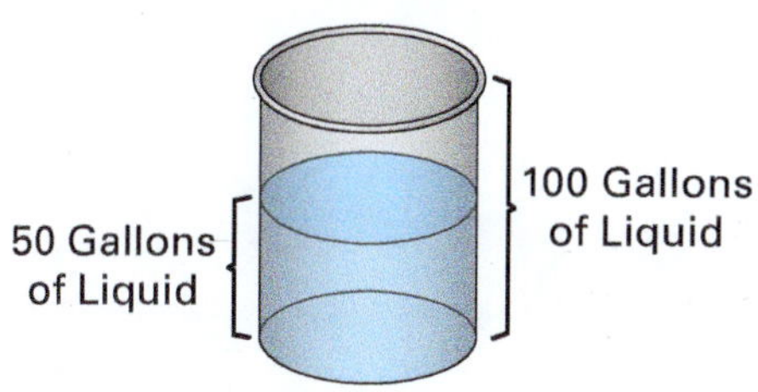

Figure 5 100-gallon tank.

Measuring Coating Thickness

Coating thickness is important on many parts and surfaces because too little or too much can cause problems. A paint coating must be thick enough to prevent corrosion, withstand abrasion, and look nice. But too thick a coating might not cure properly or could crack, flake, or blister.

Customers often specify the desired coating thickness. A painter can use a wet-film thickness gauge to check a coat of paint. Typically, these gauges measure in mils or microns. The gauge shown here uses the mil, which equals one one-thousandth of an inch (0.001"). This value can also be written as 10^{-3}. A mil equals 0.0254 millimeters.

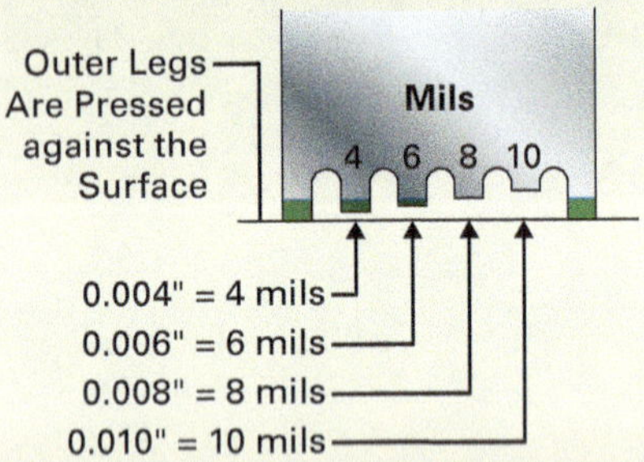

Sometimes, workers must convert decimals to percentages. Suppose a farm worker needs to prepare one gallon of insect control solution. The mixture must contain 10% to 15% of the solution's concentrate. The rest is water. There is 0.12 gallon of concentrate available. Will this be enough to make one gallon? To answer the question, convert the decimal value of the concentrate (0.12) to a percentage. Follow these steps to convert to a percentage:

Step 1 Multiply the decimal value by 100. (Tip: to multiply by 100, simply move the number's decimal point two places to the right.)

$$0.12 \times 100 = 12$$

Step 2 Add a % sign.

12%

The mixture must contain 10% to 15% of the concentrate. Since there is 12% of a gallon available, there is enough concentrate to make one gallon of working solution.

Converting percentages to decimals is simple too. Suppose a one-pound mixture must contain 22% of a certain fertilizer by weight. A digital scale, which reads in decimal values, weighs the ingredients. To determine the correct amount of chemical in decimal form, convert the percentage (22%) to a decimal by following these steps:

Step 1 Drop the % sign.

22

Step 2 Divide the number by 100. (Tip: to divide by 100, simply move the number's decimal point two places to the left.)

$$22 \div 100 = 0.22$$

Make the mixture by weighing out 0.22 pounds of the fertilizer. Add it to 0.78 pounds of the other ingredients, giving a total of 1 pound.

$$0.22 + 0.78 = 1$$

Decimals at Work

Several craft professions, including millwrights, use decimal measurements daily. Millwrights often align pump and drive motor shafts. Extremely accurate alignment eliminates vibration and prevents damage to couplings and seals. Laser technology helps align this pump and motor. The pump manufacturer specifies the alignment, often to within 0.002 or 0.003 of an inch. The millwright places shims of various thicknesses under the pump and/or motor feet to tweak the alignment.

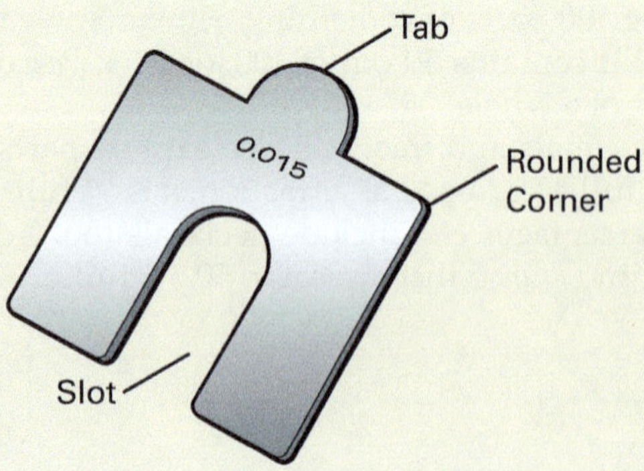

Source: LUDECA, INC.

3.3.2 Converting Fractions to Decimals

Changing a fraction into a decimal requires nothing more than division. Remember, a fraction is already a kind of division problem. Changing a fraction into a decimal value requires dividing the numerator by the denominator. The following steps show how to convert $^3/_4$ of a liter into a decimal value:

Step 1 Divide the fraction's numerator by the denominator.

$$4\overline{)3.0}$$

- For this example, place a decimal point and a 0 after the number 3. Since 4 will not divide into 3, the extra digit is necessary.

Step 2 Place the answer's decimal point directly above the decimal in the dividend.

$$\begin{array}{r} . \\ 4\overline{)3.0} \end{array}$$

Step 3 Divide as usual.

$$\begin{array}{r} 0.75 \\ 4\overline{)3.00} \\ -\ 2.8 \\ \hline 0.20 \\ -\ 0.20 \\ \hline 0.00 \end{array}$$

The decimal equivalent of $^3/_4$ is 0.75. If required, change this value into a percentage by moving the decimal point two positions to the right and adding a % sign. So, $^3/_4$ of a liter is also 0.75 liters or 75% of a liter.

3.3.3 Converting Decimals to Fractions

Converting a decimal to a fraction is also simple. Remember that both decimals and fractions express the same thing in different ways. Saying a decimal value aloud sounds like a fraction. Writing the spoken form automatically translates it.

For example, drill bits come in fractional sizes, decimal sizes, and even sets identified by special codes. A carpenter must drill a 0.125" hole. However, the available drill bits have fractional size markings on them. One of those sizes might be equivalent to 0.125". To find out, the carpenter must follow these steps to convert the decimal to a fraction:

Step 1 Say the decimal value aloud.

0.125 is one hundred twenty-five thousandths

Step 2 Write the words down as a fraction.

0.125 written as a fraction is $^{125}/_{1,000}$

Step 3 Reduce the fraction to its lowest terms.

$$\frac{125}{1,000} = \frac{125 \div 125}{1,000 \div 125} = \frac{1}{8}$$

Clearly, a $^1/_8$" drill bit will drill a 0.125" hole.

3.3.4 Converting Inches to Decimal Equivalents in Feet

Unlike the United States, most other countries use the metric system of measurement. The metric system perfectly matches the decimal system since both rely on units of 10. However, for those working with the US standard system, measurements may require conversion to decimal. For example, what decimal part of a foot does 3" represent?

First, express the inches as a fraction with 12 as the denominator. Remember, 1" equals $\frac{1}{12}$ of a foot. The fraction for 3" is $\frac{3}{12}$'.

Reducing the fraction gives $\frac{1}{4}$. Convert this to a decimal by dividing the 4 into 1.00:

$$\begin{array}{r} 0.25 \\ 4\overline{)1.00} \\ -\,0.8 \\ \hline 0.20 \\ -\,0.20 \\ \hline 0.00 \end{array}$$

So, 3" equals 0.25'.

3.3.5 Practice Problems: Converting Different Values

Convert these decimals to percentages.

1. 0.62 = _____
2. 0.475 = _____
3. 0.7 = _____

Convert these percentages to decimals.

4. 72% = _____
5. 12.5% = _____

Without using a calculator, convert the following fractions to their decimal equivalents.

6. $\frac{1}{4}$ = _____
7. $\frac{3}{4}$ = _____
8. $\frac{1}{8}$ = _____
9. $\frac{5}{16}$ = _____
10. $\frac{20}{64}$ = _____

Without using a calculator, convert the following decimals to fractions. Reduce them to their lowest terms.

11. 0.5 = _____
12. 0.12 = _____
13. 0.125 = _____
14. 0.8 = _____
15. 0.45 = _____

Convert the following measurements to a decimal value in feet. Round to the nearest hundredth.

16. 9 inches = _____ feet
17. 10 inches = _____ feet
18. 2 inches = _____ feet
19. 4 inches = _____ feet
20. 17 inches = _____ feet

Practical Math Application

When an agricultural craftworker calculates the expenses for building a new chicken coop, they must pay close attention to the percentage of money spent on various expenses. Of course, they will want to maximize profits from the chickens and eggs as well. For example, if they plan to sell $175,000 worth of poultry and eggs in one year, what will the profit be with the following percentage breakdown?

- Profit = ?%
- Overhead = 27%
- Materials = 35%
- Labor = 24%

Solution: Since $175,000 represents 100% of the revenue, the total cost and profit together must equal 100%. Adding the known percentages reveals that the total job cost is 86% of the revenue. The profit is therefore 14% of the revenue (100% – 86% = 14%). Find the profit in dollars with this method:

$$\text{Profit} = \$175{,}000 \times 14\% = \$175{,}000 \times 0.14 = \$24{,}500$$

Calculators make working with percentages easy. In fact, some calculators have a % key to make the task even easier.

For calculators with a % button, do the following:

Step 1 Enter 175000 into the calculator.
Step 2 Press the × key.
Step 3 Enter 14 into the calculator.
Step 4 Press the % key.
Step 5 Press the = key.

For calculators without a % button, do the following:

Step 1 Enter 175000 into the calculator.
Step 2 Press the × key.
Step 3 Enter 0.14 (the decimal equivalent of 14%) into the calculator.
Step 4 Press the = key.

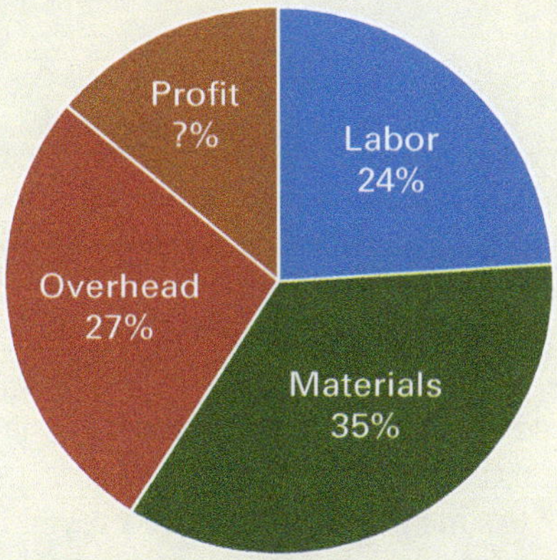

Did You Know?

Direct and Indirect Costs

When farmers and ranchers determine a percentage of profit, they must consider many different costs. The costs divide into two broad categories: direct costs and indirect costs.

Direct costs are expenses directly associated with the investment of a new building or operation. They include materials (paint, lumber, steel) and labor hours. Large-scale construction investments may also include subcontracted demolition and cleanup crews, cranes, earth-moving machinery, and jobsite safety requirements.

Indirect costs (or *overhead costs*), on the other hand, are expenses required to keep the operations running during construction. They include insurance, administrative activities (for example, utility payments), supplies, and property taxes. Try to think of some other indirect costs!

3.3.6 Practical Applications

The following are applications that craftworkers encounter daily. Use a calculator and the information provided to solve each conversion problem. Be sure to show all your work.

1. Find the costs of baseboard needed for the office building shown by the floor plan in *Figure 6*. While all door widths are a standard 36", do *not* subtract the door openings from the room measurements. Wise craftworkers always allow some extra material in case somebody makes a mistake. Round the footage number of baseboard up to the nearest linear foot. The lumber company charges $1.19 per linear foot of baseboard and gives a 12% discount. Add 6% tax after subtracting the discount amount.
   a. Total linear feet needed = _____ linear feet
   b. Initial baseboard cost = $_____
   c. Discount amount = $_____ (round to the nearest hundredth)
   d. Baseboard cost after discount = $_____
   e. Amount added for 6% tax = $_____ (round to the nearest hundredth)
   f. Total baseboard cost = $_____

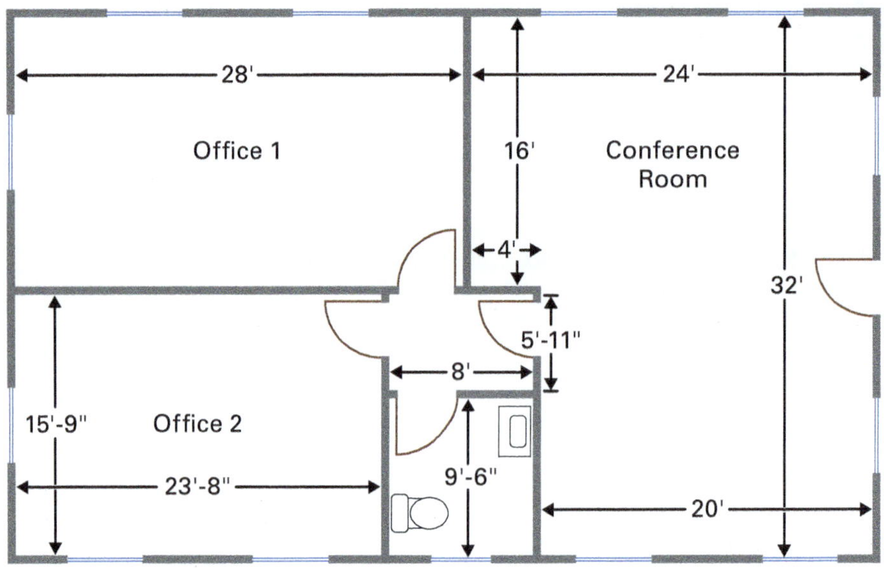

Figure 6 Office floor plan.

Going Green

City Governments Thinking Green

Governing bodies around the world come up with innovative ways to cut energy costs and lower greenhouse gas emissions. Ideas include creating energy by harnessing wave power, air-conditioning a building using rooftop gardens or lake water, and integrating wind turbines into a building's construction to generate energy.

Tuscola County, Michigan: In 2012, the Tuscola Bay Wind Farms were established in two Michigan cities, Fairgrove and Tuscola. In cooperation with the farming community, the wind turbine sites were built on farmland within the local area to operate 134 wind turbines in these two established locations. Together, the wind turbine sites generate about 220 megawatts of energy, enough to power 60,000 homes every day.

Aspen, Colorado: In 1885, Aspen became the first municipality located west of the Mississippi River to make use of hydroelectric power—a renewable energy resource. After years of work, Aspen achieved its goal of powering the city with nothing but renewable energy. Today, 100% of the city's power comes from renewable sources of energy—46% hydroelectric, 53% wind, and 1% landfill gas.

2. Use the site plan shown in *Figure 7* to determine the percentage of the parking lot that will be used for heavy equipment parking, the storage shed building, and walkways. Round off the percentages to the nearest tenth.
   a. Percentage for parking = _____%
   b. Percentage for building = _____%
   c. Percentage for walkways = _____%

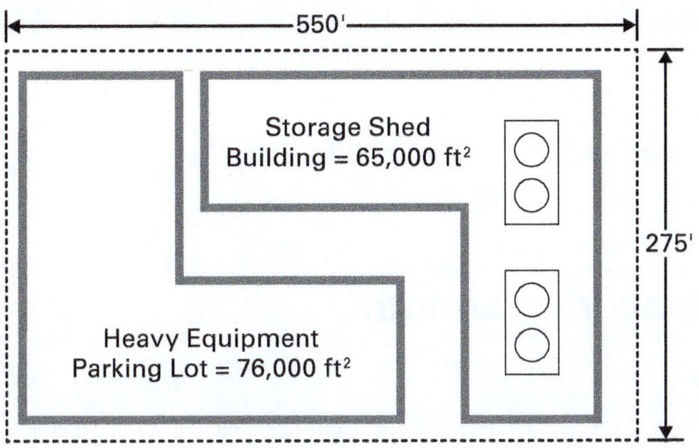

Figure 7 Site plan.

3.0.0 Section Review

1. In the number 135.792, what value does the 9 represent?
   a. 90
   b. nine tenths
   c. nine one-hundredths
   d. nine one-thousandths

2. The number 0.960 is larger than the number 0.0962.
   a. True
   b. False

3. 3.625 + 4.9 = _____.
   a. 7.525
   b. 8.525
   c. 41.15
   d. 52.63

4. 42.58 – 7.577 = _____.
   a. 35.003
   b. 35.523
   c. 36.523
   d. 50.157

5. 9.64 × 12 = _____.
   a. 11.568
   b. 21.64
   c. 115.2
   d. 115.68

6. 123.82 ÷ 6.5 = _____.
   a. 19.04923
   b. 18.76
   c. 1.905
   d. 0.19

7. Express the number 0.479 as a percentage.
   a. 0.00479%
   b. 0.479%
   c. 47.9%
   d. 479%

8. The decimal equivalent of the fraction $^7/_8$ is _____.
   a. 0.0875
   b. 0.75
   c. 0.875
   d. 8.75

9. If a fuel tank is $^2/_3$ full, then it is more than 65% full.
   a. True
   b. False

4.0.0 Measuring Length

Performance Tasks

1. Using a measuring tape, measure lumber pieces in both US standard and metric units.
2. Using a measuring tape, measure a room-sized space in both US standard and metric units.
3. Using a measuring tape, determine a short inside measurement in both US standard and metric units.
4. Add US standard measurements that include fractions.

Objective

Name common length-measuring tools and use them to measure lengths accurately.

a. Describe US standard and metric rulers, using them correctly to measure lengths.
b. Describe US standard and metric measuring tapes, using them correctly to measure lengths.

All construction workers use measuring tools to determine the dimensions of various objects. Tape measures and rulers, both US standard and metric, are the most common (*Figure 8*).

(A) US Standard Tape Measure

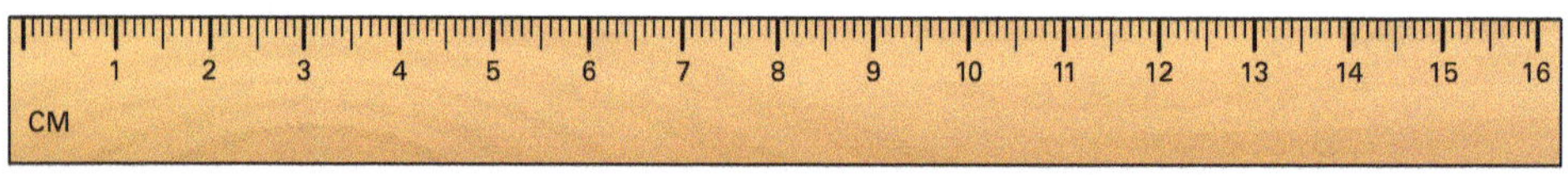

(B) Metric Ruler

Figure 8 Various measuring tools.

CAUTION

Most rulers, especially wooden ones (like the metric ruler shown in *Figure 8*) have extra material at the ends to maintain accuracy if the ends get damaged. Always start a measurement at the first marked line rather than at the ruler's physical end.

4.1.0 Reading US Standard and Metric Rulers

Reading a ruler is easy (*Figure 9*). But those who use them only occasionally may struggle to remember the values of the small marks between each major **unit**. Misreading a ruler leads to errors, expensive mistakes, and wasted material.

NOTE

In a measurement setting, the terms *US standard*, *inch-pound*, *English*, and *Imperial* mean the same thing. They are names for the measuring system used in the United States, Canada, and a few other countries. This module prefers US standard.

Unit: A standardized quantity used for a particular kind of measurement.

4.1.1 US Standard Ruler

US standard rulers provide measurements in inches. Marks divide US standard rulers into whole inches and then halves, fourths, eighths, and sixteenths. Some rulers may have marks for thirty-seconds or even sixty-fourths. Except for the whole-inch marks, everything else represents fractions of an inch. This section explains the US standard ruler and standard fractions. In *Figure 8*, the US standard ruler has two sets of marks. The top row's marks are $^1/_8$" apart. The bottom row uses marks $^1/_{16}$" apart. Other rulers may have different marking systems. For example, *Figure 10* shows a ruler with just $^1/_{16}$" marks.

Figure 9 Use a ruler to measure before cutting.
Source: National FFA Organization

In *Figure 10*, the distance between the $\frac{1}{8}$" and $\frac{1}{4}$" positions is two-sixteenths of an inch ($\frac{2}{16}$"). Similarly, the mark immediately after the $\frac{3}{4}$" position represents thirteen-sixteenths of an inch ($\frac{13}{16}$").

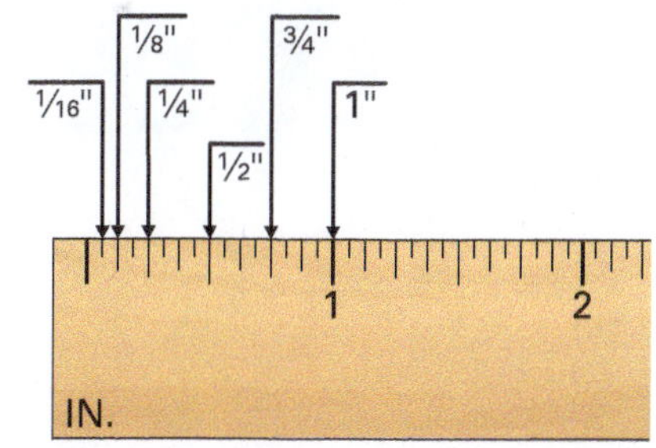

Figure 10 Standard ruler showing $\frac{1}{16}$" marks and larger values.

4.1.2 Metric Ruler

Metric rulers provide measurements in centimeters (cm) and millimeters (mm). Marks on metric tape measures and rulers (*Figure 11*) represent centimeters and millimeters. The larger numbered lines are centimeters. The smaller lines represent millimeters. The metric system is *base 10*, since all measurements use some form of the number 10. For example, each millimeter is $\frac{1}{10}$ of a centimeter.

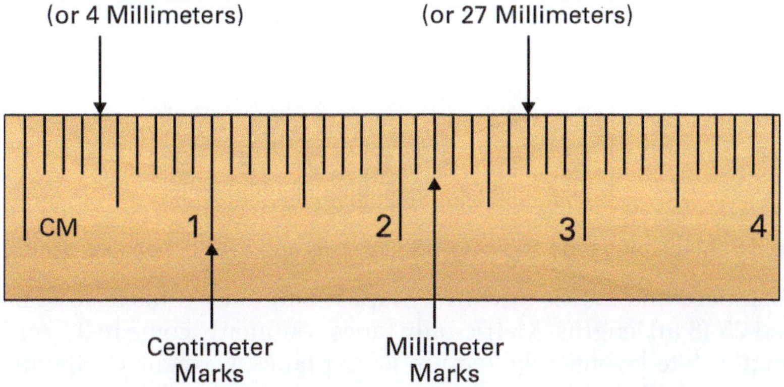

Figure 11 Marks on a metric ruler.

A measurement that is seven marks after 2 cm is 2.7 (two point seven) cm or 27 mm (*Figure 11*). Both numbers represent the same distance but use different units.

The measurement that is six marks behind the 1 cm mark is 0.4 (point four) cm or 4 mm (*Figure 11*).

4.1.3 Practice Problems: Reading Rulers

Use the information from *Figure 11* to help identify the marked lengths numbered 1 through 10 in *Figure 12* and *Figure 13*. Label each length in the space provided or on a separate sheet of paper.

Identify the marked lengths in *Figure 14*. Record the correct answers for 11, 12, and 13 in centimeters. Record the correct answers for 14 and 15 in millimeters.

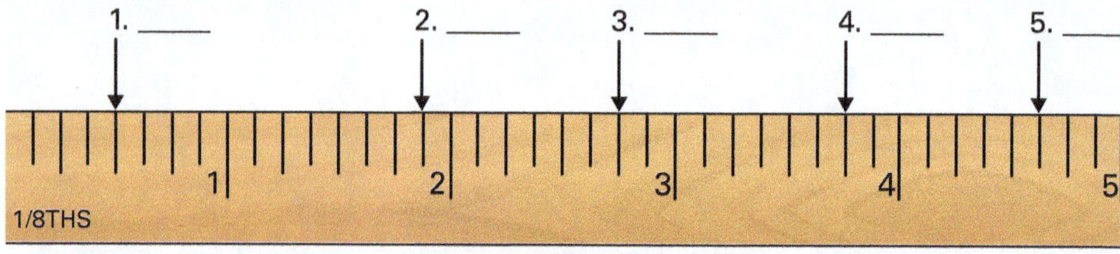

Figure 12 Practice with a ⅛" mark ruler.

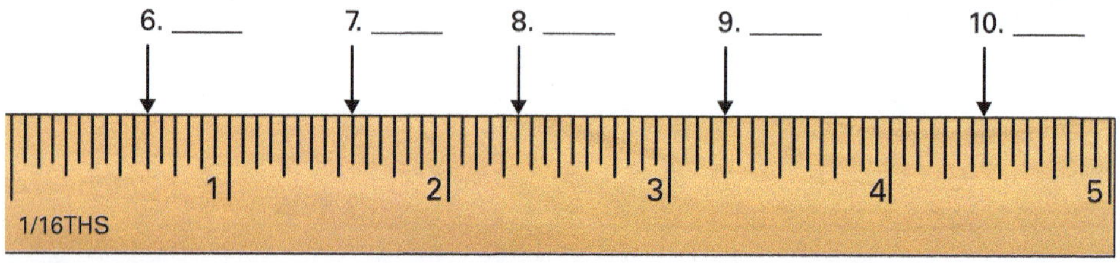

Figure 13 Practice with a $\frac{1}{16}$" mark ruler.

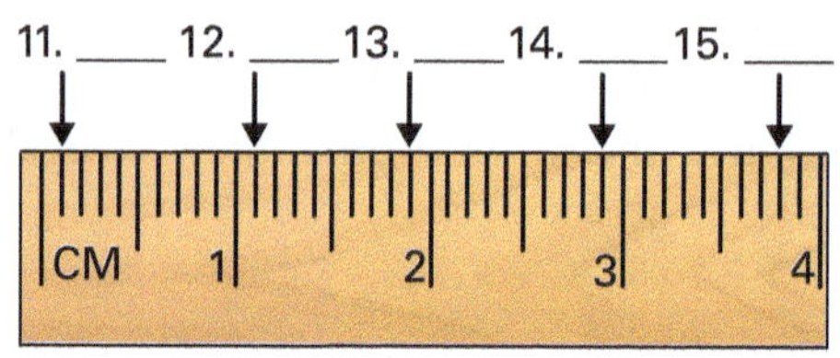

Figure 14 Practice with a metric ruler.

4.2.0 Measuring Tape

Measuring tapes (also known as *tape measures*) may have US standard or metric markings along the blade. Some, such as the one in *Figure 15*, have both. Tapes using just one system often have identical markings along both sides to make measuring easier.

Figure 15 Measuring tape showing US standard and metric measurements.

US standard and metric measuring tapes commonly come in 16' (5 meters, or 5 m) and 25' (8 m) lengths. Metric-only tapes commonly come in 3.5 m, 5 m, and 8 m lengths. Site layout tasks require longer tapes. Carpentry, pipefitting, and other crafts usually prefer short, compact ones.

4.2.1 US Standard Measuring Tape

US standard measuring tapes have markings like rulers but may also have additional markings. These marks make tasks like wall framing easier.

As *Figure 16* (A) shows, every 24" is marked with a black background. Carpenters usually use 24" spacing for wall studs that do not carry a heavy load from above. Marks 16" apart have a red background as seen in *Figure 16* (B). These help carpenters space studs 16" apart for **load-bearing** walls.

Figure 16 shows common **stud** spacing. With the measuring tape hook pulled tightly against the left side of one stud, the left side of the next stud is 16" away. Therefore, these studs are placed on 16" centers. Small black diamonds, like those in *Figure 16* (C), identify 19.2" spacings for specialized studs and **joists**.

Figure 17 shows other markings that may appear on a standard tape measure. The red foot number (1F) works with the red inch number (2) to provide quick measurements. In this case, the numbers indicate 1'-2". The number below it (the black 14) indicates 14". Adding the top scale numbers together gives the bottom scale number: 1' (or 12") + 2" = 14". The red numbers appear along the length of the blade.

Load-bearing: Carrying a significant amount of weight and/or providing necessary structural support. A load-bearing wall typically carries some portion of the roof's weight and cannot be removed without risking structural failure or collapse.

Stud: A vertical support inside a wall to which the wall finish material attaches. The base of a stud rests on a horizontal baseplate, and a horizontal cap plate rests on top of a series of studs.

Joists: Lengths of wood or steel that usually support floors, ceilings, or a roof. Roof joists will be at the same angle as the roof itself, while floor and ceiling joists are usually horizontal.

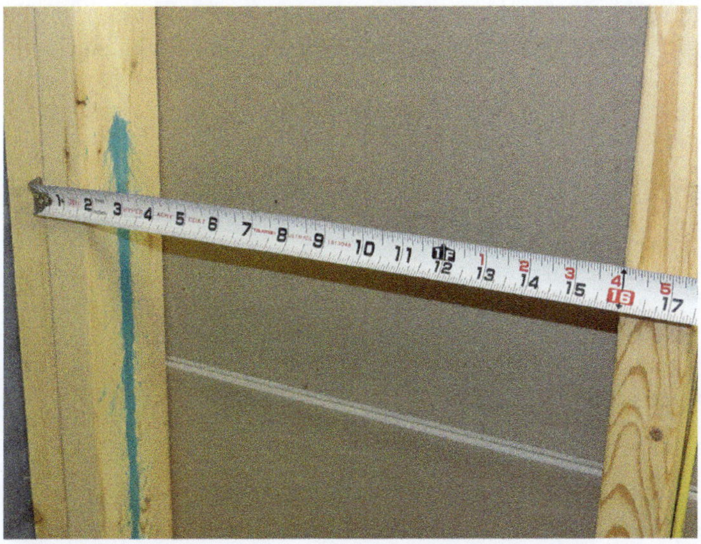

(A)

Every 24" is marked with a contrasting black background. 24" spacing on center is used most commonly for nonbearing walls.

(B)

Every 16" is marked with a red background. 16" spacing on center is used most commonly for load-bearing walls.

(C)

Every 19.2" is marked with a small black diamond. 19.2" spacing on center is an alternate, less commonly used spacing scheme for load-bearing walls and special joists.

Figure 16 Stud spacing and wall framing markings.

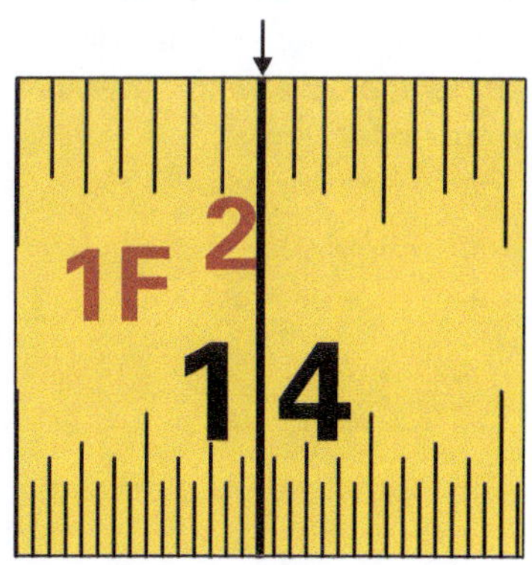

Figure 17 Other markings on a standard tape measure.

Around the World

Nominal and True Dimensions

A 2 × 4 board used in the United States is not really 2" × 4". Only the rough board cut from the log had those dimensions. After drying and planning, the finished board is $1\frac{1}{2}$" × $3\frac{1}{2}$". The following are examples of the true dimensions of common US board sizes:

Nominal Size	Actual Measure	Actual Metric Measure
1" × 10"	$\frac{3}{4}$" × $9\frac{1}{4}$"	19 mm × 235 mm
2" × 6"	$1\frac{1}{2}$" × $5\frac{1}{2}$"	38 mm × 140 mm
2" × 8"	$1\frac{1}{2}$" × $7\frac{1}{4}$"	38 mm × 184 mm
2" × 10"	$1\frac{1}{2}$" × $9\frac{1}{4}$"	38 mm × 235 mm
4" × 4"	$3\frac{1}{2}$" × $3\frac{1}{2}$"	89 mm × 89 mm

To convert to equivalent metric lumber sizes, use 25 mm per inch as a conversion. This method is not very precise, but it works. A 2" × 4" is 50 mm × 100 mm in the metric world. However, the actual dimensions are typically 40 mm × 90 mm. Over the years, actual lumber size has steadily decreased compared to nominal size. This trend may provoke changes in building codes to keep structures safe.

4.2.2 Metric Measuring Tape

Metric measuring tapes look like standard measuring tapes. In the United States, metric-only tapes are rare. In most cases, a combination tape will be the only option. Outside the United States, however, metric-only tapes are common.

Metric tapes have markings to help with framing tasks too. But they will be spaced differently to accommodate other building codes and lumber dimensions.

4.2.3 Using a Tape Measure

Most people learn to use a measuring tape easily. Measuring tapes curve slightly to make them stiffer (*Figure 18*). This feature helps them stay straight when extended. Some tapes have thicker blades and a greater curve for longer extension. Manufacturers call the amount of extension the tape's *standout*. The following steps outline the correct way to measure with a tape:

Step 1 Pull the tape straight out of its case with one hand.

Step 2 Hook the metal tape end to the measured object. Another worker may have to hold the tape end if it cannot be hooked onto the object. The tape hook should be a bit loose and slide slightly to compensate for its thickness. It slides in for making inside measurements and slides out for outside ones. If the hook binds instead of sliding, measurements could be off by as much as $\frac{1}{16}$" (1.6 mm). A bent or damaged hook can have the same effect.

Step 3 Extend the tape by pulling the case. Go a bit beyond the measurement point.

Step 4 After extending the tape the correct amount, slide the case's thumb lock to hold the tape in place. Not all tape measures have a thumb lock, and some tape measures lock automatically. A thumb release retracts the tape on automatically locking tape measures. Never let the tape snap back into the case, or the hook may break.

Step 5 Read the measurement from the correct location on the measured object.

Measuring between two objects can be challenging. For example, installing a shelf between two walls requires this kind of measurement. Measuring tapes usually have the case's exact length stamped on the side. Take the measurement by firmly bumping the hook against one wall and the back of the case against the other. Read the length where the tape enters the case. Then, add the case length to the value.

NOTE

Always double-check measurements before cutting a piece of material. Remember the old saying, "Measure twice, cut once."

4.2.4 Practice Problems: Reading Measuring Tapes

Identify the marked lengths numbered 1 through 5 in *Figure 19*. Record the answers in inches, using fractions as necessary. Identify the marked lengths numbered 6 through 10 in *Figure 20*. Record the answers in centimeters, using tenths of a centimeter as necessary.

Around the World

Visualizing Metric Units

The follow examples make it easier for US standard users to visualize metric measurements (approximation only):

1 millimeter = the thickness of a dime
1 centimeter = the width of a standard paper clip
1 decimeter = the length of a crayon
1 meter = the distance from a door handle to the floor (about 1.1 yards)
1 kilometer = the length of 6 city blocks (about 0.6 miles)
1 gram = the weight of a paper clip
1 kilogram = the weight of a brick (about 2.2 pounds)

Figure 18 Measuring tapes are slightly curved for strength.
Source: National FFA Organization

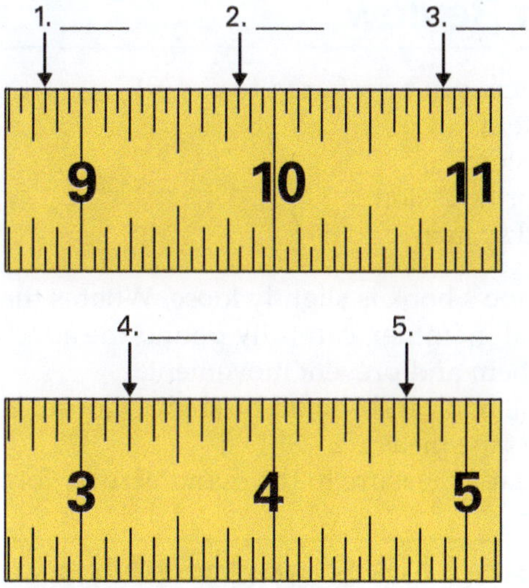

Figure 19 US standard tape measure practice.

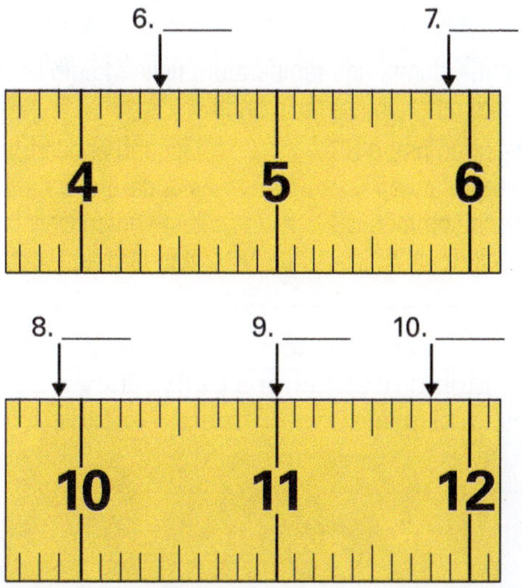

Figure 20 Metric tape measure practice.

Did You Know?

Standard Measure

Determining the distance to a neighbor's farm may have been the earliest reason for measuring. Imagine that you wanted to measure the distance to a favorite fishing pond. You could pace off the distance and count the number of steps. You might find that it was 570 paces between your house and the pond. You could then pace off the distance to another pond. Comparing paces would tell you which distance was longer, so the pace became a measurement unit. Since everyone has a different pace length, however, this method is not accurate.

Early humans used feet, arms, hands, and fingers for measuring. A few of these units survive today. We measure length in feet and horse height in hands. An inch comes close to a thumb's width, but body part sizes differ. A person with small hands could sell a horse as taller than it really is.

These problems led to standardized measurement units. Manufactured physical objects, such as special sticks or metal rods, became the basis for consistent units. Craft associations made careful copies of these so-called standards and used them to produce their measuring tools. Crafts guilds insisted on their apprentices, journeymen, and masters following the standards. A craft's reputation depended on quality and accuracy, just as it does today. Guilds often penalized those who were careless or who deliberately falsified measurements.

4.0.0 Section Review

1. The metric system can be referred to as a(n) _____.
   a. base 10 system
   b. base 100 system
   c. geometric progression
   d. open-ended system

2. A measuring tape's hook is slightly loose. What is the *correct* action?
   a. Using a small hammer, carefully pound the attaching rivets enough to tighten them and prevent movement.
   b. Nothing; the hook end should be slightly loose.
   c. Replace the tape measure.
   d. Return the tape measure to the manufacturer for repair.

5.0.0 Metric and US Standard Measurement Systems

Performance Tasks

There are no Performance Tasks in this section.

Objective

Name common length, weight, volume, and temperature units in both US standard and metric systems and convert them into other comparable units.

a. List and convert between common US standard and metric length units.
b. List and convert between common US standard and metric weight units.
c. List and convert between common US standard and metric volume units.
d. List and convert between common US standard and metric temperature units.

Most of the world uses the metric system of measurement. Companies manufacture and ship goods to global destinations daily. They may provide dimensions, weights, temperatures, and pressures in metric values, US standard values, or both.

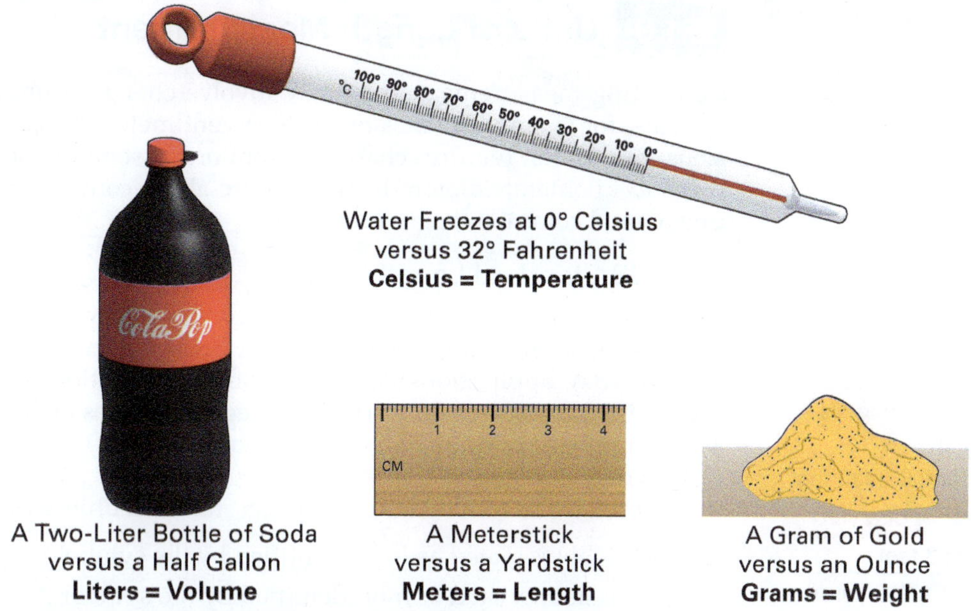

Figure 21 Common US standard and metric measured values.

Companies within the United States have started using the metric system. Many manufacturers use both US standard and metric values so they can support customers from any country. Many crafts require its workers to use both systems and convert between them. *Figure 21* shows some common metric and US standard system examples.

Science, engineering, and many crafts require exact measurements of physical quantities. Measurements involve comparing the quantity being measured to a standard dimension called a unit. Measurements must always include the appropriate unit (such as foot, liter, or pound). A number alone is not a measurement.

Measuring with the metric system is simpler than measuring with the US standard system. Scaling metric units always involves multiplying or dividing by 10s. On the other hand, the US standard system requires many different conversion factors. For example, 1 mile is 5,280 feet, and 1 inch is $^1/_{12}$ of a foot. On the other hand, 1 centimeter is $^1/_{100}$ of a meter, and 1 kilometer is 1,000 meters. Metric calculations usually take place in decimal form, eliminating the need to add and subtract fractions.

Scaling metric units up or down involves attaching a prefix to the unit. *Table 1* shows the common metric scaling prefixes. For example, the prefix *milli-* scales a unit by 0.001 (one one-thousandth). One millimeter is therefore 0.001 meter. The prefix *kilo-* scales a unit by 1,000. One kilometer is therefore 1,000 meters. While the table may look intimidating, it is quite simple. Just find the prefix and read across to find the scaling factor.

NOTE

Farmland is often measured in *acres*, a US standard unit equivalent to 43,560 square feet or 4,047 square meters. The metric unit for measuring large areas like farmland is a *hectare* (*ha*), which is equivalent to 10,000 square meters.

TABLE 1 Metric System Prefixes

Unit Prefix	Expressed Quantity as Powers of 10			Decimal System Expression
Micro- (μ)	1/1,000,000	0.000001	10^{-6}	One-millionth
Milli- (m)	1/1,000	0.001	10^{-3}	One-thousandth
Centi- (c)	1/100	0.01	10^{-2}	One-hundredth
Deci- (d)	1/10	0.1	10^{-1}	One-tenth
Deka- (da)	10	10.0	10^{1}	Tens
Hecto- (h)	100	100.0	10^{2}	Hundreds
Kilo- (k)	1,000	1,000.0	10^{3}	Thousands
Mega- (M)	1,000,000	1,000,000.0	10^{6}	Millions
Giga- (G)	1,000,000,000	1,000,000,000.0	10^{9}	Billions

5.1.0 Units of Length Measurement

Converting measurement units may involve changing units within the same system—for example, inches to yards or centimeters to meters. On other occasions, conversion requires changing from one system to another—for example, miles to kilometers. Before learning more about conversion, it is important to understand common length units.

5.1.1 US Standard Length Units

Table 2 shows the US standard system's common length units, along with their relationships. Many applications require breaking inches down into fractions. For everyday applications, $\frac{1}{64}$" is the smallest fraction required. However, a machinist doing precision work might need to express inches in decimal values. These numbers may be as small as thousandths (0.001") or ten-thousandths (0.0001").

Abbreviate the following common US standard units as shown:

- Inch is abbreviated as *in* or identified by the symbol ".
- Foot is abbreviated as *ft* or identified by the symbol '.
- Yard is abbreviated as *yd*.
- Mile is abbreviated as *mi*.

TABLE 2 Common US Standard Length Units

Original Weights	Equivalent Weights
1 inch	$\frac{1}{12}$ of a foot; 0.0833 feet
1 foot	12 inches; $\frac{1}{3}$ of a yard
1 yard	36 inches; 3 feet
1 mile	5,280 feet; 1,760 yards

Converting one unit to another requires multiplying or dividing by a conversion factor. A calculator simplifies unit conversions. For example, to change from inches to yards, divide the number of inches by 12 (the number of inches in a foot) to convert to feet first. Then, divide the result by 3 (the number of feet in a yard) to convert to yards. Alternatively, just divide the number of inches by 36 (*Figure 22*) to convert directly to yards.

Many craft professionals convert between units with smartphone apps. These apps contain conversion factors for many US standard and metric units. Search engines can also convert between units or suggest online calculators. Simply type in the two units as a search term, such as "inches to meters."

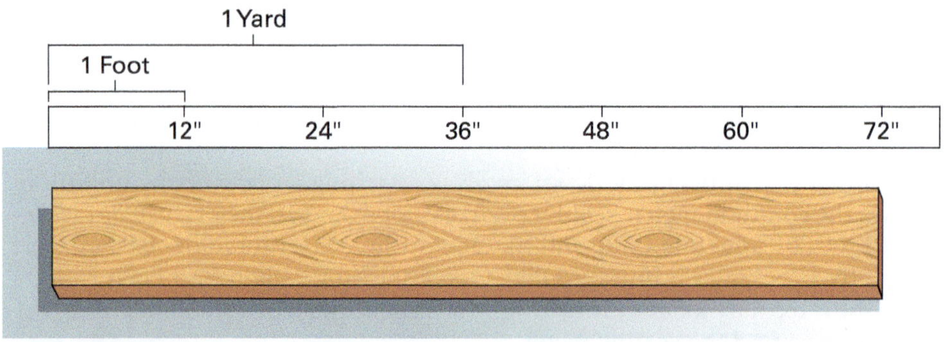

Figure 22 Converting inches to yards.

5.1.2 Metric Length Units

Table 3 shows the metric system's common length units, along with their relationships. Notice that all the units relate to each other by 10s.

TABLE 3 Common Metric Length Units

Original Weights	Equivalent Weights
1 millimeter	0.1 centimeters; 0.001 meters
1 centimeter	10 millimeters; 0.01 meters
1 meter	100 centimeters; 0.001 kilometers
1 kilometer	1,000 meters

Abbreviate the common metric length units as shown:

- Millimeter is abbreviated as *mm*.
- Centimeter is abbreviated as *cm*.
- Meter is abbreviated as *m*.
- Kilometer is abbreviated as *km*.

Converting between metric units requires multiplying or dividing by some value based on 10. The easiest way to do this operation is to move the decimal point to the left or right. Which way to move the decimal point depends on size. To convert a smaller unit to a larger one, move the decimal point to the left. To convert a larger unit to a smaller one, move the decimal point to the right. *Table 1* helps you determine how many places to move.

For example, to convert 72 cm to meters (*Figure 23*), remember that 1 m is bigger than 1 cm. The third row of *Table 1* indicates that the *centi-* prefix stands for 0.01. Therefore, move the decimal point two places to the left. As a result, 72 cm becomes 0.72 m.

Remember that moving decimal points is identical to multiplying or dividing. Dividing 72 cm by 100 also gives 0.72 m.

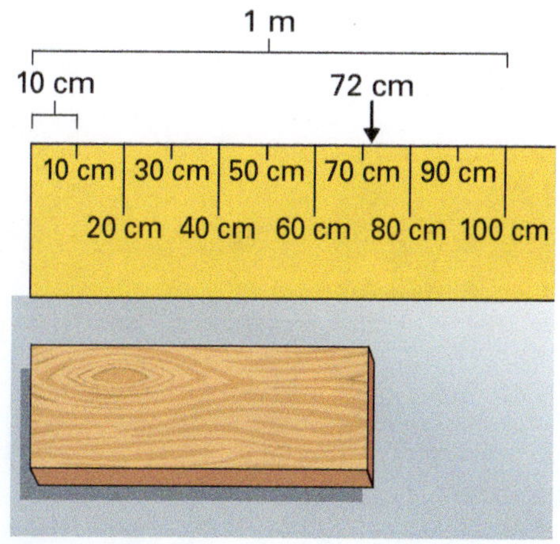

Figure 23 Converting centimeters to meters.

5.1.3 Converting Length Units between Systems

Converting measurements between the US standard and metric systems always involves multiplying or dividing by a conversion factor. Reference books, dictionaries, and textbooks all contain conversion factor charts. *Appendix 93103A* contains an example. Online tools and smartphone apps perform conversions automatically using the same conversion factors.

Converting between units requires accurate math and correct conversion factors. Choosing the wrong factor or making a calculation mistake makes the result worthless. *Table 4* provides common factors for converting length measurements from one system to another. After converting, the answer may require rounding.

According to *Table 4*, 1 cm is equal to 0.3937". To convert 13 cm to inches, multiply 13 by this factor. The result is 5.1181". For most applications, round to 5.12", or even to 5.1". Rounding decisions depend on the application. Locating the center of a steel plate to drill a hole requires high precision. In comparison, placing a mailbox post the correct distance from the end of a driveway is less demanding.

TABLE 4 US Standard and Metric Length Conversion Factors

Unit	To millimeter	To centimeter	To inch	To foot	To yard	To meter	To kilometer	To mile
From millimeter	1	0.1	0.03937	0.003281	0.001094	0.001	0.000001	0.0000006214
From centimeter	10	1	0.3937	0.03281	0.01094	0.01	0.00001	0.000006214
From inch	25.4	2.54	1	0.08333	0.02778	0.0254	0.0000254	0.00001578
From foot	304.8	30.48	12	1	0.3333	0.3048	0.0003048	0.0001894
From yard	914.4	91.44	36	3	1	0.9144	0.0009144	0.0005681
From meter	1,000	100	39.37	3.281	1.094	1	0.001	0.0006214
From kilometer	1,000,000	100,000	39,370	3,281	1,094	1,000	1	0.6214
From mile	1,609,340	160,934	63,360	5,280	1,760	1,609.3	1.6093	1

5.1.4 Practice Problems: Converting Measurements

Without using a calculator, find the answers to the following conversion problems. Round your answers to the nearest hundredth when necessary.

1. 0.45 meters = _____ centimeters
2. 3 yards = _____ inches
3. 36 feet = _____ yards
4. 90 inches = _____ yards
5. 1 centimeter = _____ meters
6. 66 inches = _____ centimeters
7. 47 feet = _____ meters
8. 54.5 centimeters = _____ feet
9. 19 yards = _____ meters
10. 4.7 meters = _____ inches

5.2.0 Units of Weight Measurement

This section focuses on US standard and metric units used to measure weight.

5.2.1 US Standard Weight Units

Force: A push or pull on a surface. In some contexts, a force is the weight of an object or a fluid.

Mass: The quantity of matter present.

Weight is the **force** that an object exerts on a surface due to its **mass** and the pull of the Earth's gravity. In the US standard system, common units for weight include ounce, pound, and ton. *Table 5* shows the relationship between these three units.

TABLE 5 Common US Standard Weight Units

Original Weights	Equivalent Weights
1 ounce	$\frac{1}{16}$ of a pound; 0.0625 pounds
1 pound	16 ounces; 0.0005 tons
1 ton	2,000 pounds

Abbreviate the common US standard weight units as shown:

- Ounce is abbreviated as *oz*.
- Pound is abbreviated as *lb*.
- Ton is abbreviated as *t*.

Converting from one US standard weight unit to another requires multiplying or dividing by an appropriate conversion factor. For example, to convert 134 oz to pounds, divide by 16 (the number of ounces in one pound). The result is 8.375 lb. Round the number to an appropriate level.

5.2.2 Metric Weight Units

The metric system uses milligram, gram, and kilogram as its weight units. As with length, these metric units relate to each other by 10s. *Table 6* shows their relationships.

TABLE 6 Common Metric Weight Units

Original Weights	Equivalent Weights
1 milligram	0.001 grams; 0.000001 kilograms
1 gram	1,000 milligrams; 0.001 kilograms
1 kilogram	1,000 grams; 0.001 metric tons
1 metric ton	1,000 kilograms

Abbreviate each metric weight unit as shown:

- Milligram is abbreviated as *mg*.
- Gram is abbreviated as *g*.
- Kilogram is abbreviated as *kg*.

Converting metric weight units requires multiplying or dividing by 10s. Alternatively, just shift the decimal point to the left or right. *Table 6* shows that 1 g is equal to 1,000 mg. To convert 6,439 mg to grams, divide 6,439 mg by 1,000—the number of milligrams in a single gram. A milligram is 0.001 g, so an easier method is to move the decimal point three places to the left. Regardless of the method chosen, the result is the same: 6.439 g.

Did You Know?

Common Metric Terms

The millimeter, centimeter, meter, and kilometer are the most common metric length units. Length units longer than a kilometer are uncommon.

Scientists and precision machinists may use units smaller than the millimeter (0.001 m). The micrometer (0.000001 m), nanometer (0.000000001 m), and the picometer (0.000000000001 m) are very specialized units.

Around the World

Metric System as a Modern Standard

As modern scientific thinking started to develop during the 1500s, scientists realized that they needed to agree on measurement standards. Every country had its own units, so scientists had difficulty communicating their measurements to each other.

In the 1790s, French scientists started developing what later became the metric system. They began with a standard length, called the *meter*. Over the next century, the global scientific community adopted the metric system. Eventually, most of the world successfully made the change. However, not until the 1970s did the United States and Canada truly acknowledge the metric system. Increasing global trade stimulated this shift, but many people resisted learning a new measurement system. Even today, many in the United States do not understand metric units.

5.2.3 Converting Weight Units between Systems

Table 7 shows the relationship between weight units in the US standard and metric systems. The conversion factors may be more precise than required. Always round answers to the values appropriate to the application.

An HVAC technician might have to convert air-conditioning refrigerant weight from one system to the other. Or a crane operator might need to verify that his US standard rigging could handle a particular metric-specified load. Imagine that the load weighs 1,200 kg. How many pounds is this?

Table 7 shows that 1 kg = 2.205 lb, so multiplying 1,200 kg by 2.205 yields 2,646 lb. The slings or lifting equipment must have a rated capacity greater than 2,646 lb.

When working with smaller weights, ounces or grams may be more appropriate. One ounce is equal to 28.35 g, so the gram is the smaller unit. Convert ounces to grams by multiplying the number of ounces by 28.35. Convert grams to ounces by dividing by 28.35.

Water and the Gram

The milliliter is a popular metric unit for measuring liquids. It equals $^1/_{1,000}$ (0.001) of a liter. One milliliter of pure water weighs exactly one gram. This is yet another example of the metric system's practicality. Remember, however, that other liquids of one milliliter have different weights.

TABLE 7 US Standard and Metric Weight Conversion Factors

Unit	To milligram	To gram	To ounce	To pound	To kilogram	To ton	To metric ton
From milligram	1	0.001	0.00003527	0.000002205	0.000001	0.000000001102	0.000000001
From gram	1,000	1	0.03527	0.002205	0.001	0.000001102	0.000001
From ounce	28,349.5	28.35	1	0.0625	0.02835	0.00003125	0.00002835
From pound	453,592	453.6	16	1	0.4536	0.0005	0.000454
From kilogram	1,000,000	1,000	35.27	2.205	1	0.0011	0.001
From ton	907,200,000	907,185	32,000	2,000	907.185	1	0.9072
From metric ton	1,000,000,000	1,000,000	35,274	2,204.62	1,000	1.102	1

5.2.4 Practice Problems: Converting Weight Units

Convert these weights from US standard to metric weight units, or vice versa. Round answers to the nearest hundredth.

1. 50 pounds = _____ kilograms
2. 50 kilograms = _____ pounds
3. 15.9 ounces = _____ grams
4. 94 grams = _____ ounces

5.3.0 Units of Volume Measurement

Volume: The amount of space contained within a three-dimensional object.

Cube: A three-dimensional object whose sides are all the same length.

This section examines units of **volume** measurement in the US standard and metric systems. Volume applies to three-dimensional objects such as a **cube** or a cylinder.

An object's volume refers to the amount of space it occupies. For example, a barrel might take up 7.35 cubic feet of space. An engine's cylinders might have a total volume of 250 cubic centimeters. Calculating volume depends on the object. The appropriate calculations appear later in this module.

A cube is the simplest object since its sides are all the same length. For example, if a cube measures 1" × 1" × 1", it has a volume of 1 cubic inch. Notice that volume units share part of the length units, with the word "cubic" added in front of the length units.

5.3.1 US Standard Volume Units

Table 8 shows the three most common volume units in the US standard system. Note that these units use the familiar length measurement units.

Abbreviate each of the US standard volume units as shown:

- Cubic inch is abbreviated as *cu in* or in^3. Sometimes *CI* is used, but this is not a standard abbreviation.
- Cubic foot is abbreviated as *cu ft* or ft^3.
- Cubic yard is abbreviated as *cu yd* or yd^3.

Convert between these units in the usual way. For example, to convert 864 in^3 to cubic feet, multiply 864 by the conversion factor 0.0005787. The result is 0.4999 ft^3, or 0.5 ft^3.

TABLE 8 Common US Standard Volume Units

Original Volumes	Equivalent Volumes
1 cubic inch	0.0005787 cubic feet
1 cubic foot	1,728 cubic inches; 0.037 cubic yards
1 cubic yard	27 cubic feet

5.3.2 Metric Volume Units

The cubic centimeter and the cubic meter are the most common metric volume units. *Table 9* shows their relationship to each other. As with all metric system units, volume units relate to each other by 10s. Convert between these in the usual way for metric units. The cubic meter represents a three-dimensional object equivalent to a cube 100 cm × 100 cm × 100 cm in size. Therefore, 1 cubic meter contains 1,000,000 cubic centimeters.

Abbreviate each of the metric volume units as shown:

- Cubic centimeter is abbreviated as *cu cm* or cm^3.
- Cubic meter is abbreviated as *cu m* or m^3.

TABLE 9 Common Metric Volume Units

Original Volumes	Equivalent Volumes
1 cubic centimeter	0.000001 cubic meters
1 cubic meter	1,000,000 cubic centimeters

5.3.3 Converting Volume Units between Systems

Table 10 shows the relationship between US standard and metric volume units. A worker excavating soil to prepare a foundation might have to convert between systems. In most cases, the units in each system should be similar in size (for example, cubic yards to cubic meters). Rarely would the worker have to convert a large unit to a small one (for example, cubic yards to cubic centimeters).

TABLE 10 US Standard and Metric Volume Conversion Factors

Unit	To cubic centimeter	To cubic inch	To cubic foot	To cubic yard	To cubic meter
From cubic centimeter	1	0.061	0.00003531	0.000001308	0.000001
From cubic inch	16.387	1	0.0005787	0.000021433	0.000016387
From cubic foot	28,316.85	1,728	1	0.037	0.0283
From cubic yard	764,554.86	46,656	27	1	0.765
From cubic meter	1,000,000	61,023.7	35.315	1.308	1

A welder fabricates a box to hold a battery. The battery's specification sheet says that the battery is cube shaped, with a volume of 1,600 cm^3. The space allowed for the battery box is cube shaped and contains 112 in^3. Will the battery fit?

To convert cubic centimeters to cubic inches, multiply the number of cubic centimeters by the conversion factor. Since the conversion factor is 0.061, multiply 1,600 by 0.061. The result is 97.6 in^3. The battery box will fit the available space with some room to spare.

A concrete craftworker calculates that a slab requires 59 yd^3 of cement. However, the supplier sells cement in cubic meters. How much should the worker order?

According to *Table 10*, the conversion factor is 0.765. Multiply 59 yd^3 by 0.765 to get 45.1 m^3.

5.3.4 Practice Problems: Converting Volume Units

Convert these volumes from the US standard system to the metric system, or vice versa. Round to the nearest tenth.

1. 11,600 cubic inches = _____ cubic feet
2. 1.9 cubic meters = _____ cubic centimeters
3. 512 cubic meters = ______ cubic yards
4. 7 cubic feet = ______ cubic centimeters
5. 0.2 cubic meters = ______ cubic feet

Volume of Liquids and Liquid Conversion

Sometimes you might want to calculate volume in terms of liquid, such as liquid level measurement or storage tank capacity, instead of the volume of an object or space. The units, calculation, and conversions are different for liquid volume.

US standard units of liquid volume include gallons, quarts, pints, and fluid ounces. Metric units of liquid volume are represented by the base unit *liter* (*L*), with any volume smaller than one liter measured in *milliliters (mL)*. Abbreviations, basic conversions, and equivalencies of a few of the most common units used in agriculture are provided here.

US Standard Unit	Calculation	Equivalency	Metric Conversion
Gallon (gal)	Volume (ft^3) × 7.48	4 qt = 1 gal	1 gal = 3.79 L
Quart (qt)	Volume (in^3) × 0.017	2 pt = 1 qt	1 qt = 0.946 L
Pint (pt)	Volume (in^3) × 0.035	16 oz = 1 pt	1 pt = 0.568 L
Ounces (oz)	Volume (in^3) × 0.554	1 gal = 128 oz	1 oz = 29.57 mL
Metric Unit	**Calculations**	**Equivalency**	**US Standard Conversion**
Liter (L)	Volume (m^3) × 1,000	1 L = 1,000 mL	1 L = 0.264 gal
Milliliter (mL)	Volume (mm^3) × 0.001	1 mL = 0.001 L	1 mL = 0.034 oz

5.4.0 Temperature Units

Temperature is a measurement of how fast the atoms or molecules in a solid, liquid, or gas are moving. If they move rapidly, the temperature is high. When they slow down, the temperature is low. People often use the terms *hot* and *cold* to describe the way they sense temperature. Many crafts regularly work with temperatures. A welder running an arc welding machine makes the metal atoms in two pieces of steel move rapidly. The metal gets hot, softens, and joins. A refrigerator chills the water in a water bottle by slowing down the water molecules.

There are many different units of temperature. Converting between units is frequently necessary in the crafts, as well as in science, engineering, and industry. This section examines common temperature units and the proper ways to convert them.

Temperature units come from a measuring scale based on one or more standard temperature points. These points usually come from water's behavior. For example, water freezes and boils at specific temperatures at sea level. Several temperature scales use these points as references.

Most temperature measurements are in degrees, represented by the degree symbol (°). A complete temperature unit consists of the degree symbol, followed by a single letter that identifies the scale. The four most common temperature scales are Fahrenheit (°F), Celsius (°C), Rankine (°R), and Kelvin (K). The Kelvin temperature scale does not use degrees.

Everyday temperature measurements are usually in Fahrenheit or Celsius. On the Fahrenheit scale, water freezes at 32°F and boils at 212°F. On the Celsius scale, water freezes at 0°C and boils at 100°C. *Figure 24* shows the freezing and boiling points for each unit.

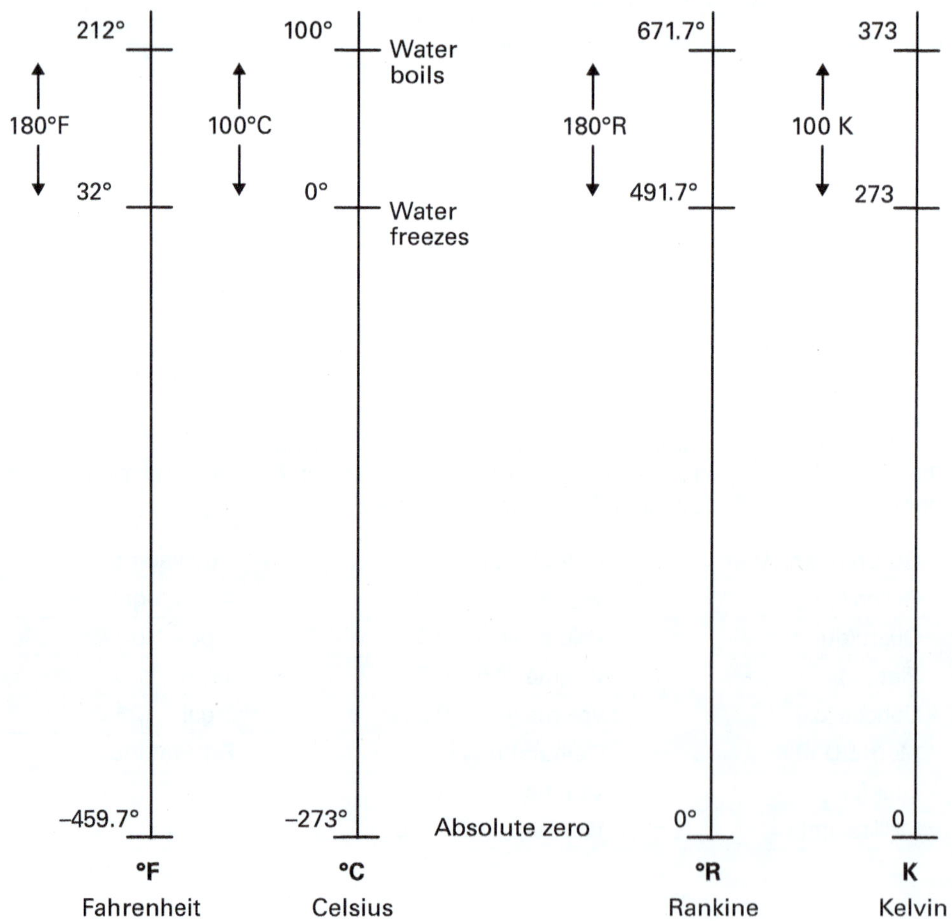

Figure 24 Comparing temperature scales.

In the 1700s, scientists speculated that there must exist a minimum temperature that nothing could be colder than. Today, we call this point *absolute zero*. At this temperature, atoms and molecules barely move. Two temperature scales, the Rankine scale and the Kelvin scale, use absolute zero as one endpoint.

On the Rankine scale, water freezes at 491.7°R and boils at 671.7°R—a 180° range. This range is identical in size to that of the Fahrenheit scale (212° – 32° = 180°). Degrees on the Rankine scale are therefore equivalent to those on the Fahrenheit scale. For this reason, some call the Rankine scale the *absolute Fahrenheit scale*.

On the Kelvin scale, water freezes at 273 K and boils at 373 K—a 100-unit range. This range is identical to that of the Celsius scale (100° – 0° = 100°). Values on the Kelvin scale are therefore equivalent to those on the Celsius scale. For this reason, some call the Kelvin scale the *absolute Celsius scale*. Note that the unit identifier for the Kelvin scale is the letter *K* without a degree symbol in front of it. In fact, the term *degree* is not used when referring to temperatures on the Kelvin scale. To express room temperature on the Celsius and Kelvin scales, say "Room temperature is around 22 degrees Celsius or 295 Kelvins." Both the Kelvin and Celsius scales are part of the metric system of measurement.

Most crafts use the Fahrenheit and Celsius scales. Scientific and some engineering applications require the Rankine or Kelvin scale.

Charts, tables, and smartphone apps make converting between temperature scales simple. However, craftworkers benefit from understanding the underlying math.

Around the World

Coldest Recorded Temperature

Scientists routinely produce extremely low temperatures using a variety of exotic laboratory techniques. To date, the coldest temperature ever achieved came to within less than a billionth of a degree of absolute zero (0 K).

In 1983, Russian scientists at the Vostok Station in Antarctica measured the lowest natural temperature on the Earth, a chilly –128.6°F (–89.2°C)!

Digital Thermometers

Thermometers with digital (numeric) readouts can provide temperature readings in both Celsius and Fahrenheit. Many analog thermometers have dual scales as well. The unit shown in the first image is a digital temperature probe. The one in the second image is a noncontact infrared thermometer. It determines temperature by measuring how much invisible infrared radiation an object emits. Both provide their readings directly on a numeric display.

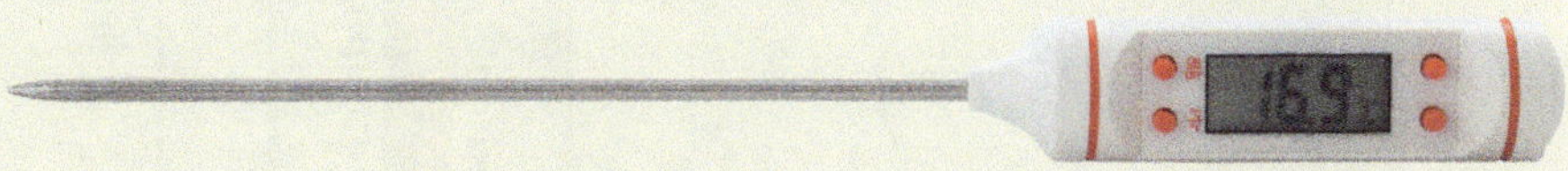

Source: iStock

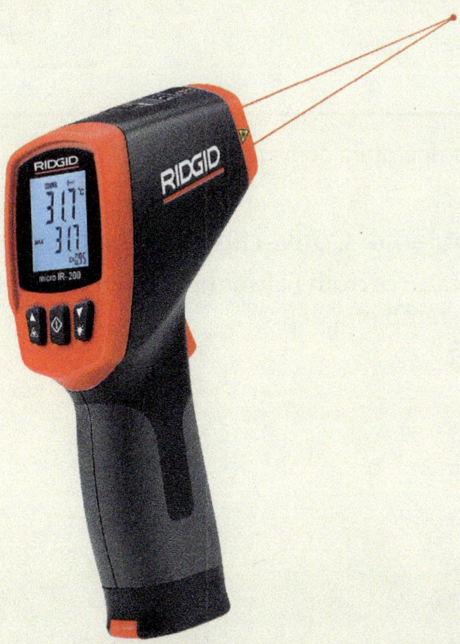

Source: Ridge Tool Company

The Fahrenheit scale has 180° between water's boiling and freezing points (212° – 32° = 180°). The Celsius scale has 100° between these two points (100° – 0° = 100°). The following fraction shows the relationship between the scales:

$$\frac{\text{Fahrenheit range}}{\text{Celsius range}} = \frac{180°\text{F}}{100°\text{C}} = \frac{9}{5}$$

One degree Fahrenheit is $\frac{5}{9}$ of one degree Celsius, and one degree Celsius is $\frac{9}{5}$ of one degree Fahrenheit. 32°F corresponds to 0°C. To convert Fahrenheit to Celsius, subtract 32° and then multiply by $\frac{5}{9}$. To convert Celsius to Fahrenheit, multiply by $\frac{9}{5}$ and then add 32°. The following equations show this process:

$$°\text{C} = \frac{5}{9} \times (°\text{F} - 32°)$$

$$°\text{F} = \left(\frac{9}{5} \times °\text{C}\right) + 32°$$

Practice these calculations to become more comfortable with them. *Figure 25* shows two examples.

Example: 77°F to °C

$$°\text{C} = \frac{5}{9}(77° - 32°)$$

$$°\text{C} = \frac{5}{9}(45°)$$

$$°\text{C} = 25°\text{C}$$

Example: 90°C to °F

$$°\text{F} = \left(\frac{9}{5} \times 90°\right) + 32°$$

$$°\text{F} = 162° + 32°$$

$$°\text{F} = 194°\text{F}$$

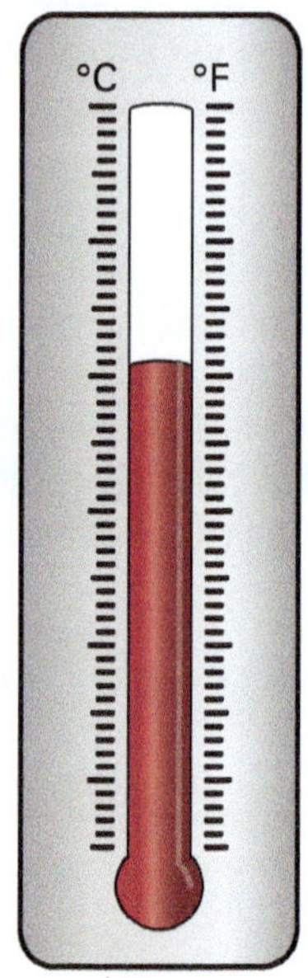

Figure 25 Sample temperature conversions.

5.4.1 Practice Problems: Converting Temperatures

Convert these temperatures from Fahrenheit to Celsius, or Celsius to Fahrenheit. Round to the nearest tenth of a degree.

1. 180°F = _____°C
2. 66°F = _____°C
3. –26°C = _____°F
4. 71°C = _____°F

5.0.0 Section Review

1. When converting from a larger metric unit to a smaller one, the decimal point must move _____.
   a. to the left
   b. to the right
   c. all the way to the left end of the number
   d. all the way to the right end of the number

2. Weight is the force that an object exerts on a surface due to its mass and _____.
   a. density
   b. volume
   c. the current temperature
   d. the pull of the Earth's gravity

3. The cubic meter represents a three-dimensional object equivalent to a cube that measures _____.
   a. 10 cm × 10 cm × 10 cm
   b. 100 mm × 100 mm × 100 mm
   c. 100 cm × 100 cm × 100 cm
   d. 10 m × 10 m × 10 m

4. On the Celsius temperature scale, the difference between the freezing and boiling points of water is _____.
   a. 32°C
   b. 100°C
   c. 180°C
   d. 212°C

6.0.0 Introduction to Geometry

Objective

Classify angles and geometric shapes, as well as calculating their areas or volumes.

a. List each angle type.
b. Name common geometric shapes and summarize their qualities.
c. Calculate the area of two-dimensional shapes.
d. Calculate the volume of three-dimensional shapes.

Performance Tasks

There are no Performance Tasks in this section.

Geometry might sound complicated, but it is a very practical kind of math. It deals with many familiar shapes including the **circle**, **triangle**, **square**, and **rectangle**. The construction industry uses geometry constantly as it works with measurements and shapes. Understanding practical geometry helps every craftworker to carry out daily tasks.

The first part of this section focuses on **plane geometry**. In plane geometry, shapes are two-dimensional (2D). They have length and width only. In contrast, **solid geometry**, covered in the second part, works with three-dimensional (3D) shapes that have three dimensions—length, width, and height.

Circle: A closed curved line drawn at a constant distance around a central point. A circle measures 360°.

Triangle: A closed shape that has three sides and three angles.

Square: A special type of rectangle with four equal sides and four 90° angles.

Rectangle: A four-sided shape with four 90° angles. Opposite sides of a rectangle are always parallel and the same length. Adjacent sides are perpendicular and may or may not be equal in length.

Plane geometry: The mathematical study of two-dimensional (flat) shapes.

Solid geometry: The mathematical study of three-dimensional shapes.

6.1.0 Angles

Angle: The shape made by two straight lines coming together at a point. The space between those two lines is measured in degrees.

Vertex: The point at which two or more lines or curves come together.

Degree: A unit of measurement for angles. For example, a right angle measures 90°.

Acute angle: Any angle greater than 0° and less than 90°.

Right angle: An angle that measures 90°. The two lines that form a right angle are perpendicular to each other. This is the angle most used in the crafts.

Obtuse angle: Any angle greater than 90° and less than 180°.

Straight angle: A 180° angle or flat line.

Adjacent angles: Angles that have the same vertex and one side in common.

Opposite angles: Two angles that are formed by two straight lines crossing. They are always equal.

Construction crafts use the term **angle** frequently. It describes the shape made by two straight lines that meet at a point, called the **vertex**. Angle measurements use **degree** as the unit, often represented by the ° symbol. Degrees describe the relationship between the two lines. A protractor measures angles. The following are typical angles used in construction (*Figure 26*):

- **Acute angle** — An angle that measures more than 0° and less than 90°. The most common acute angles are 30°, 45°, and 60°.
- **Right angle** — An angle that measures exactly 90°. The two lines that form the right angle are perpendicular to each other, like the shape of a letter "L." The construction industry uses this angle frequently. Plans and drawings indicate right angles with a square symbol at the vertex, as *Figure 26* (C) shows.
- **Obtuse angle** — An angle that measures more than 90° and less than 180°.
- **Straight angle** — An angle measuring exactly 180° (a flat line).
- **Adjacent angles** — These angles have the same vertex and one side in common. Adjacent refers to objects that are next to each other.
- **Opposite angles** — Angles formed by two straight lines that cross. Opposite angles are always equal.

6.2.0 Shapes

Figure 27 shows examples of rectangles, squares, triangles, and circles—shapes common to many crafts.

6.2.1 Rectangle

Diagonal: Line drawn from one corner of a rectangle or square to the farthest opposite corner.

Right triangles: A type of triangle that includes one 90° angle.

A rectangle is a four-sided shape with four 90° angles. The four angles always add up to 360°. A rectangle has two pairs of equal sides that are parallel to each other. The **diagonal** lines of a rectangle are always equal too. Diagonals are lines connecting opposite corners. Cutting a rectangle on the diagonal gives two identical **right triangles**, as *Figure 28* shows. All right triangles have one 90° angle.

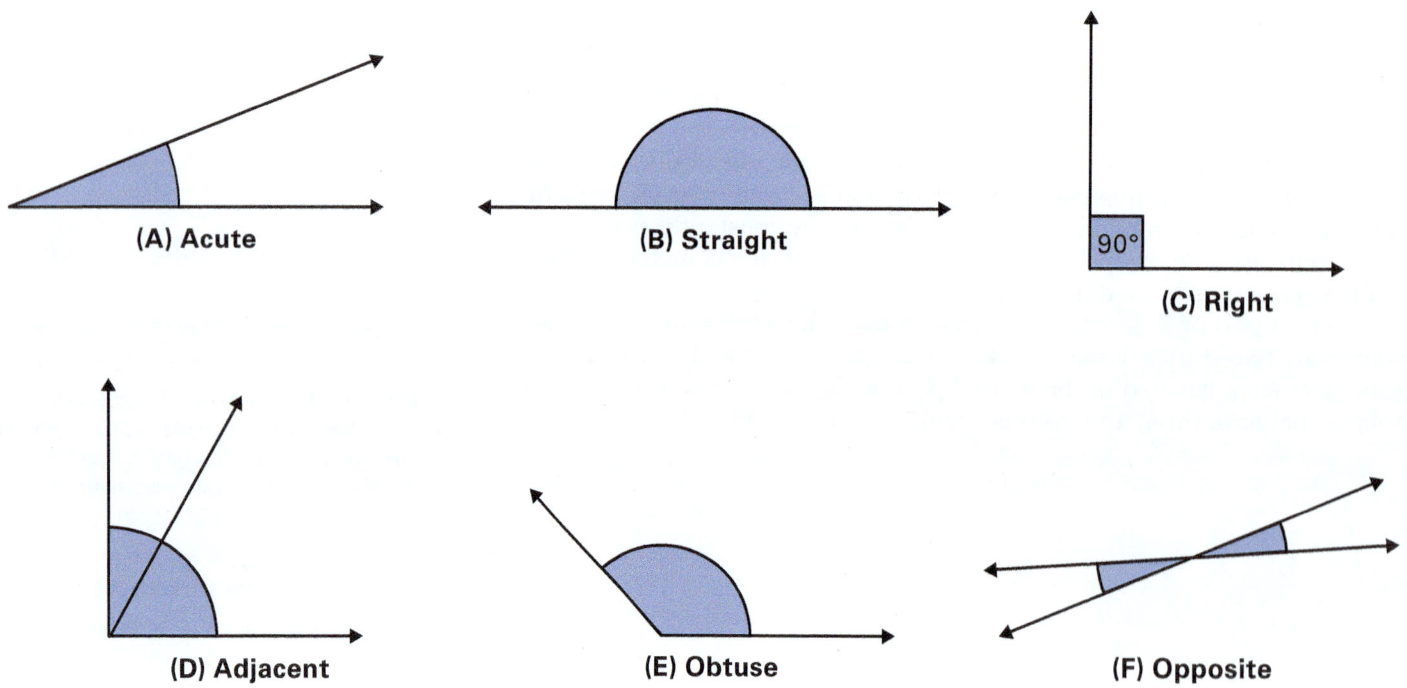

Figure 26 Angle types.

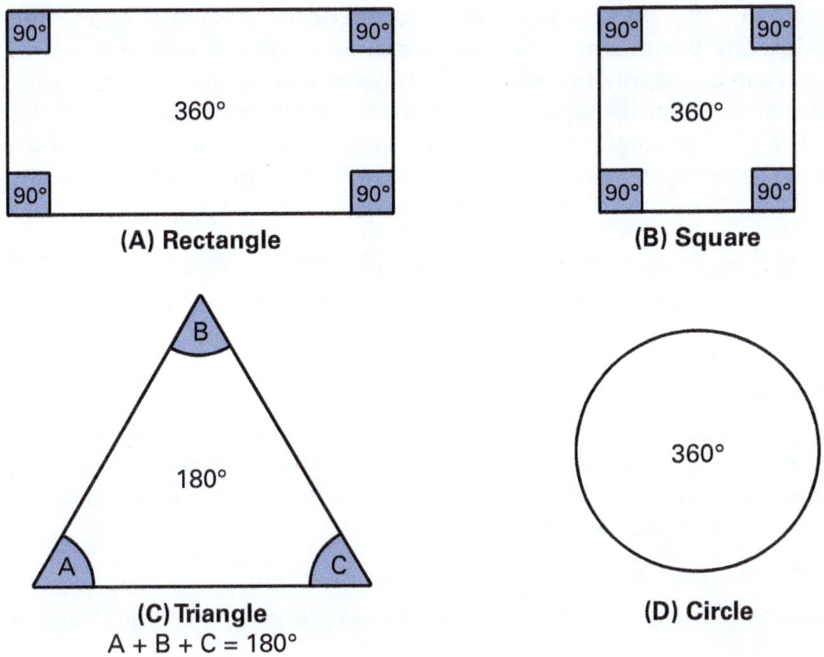

Figure 27 Common plane geometry shapes.

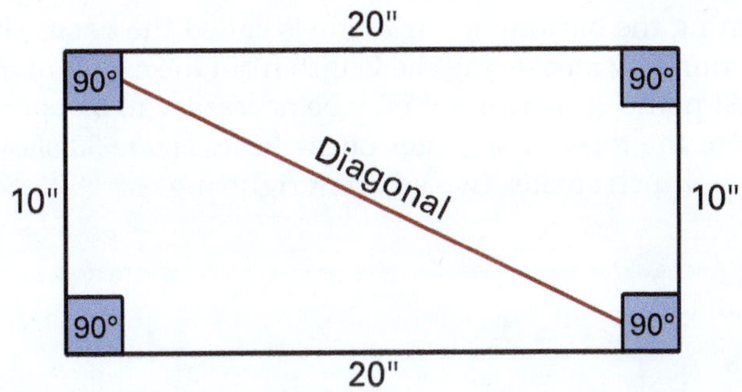

Figure 28 Cutting a rectangle on the diagonal produces two right triangles.

6.2.2 Square

A square is a special rectangle with four equal sides. It has four 90° angles that add up to 360°. Cutting a square on the diagonal gives two right triangles that have two 45° angles and one 90° angle, as *Figure 29* shows. Outside of geometry, the word **square** can also refer to the product of a number multiplied by itself. For example, 25 is the square of 5, and 16 is the square of 4.

Square: A special type of rectangle with four equal sides and four 90° angles.

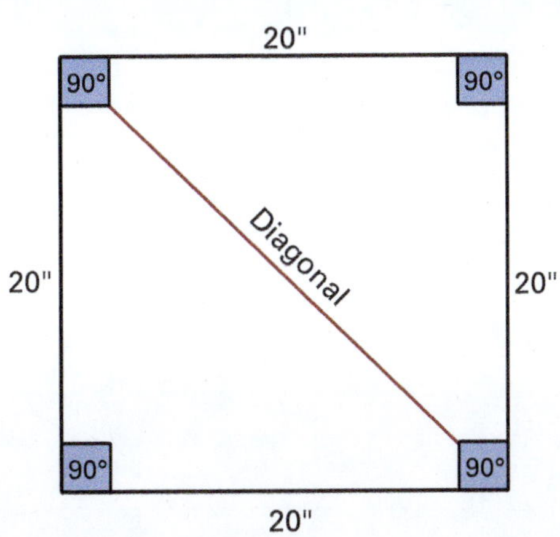

Figure 29 Cutting a square on the diagonal produces two right triangles.

Perimeter: The distance around the outside of a closed shape, such as a rectangle, circle, square, or any irregular shape.

Measuring the outside lines of a rectangle or a square determines the **perimeter**. The perimeter is the distance around any two-dimensional figure. The term can also apply to some three-dimensional figures. Craftworkers often measure or calculate a shape's perimeter when estimating material or planning a cut. For example, before installing molding along all four walls of a room, measure the perimeter first. If the room is 14' by 12', purchase 52' of molding (14' + 12' + 14' + 12' = 52'). Another way to calculate it is:

$$(2 \times 14') + (2 \times 12') = 52'$$

Since a square has four equal sides, calculate its perimeter by multiplying one side's length by four.

6.2.3 Triangle

Equilateral triangle: A triangle that has three equal sides and three equal angles.

Isosceles triangle: A triangle that has two equal sides and two equal angles.

Scalene triangle: A triangle with sides of unequal lengths.

A triangle is a closed shape that has three sides and three angles. Although the angles in a triangle can vary, they always add up to 180° (*Figure 30*). The types of triangles used in construction include the right triangle, **equilateral triangle**, **isosceles triangle**, and **scalene triangle**:

- *Right triangle* — A right triangle has one 90° angle.
- *Equilateral triangle* — An equilateral triangle has three equal angles and three equal sides.
- *Isosceles triangle* — An isosceles triangle has two equal angles and two equal sides.
- *Scalene triangle* — A scalene triangle has three sides of unequal lengths.

Base: As it relates to triangles, the base is the line forming the bottom of the triangle.

Bisect: To divide into two equal parts. For example, when an angle is bisected, the two resulting angles are equal.

The side forming the bottom of a triangle is called the **base**. The height of a triangle can be found by measuring the length from the center of the base to the triangle's highest point. Sometimes it may be necessary to **bisect** a triangle—to cut it in half from the top to the center of the base. *Figure 30* shows a bisected isosceles triangle, which creates two adjacent right angles.

Did You Know?

Rope Stretchers

The word *geometry* comes from two Greek words meaning "earth measuring." One of geometry's earliest applications was surveying (measuring land).

Ancient Egyptians strategically located their farms on the banks of the Nile River. Every year, the Nile flooded, depositing rich silt onto the surrounding farmland. However, the annual flooding destroyed property markers too. So professional surveyors, called *rope stretchers*, had to measure the land and replace the markers in the correct locations.

The rope stretchers calculated distances using ropes with equally spaced knots tied along their length. Measuring flat land was easy, but dealing with hills and other obstacles was more difficult. Solving these challenges laid the foundations of geometry. Today, geometry remains essential to many crafts and professions.

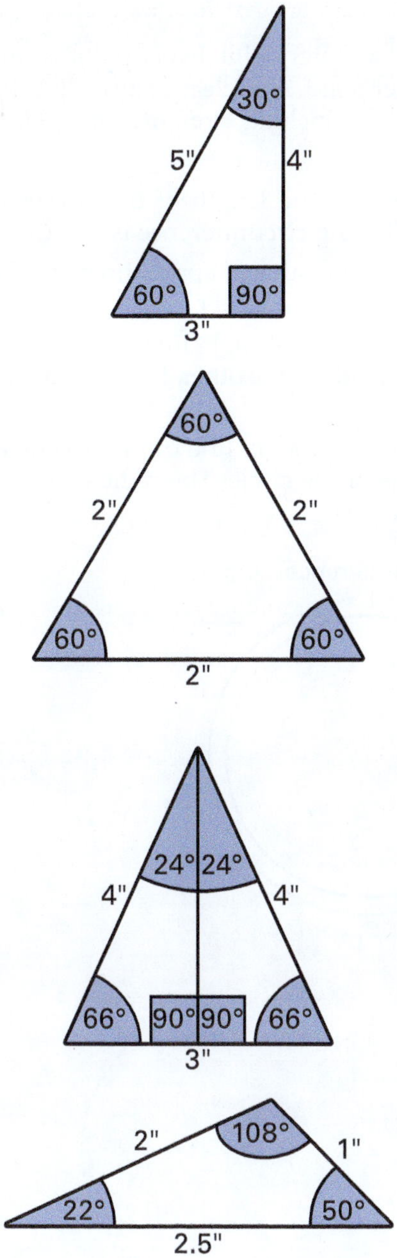

Right Triangle
30° + 60° + 90° = 180°

Equilateral Triangle
60° + 60° + 60° = 180°

Isosceles Triangle
48° + 66° + 66° = 180°

Scalene Triangle
108° + 22° + 50° = 180°

Figure 30 The sum of a triangle's three angles always equals 180°.

6.2.4 Circle

Circumference: The distance around the curved line that forms the circle.

Pi: A mathematical constant value approximately equal to 3.14 (or $^{22}/_{7}$) used to determine the area and circumference of circles. It is sometimes symbolized by the Greek letter π.

Diameter: The length of a straight line that crosses from one side of a circle, through the center point, to a point on the opposite side. The diameter is the longest straight line that can be drawn inside a circle.

Radius: The distance from a circle's center point to any point on the curved line, or half the width (diameter) of the circle.

Formula: A mathematical process used to solve a problem. For example, the formula for finding the area of a rectangle is Side A × Side B = Area, or A × B = Area.

A circle is a closed curved line around a center point. Every point on the curved line is the same distance from the center point. A circle measures 360°. Important measurement terms that apply to circles include **circumference**, **pi**, **diameter**, and **radius** (*Figure 31*):

- *Circumference* — The circumference is the length of the curved line that forms the circle. The **formula** for finding circumference is pi × diameter.
- *Pi or π* — Pi is a mathematical constant value of approximately 3.14 (or $^{22}/_{7}$) used to determine the area and circumference of circles.
- *Diameter* — The diameter is the length of a straight line that crosses through the center point of a circle, from one side to the other. It is the longest straight line possible inside a circle.
- *Radius* — The radius is the length of a straight line from the center point of the circle to any point on the edge of the circle. The radius equals half the diameter.

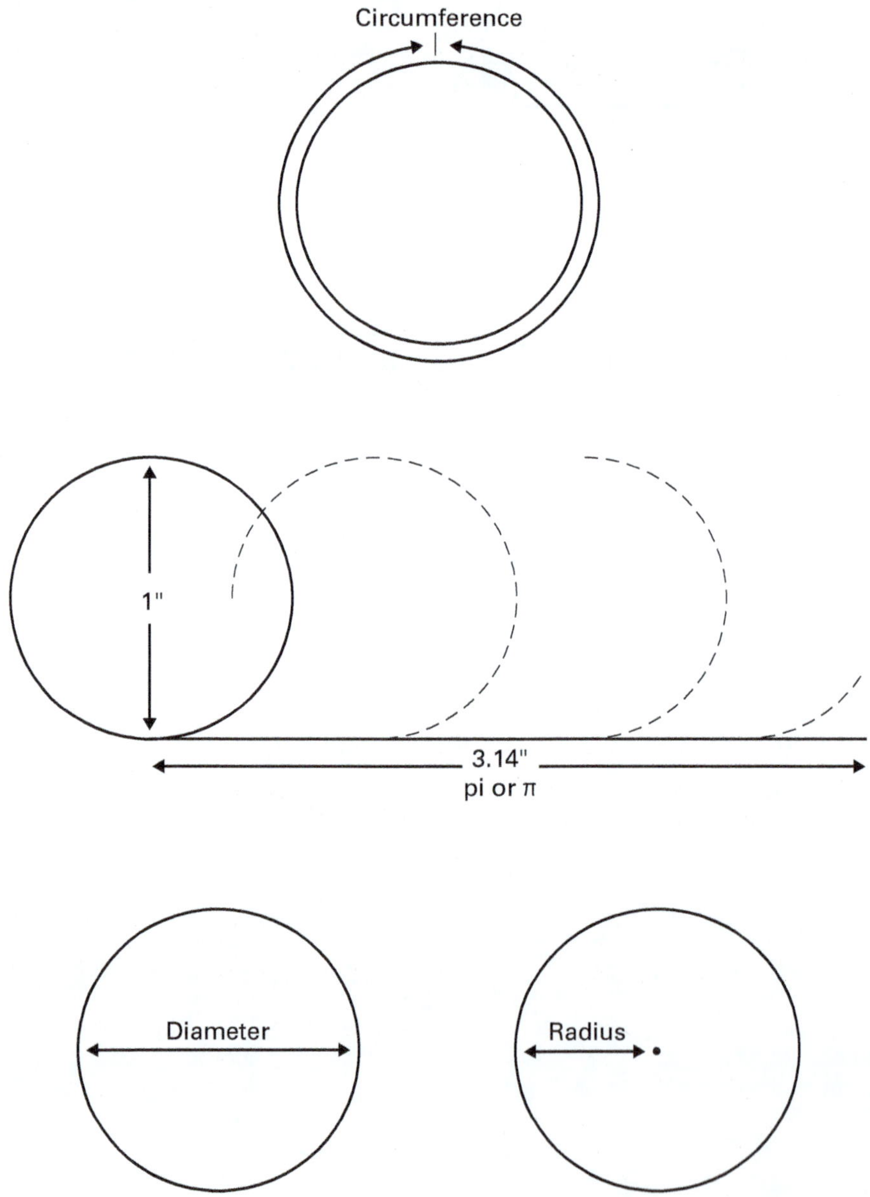

Figure 31 Measurements that apply to circles.

6.3.0 Calculating the Area of Shapes

Area refers to the space contained within a two-dimensional object's perimeter. To carpet an office floor or paint its walls, carpet layers and painters must calculate areas before ordering material. Squared measurement units specify the surface area. US standard area measurement units include square inches (*sq in* or in^2), square feet (*sq ft* or ft^2), and square yards (*sq yd* or yd^2). Metric area measurement units include square centimeters (*sq cm* or cm^2) and square meters (*sq m* or m^2). For larger areas, such as land measurement, square miles or square kilometers are common. The following equations explain the meaning of each area measurement unit:

Area: The amount of space contained in a two-dimensional object, such as a rectangle, circle, or square.

- 1 square inch = 1" × 1" = 1 in^2
- 1 square foot = 1' × 1' = 1 ft^2
- 1 square yard = 1 yd × 1 yd = 1 yd^2
- 1 square centimeter = 1 cm × 1 cm = 1 cm^2
- 1 square meter = 1 m × 1 m = 1 m^2

Craftworkers must be able to calculate the areas of basic shapes. Mathematical formulas make this task simple. *Appendix 93103A* lists the important ones. Become familiar with these now since they will come into use frequently in future modules. The following are a few significant examples:

Example 1:

Area of a rectangle = length × width

A painter measures a wall 20' long and 8' high. To calculate its area, multiply 20' × 8' = 160 ft^2.

Example 2:

Area of a square = length × width

A worker tiles a 12 m square room. Since all sides of a square are equal, the area is 12 m × 12 m = 144 m^2.

Example 3:

Area of a circle = pi × $radius^2$

A worker must seal a circular driveway with a radius of 20'. Since the value of pi is always approximately 3.14, the area is 3.14 × (20' × 20') = 1,256 ft^2.

Example 4:

Area of a triangle = $\frac{1}{2}$ × base × height

A carpenter is installing pieces of siding on a wall section. The triangle's base is 2', and its height is 4'. The area is $\frac{1}{2}$ × 2' × 4' = 4 ft^2.

Diagonals

Measuring diagonals can confirm that a surface is a true rectangle with 90° corners. For example, before applying a piece of sheathing, check if it is a true rectangle. Using a tape measure, measure the sheathing from one corner to the opposite corner. Then measure the sheathing from the other corner to its opposite corner. If the two numbers are the same, the sheathing is a true rectangle. If not, the sheathing may cause problems during installation.

6.3.1 Practice Problems: Calculating Area

1. The area of a rectangle that is 8' long and 4' wide is _____.
   a. 12 ft^2
   b. 22 ft^2
   c. 32 ft^2
   d. 36 ft^2

2. The area of a 16 cm square is _____.
   a. 256 cm^2
   b. 265 cm^2
   c. 276 cm^2
   d. 278 cm^2

3. The area of a circle with a 14' diameter is _____.
   a. 15.44 ft^2
   b. 43.96 ft^2
   c. 153.86 ft^2
   d. 196 ft^2

4. The area of a triangle with a base of 4 cm and a height of 6 cm is _____.
   a. 12 cm^2
   b. 24 cm^2
   c. 32 cm^2
   d. 36 cm^2

5. The area of a rectangle that is 14 m long and 5 m wide is _____.
   a. 60 m^2
   b. 65 m^2
   c. 70 m^2
   d. 75 m^2

6.4.0 Volume of Three-Dimensional Shapes

Volume is the amount of space contained within three-dimensional objects. Calculating volume may require measuring as many as three dimensions. For example, calculating a cube's volume requires its length, width, and height. Some people refer to height as *depth* or *thickness*. US standard volume units include cubic inch (*cu in* or in^3), cubic foot (*cu ft* or ft^3), and cubic yard (*cu yd* or yd^3). Metric volume units include cubic centimeter (*cu cm* or cm^3) and cubic meter (*cu m* or m^3). The following equations explain the meaning of each volume measurement unit:

- 1 cubic inch = 1" × 1" × 1" = 1 in^3
- 1 cubic foot = 1' × 1' × 1' = 1 ft^3
- 1 cubic yard = 1 yd × 1 yd × 1 yd = 1 yd^3
- 1 cubic centimeter = 1 cm × 1 cm × 1 cm = 1 cm^3
- 1 cubic meter = 1 m × 1 m × 1 m = 1 m^3

When calculating volumes, be sure that all measurements are in the same units. For example, a concrete slab has dimensions of 4' × 6' × 6". But multiplying these values to calculate volume will give an incorrect result. Convert feet to inches or inches to feet first. For example, 6" equals 0.5'. Therefore, 4' × 6' × 0.5' = 12 ft^3.

Geometry Practical Application

Before erecting a metal storage shed, concrete workers must pour a slab and install anchor bolts. The structure attaches to the slab by the anchor bolts. Since cement comes in cubic yards, a volume measure, the worker must calculate the slab's volume to know how much cement to use. To determine the volume of the slab shown in the image, multiply its length by its width and its depth (thickness). Alternatively, multiply the slab's area by its depth:

$$\text{Volume} = \text{area} \times \text{depth}$$
$$\text{Area of slab} = 50' \times 40' = 2{,}000 \text{ ft}^2$$
$$\text{Slab depth} = 6''$$

Convert inches to feet:

$$6'' = \frac{6}{12} = \frac{1}{2}'$$
$$\text{Volume} = 2{,}000 \text{ ft}^2 \times \frac{1}{2}' = 1{,}000 \text{ ft}^3$$
$$\text{Volume of slab} = 1{,}000 \text{ ft}^3$$

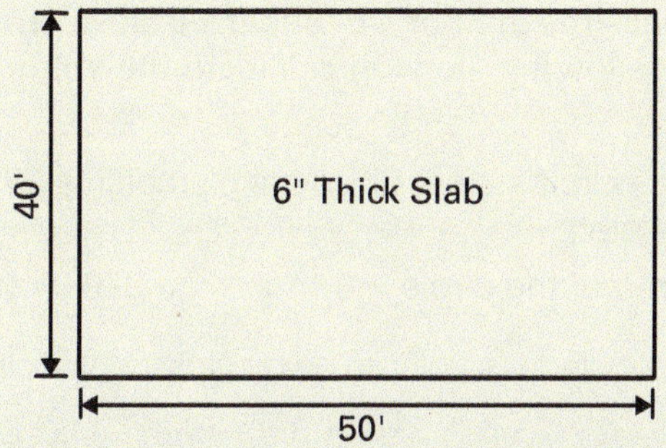

People have many names for geometric objects like the concrete slab. To make things simple, this module refers to all such objects as *three-dimensional rectangles.*

Craft professionals often calculate the volumes of three-dimensional spaces. *Appendix 93103A* contains a list of formulas for many objects. The following sections illustrate some common formulas. Do not forget that all units must be the same!

6.4.1 Three-Dimensional Rectangles

Calculate a three-dimensional rectangle's volume by multiplying length with width and depth. Perform the following steps to calculate the volume of the rectangle in *Figure 32*:

Step 1 Convert all units to a single type.

$$4'' = \frac{4}{12} = 0.33333'$$

Step 2 Determine the volume.

$$20' \times 8' \times 0.33333' = 53.3328 \text{ ft}^3$$

Step 3 If necessary, convert cubic feet to cubic yards.

$$53.3328 \text{ ft}^3 \div 27 \text{ (ft}^3 \text{ per yd}^3) = 1.98 \text{ yd}^3$$

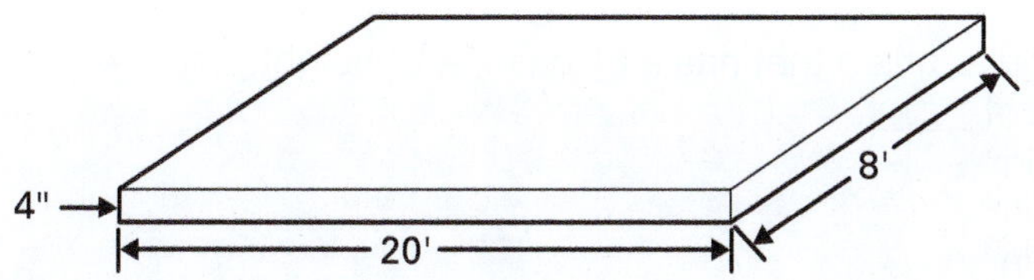

Figure 32 A three-dimensional rectangle.

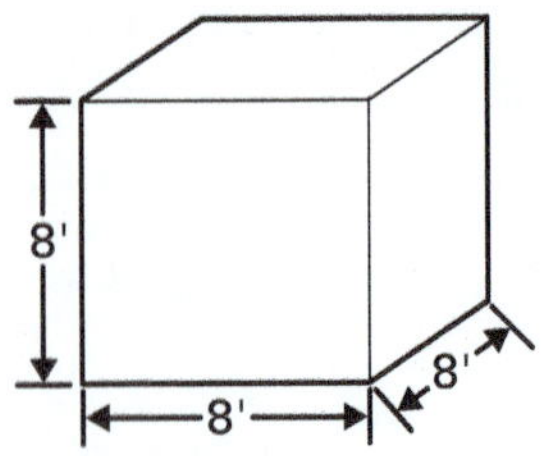

Figure 33 A cube.

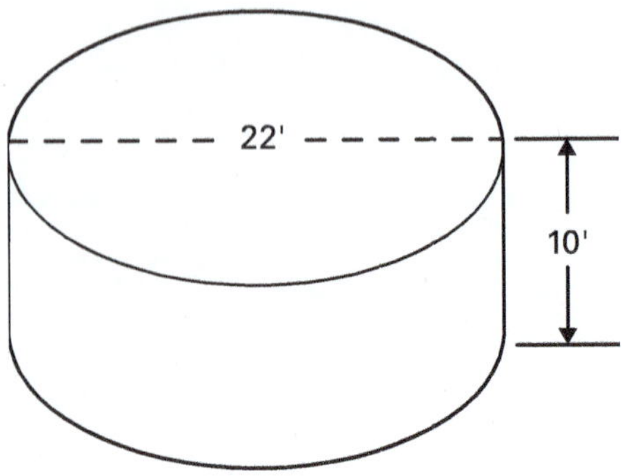

Figure 34 A cylinder.

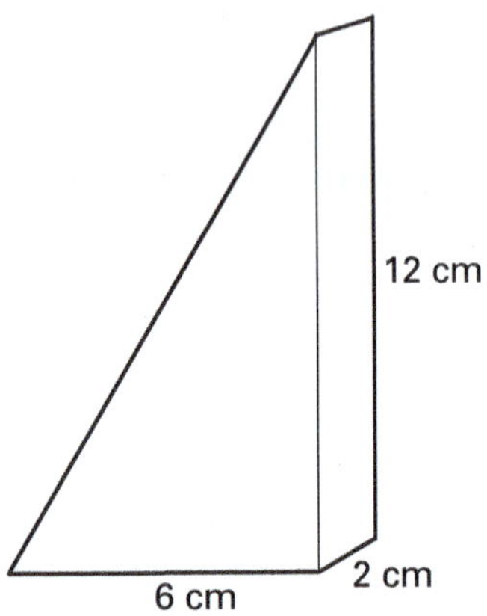

Figure 35 A triangular prism.

> **Did You Know?**
>
> **Pi**
>
> Pi is a mathematical constant that describes the relationship between a circle's circumference and its diameter. The value of pi goes on forever without any repeating patterns. In most cases, a few digits suffice for accurate calculations. Craftworkers often round pi to 3.14 or 3.14159. Mathematicians and computer scientists have developed software that calculates pi to trillions of digits. However, these calculations have little practical value beyond breaking records or testing new software techniques.

6.4.2 Cubes

A cube is special three-dimensional rectangle. Calculate its volume in the same way: length × width × depth. Since all sides are equal in length, multiply the side length by itself three times. Perform the following steps to calculate the volume of the cube in *Figure 33*:

Step 1 Determine the volume.

$$8' \times 8' \times 8' = 512 \text{ ft}^3$$

Step 2 If necessary, convert cubic feet to cubic yards.

$$512 \text{ ft}^3 \div 27 (\text{ft}^3 \text{ per yd}^3) = 18.96 \text{ yd}^3$$

6.4.3 Cylinders

Calculate a cylinder's volume using the following formula: pi × radius^2 × height. This formula multiplies the area of a cylinder's base (a circle) by the cylinder's height. Perform the following steps to calculate the volume of the cylinder in *Figure 34*:

Step 1 Determine the circle's area. Since the diameter is 22', the radius is 11' (half the diameter).

$$\text{Area of the circle} = 3.14 \times 11^2 = 379.94 \text{ ft}^2$$

Step 2 Calculate the cylinder's volume by multiplying the circle area by the height.

$$379.94 \text{ ft}^2 \times 10' = 3{,}799.4 \text{ ft}^3$$

Step 3 If necessary, convert cubic feet to cubic yards.

$$3{,}799.4 \text{ ft}^3 \div 27 (\text{ft}^3 \text{ per yd}^3) = 140.72 \text{ yd}^3$$

6.4.4 Triangular Prisms

A triangular prism is a three-dimensional triangle. Calculate its volume with the following formula: ½ × base area × height × depth (thickness). Perform the following steps to calculate the volume of the triangular prism in *Figure 35*:

Step 1 Calculate the triangle's area.

$$\tfrac{1}{2} \times 6 \text{ cm} \times 12 \text{ cm} = 36 \text{ cm}^2$$

Step 2 Calculate the prism's volume.

$$36 \text{ cm}^2 \times 2 \text{ cm} = 72 \text{ cm}^3$$

6.4.5 Practice Problems: Calculating Volume

1. A rectangular shape 5' high, 6' thick, and 13' long has a volume of _____.
   a. 24 ft^3
   b. 43 ft^3
   c. 95 ft^3
   d. 390 ft^3

2. A 3 cm cube has a volume of _____.
   a. 6 cm^3
   b. 9 cm^3
   c. 12 cm^3
   d. 27 cm^3

3. A triangular prism that has a 6" base, a 2" height, and a 4" depth has a volume of _____.
   a. 12 in^3
   b. 24 in^3
   c. 36 in^3
   d. 48 in^3

4. A cylinder that is 6 m in diameter and 60 cm high has a volume of _____.
   a. 16.96 m^3
   b. 18.23 m^3
   c. 1130.4 m^3
   d. 6782.4 m^3

5. If a work crew must spread gravel across an area that measures 17' square and the gravel layer is to be 6" thick, the volume of gravel required is _____.
   a. 3.77 yd^3
   b. 5.35 yd^3
   c. 102 yd^3
   d. 144.5 yd^3

6.4.6 Practical Applications Using Volume

The following examples illustrate practical craft applications that require geometry. Use provided information to solve each problem. Be sure to show all your work.

1. Calculate the volume of the sheet metal duct pipe shown in *Figure 36*. Round to the nearest cubic inch.

2. How many cubic feet of topsoil must a worker remove before pouring the concrete sidewalk shown in *Figure 37*? The sidewalk must be 4" thick and level with the surrounding soil. Round to the nearest cubic foot.

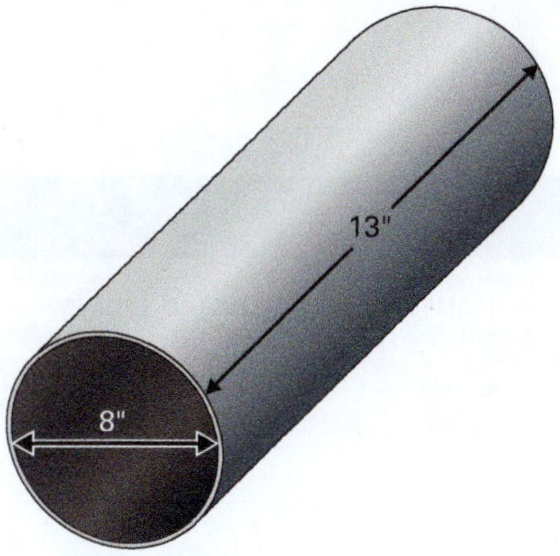

Figure 36 Finding a pipe section's volume.

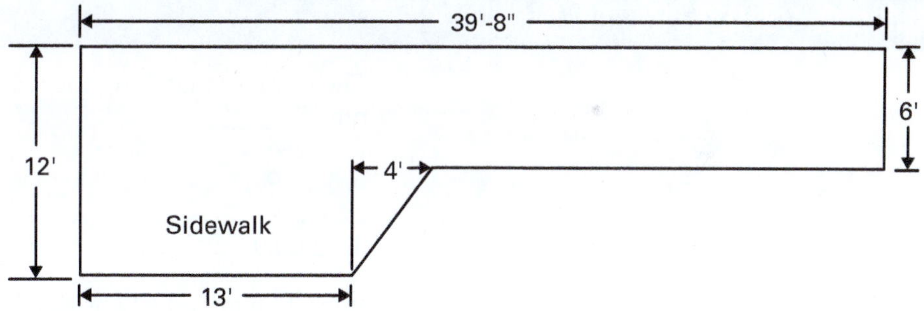

Figure 37 Removing topsoil for a sidewalk.

Did You Know?

3-4-5 Rule

The building crafts have used the 3-4-5 rule for centuries. Requiring nothing more than a tape measure, this simple layout method confirms that a rectangular space has true right-angle corners. The numbers 3, 4, and 5 represent the triangle's dimensions in any single unit. Any right triangle whose sides are multiples of 3, 4, and 5 will work. Examples include 9-12-15, 12-16-20, and 15-20-25.

To use this method, pick a 3-4-5 triangle that fits within the space being checked. For example, to check the area shown in the image, a 15-20-25 triangle works fine. Measure and mark 15'-0" down the side in one direction. Next, measure and mark 20'-0" down the adjacent side. Finally, measure the diagonal between these points. If the diagonal is not exactly 25'-0", the corners are not 90°.

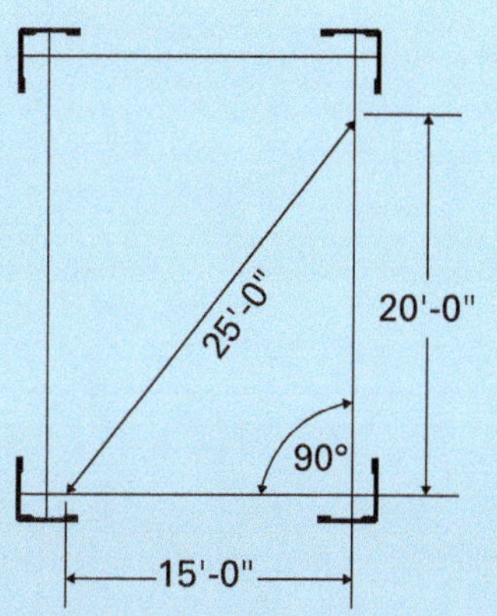

6.0.0 Section Review

1. An angle of 66° is a(n) _____.
   a. right angle
   b. straight angle
   c. obtuse angle
   d. acute angle

2. What is the difference between circumference and perimeter?
   a. Circumference applies only to shapes with angles, while perimeter applies only to circles.
   b. Perimeter applies to all shapes, while circumference applies only to circles.
   c. Perimeter equals one-half of the circumference.
   d. The circumference is never larger than the perimeter.

3. Calculate the area of a rectangle that is 27.3 m × 9.3 m.
   a. 245.7 m^3
   b. 251.1 m^2
   c. 253.89 m^2
   d. 253.89 m^3

4. Calculate the volume of a shipping container that is 26' long × 13'-8" wide × 3' deep.
   a. 355.32 ft^3
   b. 355.32 ft^2
   c. 1,065.48 ft^3
   d. 1,065.48 ft^2

Module 93103 Review Questions

1. The number matching the words "two thousand, six hundred eighty-nine" is _____.
   a. 2,286
   b. 2,689
   c. 6,289
   d. 20,689

2. A bricklayer laid 649 bricks the first day, 632 the second day, and 478 the third day. Without using a calculator, determine how many bricks the worker laid during three days.
   a. 1,759
   b. 1,760
   c. 1,769
   d. 1,770

3. A supplier delivered a total of 1,478' electrical cable for a job. Due to a delay, the electricians installed only 489'. Without using a calculator, determine how many feet of cable remain.
   a. 978
   b. 980
   c. 989
   d. 1,099

4. A worker has been asked to deliver 15 scaffolds to each of 26 different sites. Without using a calculator, determine how many scaffolds the worker will deliver.
   a. 120
   b. 240
   c. 375
   d. 390

5. A farmer has 400 rolls of hay that must be equally distributed to 5 different stables. Without using a calculator, determine how many rolls of hay will go to each site?
   a. 8
   b. 20
   c. 80
   d. 208

6. Which of the following fractions is an equivalent fraction for $^3/_8$?
   a. $^3/_{64}$
   b. $^6/_{64}$
   c. $^{24}/_{64}$
   d. $^{36}/_{64}$

7. The *lowest* common denominator for the fractions $^8/_{64}$ and $^8/_{32}$ is _____.
   a. 8
   b. 16
   c. 24
   d. 32

8. Perform the following calculation and reduce the answer to its lowest term: $^3/_8 + {}^9/_{16} =$ _____.
   a. $^3/_4$
   b. $^{12}/_{16}$
   c. $^7/_8$
   d. $^{15}/_{16}$

9. Perform the following calculation and reduce the answer to its lowest term: $^{11}/_{32} - {}^2/_8 =$ _____.
   a. $^3/_{32}$
   b. $^1/_8$
   c. $^9/_{32}$
   d. $1^9/_{32}$

10. Put the following decimals in order from *smallest* to *largest*: 0.402, 0.420, 0.042, 0.442.
    a. 0.420, 0.042, 0.442, 0.402
    b. 0.042, 0.402, 0.420, 0.442
    c. 0.442, 0.420, 0.402, 0.042
    d. 0.042, 0.402, 0.442, 0.420

11. A worker applies two coatings to a pipe. The first coating is 51.5 nanometers thick, and the second coating is 89.7 nanometers thick. How thick is the combined coating?
    a. 141.2 nanometers
    b. 142.12 nanometers
    c. 144.2 nanometers
    d. 145.02 nanometers

12. Perform the following calculation and round the answer to the nearest hundredth: $3.53 \times 9.75 =$ _____.
    a. 12.28
    b. 34.42
    c. 36.14
    d. 48.13

13. It costs a painting contractor $2.37 per square foot to paint a wall. What will it cost to paint an 864.5 ft^2 wall? Round to the nearest cent.
    a. $204.88
    b. $2,048.87
    c. $2,848.88
    d. $2,888.86

14. $89.435 \div 0.05 =$ _____.
    a. 1788.7
    b. 17.887
    c. 4.47175
    d. 447.175

15. Without using a calculator, what is the decimal equivalent of 13.9%?
    a. 0.009
    b. 0.013
    c. 0.139
    d. 1.39

16. Without using a calculator, convert 14.75 to its equivalent fraction expressed in lowest terms.
    a. $14^{75}/_{100}$
    b. $29^{5}/_{20}$
    c. $^{59}/_{4}$
    d. $14^{7}/_{4}$

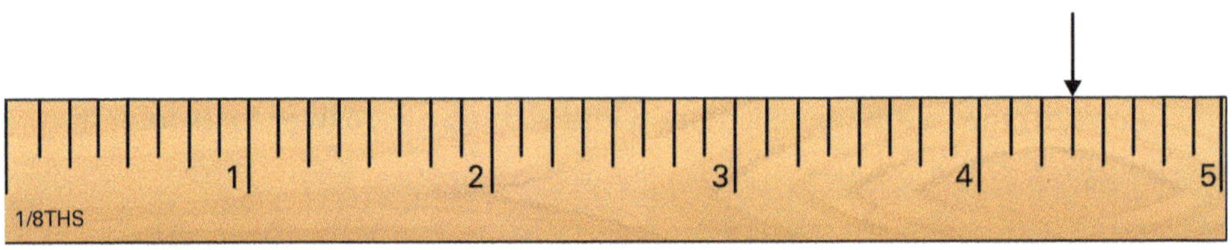

Figure RQ01

17. The arrow on the US standard ruler shown in *Figure RQ01* is pointing to _____.
    a. $4^{3}/_{8}$"
    b. $4^{1}/_{4}$"
    c. $4^{1}/_{2}$"
    d. $4^{5}/_{8}$"

18. The arrow on the metric ruler shown in *Figure RQ02* is pointing to _____.
    a. 5.2 cm
    b. 5.3 cm
    c. 5.4 cm
    d. 5.5 cm

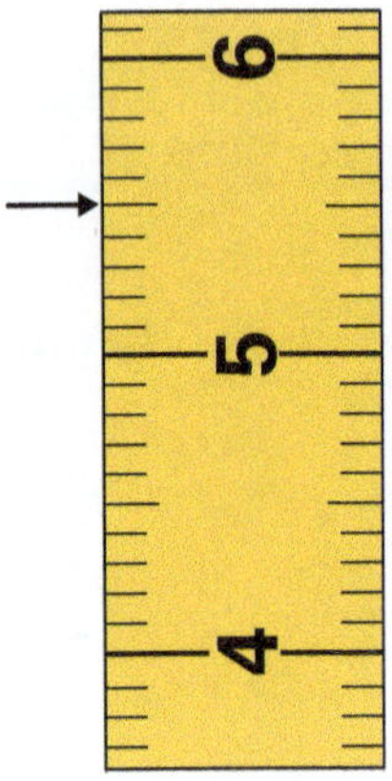

Figure RQ02

19. The red background around the number 32 in *Figure RQ03* identifies _____.
   a. that the tape measure is metric
   b. the US standard measurement equivalent to 1 m
   c. the marking for studs placed on 16" centers in a wall structure
   d. the marking for studs placed on 24" centers

Figure RQ03

20. A centimeter is $\frac{1}{100}$ of a meter, while a kilometer is equal to 1,000 meters.
   a. True
   b. False

21. To convert a measurement from centimeter to meter, move the decimal point _____.
   a. two places to the left
   b. three places to the right
   c. three places to the left
   d. four places to the right

22. What is the centimeter equivalent of 67"?
   a. 17.01 cm
   b. 26.38 cm
   c. 170.18 cm
   d. 263.80 cm

23. A pound of water weighs more than a kilogram of water.
   a. True
   b. False

24. What is the gram equivalent of 49 oz?
   a. 1.73 g
   b. 3.06 g
   c. 747.42 g
   d. 1389.15 g

25. What is the Celsius equivalent of 15°F?
   a. −9.4°C
   b. 9.4°C
   c. −59°C
   d. 59°C

26. The two angles shown in *Figure RQ04* are _____.
   a. obtuse and adjacent
   b. obtuse and opposite
   c. right and opposite
   d. acute and adjacent

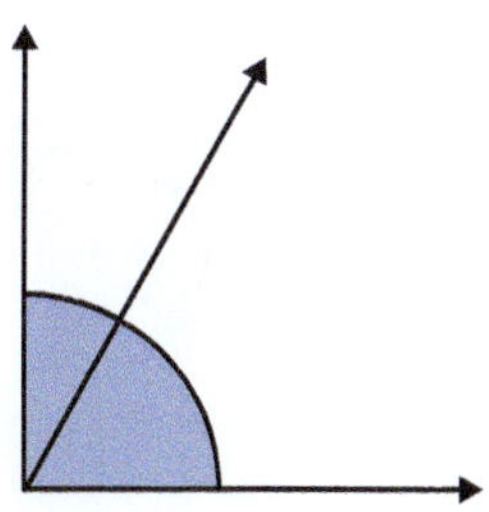

Figure RQ04

27. The sum of the four angles in a rectangle is always 180°.
   a. True
   b. False

28. The mathematical constant *pi* is needed for calculating the area of a _____.
   a. circle
   b. square
   c. rectangle
   d. cube

29. Using a calculator, calculate the volume of a cylindrical tank that is 23' high with a radius of 6.25'. Round the answer to the nearest cubic foot.
    a. 144 ft^3
    b. 898 ft^3
    c. 903 ft^3
    d. 2,821 ft^3

30. Using a calculator, calculate the volume of a three-dimensional triangular prism that has a 7 cm base, a 4 cm height, and a 30 mm depth. Remember to convert all measurements to the same unit before applying the formula.
    a. 42 cm^3
    b. 42 mm^3
    c. 82 mm^3
    d. 82 cm^3

Answers to odd-numbered questions are found in the Review Question Answer Keys at the back of this book.

Answers to Section Review Questions

Answer	Section Reference	Objective
Section 1.0.0		
1. d	1.1.0; *Figure 1*	1a
2. c	1.2.0	1b
3. b	1.3.0	1c
Section 2.0.0		
1. b	2.1.0	2a
2. b	2.2.0	2b
3. c	2.3.0	2c
4. d	2.3.0	2c
5. a	2.4.0	2d
6. c	2.4.0	2d
Section 3.0.0		
1. c	3.1.0; *Figure 4*	3a
2. a	3.1.2	3a
3. b	3.2.1	3b
4. a	3.2.1	3b
5. d	3.2.2	3b
6. a	3.2.3	3b
7. c	3.3.1	3c
8. c	3.3.2	3c
9. a	3.3.2	3c
Section 4.0.0		
1. a	4.1.2	4a
2. b	4.2.3	4b
Section 5.0.0		
1. b	5.1.2	5a
2. d	5.2.1	5b
3. c	5.3.2	5c
4. b	5.4.0	5d
Section 6.0.0		
1. d	6.1.0	6a
2. b	6.2.2; 6.2.4	6b
3. c	6.3.0	6c
4. c	6.4.1	6d

User Update

SCAN ME

Did you find an error? Submit a correction by visiting **www.nccer.org/olf** or by scanning the QR code using your mobile device.

Practice Problem Answers and Calculations

Section 1.1.1 Practice Problems: Place Values of Whole Numbers

1. a (3,964)
2. d (thousands place)
3. b (1,693)
4. b (16,500)
5. c (322,907)

Section 1.2.1 Practice Problems: Adding and Subtracting Whole Numbers

1. $1,480
   $850 + $460 + $170 = **$1,480**
2. $2,278
   $2,477 + $2,263 + $3,218 = $7,958
   $10,236 – $7,958 = **$2,278**
3. 156 hours
   Crew One: 10 + 9 + 11 + 12 + 9 = 51
   Crew Two: 9 + 12 + 12 + 9 + 9 = 51
   Crew Three: 12 + 12 + 10 + 9 + 11 = 54
   Total: 51 + 51 + 54 = **156**
4. $216
   $874 + $67 + $93 = $1,034
   $1,250 – $1,034 = **$216**
5. 891,135 ₽
   850,360 ₽ + 119,980 ₽ = 970,340 ₽
   970,340 ₽ – 79,205 ₽ = **891,135 ₽**

Section 1.3.2 Practice Problems: Multiplying and Dividing Whole Numbers

1. 12 bags

```
      12
  15)180
    – 15
      30
    – 30
       0
```

2. a. 520; b. 360; c. 200
   4 × 2 = 8 locations
   65 × 8 = **520** (2' × 4' boards)
   45 × 8 = **360** (2' × 8' boards)
   25 × 8 = **200** (2' × 10' boards)
3. 13 sections; 3 meters left over
   45 + 30 = 75
   75 ÷ 6 = 12.5, round up to **13**
   13 × 6 = 78
   78 – 75 = **3**

4. a. $2,400; b. $8,400; c. $18,100
   a. 3 × $800 = **$2,400**
   b. 2 (weeks) × $3,400 = $6,800
      2 (days) × $800 = $1,600
      $6,800 + $1,600 = **$8,400**
   c. 1 (month) x $10,500 = $10,500
      2 (weeks) × $3,400 = $6,800
      1 (day) × $800 = $800
      $10,500 + $6,800 + $800 = **$18,100**

Section 2.1.4 Practice Problems: Finding Equivalent Fractions

1. b ($\frac{4}{16}$)

$$\frac{1 \times 4}{4 \times 4} = \mathbf{\frac{4}{16}}$$

2. c ($\frac{4}{32}$)

$$\frac{2 \times 2}{16 \times 2} = \mathbf{\frac{4}{32}}$$

3. d ($\frac{6}{8}$)

$$\frac{3 \times 2}{4 \times 2} = \mathbf{\frac{6}{8}}$$

4. a ($\frac{48}{64}$)

$$\frac{3 \times 16}{4 \times 16} = \mathbf{\frac{48}{64}}$$

5. c ($\frac{6}{32}$)

$$\frac{3 \times 2}{16 \times 2} = \mathbf{\frac{6}{32}}$$

6. $\frac{1}{8}$

$$\frac{2 \div 2}{16 \div 2} = \mathbf{\frac{1}{8}}$$

7. $\frac{1}{4}$

$$\frac{2 \div 2}{8 \div 2} = \mathbf{\frac{1}{4}}$$

8. $\frac{3}{8}$

$$\frac{12 \div 4}{32 \div 4} = \mathbf{\frac{3}{8}}$$

9. $\frac{1}{2}$

$$\frac{4 \div 4}{8 \div 4} = \mathbf{\frac{1}{2}}$$

10. $^1\!/_{16}$

$$\frac{4 \div 4}{64 \div 4} = \mathbf{\frac{1}{16}}$$

11. c (12)

$$\frac{2}{6}, \frac{3}{4}$$

$$\frac{2 \times 2}{6 \times 2} = \frac{4}{12}$$

$$\frac{3 \times 3}{4 \times 3} = \frac{9}{12}$$

$$\frac{4}{12}, \frac{9}{12}$$

12. b (8)

$$\frac{1}{4}, \frac{3}{8}$$

$$\frac{1 \times 2}{4 \times 2} = \frac{2}{8}$$

$$\frac{2}{8}, \frac{3}{8}$$

13. d (8)

$$\frac{1}{8}, \frac{1}{2}$$

$$\frac{1 \times 4}{2 \times 4} = \frac{4}{8}$$

$$\frac{1}{8}, \frac{4}{8}$$

14. b (16)

$$\frac{1}{4}, \frac{3}{16}$$

$$\frac{1 \times 4}{4 \times 4} = \frac{4}{16}$$

$$\frac{4}{16}, \frac{3}{16}$$

15. d (8)

$\frac{4}{32}, \frac{5}{8}$

$\frac{4 \div 4}{32 \div 4} = \frac{1}{8}$

$\frac{1}{8}, \frac{5}{8}$

Section 2.2.1 Practice Problems: Changing Improper Fractions to Mixed Numbers

1. $4\frac{3}{8}$
   $\frac{35}{8} = 35 \div 8 =$ **4, r3** or $\mathbf{4\frac{3}{8}}$
2. $6\frac{1}{16}$
   $\frac{97}{16} = 97 \div 16 =$ **6, r1** or $\mathbf{6\frac{1}{16}}$
3. $3\frac{1}{4}$
   $\frac{13}{4} = 13 \div 4 =$ **3, r1** or $\mathbf{3\frac{1}{4}}$
4. $2\frac{4}{5}$
   $\frac{14}{5} = 14 \div 5 =$ **2, r4** or $\mathbf{2\frac{4}{5}}$
5. $3\frac{1}{16}$
   $\frac{49}{16} = 49 \div 16 =$ **3, r1** or $\mathbf{3\frac{1}{16}}$

Section 2.3.1 Practice Problems: Adding and Subtracting Fractions

1. $\frac{3}{8}$

   $\frac{1}{8} + \frac{4}{16}$

   $\frac{4 \div 2}{16 \div 2} = \frac{2}{8}$

   $\frac{1}{8} + \frac{2}{8} = \mathbf{\frac{3}{8}}$

2. $\frac{7}{8}$

   $\frac{4}{8} + \frac{6}{16}$

   $\frac{6 \div 2}{16 \div 2} = \frac{3}{8}$

   $\frac{4}{8} + \frac{3}{8} = \mathbf{\frac{7}{8}}$

3. $1\frac{1}{4}$

   $\frac{2}{4} + \frac{3}{4} = \frac{5}{4}$

   $5 \div 4 = 1$, r1 or $\mathbf{1\frac{1}{4}}$

4. 1

$$\frac{3}{4} + \frac{2}{8}$$

$$\frac{2 \div 2}{8 \div 2} = \frac{1}{4}$$

$$\frac{3}{4} + \frac{1}{4} = \frac{4}{4}$$

$$\frac{4}{4} = \mathbf{1}$$

5. $1\frac{1}{4}$

$$\frac{14}{16} + \frac{3}{8}$$

$$\frac{14 \div 2}{16 \div 2} = \frac{7}{8}$$

$$\frac{7}{8} + \frac{3}{8} = \frac{10}{8}$$

$$\frac{2 \div 2}{8 \div 2} = \frac{1}{4}$$

$$\mathbf{1\frac{1}{4}}$$

6. $\frac{1}{16}$

$$\frac{3}{8} - \frac{5}{16}$$

$$\frac{3 \times 2}{8 \times 2} = \frac{6}{16}$$

$$\frac{6}{16} - \frac{5}{16} = \mathbf{\frac{1}{16}}$$

7. $\frac{1}{16}$

$$\frac{11}{16} - \frac{5}{8}$$

$$\frac{5 \times 2}{8 \times 2} = \frac{10}{16}$$

$$\frac{11}{16} - \frac{10}{16} = \mathbf{\frac{1}{16}}$$

8. $\frac{5}{12}$

$$\frac{3}{4} - \frac{2}{6}$$

$$\frac{3 \times 3}{4 \times 3} = \frac{9}{12}$$

$$\frac{2 \times 2}{6 \times 2} = \frac{4}{12}$$

$$\frac{9}{12} - \frac{4}{12} = \mathbf{\frac{5}{12}}$$

9. $^5/_{12}$

$$\frac{11}{12} - \frac{4}{8}$$

$$\frac{11 \times 2}{12 \times 2} = \frac{22}{24}$$

$$\frac{4 \times 3}{8 \times 3} = \frac{12}{24}$$

$$\frac{22}{24} - \frac{12}{24} = \frac{10}{24}$$

$$\frac{10 \div 2}{24 \div 2} = \mathbf{\frac{5}{12}}$$

10. $^3/_{16}$

$$\frac{11}{16} - \frac{1}{2}$$

$$\frac{1 \times 8}{2 \times 8} = \frac{8}{16}$$

$$\frac{11}{16} - \frac{8}{16} = \mathbf{\frac{3}{16}}$$

11. $7^1/_4$

$8 - {}^3/_4$

$7 + 1 - {}^3/_4$

$7 + {}^4/_4 - {}^3/_4$

$7 + {}^1/_4 = \mathbf{7^1/_4}$

12. $11^3/_8$

$12 - {}^5/_8$

$11 + 1 - {}^5/_8$

$11 + {}^8/_8 - {}^5/_8$

$11 + {}^3/_8 = \mathbf{11^3/_8}$

13. d ($1^{21}/_{64}$")

Two Punches = $4^1/_{64} + 4^3/_{32}$

$$\frac{3 \times 2}{32 \times 2} = \frac{6}{64}$$

$4^1/_{64} + 4^6/_{64} = 8^7/_{64}$

$9^7/_{16} - 8^7/_{64}$

$$\frac{7 \times 4}{16 \times 4} = \frac{28}{64}$$

$9^{28}/_{64} - 8^7/_{64} = \mathbf{1^{21}/_{64}}$

14. b ($8^{11}/_{16}$")

$20\,^3/_4 - 12\,^1/_{16}$

$$\frac{3 \times 4}{4 \times 4} = \frac{12}{16}$$

$20\,^{12}/_{16} - 12\,^1/_{16} = \mathbf{8^{11}/_{16}}$

15. a ($^7/_{16}$")

$36\,^3/_8 - 35\,^{15}/_{16}$

$$\frac{3 \times 2}{8 \times 2} = \frac{6}{16}$$

$36\,^6/_{16} - 35\,^{15}/_{16}$

$$35 + 1 + \frac{6}{16}$$

$$35 + \frac{16}{16} + \frac{6}{16}$$

$$35 + \frac{22}{16}$$

$35\,^{22}/_{16} - 35\,^{15}/_{16} = \mathbf{^7/_{16}}$

Section 2.4.1 Practice Problems: Multiplying and Dividing Fractions

1. $^5/_{32}$
   $^4/_{16} \times {^5/_8} = {^{20}/_{128}} \div {^4/_4} = \mathbf{^5/_{32}}$
2. $^{21}/_{32}$
   $^3/_4 \times {^7/_8} = \mathbf{^{21}/_{32}}$
3. $3^3/_4$
   $^2/_8 \times {^{15}/_1} = {^{30}/_8} \div {^2/_2} = \mathbf{3^3/_4}$
4. 21
   $^3/_7 \times {^{49}/_1} = {^{147}/_7} \div {^7/_7} = {^{21}/_1} = \mathbf{21}$
5. $^1/_4$

$$\frac{8}{16} \times \frac{32}{64}$$

$$\frac{8 \div 8}{16 \div 8} = \frac{1}{2} \qquad \frac{32 \div 32}{64 \div 32} = \frac{1}{2}$$

$$\frac{1}{2} \times \frac{1}{2} = \frac{1}{4}$$

or $$\frac{8}{16} \times \frac{32}{64} = \frac{256 \div 256}{1{,}024 \div 256} = \mathbf{\frac{1}{4}}$$

6. $\frac{1}{8}$

   $\frac{3}{8} \div 3 = \frac{3}{8} \times \frac{1}{3} = \frac{3}{24} \div \frac{3}{3} = \mathbf{\frac{1}{8}}$

7. $1\frac{1}{4}$

   $\frac{5}{8} \div \frac{1}{2} = \frac{5}{8} \times \frac{2}{1} = \frac{10}{8} \div \frac{2}{2} = \mathbf{\frac{5}{4}}$ or $\mathbf{1\frac{1}{4}}$

8. 2

   $\frac{3}{4} \div \frac{3}{8} = \frac{3}{4} \times \frac{8}{3} = \frac{24}{12} \div \frac{12}{12} = \frac{2}{1} = \mathbf{2}$

9. a (34')

   $8\frac{1}{2} \div \frac{1}{4} = \frac{17}{2} \div \frac{1}{4} = \frac{17}{2} \times \frac{4}{1} = \frac{68}{2} \div \frac{2}{2} = \frac{34}{1} = \mathbf{34}$

10. d (8)

    $7 \div \frac{7}{8} = \frac{7}{1} \times \frac{8}{7} = \frac{56}{7} \div \frac{7}{7} = \frac{8}{1} = \mathbf{8}$

Section 3.1.3 Practice Problems: Working with Decimals

1. b (four-tenths)
2. c (five-hundredths)
3. a (two and five-tenths)
4. c (0.18)
5. d (5.8)
6. c (0.004, 0.044, 0.400, 0.404)
7. d (0.059, 0.508, 0.56, 0.567)
8. a (0.003, 0.032, 0.302, 0.320)
9. b (0.008, 0.086, 0.867, 0.870)
10. b (0.616, 0.626, 0.630, 0.641)

Section 3.2.5 Practice Problems: Decimals

1. 11.7

$$\begin{array}{r} 2.50 \\ 4.20 \\ +\ 5.00 \\ \hline \mathbf{11.70} \end{array}$$

2. 10.48

$$\begin{array}{r} \scriptstyle 1 \\ 1.82 \\ 3.41 \\ +\ 5.25 \\ \hline \mathbf{10.48} \end{array}$$

3. 92.83

$$\begin{array}{r} \scriptstyle 1 \\ 6.43 \\ +\ 86.40 \\ \hline \mathbf{92.83} \end{array}$$

4. b (0.328 cm)

$$\begin{array}{r} \scriptstyle 1 \\ 0.078 \\ +\ 0.250 \\ \hline \mathbf{0.328} \end{array}$$

5. b (4.4 tons)

$$\begin{array}{r} 6.7 \\ -\,2.3 \\ \hline \mathbf{4.4} \end{array}$$

6. c (0.9")

$$\begin{array}{r} 3 \\ \times\,0.3 \\ \hline \mathbf{0.9} \end{array}$$

7. a (5.28")

$$\begin{array}{r} 6.18 \\ -\,0.90 \\ \hline \mathbf{5.28} \end{array}$$

8. b (608 pounds)

$$\begin{array}{r} 128 \\ \times\,4.75 \\ \hline 640 \\ 8960 \\ +\,51200 \\ \hline \mathbf{608.00} \end{array}$$

9. 2.52

$$\begin{array}{r} \mathbf{2.52} \\ 18\overline{)45.36} \\ -\,36 \\ \hline 093 \\ -\,090 \\ \hline 0036 \\ -\,0036 \\ \hline 0000 \end{array}$$

10. 0.25

$$\begin{array}{r} 0.252, \text{ rounded to } \mathbf{0.25} \\ 18\overline{)4.536} \\ -\,36 \\ \hline 093 \\ -\,090 \\ \hline 0036 \\ -\,0036 \\ \hline 0000 \end{array}$$

11. 0.03

$$\begin{array}{r} 0.0252, \text{ rounded to } \mathbf{0.03} \\ 18\overline{)0.4536} \\ -\,00 \\ \hline 045 \\ -\,036 \\ \hline 0093 \\ -\,0090 \\ \hline 00036 \\ -\,00036 \\ \hline 00000 \end{array}$$

12. 0.41

```
     0.408, rounded to 0.41
25)10.200
  – 100
   0020
  – 0000
   00200
  – 00200
   00000
```

13. 5.2

```
   5.2
6)31.2
– 30
  012
– 012
  000
```

14. 20

```
                      20
14.1)282  =  141)2820
                – 282
                   00
                 – 00
                    0
```

15. 216.88

```
                216.875, rounded to 216.88
3.2)694  =  32)6940.000
             – 64
               054
             – 032
               0220
             – 0192
               00280
             – 00256
               000240
             – 000224
               0000160
             – 0000160
               0000000
```

16. 220

```
                  220
0.45)99  =  45)9900
             – 90
                90
              – 90
                 00
               – 00
                  0
```

17. 40.8

$$2.5\overline{)102} = 25\overline{)1020.0}$$

40.8

- 100
0020
- 0000
00200
- 00200
00000

18. 520

$$0.6\overline{)312} = 6\overline{)3120}$$

520

- 30
12
- 12
00
- 00
0

19. 4.91

$$4.24\overline{)20.82} = 424\overline{)2082.000}$$

4.910

- 1696
03860
- 03816
000440
- 000424
0000160
- 0000000
160r

20. 10.51

$$3.7\overline{)38.9} = 37\overline{)389.000}$$

10.513, rounded to **10.51**

- 37
019
- 000
0190
- 0185
00050
- 00037
000130
- 000111
19r

21. 22

$$0.45\overline{)9.9} = 45\overline{)990}$$

22

- 90
90
- 90
0

22. 40.8

$$0.25\overline{)10.20} \;=\; 25\overline{)1020.0}\quad \mathbf{40.8}$$

- 100
0020
- 0000
00200
- 00200
00000

23. 52

$$0.6\overline{)31.2} \;=\; 6\overline{)312}\quad \mathbf{52}$$

- 30
12
- 12
0

24. 53.74

25. 0.91

26. 123.13

27. 1.98

28. 38.4

29. b (24)

24.13, rounded to **24**

$$3.75\overline{)90.5} \;=\; 375\overline{)9050.00}$$

- 750
1550
- 1500
00500
- 00375
001250
- 001125
000125r

30. d (6.6)

6.59, rounded to **6.6**

$$151.8\overline{)1001} \;=\; 1518\overline{)10010.00}$$

- 09108
009020
- 007590
0014300
- 0013662
638r

31. d (28.13)

28.127, rounded to **28.13**

4.30)120.95 = 430)12095.000
– 0860
03495
– 03440
000550
– 000430
0001200
– 0000860
00003400
– 00003010
390r

32. c (311.8)

311.83, rounded to **311.8**

0.37)115.38 = 37)11538.00
– 111
0043
– 0037
00068
– 00037
000310
– 000296
0000140
– 0000111
29r

33. b (240.4)

240.375, rounded to **240.4**

0.48)115.38 = 48)11538.000
– 096
0193
– 0192
00018
– 00000
000180
– 000144
0000360
– 0000336
00000240
– 00000240
00000000

Section 3.3.5 Practice Problems: Converting Different Values

1. 62%
   0.62 = 0.62 × 100 = **62%**
2. 47.5%
   0.475 = 0.475 × 100 = **47.5%**
3. 70%
   0.7 = 0.7 × 100 = **70%**

4. 0.72

   72% = 72 ÷ 100 = **0.72**

5. 0.125

   12.5% = 12.5 ÷ 100 = **0.125**

6. 0.25

```
   0.25
 4)1.00
 – 08
   020
 – 020
   000
```

7. 0.75

```
   0.75
 4)3.00
 – 28
   020
 – 020
   000
```

8. 0.125

```
   0.125
 8)1.000
 – 08
   020
 – 016
   0040
 – 0040
   0000
```

9. 0.3125

```
    0.3125
 16)5.0000
  – 48
    020
  – 016
    0040
  – 0032
    00080
  – 00080
    00000
```

10. 0.3125

```
    0.3125
 64)20.0000
  – 192
    0080
  – 0064
    00160
  – 00128
    000320
  – 000320
    000000
```

11. $\frac{1}{2}$

$$0.5 = \frac{5 \div 5}{10 \div 5} = \mathbf{\frac{1}{2}}$$

12. $\frac{3}{25}$

$$0.12 = \frac{12 \div 4}{100 \div 4} = \mathbf{\frac{3}{25}}$$

13. $\frac{1}{8}$

$$0.125 = \frac{125 \div 125}{1{,}000 \div 125} = \mathbf{\frac{1}{8}}$$

14. $\frac{4}{5}$

$$0.8 = \frac{8 \div 2}{10 \div 2} = \mathbf{\frac{4}{5}}$$

15. $\frac{9}{20}$

$$0.45 = \frac{45 \div 5}{100 \div 5} = \mathbf{\frac{9}{20}}$$

16. 0.75

$$\frac{9 \div 3}{12 \div 3} = \frac{3}{4}$$

$$\begin{array}{r} \mathbf{0.75} \\ 4\overline{)3.00} \\ \underline{-\ 28} \\ 020 \\ \underline{-\ 020} \\ 000 \end{array}$$

17. 0.83

$$\frac{10 \div 2}{12 \div 2} = \frac{5}{6}$$

$$\begin{array}{r} 0.833 \\ 6\overline{)5.000} \\ \underline{-\ 48} \\ 020 \\ \underline{-\ 018} \\ 0020 \\ \underline{-\ 0018} \\ 2r \end{array}$$

Rounded to the nearest hundredth = **0.83**

18. 0.17

$$\frac{2}{12}\begin{matrix}\div 2\\ \div 2\end{matrix} = \frac{1}{6}$$

```
   0.166
6)1.000
 - 06
   040
 - 036
   0040
 - 0036
      4r
```

Rounded to the nearest hundredth = **0.17**

19. 0.33

$$\frac{4}{12}\begin{matrix}\div 4\\ \div 4\end{matrix} = \frac{1}{3}$$

```
   0.333
3)1.000
 - 09
   010
 - 009
     1r
```

Rounded to the nearest hundredth = **0.33**

20. 1.42

```
      1.416
12)17.000
  - 12
    050
  - 048
    0020
  - 0012
    00080
  - 00072
        8r
```

Rounded to the nearest hundredth = **1.42**

Section 3.3.6 Practical Applications

1. a. 342 linear feet; b. $406.98; c. $48.84; d. $358.14; e. $21.49; f. $379.63
   a. 28' × 2 + 24' + 32' + 20' + 16' × 3 + 15'-9" × 3 + 23'-8" × 2 + 5'-11" × 2 + 9'-6" × 2 + 8' × 4 + (24' – 20') = 341.416', rounded up to **342'**
   b. 342' × $1.19 = **$406.98**
   c. $406.98 × 12% = $406.98 × 0.12 = **$48.84**
   d. $406.98 – $48.84 = **$358.14**
   e. $358.14 × 6% = $358.14 × 0.06 = **$21.49**
   f. $358.14 + $21.49 = **$379.63**
2. a. 50.2%; b. 43.0%; c. 6.8%
   a. 76,000 ft^2 ÷ (550' × 275') = 76,000 ft^2 ÷ 151,250 ft^2 = **50.2%**
   b. 65,000 ft^2 ÷ (550' × 275') = 65,000 ft^2 ÷ 151,250 ft^2 = **43.0%**
   c. 100% – 50.2% – 43.0% = **6.8%**

Section 4.1.3 Practice Problems: Reading Rulers

1. $\frac{4}{8}$"
2. $1\frac{7}{8}$"
3. $2\frac{6}{8}$"
4. $3\frac{6}{8}$"
5. $4\frac{5}{8}$"
6. $\frac{10}{16}$"
7. $1\frac{9}{16}$"
8. $2\frac{5}{16}$"
9. $3\frac{4}{16}$"
10. $4\frac{7}{16}$"
11. 0.1 cm
12. 1.1 cm
13. 1.9 cm
14. 29 mm
15. 38 mm

Section 4.2.4 Practice Problems: Reading Measuring Tapes

1. $8\frac{13}{16}$"
2. $9\frac{13}{16}$"
3. $10\frac{14}{16}$" (or $10\frac{7}{8}$")
4. $3\frac{4}{16}$" (or $3\frac{1}{4}$")
5. $4\frac{11}{16}$"
6. 4.4 cm
7. 5.9 cm
8. 9.9 cm
9. 11 cm
10. 11.8 cm

Section 5.1.4 Practice Problems: Converting Measurements

1. 45
   0.45 × 100 (conversion factor) = **45**
2. 108
   3 × 36 (conversion factor) = **108**
3. 12
   36 × 0.3333 (conversion factor) = **12**
4. 2.5
   90 × 0.02778 (conversion factor) = **2.5**
5. 0.01
   1 × 0.01 (conversion factor) = **0.01**
6. 167.64
   66 × 2.54 (conversion factor) = **167.64**
7. 14.33
   47 × 0.3048 (conversion factor) = **14.33**

8. 1.79
   54.5 × 0.03281 (conversion factor) = **1.79**
9. 17.37
   19 × 0.9144 (conversion factor) = **17.37**
10. 185.04
   4.7 × 39.37 (conversion factor) = **185.04**

Section 5.2.4 Practice Problems: Converting Weight Units

1. 22.68
   50 × 0.4536 (conversion factor) = **22.68**
2. 110.25
   50 × 2.205 (conversion factor) = **110.25**
3. 450.77
   15.9 × 28.35 (conversion factor) = **450.77**
4. 3.32
   94 × 0.003527 (conversion factor) = **3.32**

Section 5.3.4 Practice Problems: Converting Volume Units

1. 6.7
   11600 × 0.0005787 (conversion factor) = **6.7**
2. 1,900,000.0
   1.9 × 1,000,000 (conversion factor) = **1,900,000**
3. 669.7
   512 × 1.308 (conversion factor) = **669.7**
4. 198,218.0
   7 × 28,316.85 (conversion factor) = **198,218**
5. 7.1
   0.2 × 35.315 (conversion factor) **= 7.1**

Section 5.4.1 Practice Problems: Converting Temperatures

1. 82.2

   $°C = {}^{5}/_{9}\,(180 - 32)$

   $°C = {}^{5}/_{9}\,(148)$

   $°C = {}^{5}/_{9} \times {}^{148}/_{1} = {}^{740}/_{9}$

   °C = **82.2**

2. 18.9

   $°C = {}^{5}/_{9}\,(66 - 32)$

   $°C = {}^{5}/_{9}\,(34)$

   $°C = {}^{5}/_{9} \times {}^{34}/_{1} = {}^{740}/_{9}$

   °C = **18.9**

3. −14.8

   $°F = [\,{}^{9}/_{5} \times (-26)] + 32$

   $°F = [\,{}^{9}/_{5} \times (-{}^{26}/_{1})] + 32$

   $°F = -{}^{234}/_{5} + 32$

   $°F = -46.8 + 32$

   $°F =$ **−14.8**

4. 159.8

   $°F = ({}^{9}/_{5} \times 71) + 32$

   $°F = ({}^{9}/_{5} \times {}^{71}/_{1}) + 32$

   $°F = {}^{639}/_{5} + 32$

   $°F = 127.8 + 32$

   $°F =$ **159.8**

Section 6.3.1 Practice Problems: Calculating Area

1. c (32 ft^2)
   Area = length × width = 8 × 4 = **32 ft^2**
2. a (256 cm^2)
   Area = length × width =16 × 16 = **256 cm^2**
3. c (153.86 ft^2)
   radius = ½ diameter
   radius = ½ × 14 = 7'
   Area = pi × radius^2 = 3.14 × 7 × 7 = 3.14 × 49 = **153.86 ft^2**

```
   1
   1 3
   3.14
  × 49
  2826
+ 12560
 153.86
```

4. a (12 cm^2)
   Area = 0.5 × base × height = 0.5 × 4 × 6 = 0.5 × 24 = **12 cm^2**

```
   2
  24
 × 0.5
 12.0
```

5. c (70 m^2)
   Area = base × height = 14 × 5 = **70 m^2**

Section 6.4.5 Practice Problems: Calculating Volume

1. d (390 ft^3)
   Volume = length × width × depth = 5 × 6 × 13 = 30 × 13 = **390 ft^3**

```
   30
 × 13
   90
+ 300
  390
```

2. d (27 cm^3)
   Volume = length × width × depth = 3 × 3 × 3 = **27 cm^3**
3. b (24 in^3)
   Volume = 0.5 × base × height × depth = 0.5 × 6 × 2 × 4 = **24 in^3**
4. a (16.96 m^3)
   Radius = $\frac{1}{2}$ diameter = $\frac{1}{2} \times 6 = \frac{6}{2} \div \frac{2}{2} = \frac{3}{1}$ = 3 m
   Height = 60 cm = 0.6 m
   Volume = pi × radius^2 × h = 3.14 × 3 × 3 × 0.6 = 3.14 × 9 × 0.6

$$\begin{array}{r} 9 \\ \times\ 0.6 \\ \hline 5.4 \end{array}$$

   Volume = 3.14 × 5.4

$$\begin{array}{r} {\scriptstyle 1} \\ {\scriptstyle 2} \\ 3.14 \\ \times\ 5.4 \\ \hline 1256 \\ +\ 15700 \\ \hline 16.956 \end{array}$$

   Volume = 16.956 m^3, rounded up to **16.96 m^3**
5. b (5.35 yd^3)
   Depth = 6"
   $\frac{6}{12} = \frac{1}{2}$ = 0.5'
   Volume = length × width × depth = 17 × 17 × 0.5 =

$$\begin{array}{r} {\scriptstyle 4} \\ 17 \\ \times\ 17 \\ \hline 119 \\ +\ 170 \\ \hline 289 \end{array}$$

   Volume = 289 × 0.5

$$\begin{array}{r} {\scriptstyle 4\ 4} \\ 289 \\ \times\ 0.5 \\ \hline 144.5 \end{array}$$

   Volume = 144.5 ft^3
   1 yd = 3'
   1 yd^3 = 3 × 3 × 3 = 27 ft^3

$$144.5\ ft^3 \times \frac{1\ yd^3}{27\ ft^3} = 144.5 \div 27$$

$$\begin{array}{r} 5.35 \\ 27\overline{)144.50} \\ -\ 135 \\ \hline 95 \\ -\ 81 \\ \hline 140 \\ -\ 135 \\ \hline 5r \end{array}$$

   Volume = **5.35 yd^3**

Section 6.4.6 Practical Applications Using Volume

1. 653 in^3
   Volume = $\pi \times 4^2 \times 13"$ = **653.12 in^3**
2. 109 ft^3
   Area of upper rectangle = 39.666' × 6' = 237.996 ft^2
   Area of remaining rectangle = 6' × 13' = 78 ft^2
   Area of remaining triangle = ½ × 4' × 6' = 12 ft^2
   Total area = 237.996 ft^2 + 78 ft^2 + 12 ft^2 = 327.996 ft^2
   Total volume of topsoil = 327.996 ft^2 × 0.333' (thickness of the concrete) = **109 ft^3**

Section Review Calculations

Section 1.0.0

Question 2

46 + 8 + 8 + 30 + 8 + 10 + 20 + 10 + 8 + 10 + 8 + 10 + 10 + 38 = **224'**

Question 3

224' × 12 = 2688"

2688" ÷ 16" per block = 168 blocks per course

A 24" high foundation requires 3 courses of 8" block

Therefore, 3 × 168 blocks = **504 blocks**.

Section 2.0.0

Question 1

$^1/_2 \times {}^{32}/_{32} = {}^{32}/_{64}$

Question 2

$^{76}/_{64} = 76 \div 64 = 1$ with a remainder of $^{12}/_{64}$

Reduce $^{12}/_{64}$ by dividing both numerator and denominator by 2, giving $^6/_{32}$.

Therefore, $^{76}/_{64}$ $(1^6/_{32}) \neq 1^9/_{32}$.

Question 3

Convert $^3/_8$ to a common denominator with $^{11}/_{16}$: $^3/_8 \times {}^2/_2 = {}^6/_{16}$

$^{11}/_{16} + {}^6/_{16} = {}^{17}/_{16}$, or $1^1/_{16}$

Question 4

Convert $1^1/_4$ to a common denominator with $^7/_{16}$: $1^1/_4 = {}^5/_4 = {}^5/_4 \times {}^4/_4 = {}^{20}/_{16}$

$^{20}/_{16} - {}^7/_{16} = {}^{13}/_{16}$

Question 5

$^1/_8 \times {}^1/_{10} = (1 \times 1)/(8 \times 10) = {}^1/_{80}$

Question 6

Invert the divisor and multiply:

$^7/_{16} \div {}^7/_8 = {}^7/_{16} \times {}^8/_7 = (7 \times 8)/(16 \times 7) = {}^{56}/_{112}$

Using a factor of 56, reduce to lowest terms:

$(56 \div 56)/(112 \div 56) = {}^1/_2$

Section 3.0.0

Question 8

$^7/_8 = 7 \div 8 =$ **0.875**

Question 9

$^2/_3 = 2 \div 3 = 0.666$

Convert 0.666 to a percentage by moving the decimal 2 places to the right: 66.6%

66.6% is greater than 65%.

Section 6.0.0

Question 3

Area = length × width

Area = 27.3 m × 9.3 m

Area = **253.89 m²**

Question 4

Volume = length × width × depth

Volume = 26' × 13.66' × 3'

Volume = **1,065.48 ft^3**

MODULE 93104

Introduction to Hand Tools

Source: Shutterstock

Objectives

Successful completion of this module prepares you to do the following:

1. Name common hand tools and state how to use them.
   a. Identify various hammers and demolition tools and explain how to use them.
   b. Describe chisels and punches and how they are used.
   c. Match screwdrivers to the appropriate hardware.
   d. Differentiate between nonadjustable, adjustable, and socket wrenches.
   e. Describe various types of pliers and explain how they are used.
2. Identify common measurement and layout tools and describe how to use them.
   a. Explain how to use a variety of measuring tools.
   b. Define various types of levels and layout tools and indicate how they are used.
3. Identify and describe other hand tools common to shops and jobsites.
   a. Differentiate between various handsaws and their designated applications.
   b. Identify common clamp designs.
   c. Explain how different utility knives are used with various materials.
   d. Describe shovels and picks and the tasks for which each one is best suited.

Performance Task

Under supervision, you should be able to do the following:

1. Inspect and demonstrate the safe and proper use of the following hand tools:
   - Hammers
   - Demolition tools
   - Chisels and punches
   - Screwdrivers
   - Adjustable wrenches
   - Nonadjustable wrenches
   - Sockets
   - Pliers
   - Tape measures
   - Levels
   - Squares
   - Handsaws
   - Clamps
   - Files
   - Utility knives
   - Shovels

Digital Resources for Agricultural Technology

SCAN ME

Scan this code using the camera on your phone or mobile device to view the digital resources related to this craft.

Overview

Every profession has its tools. A surgeon uses a scalpel, an instructor uses a whiteboard, and an accountant uses a calculator. The construction and agricultural fields require a broad array of hand tools. Even if you are familiar with some tools, all craftworkers must learn to select, maintain, and use them safely. Quality hand tools are expensive, so proper use and maintenance will greatly extend their lifetimes. A true craft professional invests wisely in hand tools, and uses, maintains, and stores them with the same wisdom. Always use the right tool for the job and learn the best techniques for each task to improve your efficiency on the jobsite or farm.

NCCER **Industry-Recognized Credentials**

If you are training through an NCCER-accredited sponsor, you may be eligible for credentials from NCCER. The ID number for this module is 93104. Note that this module may have been used in other NCCER curricula and may apply to other level completions. Contact NCCER at 1.888.622.3720 or go to **www.nccer.org** for more information.

You can also show off your industry-recognized credentials online with NCCER's digital credentials. Transform your knowledge, skills, and achievements into credentials that you can share across social media platforms, send to your network, and add to your resume. For more information, visit **www.nccer.org**.

1.0.0 Common Hand Tools

Objective

Name common hand tools and state how to use them.

a. Identify various hammers and demolition tools and explain how to use them.
b. Describe chisels and punches and how they are used.
c. Match screwdrivers to the appropriate hardware.
d. Differentiate between nonadjustable, adjustable, and socket wrenches.
e. Describe various types of pliers and explain how they are used.

Performance Task

1. Inspect and demonstrate the safe and proper use of the following hand tools:
   - Hammers
   - Demolition tools
   - Chisels and punches
   - Screwdrivers
   - Adjustable wrenches
   - Nonadjustable wrenches
   - Sockets
   - Pliers

Construction craftworkers, farmhands, and ranch hands require a combination of knowledge and practical skills to be successful. While many things can be done with hands alone, every craft must use hand tools to get the job done. How craftworkers handle and care for their tools often demonstrates their level of skill and respect for the craft.

Many unique hand tools are used across the crafts and agricultural professions—far too many to present in a single module. You will likely be exposed to more hand tools in the early stages of your career. Learning their functions and capabilities will allow you to solve problems in the field and increase your output. Just think of how the first farmers used hand tools to grow and harvest a surplus of food. Then, the industrial revolution produced machines that could complete tasks that once required manual labor to complete and in a fraction of the time once needed. Even though these machines continue to advance construction and agriculture, specialized tasks still require an appropriate hand tool and a hardworking, well-trained individual wielding it.

Hand tools are often grouped by purpose. Each group has different physical characteristics that determine how each similar tool is applied to the job. This section presents common hand tools used in construction crafts, including:

- Hammers and demolition tools
- Chisels and punches
- Screwdrivers
- Wrenches

- Pliers and wire cutters
- Axes
- Shovels
- Post-hole diggers
- Hoes
- Rakes
- Seeders

Some personal protective equipment (PPE) must be worn consistently on the jobsite, such as safety glasses, safety shoes, hard hats, and gloves. This protective gear is typically required on every jobsite for your protection.

When working at height, all tools must be handled carefully so that they do not fall and injure someone below. A tool should be connected to a tool belt with a **lanyard** to prevent injury to others and damage to the work below.

Lanyard: A cord, rope, or similar material used to limit the travel of two connected objects.

1.1.0 Hammers and Demolition Tools

Hammers are made in different sizes and weights for specific types of work. The safest hammers have heads made of special alloys or drop-forged steel. Due to their light weight and durability, titanium hammers are also becoming popular.

Some of the most common hammers are the claw hammer, the **drywall** hammer, and the ball peen hammer. Demolition tools include various styles of sledgehammers and nail pullers. Several examples of each tool will be presented in this section.

Drywall: A large, flat board made of layers of fiberboard and gypsum used primarily for wall construction and finishing. Drywall, also known as *Sheetrock* and *wallboard*, is typically manufactured in 4' × 8' panels and readily accepts paint.

1.1.1 Claw Hammers

The claw hammer (*Figure 1*) typically has a steel head and a handle made of wood, steel, fiberglass, and graphite. Claw hammers are generally associated with carpentry and similar crafts. The head is used to drive a nail, wedge, or **dowel**, and the claw is used to pull nails out of wood. The nailing face of the hammer may be slightly rounded or have a milled or waffled surface. While some specialized hammers do have a flat face, such faces are rarely found on hammers designed for driving nails.

Dowel: A pin or rod, usually round, that fits into a corresponding hole to fasten or align two workpieces.

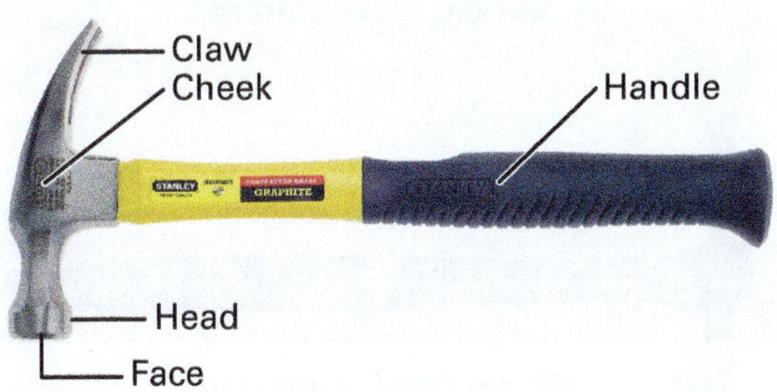

Figure 1 Claw hammer.
Source: Image property of Stanley Black & Decker. Used with permission

A flat hammer face would leave deeper depressions when you drive the head of a nail **flush** with the surface of the work. A slightly rounded (convex) face allows a skilled worker to drive a nail head flush without seriously damaging the surface of the work. A milled or waffled nailing face is another type of hammer face (*Figure 2*). The milled face is generally reserved for framing hammers, where the marks it leaves will be permanently concealed behind finish work.

Flush: As an adjective, to be even or at the same level with another object. For example, when a nail is fully seated, the top of the head is flush (even) with, or slightly beyond, the surface of the wood.

There are several common types of claw hammers. Finish hammers (*Figure 1*) are designed for lighter work such as finish carpentry, as the name implies. They are well suited for driving slender nails without **overstriking** or bending them. Most have curved claws and are around 16–20 ounces (oz), or 454–567 grams (g), in weight. Smaller versions for hammering brads or small finish nails may be as light as 7 oz (198 g).

Overstriking: When a hammer misses the head of a nail, with the head going past it and allowing the hammer handle to impact the nail head instead. Overstriking is a common cause of hammer damage.

Figure 2 Milled nailing face.
Source: Getty Images

Framing hammers (*Figure 3*) are often heavier than common finish hammers since heavier blows are often needed for the larger nails and thicker material used in wood framing. The additional weight is in the head. A framing hammer may weigh as much as 28 oz (794 g). They also have longer handles to generate more power. Framing hammers may also have a milled face, as it helps prevent the face from slipping off the nail on impact.

Hammers also differ by the design of the claws (*Figure 4*). A curved claw hammer provides some additional leverage for pulling nails. Rip claw hammers have flatter claws and are best for ripping boards and woodwork apart.

Driving a nail with a claw hammer is simple using the following technique, although practice is required to develop rhythm and accuracy:

Step 1 Don the appropriate PPE, including safety glasses.

Step 2 Hold the nail in position and rest the face of the hammer on it to give you a visual starting place.

Step 3 Draw the hammer back and lightly tap the nail to start it.

Step 4 Once the nail is started, move your fingers away.

Step 5 Hold the hammer **level** with the head of the nail, using the motion of the wrist, elbow, and shoulder to strike the nail squarely in the center of the hammer face.

Level: Used as an adjective in this context, meaning flat or even with another surface or in relation to Earth's horizon.

Figure 3 Framing hammer.
Source: Image property of Stanley Black & Decker. Used with permission

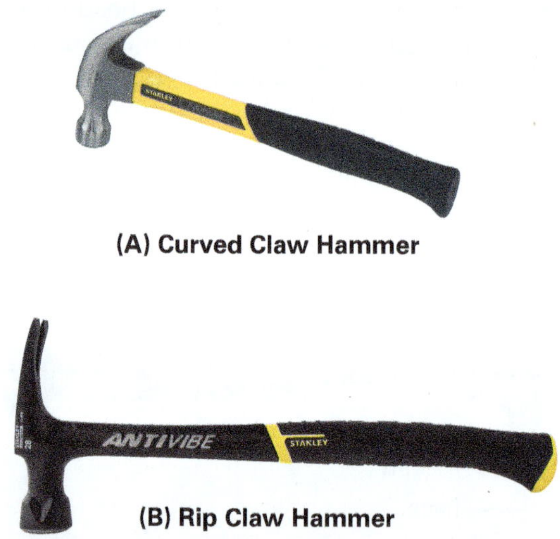

Figure 4 Curved claw and rip claw hammers.
Source: Image property of Stanley Black & Decker. Used with permission

The hammer is designed to produce a certain amount of force on the object it strikes. If you hold the hammer incorrectly, the benefits of its design are not evident. The distance between your hand and the head of the hammer affects the force you apply to drive a nail. The closer to the head you hold the hammer, the harder you will need to swing to achieve the desired force. Hold it with the end of the handle even with the lower edge of your palm. Make it easier on yourself by holding the hammer properly; it takes less effort to drive the nail.

Pulling a nail with a claw hammer is as easy as driving one:

Step 1 Slip the claw of the hammer under the nail head.

Step 2 Use the hammer's leverage to pry the nail up until the handle is nearly straight up (vertical), partially drawing the nail out of the wood.

Step 3 Try to pull the nail straight up from the wood. If it is too tight and more leverage is needed, place a small block of wood between the hammer head and the wood to elevate it, then repeat the prying motion.

Longer and larger nails require more effort to remove, depending on the material in which they are driven. Placing a small block of wood beneath the head when pulling a nail also protects the workpiece from being scarred by the hammer.

Figure 5 Drywall hammer.
Source: Image property of Stanley Black & Decker. Used with permission

1.1.2 Drywall Hammer

A drywall hammer (*Figure 5*) is used for driving nails in drywall, also referred to as *Sheetrock*. Instead of claws, the opposite end of the hammer is shaped like a small hatchet.

Drywall hammers are lighter than normal hammers, generally weighing 12–14 oz (340–397 g). This light weight helps prevent deep indentations in the drywall. The nailing face is more rounded than other hammers for the same reason. The hatchet-like tail of the hammer was originally used by plasterers to cut **lath**. However, it is equally valuable for drywall work. It can cut through small woodwork, and the U-shaped area is used to help grab and carry drywall sheets.

Lath: A thin, flat strip of wood used to form latticework or a foundation for wall plaster.

1.1.3 Ball Peen Hammer

A ball peen hammer (*Figure 6*) is primarily used in metalworking. It has a flat face for striking and a spherical or hemispherical head for **peening** (forming) metal or rivet heads. This hammer is also used with chisels and punches. In **welding** operations, the ball peen hammer is used to reduce stress in a fresh weld by peening, or repeatedly striking, the **joint** as it cools.

Peening: The process of bending, shaping, or cutting metal by striking it with a tool.

Welding: The process of heating two or more metal workpieces to their liquid state and fusing them to create a very strong joint.

Joint: The intersection of workpiece components where a physical connection is generally made. Joints are often named by the style of connection, such as butt joints and miter joints.

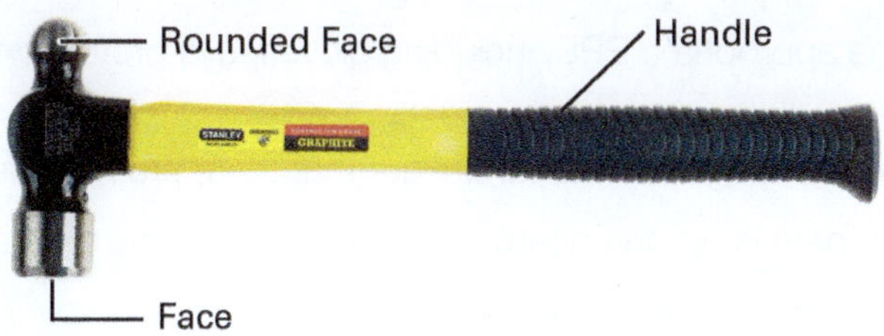

Figure 6 Ball peen hammer.
Source: Image property of Stanley Black & Decker. Used with permission

Ball peen hammers are also known as *engineer's hammers* or *machinist's hammers*. They are classified by weight, ranging from 4 oz to 2 pounds (lb), or 113 g to 0.9 kilograms (kg).

It is best to strike material with a ball peen hammer in a controlled manner. Heavy blows are not often required. Never use a ball peen hammer to drive nails, because the head of the hammer is made from a milder steel. Continual pounding on nails can deform or damage the face. Ball peen hammers and hammers for woodworking are not interchangeable.

1.1.4 Sledgehammers

A sledgehammer is a heavy-duty tool used to drive posts or take on demolition work. You can also use it to break up cast iron or concrete. The head of the sledgehammer is made of high-carbon steel and weighs 2–20 lb (0.9–9 kg).

The shape of the head depends on the job the sledgehammer is designed to do. Sledgehammers can be either long-handled or short-handled. *Figure 7* shows three styles of sledgehammers.

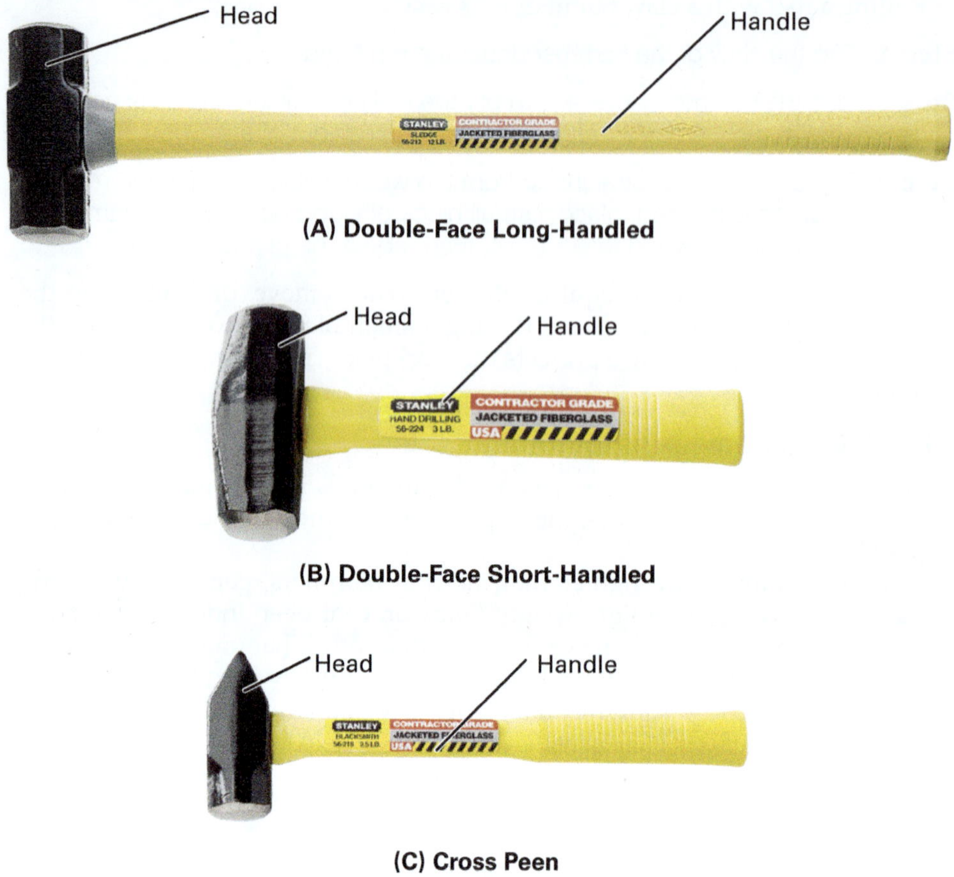

Figure 7 Sledgehammers.
Source: Image property of Stanley Black & Decker. Used with permission

A sledgehammer can cause injury to you or anyone working near you. Follow these steps to use a sledgehammer properly and safely:

Step 1 Wear the appropriate PPE, including gloves, eye protection, and safety shoes.

Step 2 Inspect the sledgehammer to ensure that there are no defects.

Step 3 Be sure there is no one nearby who could be struck on the backswing.

Step 4 Hold the sledgehammer with both hands apart (hand over hand).

Step 5 Stand directly in front of the object to be driven.

Step 6 Lift the sledgehammer straight up above the target.

Step 7 Set the head of the sledgehammer on the target.

Step 8 Begin delivering short blows to the target and gradually increase the length and force of the stroke.

WARNING!

Do not swing a sledgehammer beyond vertical and over your head. Doing so may cause a back injury and significantly reduce control over the tool.

Hold the sledgehammer with both hands. Never use your hands to hold an object while someone else drives with a sledgehammer. Doing so could result in serious injury, such as crushed or broken bones.

1.1.5 Nail Pullers and Staple Removers

An existing wooden structure often needs to be disassembled. Nail pullers and wrecking bars (*Figure 8*) are tools for the task.

The three most common nail-pulling tools are the cat's paw, the chisel bar, and the wrecking bar. Craftworkers use many other names to describe these tools. The cat's paw (*Figure 8* [A]) is a straight steel rod with a curved claw at one end. It is used to pull nails that have been driven flush with the surface of the wood or slightly below it. The cat's paw pulls the nail head just above the surface so another tool can completely pull it out. A cat's paw may lack the leverage needed to completely remove a long nail due to the tight curve at the head.

A chisel bar often has a claw at one end, like a cat's paw, but the other end is flattened into a broad chisel-like head with a long **bevel** (*Figure 8* [B]). The chisel bar can pull a nail like a claw hammer. The chisel-shaped end can be driven into wood joints to split and rip apart the pieces, or to get beneath a nail head.

Bevel: The inclination of a surface that intersects another surface at any angle less than 90°; also used as a verb, meaning to cut or trim an edge to an angled profile.

Wrecking bars (*Figure 8* [C]), also called *ripping bars* or *wonder bars*, have a nail slot at the end to pull nails. An angled, wedge-shaped face at the other end is used like a chisel and a prying tool. They are used for dismantling woodwork, such as demolishing wooden framing, opening crates, and disassembling pallets. Use the straighter end to force wood joints apart and use the claw to pull large nails and spikes. Wrecking bars are generally 12" to 42" (31 centimeters [cm] to 107 cm) long. Longer bars provide more leverage and pulling power.

(A) Cat's Paw

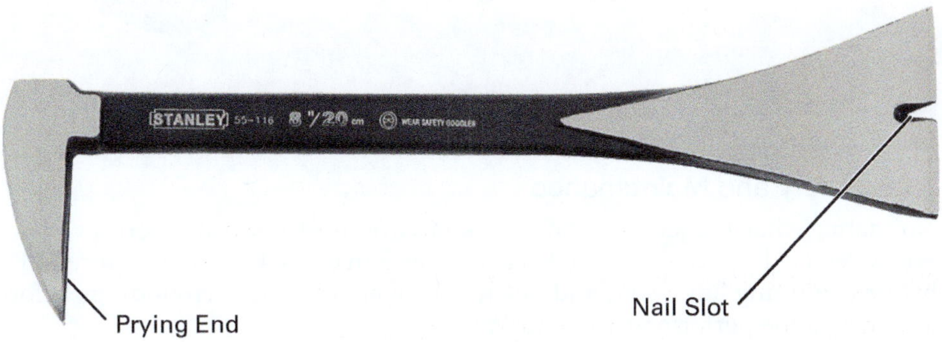

(B) Chisel Bar

(C) Wrecking Bar

Figure 8 Nail pullers and wrecking bars.
Source: Image property of Stanley Black & Decker. Used with permission

Before using a ripping bar or similar tool, make sure that you are wearing the appropriate PPE. Eye protection is extremely important due to the risk of flying debris. Before prying on wood or a nail, ensure that you are in a stable position and prepared for the release of resistance against your effort.

WARNING!

Wear a hard hat, safety glasses, and gloves for protection when using these tools. When manipulating the bar, never pull the bar toward your face. Instead, pull the bar toward your shoulder. Do not push hard away from your body, as this may cause you to lose your balance.

There are also several handheld staple pullers that are designed to remove staples from fencing posts. You could use pliers to complete this task but models like the Caldwells Quick Staple Puller or the Stockade Staplemate allow a worker to quickly detach fasteners one-handed without damaging the wire fence (*Figure 9*).

Figure 9 Fence staple puller.
Source: Shutterstock

1.1.6 Safety and Maintenance

No matter what the job is, a safe work environment and safe work practices are extremely important to avoid accidents and injuries. Wear the required PPE before beginning any work, and always be aware of your surroundings before and during the performance of a task.

Consider the following guidelines for the safe use and maintenance of hammers and similar tools:

- Make sure the handle is set securely in the head of a hammer.
- Replace cracked or broken handles.
- Make sure the face of the hammer is clean.
- Hold the hammer properly. Grasp the handle firmly near the end and hit the nail squarely.
- Do not hit with the cheek or side of the hammerhead unless the hammer has been designed for that purpose.
- Do not use hammers with chipped or otherwise damaged heads.
- Do not use a hammer with a cast head.
- Never strike hammerheads together.

The following are additional considerations for safety and maintenance when working with sledgehammers:

- Replace cracked or broken handles before you use a sledgehammer.
- Use the right amount of force for the job.
- Keep your hands away from the object being driven.
- Never swing until you have checked behind you to make sure you have plenty of clearance from both people and objects.

The following are safety guidelines for ripping and nail pulling:

- Use two hands when ripping to enhance control.
- To prevent injury, be sure the material holding the nail is braced securely before pulling.

Many accidents with prying tools occur when a pry bar slips and the worker falls. Be sure to maintain your balance and keep a firm grip on the tool. This technique also helps reduce damage to materials that must be reused, such as concrete forms.

1.2.0 Chisels and Punches

Chisels are used to cut and shape wood, stone, or metal. Punches are used to indent metal, drive pins, and align holes.

1.2.1 Chisels

A chisel is a metal tool with a sharpened, beveled (sloped) edge. Wood chisels and cold chisels (*Figure 10*) are presented in this section. Both types are made from heat-treated steel to make them harder. Technically, a chisel can cut any softer material than the chisel's steel. However, wood chisels are used exclusively on wood, while cold chisels are used on metal. The term *cold chisel* stems from using the tool with cold metal instead of metal heated by a torch or furnace. Both cold and wood chisels are manufactured in various shapes and sizes to support specific tasks.

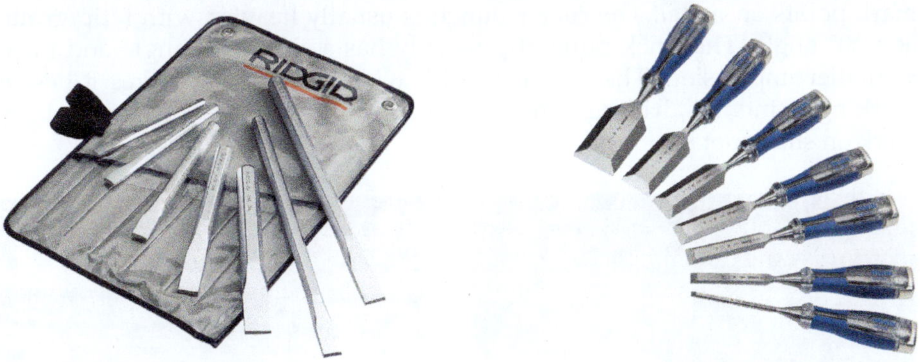

(A) Cold Chisels **(B) Wood Chisels**

Figure 10 Cold and wood chisels.
Sources: Ridge Tool Company (10A); Image property of Stanley Black & Decker. Used with permission (10B)

Cold chisels are designed for hard materials, such as steel and stone. The flat cold chisel is the most common type. Its tip is ground to a 60° angle. Flat cold chisels are often used to cut through bolts and similar hardware that are rusted or damaged and cannot be removed with wrenches.

Before using a cold chisel, it is important to wear the appropriate PPE, especially proper eye protection. The first step to using a cold chisel is to secure the object to be cut in a vise. Place the blade of the chisel at the spot where the material is to be cut. Strike the chisel with a ball peen hammer to force the chisel into and through the material. Repeat as necessary.

There are also several different styles of wood chisel. Bevel-edged chisels are easy to push into corners because of the slightly angled edges. This chisel is commonly used for completing or cleaning up dovetail joints. Stronger wood chisels have a rectangular cross-sectional blade and are typically used for heavier work requirements due to their strength.

Paring chisels are longer and thinner than standard wood chisels. They can reach deeper into long joints to clean them for a precise fit. Note that most paring chisels are designed for hand use only; they should not be struck with a hammer and must be kept very sharp to work properly. If the handle is wooden or plastic with no obvious metal striking surface on the end, it should not be struck.

Before using a wood chisel, it is important to wear the appropriate PPE, especially proper eye protection. First, outline the area to be chiseled out. Next, set the chisel at one end of the outline, with its edge on the cross-grained line and the bevel facing the recess to be made. Lightly strike the chisel head with a mallet. Repeat this process at the other end of the outline, again with the bevel of the chisel blade toward the recess. Then make a series of cuts approximately $\frac{1}{4}$" apart from one end of the recess to the other.

To pare (trim) away the notched wood, hold the chisel bevel-side down to slice inward from the end of the recess (*Figure 11*). Remember that most paring chisels should not be struck with a hammer. However, there are times you can use a hammer to lightly bump the chisel ahead.

1.2.2 Punches

A punch (*Figure 12*) uses the impact of a hammer to indent metal, start a drill bit, drive pins, and align holes in mated parts. Punches are made of hardened and **tempered** steel.

Tempered: Treated with heat to enhance or restore steel hardness.

Three common types of punches are the center punch, the prick punch, and the tapered, or straight, punch. The center and prick punches are both used to mark points on metal. The center punch is usually heavier, with a tip ground at a 60° angle. The prick punch tip usually has a shallower angle and makes a smaller impression. The center punch's wider depression makes it best for marking points for drilling, while the prick punch is best for general layout work on sheet metal.

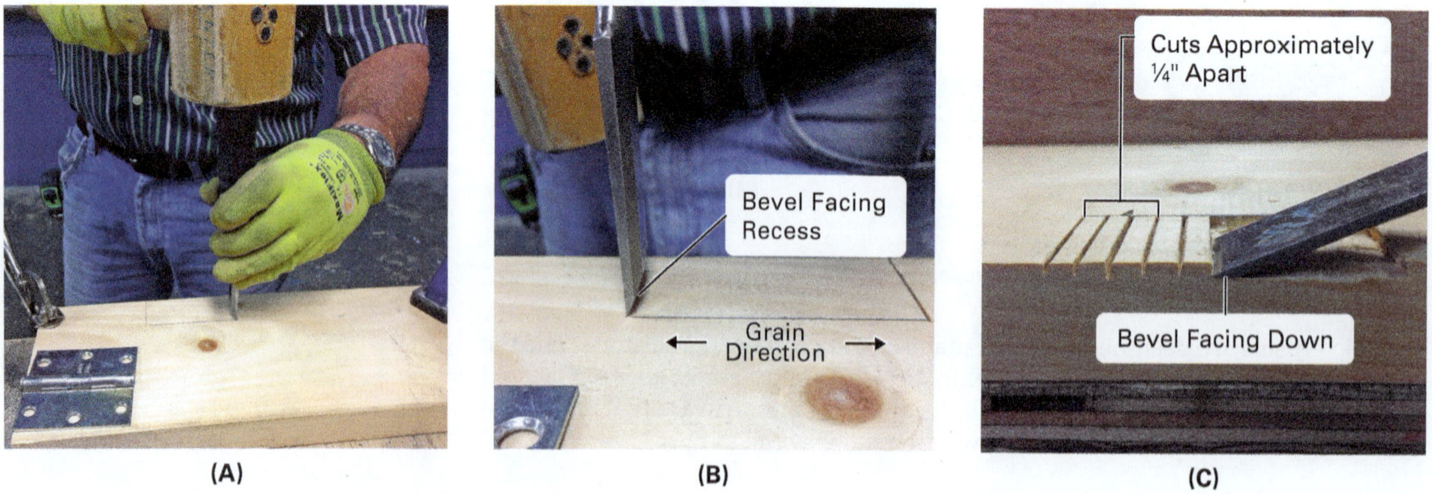

Figure 11 Proper use of a wood chisel.
Source: Courtesy of Cianbro

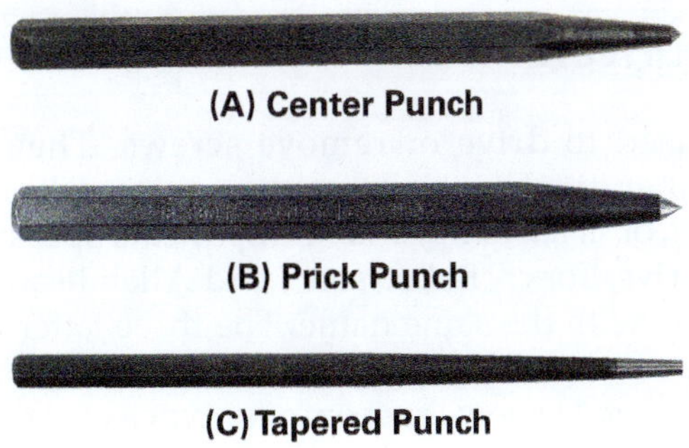

Figure 12 Punches.
Source: Image property of Stanley Black & Decker. Used with permission

The tapered punch (*Figure 12* [C]) has a slight taper along much of its length. Tapered punches typically have a blunt end rather than a point and are used to drive out pins or align the holes of mating parts for assembly.

To use a punch, hold it straight with one hand in the desired position and strike it squarely with a hammer.

1.2.3 Safety and Maintenance

Follow these guidelines for the safe use and proper maintenance of chisels and punches:

- Always wear the appropriate PPE, including work gloves and safety goggles.
- Make sure the wood chisel blade is beveled at a precise 25° angle so it will cut well.
- Ensure a cold chisel blade is beveled at a 60° angle.
- Sharpen the cutting edge of a chisel on an oilstone to produce a keen edge.
- Inspect the punch's point to ensure it is sharp and not damaged. A proper bench grinder can easily sharpen punches.
- Do not use a chisel, punch, or hammer with a head that has become flattened or **mushroomed**, as shown in *Figure 13*. Also shown is a properly dressed head, showing the bevel around the top. Repairing mushroomed heads is best done with a bench grinder.
- Ensure the right tool for the job is used.

Mushroomed: Refers to the head of a tool that is struck or used for striking when the head becomes wider and distorted, like the head of a mushroom.

WARNING!

Striking a chisel or punch with a mushroomed head can cause metal chips to break off. These flying chips can cause serious eye injuries. If a chisel has a mushroomed head, remove the tool from service until it is repaired or replaced according to company policy.

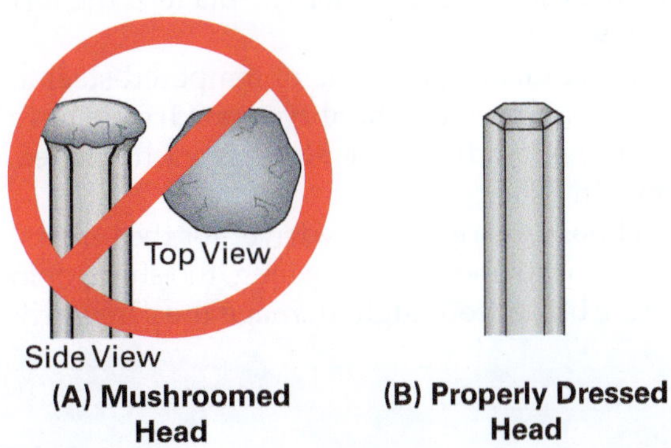

Figure 13 Chisel damage.

1.3.0 Screwdrivers

Screwdrivers are used to drive or remove screws. They are identified by the type of screw they fit. The most common screwdrivers are slotted (also known as *straight*, *flat*, or *standard* tip) and Phillips head. Specialized screwdrivers include the clutch-drive, Torx®, Robertson®, and Allen head, all designed to fit the heads of fasteners with the same name. The three latter heads offer greater grip and allow more **torque** to be applied than slotted or Phillips screws without **stripping** the head. *Figure 14* shows six common types of screws.

Torque: The rotating or twisting force applied to an object such as a fastener or the shaft of a motor.

Stripping: Damaging or distorting the head or threads of a fastener.

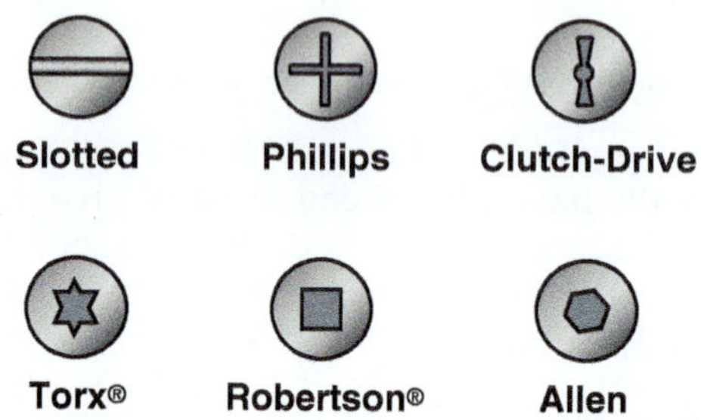

Figure 14 Common screwhead types.

The screwdrivers designed to fit common screws are described here:

- *Slotted* — A very common type of screwdriver, but it also provides the least amount of grip in the screwhead, limiting the amount of torque that can be applied.
- *Phillips* — The most common type of crosshead screwdriver, offering better grip over the slotted design.
- *Clutch-drive* — Has an hourglass-shaped head that provides good grip; common where security from vandalism is desirable.
- *Torx®* — Has a star-shaped tip, widely used for mechanical assembly and repair work; also popular in residential construction.
- *Robertson® (square)* — Has a square drive that provides extra grip to apply torque. Like the Torx®, the Robertson® drive has also become popular in residential construction, for both interior and exterior applications.
- *Allen (hex)* — Works with socket-head screws that have a female, hex-shaped (six-sided) recess. Allen wrenches, which are typically L-shaped steel bars, are generally referred to as wrenches instead of screwdrivers.

Screwdrivers may be color coded according to their size, to make visual identification easier.

To choose the right screwdriver and use it correctly, it helps to know something about screwdriver construction. Each section has a name, as shown in *Figure 15*. The handle is designed to give you a firm grip. The shank is the metal portion between the handle and the blade. The blade is the formed end that fits into the head of a screw.

Industrial screwdriver blades are made of tempered steel to resist wear and prevent bending and breaking. Good quality screwdrivers have shanks that run through the length of the handle. Some also have a hardened tip, resulting in the darker tips shown in *Figure 15*.

It is important to choose the right screwdriver for the screw. The blade should fit snugly into the screwhead and not be too long, too short, or loose (*Figure 16*). If you use the wrong size blade, you might damage the screwdriver or screwhead.

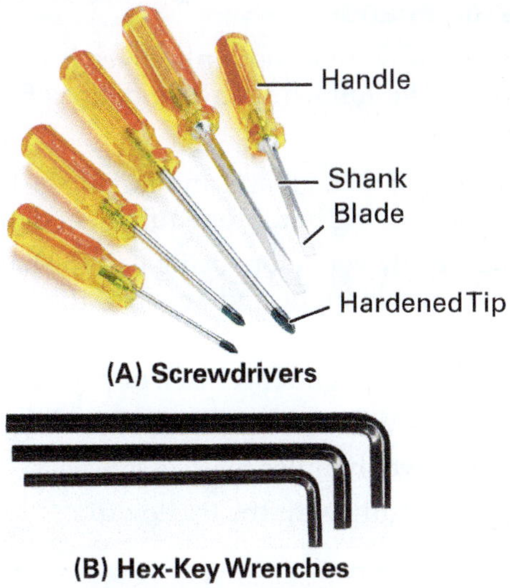

Figure 15 Screwdrivers and hex-key wrenches.
Sources: Ridge Tool Company (15A); Klein Tools (15B)

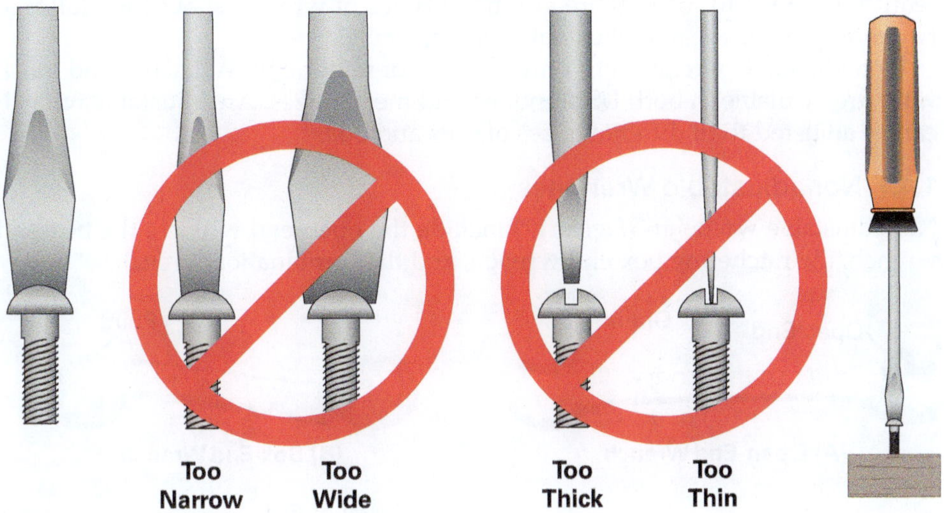

Figure 16 Proper use of a screwdriver.

Note that there are different sizes of each screw type. In addition, there are several styles of screws that are commonly called Phillips screws, but with slight differences. A Reed & Prince (R&P) screw, for example, looks like a Phillips head but comes to a sharper point at the socket base. The base of a Phillips head screw slot is slightly rounded. Using the wrong screwdriver on these two styles will often result in slippage and damage to the screw.

It is very important to use a screwdriver correctly. Using one incorrectly can damage the screwdriver, screwhead, or both. Wear work gloves to protect your hands. Once the right size and type of blade for the screwhead is chosen, position the shank perpendicular (at a right angle) to the work. Apply firm, steady pressure to the screwhead and turn clockwise to tighten, or counterclockwise to loosen. Do not allow the screwdriver bit to rotate inside the screwhead, as it will damage the head.

Screws vs. Nails

Screws hold better than nails in most situations. The spiral ridges, called *threads*, help hold the screw tightly inside the material, unlike the smooth surface of most nails. Self-tapping screws end in a sharp point and have sharper threads. These screws cut their own threads in the material, eliminating the need to drill a starter hole. However, making a small starter hole with a drill helps keep the wood from splitting. This is especially important in cabinetry, trim work, and furniture construction.

1.3.1 Safety and Maintenance

Properly maintaining a screwdriver not only gives it a longer life, but it also helps to prevent injuries. The following are guidelines for the proper use and maintenance of screwdrivers:

- Keep screwdrivers free of dirt, grease, and grit so the blade will not slip out of the screwhead slot, causing injury or equipment damage.
- Visually inspect a screwdriver before using it. If the handle is worn or damaged, or if the tip is not straight and smooth, the screwdriver should be repaired or replaced.
- Never use a screwdriver as a punch, chisel, or pry bar.
- Do not use or lay a screwdriver near electrically energized conductors.
- Do not expose screwdriver handles to excessive heat.
- Carry screwdrivers carefully with the tip down.

1.4.0 Wrenches

Wrenches are used to hold and turn nuts, bolts, pipes, and similar objects that require leverage to turn. There are many types of wrenches, but they fall into two main categories: nonadjustable and adjustable.

Nonadjustable wrenches fit only one size of nut or bolt. Wrenches and hardware are available in both US standard and metric sizes. An adjustable wrench can be adjusted to fit different sizes of nuts and bolts.

1.4.1 Nonadjustable Wrenches

Nonadjustable wrenches (*Figure 17*) include the open-end wrench, the box-end wrench, the ratcheting box-end wrench, and the combination wrench.

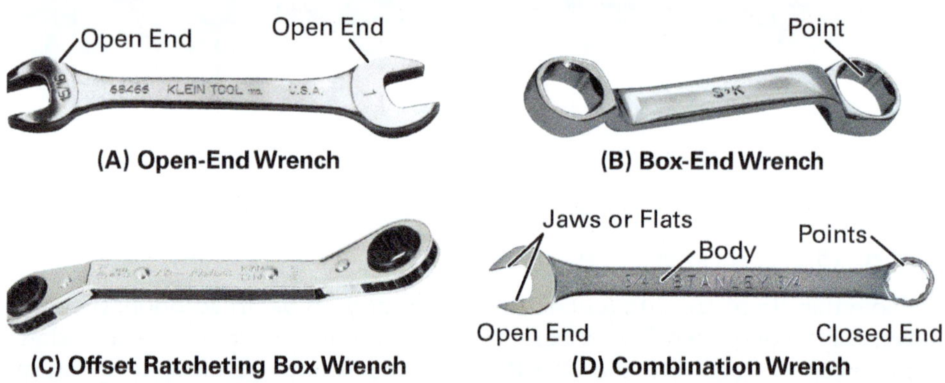

Figure 17 Nonadjustable wrenches.

Sources: Klein Tools (17A); Courtesy of SK Tools USA, LLC (17B); Ritchie Engineering Company, Inc. / YELLOW JACKET (17C); Image property of Stanley Black & Decker. Used with permission (17D)

The open-end wrench is one of the easiest wrenches to use. The open end allows you to position the tool around the fastener easier than a box-end wrench, but with less grip. Most open-end wrenches offer two different sizes in a single wrench, such as $^{7}/_{16}$" and $^{1}/_{2}$". Note that metric equivalents are not offered in this context, as metric wrenches are sized by the millimeter (mm), and there is no functional equivalency between metric and US standard wrenches. These sizes refer to the distance between the **flats**—the straight sides of the wrench opening.

Flats: The flat, straight sides of a wrench opening; also, the sides on a nut or bolt head.

The wrench size must match the distance across the head of a fastener. For example, a ½" bolt head requires a ½" wrench. Note, however, that this does not refer to the size of the bolt itself. Bolts are sized according to the nominal diameter of the threaded shaft, not the head. A 5⁄16" bolt typically has a ½" head, while a 3⁄8" bolt has a 9⁄16" head. You will learn more about hardware later in your training. If you work with hardware often in your chosen craft, their sizes and specifications will quickly be committed to memory.

Box-end wrenches form a continuous perimeter around the head of a fastener. The wrench opening typically has 6 or 12 flats. A hexagonal shape, typical of bolts and nuts, has 6 sides. Wrenches with 6 flats are less likely to round off and damage the flats of a bolt or nut. Although wrenches with 12 flats offer slightly less security, the advantage is that there are 12 possible positions for the wrench to settle onto the fastener instead of just 6. In tight quarters, this is a significant advantage.

Like open-end wrenches, box-end wrenches typically fit two different sizes of hardware. They offer a firmer grip and are safer because they will not slip off the fastener quite so easily. Box-end wrenches with a ratcheting feature are popular, but they are generally shorter in length, reducing the leverage. Ratcheting box-end wrenches are excellent for assembly work but are not the best choice for freeing corroded or tight hardware assemblies.

As the name implies, combination wrenches are a combination of two types of wrenches. One end is open, and the other is a box-end. Each combination wrench generally fits only one size of fastener; both ends are the same size.

Use the correct wrench size for a fastener and learn to pull the wrench toward your body to apply torque. Pushing the wrench away from your body is more likely to result in an injury.

Be sure that the wrench fits snugly on the nut, bolt, or other fastener. If the fit is too loose, the nut or bolt head may slip and round off. A damaged head makes a fastener much more difficult to remove.

1.4.2 Adjustable Wrenches

Adjustable wrenches, often called Crescent® wrenches due to name recognition, are used to loosen or tighten nuts and bolts, as are nonadjustable models. Although a fixed-jaw wrench is usually better and safer, an adjustable wrench offers significant flexibility in a single tool. In many cases, the two types are used together. For example, while a nut is being tightened with a box-end wrench, the adjustable wrench might be used to prevent the bolt head from rotating with the nut.

Adjustable wrenches have one fixed jaw and one movable jaw. The adjusting nut on the wrench interlocks with the teeth in the body of the wrench and positions the movable jaw. These wrenches come in lengths from 4" to 30" with openings several inches wide. Adjustable wrenches include the pipe wrench, spud wrench, and common adjustable wrench (*Figure 18*).

To use an adjustable wrench properly, set the jaws to the rough size for the nut, bolt, or threaded pipe. Ensure that pressure, regardless of whether the hardware is being tightened or loosened, is applied to the fixed jaw rather than the movable jaw. Then fully tighten the jaws on the hardware; it helps to lightly shake the wrench as you use your thumb and forefinger to tighten the adjusting nut. Pull on the wrench as needed to tighten or loosen the component. Pulling toward your body is safer and can develop more power than pushing away from your body (*Figure 19*).

WARNING!

Improperly adjusted jaws could cause the wrench to slip, resulting in a loss of balance and serious injury. Whenever possible, pull the wrench toward you. Pushing the wrench away from your body is more likely to result in a loss of balance and an injury. If you must push on the wrench, brace yourself with your free hand. Gloves should always be worn when working with wrenches.

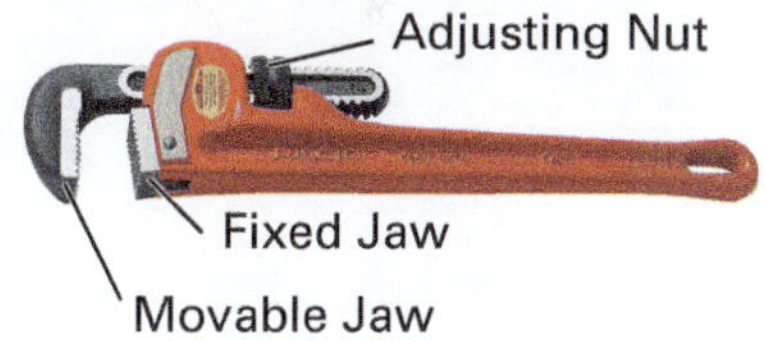

(A) Pipe Wrench

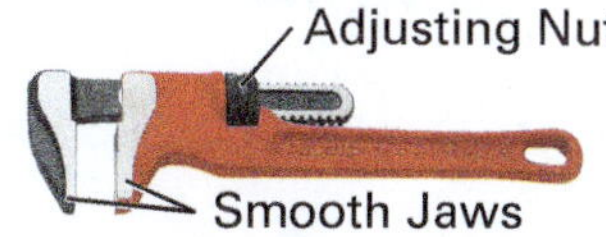

(B) Spud Wrench

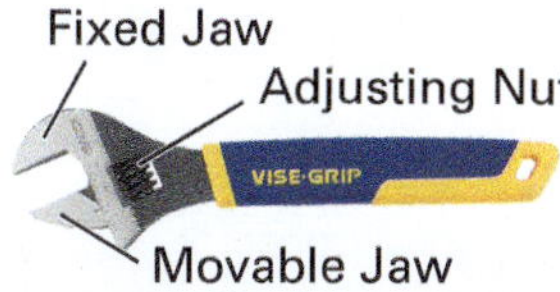

(C) Adjustable Wrench

Figure 18 Adjustable wrenches.
Sources: Ridge Tool Company (18A–18B); Image property of Stanley Black & Decker. Used with permission (18C)

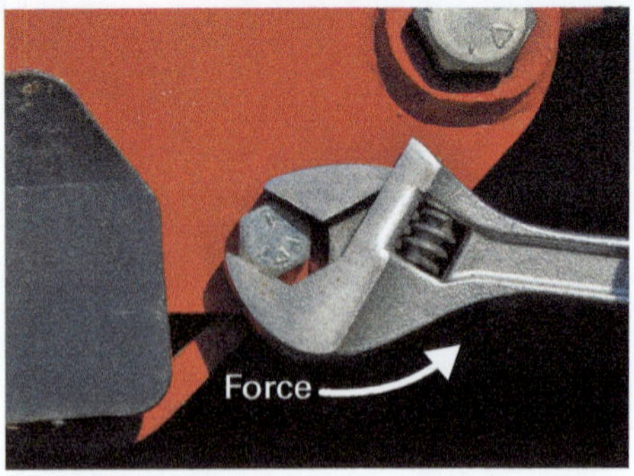

Figure 19 Loosening a fastener with an adjustable wrench.

Pipe wrenches are used to tighten and loosen threaded pipe. The upper jaw of the wrench is adjusted by turning the adjusting nut. Both jaws have serrated teeth for gripping power on smooth surfaces. The jaw is spring-loaded and slightly angled so you can release the grip and reposition the wrench without having to readjust the jaw.

The spud wrench operates in the same manner as a pipe wrench. However, the jaws are smooth, so it has no value in gripping smooth, round pipe. Instead, the wrench is used on large nuts common to plumbing and piping. Large, yet thin nuts might be found on the bottom of a sink drain, for example.

Although some hardware has left-hand threads, most have right-hand threads. Right-hand threads are turned clockwise to tighten them and counterclockwise to loosen them. This is very important to remember, as it is easy to turn a fastener the wrong way when attempting to remove it, which will only make it tighter.

1.4.3 Socket Wrenches

Socket wrenches typically use a ratcheting mechanism that forces the socket to rotate in one direction, while it can freely spin when turned in the opposite direction. This enables the user to quickly tighten or loosen a fastener without removing and refitting the wrench after each turn or when a complete revolution cannot be made because of poor accessibility.

Socket wrench sets include different combinations of sockets and ratcheting handles used to turn the sockets. The handle's ratcheting mechanism attaches to various socket sizes using a short square shaft. The size of the shaft is referred to as the drive size. Common drive sizes are $\frac{1}{4}$", $\frac{3}{8}$", and $\frac{1}{2}$". Note, though, that these sizes refer only to the drive size, not the size of the hardware they fit. Much larger sizes are available—up to $2\frac{1}{2}$" for heavy industrial work. Oddly enough, ratchet drive sizes are the same globally; no metric drive sizes exist. This is reportedly because the ratchet was largely an American invention. However, metric sockets, designed to fit metric hardware, are very common today.

Socket sets are manufactured to fit many US standard and metric hardware sizes. Most sockets (*Figure 20*) have 6 or 12 gripping flats, like box-end wrenches, and come in different lengths. Like box-end wrenches, those with 6 flats offer better grip. Deep sockets, which are longer than standard sockets, allow the socket to accommodate a bolt that protrudes through the nut. A short extension is also shown in *Figure 20*. Extensions are available in various lengths and are used to lengthen the socket and reach into tight places.

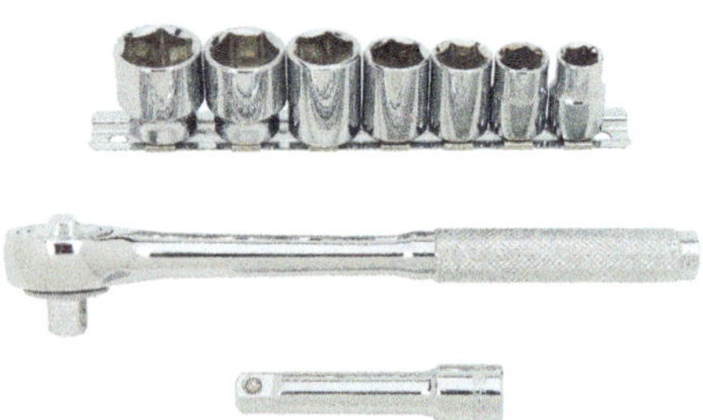

Figure 20 Socket set.
Source: Image property of Stanley Black & Decker. Used with permission

Spud Wrenches

The definition of a spud wrench varies among the crafts. Ridgid® and others refer to the example shown in *Figure 18* as a spud wrench. However, another type of spud wrench, shown here, looks like a common adjustable wrench on one end, while the opposite end comes to a blunt point shaped like a tapered punch. This wrench was developed for structural steel workers who need a wrench to tighten large bolts and nuts. However, another tool is required to align the holes before the bolt can be inserted through two pieces of structural steel or pipe flanges. The spud wrench handles both tasks without changing tools. In some cases, the wrench is not adjustable, but an open-end or box-end style instead.

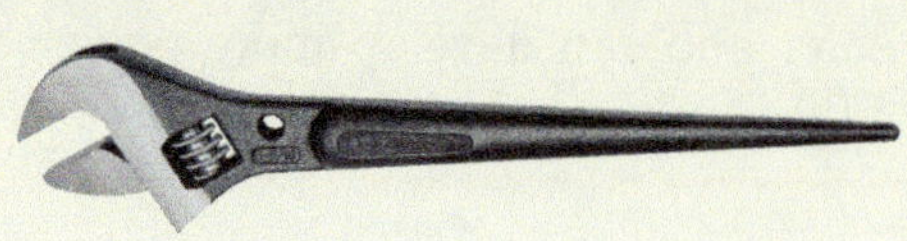

Source: Klein Tools

Socket sets may contain different types of handles. The ratchet handle shown in *Figure 20* has a knob at the top, used to change the direction of free rotation. Ratchet handles come in a variety of lengths. The longer the handle, the more leverage that can be applied. Larger hardware requires more torque, so handles used with the largest sockets are longer and stronger. In addition to ratchet handles, hinged handles are available (*Figure 21*). A hinged socket handle, or a breaker bar, can be used when a ratcheting action is unnecessary. Hinged handles are often longer and more durable, designed to break a difficult fastener loose.

Figure 21 Hinged socket handle, or breaker bar.

Source: Image property of Stanley Black & Decker. Used with permission

First, select a socket that fits the fastener head to use sockets and ratchets properly. Then select a ratchet that matches the socket drive. Push the square end of the socket over the spring-loaded ball on the drive shaft. Place the socket over the fastener and pull the handle in the appropriate direction to turn the fastener. Reverse the direction of free rotation by using the knob or lever located on the head of the ratchet handle.

Torque Wrenches

Torque wrenches are designed to apply specific torque to a fastener. Torque is measured in units of distance by force. In the US standard system, torque values are stated in inch-pounds or foot-pounds. Inches and feet represent the distance, and pounds represent the force. In the metric system, torque is measured in Newton-meters. Newtons measure force, and meters represent the distance. One inch-pound is equal to 0.113 Newton-meters. Metric system users commonly refer to torque as *moment*.

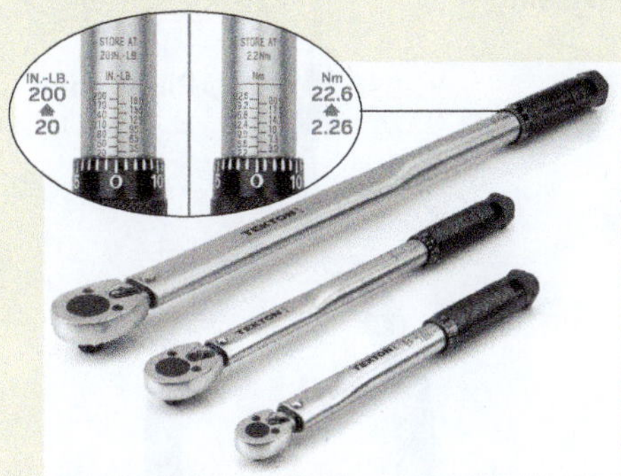

Source: TEKTON

1.4.4 Safety and Maintenance

Note the following safety and maintenance guidelines for working with wrenches:

- Pull the wrench toward your body, not toward your face or away from your body. Leaning into the wrench and pushing away can result in serious injury if the wrench slips off.
- Check the condition of the handle and jaw working surfaces before use.
- Keep adjustable wrenches clean. Do not allow mud or grease to clog the adjusting screw and slide.
- Do not use a wrench as a hammer.
- Do not use any wrench beyond its capacity. Never use a **cheater bar** or pipe on a wrench. This could snap the tool or break the head off the bolt.
- Thoroughly clean sockets and ratchets after each use. Grease and grit can collect inside the socket and on the ratchet shaft, interfering with the wrench's operation.

Cheater bar: A length of pipe or rod used to increase the leverage on a wrench or similar tool. The increased leverage provided by a cheater bar often exceeds the tool's capacity and strength, resulting in a damaged tool and workpiece.

1.5.0 Pliers and Wire Cutters

Pliers are hinged tools that hold, cut, and bend wire and soft metals. Pliers should not be used for tightening or loosening hardware such as bolts and nuts. If any significant torque is applied, they will round off the corners of the bolt or nut. Wrenches may no longer fit the damaged hardware properly, creating a larger problem.

High-quality pliers are made of hardened tool steel and come in many different styles, depending on their use (*Figure 22*). The following types of pliers are the most common:

- Slip-joint pliers
- Long-nose pliers
- Lineman's pliers
- Tongue-and-groove pliers
- Locking pliers

It is important to note that these styles have many variations, and some have alternate names.

Tool Lanyards

Tools are as important when working at heights as on the ground. However, a dropped tool becomes a serious hazard when working at heights. If the tool has moving parts, such as a battery-powered drill, the fall may destroy it. Almost any dropped tool can cause serious injury or even death, especially when the impact is unexpected.

Many companies have introduced tool lanyards specifically designed for work in an elevated environment. Tethering tools to the tool belt or wrist when working at height can prevent injuries, protect the tools, and eliminate the time lost in retrieving a dropped tool.

Source: Image property of Stanley Black & Decker. Used with permission

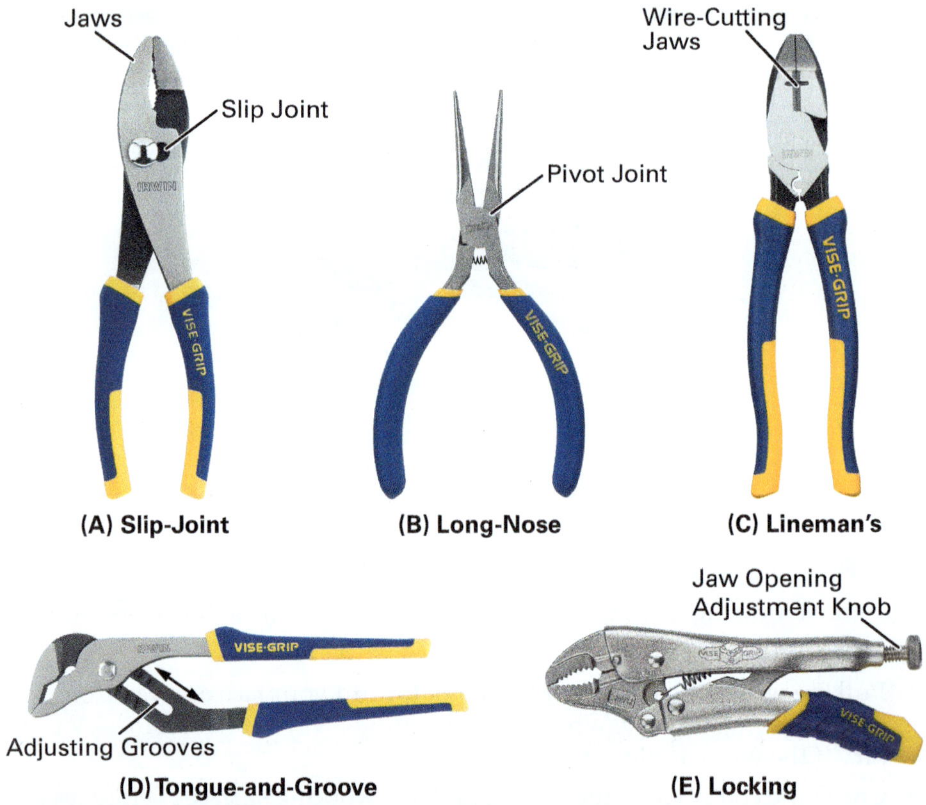

Figure 22 Common types of pliers.
Source: Image property of Stanley Black & Decker. Used with permission

1.5.1 Slip-Joint Pliers

Slip-joint pliers are used to hold and bend wire and to grip and hold objects during assembly. They have adjustable jaws with two or three possible positions. To use slip-joint pliers properly, place the jaws on the object to be held and squeeze the handles until the pliers grip the object. Adjust the jaw position to improve the grip if necessary.

1.5.2 Long-Nose Pliers

Long-nose pliers, also known as *needle-nose pliers*, are used to get into tight places where other pliers cannot reach or to grip parts too small to hold with fingers. They are also useful for bending angles or loops in electrical wire. Most have a set of wire-cutting jaws near the pivot. Like many other types of pliers, long-nose pliers are available with spring openers—a spring between the handles that keeps them apart; the jaws are open until pressure is applied to the handles. Otherwise, the tips of your fingers must push the handles apart to open the jaws.

1.5.3 Lineman's Pliers

Lineman's pliers, also known as *side cutters*, have wider, heavier jaws than slip-joint pliers. They are typically used to cut electrical wire and to hold workpieces. The wedged jaws reduce the chance that wires will slip, and the hook bend at the base of both handles allows for a better grip.

To properly cut wire with lineman's pliers, always point the loose end of the wire down. Squeeze the handles as you cut at a right angle to the wire.

1.5.4 Tongue-and-Groove Pliers

Tongue-and-groove pliers have serrated teeth that grip flat, square, round, or hexagonal objects. Slipping the curved tongue into the desired groove lets you set the jaws in several positions. Large tongue-and-groove pliers are often used to grip pipes because the longer handles provide good leverage.

They are available with several basic jaw designs: straight, smooth, V-shaped, or curved. V-shaped or curved jaws work best with round objects. Straight jaws, as shown on the pliers in *Figure 23*, are best for working with flat material. Smooth jaws are chosen when serrated jaws may damage a delicate surface, but the grip on the object is compromised due to the lack of teeth.

Use the pliers by opening them to the widest position and placing the jaws on the object. Determine which jaw position provides the best and most comfortable grip, and then squeeze the handles until the pliers grip the object.

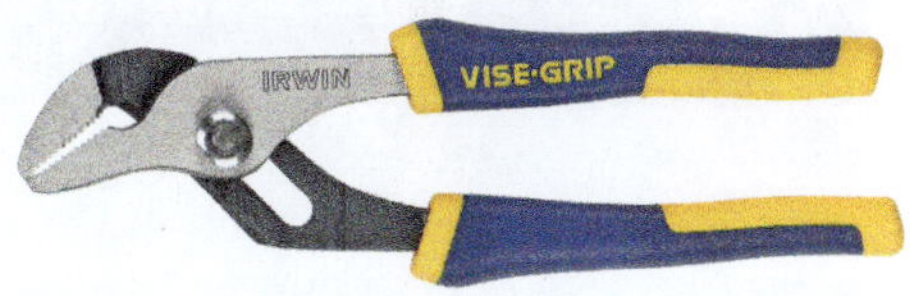

Figure 23 Straight jaw tongue-and-groove pliers.
Source: Image property of Stanley Black & Decker. Used with permission

1.5.5 Locking Pliers

Locking pliers clamp firmly onto objects like a vise. Due to brand recognition, they are often called Vise-Grips®. They are especially handy for holding metal components together for welding or assembly and gripping a damaged nut or bolt.

A knurled knob on the handle is used to adjust jaw spacing. Close the handles to lock the pliers and release the lock by depressing the lever. Many different styles of locking pliers have been developed for unique applications. In *Figure 24*, large jaw locking pliers are used in a fabrication shop. Note the unusual construction of the tool.

To use locking pliers, first place the jaws on the object to be held. Turn the adjusting screw in the handle until the jaws contact the workpiece, and then squeeze the handles together to lock the pliers. Readjust the jaws if the locking tension is too loose or too tight. Squeeze the release lever to unlock and remove the pliers.

Figure 24 Using large jaw locking pliers while operating a plasma arc cutter.
Source: Photo courtesy of The Lincoln Electric Company, Cleveland, OH, U.S.A.

1.5.6 Safety and Maintenance

Pliers may not seem to be a dangerous tool, but they can cause an injury if misused. Remember the following guidelines when using pliers:

- Wear the appropriate PPE, especially if you cut wire.
- Hold pliers close to the end of the handles to avoid pinching your fingers in the hinge.
- Do not extend the length of the handles for greater leverage; use a larger pair of pliers instead.
- Hold the scrap end of cut wire to avoid flying metal bits when you cut.
- Always cut at right angles. Do not rock the pliers from side to side or bend the wire back and forth against the cutting blades.
- Lubricate pliers regularly to prevent rust and to keep them working smoothly.
- Never use pliers around energized electrical circuits. Although the handles may be plastic-coated, they are not always rated to prevent electrical shock.
- Do not expose plier handles to excessive heat.
- Do not use pliers to turn nuts or bolts. They are poor substitutes for wrenches.
- Never use pliers as hammers.

1.0.0 Section Review

1. The type of hammer that is the *best* choice for pulling a nail is the _____.
   a. sledgehammer
   b. drywall hammer
   c. curved claw hammer
   d. ball peen hammer

2. What is the bevel angle of a flat cold chisel?
   a. 60°
   b. 50°
   c. 40°
   d. 30°

3. The type of screwhead shown in *Figure SR01* is the _____.
   a. Torx®
   b. Robertson®
   c. Phillips
   d. clutch-drive

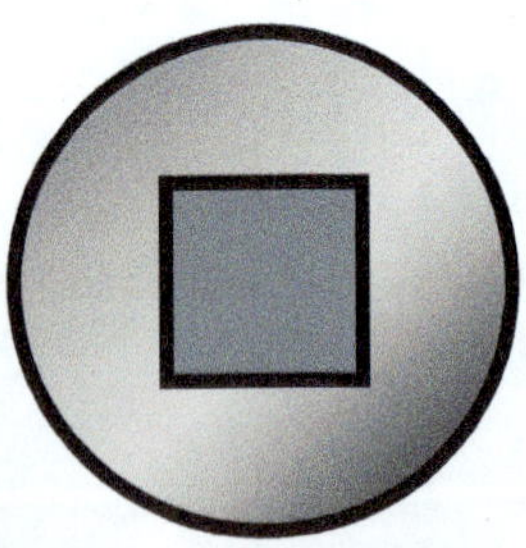

Figure SR01

4. When using an adjustable wrench, pressure to tighten or loosen should *always* be applied _____.
   a. clockwise
   b. counterclockwise
   c. to the fixed jaw
   d. to the adjustable jaw

5. Locking pliers are also commonly referred to as _____.
   a. tongue-and-groove pliers
   b. clutch grips
   c. Robertsons®
   d. Vise-Grips®

2.0.0 Measurement and Layout Tools

Objective

Identify common measurement and layout tools and describe how to use them.

a. Explain how to use a variety of measuring tools.

b. Define various types of levels and layout tools and indicate how they are used.

Performance Task

1. Inspect and demonstrate the safe and proper use of the following hand tools:
   - Tape measures
   - Levels
   - Squares

Knowing how to collect precise measurements, create straight vertical lines, and level accurately are foundational skills for a craftworker. Incorrect measurements and crooked components have adverse effects on the job. It is important to ensure that measurements are precise, and components are set **plumb** and level.

Plumb: Perfectly vertical, meaning a surface is at a right angle (perpendicular) to the horizon or other surface used for comparison.

2.1.0 Rules and Other Measuring Tools

This section reviews measuring tools and guidelines for their use. Craftworkers most often use four basic types of measuring tools:

- Flat steel rule
- Tape measure
- Wooden folding rule
- Laser measuring tool

Several factors influence the choice of a measuring tool, including accuracy, ease of use, durability, and readability. Of course, the chosen tool must also be suitable for the measurement length.

2.1.1 Steel Rule

The flat steel rule (*Figure 25*) is a simple measuring tool. Flat steel rules are usually 6" or 12" long, with metric versions available at 15 cm and 30 cm. Longer lengths, generally for shop use, are 1 yard (yd) or 1 meter (m) long. Steel rules can be thin and flexible or thicker and rigid. While steel rules are accurate measuring tools, they can also be used as straightedges for layout work.

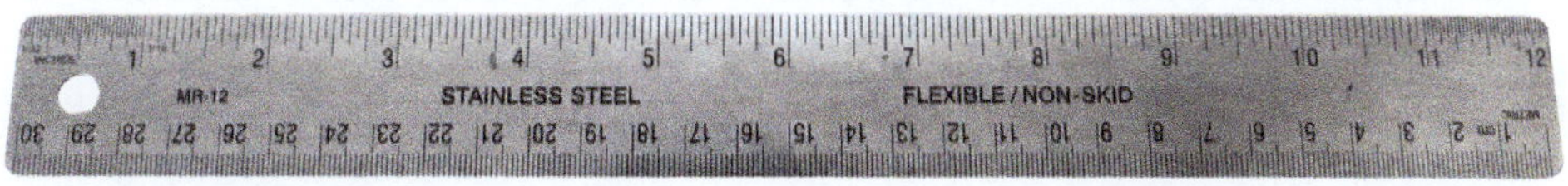

Figure 25 Steel rule.

2.1.2 Tape Measure

The blade of a tape measure (*Figure 26*), or measuring tape, is usually marked in $^1/_{16}$" or 1 mm increments, although some tapes offer smaller US tandard increments. For versatility, a tape measure may include both US standard and metric markings. Tape measures are also available with special markings for residential and light commercial construction wall framing.

Figure 26 Tape measure.
Source: Image property of Stanley Black & Decker. Used with permission

Concave: Having a shape that curves inward, such as the interior of a circle. *Convex* is an antonym, meaning a shape that curves outward.

The **concave** (curved, as viewed from the top) shape of a tape measure blade is designed to strengthen the blade when extended, helping it remain rigid. Once the blade tip is secure and the proper measurement is established, move the blade edge until it lies flat on the surface, then mark the material. This method makes it easier to read the tape and mark the material accurately.

NCCER Module 93103, *Introduction to Construction Math*, explains how to effectively use and read a tape measure. Remember that the hook on the end of the tape is designed to slide in and out slightly, allowing for accurate inside and outside measurements. The movement compensates for the thickness of the hook. If the hook becomes damaged or distorted, it is best to replace the tape measure.

Remember to check and add the dimensions stamped or printed on the housing when taking an inside measurement. This represents the exact width of the housing.

2.1.3 Wooden Folding Rule

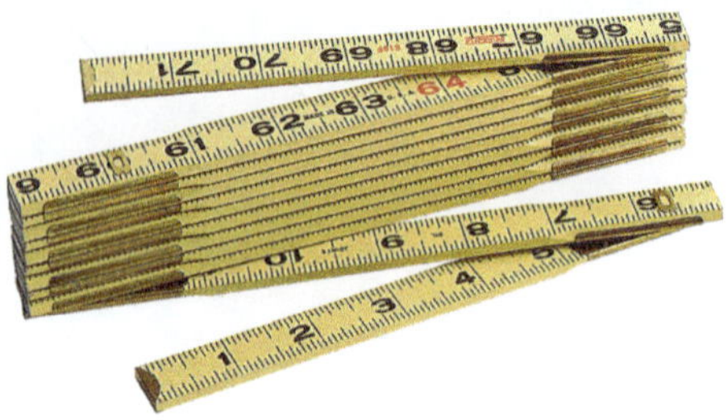

Figure 27 Wooden folding rule.
Source: Ridge Tool Company

A wooden folding rule (*Figure 27*) is usually marked in sixteenths of an inch on both edges of each side. Folding rules with US standard markings come in 6' and 8' lengths, with 1 m and 2 m models available for metric users. Because it is rigid, a folding rule is better than a cloth or steel tape in some situations, such as vertical measurement. Like a tape measure, the folding rule can also have special marks at the 16", 19.2", and 24" increments to make wall framing easier. These markings will differ on metric versions due to differences in construction standards.

2.1.4 Laser Measuring Tools

Figure 28 Laser measuring tool.
Source: Image property of Stanley Black & Decker. Used with permission

A laser measuring tool (*Figure 28*) is a battery-powered, electronic version of a tape measure. The tool works by pointing it at a specific object and pressing a button. When the button is pressed, a laser beam shoots toward the object and returns to the instrument. The travel time of the beam determines the distance, which is displayed on the screen. Measurements are reported very quickly. Laser measuring tools are designed to report and record in both US standard and metric units.

These tools are precision instruments that are handled carefully and stored appropriately. The following are some of the advantages a laser measuring tool has over a traditional tape measure:

- Long vertical measurements can be taken from the ground.
- Longer measurements can be taken. Some laser tools can measure up to 600' (183 m). A special target plate may be required for long distances where more powerful lasers are required.
- Some laser measuring tools have built-in options that make length-related calculations (such as addition, subtraction, and area) easier to perform.
- Measurements can be electronically stored on the tool so that measurements can be written down immediately after the task is completed. In addition, many are equipped with Bluetooth® wireless technology so that measurements can be sent to a smartphone or tablet.
- Some laser measuring tools can have built-in electronic levels or spirit levels.

The disadvantages of a laser measuring tool may include accuracy, especially over long distances. This is not to say that they fail to meet their specifications for accuracy, but some craftworkers struggle to place their trust in electronic measurement, regardless of reported accuracy. In construction, absolute precision is rarely required for very long measurements. It may also be difficult to take accurate measurements in very hot or smoky environments because the beam bounces from particle to particle in the atmosphere, causing the light to scatter.

Note that highly reflective targets may also interfere with accurate measurement. In these cases, apply a piece of tape or similar material with a matte finish to the target.

2.1.5 Safety and Maintenance

Follow these guidelines to ensure measuring tools are safely and properly used:

- Occasionally apply a few drops of lubricant on the spring joints of a wooden folding rule and steel tape.
- Wipe moisture from steel tapes to help prevent rust in areas where the coating has become worn or damaged.
- Do not kink or twist steel tapes. Once a tape is significantly distorted, the damage is permanent.
- Do not use a steel measuring device near energized electrical components.
- Do not allow laser measuring tools to get wet or be exposed to excessive heat.

2.2.0 Levels and Layout Tools

A level is a tool used to determine how close a surface is to a true horizontal plane (level) or vertical plane (plumb). Layout tools include the carpenter's square and combination square, which are also introduced here. Some squares are designed specifically for a given craft. For example, although a carpenter's square and a pipefitter's square may appear identical from a distance, they have different formulas and other information important to their craft. Other layout tools include chalk lines and plumb bobs.

2.2.1 Spirit Levels

The standard spirit level (*Figure 29*) is the most common leveling instrument in the construction trade. The spirit level got its name because its vials are often filled with colored ethanol, a type of alcohol, and alcohol is often referred to as spirits. Alcohol is used because it does not freeze like water and moves more freely inside the vial.

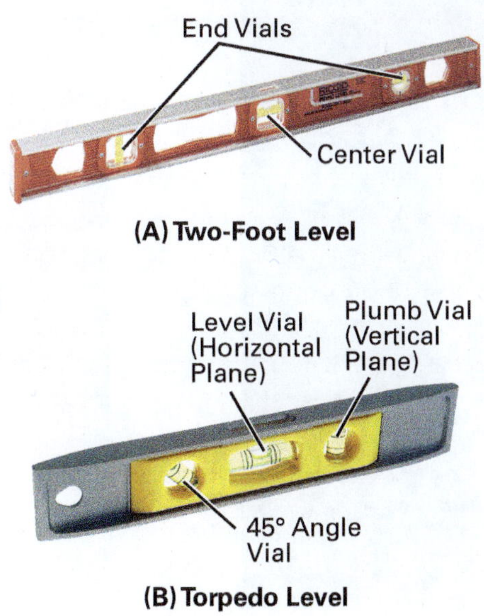

Figure 29 Spirit levels.
Sources: Ridge Tool Company (29A); Image property of Stanley Black & Decker. Used with permission (29B)

Most spirit levels are made of tough, lightweight metals such as magnesium or aluminum, or types of plastic that can reliably maintain their true form. They typically have three vials filled with alcohol. The center vial checks for level, and the two end vials are used to check for plumb. Some spirit levels, like the torpedo level shown in *Figure 29* (B), will also show a true 45° angle. Torpedo levels usually have a magnetic and/or V-shaped base, making it useful for pipefitting work. Spirit levels are available in a variety of lengths.

The limited amount of liquid in each vial is intentional, so there is always a bubble. When the bubble is centered precisely between the lines on the vial, the surface is either level (perfectly horizontal) or plumb (perfectly vertical) as shown in *Figure 30*. Before placing a level against a surface, be sure no debris on the surface will prevent an accurate measurement.

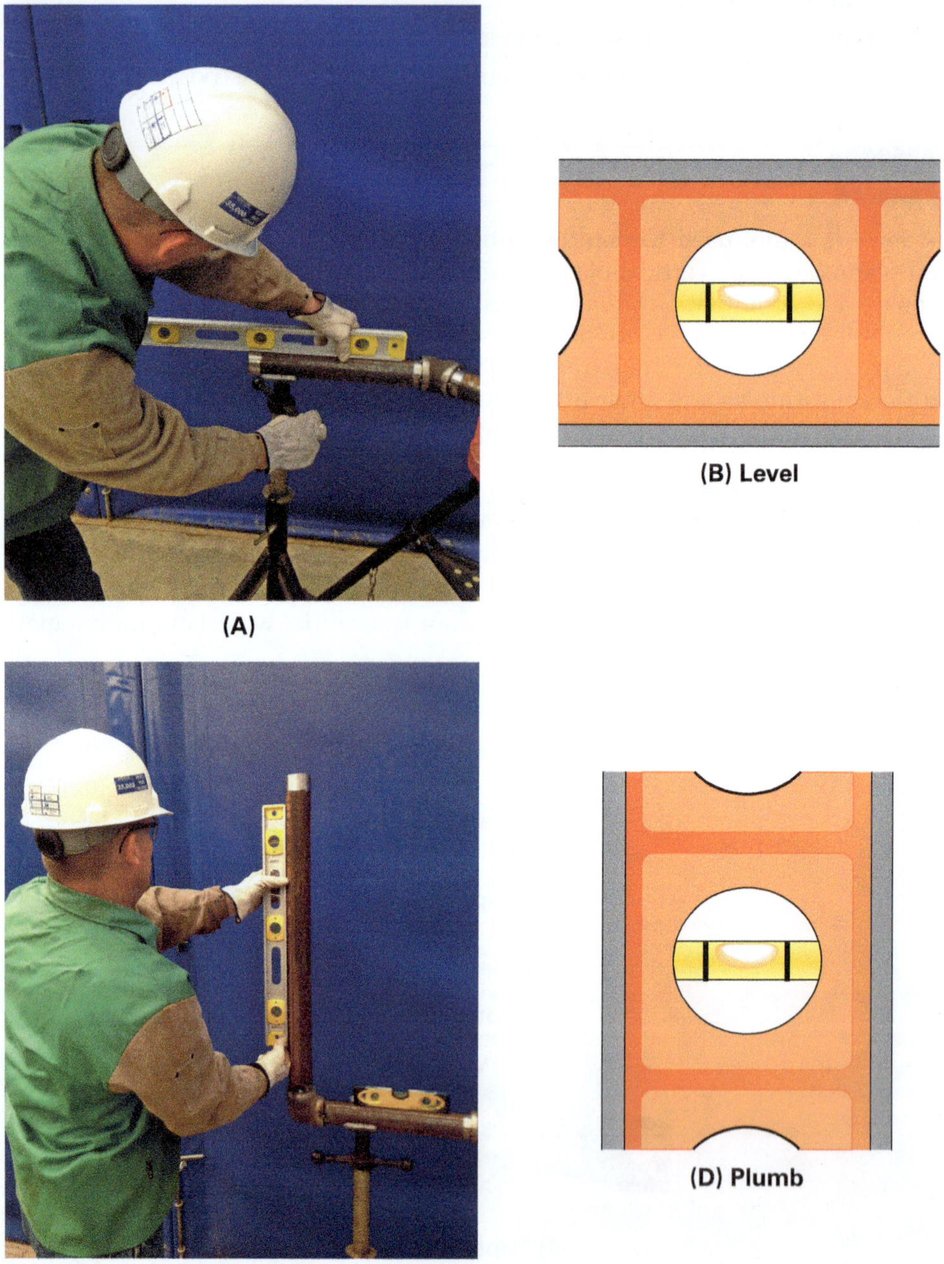

(A) **(B) Level**

(C) **(D) Plumb**

Figure 30 Levels showing level and plumb.
Source: Courtesy of Cianbro (30A, 30C)

When setting posts, you must be sure they are plumb in two different planes. A post level (*Figure 31*) is designed to be strapped to a post, providing a view of the condition in both planes. Many also have magnets that allow them to be used on steel posts without a strap.

Figure 31 Post level.
Source: Image property of Stanley Black & Decker. Used with permission

2.2.2 Digital Levels

Digital levels provide a digital readout of degrees of slope and rate of rise to lay out stairs and roofs or determine the fall of a drain line. Some may also use a simulated bubble display for a more traditional feel. Due to their versatility, digital levels, like the one shown in *Figure 32*, are becoming more common on construction sites and farms. Always handle and store a digital level carefully to avoid damaging it, following the manufacturer's recommendations for its use and care.

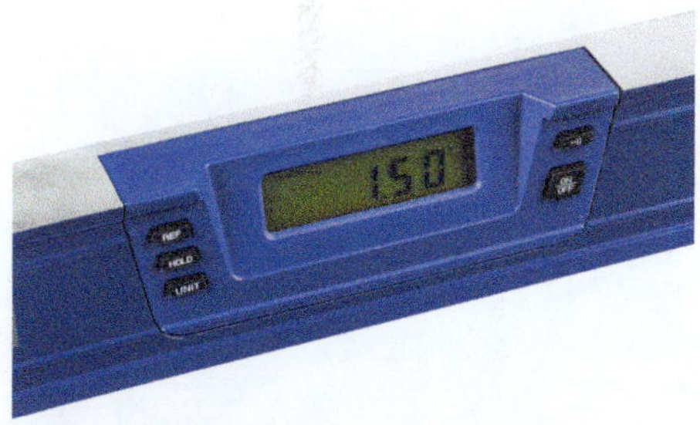

Figure 32 Digital level display.
Source: Image property of Stanley Black & Decker. Used with permission

2.2.3 Laser Levels

With a laser level (*Figure 33*), a single worker can accurately establish plumb, level, or **square** measurements. Laser levels help establish foundation elevations and drainage slopes, square framing, align plumbing and electrical lines, and hang suspended ceilings. A laser level may be mounted on a tripod, fastened onto pipes or framing studs, or suspended from a ceiling. They are used for leveling over a significant distance well beyond the reach of a conventional level.

Square: Used as an adjective, to have the shape of a square where two components are at precise 90° angles (perpendicular) to each other.

Laser levels come in a variety of sizes and weights. They are generally housed in sturdy casings designed to withstand the construction environment. Always handle and store laser levels carefully, following the manufacturer's recommendations for care and use. Although they are constructed for the environment, they are instruments that require care in their use and movement from place to place.

Figure 33 Laser level.
Source: Image property of Stanley Black & Decker. Used with permission

WARNING!

Never look directly into a laser beam. It can permanently damage your eyes and impair your vision.

2.2.4 Level Safety and Maintenance

All levels are precision instruments that must be handled with care. Unless the user looks directly into the laser beam, there is little risk of personal injury when working with this tool. Remember these guidelines when working with levels:

- Never look directly into a laser beam or its source. Serious and permanent eye injury can result.
- Replace a level if a crack or break appears in any of the vials, or if they appear loose and insecure.
- Keep levels clean and dry.
- Do not bend or otherwise apply excessive pressure to a level.
- Do not drop or throw any level.

2.2.5 Squares

Rafter: One of a series of beams that extend from the perimeter of a building to the high point of a roof, providing structural support for the roof above.

Squares (*Figure 34*) are used to mark, measure, and check if a joint is square. The type of square you use depends on your job type and preference. Common squares are the carpenter's square, the **rafter** angle square (also called a *speed square*), the try square, and the combination square.

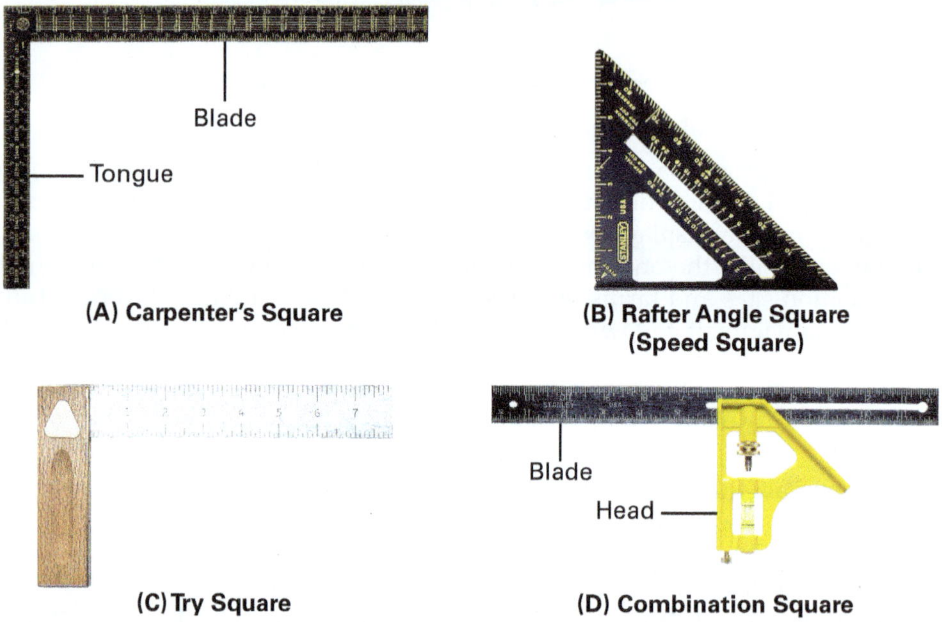

Figure 34 Types of squares.
Source: Image property of Stanley Black & Decker. Used with permission

The carpenter's square, also called a *framing square*, is L-shaped and is used for squaring up work sections such as baseplates and walls to ensure components are perpendicular to each other. As shown in *Figure 35*, it is also used to mark square cutting and matching lines, especially when working with wide framing lumber. The carpenter's square has a 24" blade and a 16" tongue, forming a right angle. Metric versions have a 600 mm blade with a 400 mm tongue. The blade and tongue are marked with inches and fractions of an inch, or metric centimeters and millimeters. Both the blade and tongue can be used as a measuring device or straightedge. Tables and formulas relevant to carpentry are usually printed on the blade as a convenience.

Figure 35 Marking a cutting line.
Source: Zachary McNaughton, River Valley Technical Center

The carpenter's square is virtually identical to a pipefitter's square, except for one feature. The carpenter's square has formulas and other information relevant to carpentry work. The pipefitter's square has formulas and other helpful information relevant to the pipefitting craft.

The rafter angle square is another popular square in carpentry. Frequently made of cast aluminum or tough plastics, it combines a protractor, try square, and framing square. It is marked with degree gradations for a fast, easy rafter layout. The square is small, so it is easy to store and carry.

The try square has a fixed 90° angle and is used mainly for woodworking. Like other squares, it can be used to lay out 90° cutting lines, check a joint to ensure it is square, and check if a piece of lumber is warped or cupped (bowed).

The combination square has a ruled blade that slides back and forth in a head. The position of the ruled blade can be changed as needed. The head is permanently formed into 45° and 90° angles. Many combination squares also contain a small spirit level and a metal scribe for marking metal. Because of its versatility, the combination square is one of the most useful squares for layout work. Good combination squares have all-metal parts, with the head often made from cast iron and the blade from tempered or stainless steel.

To mark a 90° angle using a combination square (*Figure 36*), set the blade so that it is long enough to reach fully across the lumber or metal. Position the front edge of the head firmly against the material to be marked. Start at the edge of the material and follow the blade with a pencil to make the mark.

To mark a 45° angle using a combination square (*Figure 37*), again set the blade so that it is long enough to reach fully across the lumber or metal. Position the back of the head firmly against the edge of the material to be marked and use the pencil to follow the blade at a 45° angle to make the mark.

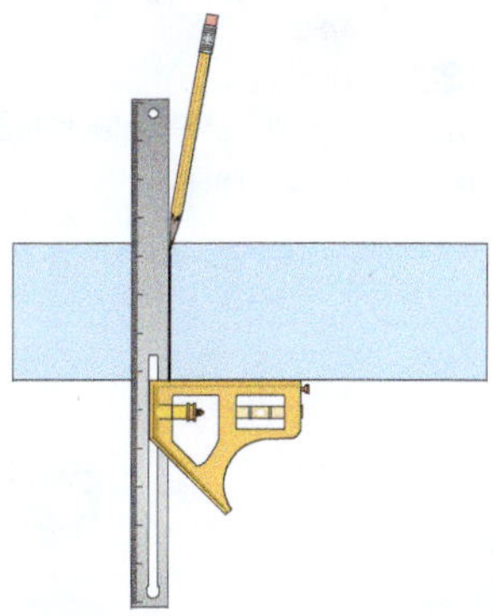

Figure 36 Using a combination square to mark a 90° angle.

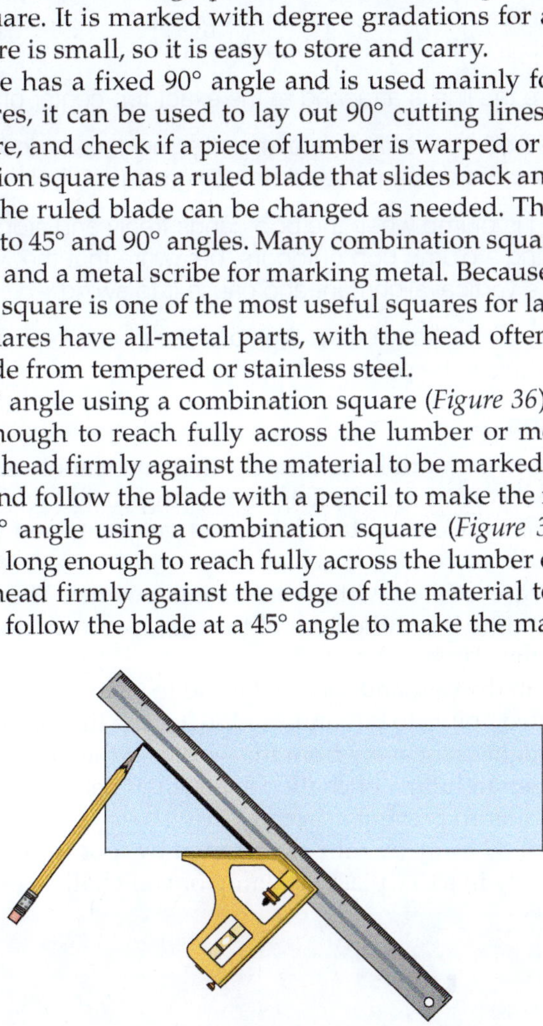

Figure 37 Using a combination square to mark a 45° angle.

2.2.6 Use and Maintenance

Follow these guidelines for using squares:

- Do not use a square for something it was not designed for, especially prying or hammering. It is especially important to protect the edges from damage and avoid bending the blades.
- Keep squares dry to prevent rusting. For protection, apply a light coat of silicone spray or lubricant to the blade, and occasionally clean and lubricate the moving parts of combination squares.
- Do not drop or strike the square hard enough to change the angle between the blade and the head. A try square can be more easily damaged by a drop than a carpenter's square because of the differences in construction.

A square can be checked to ensure that it has not become distorted. To scribe a line, use it on a straight and true piece of lumber. Then, turn the square over and scribe a line in the same location. The two lines should be identical and true to each other.

2.2.7 Plumb Bob

Figure 38 Plumb bob.
Source: Image property of Stanley Black & Decker. Used with permission

The plumb bob is a pointed weight attached to a string (*Figure 38*). It uses the force of gravity to make the string hang perfectly vertical or plumb. Heavier weights are generally better for longer drops to ensure the plumb bob can hold its position.

The string is plumb when the bob hangs freely and is motionless. For example, a plumb bob can ensure a wall or a doorjamb is vertical. When installing a support post under a beam, a plumb bob can show the point on the floor directly under the beam section requiring support.

Follow these steps to use a plumb bob properly:

Step 1 Make sure the line is attached at the exact top center of the plumb bob.

Step 2 Hang the bob from a horizontal member, such as a doorjamb, joist, or beam. To avoid damage, avoid dropping a plumb bob on its point.

Step 3 The string is plumb when the bob hangs freely and stops swinging. When using a plumb bob outdoors, be aware that the wind may blow it out of true vertical alignment and cause it to swing.

Step 4 Mark the point directly below the tip of the plumb bob. This point is precisely below the point of attachment above.

2.2.8 Chalk Lines

(A) **(B)**

Figure 39 Chalk line and chalk.
Source: Image property of Stanley Black & Decker. Used with permission

Chalk lines are used to mark long, straight lines on smooth surfaces. A chalk line is a simple tool consisting of a length of string coated with chalk on a reel (*Figure 39*). The line is stretched taut between two points and then snapped to mark a temporary line.

The chalk line case is filled with colored chalk powder. The line is automatically chalked each time it is pulled out of the case. Some models can also serve as a lightweight plumb bob.

Pull the line from the case and secure the end to snap a chalk line. Stretch the line between the two points to be connected. After the line has been pulled taut, pull the string straight up or away from the work and then release it. This marks the surface with a straight line of chalk (*Figure 40*). If necessary, spray the chalk line with clear lacquer to prevent it from wearing away quickly. Otherwise, the chalk is easily washed away by rain or distorted by foot traffic. Store the chalk line and chalk supply in a dry place, as damp or wet chalk is unusable.

Figure 40 Using a chalk line.
Source: Image property of Stanley Black & Decker. Used with permission

2.0.0 Section Review

1. Folding rules with US standard markings come in _____.
   a. 2' and 4' lengths
   b. 2' and 6' lengths
   c. 3' and 8' lengths
   d. 6' and 8' lengths

2. The carpenter's square is also called a _____.
   a. framing square
   b. speed square
   c. try square
   d. combination square

3.0.0 Other Common Hand Tools

Objective

Identify and describe other hand tools common to shops and jobsites.

a. Differentiate between various handsaws and their designated applications.
b. Identify common clamp designs.
c. Explain how different utility knives are used with various materials.
d. Describe shovels and picks and the tasks for which each one is best suited.

Performance Task

1. Inspect and demonstrate the safe and proper use of the following hand tools:
   - Handsaws
   - Clamps
   - Files
   - Utility knives
   - Shovels

Many different saws are on the market. Some have extremely specific uses, while others are more versatile. Handsaws were used before 1500 BC, and they remain on the jobsite today. However, smaller, lighter, and more powerful power tools have replaced much of the work previously done by hand.

Clamps and vises have been in use for years, and a power tool cannot replace them. Today's clamps are convenient and can be used in various situations. This section will introduce these basic tools and a few others you will encounter on the jobsite.

3.1.0 Saws

Hand sawing can be hard work. Using the right saw for the job makes sawing easier. Saws come in various shapes and sizes, yet all share one important feature—cutting teeth. The shape and size of the teeth are related to cut texture and speed.

Differences in blade and tooth design make cutting straight or curved lines through wood, metal, plastic, and many other materials possible. Generally, the fewer teeth per inch (tpi) of length found on a saw blade, the coarser and faster the cut will be. More teeth per inch, which means smaller teeth, results in a slower but smoother cut.

Figure 41 shows several types of saws. The following are brief descriptions of these saws:

WARNING!

Saw teeth are very sharp. Use gloves and do not handle saw teeth with bare hands. When cutting with a saw, ensure that your fingers always remain clear of the teeth.

- *Backsaw* — The standard blade of this saw is 8" to 14" (20 cm to 36 cm) long with 11 tpi to 14 tpi. A backsaw has a broad, flat blade and a reinforced back edge. The reinforced back keeps the blade straight and true. It is used for cutting finish lumber and **miter joints**, as the blade is designed to produce smooth cuts.
- *Keyhole saw* — The standard blade of this saw is 12" to 14" (30 cm to 36 cm) long with 7 tpi to 8 tpi. This saw cuts curves quickly in wood, plywood, or wallboard. It cuts holes for large-diameter pipes, vents, and electrical boxes. The name, however, comes from its original use of cutting holes in doors for door hardware.
- *Coping saw* — This saw has a narrow, flexible blade attached to a U-shaped frame. The blade can be rotated in the frame to any angle. The blade holders also maintain proper blade tension. Standard blades range from 10 tpi to 20 tpi. The coping saw is used for tight curves and circles in finish carpentry. Generally, coping saw blades are mounted with the teeth pointing back toward the handle, allowing it to cut on the pull stroke. However, the blade can also be placed so the cutting action occurs on the push stroke, like most handsaws.

Miter joints: Joints made by fastening parts together with the ends cut at a similar angle, usually 90°.

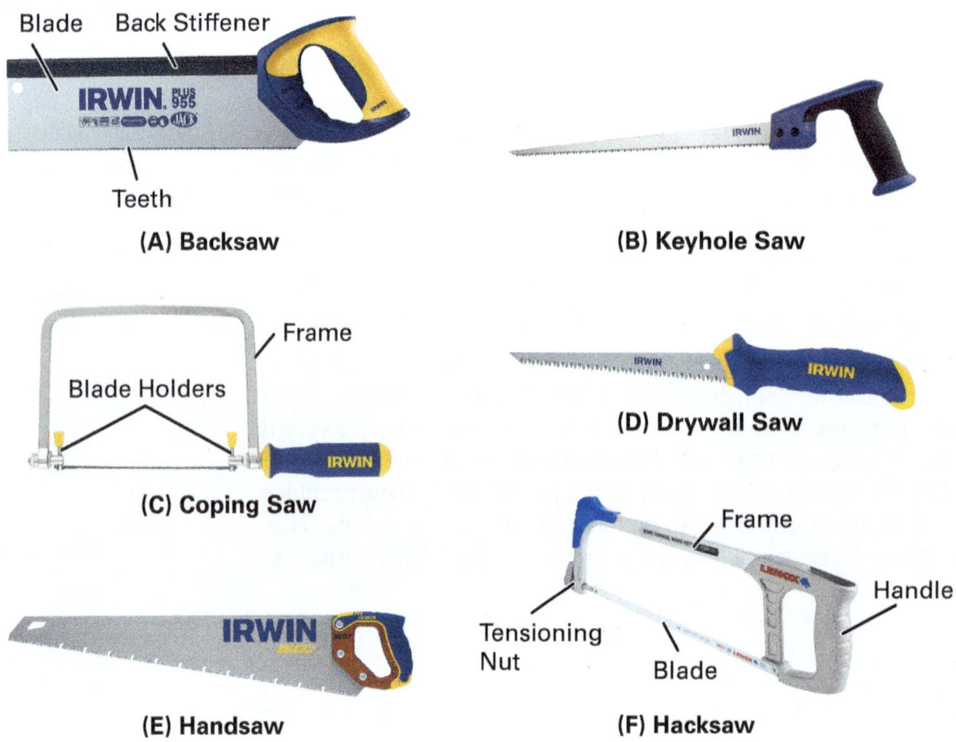

Figure 41 Common saws.
Source: Image property of Stanley Black & Decker. Used with permission

- *Drywall saw* — A drywall saw is a long, narrow saw used to cut softer building materials quickly. It can have a fixed or retractable blade mounted in a wood or plastic handle. The blade on a drywall saw has a very sharp point, making it easy to make a hole in soft material and start a cut.
- *Handsaw* — The standard blade of this saw is 26" (66 cm) long with 8 tpi to 14 tpi for a crosscut saw and 5 tpi to 9 tpi for a ripsaw. A crosscut saw, as the name implies, is designed to cut across the grain of the wood. A ripsaw is designed to cut with the grain.
- *Hacksaw* — The standard blade of this saw is 8" to 16" (20 cm to 41 cm) long with 14 tpi to 32 tpi. The hacksaw is used to cut through metals such as bolts and pipe. For this reason, the blades have more and finer teeth than most handsaws. It has a sturdy frame and a pistol grip handle. The blade is tightened using a tensioning nut or similar mechanism. When installing a hacksaw blade, ensure the teeth are facing away from, not toward, the saw handle. Hacksaws are designed to cut on the push stroke, not the pull stroke.

3.1.1 Handsaws

The handsaw's blade is made of tempered steel, so it will stay sharp and not buckle. When a handsaw is not handled properly, and the blade does buckle, it is virtually useless. Handsaws are classified mainly by the characteristics of the teeth. Saw teeth are set or angled alternately in opposite directions to make a **kerf** slightly wider than the thickness of the saw blade body.

Kerf: A cut taken by a saw blade with a width equal to that of the blade teeth.

Two common types of handsaws are the crosscut saw and the ripsaw. The crosscut saw, which has 8 tpi to 14 tpi, is designed to cut across the grain of wood. It cuts slower than a ripsaw but leaves a smoother surface. Blade lengths range from 20" to 28" (51 cm to 71 cm). A 24" or 26" (61 cm or 66 cm) blade is good for general use. Longer blades allow longer cutting strokes, contributing to a better, more consistent cut.

Follow these steps to use a crosscut saw properly:

Step 1 Wear safety glasses and work gloves before beginning. Mark the cut to be made with a square. Also place an X or similar mark on one side of the squared line to show on which side of the line the cut should be made. The kerf should always be taken from the scrap side of the mark.

Step 2 Make sure the piece to be cut is well supported on a sawhorse or similar support. Secure the material with one or more clamps. Plan to support the scrap end as well as the main part of the lumber to keep it from breaking away as the blade nears the end of the cut. With short pieces of lumber, you can support the scrap end of the piece with a free hand. With longer pieces, additional support or assistance is required.

Step 3 Place the saw teeth on the edge of the wood farthest from you, just at the outside edge of the mark. Remember that the blade's edge should be aligned with the line on the scrap side of the lumber, so that the kerf is removed from the scrap side of the cut.

Step 4 Start the cut with the blade touching the lumber near the saw's handle, pulling toward your body on the first stroke. Place the saw at about a 45° angle to the wood and pull the length of the blade to make a small kerf.

Step 5 Push the blade in the opposite direction to cut and deepen the kerf. Start sawing slowly, increasing the length of the stroke as the kerf deepens. Remember that the cutting is done on the push stroke, which is more forceful and faster than the pull stroke.

Step 6 Use the thumb of the hand that is not sawing to guide the saw as the cut progresses, so the blade stays perpendicular to the workpiece.

Step 7 Do not push too hard or ride the saw into the wood. Let the weight of the saw set the cutting rate. This will make it easier to control the saw and is less tiring.

Step 8 Continue to saw with the blade at a 45° angle to the wood. Work the blade back toward the saw if it wanders from the line. If the saw blade binds in the kerf, place a wooden wedge into the kerf to hold it open or reposition the support under the scrap end to prevent binding.

Figure 42 Crosscut saw kerf.
Source: Courtesy of Cianbro

Figure 42 shows the kerf that results from a crosscut. Note that the kerf is parallel with the line on one side, with the left side of the line being the scrap side. This ensures the length of the cut lumber remains accurate.

The ripsaw has 5 tpi to 9 tpi and is designed to cut with the grain (parallel to the wood fibers), meeting less resistance than a crosscut saw. Because it has fewer teeth than the crosscut saw, it will leave a coarser surface but cuts faster.

To use a ripsaw properly, mark and start a cut the same way you would with a crosscut saw. Once the cut is underway, saw with the blade at a steeper 60° angle to the wood.

3.1.2 Safety and Maintenance

It is important to focus on the blade when sawing. Saw teeth are sharp enough to cause a serious injury. Here are some basic guidelines for caring for and working with handsaws:

Emery cloth: A material made with a thin cloth backing, coated with an abrasive tightly bonded to the backing; typically sold in sheets or narrow rolls.

- Brace yourself when sawing so you are not thrown off balance when the scrap separates from the workpiece.
- Clamp or otherwise secure the workpiece to prevent it from moving while sawing.
- Clean saw blades occasionally with fine **emery cloth** or steel wool and apply a coat of silicone lubricant if it starts to rust.
- Always lay a saw down gently. It cannot be repaired if the blade develops a permanent bend or buckle.
- Craftworkers, farmhands, and ranch hands can sharpen good-quality saws with the proper equipment and skill.
- Do not allow the handsaw blade to saw into stone, concrete, or metal, as these materials can easily damage the teeth.

3.2.0 Clamps

There are many types and sizes of clamps (*Figure 43*). Clamps are sized by the maximum opening of the jaws. The depth, or throat, of the clamp determines how far from the edge of the workpiece the clamp can be placed. The following descriptions are related to the clamps shown in *Figure 43*:

- *C-clamp* — This multipurpose clamp is named for the C-shaped frame. The clamp has a metal shoe at the end of a screw. Using a sliding T-bar handle, the clamp is tightened to hold material between the frame's metal jaw and the shoe. C-clamps are strong and durable, usually made of cast iron or steel.
- *Locking C-clamp pliers* — Technically, this clamp also falls into the pliers category. However, the locking feature of these pliers also makes them a versatile clamp. A knob in the handle controls the width and tension of the jaws. The handles are closed to lock the clamp, and a lever is pressed to unlock the jaws.

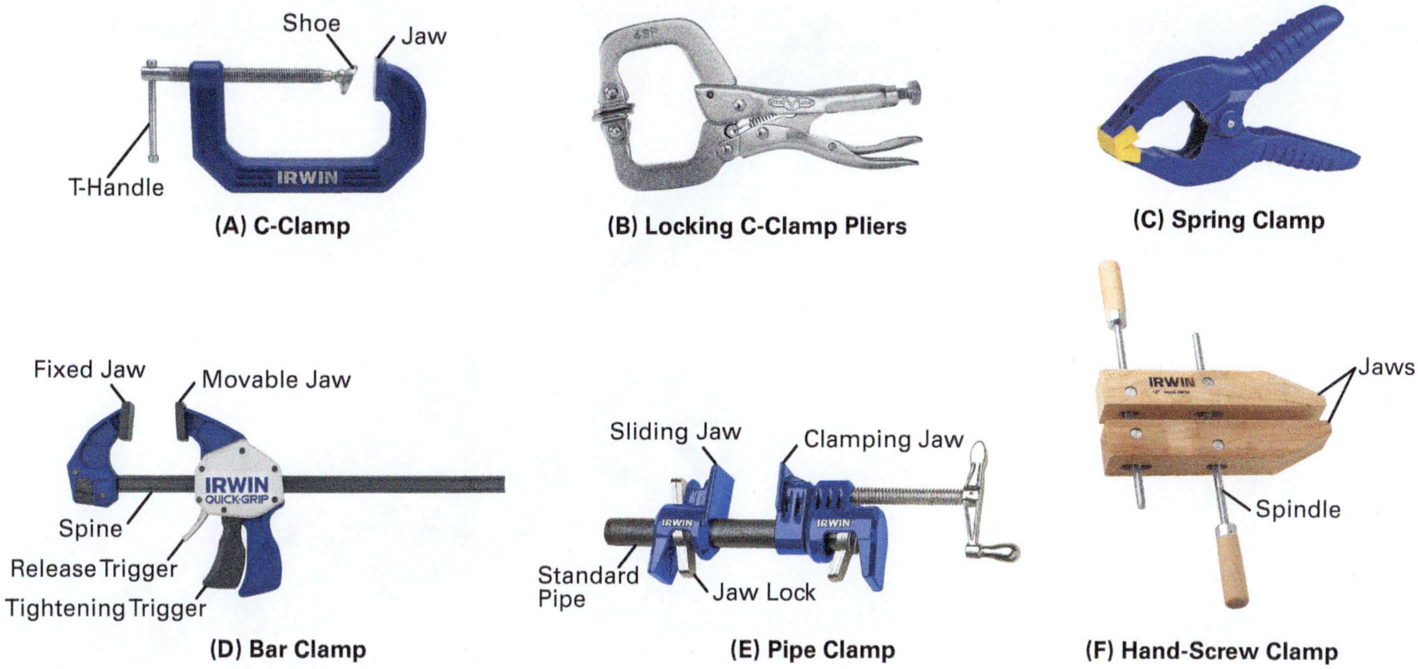

Figure 43 Types of clamps.
Source: Image property of Stanley Black & Decker. Used with permission

- *Spring clamp* — These simple clamps are spring-operated. When the handles are released, the spring holds the clamp tightly shut. The jaws are usually cushioned with plastic or rubber to protect the workpiece surface.
- *Bar clamp* — A rectangular piece of steel or aluminum forms the spine of the bar clamp. It has a fixed jaw at one end and a sliding jaw that moves along the bar. Small grooves in the spine lock the moving head into position. Pads on the jaws prevent it from damaging delicate surfaces. It is also designed to release the material quickly and easily by simply squeezing the release trigger. Another feature of this tool is that you can change the direction of the jaw to turn it into a spreader bar.
- *Pipe clamp* — The spine of a pipe clamp is a selected length of pipe. It has one fixed jaw and a movable jaw that slides along the pipe. The movable jaw has a lever mechanism that is squeezed to slide it along the spine. The pipe connecting the two jaw assemblies can be any length required; even a full length of pipe can be used. As a result, the pipe clamp is popular for large and wide clamping tasks.
- *Hand-screw clamp* — This clamp has broad wooden jaws that spread the pressure over a wide area. Each jaw works independently along two threaded rods. The jaws can be angled toward or away from each other or remain parallel.
- *Web (cargo, band) strap* — Although not technically a clamp, the web strap (*Figure 44*) uses a belt-like canvas or nylon strap or band to apply even pressure around a bundle of material or to secure a load. After looping the band around the work, the clamp head tightens the band with a ratcheting action. A quick release lever loosens the band when finished.
- *Come-along* — A come-along, sometimes called a power puller, is an excellent tool for both construction and agricultural applications. It can pull heavy loads or be a tensioned strap for holding items in place. In agriculture, these hand-operated ratchet levers are commonly used to tighten wire fences, resulting in a strong, taut barrier for keeping animals in and potential predators and pests out (*Figure 45*).

Figure 44 Web strap.
Source: Image property of Stanley Black & Decker. Used with permission

Figure 45 Come-along.
Source: iStock

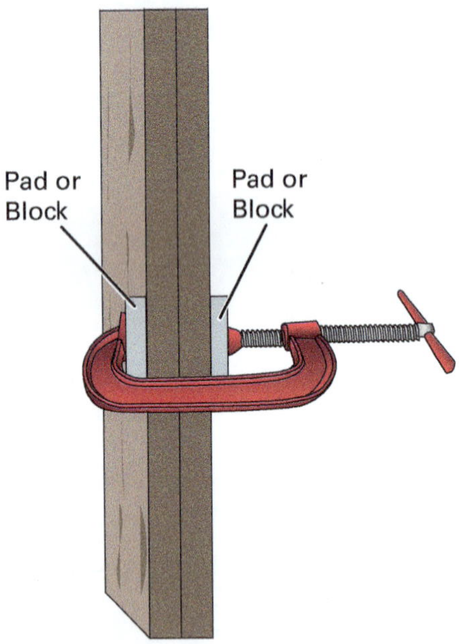

Figure 46 Placing cushions for clamp jaws.

When clamping wood or other soft material, place rubber pads or thin woodblocks between the workpiece and the jaws to protect the workpiece, as shown in *Figure 46*. When wrapping heavy web straps around bundles of finish lumber, it is also a good idea to protect the corners of the wood with cushions, preventing the straps from digging into the edges.

3.2.1 Safety and Maintenance

The following are guidelines for using clamps:

- Store clamps by lightly clamping them to a rack.
- Discard clamps with bent frames or spines. The spine of a pipe clamp is easily replaced.
- Clean and lubricate the moving parts occasionally, but do not overdo it, as oil attracts and holds dirt and dust.
- Never use any of the clamps shown for hoisting work.
- Do not use pipe or other tools to add leverage when tightening a clamp. Doing so allows you to exceed the clamp's capacity, likely leading to permanent damage.

3.3.0 Utility Knives

As the name implies, utility knives are general-purpose tools. Because of their versatility, they are common among all crafts. However, they are also hazardous and have been the cause of many injuries. Many companies have specific policies regarding their use. Some companies and jobsites prohibit the use of utility knives without a self-retracting blade.

WARNING!

All types of files and utility knives have sharp points and edges that can cause serious injury. Handle knives and files with extreme caution and the greatest of care. Never use a file or rasp without first installing a handle.

3.3.1 Utility Knife Basics

A utility knife (*Figure 47*) is used to cut various materials, including roofing felt, fiberglass or asphalt shingles, vinyl or linoleum floor tiles, fiberboard, and drywall. Utility knives can also trim insulation and open cartons.

The utility knife has a replaceable, thin, razor-like blade. The handle, which is approximately 6" (15 cm) long and is made of die-cast metal or plastic, holds the blade. The least expensive, simpler models are made in two halves, held together with a screw for blade replacement. Others fold to a more compact size. Some utility knives automatically retract the blade when thumb pressure is taken off the blade control for increased safety. Some jobsites or employers may require self-retracting knives to maximize safety.

Note the different blades used in the carpet knife. Carpet is highly abrasive and hard on blades. Carpet knife blades have two sharpened edges, so they can be reversed to extend their usable life.

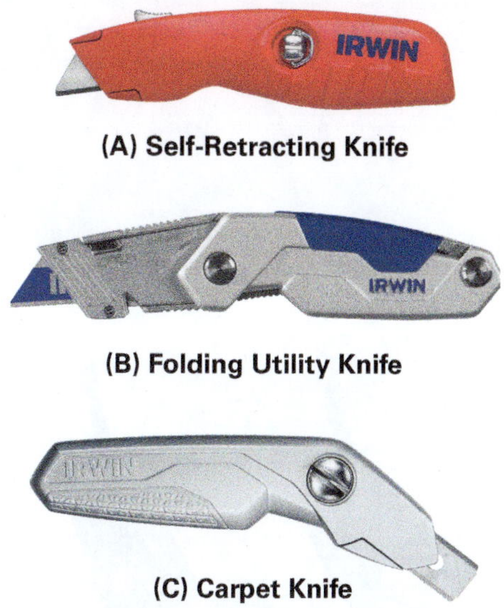

(A) Self-Retracting Knife

(B) Folding Utility Knife

(C) Carpet Knife

Figure 47 Utility knives.
Source: Image property of Stanley Black & Decker. Used with permission

WARNING!

Utility knife blades are extremely sharp. Handle these knives carefully and ensure the blade is safely secured after each use.

To use a retractable blade utility knife safely and properly, place a protective barrier, such as scrap wood, under the object to be cut. Don an appropriate pair of gloves. Unlock and push the blade out by pushing on the lever and sliding it forward. Release pressure on the lever to lock the blade open or hold the lever if the knife automatically retracts. Always cut away from your body or perpendicular to it. Once cutting is complete, unlock and pull the blade in by pushing on the lever and sliding it back. Folding models have a blade lock that must be depressed to fold and unfold.

These guidelines apply to utility knives:

- Replace blades when they stop cutting and start tearing the material. Sharp blades are safer than dull ones.
- Always keep the blade closed and locked when the knife is not used.
- Be sure to position yourself properly, keeping your free hand out of the line of the cut. Cut perpendicular to your body (left to right or right to left) or away from your body.
- Utility knife blades may be somewhat brittle and snap off easily, leaving a dangerous jagged edge.

3.4.0 Earth-Moving Tools

This section reviews various earth-moving tools and guidelines for their use. Craftworkers, farmhands, and ranch hands most commonly use the following types of earth-moving equipment:

- Shovels
- Picks and mattocks
- Post-hole diggers
- Hoes
- Rakes

3.4.1 Shovels

Many different construction trades use shovels. An electrician running underground wiring may dig a trench, and a mason may dig footers for a foundation. Carpenters, plumbers, and many other craftworkers must occasionally use a shovel.

Figure 48 Round and square shovels.
Source: Ridge Tool Company

Figure 49 Spade.
Source: Shutterstock

The three basic shapes of shovel blades are round, square, and spade. Round shovels (*Figure 48* [A]) are used to dig holes or move soil. A square shovel (*Figure 48* [B]) is used to move gravel or clean up debris. A traditional spade (*Figure 49*) is used to shape trenches or holes that need smooth, straight sides. It is flatter than the others and does not have curved sides. However, many variations of the traditional spade are available today.

Round shovels and spades are best for digging fresh earth. When using them, place the tip of the shovel where digging or soil removal will begin. Balance a foot on the step at the top of the blade, shift some of your weight to it, and press down to cut into the soil.

Square shovels are best for collecting debris from a hard surface, or for moving sand and gravel. When using a square shovel, place the leading edge of the shovel blade against the gravel or debris and push until the shovel is loaded.

All types of shovels can have wooden or fiberglass handles. They generally come in two lengths. A long handle is usually 47" to 48" (1.2 m) long; a short handle is usually 27" (69 cm) long. Short handles usually have a D-ring on the end to provide a better grip. Generally, the longer handles provide better leverage, but a shovel is not meant to be used as a lever. Roots and other obstructions should first be cleared with other tools.

3.4.2 Picks and Mattocks

A pick (*Figure 50* [A]) consists of a wooden or synthetic handle and a forged steel head with a thin pick on one end and a flat cutting head on the other. Depending on the size and strength of the pick, it can be used to break hardened or rocky soil, to level out stones and pavers, to loosen soil, and to break up stones and concrete. Picks with the longest handles are used when a normal amount of swing force is needed and there is clearance. Short-handled picks are especially useful when working below the surface and down on your knees.

Like a pick, a mattock is also used to break hard, rocky soil and dig trenches. Mattocks are much better than picks for clearing tree roots. One side of the mattock head has a wide cutting blade, as shown in *Figure 50* [B]. The other end may be like a pick or another blade turned 90° from the mattock blade. Mattock handles are rarely longer than 36" (91 cm).

The first step to safely and effectively using a pick or mattock is to select the length appropriate for your size and the task at hand. You should always wear gloves and steel-toed work boots or shoes.

Figure 50 Pick and mattock.
Source: A.M. Leonard, Inc.

Place one hand at the end of the handle and place the dominant hand about two-thirds up the handle. To strike hard with a short-handled pick, raise the pick to a near vertical position. With your knees bent, plunge the tool into the ground. Raise the pick again and then swing it back toward the ground, using the weight of the toolhead and the leverage of the handle to deliver a solid blow.

WARNING!

Do not swing a pick or mattock beyond a vertical position and over your head. Doing so may cause a back injury and significantly reduce control over the tool.

3.4.3 Post-Hole Diggers

Although posts are typically dug with an auger attached to a tractor in most agricultural settings, a handheld post-hole digger is more efficient in certain situations (*Figure 51*).

This tool is essentially two small shovels connected by a hinging bolt. As you push the long wooden handles toward each other, the shovels separate. The sharp edges can be used to carve the earth and define the hole diameter. Once you have broken up enough soil, you can pull the handles together to scoop up the loose dirt.

If you plan on installing steel posts, you can use T-post drivers once the hole is completed (*Figure 52*). This specialized tool is a sturdy, hollow steel tube that slips over the top of a fence post. The installer then lifts the driver by the welded handles on the sides to slam the post downward until it is set firmly into the ground. When using a post driver, ensure your hands never leave the handles to avoid pinch points and smashing your fingers.

Figure 51 Post-hole digger.
Source: iStock

Figure 52 T-post driver.
Source: Shutterstock

3.4.4 Hoes and Rakes

Hoes and rakes are some of the first agricultural tools ever invented for moving dirt and debris (*Figure 53*).

Both tools have a long, lightweight handle. However, the heads differ slightly. A rake has a fan of flexible claws designed to grab loose leaves, mulch, and other materials, while a hoe is built with a sturdy, flat metal face for moving small sections of dirt for crops and gardens.

(A) **(B)**

Figure 53 Rake and hoe.
Sources: Shutterstock (53A); iStock (53B)

3.4.5 Axes

Axes are multifunctional tools with long wooden or synthetic handles and a heavy metal head with one or two sharp edges called *bits*. Although chainsaws have taken the job of felling axes, hatchets and splitting mauls are still common tools for harvesting firewood (*Figure 54*).

Hatchets are lighter and have a short handle similar to a framing hammer that accommodates one-handed use while trimming branches off the main logs or shaving off smaller logs into kindling. The splitting maul has a much heavier head with a thick, wedge-shaped blade for improved driving force for splitting larger logs and stumps.

(A) **(B)**

Figure 54 Splitting maul and hatchet.
Source: iStock

Safety requirements for axes match many long-handled earth-moving tools like the pick, although the sharp blade can cause more damage if used improperly. Always check your backswing, use minimal force during your swing, and ensure all body parts are clear of the cutting path. Finally, sharpen your blade regularly, check the handle to maintain tool function, and guarantee that the axe-head won't loosen or fly off the handle during woodcutting operations.

3.4.6 Seeders

Due to the scale of commercial farming, many seeders are now attached to heavy equipment to disperse millions of seeds across multiple acres of land. However, hand-operated seeders are still used in smaller settings such as gardens and homesteads (*Figure 55*).

These wheeled seed dispersal tools are often used in residential settings to spread grass seed and other chemicals to improve the aesthetic quality of home lawns. The same function can help farmhands and ranch hands improve the efficiency of planting produce compared to traditional hand-seeding methods.

Figure 55 Using a lawn seeder.
Source: iStock

3.4.7 Safety and Maintenance

The following are guidelines for working safely with earth-moving tools:

- Always ensure that the blade is fixed firmly to the handle and that no cracks or splits are present.
- Wear the appropriate PPE, including safety glasses and gloves, when digging, trenching, or clearing debris. Wear safety shoes or boots to protect your feet from dropped materials and tool blades.
- Do not leave dirt or debris on the blade. Rinse or brush off any blades after use.
- Always ensure that the head is firmly fixed to the handle and that no cracks or splits are present.
- Ensure no other workers are in the swing path before beginning the work.
- Always use a pick of the appropriate length and weight for your physical size.
- Only use maximum force swings when necessary. Hard swings place more strain on your back and shoulders.

- Do not swing a pick or mattock beyond a vertical position and over your head. Doing so may cause a back injury and significantly reduce control over the tool.
- Be sure to wear appropriate eye protection, gloves, and safety shoes or boots.

3.0.0 Section Review

1. To cut through a metal pipe, the correct saw to use is a _____.
   a. coping saw
   b. crosscut saw
   c. drywall saw
   d. hacksaw

2. Which type of clamp is *best* suited for a very wide clamping application, such as an 8' (2.4 m) span across the jaws?
   a. C-clamp
   b. Bar clamp
   c. Spring clamp
   d. Pipe clamp

3. Which utility knife type has a double edge that can be flipped to improve the blade's life?
   a. Folding
   b. Self-retracting
   c. Carpet
   d. Serrated

4. Which tool is *best* for moving gravel?
   a. Square shovel
   b. Round shovel
   c. Mattock
   d. Spade

Module 93104 Review Questions

1. A slightly rounded face on a hammer _____.
   a. helps deliver a harder blow
   b. prevents overstriking
   c. is only found on drywall hammers
   d. helps avoid significant damage to the material when nailing

Figure RQ01

Source: Image property of Stanley Black & Decker. Used with permission

2. What type of hammer is shown in *Figure RQ01*?
   a. Curved claw hammer
   b. Rip claw hammer
   c. Drywall hammer
   d. Ball peen hammer

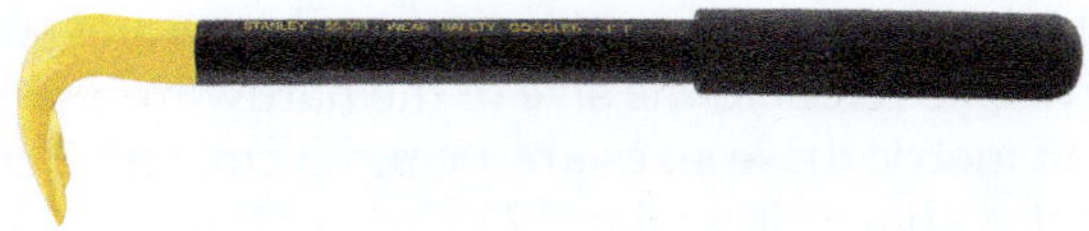

Figure RQ02

Source: Image property of Stanley Black & Decker. Used with permission

3. The tool shown in *Figure RQ02* is a _____.
   a. cat's paw
   b. wrecking bar
   c. wood chisel
   d. cape chisel

4. What type of chisel is longer and thinner than a standard wood chisel?
   a. Beveled
   b. Dovetail
   c. Paring
   d. Cold

5. Which type of chisel is typically designed to be used without a hammer?
   a. Cold chisel
   b. Paring chisel
   c. Cape chisel
   d. Diamond-pointed chisel

6. The *best* punch used to mark a spot for drilling a hole is the _____.
   a. center punch
   b. tapered punch
   c. prick punch
   d. angle punch

7. The *most* common type of crosshead screwdriver is the _____.
   a. Robertson® screwdriver
   b. Torx® screwdriver
   c. slotted screwdriver
   d. Phillips screwdriver

8. The wrench that is *most* likely to fit only one size of hardware is the _____.
   a. open-end wrench
   b. box-end wrench
   c. combination wrench
   d. adjustable wrench

Figure RQ03

9. In *Figure RQ03*, the fastener is likely being tightened.
   a. True
   b. False

10. Which of the following statements about ratchets and sockets is *correct*?
    a. The drive size refers to the size of the hardware a ratchet handle fits.
    b. Common metric drive sizes are 1 cm, 1.5 cm, and 2 cm.
    c. Most sockets have either 6 or 12 gripping flats.
    d. Socket extensions are available in only one length for each drive size.

11. Pliers should *not* be used to _____.
    a. tighten or loosen hardware
    b. cut wire
    c. bend wire
    d. grip pipe

12. A disadvantage of a laser measuring tool is that it _____.
    a. can only measure up to about 100' (30 m)
    b. does not work well in smoky conditions
    c. must be powered with a drop cord
    d. requires several minutes to complete a measurement

13. When something is plumb, it is _____.
    a. precisely vertical
    b. level to the horizon
    c. parallel to the ground
    d. at some angle other than 90°

Figure RQ04
Source: Image property of Stanley Black & Decker. Used with permission

14. The tool shown in *Figure RQ04* is called a _____.
    a. plumb level
    b. post level
    c. torpedo level
    d. laser level

15. The try square is made at a fixed _____.
    a. 45° angle
    b. 90° angle
    c. 180° angle
    d. 360° angle

16. Accurate and smooth miter joints in wood can be cut with a _____.
    a. keyhole saw
    b. hacksaw
    c. drywall saw
    d. backsaw

17. A hand-screw clamp has _____.
    a. metal jaws
    b. nylon jaws
    c. wooden jaws
    d. fiberglass jaws

18. How is a carpet knife different from a common utility knife?
    a. The blade has a single sharpened edge.
    b. The blade has two sharpened edges.
    c. A carpet knife is much longer than a utility knife.
    d. A carpet knife has a serrated blade.

19. A spade is used to _____.
    a. clear tree roots
    b. move gravel or clean up construction debris
    c. move large volumes of soil
    d. shape trenches and holes requiring straight sides

Answers to odd-numbered questions are found in the Review Question Answer Keys at the back of this book.

Answers to Section Review Questions

Answer	Section Reference	Objective
Section 1.0.0		
1. c	1.1.1	1a
2. a	1.2.1	1b
3. b	1.3.0; *Figure 14*	1c
4. c	1.4.2	1d
5. d	1.5.5	1e
Section 2.0.0		
1. d	2.1.3	2a
2. a	2.2.5	2b
Section 3.0.0		
1. d	3.1.0	3a
2. d	3.2.0	3b
3. c	3.3.1	3c
4. a	3.4.1	3d

MODULE 93105

Introduction to Power Tools

Source: iStock

Objectives

Successful completion of this module prepares you to do the following:

1. Identify and explain how to use various types of power drills and impact wrenches.
   a. Summarize basic power tool safety guidelines.
   b. Identify common power drills and bits and explain how to use them.
   c. Describe the difference between hammer drills and impact drivers.
   d. Identify pneumatic drills and impact wrenches and explain how to use them.
2. Identify and explain how to use various types of power saws.
   a. Explain how to use a circular saw and identify different types of blades.
   b. Differentiate between jigsaws and reciprocating saws and explain how to use them.
   c. Explain how to use a portable band saw.
   d. Describe the difference between miter saws and cutoff saws.
   e. Explain how to use table saws and describe the jobs they are best suited for.
   f. Describe the proper use of chainsaws and electric pruning shears.
3. Describe the jobs best suited to grinders and oscillating multi-tools.
   a. Explain how to use various types of grinders.
   b. Identify grinder accessories and the jobs for which they are used.
   c. List the jobs that can be performed using an oscillating multi-tool.
4. Identify and explain how to use miscellaneous power tools.
   a. Discuss the hazards of using power nailers.
   b. Describe jobs that can be performed with hydraulic jacks and wrenches.

Performance Task

Under supervision, you should be able to do the following:

1. Safely and properly demonstrate the use of the following tools:
   - Electric drill (corded and cordless)
   - Hammer drill
   - Impact driver
   - Circular saw
   - Jigsaw
   - Reciprocating saw
   - Portable band saw
   - Miter or cutoff saw
   - Table saw
   - Portable or bench grinder
   - Oscillating multi-tool
   - Power nailer

Digital Resources for Agricultural Technology

SCAN ME

Scan this code using the camera on your phone or mobile device to view the digital resources related to this craft.

Overview

Power tools play an important role in the construction and agriculture industries. Regardless of their specialization, any worker will face a situation where power tools can simplify and expedite a challenging task. This module provides an overview of the common types of power tools, how they function, and the proper techniques required for safe and efficient operation.

NCCER **Industry-Recognized Credentials**

If you are training through an NCCER-accredited sponsor, you may be eligible for credentials from NCCER. The ID number for this module is 93105. Note that this module may have been used in other NCCER curricula and may apply to other level completions. Contact NCCER at 1.888.622.3720 or go to **www.nccer.org** for more information.

You can also show off your industry-recognized credentials online with NCCER's digital credentials. Transform your knowledge, skills, and achievements into credentials that you can share across social media platforms, send to your network, and add to your resume. For more information, visit **www.nccer.org**.

1.0.0 Power Drills/Drivers

Performance Task

1. Safely and properly demonstrate the use of the following tools:
   - Electric drill (corded and cordless)
   - Hammer drill
   - Impact driver

Objective

Identify and explain how to use various types of power drills and impact wrenches.

a. Summarize basic power tool safety guidelines.
b. Identify common power drills and bits and explain how to use them.
c. Describe the difference between hammer drills and impact drivers.
d. Identify pneumatic drills and impact wrenches and explain how to use them.

Alternating current (AC): The common power supplied to most all wired devices, where the current reverses its direction many times per second. AC power is the type of power generated and distributed throughout settled areas.

Direct current (DC): An electric power supply where the current flows in one direction only. DC power is supplied by batteries and transformer-rectifiers that change AC power to DC.

This module introduces four categories of power tools:

- *Electric tools (corded and cordless)* — These tools are powered by one of two types of electricity: **alternating current (AC)** or **direct current (DC)**. Tools powered by AC plug into a wall receptacle using an integrated power cord. DC-powered tools, known as *cordless tools*, operate on rechargeable batteries and make up many power tools used on today's construction and agricultural sites.
- *Pneumatic tools* — These tools are powered by air. Electric or gasoline-powered compressors produce air pressure. Air hammers and pneumatic nailers are examples of pneumatic tools.
- *Specialty tools* — These tools include powder-actuated fastening tools that use gunpowder to drive pins and studs into concrete and steel. Powder-actuated tools require special training and certification to use.
- *Hydraulic tools* — These tools are powered by fluid pressure. Hand pumps or electric pumps are used to produce fluid pressure. Pipe benders, jackhammers, and wrenches are examples of hydraulic tools.

Before reviewing specific types of tools from each of these categories, it is important to address general safety guidelines that apply to all power tools, regardless of category.

1.1.0 Basic Power Tool Safety Guidelines

Always remember, if not used properly, power tools have the potential to be dangerous. Safety guidelines are covered for each power tool in this module, but general safety issues—safety in the work area, safety equipment, and working with electricity—are covered in NCCER Module 93102, *Basic Safety*. This information is vital for working with power tools.

WARNING!

NCCER Module 93102, *Basic Safety*, must be completed before studying this module. Remember that appropriate personal protective equipment (PPE) must be worn when operating any power tool or when near someone else who is operating a power tool.

Basic safety guidelines that must be followed when working with all power tools include the following:

- Always read and follow guidelines provided in the power tool's operating instructions before attempting to use it.
- Never operate power tools without proper instruction and supervision.
- Confirm you have the right power tool for the job.
- Disconnect the tool from its energy source when it is not in use and before performing maintenance or replacing parts, such as bits, blades, or discs.
- Regularly inspect power tools, including power cords and hoses, for damage. Report any damaged tools to your supervisor.
- Proper PPE, including safety glasses, is required regardless of what power tool is being used. Some tools require the use of a face shield in addition to safety glasses. Safety shoes should also be worn. Tight-fitting gloves are required for most power tools, as loose gloves can be a safety hazard of their own.
- Keep other craftworkers who are not directly involved with the work at a safe distance.
- Do not wear loose-fitting clothing that could be caught in the mechanism of your power tool.
- Never use electric power tools in wet conditions unless they are approved for this type of application. When a power tool has a three-pronged plug, make certain it is plugged into a grounded receptacle. Additionally, before plugging into a receptacle, confirm that the power switch on the tool is set to the Off position.
- If a power tool is equipped with a **trigger lock**, do not use the lock. A trigger lock is a small lever, switch, or part that can be used to activate a locking catch or spring to hold a power tool trigger in the operating mode even when the trigger is released. Locking any power tool in the On position can be very dangerous.

Trigger lock: A small lever, switch, or part that can be used to activate a locking catch or spring to hold a power tool trigger in the operating mode without finger pressure.

WARNING!

An employer or jobsite may prohibit the use of trigger locks. If a power tool has a trigger lock, do *not* activate it.

- Never carry an electric or pneumatic tool by its cord or hose.
- Do not allow yourself to be distracted while using a power tool. If you are distracted or interrupted while operating the tool, turn it off.

1.2.0 Types of Power Drills/Drivers

Various types of power drills, also known as *drivers*, are used in the construction industry. A power drill is an efficient tool for making holes by spinning drill bits into wood, metal, plastic, or other materials. With different attachments and accessories, the power drill can also be used as a sander, polisher, screwdriver, grinder, or **countersink**.

Countersink: A bit or drill used to set the head of a screw at or below the surface of the material.

The most common types of power drivers include the following:

- Electric corded drills
- Electric cordless drills
- Electromagnetic (mag drill press) drills
- Hammer drills and impact drivers
- Pneumatic drills and wrenches
- Electric screwdrivers

1.2.1 Power Drills and Bits

Most power drills have several common features. For example, most have a pistol grip with a trigger switch for controlling power (*Figure 1*).

The harder a variable-speed drill trigger is pulled, the faster the drill spins. Drills also have reversing switches that enable the drill to spin backward to back the drill bit out if it becomes stuck in the material while drilling.

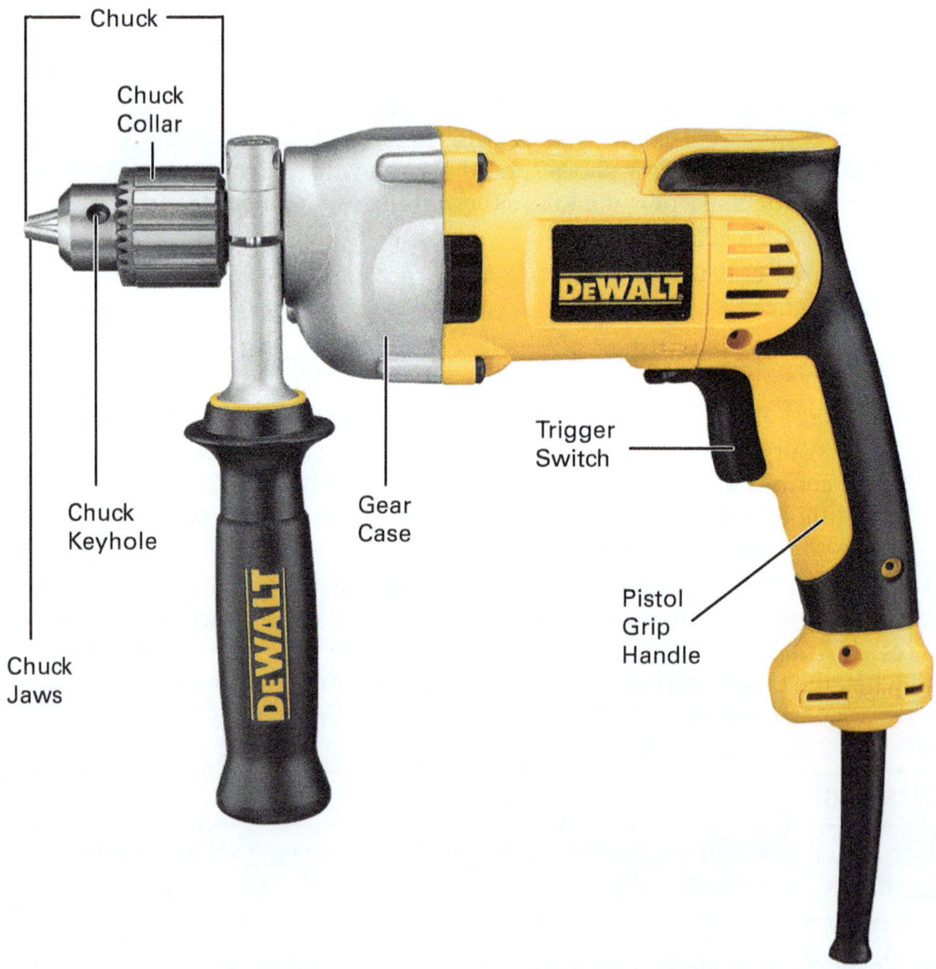

Figure 1 Parts of power drill.
Source: Image property of Stanley Black & Decker. Used with permission

Power drills use replaceable bits for different drilling tasks (*Figure 2*). On variable-speed power drills, a screwdriver bit can be used in place of a drill bit so the drill can be used as a screwdriver. Only use screwdriver bits designed for use in a power drill.

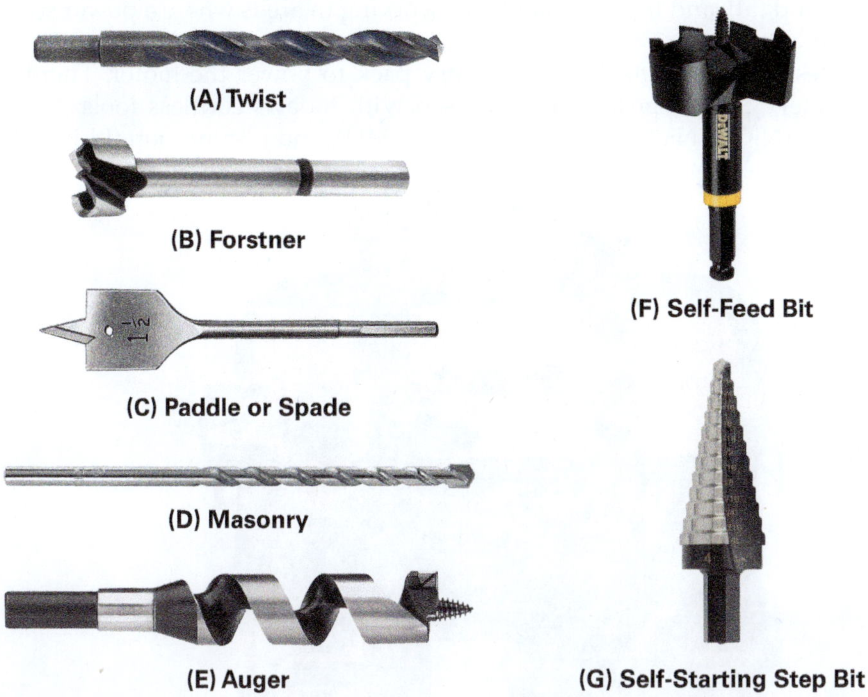

Figure 2 Drill bits.
Sources: Image property of Stanley Black & Decker. Used with permission (2A, 2D–2G); Courtesy Milwaukee Tool (2B–2C)

Twist drill bits are used to drill wood and plastic at high speeds or metal at lower speeds. A **Forstner bit** is a common accessory for wood and is particularly good for boring a flat-bottom hole. A paddle bit or spade bit is also used in wood. The bit size is measured by the paddle's diameter, which generally ranges from $\frac{1}{2}$" to $1\frac{1}{2}$" (roughly 13 mm to 38 mm).

A **masonry bit** with a carbide tip is used in concrete, stone, slate, and ceramic materials. The **auger bit**, which is designed to turn at low speeds, is used for drilling wood and other soft materials but not for drilling metal. A **self-feed bit** uses a lead screw that pulls the larger rear portion of the bit into the wood, making it simple to drill larger holes. The **step drill bit** is perfect for drilling holes in metal up to $\frac{1}{4}$" (roughly 6 mm) thick. As the name implies, this bit includes different 'steps' that allow the user to cut a hole to a desired diameter.

Drill bit sizes can vary a great deal. For most applications, bits are available in both fractional inch sizes and metric sizes. Common fractional inch sizes range from $\frac{1}{64}$" to 1". Common metric sizes for twist drill bits range from 0.5 mm to 25 mm. It is important to note that fractional inch bits and metric bits are not directly interchangeable. In other words, no fractional inch bit has a corresponding metric bit, and vice versa.

However, twist drill bits are also available in numbered and lettered bits. Numbered sets range from #80 up to #1, the largest of the set. The next bit size larger is the A bit, with the Z bit being the largest of the lettered set. These sets correspond to decimal fractions and metric sizes. For example, a #8 drill bit is sized at 0.199" and 5.055 mm. Number and letter drills are used or specified when the precise size of the hole is important.

Bits are held in a drill by the drill **chuck**. Keyed chucks are opened and closed using a **chuck key** (*Figure 3*). Chuck keys are typically interchangeable in design, but several sizes are different. Keyless chucks are typically found on cordless drills. Remember that the size of the chuck limits the size of the bit or tool a drill can accommodate.

Forstner bit: A bit designed for use in wood or similar soft material. The design allows it to drill a flat-bottom blind hole in the material.

Masonry bit: A drill bit with a carbide tip designed to penetrate materials such as stone, brick, or concrete.

Auger bit: A drill bit with a spiral cutting edge for boring holes in wood and other materials.

Self-feed bit: A drill bit that uses a lead screw to pull the larger rear portion of the bit into the wood, making it simple to drill larger holes.

Step drill bit: A drill bit that includes steps that enable the user to cut holes in the metal to a desired diameter.

Chuck: A clamping device that holds an attachment; for example, the drill's chuck holds the drill bit.

Chuck key: A small, T-shaped steel piece used to open and close the chuck on power drills. Cordless drills typically do not have a chuck key.

1.2.2 Cordless Tools and Batteries

Most tools on today's construction sites and farms are cordless. Cordless tools offer two important benefits: safety and mobility. They are safer and easier to handle because they lack inconvenient power cords that can cause people to stumble and fall, and they are handy for working in areas where a power source is unavailable.

Cordless drills (*Figure 4*) use a battery pack to power the motor. There are three battery pack types commonly used with today's cordless tools: Nickel-Cadmium (NiCd), Nickel-Metal Hydride (NiMH), and Lithium-ion (Li-ion).

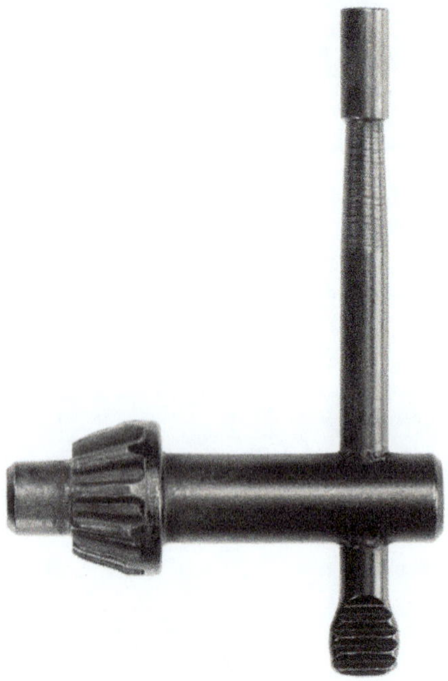

Figure 3 Chuck key.
Source: Image property of Stanley Black & Decker. Used with permission

Figure 4 Cordless drill.
Source: Image property of Stanley Black & Decker. Used with permission

The battery pack can be detached and plugged into a charger whenever the tool is idle. Some chargers can recharge the battery pack quickly, while others take longer to regain a full charge.

The quality of charge and the battery's lifecycle must be considered when determining how to best charge the battery. Li-ion batteries have several advantages over the other NiCd and NiMH batteries, including the highest number of **charge cycles** and the longest runtime on a full charge. The charge cycle describes the rechargeable battery's complete charge and discharge cycle.

Amp hour (Ah) describes the maximum amperes (amps or A) a battery can continuously provide power for up to 60 minutes. Drawing more amps reduces the time before the battery is drained and must be recharged. For example, under ideal conditions, a 4 Ah battery can continuously provide up to 4 amps for one hour before draining the battery. Drawing 8 amps from the same battery reduces the tool's runtime by up to 30 minutes.

Another critical factor to consider is the battery's **voltage rating**. A voltage rating measures a cordless tool's power. The higher the voltage, the more power the tool can deliver.

Regardless of the battery type, every manufacturer provides information about battery strength and expected lifetime in their product literature. Craftworkers, farmhands, and ranch hands using cordless tools should always carry extra battery packs.

Charge cycles: Describe one complete charge and discharge cycle for a rechargeable battery.

Amp hour (Ah): A rating that describes the maximum amps a battery can provide continuously for up to 60 minutes. Drawing more amps than the battery's rating is possible, but it reduces the time before the battery is drained.

Voltage rating: A measurement of the power a tool can deliver.

Some cordless drills have adjustable clutches so the drill motor can serve as a power screwdriver without applying too much power to the screw. Only heavy-duty cordless drills use keyed chucks, while most are keyless.

1.2.3 Power Drill Safety and Maintenance Guidelines

In addition to following basic power tool safety guidelines, familiarize yourself with these power drill safety rules before operating a power drill:

- Always wear the appropriate PPE when working with drills, especially safety glasses.
- Disconnect the power source (power cord or battery) before performing maintenance or changing drill bits.
- Keep your hands away from the drill bit and chuck.

WARNING!

Keep your hands away from a spinning drill bit. The spinning bit will cut your hands. Apply even pressure to the drill to prevent it from twisting or binding.

- When using a corded drill, operate only double-insulated electric power tools with proper **ground fault protection**. A **ground fault circuit interrupter (GFCI)** device protects equipment from continued electrical current in case of a circuit fault. The GFCI monitors the current flow and opens the circuit (which stops the flow of electricity) if it detects a difference between positive and negative flow.
- When using power drills with two-prong plugs, confirm that they are double-insulated. Tools that are not double-insulated must have a third prong to provide a ground circuit. Do not use a power drill if the ground prong on the plug is broken off.
- Before connecting a drill to its power source, confirm the trigger is NOT turned on.
- Before drilling through a wall or partition, verify what is inside the wall or on the other side of the work surface to avoid hitting water lines or electrical wiring.

Ground fault protection: Protection against short circuits; a safety device cuts power off as soon as it senses any imbalance between incoming and outgoing current.

Ground fault circuit interrupter (GFCI): A device designed to protect people from electric shock and equipment from damage by interrupting the flow of electricity if a circuit fault occurs.

WARNING!

Do not drill into or through a wall before confirming what is on the other side. Take steps to avoid hitting anything that would present a safety hazard or cause damage. Spaces between studs (upright pieces in the walls of a building) often contain electrical wiring, plumbing, or insulation. Care must be taken to avoid drilling directly into the wiring, pipes, or insulation.

- Before drilling, select the proper bit for the job and ensure it is sharp. Then, ensure the drill bit is securely tightened in the chuck before using it.
- When using drills with keyed chucks, attach the key to the power cord to ensure the key is available when tightening or loosening.
- Hold the drill with both hands and apply steady pressure.
- Never ram the drill while drilling because the force can chip the cutting edge and damage the bearings.
- As noted earlier, never use a drill's trigger lock to hold the trigger in the operating mode.

- Do not overreach when using a power drill; maintain a stable and balanced stance.
- Always keep your power drill clean.

WARNING!

If power tools are shared between workers, it is important *not* to grab the tool by the bit or any other moving part. Power tools should be disconnected from their power source whenever they are passed between workers to avoid accidental startups. Hand off tools by their handles and/or bodies only.

CAUTION

Power drills can be dangerous if not used properly. Always wear the proper PPE, including appropriate eye, head, and hand protection.

1.2.4 Preparing to Use a Power Drill

Before using a drill, make sure it is designed to perform the task at hand. Some electric power drills are made to be used in tight spaces, such as between studs and joists. Drills like those shown in *Figure 5* are referred to as right-angle drills. The drill on the right is larger and develops more power for larger holes.

(A) (B)

Figure 5 Right-angle drills.
Source: Courtesy Milwaukee Tool

Preparing Drills with Keyed Chucks

Once the proper drill has been selected, it is time to prepare it for use. Follow the steps below to prepare the power drill:

Step 1 Disconnect the drill from its power source.

Step 2 Turn the chuck counterclockwise (to the left) until the chuck opening is large enough to insert the bit **shank**. The shank is the smooth part of the bit.

Step 3 Insert the bit shank into the chuck opening (*Figure 6*). Keeping the bit centered in the opening, turn the chuck by hand until the jaws grip the bit shank.

NOTE

The chuck key has a grooved ring called a gear. Ensure that the proper size chuck key is used so that the key's gear meshes with the matching gears on the end of the chuck.

Shank: The smooth part of a drill bit that fits into the chuck.

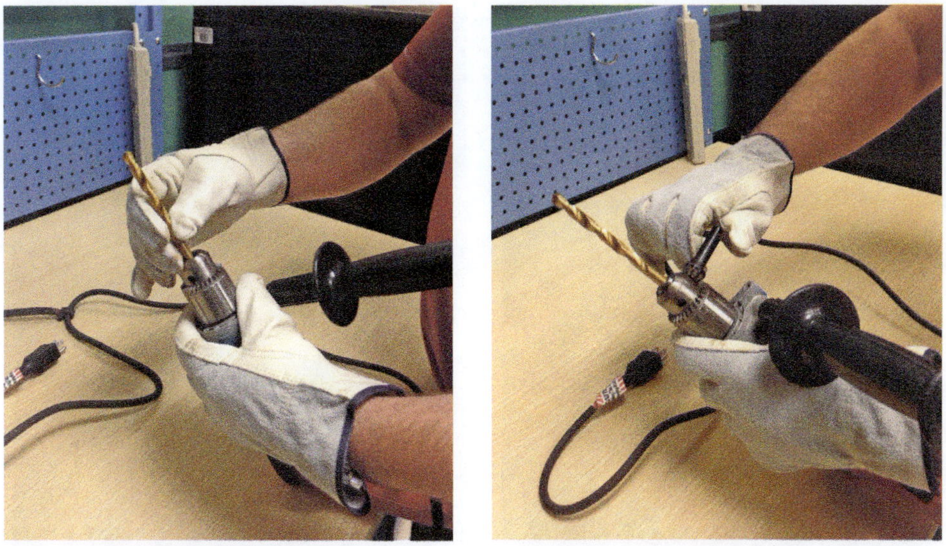

(A) Insert the Bit Shank Into the Chuck Opening **(B) Tighten with the Chuck Key**

Figure 6 Loading the bit on a keyed chuck.
Source: Courtesy of Cianbro

Step 4 Insert the chuck key (*Figure 6*) into one of the holes on the side of the chuck and turn the key clockwise (to the right) to tighten it. With larger chucks, you can tighten the individual jaws of the chuck uniformly by inserting and tightening the key in each hole on the chuck. Remove the key from the chuck once it is tightened. You are now ready to use the electric drill.

WARNING!

Remember to remove the key from the chuck after it has been tightened. Otherwise, the key could fly out and injure you or a coworker when you start the drill.

Preparing Drills with Keyless Chucks

Most cordless drills use a keyless chuck. While the steps for preparing a cordless drill are similar, there are some small differences. Follow the steps below when preparing to use drills with keyless chucks:

Step 1 Disconnect the drill from its power source by removing the battery pack before loading a bit.

Step 2 Open the chuck by turning it counterclockwise until the jaws are wide enough to insert the bit shank (*Figure 7*).

Step 3 Insert the bit shank into the chuck opening (*Figure 7*). Keeping the bit centered in the opening, turn the chuck by hand until the jaws grip the bit shank.

Step 4 Tighten the chuck securely with your hand so the bit does not move (*Figure 7*). You are now ready to use the cordless drill.

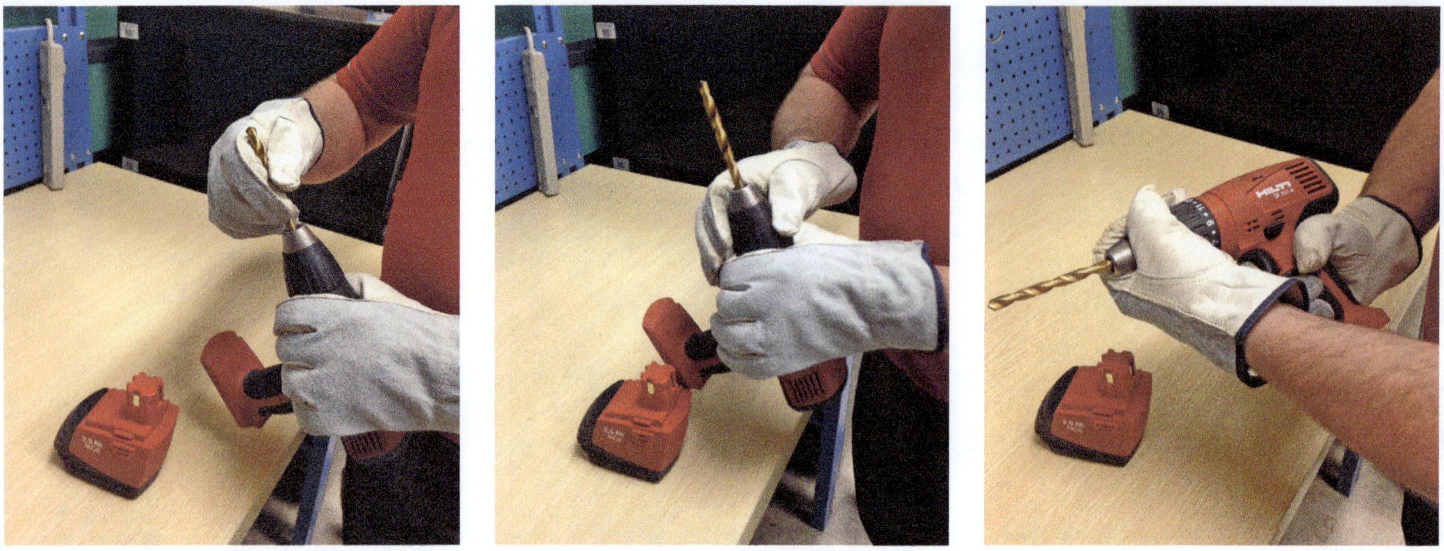

(A) Insert the Bit Shank

(B) Keep Bit Straight and Partially Tighten the Chuck

(C) Tighten the Chuck Securely

Figure 7 Loading the bit on a keyless chuck.
Source: Courtesy of Cianbro

1.2.5 Operating Power Drills

After securing the bit, connect the drill to its power source. Use the following steps to operate the power drill:

Step 1 Put on the appropriate PPE.

Step 2 Make a small indent or mark on the material exactly where the hole needs to be drilled. For wood, use a small punch to make the indent; for metal, use a center punch.

Step 3 Firmly clamp or support the drilled material, then hold the drill perpendicular (at a right angle) to the material surface and start the drill motor. With a variable-speed drill, start the bit slowly and verify the drill is rotating in the right direction. (When the bit is faced away, it should turn clockwise.)

Revolutions per minute (rpm): The rotational speed of a motor or shaft based on the number of times it rotates each minute.

Step 4 Place the tip of the bit on the indent or mark and hold the drill with both hands. Apply only moderate pressure when drilling. *Figure 8* shows the proper way to hold the drill. The drill motor should operate at approximately the same **revolutions per minute (rpm)** as it does when pressure is not exerted. If the sound of the drill indicates it is slowing considerably, apply less pressure.

Step 5 Reduce the pressure when the bit is about to emerge from the other side of the work, especially when drilling metal. If the drill bit gets stuck in the material during drilling, release the trigger, use the reversing switch to change the direction of the drill, and gently back it out of the material.

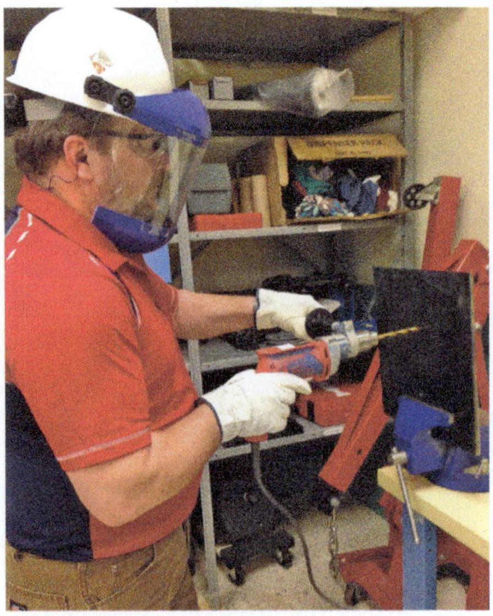

Figure 8 Properly holding a drill.
Source: Courtesy of Cianbro

WARNING!

As the drill bit emerges through the opposite side of the material, be prepared for it to grab. With the power still on, the drill motor tends to rotate the operator, not the bit. Be sure to maintain firm control if this happens.

WARNING!

To avoid injury, adjust the drill chuck properly prior to each use. Do not use the tool motor to tighten the chuck while gripping it.

Going Green

Recycling Rechargeable Batteries

Recycling rechargeable batteries once their power supply has become exhausted is no longer a luxury; it has become a matter of necessity. In fact, federal and state laws now regulate the disposal of some types of rechargeable batteries.

The US Environmental Protection Agency (EPA) estimates that more than 350 million rechargeable batteries are purchased annually in the United States. Likewise, hundreds of millions of rechargeable batteries and cell phones are retired each year. Rechargeable batteries use heavy metals and elements such as nickel, lithium, cadmium, mercury, and lead. All these elements can be toxic to people and the environment if not disposed of properly. Most landfills are not designed to handle the toxic metals that will eventually leak out of the batteries.

Rechargeable batteries are commonly found in the following:

- Cordless power tools
- Cellular and cordless phones
- Two-way radios
- Laptops and computers
- Digital cameras
- Camcorders

If a battery is rechargeable, it is recyclable. Rechargeable batteries are not only better for the environment if they are discarded properly, but they are also more cost-effective. Using rechargeable batteries can help save money and protect the environment at the same time.

Recently, most cities have added hazardous waste collection centers that collect rechargeable and regular batteries, paint, oil, refrigerant, and other hazardous wastes.

Drilling Metal

When drilling metal, it is best to lubricate the bit to help cool the cutting edges and produce a smoother finished hole. A small amount of noncombustible cutting oil or tapping fluid makes a good lubricant for drilling softer metals. Lubrication is *not* needed for drilling wood. When drilling deep holes, pull the drill bit partly out of the hole occasionally. Doing so clears the hole of shavings.

1.2.6 Electromagnetic Drill Presses

An electromagnetic drill press, commonly known as a *mag drill press*, is a portable drill mounted on an electromagnetic base (*Figure 9*). This type of tool is designed for drilling holes in thick metal. Once the drill is placed on a metal surface, the power is turned on, and the magnet is energized. The magnetic base will hold the drill in place for drilling. Some drills can also be rotated while the base remains stationary.

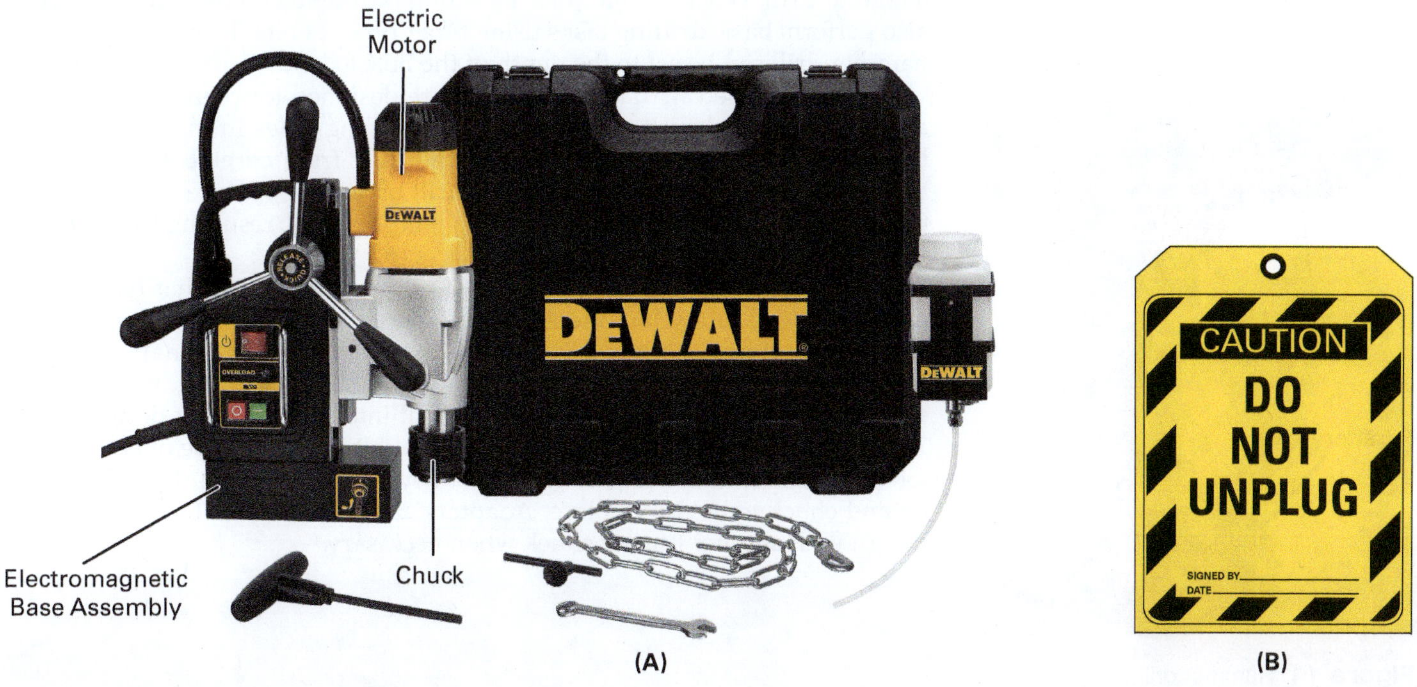

Figure 9 Electromagnetic drill press.
Source: Image property of Stanley Black & Decker. Used with permission (9A)

A switch on the junction box controls the electromagnetic base. When the switch is turned on, the magnet holds the drill in place on any surface with magnetic characteristics, such as carbon steel. The drill base must be clean, and the surface must be flat. The switch on the top of the drill turns the drill motor on and off. A depth gauge can be used to set the depth of the hole being drilled. It operates like a drill press, with the operator turning a hand wheel to raise and lower the drill against the workpiece. Workers who use electromagnetic drills should be properly trained to safely operate the specific drill being used.

There are several safety rules specific to electromagnetic drill presses, including the following:

- Securely clamp the material being drilled by the press. Unsecured materials can become dangerous flying or spinning objects.
- Since electromagnetic drill presses are mounted onto an electromagnetic base, it is important to maintain electrical power to the drill's electromagnet. To prevent the electrical power supply from being interrupted, attach a Do Not Unplug tag (*Figure 9 [B]*) to the cord.

WARNING!

Removing power from an electromagnetic drill while it is in use can cause the drill to fall, potentially harming the operator.

- Use a safety chain to secure the electromagnetic drill in case power is lost or shut off (*Figure 10*).
- Before using an electromagnetic drill press on a construction site, confirm whether safety attachments must be used. Some sites require attaching shields and/or safety lines before an electromagnetic drill press can be operated.

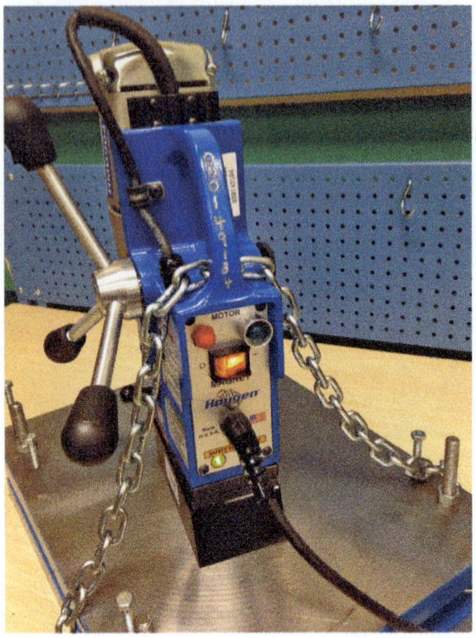

Figure 10 Electromagnetic drill with safety chain.
Source: Courtesy of Cianbro

1.3.0 Hammer Drills and Impact Drivers

Carbide: A very hard material made of tungsten carbide and heavy metals such as cobalt. Commonly baked on to the tips of masonry bits and to the edges of saw blades.

A hammer drill (*Figure 11*) has a light pounding action that enables it to drill into concrete, brick, or tile. The bit rotates and hammers at the same time, allowing faster drilling in these types of hard materials than would be possible with a regular drill. When the hammering action is disabled on these drills, they can also perform basic drilling tasks using other types of bits. The depth gauge on a hammer drill can be set to the depth of the hole to be drilled.

Special bits used with hammer drills are designed for pounding while they are turning. As a result, the bits used for drilling holes in masonry and rock have very hard tips, which are frequently made from **carbide**. Carbide-tipped bits are made by coating the tip of the drill bit with a mixture of tungsten carbide and cobalt baked at extremely high temperatures, resulting in a harder tip than common bits.

The term *hammer drill* is often used to describe all tools that hammer and drill. However, rotary hammers (*Figure 12*) are designed for much heavier jobs than a typical hammer drill. They usually have slower rotational speeds than hammer drills and strike harder and less often.

Most rotary hammers require bits that fit into special chuck designs (*Figure 13*), so selected masonry bits must be compatible with the type of chuck on the tool. Chucks on rotary hammers are not keyed but rely upon the design of the bit and chuck to hold it in place. Adapters are also available to use a bit shank design with another type of chuck when necessary.

Figure 11 Hammer drill.
Source: Image property of Stanley Black & Decker. Used with permission

Figure 12 Rotary hammer.
Source: Image property of Stanley Black & Decker. Used with permission

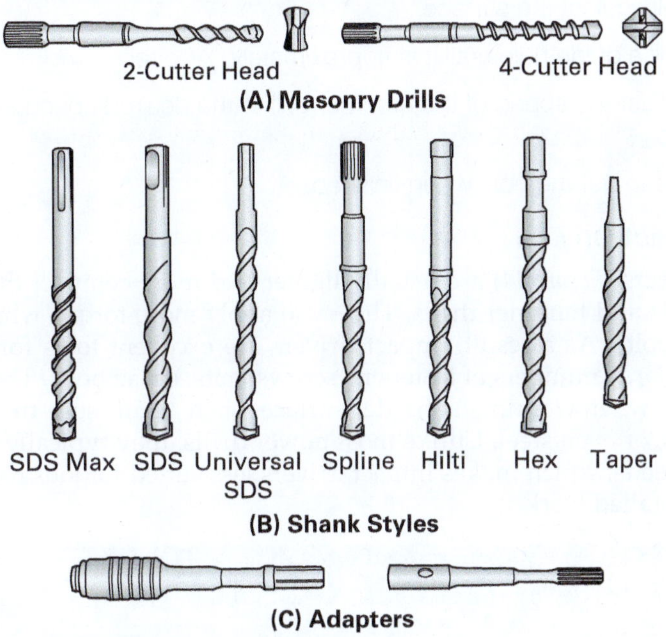

Figure 13 Rotary hammer bit designs.

1.3.1 Hammer Drill Safety and Maintenance

In addition to following basic power tool safety guidelines, before operating a hammer drill, familiarize yourself with these safety rules:

- Always wear the appropriate PPE when working with hammer drills, especially safety glasses and hearing protection.
- Never operate hammer drills or rotary hammers with one hand. Use both hands for maximum control.
- Do not operate hammer drills or rotary hammers near flammable materials. Sparks generated by the bits may ignite them.
- Use a clamp to secure smaller or lighter workpieces, and never hold a workpiece with any part of your body.
- Always disconnect the hammer drill from its power source before changing bits or making any adjustments.
- Check the hammer drill and bit for damage prior to using it on a job.
- Make sure you use the correct bit for the job. Using regular, non-masonry bits on concrete can damage or destroy the bit.

1.3.2 Operating Hammer Drills

If the hammer drill is being used on concrete or similar materials that produce airborne dust, there are important silica exposure standards that must be followed. Silica exposure standards are outlined in NCCER Module 93102, *Basic Safety.*

Use the following steps to operate a hammer drill:

Step 1 Wear appropriate PPE.

Step 2 Ensure you have selected the appropriate drill bit for the job. If you need to replace the bit, disconnect the drill from its power source prior to doing so. Once the bit has been replaced, reconnect the drill to its power source.

Step 3 Make a small indent or mark on the material exactly where the hole needs to be drilled.

Step 4 Set the number of blows the bit will exert each minute.

Step 5 Set the drill to turn slowly.

Step 6 Place the tip of the bit on the indent or mark, then apply pressure on the drill against the surface.

Step 7 Drill a guide hole until it is approximately $\frac{1}{4}$" deep.

Step 8 Adjust the speed of the hammer drill to the desired speed and finish the job.

Step 9 Clean out the hole when finished.

1.3.3 Impact Drivers

Impact drivers (*Figure 14*) are usually lighter and more compact than common power drills and hammer drills. They also apply more torque when fastening screws or bolts. As a result, impact drivers are excellent tools for driving (or loosening) large numbers of fasteners, screws, nuts, or lag bolts. They are also a good choice when working on harder surfaces such as pressure-treated lumber, hardwoods, or even steel. Unlike most power drills, they typically do not offer variable speeds, which makes impact drivers less suited for jobs requiring precision or detailed work.

Figure 14 Impact driver.
Source: Image property of Stanley Black & Decker. Used with permission

Another more subtle difference between hammer drills and impact drivers is their function. Hammer drills exert downward, pounding force on the bit while the chuck is turning. Impact drivers use a spring to exert perpendicular force on the bit. When the bit detects resistance as it turns, the driver exerts greater force, called **torque**, on the turning bit. Torque is the force produced by the drill as it turns the bit.

Finally, impact drivers have a hexagonal-shaped collet, or clamp, instead of a chuck. The hexagonal design is needed because of the extra torque exerted on the bit by the impact driver.

Torque: The turning force produced by the drill. As the bit's turning speed increases, torque decreases, and vice versa.

CAUTION

Impact drivers generate much more torque than regular drills. As a result, they can split, break, or rip through softer or more brittle surfaces. Be extra careful when using an impact driver to attach fasteners to wood or softer materials.

1.3.4 Impact Driver Safety and Maintenance

In addition to following basic power tool safety guidelines, familiarize yourself with these impact drill safety rules before operating the tool:

- Always wear the appropriate PPE when working with impact drivers.
- Read and understand the manufacturer's manual.
- Clean the impact driver before using it.
- Ensure the proper accessory is used for the job.
- Confirm that the fastener and material you are drilling into will not be damaged by the torque generated by the impact driver.

1.3.5 Operating Impact Drivers

Use the following steps to operate an impact driver:

Step 1 Wear appropriate PPE.

Step 2 Select the appropriate bit or attachment for the job.

Step 3 Disconnect the driver from its power source.

Step 4 Pull the collet forward and insert the bit. Once the collet snaps back in place, the bit is attached.

Step 5 Reconnect the driver to its power source.

Step 6 Set the direction you want the bit to turn (tighten or loosen).

Step 7 Place the bit or socket squarely on the fastener.

Step 8 Start the tool slowly and release the trigger when the fastener is attached or removed. Driving the fastener too hard can damage it.

1.4.0 Pneumatic Drills and Impact Wrenches

Pneumatic tools are powered by compressed air. An air hose transfers the compressed air from air compressor to the tool. Pneumatic tools tend to have more power for their weight than comparable electric tools. Two pneumatic power tools commonly used by construction workers are pneumatic drills and impact wrenches.

1.4.1 Pneumatic Drills

Pneumatic drills have many of the same parts, controls, and applications as electric drills. Since there is no motor, they are generally more compact in size. These drills are typically used when there is no available source of electricity or when a high rate of production is necessary. Like electric drills, they can be used as power screwdrivers when fitted with the proper attachments. The pneumatic drill in *Figure 15* has a keyless chuck. They are also equipped with keyed chucks.

Common sizes of pneumatic drills are $\frac{1}{4}$", $\frac{3}{8}$", and $\frac{1}{2}$". The size refers to the diameter of the largest bit shank that can be gripped in the chuck, not the drilling capacity. Some common metric sizes are 8 mm, 10 mm, and 13 mm.

Figure 15 Pneumatic drill.
Source: Image property of Stanley Black & Decker. Used with permission

Figure 16 Pneumatic impact wrench.
Source: Asset: Atlas Copco

1.4.2 Impact Wrenches

Impact wrenches are available in pneumatic (*Figure 16*) and electric, corded and cordless, (*Figure 17*) models. Both types are used to fasten, tighten, and loosen nuts and bolts. Like pneumatic drills, pneumatic impact wrenches must be connected with a hose to an air compressor. The speed and torque of these wrenches can be adjusted based on requirements of the job. Electric models lack the strength of pneumatic wrenches, but they are easier to maintain, especially since there is no need to service hoses and compressors.

Figure 17 Cordless impact wrench.
Source: Image property of Stanley Black & Decker. Used with permission

While pneumatic impact wrenches have traditionally been more popular than their electric cousins, advances in battery-powered technology make cordless wrenches a more viable option today. Some of the most advanced cordless impact wrenches can generate up to 1,500 foot-pounds (ft-lb) of breakaway torque. Although this is nowhere near the forward torque available on some larger pneumatic wrenches, it still provides a convenient and strong alternative for smaller jobs.

1.4.3 Pneumatic Tool Safety and Maintenance Guidelines

In addition to following basic power tool safety guidelines, familiarize yourself with the following pneumatic drill and impact wrench safety rules before operating either type of tool:

- Always wear the appropriate PPE, including eye, hand, and ear protection.
- Review the manufacturer's instructions provided with the pneumatic tool.
- Before changing attachments or performing any maintenance on a pneumatic drill or impact wrench, make sure the air supply is turned off and the tool is physically disconnected from the supply hose.
- Make sure the workpiece is secure, and always maintain a balanced body stance when operating the tool.
- Regularly check the air hose for damage.
- Keep your hands away from the working end of the tool.
- If you are using a pneumatic wrench to tighten a bolt-and-nut combination, use a backup wrench to keep the bolt or nut from spinning.
- Confirm the air supply is clean, dry, and at the proper pressure.
- Keep the drill or wrench lubricated.
- Only use drill bits and impact sockets designed for use with the tool.

WARNING!

The amount of air pressure used to power a pneumatic drill is higher than the Occupational Safety and Health Administration (OSHA) recommendation for cleaning work areas or similar activities. Improper use of pressurized air can lead to injuries and damage to the tool or materials.

1.4.4 Operating Pneumatic Drills and Impact Wrenches

Use the following steps to operate a pneumatic drill or impact wrench:

Step 1 Wear the appropriate PPE.

Step 2 Before operating the pneumatic drill or impact wrench, ensure there is an oiler if the tool requires one, either at the air source or at the tool (*Figure 18*). Confirm the oiler is set properly.

Step 3 Attach the appropriate bit or socket to the tool.

Step 4 Ensure the pneumatic connection between the tool and the supply hose is secure, and install a **whip check** as required (*Figure 19*). A whip check is a safety attachment used to prevent whiplashing in hoses that are inadvertently uncoupled.

WARNING!

The air hose must be connected properly and securely. An unsecured air hose can come loose and whip around violently, causing serious injury. Some fittings require the use of whip checks to keep them from coming loose.

Step 5 Confirm the compressor's regulator is set to the appropriate pounds per square inch (psi). Also, if tightening a bolt or nut, check the recommended torque specifications prior to using the tool.

WARNING!

Only use impact sockets made for pneumatic impact wrenches. Using handheld sockets can damage property and cause injury.

Step 6 Grip the drill or wrench firmly before operating the tool.

Step 7 When you finish using the tool, disconnect it from the air hose.

Whip check: A safety attachment used to prevent whiplashing in hoses when they are inadvertently uncoupled.

Figure 18 Inline pneumatic oiler reservoir.
Source: C.R. Laurence Co., Inc.

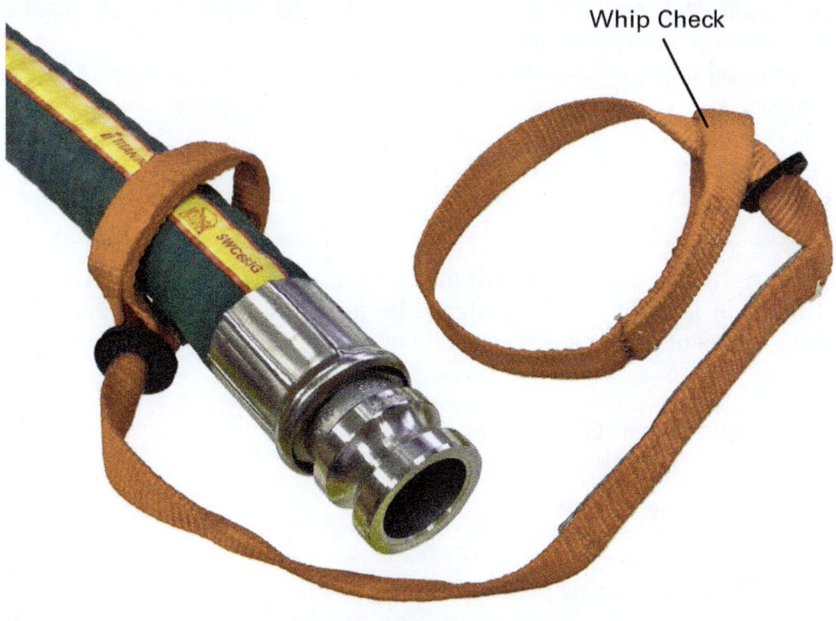

Figure 19 Connect the pneumatic tool and hose using the whip check.
Source: Courtesy of Capital Rubber Corporation -Bensenville IL

1.0.0 Section Review

1. The trigger lock on power tools should *never* be used to lock the trigger in the On position.
   a. True
   b. False

2. The part on a power drill that enables a user to back out a drill bit when it is stuck in the work material is called a(n) _____.
   a. chuck key
   b. reversing switch
   c. whip check
   d. auger switch

3. An important difference between hammer drills and rotary hammers is _____.
   a. rotary hammers turn faster
   b. hammer drills are designed for heavy duty jobs
   c. rotary hammers have keyed chucks
   d. rotary hammers have specially designed chucks

4. A pneumatic power tool that is *best* suited for fastening, tightening, and loosening nuts and bolts is a _____.
   a. cordless screwdriver
   b. hydraulic ratchet
   c. pneumatic impact wrench
   d. right angle drill

2.0.0 Power Saws

Performance Task

1. Safely and properly demonstrate the use of the following tools:
   - Circular saw
   - Jigsaw
   - Reciprocating saw
   - Portable band saw
   - Miter or cutoff saw
   - Table saw

Objective

Identify and explain how to use various types of power saws.

a. Explain how to use a circular saw and identify different types of blades.
b. Differentiate between jigsaws and reciprocating saws and explain how to use them.
c. Explain how to use a portable band saw.
d. Describe the difference between miter saws and cutoff saws.
e. Explain how to use table saws and describe the jobs they are best suited for.
f. Describe the proper use of chainsaws and electric pruning shears.

Using the right saw for the job will make your work much easier. Always make sure the blade is right for the material being cut. This section focuses on the following types of power saws:

- Circular saws
- Jigsaws and reciprocating saws
- Portable handheld band saws
- Power miter saws
- Table saws
- Chainsaws and electric pruning shears

2.1.0 Circular Saws

Many years ago, a company named Skil® made power tool history by introducing the portable circular saw. Today, many companies make dozens of models, but a lot of people still refer to portable circular saws as Skilsaws. Other names include utility saw, electric handsaw, and builder's saw. The portable circular saw (*Figure 20*) is designed to cut lumber and boards to size, and it is available in corded and cordless models.

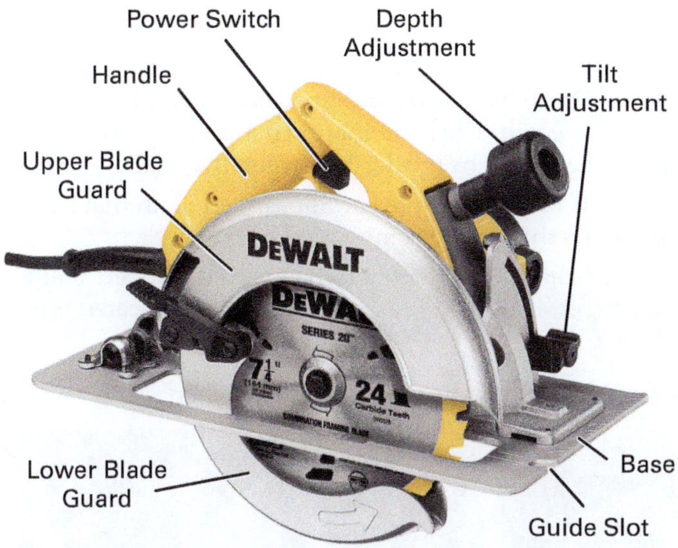

Figure 20 Portable circular saw.
Source: Image property of Stanley Black & Decker. Used with permission

The size of a circular saw is based on the diameter of the circular blade. Circular saws in the United States typically use fractional inch measurements with blade diameters ranging from $3\frac{3}{8}$" to $16\frac{1}{4}$". The most popular blade size for corded saws is $7\frac{1}{4}$". Many smaller cordless circular saws use a $6\frac{1}{2}$" blade. The hole in the center of a circular saw blade fits onto the **arbor**, or *shaft*, of the saw. The most common arbor size for a circular saw blade is $\frac{5}{8}$".

Arbor: The end of a circular saw shaft where the blade is mounted.

Circular saws are also available in metric sizes. Metric circular saws typically have circular blades of 165 mm, 190 mm, or 235 mm. Most metric circular saws have a 20 mm arbor, although some have a 25.4 mm arbor. It is important to note that standard saws with fractional inch measurements are not interchangeable with metric saws. For instance, a $7\frac{1}{4}$" blade with a $\frac{5}{8}$" arbor hole is different from a 190 mm blade with a 20 mm arbor hole. Only metric blades can be used on metric saws, and only fractional inch blades can be used on standard saws.

Circular saw weights can vary, but most of them weigh between 7 lb and 14 lb (3.175 kg and 6.35 kg). The handle of the circular saw has a trigger switch that starts the saw. The motor is protected by a rigid housing. Blade speed is stated in rpm when the blade is not engaged in cutting. The teeth of the blade point in the direction of rotation.

Two guards protect the blade. The upper blade guard protects workers from flying debris and from accidentally touching the spinning blade. The lower guard is spring-loaded; when the saw is pushed forward, it retracts up and under the top guard to allow the saw to cut.

WARNING!

Never use a saw unless the lower blade guard is properly attached. The top guard protects the user from the blade and from flying particles.

Saw blades fall into two categories: standard steel and carbide-tipped. Standard steel blades must be sharpened regularly, while carbide-tipped blades remain sharp for a much longer period. When using carbide-tipped blades, however, it is important to select the right blade for the job. *Figure 21* shows some of the most common types of saw blades, including the following:

- *Rip* — These blades are designed to cut with the grain of the wood. The square chisel teeth cut parallel with the grain and are generally larger than other types of blade teeth.
- *Crosscut* — These blades are designed to cut across the grain of the wood, that is, at a 90° angle. Crosscut teeth cut at an angle and are finer than rip blade teeth.
- *Combination* — These blades are designed to cut hard or soft wood, either with or across the grain. The combination blade features both rip and crosscut teeth with deep troughs (gullets) between the teeth.
- *Nail cutter* — This blade has large carbide-tipped teeth that can make rough cuts through nails that may be embedded in the work.
- *Nonferrous metal cutter* — This blade has carbide-tipped teeth for cutting aluminum, copper, lead, and brass. It should be lubricated with oil or wax before each use.

Always follow the manufacturer's instructions when using saw blades.

Figure 21 Circular saw blades.
Source: Image property of Stanley Black & Decker. Used with permission

2.1.1 Circular Saw Safety Guidelines and Maintenance

There are numerous factors to consider in order to use a circular saw safely and effectively. In addition to following basic power tool safety guidelines, familiarize yourself with the following circular saw safety rules before operation:

- Always wear the appropriate PPE when working with saws, especially safety glasses.
- Always be aware of the blade's location as it relates to your hands or any other part of your body.
- Before connecting the saw to its power source, ensure the blade is tight and the blade guard is working correctly. The chosen blade should have a maximum rpm equal to or higher than the speed of the saw.
- To avoid hitting water lines or electrical wiring, find out what is inside the wall or on the other side of a partition before cutting through a structure.

- During operation, keep both hands on the saw grips.
- Never force the saw through the work. This causes binding and overheating and may cause injury.
- Never reach underneath the work while operating the saw, and never stand directly behind the work. Always stand to one side of the work.
- Use clamps to secure small pieces of material to be cut.
- Know where the saw's power cord is located at all times. Accidentally cutting through the power cord can cause electrocution.

WARNING!

When using a circular saw, workers should never hold material to be cut with their hands; always use a clamp instead.

The most important maintenance on a circular saw is at the lower blade guard. Sawdust builds up and causes the guard to stick. If the guard sticks and does not move quickly over the blade after it makes a cut, the bare blade may still be turning when the saw is set down and may cause damage. Remove sawdust from the blade guard area. Remember to always disconnect the power source before performing any maintenance.

To avoid personal injury and damage to materials, check often to make sure the guard quickly and smoothly snaps shut. To ensure smooth operation of the guard, disconnect the saw from its power source, allow it to cool, and clean foreign material from the track. Be aware of fire hazards when using cleaning liquids such as isopropyl alcohol. Do not lubricate the guard with oil or grease since this could cause sawdust to stick in the mechanism. Always keep blades clean and sharp to reduce friction and kickback. Blades can be cleaned with hot water or mineral spirits. Be careful with mineral spirits as they are very flammable.

2.1.2 Operating Circular Saws

Use the following steps to operate a circular saw:

Step 1 Wear appropriate PPE.

Step 2 Secure and support the material you are cutting. If the work isn't heavy enough to stay in position without moving, it should be clamped down.

Step 3 Mark the cut using a pencil or marking tool.

Step 4 Unplug or remove the battery pack from the saw before adjusting the blade depth to the thickness of the wood being cut plus $\frac{1}{4}$" (roughly 6 mm). This prevents the blade from protruding through the material farther than necessary.

Step 5 Place the front edge of the baseplate on the work so the guide notch and the blade are in line.

Step 6 After the saw has been started and is up to full speed, slowly move it forward to begin cutting. As the blade begins to cut, it will leave a **kerf** (*Figure 22*) that measures roughly $\frac{1}{8}$" (3.2 mm) wide. It is important to consider the width of the kerf when making a cut with a circular saw. Be sure to align the blade with the waste side of the cutting line, or the finished piece will be short. When marking the cut, mark an X on the waste side of the cut mark as a reminder of which side of the mark to cut.

CAUTION

Make sure the blade is appropriate for the material being cut.

Kerf: The channel created by a saw blade passing through the material, which is equal to the width of the blade teeth.

Figure 22 Saw kerf.
Source: Courtesy of Cianbro

Step 7 The lower blade guard will automatically rotate up and under the top guard when the saw is pushed forward. While cutting with the saw, grip the saw handles firmly with two hands, as shown in *Figure 23*. As the cut nears completion, the guide notch on the baseplate will move off the end of the work. At that point, use the blade as a guide.

Step 8 Release the trigger switch once the cut is complete. The blade will stop rotating. Make sure the blade has stopped before setting the saw down.

NOTE

If the saw cuts off the line, stop the saw, back up slightly, and restart the cut. Do not force the saw.

Figure 23 Proper use of a circular saw.

The Worm Drive Saw

The worm drive saw is a heavy-duty type of circular saw. Most circular saws have a direct drive. The blade is mounted on a shaft that is part of the motor. With a worm drive saw, the motor drives the blade from the rear through two gears. One gear (the worm gear) is cylindrical and threaded like a screw. The worm gear drives a wheel-shaped gear (the worm wheel) that is directly attached to the shaft to which the blade is fastened. This setup delivers much more rotational force (torque), making cutting lumber with double thickness easier. The worm drive saw is almost twice as heavy as a conventional circular saw and should only be used by an experienced craftworker.

Source: Ridge Tool Company

2.2.0 Jigsaws and Reciprocating Saws

Jigsaws and **reciprocating** saws are two types of saws capable of making straight and curved cuts. Both saws have a blade that moves back and forth to enable the cutting action.

Reciprocating: Moving backward and forward on a straight line.

2.2.1 Jigsaws

Jigsaws have very fine blades, which makes them effective tools for doing delicate and intricate work. They are commonly used to cut patterns or irregular shapes from wood or thin metals.

Today's cordless jigsaw (*Figure 24*) is an extremely useful portable power tool. It can make straight or curved cuts in wood, metal, plastic, wallboard, and other materials. The jigsaw cuts with a blade that moves up and down, unlike the spinning blade of a circular saw. This means that each cutting stroke (upward) is followed by a return stroke (downward), so the saw cuts only half the time it is operating. This is called *up-cutting* or *clean-cutting.*

The baseplate (shoeplate or footplate) is an important part of the jigsaw. Its broad surface helps to keep the blade lined up, prevents the work from vibrating, and allows the blade teeth to bite into the material.

Many models are available with tilting baseplates for cutting beveled edges. They come with a top handle or a barrel handle, and some cordless models are available.

The jigsaw has changeable blades that enable it to cut many different materials, from wood and metal to wallboard and ceramic tile. Fine-toothed blades are used for thin materials and smoother cuts. Coarse blades are used for faster cutting of thicker materials and when smooth cuts are not a concern. Blades are rated by the number of teeth per inch (TPI) or teeth per centimeter (TpCM).

Most jigsaws can be operated at various blade speeds. Types of jigsaws include single-speed, two-speed, and variable-speed. The speed of a variable-speed jigsaw is controlled by how far the trigger is depressed. The low-speed range is for cutting hard materials, and the high-speed range is for soft materials.

Top Handle

Figure 24 Jigsaw.
Source: Image property of Stanley Black & Decker. Used with permission

CAUTION

Do not lift the blade out of the work while the saw is still running. If the blade is lifted out, the tip of the blade may hit the wood surface, marring the work and possibly breaking the blade.

2.2.2 Reciprocating Saws

A reciprocating saw, regardless of the manufacturer, is often referred to as a SawZall® (a trademark of Milwaukee Tool) in the United States because it was the first saw designed to serve as an electric hacksaw. Both the jigsaw and the reciprocating saw can make straight and curved cuts. They are used to cut irregular shapes and holes in plaster, plasterboard, plywood, studs, metal, and most other materials that can be cut with a saw.

Both saws have straight blades that move up and down along a straight line as they are guided along the cut. The reciprocating saw (*Figure 25*) is designed for more heavy-duty jobs than the jigsaw. It also uses longer and tougher blades than a jigsaw. In fact, the reciprocating saw is used for jobs that require brute strength, which makes it an excellent choice for general demolition work. It can saw through walls or ceilings and create openings for windows, plumbing lines, and more.

Jigsaw and reciprocating saw blades are available for various cutting tasks and materials. The blades are often labeled for the material or specific use to make selection easier. Generally, metal cutting blades have more TPI (or TpCM) than wood-cutting blades, and the teeth are also smaller.

Like the jigsaw, reciprocating saws come in single-speed, two-speed, and variable-speed models. The low-speed setting is best for metalwork, while the high-speed setting is best for sawing wood and other soft materials.

The baseplate (shoeplate or footplate) may have a swiveling action or be fixed. Whatever the design, the baseplate provides a brace or support point for the sawing operation.

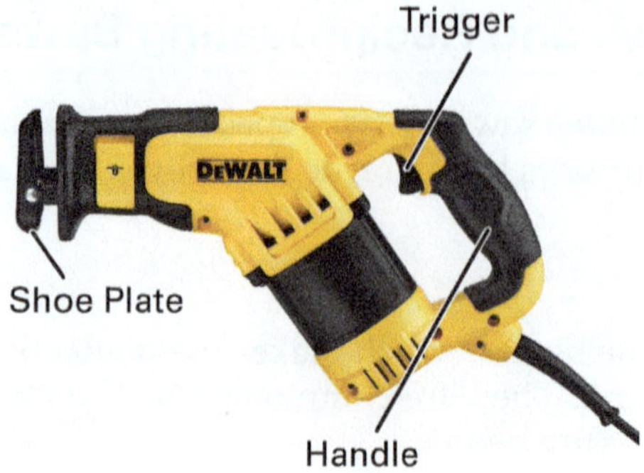

Figure 25 Reciprocating saw.
Source: Image property of Stanley Black & Decker. Used with permission

2.2.3 Jigsaw and Reciprocating Saw Safety and Maintenance

As with any power tool, jigsaws and reciprocating saws can be dangerous to operate if safety guidelines are ignored. In addition to following basic power tool safety guidelines, familiarize yourself with the following safety rules before operating a jigsaw or reciprocating saw:

- Always wear the appropriate PPE when working with saws, especially safety glasses.
- Before cutting through a wall or partition, find out what is inside the wall or on the other side of the partition. This will prevent the accidental cutting of water lines or electrical wiring.
- Make sure the saw is disconnected from its power source before installing or changing blades or performing any maintenance on the saw.
- When installing a blade in the saw, ensure it is in the collar as far as it will go before securely tightening the setscrew.
- When replacing a broken blade, look for any pieces of the blade that may be stuck inside the collar.
- Verify that the switch is in the Off position before it is plugged into a power source.
- Always use a sharp blade, and never force the blade through the work. Forcing or leaning into the blade can cause you to lose balance and possibly even be cut by the saw.
- When cutting metal, use a metal cutting blade.
- Lubricate the blade with an agent, such as beeswax, to help make tight turns and to reduce the chance of breaking the blade.

2.2.4 Operating Jigsaws and Reciprocating Saws

Many of the steps involved in the safe and efficient use of jigsaws and reciprocating saws are the same. Use the following steps to operate both types of saws:

Step 1 Wear the appropriate PPE.

Step 2 Clamp the material being sawed to a pair of sawhorses or secure it in a vise to reduce vibration.

Step 3 Disconnect the saw from its power source.

Step 4 Confirm that the blade you have chosen suits the material you are cutting.

Step 5 Check the blade for dulling or damage.

Step 6 Measure the material and mark it before cutting.

NOTE

If cutting from the edge of a board or panel with a jigsaw, ensure the front of the baseplate rests firmly on the work's surface before starting the saw. The blade should not be touching the work at this stage.

Step 7 If a cut must be made from the middle of a board, drill a hole at the starting point large enough for the blade to pass through.

Step 8 Reconnect the saw to its power source.

Step 9 Squeeze the trigger to start the saw and move the blade gently but firmly into the material. Continue feeding the saw into the work at a reasonable pace without forcing it. Never force the blade into the work.

Step 10 When the cut is finished, release the trigger and let the blade come to a stop *before* removing it from the work.

Set the saw to the desired speed before cutting the material with a reciprocating saw. Use lower speeds for sawing metal; use higher speeds for sawing wood and other soft materials. Grip the saw with both hands (*Figure 26*) and place the baseplate firmly against the workpiece. Once the trigger is squeezed, the blade moves back and forth, cutting the material on the backstroke.

WARNING!

Use both hands to grip the reciprocating saw firmly. Otherwise, the pull from the blade's grip may jerk the saw out of your grasp.

NOTE

When a reciprocating saw blade first strikes the surface, it may jump back. Keep a steady hand; the cutting action will eventually allow the blade to enter the workpiece. Forcefully plunging the blade into the surface is a common cause of broken blades.

Figure 26 Proper use of a reciprocating saw.
Source: Image property of Stanley Black & Decker. Used with permission

2.3.0 Portable Band Saws

The portable band saw (*Figure 27*) can cut pipe, metal, plastics, wood, and irregularly shaped materials. It is especially good for cutting heavy metal but performs equally as well on fine-cutting jobs. Although it can cut wood, it is used almost exclusively to cut metal products on the jobsite.

The band saw has a one-piece blade that runs in one direction around guides at either end of the saw. The blade is a thin, flat piece of steel that must be of the proper length to fit the revolving pulleys that drive and support it. Proper blade length is determined by referencing the manufacturer's documentation. Like most blades, its coarseness is rated in TpCM or TPI. Generally, higher TPIs produce smoother cuts. Thicker materials require coarser blades. However, if the blade is too coarse for the material, the individual teeth may begin to break off. *Figure 28* shows how the blade is routed around the pulleys and through the blade guides.

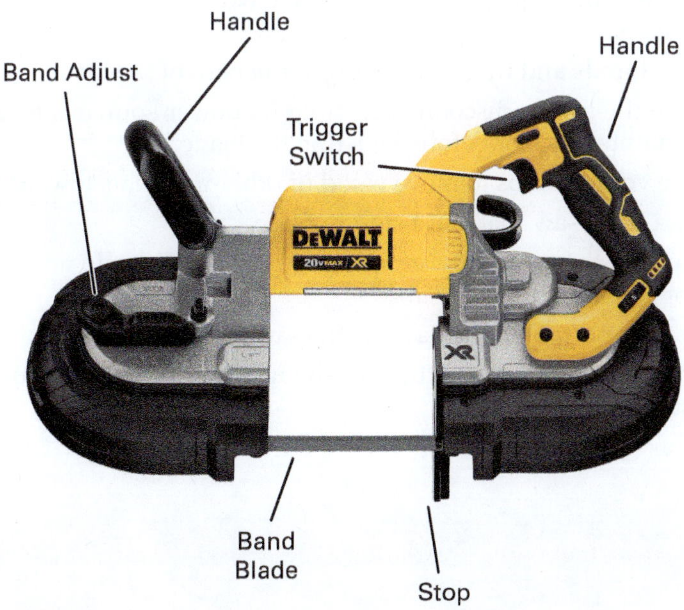

Figure 27 Portable handheld band saw.
Source: Image property of Stanley Black & Decker. Used with permission

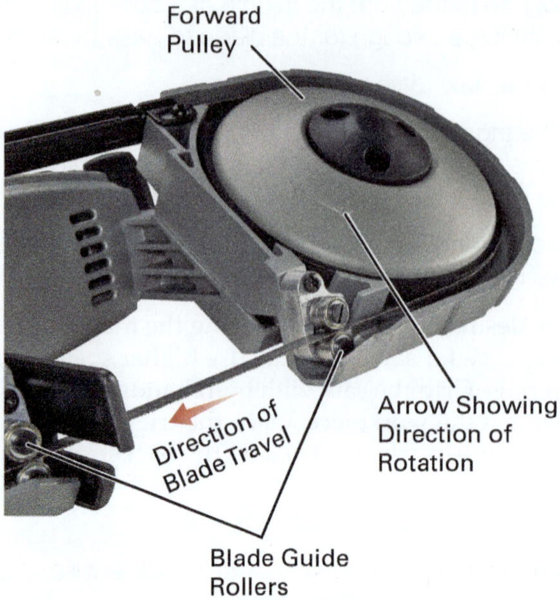

Figure 28 Band saw pulley and blade guide rollers.
Source: Courtesy of Milwaukee Tool

While some band saws have multiple speeds, most do not. The portable band saw generally cuts best at a low speed since higher speeds cause the blade's teeth to rub rather than cut. This can create heat through friction, which reduces the blade's life.

WARNING!

A portable band saw always cuts in the direction of the user. Therefore, workers must be especially careful to avoid injury when using this saw type. Always wear appropriate PPE and stay focused on the work.

2.3.1 Band Saw Safety and Maintenance

In addition to following basic power tool safety guidelines, familiarize yourself with the following band saw safety rules before operation:

- Always wear the appropriate PPE when working with saws, especially safety glasses.
- Keep your hands and fingers away from the path of the blade.
- Make sure the saw is disconnected from its power source before performing any maintenance or before adjusting the blade.
- Never use your thumbs to move wood or other materials toward the blade.
- Use only a band saw that has a stop in place.
- Before cutting through lines or pipes, confirm that they do not contain hazardous or combustible materials.
- Never force a portable band saw; let the saw do the cutting.
- Before backing out of a cut, turn off the machine, then back out the blade after it stops moving.

2.3.2 Operating a Portable Band Saw

Use the following steps to operate a portable band saw:

Step 1 Wear the appropriate PPE.

Step 2 Disconnect the portable band saw from its power source.

Step 3 Verify you are using the right blade for the job.

Step 4 Check the blade tension and adjust, if necessary.

Step 5 Confirm that both blade guides are close to the blade.

Step 6 Reconnect the portable band saw to its power source.

Step 7 Place the stop firmly against the object to be cut.

Step 8 Plan saw cuts to avoid backing out of curves in the cut object.

Step 9 Place marks on the wood to guide you while cutting. Remember to use pencil since ink will permanently stain most wood.

Step 10 Hold the material you intend to cut flat on the table before starting the cut.

Step 11 Turn on the band saw and allow the blade to come to full speed.

Step 12 Feed the material so that the blade follows the path of the marks. Little or no downward pressure is needed to make a good cut since the saw's weight provides pressure for cutting.

Step 13 When finished cutting, turn off the bandsaw.

NOTE

The direction of blade rotation can pull the saw toward the stop. If the stop is not firmly against the workpiece, the saw will jump when started until the stop slams against the material being cut.

2.4.0 Miter and Cutoff Saws

Miter saws and cutoff saws are similar as both are used to make straight or miter cuts. Both types of saws can be mounted permanently, but portable and cordless versions of the saws allow users to move them from a workshop to a worksite.

2.4.1 Power Miter Saws

The power miter saw combines a miter box with a circular saw, allowing it to make straight and miter cuts. There are two types of power miter saws: standard miter saws and sliding compound miter saws.

The saw blade of a standard miter saw pivots horizontally from the rear of the table and locks it in position to cut angles from 0° to 45° right and left. Stops are preset for common angles. The difference between the standard miter saw and the compound miter saw (*Figure 29*) is that the blade on the compound miter saw can be tilted vertically, allowing the saw to be used to make a compound cut (combined bevel and miter cut).

The sliding compound miter saw is like a miter saw and compound miter saw (*Figure 29*). It has a rail in the table that allows the motor and blade assembly to slide forward and backward. This sliding capability enables the tool to cut wider material than a standard miter saw can.

2.4.2 Abrasive Cutoff Saws

An abrasive cutoff saw (*Figure 30*), also known as a *chop saw* or *cutoff saw*, can be used to make straight or angular cuts through materials such as angle iron, flat bar, and channel. Like miter saws, cutoff saws can be stationary or portable.

The abrasive blade on a cutoff saw can be between 10" and 18" in diameter. Metric blades are commonly 250 mm to 350 mm in diameter. When the saw is in operation, the blade spins so quickly that the resulting friction is hot enough to burn through the material. Like all rotating blades and stones, the maximum rpm of the blade must be equal to or greater than that of the saw.

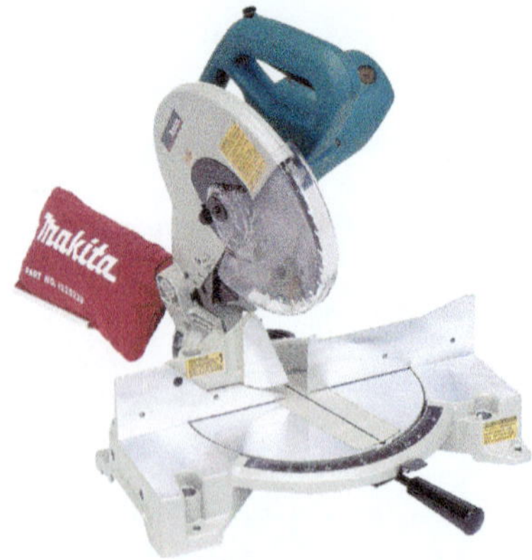

(A) Compound Miter Saw

(B) Compound Sliding Miter Saw

Figure 29 Miter saws.
Sources: Ridge Tool Company (29A); Image property of Stanley Black & Decker. Used with permission (29B)

Figure 30 Abrasive cutoff saw.
Source: Image property of Stanley Black & Decker. Used with permission

2.4.3 Miter and Cutoff Saw Safety and Maintenance

Several basic safety guidelines should be followed when operating a miter or cutoff saw to ensure worker safety and equipment protection. If either saw is being used on materials that produce airborne dust, important silica exposure standards must be followed. These standards are outlined in the NCCER Module 93102, *Basic Safety*.

In addition to following basic power tool safety guidelines, familiarize yourself with the following safety rules before operating a miter or cutoff saw:

- Always wear the appropriate PPE when working with saws, especially safety glasses.
- Never wear a watch or jewelry while operating the saw because they can get caught in the machinery.
- Securely lock the blade at the correct cutting angle.
- While operating the saw, keep your hands and fingers clear of the blade and never attempt to adjust the saw while it runs.
- Confirm the blade is in good condition and secure before using the saw.
- Do not allow other workers to stand nearby while operating a saw.
- Make sure the work area is clear of flammable materials, such as chemicals and rags, that could ignite from sparks. Abrasive cutoff saws tend to produce streams of sparks.
- Disconnect the saw from its power source before changing the blade or performing any sort of maintenance or setup.
- Verify that the rpm rating of the blade meets or exceeds the saw's spindle speed.
- Check all saw guards to ensure they are in place and working properly.
- Never retract a safety guard to view the cut material while the saw is in use.
- If cutting long material, ensure the other end of the material is supported.

2.4.4 Operating Miter and Cutoff Saws

Using a power miter saw and an abrasive cutoff saw safely and efficiently involves many of the same steps used for other types of power saws. Use the following steps to operate a power miter or cutoff saw:

Step 1 Wear the appropriate PPE.

Step 2 Ensure the power supply is disconnected from the saw.

Step 3 Set the angle or tilt of the blade according to the cut that needs to be made.

Step 4 Reconnect the power supply to the saw.

Step 5 Place the material to be cut firmly against the fence.

Step 6 With the blade arm raised, turn on the saw and allow it to reach maximum speed.

Step 7 Firmly grasp the handle of the rotating saw arm and lower the blade onto the material. Some compound miters allow the rotating arm to slide while cutting.

Step 8 Once the cut has been made, turn off the saw and raise the rotating saw arm.

2.5.0 Table Saws

Table saws are designed with a circular saw blade that protrudes up through a slot on the bench. They are typically distinguished by the size of the blade they can accommodate. Table saws using 10" blades are excellent options for most carpentry jobs, while saws with 8" blades are designed for cutting smaller materials. Those with 12" blades are the best choice for larger jobs.

Table saws are typically used for ripping and crosscutting lumber. Ripping involves cutting lumber to a specific width, while crosscutting involves cutting it to a specific length. Since the table saw has an exposed blade and a fence, users can feed lumber into the blade in a controlled manner. The result is a saw that is excellent for making precise and accurate cuts, which is one reason table saws are a favorite for finish carpenters.

While many older table saws were large, stationary tables that were plugged into a receptacle for power, many of today's table saws are cordless and/or portable with folding legs (*Figure 31*).

The table part of this tool is where wood is laid before it is pushed against the rip fence and into the saw blade. The saw part of this tool is mounted under the table and can be adjusted up and down to expose more or less of the blade. Some table saws also allow the blade to be tilted up to 45°.

Figure 31 Portable table saw.
Source: Image property of Stanley Black & Decker. Used with permission

2.5.1 Table Saw Safety and Maintenance

In addition to following basic power tool safety guidelines, familiarize yourself with the following table saw safety rules before operating it:

- Always wear the appropriate PPE when working with saws, especially safety glasses.
- Use blades designed for the job.
- Confirm the blade is sharp before cutting.
- Verify that the blade guard moves freely before connecting to power.
- Use a push stick or board to push any material requiring your hand to pass near the blade.
- Do not use the table saw fence during a crosscutting operation without placing a block between the cut material and the fence. This prevents the cutoff piece from being trapped between the blade and the fence.

- Keep the table saw bench clear of all debris, especially when it is in use.
- Disconnect the table saw from its power source when adjusting the blade or performing any maintenance.
- Always anticipate kickback when ripping any type of material. This means you should not stand in line with the blade while it is cutting. Instead, stand to one side until cutting is finished.
- Set the blade to the correct height.
- Make sure any lumber you are cutting is flat on the bench and the board is pressed level against the rip fence when cutting. Otherwise, binding and kickback may occur.

2.5.2 Ripping Wood with a Table Saw

CAUTION

Do not push lumber or any other material against the table saw's blade before the blade reaches full speed. Contacting the blade any earlier may cause kickback and/or serious injury.

Use the following steps to rip wood or other material on a table saw:

Step 1 Wear the appropriate PPE.

Step 2 Confirm the table saw is disconnected from any power source.

Step 3 Ensure the table saw is fitted with a rip blade.

Step 4 Adjust the blade so that it is no more than $\frac{1}{4}$" above the thickness of the material to be cut.

Step 5 Adjust the rip fence so that its inner edge corresponds with the desired width of the material once it is cut.

Step 6 Connect the power source and place the material you want to cut on the bench. Ensure it is aligned against the rip fence but not touching the blade.

Step 7 Slowly push the material along the edge of the rip fence, ensuring it is held snugly against the rip fence while it is pushed.

Step 8 Use a push stick to guide the final few inches of the material past the blade. Push sticks should always be used if there is any chance that pushing the material will put your fingers within a few inches of the blade.

Step 9 When finished ripping, turn off the table saw.

2.6.0 Chainsaws and Electric Pruning Shears

Few tools are more versatile and widely used in the world of agriculture than chainsaws and electric pruning shears (*Figure 32*).

(A) (B)

Figure 32 Chainsaw and electric pruning shears.
Sources: iStock (32A); Shutterstock (32B)

Chainsaws are powerful tools primarily used for felling trees, cutting firewood, and pruning large branches. They are particularly valuable in orchards and timber production, where their ability to cut wood swiftly saves time and labor compared to manual saws and axes. The development of lighter, more ergonomic models has made chainsaws more accessible to a wider range of users, and battery-operated models provide the same mobility as gas-operated options with reduced maintenance.

On the other hand, electric pruners are designed for more precise and delicate tasks (*Figure 33*). These tools are ideal for trimming vines, small branches, and shrubs. In vineyards and fruit orchards, electric pruners allow for quick, clean cuts, which are crucial for the plant's health and the produce's quality. Their lightweight design and ease of use reduce hand fatigue and the risk of repetitive strain injuries, making them a favorite among workers who spend long hours pruning.

Some electric pruners are equipped with a telescopic pole, like the one shown in *Figure 32*. This pole greatly extends the worker's reach, allowing them to tackle hard-to-reach areas.

Both chainsaws and electric pruners contribute to better crop management and yield. Chainsaws facilitate the removal of old or diseased trees, creating space for healthier growth, while electric pruners ensure that plants are properly maintained and shaped for optimal sunlight exposure and air circulation. This maintenance is essential for preventing diseases and promoting robust plant growth.

Figure 33 Electric pruners.
Source: Shutterstock

2.6.1 Maintenance and Safety Tips

Strong maintenance and safety habits can help reduce the risk of injury and improve the usable lifetime of your tools. Familiarize yourself with the following chainsaw safety guidelines prior to operating it:

- Read and follow the manufacturer's instructions and safety guidelines before operating the tools.
- Sharpen chainsaw blades and pruner blades regularly to ensure clean cuts and prevent strain on the tools.
- Check and tighten all screws and bolts to avoid operational hazards. For instance, a saw chain should be taught so that it doesn't come loose and fall off the bar, but it shouldn't be too tight to limit function (*Figure 34*).

- Ensure the chainsaw is in the Off position, and slide the chain back and forth before the initial operation begins. Wear gloves while you complete this task to avoid cutting your hand from the sharp teeth.
- Clean the tools after each use to remove sap, sawdust, and debris that can clog moving parts. To avoid damaging the tool, only use manufacturer-approved cleaners.
- Lubricate moving parts to reduce friction and wear.
- Inspect electrical components and cables of electric pruners for damage.
- Wear appropriate PPE such as protective leg chaps or pants, a chainsaw-protective shirt, gloves, goggles, ear protection, and sturdy footwear.
- Ensure the work area is clear of obstacles and other people, maintaining a safe distance.
- Use both hands to operate chainsaws, maintaining a firm grip to control the tool.
- Avoid using chainsaws and electric pruners to prevent accidents in wet or slippery conditions.

Figure 34 Chainsaw maintenance.
Source: iStock

2.0.0 Section Review

1. The proper way to start cutting material with a circular saw is to _____.
   a. power up the saw to full speed then slowly move it forward into the material
   b. hold the lower blade guard up to position the blade on the cut mark
   c. press the blade against the material being cut and set the saw rpm to low
   d. tilt the front edge of the baseplate upward and push the saw forward

2. A jigsaw is an effective tool for _____.
   a. drilling holes in concrete or pavement
   b. making long straight cuts through thick metal
   c. cutting through walls in demolition jobs
   d. doing delicate work on thin materials

3. The blade on a portable band saw _____.
   a. moves up and down through a shoeplate
   b. spins in a circular motion on an arbor
   c. runs in one direction around guides
   d. reciprocates in and out from a guard

4. A type of miter saw in which the blade can be pivoted horizontally and vertically is called a _____.
   a. sliding jigsaw
   b. compound miter saw
   c. reciprocating saw
   d. sliding abrasive saw

5. Pushing a board against a table saw's blade before it reaches full speed _____.
   a. is the desired approach when ripping thin lumber
   b. is the best approach when using larger blades
   c. can cause kickback and serious injury
   d. can increase efficiency and speed

3.0.0 Grinders and Oscillating Multi-Tools

Objective

Describe the jobs best suited to grinders and oscillating multi-tools.

a. Explain how to use various types of grinders.
b. Identify grinder accessories and the jobs for which they are used.
c. List the jobs that can be performed using an oscillating multi-tool.

Performance Task

1. Safely and properly demonstrate the use of the following tools:
   - Portable or bench grinder
   - Oscillating multi-tool

Grinding tools can power all kinds of **abrasive** wheels, brushes, buffs, drums, bits, saws, and discs. These wheels come in a variety of materials and **grit**. They can drill, cut, smooth, polish, shape or sand wood or metal, mark steel and glass, and sharpen or engrave. They can even be used on plastics.

WARNING!

Always wear safety goggles and a face shield when working with grinders. Make sure the work area is free of combustible materials such as rags or flammable liquids and that a fire extinguisher is easily accessible. Clothes should be snug, comfortable, and free of cuffs at the wrists and ankles. Wearing excessively loose clothing on the worksite can be extremely dangerous.

Abrasive: A substance, such as sandpaper, that is used to wear away material.

Grit: A granular, sand-like material used to make sandpaper and similar materials abrasive. Grit is graded according to its texture. The grit number indicates the number of abrasive granules in a standard size (inches or centimeters). The higher the grit number, the more particles in a given area, indicating a finer abrasive material.

3.1.0 Grinders

Grinders are available in various configurations. Common handheld grinders include angle grinders (also called *side grinders* or *right-angle grinders*), end grinders, and detail grinders. Stationary grinders, called *bench grinders*, are permanently mounted on a worktable or bench.

Angle grinders are used to grind away hard, heavy materials and to grind surfaces such as pipes, plates, or welds (*Figure 35*). The angle grinder has a rotating grinding disc set at a right angle to the motor shaft.

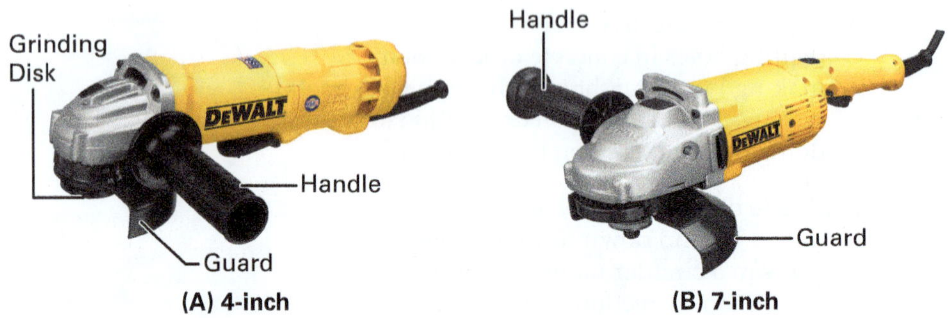

Figure 35 Angle grinders.
Source: Image property of Stanley Black & Decker. Used with permission

Figure 36 End grinder.
Source: Courtesy of Milwaukee Tool

End grinders, sometimes called *horizontal grinders* or *pencil grinders*, are used to smooth the inside of materials, such as pipe (*Figure 36*). The grinding disc on the end grinder rotates in line with the motor shaft. Grinding is also done with the outside of the grinding disc.

Like end grinders, detail grinders (*Figure 37*), also known as *die grinders*, have an arbor that extends from the motor shaft onto which small attachments, called points, can be mounted to smooth and polish intricate metallic work. These attachments, as shown in *Figure 37*, are commonly made in shaft sizes ranging from $\frac{1}{16}$" to $\frac{1}{4}$". The shaft of these points is called the spindle. Metric points typically come in spindle sizes of 3 mm and 6 mm. A tremendous variety of point shapes are available to suit the grinding task.

The primary difference between an end grinder and a detail grinder is power. End grinders offer more power than a detail grinder.

Bench grinders (*Figure 38*) are electrically powered stationary grinding machines. They usually have two grinding wheels used for grinding, rust removal, and metal buffing. They are also great for renewing worn edges and maintaining the sharp edges of cutting tools. For example, the bench grinder can be used to smooth the mushroomed heads of cold chisels.

Figure 37 Detail grinder and point.
Sources: Image property of Stanley Black & Decker. Used with permission (37A); Courtesy of Dremel (37B)

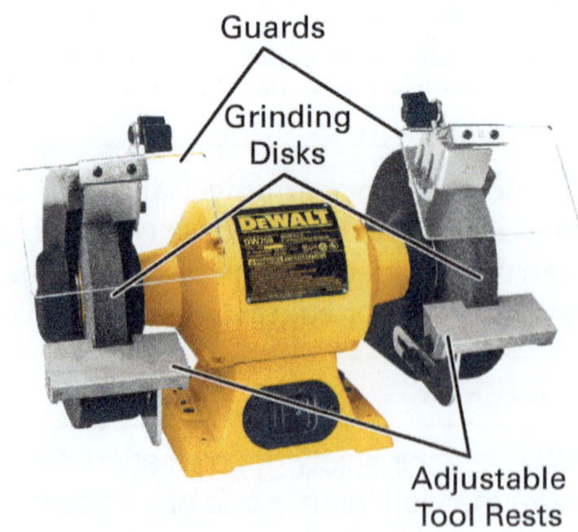

Figure 38 Bench grinder.
Source: Image property of Stanley Black & Decker. Used with permission

Heavy-duty grinder wheels range from 6¾" to 10" (150 mm to 250 mm) in diameter. Each wheel's maximum speed is given in rpm. Never use a grinding wheel above its rated maximum speed. Its rated speed must be equal to or faster than the maximum speed of the power tool.

Bench grinders come with an adjustable tool rest. This is the surface on which you position the material you are grinding, such as cold chisel heads. There should be only ⅛" (about 3 mm) between the tool rest and the wheel.

WARNING!

Never adjust tool rests when the grinder is on or when the grinding wheels are spinning. Doing so may damage the work or cause injuries. Disconnect the power source before adjusting any type of grinder.

3.1.1 Grinder Safety and Maintenance

In addition to following basic power tool safety guidelines, familiarize yourself with the specific grinder safety rules before operating any type of grinder.

Before operating angle, end, and detail grinders, familiarize yourself with these safety rules:

- Always wear the appropriate PPE when working with grinders, especially gloves and safety glasses. Wearing PPE is critically important since grinders generate a lot of high-speed sparks and flying debris.
- Never wear loose clothing or jewelry that can get caught in a grinder's wheels.
- Always disconnect the power source before performing maintenance.
- Keep your hands away from the grinding wheel.
- Direct sparks and debris (*Figure 39*) away from coworkers and any flammable materials. To keep sparks from flying into the path of other workers or from igniting flammable materials, use standing screens when grinding near walkways, other workers, or storage areas.
- Never use an angle grinder, end grinder, or detail grinder unless equipped with the manufacturer-provided guard surrounding the grinding wheel.
- Choose a grinding disc that is appropriate for the type of work being performed.
- Before installing, ensure the new grinding wheel's maximum rpm markings are equal to or greater than the grinder's maximum speed.
- Remember that individual wires from a wire wheel can separate from the wheel at a high rate of speed and cause injury.
- Before starting a grinder, ensure the grinding disc is properly secured and in good condition.
- Establish firm footing and maintain a firm grip on the grinder with both hands to avoid being pulled off balance.
- Let the grinder come up to full speed before grinding.
- When grinding on a platform, use a flame-retardant blanket to catch falling sparks.
- After shutting off the power, do not leave the tool until the grinding disc has completely stopped.

WARNING!

Grinding discs can instantaneously disintegrate if used when they are cracked. Inspect the disc for cracks before using the grinder.

Figure 39 Be aware of sparks created by grinders.
Source: Asset: Atlas Copco

Before operating bench grinders, familiarize yourself with these safety rules:

- Always wear the appropriate PPE when working with a bench grinder, especially gloves and safety glasses. Wearing PPE is critically important since bench grinders generate a lot of high-speed sparks and flying debris.
- Grinding metal on a bench grinder creates sparks, so keep the area around the grinder clean.
- Always adjust the tool rests to be within $^1/_8$" (roughly 3 mm) of the wheel. This reduces the chance of getting the work wedged between the rest and the wheel.
- Never use a grinding wheel above its rated maximum speed.
- After finishing a job with the bench grinder, shut it off.
- Before mounting a grinding wheel onto a bench grinder, inspect the wheel for chipped edges and cracks.

WARNING!

Never grind soft metals or wood on a grinding wheel. The material will wedge itself between the grit, creating stresses that may cause the wheel to shatter.

3.1.2 Operating Grinders

Be sure to follow and familiarize yourself with the steps to safely operate a grinder.

Before operating angle, end, and detail grinders, familiarize yourself with these steps:

Step 1 Wear appropriate PPE, especially a face shield.

Step 2 Secure the material in a vise or clamp it to the workbench.

Step 3 To use an angle grinder, place one hand on the handle of the grinder and one on the trigger. To use an end grinder or detail grinder, grip the grinder handle at the shaft end with one hand and cradle the opposite (motor) end of the tool in your other hand.

Step 4 Always allow the grinder to reach its maximum speed before grinding.

Step 5 Turn off the grinder and ensure the wheel has stopped rotating before putting it down.

Step 6 Complete the work by removing any loose material with a wire brush.

CAUTION

Do not force the tool or overload the motor. If too much pressure is applied, there will be a significant reduction in the grinder's rpm.

Before operating bench grinders, familiarize yourself with these steps:

Step 1 Wear appropriate PPE, especially a face shield.

Step 2 Always use the adjustable tool rest as a support when grinding or beveling metal pieces. There should be a maximum gap of $\frac{1}{8}$" (roughly 3 mm) between the tool rest and the wheel and $\frac{1}{4}$" (roughly 6 mm) between the wheel guard (also called a spark arrestor) and wheel.

Step 3 Confirm that the bench grinder is firmly secured to a stable surface.

Step 4 Always allow the grinder to reach its maximum speed before grinding.

Step 5 Keep the metal cool while it is being ground. If the metal gets too hot, it can destroy the temper (hardness) of the material.

Step 6 Always work on the face of the wheel, not on the side. *Figure 40* shows the proper way to use a bench grinder.

Did You Know?

Grinding Tungsten Can Be Hazardous to Your Grinder

Workers who use power tools are almost certain to encounter objects that contain tungsten. As mentioned earlier, tungsten is an extremely hard and brittle metal with a very high melting point. Attempting to grind tungsten carbide tools with conventional grinding discs and wheels is a bad idea. Rather than sharpening the tool, the most likely outcome of such an attempt is a damaged and/or weakened grinding wheel.

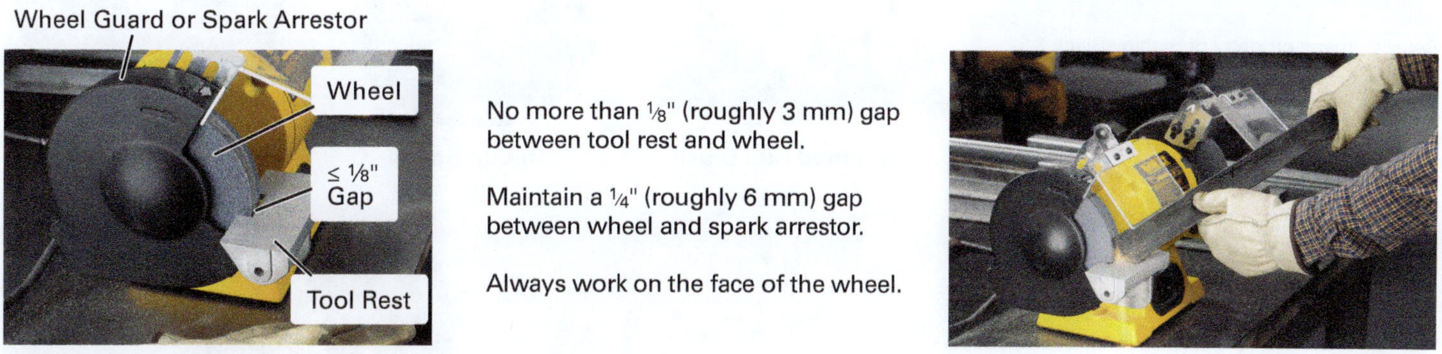

Figure 40 Proper use of a bench grinder.
Source: Image property of Stanley Black & Decker. Used with permission

3.2.0 Grinder Attachments and Accessories

Grinders are versatile pieces of equipment. While the primary uses of a typical grinder are to grind, cut, sand, and polish materials, numerous attachments and accessories can extend a grinder's capability.

Accessories, such as guards (*Figure 41*) and shields, can be added to a portable grinder to provide additional protection when special grinding discs are used. For example, there are specialized guards for use with diamond discs, flap discs, and cup brushes. Side handles can be added to some grinders to improve stability and grip during use. There are also shields available to help protect against dust. Some dust shields are even equipped with a vacuum hose outlet so that dust can be vacuumed away when the grinder is in use.

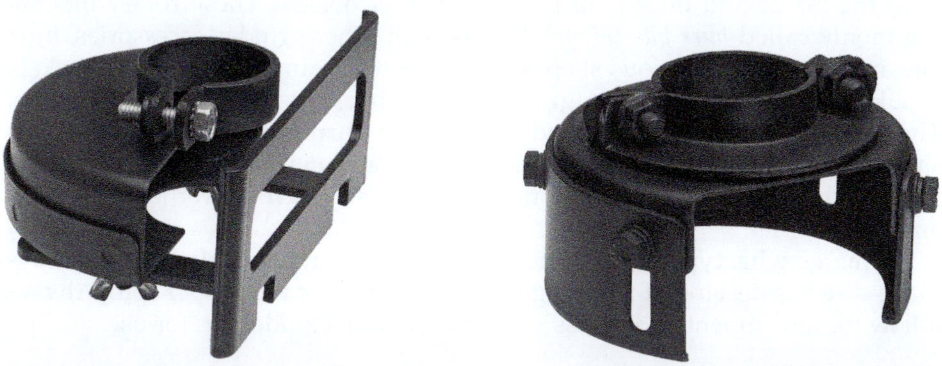

(A) Diamond Blade Guard **(B) Flaring Cup Guard**

Figure 41 Grinder accessories.
Source: Image property of Stanley Black & Decker. Used with permission

Figure 42 Backing pad and adapter nut.
Source: Image property of Stanley Black & Decker. Used with permission

Adapters and backing pads (*Figure 42*) can be installed on a grinder's arbor to accommodate different types of sanding discs, wheels, and buffing pads. The backing pad provides a stable surface for flexible discs to rest against.

An almost endless variety of grinder attachments are available, and most are available in various sizes. Some of the more common accessories are shown in *Figure 43*. They include wire brushes, cup brushes and stones, flap discs, cutting wheels, and polishing and buffing wheels. Some grinders can also be equipped with a cutoff wheel. These are best described as smaller, thinner blade versions used on an abrasive saw. Because of their tendency to break or fly apart when too much stress is applied, many companies require workers to have specific permission and documentation before they can be used.

Figure 43 Grinder attachments.
Sources: Courtesy of Milwaukee Tool (43A–43D); Camel Grinding Wheels Works Sarid LTD (43E)

Wire brushes and cup brushes are good for removing rust, scale, and file marks from metal surfaces. Flap discs are used for similar jobs but tend to last longer than fiber discs. Cutting discs are designed for cutting through material. Cloth buffing wheels enable a grinder to polish and buff metal surfaces. Coring bits are sometimes used to cut holes in stone or composite materials, such as countertops.

Special bits can be used on some grinders to remove rough edges and burrs from the inside and outside of pipes and other objects. These rotary bits are commonly called *burr bits* (*Figure 45*). As with other grinder accessories, burr bits are available in various shapes and sizes to fit almost any need. Workers should be aware that using a grinder with a burr bit requires steady hands. These bits operate at high rpm and can catch on burrs and cause the grinder to shift around quite a bit. They can also heat up the material being smoothed and possibly damage it. It is best to work in short bursts to avoid overheating the bit and the material.

No matter what type of attachment is used on a grinder, workers must always make sure it is designed to fit the grinder and rated for the proper rpm. Always follow the attachment and accessory manufacturer's guidelines for use.

Figure 44 Burr bit assortment.
Source: Asset: Atlas Copco

3.3.0 Oscillating Multi-Tools

An oscillating multi-tool (*Figure 45*), also called an *OMT*, looks like a handheld grinder, but the types of jobs it can perform are much more varied. This variety results from arbors that accept many different attachments and accessories. OMTs operate by moving these attachments back and forth at speeds of up to 21,000 oscillations per minute. Unlike saws or grinders, where you can see the blade or wheel turn, multi-tools move attachments so fast that their movement is hard to detect. This high-speed motion, combined with the smaller design of a multi-tool, makes the tool perfect for small cutting and sanding jobs. OMTs can also be used for scraping grout, cutting metal and PVC piping, removing adhesives and paint, or sanding and preparing different types of surfaces.

One of the most useful characteristics of a multi-tool is its ability to fit into tight, confined places. As a result, they are an excellent choice for undercutting low baseboards, cutting small openings in drywall, or cutting pipe where there is little room for larger tools. Like most other basic power tools, OMTs are available in corded versions, but cordless models make up the bulk of those found on today's jobsites.

Figure 45 OMT.
Source: Image property of Stanley Black & Decker. Used with permission

3.3.1 Oscillating Multi-Tool Safety and Maintenance

Follow these guidelines for the safe use and proper maintenance of OMTs:

- Always wear the appropriate PPE when working with OMTs, especially gloves and safety glasses.
- Disconnect the OMT from its power source when changing the attachment or performing maintenance.
- Confirm that the attachment on the tool is designed for the job being performed.
- Use clamps or vices to securely hold the workpiece.
- Never wear loose clothing or jewelry that can get caught on the OMT attachment.
- Keep your hands away from the attachment when the OMT is turned on.

3.0.0 Section Review

1. A stationary grinder that has two wheels for grinding, removing rust, and buffing metal is a(n) _____.
   a. angle grinder
   b. tandem grinder
   c. bench grinder
   d. end grinder

2. Special rotary bits that can be used on grinders to remove rough edges from the inside and outside of pipes and other objects are commonly called _____.
   a. burr bits
   b. backing pads
   c. coring bits
   d. buffing wheels

3. One characteristic that makes an OMT so useful is its ability to _____.
   a. fit into tight places
   b. drill large holes in concrete
   c. cut through heavy steel
   d. handle large carpentry jobs

4.0.0 Miscellaneous Power Tools

Performance Task

1. Safely and properly demonstrate the use of the following tool:
   - Power nailer

Objective

Identify and explain how to use miscellaneous power tools.

a. Discuss the hazards of using power nailers.

b. Describe jobs that can be performed with hydraulic jacks and wrenches.

Power tools are used for many typical construction or agricultural applications. In addition to drilling, cutting, and grinding, some of the more common applications include fastening and jacking. This section covers the following power tools:

- Power nailers
- Pneumatic impact wrenches
- Hydraulic jacks

4.1.0 Power Nailers

Figure 46 Pneumatic nailer.
Source: Image property of Stanley Black & Decker. Used with permission

Power nailers (*Figure 46*), often called *nail guns*, are common on construction sites. They greatly speed up the installation of materials such as wallboard, molding, framing members, and shingles.

Pneumatic nailers are driven by compressed air traveling through air lines connected to an air compressor. They are designed for specific purposes, such as roofing, framing, siding, flooring, sheathing, trimming, and finishing. Nailers use specific types of nails, depending on the material to be fastened. Nails come in coils and in strips and are loaded into the nail gun.

WARNING!

Never exceed the maximum specified operating pressure of a pneumatic nailer. Doing so will damage the pneumatic nailer and may cause injury.

Pneumatic nailers are designed to fire when the tool is pressed against the material being fastened and the trigger is pressed. An important safety feature of all pneumatic nailers is that the nailer will not fire unless it is pressed against the material.

The use of powder-actuated anchor or fastening systems has increased rapidly in recent years. They are used for anchoring static loads to steel and concrete beams, walls, etc.

Power Screwdrivers

Power screwdrivers use a power source (the model shown here uses a battery) to speed production in various applications, such as drywall installation, floor sheathing and underlayment, decking, fencing, and cement board installation. A chain of screws feeds automatically into the firing chamber. Most models incorporate a back out feature to drive out screws as well as a guide that keeps the screw feed aligned and tangle-free. Power screwdriver tools can accept Phillips or square slot screws and weigh an average of six pounds.

Source: Image property of Stanley Black & Decker. Used with permission

Powder-actuated fasteners are used to drive steel pins or threaded steel studs directly into masonry and steel (*Figure 47*). An advantage of these tools is that they eliminate the need for compressed air to operate the tool. However, these tools can be extremely dangerous. They look and fire like a gun and use the force of a gunpowder load (typically .22, .25, or .27 caliber) to drive the fastener into the material. The depth to which the pin or stud is driven is controlled by the density of the base material and the power level or strength of the powder load.

WARNING!

Operators of powder-actuated tools must be trained and certified by the manufacturer of the tool being used.

Avoid firing a powder-actuated tool into easily penetrated materials. The fastener may pass through the material and become a flying missile on the other side.

If a powder load fails to fire, wait 10 seconds, remove it from the installation tool, and then dispose of it in a bucket of water or oil.

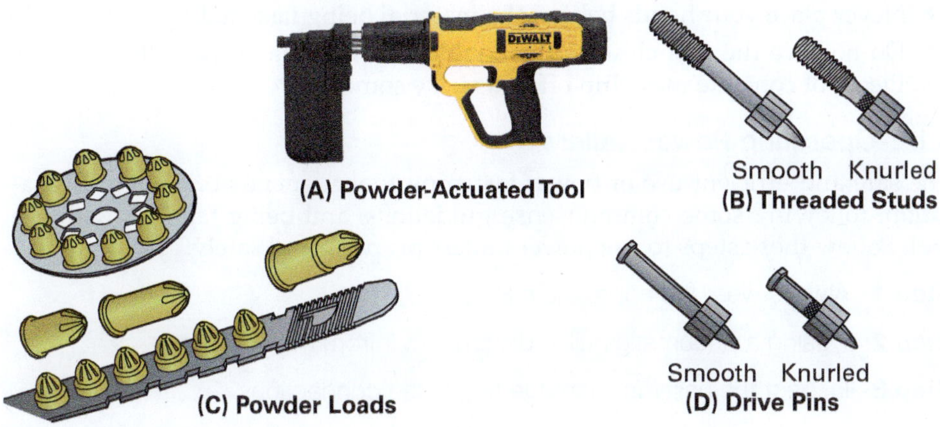

Figure 47 Powder-actuated fastening system.
Source: Image property of Stanley Black & Decker. Used with permission (47A)

While pneumatic and powder-actuated tools have traditionally been the choice of craftworkers nailing into concrete and steel, cordless electric nailers (*Figure 48*) are another alternative. Cordless nailers are battery-operated and have enough power to fasten pins, straps, and clips to concrete and steel. Although they may not have the power that a powder-actuated tool can generate, they are strong enough to handle most concrete fastening jobs. Cordless nailers have several advantages, including the lack of a hose that must be connected to pneumatic nailers. Using cordless nailers also allows craftworkers to avoid the training and certification required to use powder-actuated tools.

Figure 48 Cordless electric nailer.
Source: Image property of Stanley Black & Decker. Used with permission

4.1.1 Power Nailer Safety and Maintenance

There are numerous safety and maintenance guidelines associated with pneumatic, powder-actuated, and electric concrete fastening tools. In addition to following basic power tool safety guidelines, familiarize yourself with these safety rules before operating a power nailer:

- Always wear the appropriate PPE, especially safety glasses.
- Read and understand the manufacturer's manual.
- When using a pneumatic nailer, pay close attention to where you are pointing the nail gun. Never point the nail gun toward another worker or toward any part of your body.
- Be sure you select the correct nail gun and fastener for the job, and never load the tool until you are ready to fire it.
- Never try to operate a powder-actuated tool without the proper training.

- Keep pneumatic nailers oiled according to the manufacturer's instructions.
- Disconnect the power source from the nailer before attempting to load or repair it.
- Always treat nailers as if they are loaded and ready to fire.
- Check for pipes, electrical wiring, vents, and other materials behind wallboard before nailing.
- Be careful about nailing into paneling or any other thin surface. The fastener may go through it and strike another worker.
- Never leave a pneumatic nailer connected to a compressor hose or power when not used.

WARNING!

The use of powder-actuated tools may be prohibited by local codes because of safety issues, seismic concerns, or because the building is occupied. Check with your local authorities before using these tools.

- When loading the driver, install the drive pin before loading the charge.
- Never place your hands behind the material being fastened.
- Do not fire the tool close to the edge of the material, especially concrete. Pieces of concrete may chip off and strike someone.

4.1.2 Operating Power Nailers

The safe and efficient use of power fastening tools depends on careful preparation, following some commonsense guidelines, and being familiar with the tool. Follow these steps to use power nailers properly and safely:

Step 1 Always wear the appropriate PPE.

Step 2 If using a pneumatic nailer, disconnect it from the air hose.

Step 3 Inspect the nailer for damage and loose connections.

Step 4 Load the fastener into the nailer. When loading a powder-actuated tool, feed the pin or stud into the piston, followed by the gunpowder cartridge.

Step 5 If you are using a pneumatic nailer, connect it to the air hose. Also, check the air compressor and adjust the pressure to the recommended level. Most nailers operate at pressures of 70 psi to 120 psi (roughly 480 kPa to 830 kPa).

Step 6 Test the tool's nailing ability using a piece of scrap material. If the nail penetration is not deep enough, follow the manufacturer's instructions for increasing the nailer's power.

Step 7 Hold the pneumatic or battery-powered nailer firmly against the material to be fastened, then press the trigger (*Figure 49*). If you use a powder-actuated nailer, position the tool in front of the item to be fastened and press firmly against the mounting surface (*Figure 50*).

Step 8 After finishing the job, disconnect the power source from the nailer. Turn off the air supply and disconnect the air hose from the nailer. Remove the battery from cordless nailers. Remove the powder charge(s) from powder-actuated fastening tools.

WARNING!

A nail gun is not a toy. Playing with a nail gun can cause serious injury or death. Nails can easily pierce a hand, leg, or eye. Never point a nail gun at anyone; use it only as the manufacturer directs.

Figure 49 Proper use of a nailer.
Source: Zachary McNaughton, River Valley Technical Center

Figure 50 Using a powder-actuated fastening tool.
Source: Photo courtesy of Simpson Strong-Tie Company Inc.

4.2.0 Hydraulic Jacks

Hydraulic tools are used when an application calls for extreme force to be applied in a controlled manner. These tools do not operate at high speed, but you should exercise great care when operating them. The forces generated by hydraulic tools can easily damage equipment or cause personal injury if the manufacturer's procedures are not strictly followed.

Hydraulic jacks are portable devices used for a wide variety of purposes. They can move or lift heavy equipment and other heavy material, position heavy loads precisely, and straighten or bend frames. Hydraulic jacks have two basic parts: a pump and a cylinder (sometimes called a ram). There are various hydraulic jacks, including those with internal pumps and those that use a lever-operated pump. The latter type is often called a Porta-Power®, the name of one common brand.

A hydraulic jack with an internal pump is a general-purpose jack available in many capacities (*Figure 51*). The pump inside the jack applies pressure to the hydraulic fluid when the handle is pumped. The pressure on the hydraulic fluid applies pressure to the cylinder, which lifts or moves the load.

A typical lever-operated pump kit (*Figure 52*) includes a length of hydraulic hose, a hydraulic hand pump cylinder, and a variety of attachments made for different jobs. Lever-operated pumps are available in different capacities. Cylinders are available in many sizes; they are rated by the weight (in tons or metric tonnes) they can lift and the distance they can move it. This distance is called stroke and is measured in inches or millimeters. Hydraulic cylinders can lift more than 500 tons (about 454 metric tonnes). Strokes range from $\frac{1}{4}$" (roughly 6 mm) to more than 48" (roughly 122 cm). Different cylinder sizes and ratings are used for different jobs. Lever-operated pumps are especially useful for horizontal jacking.

Figure 51 Portable hydraulic jack.
Source: Courtesy of JPW Industries, Wilton Tools, JET Tools, Baileigh Industrial

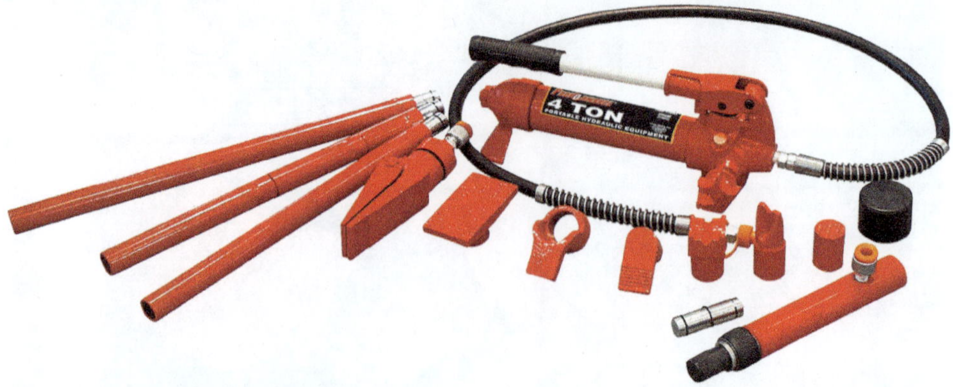

Figure 52 Lever-operated hydraulic pump kit.
Source: Torin Inc.

4.2.1 Hydraulic Jack Safety and Maintenance

Using hydraulic jacks safely and effectively requires an awareness of the area surrounding the load, the load itself, and the jack. In addition to following basic power tool safety guidelines, familiarize yourself with these safety rules:

- Always wear the appropriate PPE, especially safety glasses.
- Operate the jack according to the manufacturer's guidelines.
- Check the area before jacking a load to ensure that other workers are safely out of the way and that the load will clear all obstacles.
- Make sure you have the appropriate jack for the job and never exceed its lifting capacity.
- Check the fluid level in the jack before using it and watch for any fluid leaks during use. If a Porta-Power® is being used, ensure the hydraulic hose is not twisted or kinked. Do not move the pump if the hose is under pressure.
- Make sure the jack's base can be placed on a solid, even, and level surface. Never place it on bare soil or any other surface that could compact or shift under the load.
- If the load could move while jacking, make sure it is chocked and restrained. Never jack metal against metal; use wood softeners as a buffer between metal surfaces.
- Place the jack under the load so the load is centered and the weight is uniformly distributed.
- Stay clear of the lifted object to avoid injury if the load slips off the jack.
- Do not use an extension bar, cheater, or step on the pump handle to gain more leverage.
- Never leave a jack under a load as a support.
- Block the load up as you progress through the lift so it will only fall a short distance if the jack fails.
- Once at the proper height, add blocking as needed to support the load while the jack is removed.
- To reduce tripping hazards, remove the jack's lever any time the jack is not being pumped.

4.2.2 Portable Generators and Air Compressors

Portable generators and portable air compressors (*Figure 53*) play a vital role in modern agriculture by providing essential power and utility on farms and in remote locations. Portable generators are indispensable for supplying electricity to farm equipment and tools, especially in areas where the electrical grid is unreliable or in rural regions where the power grid is vulnerable to extreme weather events. They ensure that essential operations, such as irrigation systems, heating, cooling, and lighting, can continue uninterrupted to protect crops and livestock.

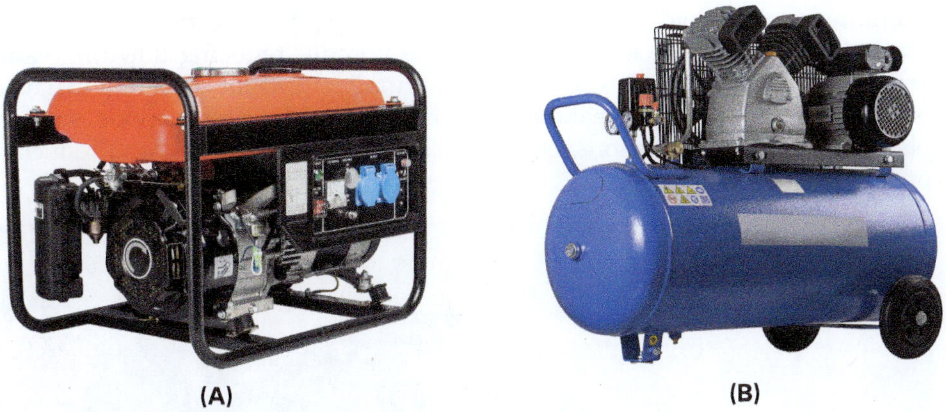

Figure 53 Portable generator and air compressor.
Source: iStock

In addition to providing backup power, portable generators are used to power various mobile and stationary equipment. For instance, they can supply electricity to mobile milking machines (*Figure 54*), electric fences, and grain dryers. This versatility allows farmers to operate essential machinery regardless of their proximity to a permanent power source. Furthermore, the mobility of portable generators makes them ideal for use in diverse agricultural settings, from large-scale crop fields to small livestock farms, ensuring operational efficiency across different environments.

Figure 54 Mobile milking machine.
Source: iStock

Portable air compressors are equally valuable in agricultural settings, offering various applications that enhance productivity and maintenance. One of their primary uses is powering pneumatic tools, such as nail guns, staple guns, and sprayers, used for tasks like building and repairing fences, constructing storage facilities, and applying pesticides or fertilizers. The efficiency and power of pneumatic tools enable farmers to complete these tasks quickly and with less physical strain, improving overall work efficiency.

Another significant application of portable air compressors is maintaining farm equipment and machinery. They are used to inflate tires (*Figure 55*), clean equipment with compressed air, and provide the necessary pressure for operating various hydraulic systems. Regular maintenance facilitated by air compressors ensures that machinery runs smoothly, reducing downtime and extending the lifespan of expensive agricultural equipment. This maintenance is crucial for preventing breakdowns during critical farming periods, such as planting and harvest seasons, ensuring consistent productivity, and reducing operational costs.

Figure 55 Tractor tires.
Source: iStock

4.0.0 Section Review

1. Pneumatic nailers are designed to safely fire when the trigger is squeezed and the tool is _____.
   a. filled with a charge of compressed air
   b. pressurized with hydraulic fluid
   c. connected to its battery pack
   d. pressed against the material being fastened

2. When using a portable hydraulic jack, be sure to _____.
   a. avoid twisting or kinking the air lines
   b. leave the jack under the load as a support
   c. use an extension bar on the pump handle
   d. place the base of the jack on a solid, level surface

Module 93105 Review Questions

1. Pneumatic tools get their power from _____.
   a. air pressure
   b. fluid pressure
   c. hand pumps
   d. AC power sources

2. When operating a power tool, it is important to *always* _____.
   a. wear the proper PPE
   b. notify a coworker
   c. wear rubber-soled footwear
   d. ask a coworker how to operate the tool

3. Using a trigger lock to lock a tool's trigger in an On position _____.
   a. can increase productivity
   b. ensures the tool will have enough power
   c. can be dangerous and cause injury
   d. is a great way to keep your hand from getting fatigued

4. The most common use of the power drill is to _____.
   a. cut wood, metal, and plastic
   b. drive nails into wood, metal, and plastic
   c. make holes in wood, metal, and plastic
   d. carve letters in wood, metal, and plastic

5. A masonry bit is able to drill into concrete and similar material because it has a _____.
   a. countersink shank
   b. ceramic core
   c. whip check
   d. carbide tip

6. An example of an electric power drill that is designed to be used in tight spaces is a(n) _____.
   a. electromagnetic drill
   b. right-angle drill
   c. hammer drill
   d. keyless chuck drill

7. The electromagnetic drill is a _____.
   a. handheld drill used on wood
   b. cordless drill used on masonry and tile
   c. portable drill used on thick metal
   d. pneumatic drill that has a pounding action

8. Hammer drills are designed to drill into _____.
   a. wood, metal, and plastic
   b. concrete, brick, and tile
   c. drywall, fiberglass, and wood
   d. roofing shingles, plastic, and wood

9. Compared to a common drill, an impact driver has _____.
   a. more torque
   b. more weight
   c. more variable speed options
   d. more bits for precision drilling

10. A pneumatic impact wrench requires the use of _____.
    a. impact sockets designed for the applicable tool
    b. an adapter so that handheld sockets will fit
    c. shear pins between the wrench and the socket
    d. a trigger lock to prevent accidental starting

11. When cutting with a circular saw, grip the saw handles _____.
    a. and pull the saw toward you
    b. firmly with two hands
    c. firmly with one hand
    d. and engage the trigger lock

12. The high-speed setting on a reciprocating saw is used for _____.
    a. cutting through drywall
    b. metal work
    c. grinding surfaces
    d. sawing wood and other soft materials

13. Before using a reciprocating saw to cut through a wall or partition, *always* _____.
    a. find out what is on the other side
    b. remove the lower blade guard
    c. increase the rpm
    d. lubricate the guard with oil or grease

14. When using a jigsaw, avoid vibration by _____.
    a. using a low-speed setting
    b. using a clamp or vise to hold the workpiece
    c. setting a heavy object on the workpiece
    d. holding the workpiece down with your free hand

15. Only use a band saw that has a _____.
    a. stop
    b. breastplate with a broad surface
    c. battery pack
    d. thick, three-piece blade

16. A sliding compound miter saw has a rail that allows the blade to slide forward and backward, which enables the saw to _____.
    a. use much thinner blades than a standard miter saw
    b. produce much less dust than a standard miter saw
    c. cut wider material than a standard miter saw
    d. cut harder material than a standard miter saw

17. The blade of an abrasive cutoff saw spins at such a high speed that _____.
    a. the resulting friction is hot enough to burn through the material
    b. it can only be used for straight cuts
    c. the abrasive particles will melt into some metals
    d. it can never be more than 8" in diameter

18. The end grinder is used to _____.
    a. polish intricate work
    b. grind surfaces
    c. smooth the inside of materials, such as pipe
    d. smooth the work before painting

19. A detail grinder smoothes and polishes intricate metallic work by using attachments called _____.
    a. points
    b. rollers
    c. pins
    d. studs

20. Powder-actuated fastening systems are used to _____.
    a. penetrate drywall and treated wood
    b. anchor static loads
    c. hammer nails into metal
    d. remove nails

21. Porta-Power® cylinders are rated by how much weight they can lift and by _____.
    a. their torque
    b. the amount of electromagnetic material they have
    c. the distance they can move the weight
    d. how much they weigh

22. Hydraulic jacks are used when the application calls for _____.
    a. operation at high speed
    b. quiet operation
    c. extreme force to be applied
    d. manually assisted lifting

Answers to odd-numbered questions are found in the Review Question Answer Keys at the back of this book.

Answers to Section Review Questions

Answer	Section Reference	Objective
Section 1.0.0		
1. a	1.1.0	1a
2. b	1.2.5	1b
3. d	1.3.0	1c
4. c	1.4.2	1d
Section 2.0.0		
1. a	2.1.2	2a
2. d	2.2.1	2b
3. c	2.3.0	2c
4. b	2.4.1	2d
5. c	2.5.2	2e
Section 3.0.0		
1. c	3.1.0	3a
2. a	3.2.0	3b
3. a	3.3.0	3c
Section 4.0.0		
1. d	4.1.0	4a
2. d	4.2.1	4b

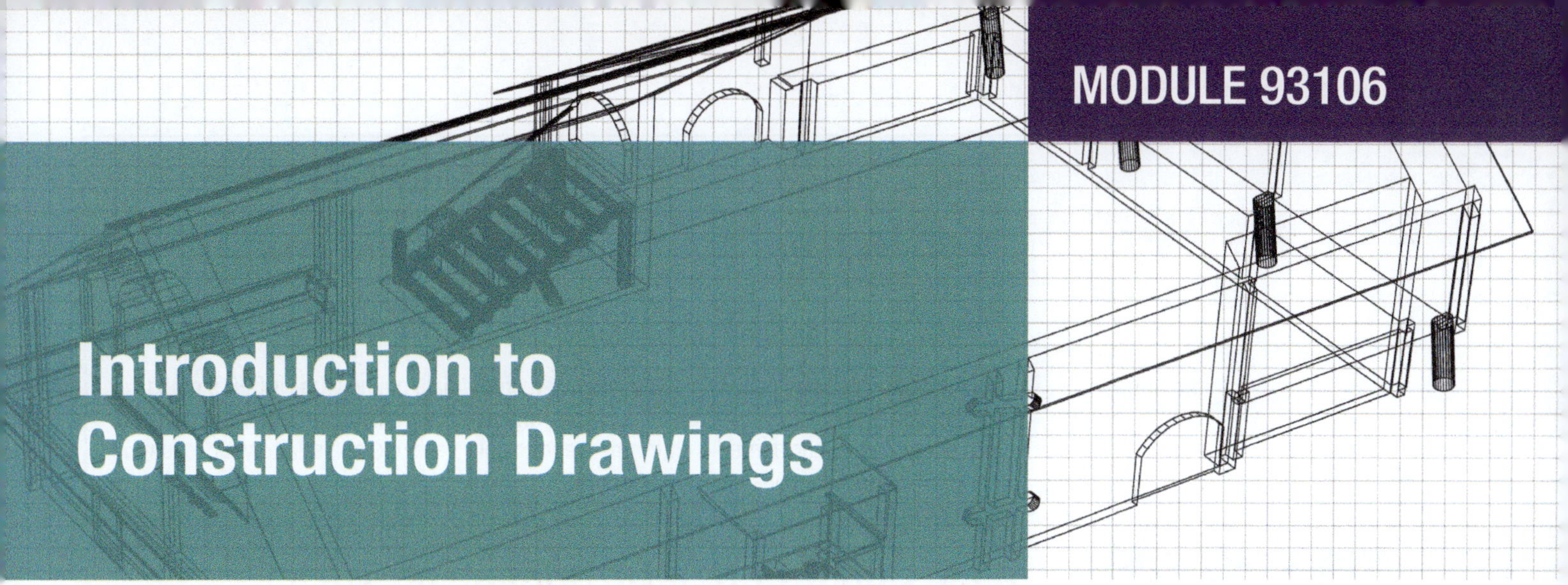

MODULE 93106

Introduction to Construction Drawings

Source: iStock

Objective

Successful completion of this module prepares you to do the following:

1. Describe components and features used in construction drawings and identify how the drawings are different.
   a. Summarize the purpose of the six basic construction drawing components.
   b. List and explain the significance of various drawing elements, such as lines of construction, symbols, and grid lines.
   c. Explain how dimensions relate to various drawing scales.
   d. Demonstrate how to use engineer's and architect's scales.
   e. Identify the six types of construction drawings.

Performance Task

Under supervision, you should be able to do the following:

1. Using the residential floor plan and elevation drawings supplied in the Performance Profile Sheet:
   - Locate the wall common to bedroom #3 and the garage.
   - Determine the overall width of the home.
   - Calculate the distance from the outside northwest corner of the house to the center of the window in bedroom #2.
   - Determine the overhang of the eaves.
   - Using an architect's scale, determine the width of the garage.

Overview

Various types of construction drawings are used to represent actual components of a building project. The drawings provide specific information about locations of parts of a structure, the types of materials to be used, and the correct layout of the building. Knowing the purposes of different types of drawings and interpreting drawings correctly are important skills for anyone who works in the construction trades. This module introduces common types of construction drawings, their basic components, standard drawing elements, and measurement tools that are typically used when working with construction drawings.

Digital Resources for Agricultural Technology

SCAN ME

Scan this code using the camera on your phone or mobile device to view the digital resources related to this craft.

NCCER Industry-Recognized Credentials

If you are training through an NCCER-accredited sponsor, you may be eligible for credentials from NCCER. The ID number for this module is 93106. Note that this module may have been used in other NCCER curricula and may apply to other level completions. Contact NCCER at 1.888.622.3720 or go to **www.nccer.org** for more information.

You can also show off your industry-recognized credentials online with NCCER's digital credentials. Transform your knowledge, skills, and achievements into credentials that you can share across social media platforms, send to your network, and add to your resume. For more information, visit **www.nccer.org**.

1.0.0 Construction Drawings and Their Components

Performance Task

1. Using the residential floor plan and elevation drawings supplied in the Performance Profile Sheet:
   - Locate the wall common to bedroom #3 and the garage.
   - Determine the overall width of the home.
   - Calculate the distance from the outside northwest corner of the house to the center of the window in bedroom #2.
   - Determine the overhang of the eaves.
   - Using an architect's scale, determine the width of the garage.

Objective

Describe components and features used in construction drawings and identify how the drawings are different.

a. Summarize the purpose of the six basic construction drawing components.
b. List and explain the significance of various drawing elements, such as lines of construction, symbols, and grid lines.
c. Explain how dimensions relate to various drawing scales.
d. Demonstrate how to use engineer's and architect's scales.
e. Identify the six types of construction drawings.

Construction drawings are architectural or working drawings used to represent a structure or system. These were traditionally referred to as **blueprints** because years ago the process for reproducing these drawings resulted in white lines on a blue background. With the arrival of **computer-aided drafting (CAD)**, and most recently building information modeling (BIM), drawings are created on a computer and reproduced using large-format plotters, printers, and copiers. Today, construction drawings are simply called drawings or prints.

Various kinds of construction drawings, including residential drawings, commercial drawings, landscaping plans, agricultural plans, shop drawings, and industrial drawings, are used in construction. In this module, several of the most common types of drawings will be introduced.

Construction drawings, together with the set of **specifications** (often abbreviated as specs), detail what is to be built and what materials are to be used. Specifications are written statements that the architectural and engineering firm provides to the general contractors. The specs define the quality of work to be done and describe the materials to be used.

The set of construction drawings forms the basis of agreement and understanding that a building will be built, or an agricultural plan laid out, as detailed in the drawings. Therefore, everyone involved in planning, supplying, and building any structure should be able to read construction drawings. For any building project, also consult the civil engineering plans for that location, including sewer, highway, and water installation plans.

Blueprints: A traditional method of reproducing technical drawings using a contact print process and light-sensitive paper that resulted in white lines on a blue background. Also, the traditional name used to describe construction drawings.

Computer-aided drafting (CAD): The making of a set of construction drawings with the aid of a computer.

Specifications: Precise written presentation of the details of a plan.

Before the 1950s, construction drawings were all drawn by hand. This was a long and tedious process. Many times, if there was a mistake or a drawing required a change, the drawing had to be recreated from scratch. The evolution from hand drafting to CAD started in 1950 when Dr. Paul J. Hanratty introduced a numerically controlled program that allowed users to draw simple lines utilizing a computer. Years later, in 1972, Dr. Hanratty invented a program called Automated Drafting and Machining. The program could draw more complex objects and included many of the commands and features still used in modern CAD programs today. By the 1980s, CAD programs were becoming commonplace with designers, **architects**, and **engineers**.

Architects: Qualified, licensed people who create and design drawings for a construction project.

Engineers: People who apply scientific principles in design and construction.

Using CAD is a cost-effective way to increase drafting productivity because the computer program automates much of the repetitive work related to generating drawings. Ultimately, CAD has the following advantages over hand-drawn construction drawings:

- It is automated.
- The computer performs calculations quickly and easily.
- Changes can be made easily.
- Commonly used **symbols** can be retrieved without difficulty.
- CAD can produce 3D modeling of the structure.

Symbols: Drawings that represent a material or component on a plan.

While CAD is still widely used, more and more drawings are created today using BIM. Traditionally, BIM was a system of paper-based documents and construction drawings designed to monitor and track specific information about a building.

Think About It

Planning for Success

Developing a blueprint to plan agricultural structures and production areas not only identifies the physical space for the operation, it also establishes the mission, vision, and management of the agricultural activities that will operate on the farm. Planning helps answer important questions, such as what equipment you will need, what resources you have access to, and what conservation practices can sustain your operation.

Source: iStock

In the last decade, BIM has increasingly moved toward computer technology, which allows the various trades and the building owner to track the life cycle of the building, from the start of planning, through construction, to completion. BIM allows building owners and trades to easily track and identify the various components and assemblies of individual systems (such as electrical, **HVAC**, and plumbing) and building materials within a specific location or locations of a building.

HVAC: Heating, ventilating, and air conditioning.

BIM also has the capabilities of simulating the life cycles of various systems within the building to provide information on determining the wear on, and lifespan of, individual components. Additionally, BIM software allows the craftworkers and the building owner to input and save information on the building as repairs, updates, or renovations are made to the building. Information such as part sizes, manufacturers, part numbers, related drawings, or any other information researched in the past can be stored in a BIM system.

Another advantage of BIM is that it allows documentation and information on a building to be easily transferred and communicated between all the members of a project.

Regardless of how drawings are created, there are almost always changes that occur during the design and construction phases. This means that the drawing needs to be updated to reflect those changes. Once there are multiple versions of a drawing, it is important to ensure that everyone is working from the most recent, up-to-date drawing. This is accomplished by using **drawing revision control**.

Drawing revision control: A system for recording, documenting, and identifying drawing revisions by using a numeric or alphabetic sequence.

Drawing revision control is a system for recording, documenting, and identifying drawing revisions by using a numeric or alphabetic sequence. Drawing revisions are recorded in the revision history block of the drawing. Some revision history blocks are included as part of the title block, while some are inserted as a separate block. A revision history block typically contains columns for the letter or number designating the revision, the drafter's initials, the date of the revision, and the description of the revision. Items revised on the drawing are typically highlighted with a revision cloud around them. A numbered or lettered triangle next to the cloud identifies its sequence in the revision history. The revised drawing is saved as a new drawing file to preserve the revision history.

1.1.0 Basic Components of Construction Drawings

Most construction drawings are laid out in a standardized format. This section describes the following six parts of a construction drawing:

- **Title block**
- Border
- Drawing area
- Revision block
- **Legend**
- North arrow

Title block: A part of a drawing sheet that includes some general information about the project.

Legend: A description of the symbols and abbreviations used in a set of drawings.

1.1.1 Title Block

The first thing to look at on any drawing is the title block. The title block is normally in the lower right-hand corner of the drawing or across the right edge of the paper (*Figure 1*) and serves two primary purposes. First, it provides information about the structure or assembly. Second, it is numbered so the print can be filed easily.

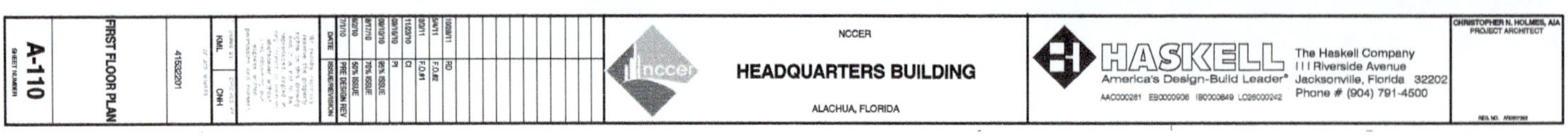

Figure 1 The title block of a construction drawing.

Different companies put different information in the title block. Generally, it contains the following:

- *Company logo* — It is usually preprinted on the drawing.
- *Sheet title* — This identifies the project.
- *Date* — The date the drawing was checked and readied for seal or issued for construction.
- *Drawn by* — The initials of the person who drafted the drawing.
- *Drawing number* — The code numbers assigned to a project.
- **Scale** — The ratio of the size of the object as drawn to the object's actual size.
- *Revision blocks* — This details information on revisions, including (at a minimum) the date and the initials of the person making the revision. Other information may include descriptions of the revision and a revision number.

Scale: The ratio between the size of a drawing of an object and the size of the actual object.

Every company has its own system for such things as project numbers and departments. It also has its own placement locations for the title and revision blocks. Your supervisor should explain the company's standards.

1.1.2 Border

The border is a clear area of approximately ½" around the edge of the drawing area. It is there so that everything in the drawing area can be printed or reproduced on printing machines with no loss of information.

1.1.3 Drawing Area

The drawing area presents the information for constructing the project, such as the **floor plan**, elevations of the building, sections, and details.

Floor plan: A drawing that provides an aerial view of the layout of each room.

CAUTION

It is essential to note the revision designation on a construction drawing and to use only the latest version. Otherwise, costly mistakes may result.

1.1.4 Revision Block

A revision block is located in the drawing area, usually in the lower right corner inside or near the title block. Different companies put the revision block in different places. This block is used to record any changes (revisions) to the drawing. It typically contains the revision number, a brief description, the date, and the initials of the person who made the revisions (*Figure 2*). All revisions must be noted in this block and dated and identified by a letter or number.

1.1.5 Legend

Each line on a construction drawing has a specific design and thickness that identifies it. Note that some of the lines may be used to identify off-site utilities. The identification of these lines and other symbols is called the legend. Although a legend doesn't automatically appear on every construction drawing, when it does, it explains or defines symbols or special marks used in the drawing (*Figure 3*). Be aware that legends are specific only to the set of drawings in which they are contained.

Proj.	No.	Revision	Rsvd.	Chkd.	Appd.	Date
3483	01	Released for Construction		APD	NWS	July 21
3483	02	Deleted Part of Line 12037		APD	NWS	Aug 21
3483	03	△A Added Welding Symbol		APD	NWS	Aug 21

Figure 2 The revision block of a construction drawing.

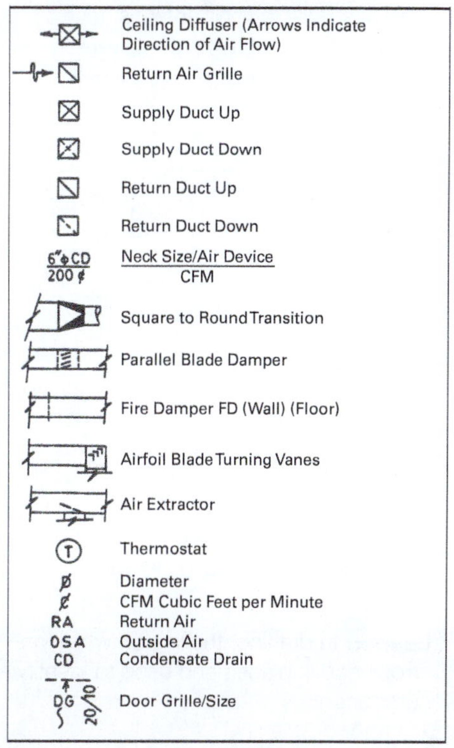

Figure 3 Sample legend.

Did You Know?

Plan North

When true north is not noted on a drawing and cannot be verified, plan north is used to describe direction on the drawing. This helps to communicate the location of features on a drawing. Plan north is toward the top of the page, east is to the right, south to the bottom, and west to the left.

Agricultural Symbols

Architects and engineers use many common symbols in drawings and specifications, but some industries, such as the agricultural industry, may use unique symbols that only pertain to their processes, operations, or requirements. Always verify the meaning of symbols on a drawing with the legend.

Source: iStock

1.1.6 North Arrow

Most construction drawings have an arrow indicating north. It may be located near the title block or in some other conspicuous place. The north arrow is a reference point that helps to verify the locations of objects, walls, and building parts shown on the construction drawing.

Regional and Company Differences

Although most symbols are standard, there can be slight variations in different regions of the country. Always check the title sheet or other introductory drawing to verify the symbols on the project drawings.

Your company may also use some special symbols or terms. Your instructor or supervisor should explain the conventions that are unique to the company.

Did You Know?

Legality of Construction Drawings

Construction drawings are incorporated into building contracts by reference, making them part of the legal documents associated with a project. When describing the project to be completed, the legal contract between the builder and the owner refers to the accompanying construction drawings for details that would be too lengthy to write out. That makes tracking changes or revisions to construction drawings over the course of a project vitally important. If an error is made along the way, either the owner or the builder must be able to find the discrepancy by reading the drawings. Taking care of construction drawings over the course of the project makes good business sense.

1.2.0 Drawing Elements

Various drawing elements make it easier and quicker for people to read and understand construction drawings. This section covers several of the most common drawing elements.

1.2.1 Lines of Construction

It is very important to understand the meanings of lines on a drawing. The lines commonly used in a drawing are sometimes called the Alphabet of Lines. Here are some of the more common types of lines (*Figure 4*):

Dimension line: A line on a drawing with a measurement indicating length.

Leader: In drafting, the line on which an arrowhead is placed and used to identify a component.

- **Dimension line** — Establishes the dimensions (sizes) of parts of a structure. These lines end with arrows (open or closed), dots, or slashes at a termination line drawn perpendicular to the dimension line.
- **Leader** *and arrowhead* — Identify the location of a specific part of the drawing. They are used with words, abbreviations, symbols, or keynotes.
- *Property line* — Indicates land boundaries.

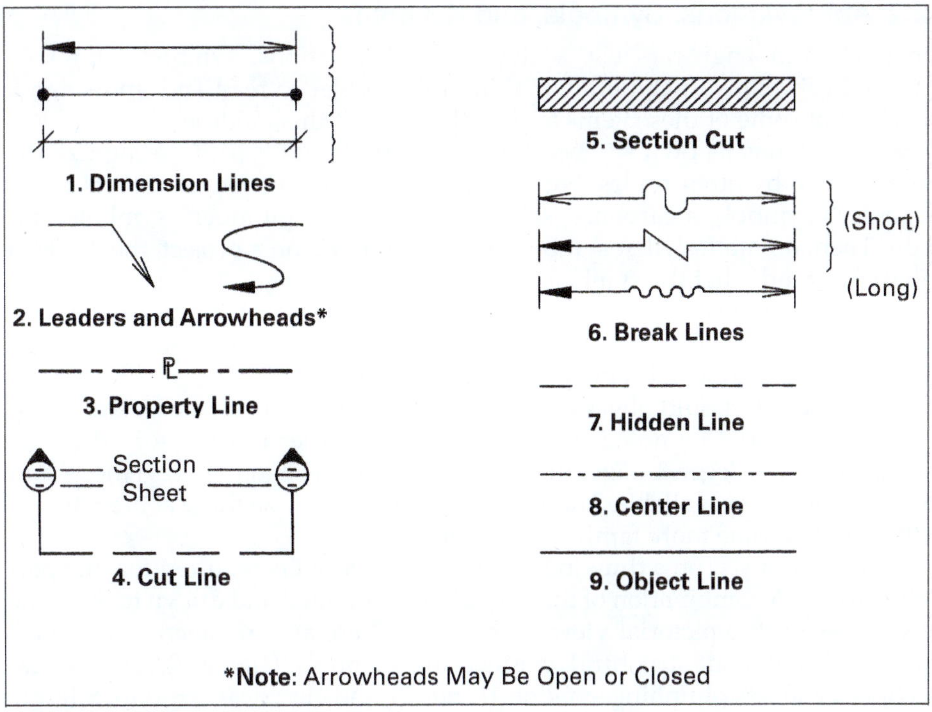

Figure 4 Lines of construction.

- *Cut line* — Lines around part of a drawing that are to be shown in a separate cross-sectional view.
- *Section cut* — Shows areas not included in the cutting line view.
- *Break line* — Shows where an object has been broken off to save space on the drawing.
- **Hidden line** — Identifies part of a structure that is not visible on the drawing. You may have to look at another drawing to see the part referred to by the lines.
- *Center line* — Shows the measured center of an object, such as a column or fixture.
- *Object line* — Identifies the object of primary interest or the closest object.

Hidden line: A dashed line showing an object obstructed from view by another object.

Check All the Plans

Always double-check all plans for a project. Be familiar with other trade work that may affect your job. Determine whether an electrical conduit is close to the plumbing pipes or whether a framing member is placed too close to an HVAC duct. By knowing all the work planned for a particular area, errors can be prevented, saving time and money.

If a particular dimension is missing on the plan, check the other plans to see if it is included there as that part of the building will also be shown on other plans.

When taking measurements from a set of plans, make sure that the dimensions of each section add up to the total measurement. For instance, when looking at a drawing of a wall with one window, add up the measurement from the left end of the wall to the left side of the window, the width of the window, and the measurement from the right side of the window to the right end of the wall. Make sure this total matches the measurement of the entire wall.

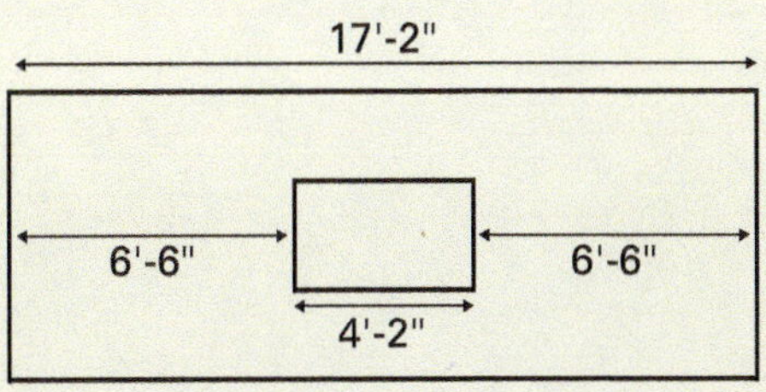

1.2.2 Abbreviations, Symbols, and Keynotes

Architects and engineers use systems of abbreviations, symbols, and keynotes to keep plans uncluttered, making them easier to read and understand. Examples of some of these items appear throughout this module.

Each trade has its own symbols, and workers should learn to recognize the symbols used by other trades. For example, an electrician should understand a carpenter's symbols, a carpenter should understand a plumber's symbols, and so on. Then, no matter what symbols are encountered on a project, the workers will understand what the symbols mean.

Abbreviations used in construction drawings are short forms of common construction terms. For example, the term *on center* is abbreviated OC. Some common construction abbreviations are listed in *Figure 5*.

Abbreviations should always be written in capital letters. Abbreviations for each project should be noted on the title sheet or other introductory drawing page such as the legend page. Books that list construction abbreviations and their meanings are available. There is no need to memorize these abbreviations, as they will become more familiar with time.

Symbols are used on a drawing to tell what material is required for that part of the project. A combination of these symbols, expanded and drawn to the same size, makes up the pictorial view of the plan. There are architectural symbols (*Figure 6*), civil and structural engineering symbols (*Figure 7*), mechanical symbols (*Figure 8*), plumbing symbols (*Figure 9*), and electrical symbols. Slightly different symbols may be used in different parts of the country or by different companies. The symbols used for each set of plans should be indicated on the title sheet or other introductory drawing. Many codebooks, manufacturers' brochures, and specifications include symbols and their meanings.

Some plans use keynotes (*Figure 10*) instead of symbols. A keynote is a number or letter (usually in a square or circle) with a leader and arrowhead that is used to identify a specific object. Part of the drawing sheet (usually on the right-hand side) lists the keynotes with their numbers or letters. The keynote descriptions normally use abbreviations.

> **Request for Information (RFI)**
>
> A request for information (RFI) is used to clarify any discrepancies in the plans. If there is a discrepancy, notify the foreman. The foreman will write up an RFI, explaining the problem as specifically as possible and putting the date and time on it. The RFI is submitted to the superintendent, who passes it to the general contractor, who passes it to the architect or engineer, who then resolves the discrepancy.
>
> Always refer to specifications and the RFI when deciding how to interpret drawings.

1.2.3 Using Gridlines to Identify Plan Locations

Think of using a road map to find a particular street. The map may use a grid to make locating a detailed area easier. For example, the index might indicate that the street is in section B-3, referring to B along the side of the map and 3 along the top. The street is located where B and 3 intersect.

The grid line system shown on a plan (*Figure 11*) is used like the grid on a map. On a drawing such as a floor plan, a grid divides the area into small parts called bays.

The numbering and lettering system begins in the upper left-hand corner of the floor plan. The numbers are normally across the top and the letters are along the side. To avoid confusion, certain letters and the symbol for zero are not used. Omitted from the grid line system are the letters I, O, and Q and numbers 1 and 0.

A grid line system makes it easy to refer to specific locations on a plan. A single outlet can be pointed out on a plan, even if there are a dozen shown. For example, referring to an outlet in bay C-8 helps to distinguish a particular outlet.

Abbreviations

A.B.	– ANCHOR BOLT	FDN.	– FOUNDATION	RM.	– ROOM
ADD'L	– ADDITIONAL	FIN.	– FINISH	SCHED.	– SCHEDULE
ADJ.	– ADJACENT	FLR.	– FLOOR	SECT.	– SECTION
A.I.S.C.	– AMERICAN INSTITUTE OF STEEL CONSTRUCTION	F.O.B.	– FACE OF BRICK	SHT.	– SHEET
		F.O.CONC.	– FACE OF CONCRETE	SIM.	– SIMILAR
ALT.	– ALTERNATE	F.O.W.	– FACE OF WALL	S.L.V.	– SHORT LEG VERTICAL
ARCH.	– ARCHITECTURAL	FS.	– FLAT SLAB	SPC.	– SPACE
A.S.T.M.	– AMERICAN SOCIETY FOR TESTING & MATERIALS	FT.	– FOOT	SPEC.	– SPECIFICATION
		FTG.	– FOOTING	SQ.	– SQUARE
BLDG.	– BUILDING	F.W.	– FILLET WELD	STD.	– STANDARD
BM.	– BEAM	GA.	– GAUGE	STIFF.	– STIFFENER
B.O.	– BOTTOM OF	GAL.	– GALVANIZED	STL.	– STEEL
BOT.	– BOTTOM	G.L.	– GLU-LAM BEAM	STOR.	– STORAGE
BSMT.	– BASEMENT	GR.	– GRADE	SYM.	– SYMMETRICAL
BTWN.	– BETWEEN	GR. BM.	– GRADE BEAM	T.&B.	–TOP AND BOTTOM
CANT.	– CANTILEVER	H.A.S.	– HEADED ANCHOR STUD	THK.	–THICKNESS
CB.	– CARDBOARD	HORIZ.	– HORIZONTAL	T.O.	–TOP OF
CH.	– CHAMFER	H.S.B.	– HIGH STRENGTH BOLT	TYP.	–TYPICAL
C.J.	– CONTROL/CONSTRUCTION JOINT	I.D.	– INSIDE DIAMETER	U.N.O.	– UNLESS NOTED OTHERWISE
CLR.	– CLEAR, CLEARANCE	IN.	– INCH	VAR.	–VARIES
C.M.U.	– CONCRETE MASONRY UNIT	INT.	– INTERIOR	VERT.	–VERTICAL
COL.	– COLUMN	JNT.	– JOINT	V.I.F.	–VERIFY IN FIELD
CONC.	– CONCRETE	LB.	– POUND	WT.	–WEIGHT
CONN.	– CONNECTION	LIN. FT.	– LINEAL FEET		
CONST.	– CONSTRUCTION	L.L.V.	– LONG LEG VERTICAL		
CONT.	– CONTINUOUS	MAT'L.	– MATERIAL		
CONTR.	– CONTRACTOR	MAX.	– MAXIMUM		
CTRD.	– CENTERED	MECH.	– MECHANICAL		
DET.	– DETAIL	MID.	– MIDDLE		
DIAG.	– DIAGONAL	MIN.	– MINIMUM		
DIAM.	– DIAMETER	MISC.	– MISCELLANEOUS		
DIM.	– DIMENSION	MTL.	– METAL		
DISCONT.	– DISCONTINUOUS	N.I.C.	– NOT IN CONTRACT		
DWG.	– DRAWING	NO.	– NUMBER		
EA.	– EACH	NOM	– NOMINAL		
E.F.	– EACH FACE	N.T.S.	– NOT TO SCALE		
EL.	– ELEVATION	OC	– ON CENTER		
ELECT.	– ELECTRICAL	OD	– OUTSIDE DIAMETER		
ELEV.	– ELEVATOR	O.H.	– OPPOSITE HAND		
EQ.	– EQUAL	OPNG.	– OPENING		
E.W.B.	– END WALL BARS	⅊	– PLATE		
E.W.	– EACH WAY	P.S.F.	– POUND PER SQUARE FOOT		
EXIST.	– EXISTING	P.S.I.	– POUND PER SQUARE INCH		
EXP. JNT.	– EXPANSION JOINT	R.	– RADIUS		
EXT.	– EXTERIOR	REINF.	– REINFORCEMENT		
F.D.	– FLOOR DRAIN	REQ'D.	– REQUIRED		

SYMBOLS

℄	CENTER LINE
ϕ	DIAMETER
	ELEVATION
&	AND
W/	WITH
⅊	PLATE
X	BY
#	NUMBER
O	AT
⌑	SQUARE
L	ANGLE

Figure 5 Abbreviations.

Care of Construction Drawings and Mobile Devices

Construction drawings and mobile devices used to store drawings and other supplemental documents are valuable records and should be protected. Follow these rules when handling construction drawings and mobile devices:

- Never write on a construction drawing without authorization.
- Keep drawings clean. Dirty drawings are hard to read and can cause errors.
- Fold drawings so that the title block is visible.
- Fold and unfold drawings carefully to avoid tearing.
- Do not lay sharp tools or pointed objects on construction drawings.
- Keep drawings away from moisture.
- Make copies for field use; don't use originals.
- Use a case when transporting or storing a mobile device.
- Keep the mobile device charged and make sure all data is backed up.
- Protect the mobile device display with a screen cover.
- Avoid prolonged exposure of the mobile device to direct sunlight and extreme heat/cold.
- Keep the mobile device away from water, food, and drink.

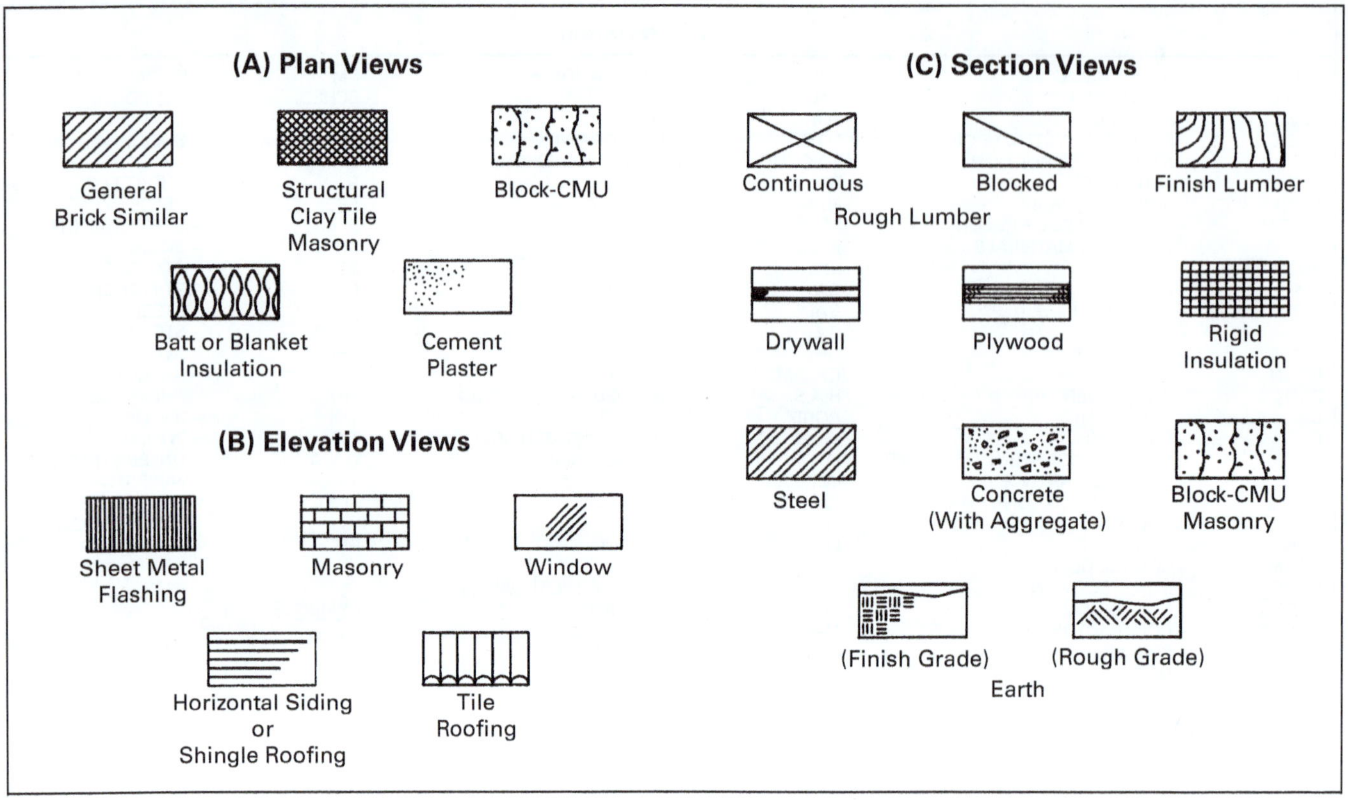

Figure 6 Architectural symbols.

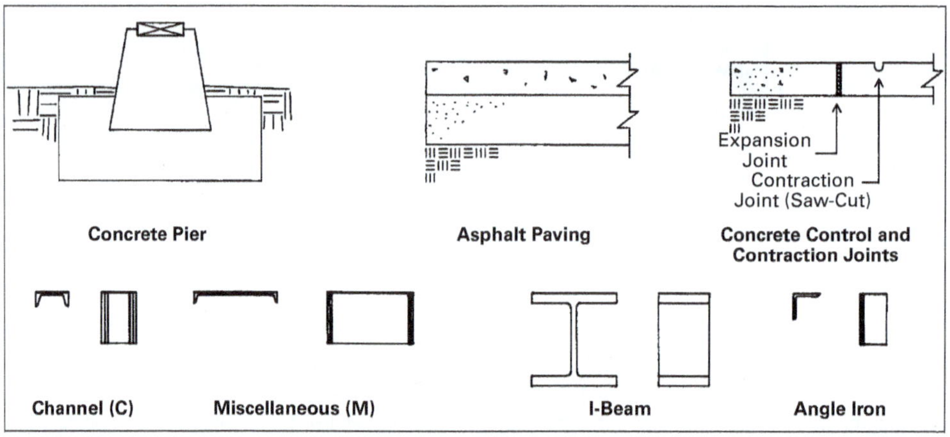

Figure 7 Civil and structural engineering symbols.

1.3.0 Dimensions and Drawing Scale

Dimensions and their associated drawing scales provide information that is essential for translating construction drawings into actual buildings correctly. This section describes basic features and the purpose of dimensions and drawing scales. How to read and use different types of drawing scales will be discussed later.

1.3.1 Dimensions

Dimensions are the parts of the drawings that show the size and placement of the objects that will be built or installed. Dimension lines can have arrowheads or slashes at both ends, with the dimension itself written near the middle of the line. The dimension is a measurement written as a number, and it may be written in inches with fractions ($6\frac{1}{2}$"), in feet with inches (1'-2"), or in inches with decimals (3.2"). When the metric system is used, the dimensions are usually expressed in meters (m), centimeters (cm), or millimeters (mm).

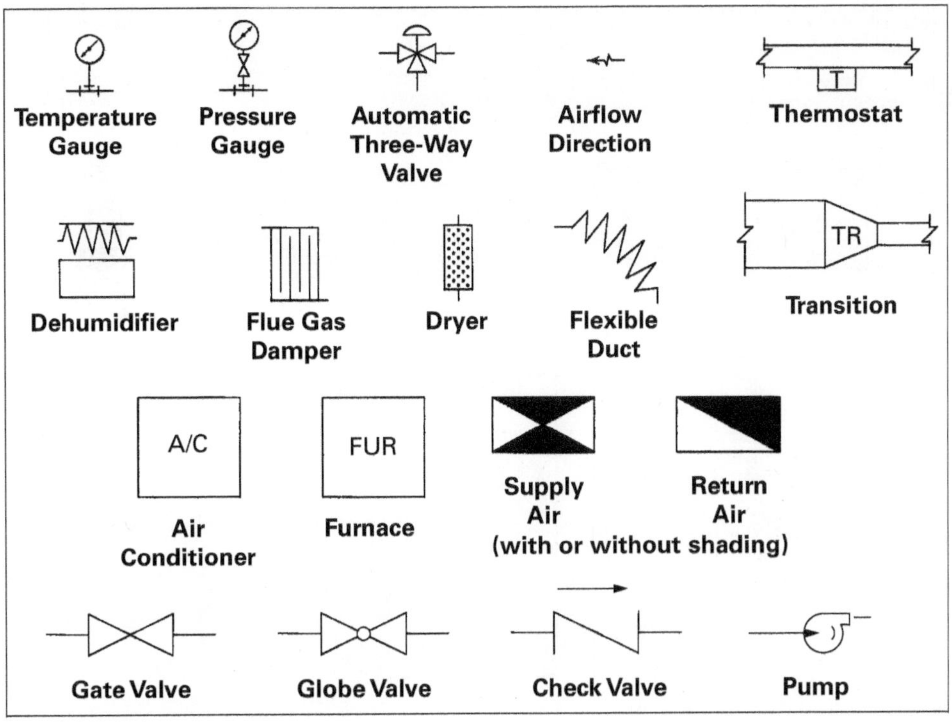

Figure 8 Mechanical symbols.

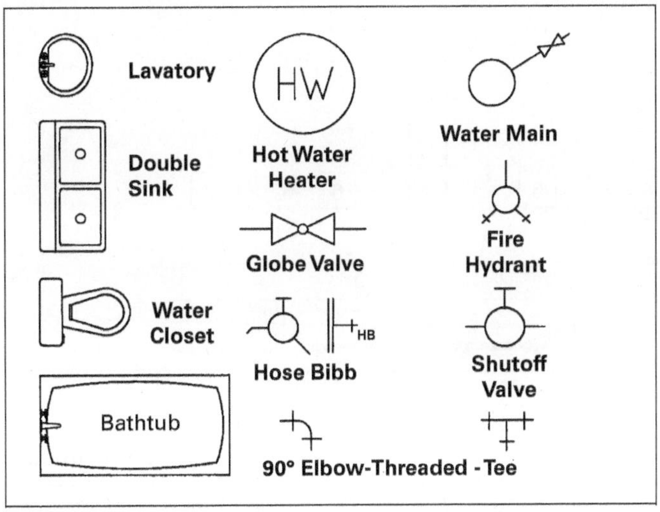

Figure 9 Plumbing symbols.

To do accurate work, workers need to know how to read dimensions on construction drawings. This means they need to know whether the dimensions measure to the exterior or the interior of an object. To understand the difference, look at *Figure 12*, which shows a piece of pipe. There are two measurements that could be taken to get the pipe's dimensions.

The first measurement is taken from the pipe's exterior edge on one side directly across to its exterior edge on the other side. The second measurement is from the pipe's inside (interior) edge on one side directly across to its interior edge on the other side. Even though the difference between these two dimensions may be only a fraction of an inch (the thickness of the pipe), they are still two completely different dimensions. This is important to remember because any dimensioning inaccuracy or miscalculation in one place will affect the accuracy of calculations in other places.

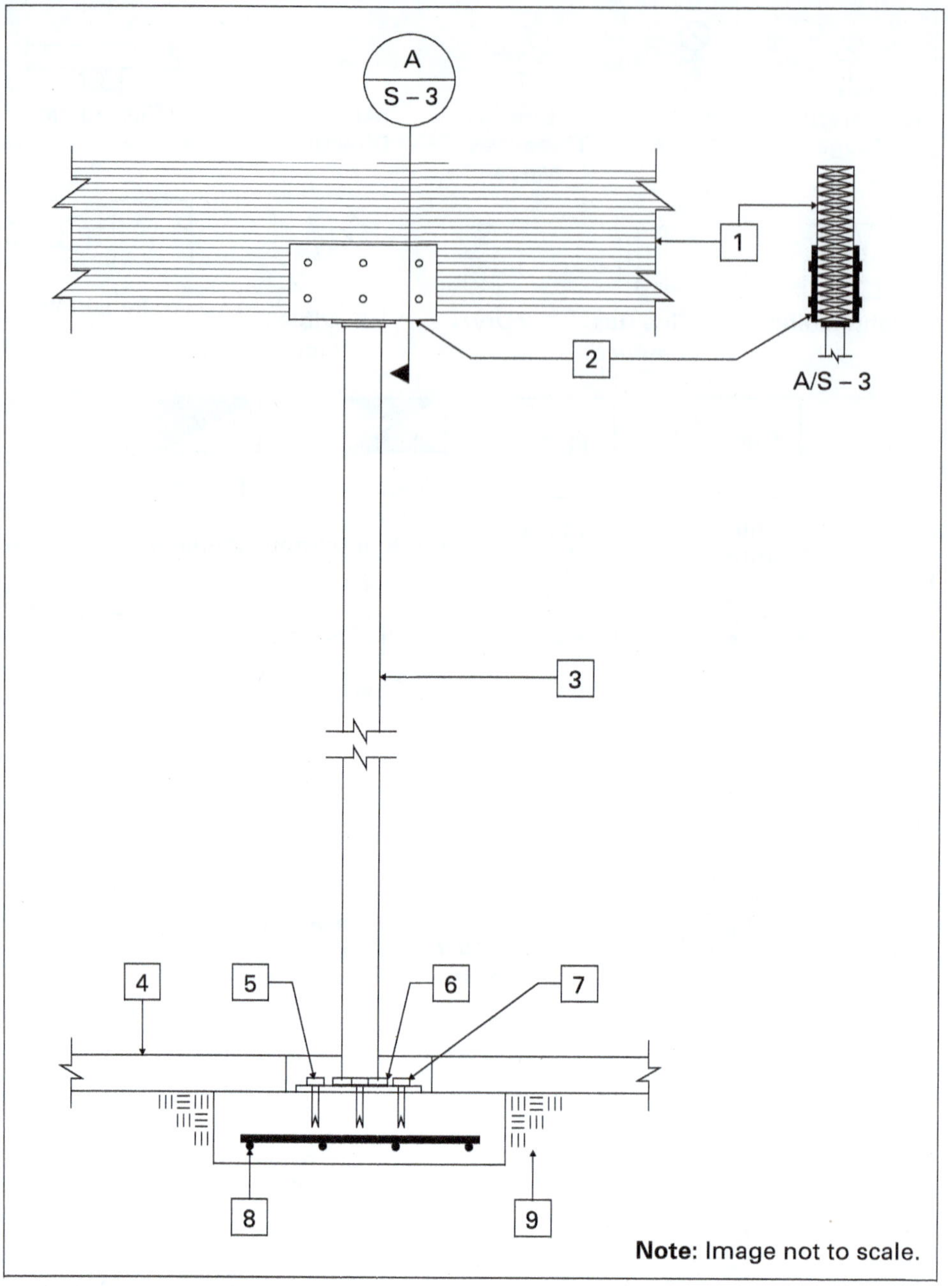

Figure 10 Keynotes.

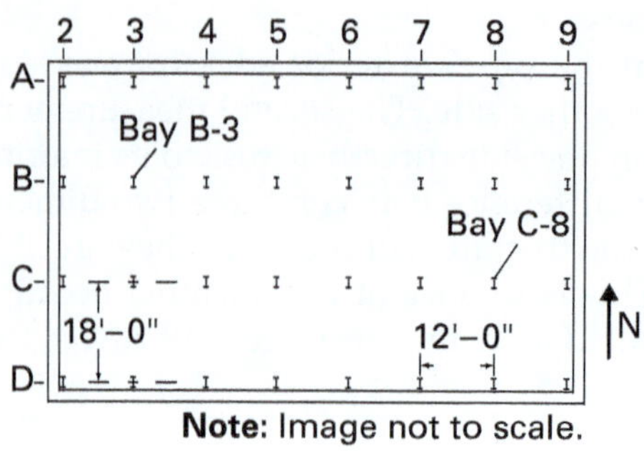

Figure 11 Grid.

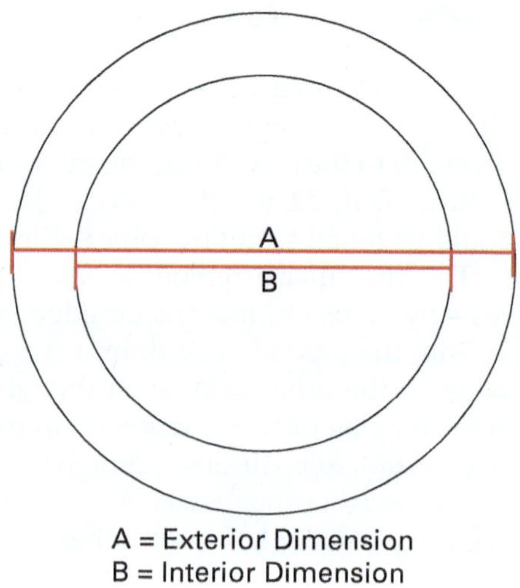

Figure 12 Exterior and interior dimensions on pipe.

Going Green

Green Construction

Green construction refers to a method of designing and building structures using materials and techniques that help minimize the stress on our natural resources and the environment.

With just under 5% of the world's population, the United States manages to consume about 19% of the world's energy (buildings account for 40% of this consumption) and produces 170 million tons (154,221,406 metric tons) of construction and demolition debris a year.

In response to these numbers, and a general increase in environmental awareness, organizations such as the US Green Building Council (USGBC) created environmental assessment systems, such as the LEED (Leadership in Energy and Environmental Design) Green Building Rating System. Systems like LEED provide green standards for the construction industry to follow, and they officially certify nonresidential structures that meet USGBC's strict criteria.

The LEED program was launched in 1988. It is a voluntary national standard that awards points for incorporating green strategies into areas such as site planning, safeguarding water quality and efficiency, efficiency in energy use and recycling, conservation of resources and materials, and the design, quality, and efficiency of the indoor environment.

Going Green

Recycling at Rinker Hall, University of Florida in Gainesville

An example of a LEED Gold-rated building is Rinker Hall on the campus of the University of Florida in Gainesville, Florida. During the planning phase of the building, special consideration was given to land use and the use of recycled and recyclable materials. Special attention was also given to planning the incorporation of the most recent green technologies in the areas of water consumption and conservation, heating and cooling energies, and other power needs required to run the various systems of a building.

A few of the green strategies incorporated into the construction of Rinker Hall include:

- Orienting the structure on a pure north-south axis to allow it to use low-angle light for daytime lighting.
- Using rooftop skylights and a central skylight-covered atrium to light specific areas using natural sunlight.
- Using shade walls on the east and west sides of the building.
- Landscaping with native trees and flora that require minimal watering.
- Installing water-free urinals and fixtures that use 20% less water than mandated.

With the various green strategies, materials selection, and building techniques employed, Rinker Hall and all its systems use up to 57% less energy than a similar structure designed in minimal compliance with the American Society of Heating, Refrigerating, and Air Conditioning Engineers (ASHRAE).

Engineer's scale: A straightedge measuring device divided uniformly into multiples of 10 divisions per inch so that drawings can be made with decimal values. Used mainly for land measurements on site plans.

Architect's scale: A specialized ruler used in making or measuring reduced scale drawings. The ruler is marked with a range of calibrated ratios for laying out distances, with scales indicating feet, inches, and fractions of inches. Used on drawings other than site plans.

1.3.2 Drawing Scale

The scale of a drawing tells the size of the object drawn compared with the actual size of the object represented. The scale is shown in one of the spaces in the title block, beneath the drawing itself, or in both places. The type of scale used on a drawing depends on the size of the objects being shown, the space available on the paper, and the type of plan.

On a site plan, the scale may read SCALE: 1" = 20'-0". This means that every 1 inch on the drawing represents 20 feet, 0 inches. The scale used to develop site plans is an **engineer's scale**.

On a floor plan, the scale may read SCALE: $\frac{1}{4}$" = 1'-0". This means that every $\frac{1}{4}$ inch on the drawing represents 1 foot, 0 inches. Floor plans are developed using an **architect's scale**. This scale is divided into fractions of an inch. Metric floor plans typically use a ratio of 1:50, indicating that each 1 mm on the drawing represents 50 m.

Some drawings are not drawn to scale. A note on such drawings reads **not to scale (NTS)**.

CAUTION

When a plan is marked NTS, workers cannot measure dimensions on the drawing and use those measurements to build the project. NTS drawings give relative positions and sizes. The sizes are approximate and are not accurate enough for construction.

Not to scale (NTS): Describes drawings that show relative positions and sizes only, without scale.

Checking Scales on Drawings

Normally the scale of a drawing is marked directly on the drawing itself. If not, however, do not guess what the scale is. Rather, compare different scales by identifying a given length on the drawing (for example, an 8'-6" pipe), and then find the scale that measures the pipe at 8'-6". Check at least two items on the drawing this way before proceeding.

Schematic Drawings

Most plumbing and electrical sketches are single-line drawings or schematic drawings. These drawings illustrate the scale and relationship of the project's components. In a single-line or schematic plumbing drawing, the line represents the centerline of the pipe. In a single-line or schematic electrical drawing, the line represents electrical wiring routing or circuit.

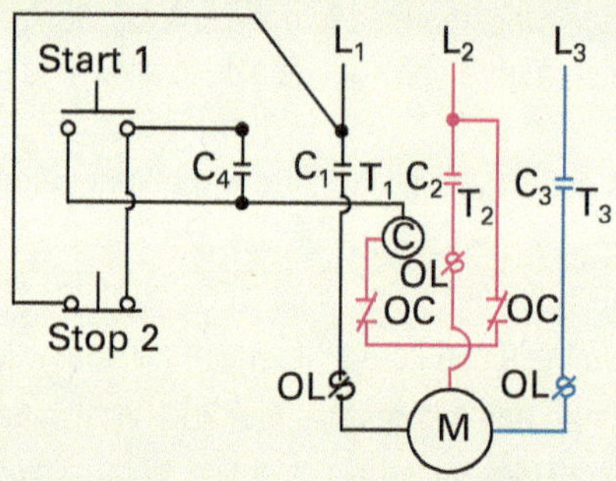

1.4.0 Measuring Scales

US standard and metric rulers and measuring tapes are usually used in a shop or in the field for a variety of measuring tasks. However, there are other types of rulers called scales used by draftspeople to produce drawings. They are also used by many workers to take scaled measurements on a drawing. These include the architect's scale, the **metric scale** (metric architect's scale), and the engineer's scale. Knowing how these scales work makes it easier to understand the information contained in a set of drawings.

Metric scale: A straightedge measuring device divided into centimeters, with each centimeter divided into 10 mm. Usually used for architectural drawings and sometimes referred to as a *metric architect's scale*.

1.4.1 Architect's Scale

The architect's scale is often used to create construction drawings. An architect's scale translates the large measurements of real structures (rooms, walls, doors, windows, ducts, etc.) into smaller measurements for drawings. Architect's scales are available in several types, but the most common include the triangular and flat scales. A flat scale is shown in *Figure 13*. The triangular architect's scale is most used because it can combine up to 12 different scales on one tool. Each side of the triangular form has two faces, and two scales are combined on each face. Architect's scales are available in 6" and 12" lengths.

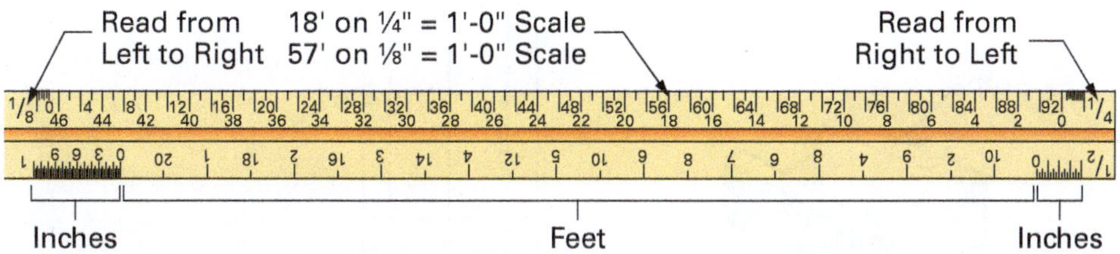

Figure 13 The architect's scale.

Each scale on an architect's scale is designated to a different fraction of an inch that equals a foot. These fraction designations appear on the right and left corners of each scale. An architect's scale is read from left to right or right to left, depending on which scale is being read.

Look at the point of measurement in *Figure 13*. It represents 57' when read from the left on the 1/8" scale and equals 18' when read from the right on the 1/4" scale.

Now refer to *Figure 14*. Using the 3/8" scale and reading from the right, it can be determined that the section of duct is 7'-5" long. Notice how the 0 point on an architect's scale is not at the extreme end of the measuring line. This is because numbers to the right of the 0 represent fractions of one foot (or inches).

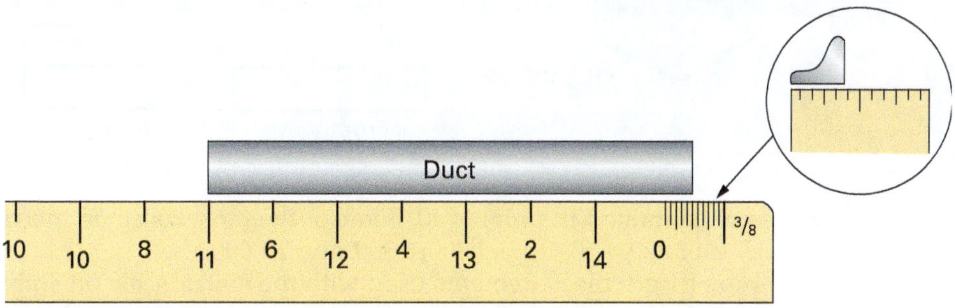

Figure 14 Measuring a section of duct with an architect's scale.

1.4.2 Metric Scale (Metric Architect's Scale)

Similar to the architect's scale is the metric scale, sometimes referred to as a metric architect's scale (*Figure 15*). Common lengths indicated on metric scales are 30 mm and 60 mm. Like the architect's scale, a metric scale can have several scales on it and is used to generate drawings. *Figure 15* shows how 10 m would be shown on a 1:200 scale. The actual length on the plans is 50 mm, equaling 10 m of real distance on the project.

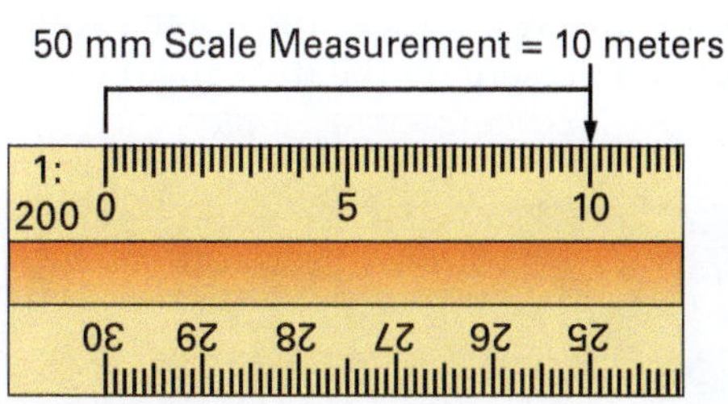

Figure 15 Metric architect's scale.

Orthographic Drawings

Orthographic drawings are used for elevation drawings. They show straight-on views of the different sides of an object with dimensions that are proportional to the actual physical dimensions. In orthographic drawings, the designer draws lines that are scaled-down representations of real dimensions. Every 12", for example, may be represented by ¼" on the drawing. Similarly, in an example using metric measurements with a ratio of 1:2, every 30 mm may be represented by 15 mm on the drawing.

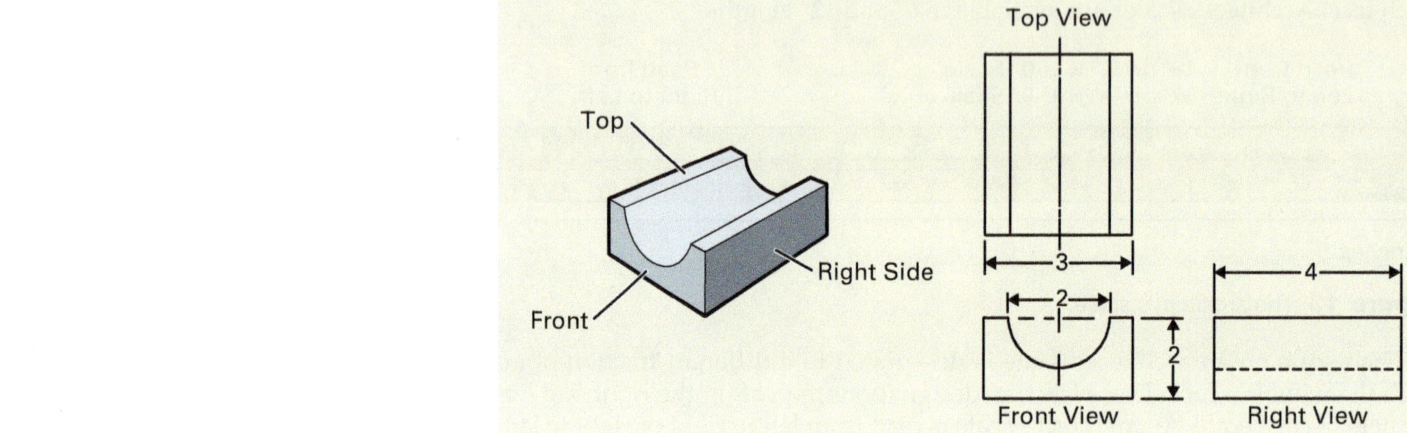

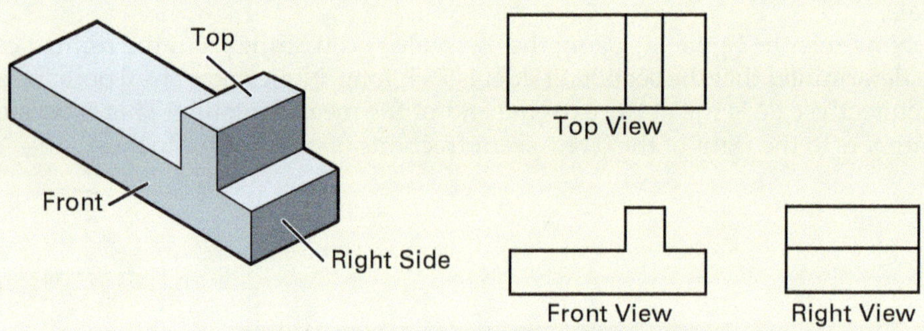

Metric scales are calibrated in units of 10. Some of the most common metric scales include 1:5, 1:10, 1:20, 1:50, 1:100, 1:200, 1:500, and 1:1,000.

The two common length measurements used with the metric scale on architectural drawings are the meter and millimeter, the millimeter being $1/1000$ of a meter. On drawings drawn to scale between 1:1 and 1:100, the millimeter is typically used. The millimeter symbol (mm) will not be shown, but there should be a note on the drawing indicating that all dimensions are given in mm unless otherwise noted.

On drawings with scales between 1:200 (*Figure 15*) and 1:2,000, the meter is generally used. Again, the meter symbol (m) will not be shown, but the drawing will have a note indicating that all dimensions are in meters unless otherwise noted. Land distances shown on site and plot plans, expressed in metric units, are typically given in m or km (1,000 m).

Reading a metric scale is easy after identifying the scale to use and the length that the scale increments represent on the drawing. Here's how it works.

The unit of length in *Figure 15* is the meter and the object on the drawing starts at 0 and extends out to 10 on the scale. At a 1:200 ratio this object would be 10 m long. This is because every millimeter on the scale represents 200 mm on the object. For example, if there are 50 mm from 0 to 10, multiply the 50 mm by 200 mm, which gives 10,000 mm (50 mm × 200 mm = 10,000 mm). Because the unit being used is the meter and there are 1000 mm in 1 m, divide the 10,000 mm by 1,000, which results in 10 m. This same process is also used to determine lengths for other ratios on the scale.

1.4.3 Engineer's Scale

The engineer's scale (*Figure 16*) is used mainly for land measurements on site plans, which means the scale must accommodate very large measurements. Each engineer's scale is set up as multiples of 10 and the measurements are taken in decimals. This is different from the architect's scale in that a unit is represented by a fraction of an inch. The most common engineer's scales are 10, 20, 30, 40, 50, and 60, which can all be combined on the triangular engineer's scale.

For each scale, the measurements can represent various units derived from that scale number and a multiple of 10. For example, on a 10 scale, 1" can represent 1' (10 × 0.10), 10' (10 × 1.0), 100' (10 × 10.0), and so on. On a 50 scale, an inch can be 5' (50 × 0.10), 50' (50 × 1.0), or 500' (50 × 10.0), and so on. This same process can also be used to determine lengths on the scales.

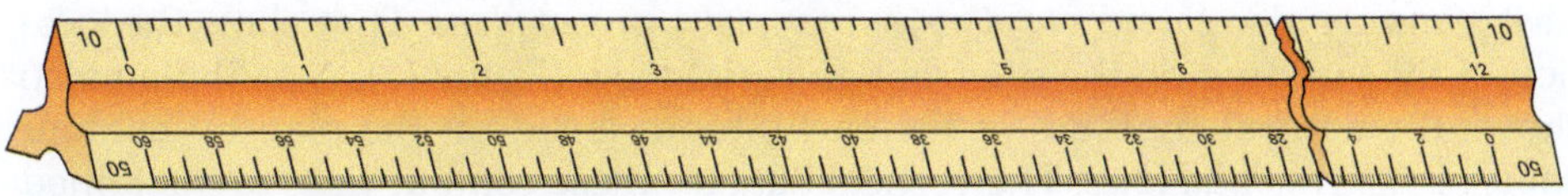

Figure 16 The engineer's scale.

1.5.0 Six Types of Construction Drawings

A set of building construction drawings typically includes six major types of drawings. Other types of drawings and documents may be used to supplement the drawing set, but their presence depends on each building's usage and installed subsystems. For example, both a fire station and a convenience store drawing package would include the six major drawing types, but the rest of the package would differ significantly. The six major drawing types (*Figure 17*) include the following:

- Civil
- Architectural
- Structural
- Mechanical
- Plumbing
- Electrical

As mentioned previously, other types of drawings and documents may be included to supplement a set of construction drawings depending on the type and use of the building. These could include:

- Fire protection
- Shop drawings
- As-built
- Change order

This section examines the characteristics of each type of construction drawing.

Blueprints

Before the introduction of large-format plotters, printers, and copiers, technical drawings were reproduced using a contact print process that utilized light-sensitive paper to produce what were called blueprints. The name *blueprint* came from the fact that the reproduction resulted in a negative of the original drawing, shown with white lines on a blue background. This technology allowed construction drawings to be reproduced quickly and accurately. The term *blueprints* were typically used to describe any set of technical or construction drawings. The process for making blueprints was developed in 1842 by an English astronomer named Sir John F. Herschel and was used widely for over a century until it was replaced by modern printing and copying technology. Today these reproductions are simply referred to as *drawings* or *prints*.

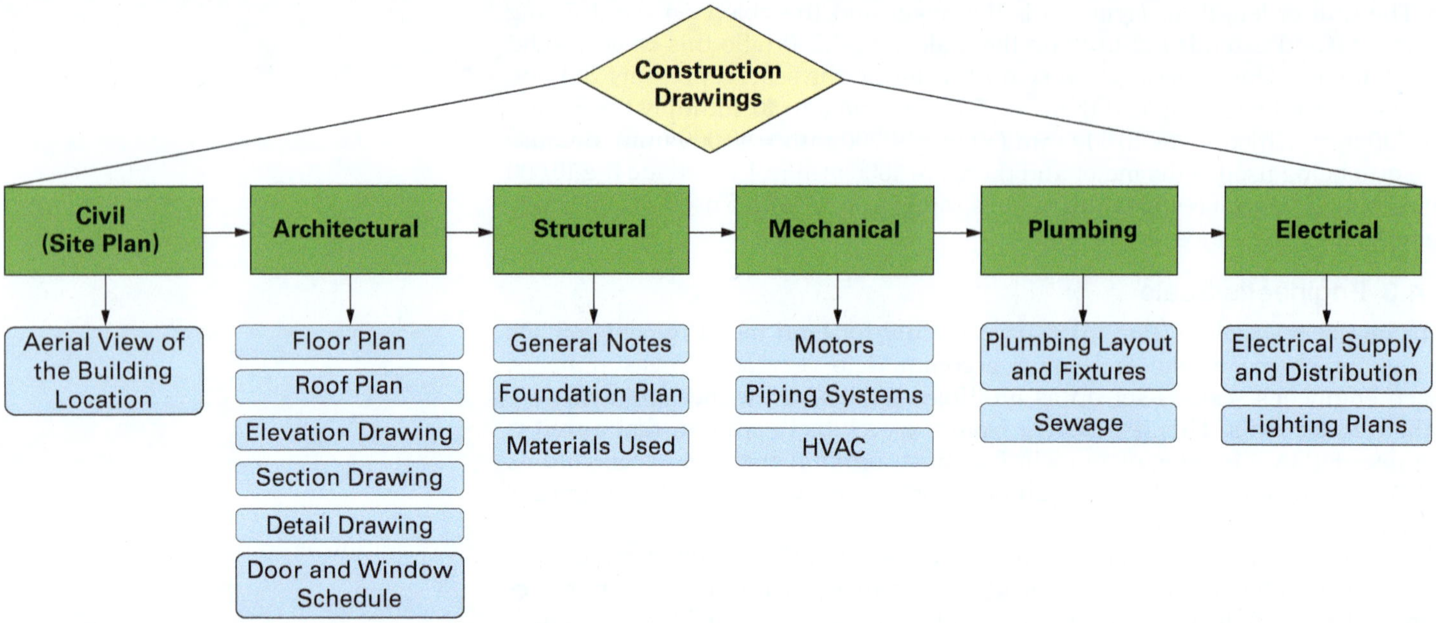

Figure 17 Types of construction drawings.

1.5.1 Civil Plans

Civil plans: Drawings that show the location of the building on the site from an aerial view, including contours, trees, construction features, and dimensions.

Contour lines: Solid or dashed lines showing the elevation of the earth on a civil drawing.

Civil plans are also called *site plans, survey plans*, or *plot plans*. They show the location of the building on the site from an aerial view (*Figure 18*). A civil plan also shows the natural contours of the earth, represented on the plan by **contour lines**, or *topographical lines*. The vertical data for these contour lines are obtained by a surveyor. Before a topographical survey can take place, a benchmark must be established. A benchmark is a fixed point with a known elevation that is used as the base point for measuring all elevation values (*Figure 19*). Surveyors also use a special coordinate system for identifying location, direction, and distance by utilizing lines of latitude and longitude (*Figure 20*). The civil plans can also include a landscape plan (*Figure 21*) that shows any trees on the property, construction features such as walks, driveways, or utilities, the dimensions of the property, and possibly a legal description of the property.

1.5.2 Architectural Plans

Architectural plans: Drawings that show the design of the project. Also called *architectural drawings*.

Architectural plans (also called *architectural drawings*) show the design of the project. Look at architectural plans first because all other drawings follow from them. Architectural plans are the most general; they show how all the parts of the project fit together. One part of an architectural plan is a floor plan, also known as a *plan view*. Any drawing made looking down on an object is commonly called a plan view. The floor plan is an aerial view of the layout of each room. It provides the most information about the project. It shows exterior and interior walls, doors, stairways, and mechanical equipment. The floor plan shows the floor as someone would see it from above if the upper part of the building were removed.

Roof plan: A drawing of the view of the roof from above the building.

Elevation (EL): Height above sea level, or other defined surface, usually expressed in feet or meters.

Elevation drawing: A side view of a building or object, showing height and width.

An architectural plan also includes a **roof plan** (*Figure 22*), which is a view of the roof from above the building. It shows the shape of the roof and the materials that will be used to finish it.

Elevation (EL) is another element of architectural drawings. An **elevation drawing** is a side view that shows height. On a building drawing, there are standard names for different elevations. For example, the side of a building that faces south is called the *south elevation*. Exterior elevations show the size of the building, the style of the building, and the placement of doors, windows, chimneys, and decorative trim.

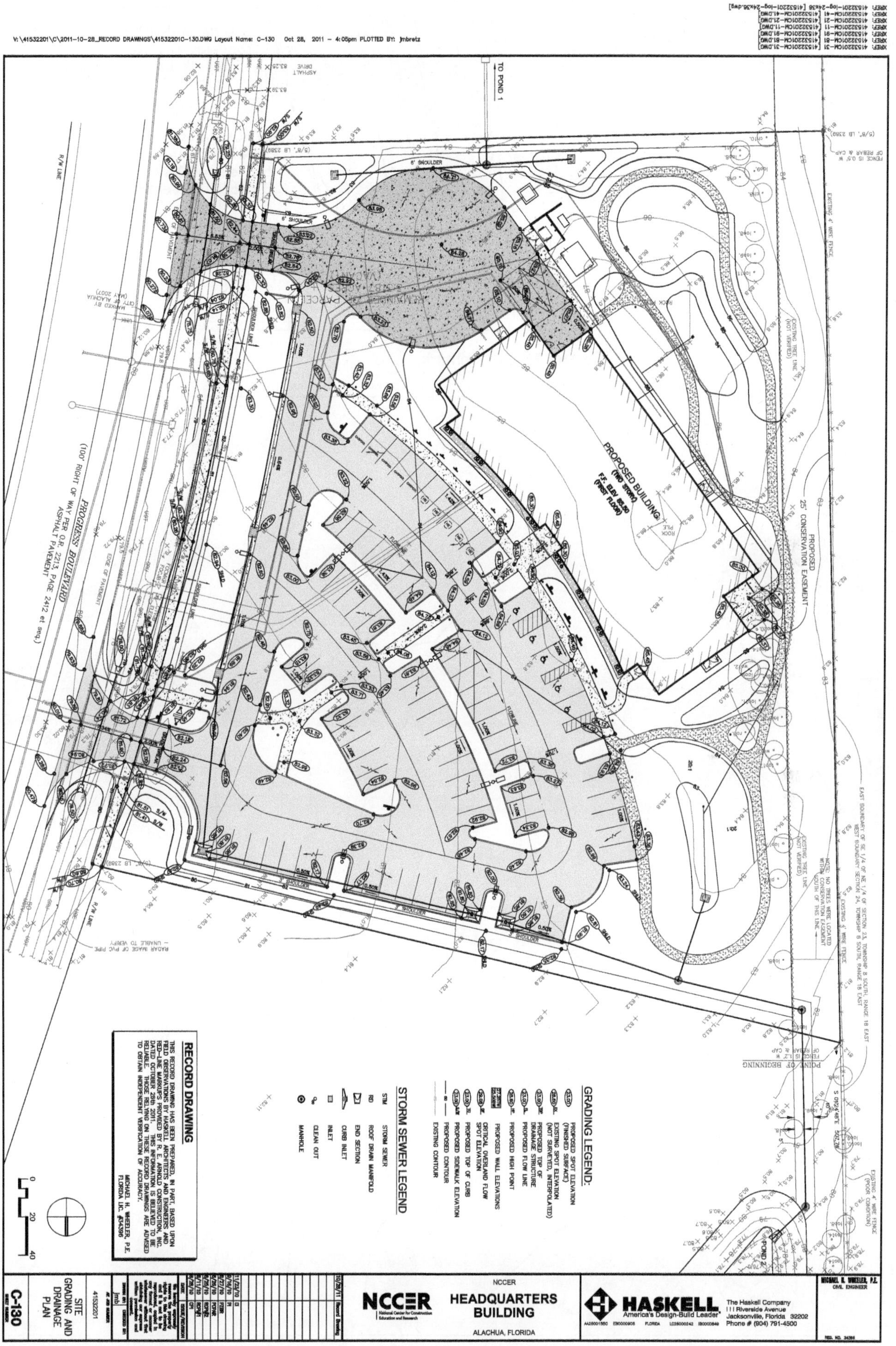

Figure 18 Civil plan, aerial view.

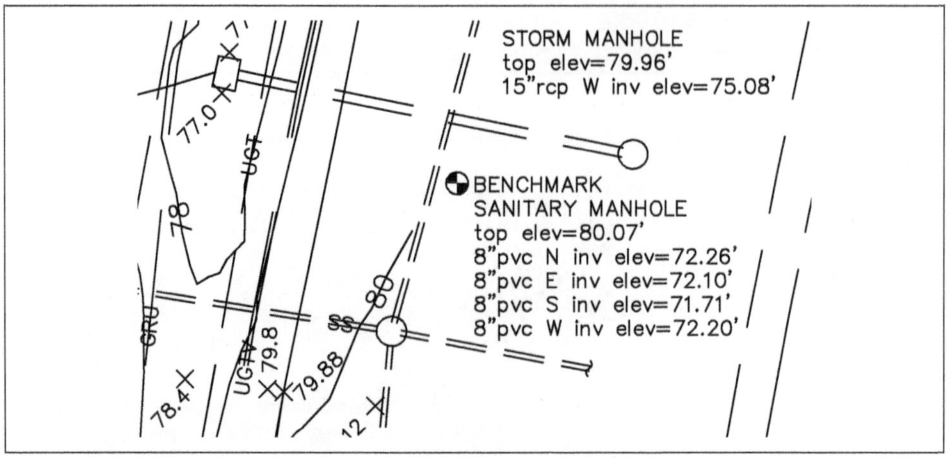

Figure 19 Benchmark.

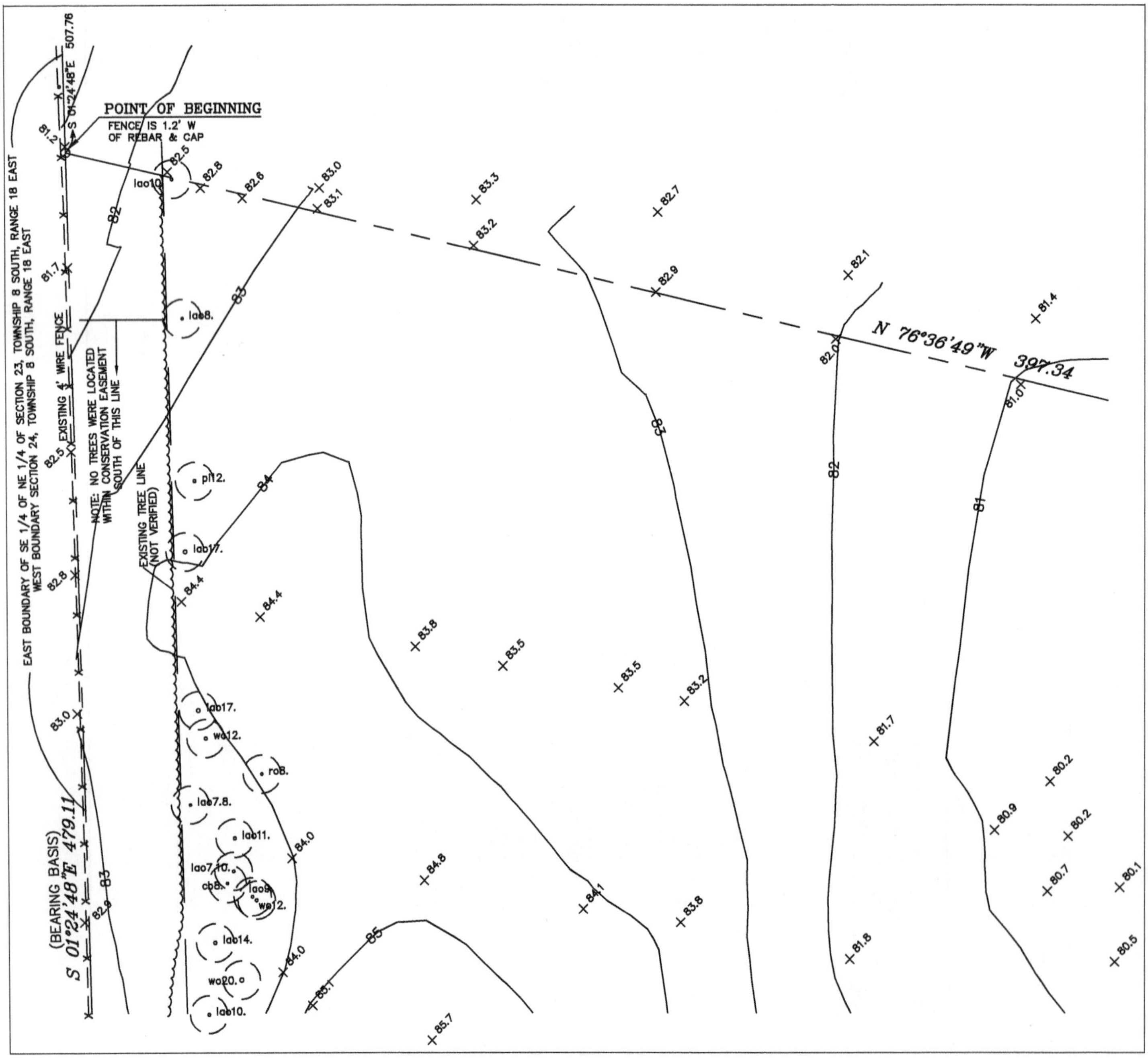

Figure 20 Coordinate system.

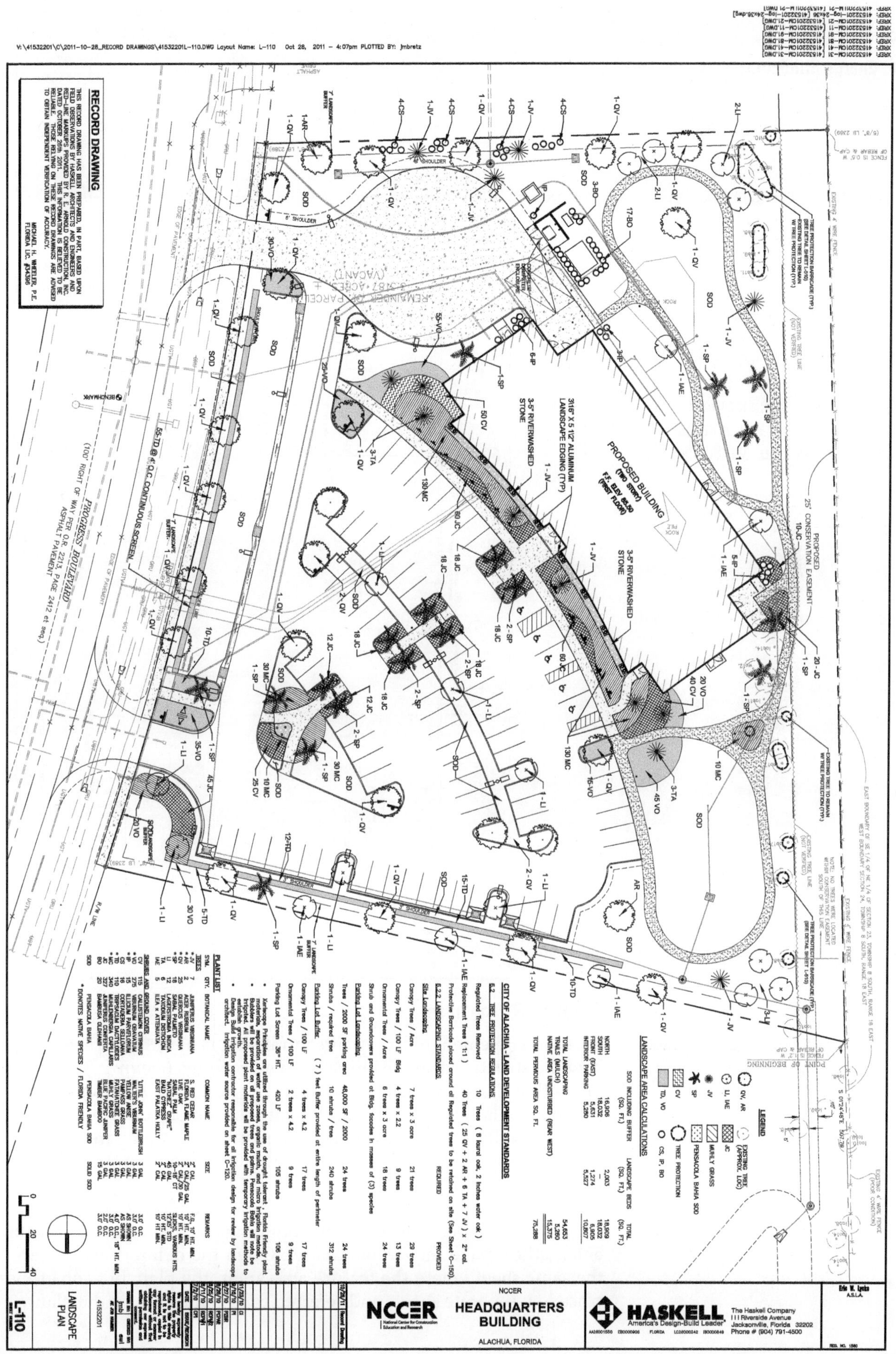

Figure 21 Landscape plan.

Did You Know?

BIM and Energy Efficient Design

Simulation tools within BIM software allow architects and designers to look at different building designs and test how energy efficient a building will be long before ground is ever broken on the project. These tools allow the user to simulate sunlight and shading throughout different seasons based on the exact location of the project along with existing surrounding landscape features, buildings, and structures. This means that the final design will make the best use of building location and direction, windows, walls, existing site features, etc., to optimize the utilization of natural sunlight and maximize energy efficiency.

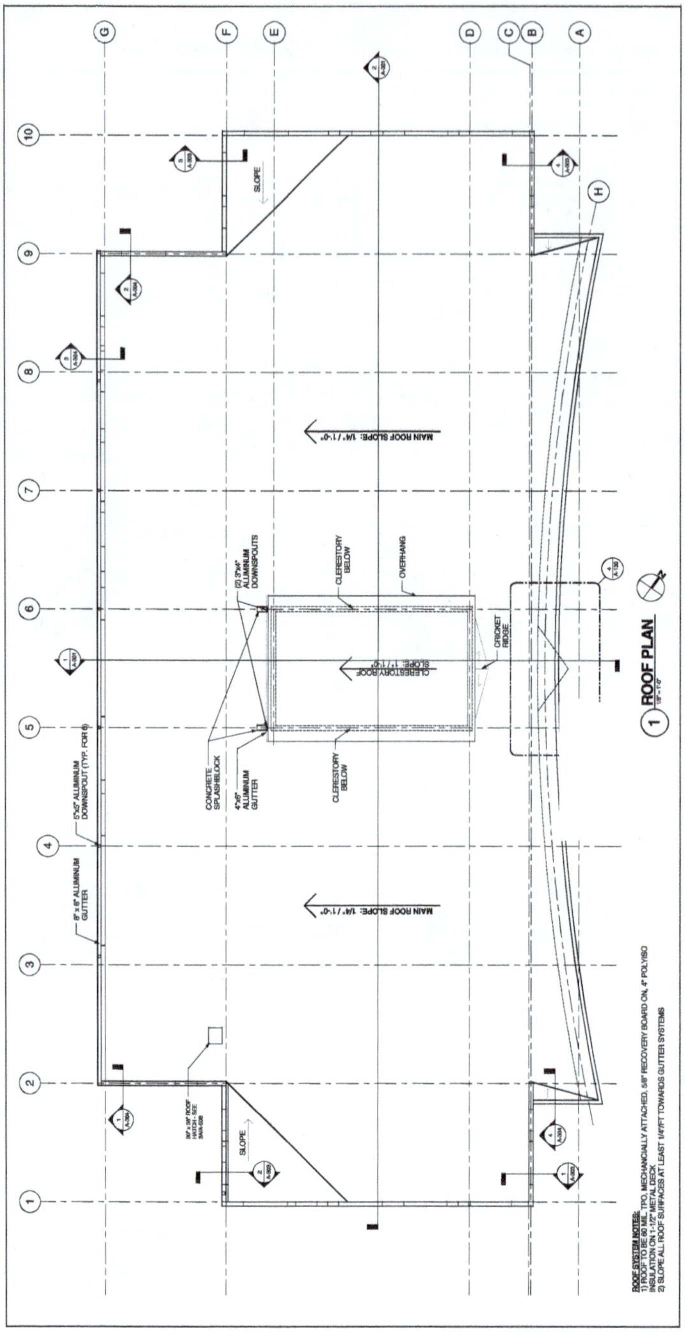

Figure 22 Roof plan.

Section drawing: A cross-sectional view of a specific location, showing the inside of an object or building.

Detail drawings: Enlarged views of part of a drawing used to show an area more clearly.

Another element of the architectural plan is a **section drawing**, which shows how the structure is to be built. Section drawings (*Figure 23*) are cross-sectional views that show the inside of an object or building. They show what construction materials to use and how the parts of the object or building fit together. They normally show more detail than plan views. Compare the information on the drawing in *Figure 23* with the information on the drawings in the Performance Profile Sheet.

Even more detail is shown in **detail drawings** (*Figure 24*), which are enlarged views of some special features of a building, such as floors and walls. They are enlarged to make the details clearer. Often the detail drawings are placed on the same sheet where the feature appears in the plan, but sometimes they are placed on separate sheets and referred to by a number on the plan view.

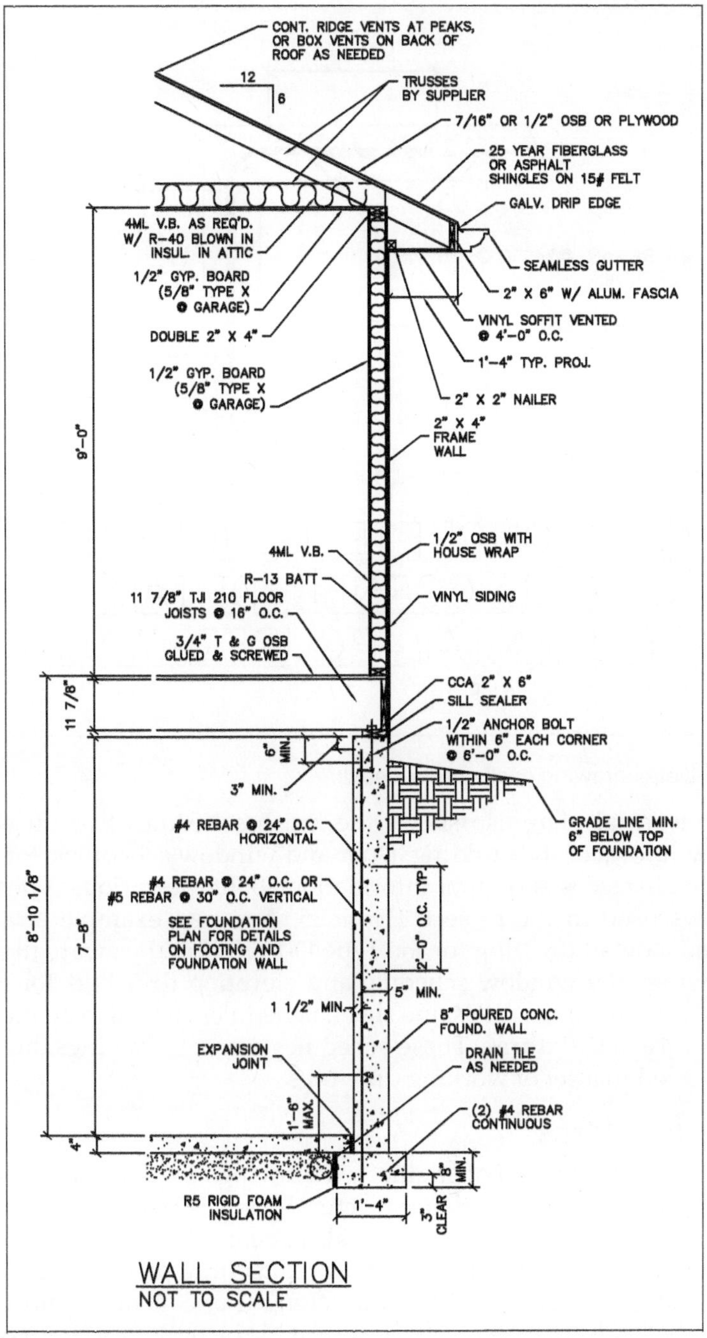

Figure 23 Residential wall section.

A change order is required when there are changes to the scope of work on a project which already has a contract in place agreed upon by all parties. These changes could come from design changes, unforeseen site conditions, or work that is added to or deleted from the scope of the project by the client. The change order document describes the changes in detail and could include the construction scope, design, project schedule, and even cost. All parties involved must agree to and sign the change order document. Depending on the change, new or revised drawings may be required as well.

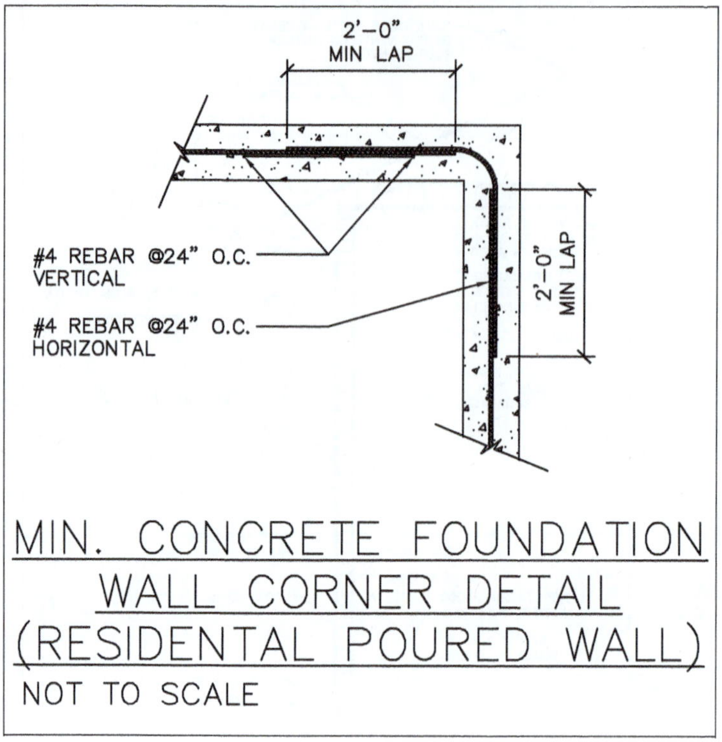

Figure 24 Detail drawing (wall corner detail).

The architectural plan also shows the finish schedules to be used for the doors and windows of the building. Door and window schedules, for example, are tables that list sizes and other information about the various types of doors and windows used in the project. *Figure 25* shows an example of a window schedule and a detail drawing for the Type D-1 window. Be aware that in some sets of drawings the window schedule and elevation drawings for each type of window may appear on the same sheet. Schedules may also be included for finish hardware and fixtures. These schedules are not drawings, but they are usually included in a set of working drawings.

1.5.3 Structural Plans

Structural plans: A set of engineered drawings used to support the architectural design.

The **structural plans** are a set of engineered drawings used to support the architectural design. The first part of the structural plans is the general notes (*Figure 26*). These notes give details of the materials to be used and the requirements to be followed to build the structure that the architectural plan depicts. The notes, for instance, might specify the type and strength of concrete required for the foundation, the loads that the roof and stairs must be built to accommodate, and codes that contractors must follow. General notes may be on a separate general notes sheet or may be part of individual plan sheets.

Foundation plan: A drawing that shows the layout and elevation of the building foundation.

Beams: Large, horizontal structural members made of concrete, steel, stone, wood, or other structural materials to provide support above a large opening.

The structural plans also include a **foundation plan** (*Figure 27*), which shows the lowest level of the building, including concrete footings, slabs, and foundation walls. They also may show steel girders, columns, or **beams**, as well as detail drawings to show where and how the foundation must be reinforced. Column schedules, spread footing schedules, and foundation notes may be included on the foundation plan. A related element is the structural floor plan, which depicts a wood or metal joist framing and the underlayment of each floor of the structure.

The structural plans show the materials to be used for the walls, whether concrete or masonry, and whether the framing is wood or steel. Structural plans also include a framing plan, showing what kinds of ceiling joists and rafters are to be used and where trusses are to be placed. Notes for the framing plan are usually found on the same sheet as the drawing.

Window Schedule				
Type	**Frame Size**	**Frame Finish**	**Glazing**	**Remarks**
A	4'-2" × 4'-2"	Clear Anodized Aluminum	Viracon VE-7-2M	
B	24'-2½" × 11'-0"	Clear Anodized Aluminum	Viracon VE-7-2M	
B-1	12'-10" × 11'-0"	Clear Anodized Aluminum	Viracon VE-7-2M	
C	15'-11½" × 8'-3½"	Clear Anodized Aluminum	Viracon VE-7-2M	
C-1	15'-11½" × 4'-1½"	Clear Anodized Aluminum	Viracon VE-7-2M	
D	15'-11½" × 11'-1½"	Clear Anodized Aluminum	Viracon VE-7-2M	
D-1	15'-11½" × 11'-1½"	Clear Anodized Aluminum	Viracon VE-7-2M	
D-2	31'-11½" × 4'-4"	Clear Anodized Aluminum	Viracon VE-7-2M	
E	14'-4" × 7'-8"	Clear Anodized Aluminum	Viracon VE-7-2M	
F	15'-11½" × 4'-4"	Clear Anodized Aluminum	Viracon VE-7-2M	
H	11'-11½" × 8'-¼"	Clear Anodized Aluminum	Viracon VE-7-2M	
H-1	11'-11½" × 10'-1¾"	Clear Anodized Aluminum	Viracon VE-7-2M	
I	15'-11½" × 4'-1½"	Clear Anodized Aluminum	Viracon VE-7-2M	
CW - 1	12'-0" × 27'-11¼"	Clear Anodized Aluminum	Viracon VE-7-2M	
CW - 3	15'-11½" × 37'-3¼"	Clear Anodized Aluminum	Viracon VE-7-2M	
CW - 4	36'-10" × 20'-9½"	Clear Anodized Aluminum	Viracon VE-7-2M	
CW - 5	28'-11⅞" × 20'-9½"	Clear Anodized Aluminum	Viracon VE-7-2M	

NOTES

1. FRAME SIZES HAVE BEEN INDICATED ASSUMING 1/2" JOINT ADJACENT TO METAL PANEL SYSTEM. ALIGNMENT OF JOINTS W/ METAL PANEL MODULE IS REQUIRED. INTERMEDIATE MULLIONS TO BE CENTERED ON ADJACENT METAL PANEL JOINTS (1)
2. "SG" INDICATES SAFETY GLAZING (TEMPERED GLASS REQUIRED)

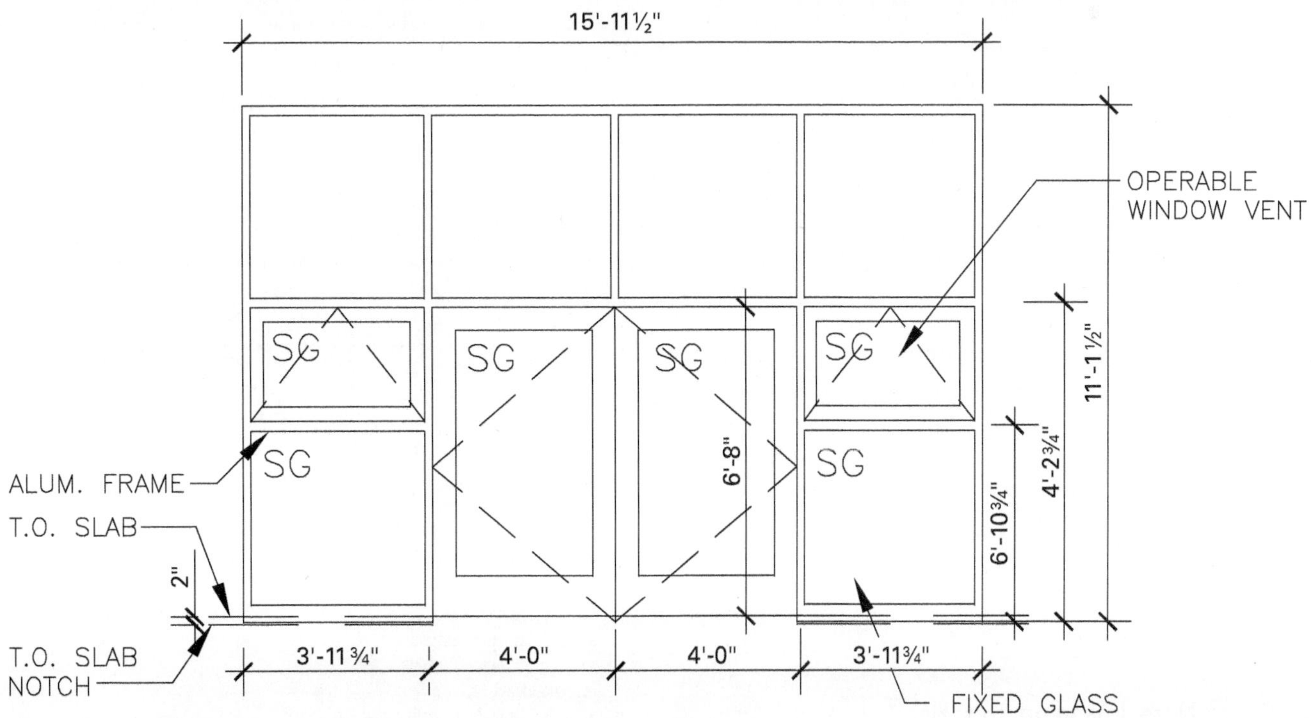

Figure 25 Window schedule and window detail.

GENERAL NOTES

I. DESIGN CRITERIA

A. GENERAL BUILDING CODE
The Contract Documents are based on the requirements of the:
1. Standard Building Code, 1997 edition.

B. DEAD LOADS
1. Partitions. An allowance of 20 PSF has been made for partitions as a uniformly distributed dead load.
2. Hanging Ceiling and Mechanical Loads. An allowance of 10 PSF has been made for hanging ceiling and mechanical equipment loads such as duct work and sprinkler pipes.

C. LIVE LOADS
1. Design live loads are based on the more restrictive of the uniform load listed below or the concentrated load listed acting over an area 2.5 feet square.

CATEGORY	UNIFORM LOAD (PSF)	CONCENTRATED LOAD (LB)
1. Roof	20	N/A
2. Elevated Floors	50	0
3. Terraces, Lobbies	100	0
4. Stairways, Exit Facilities	100	0
5. Elevator Machine Rooms	100	Assumed Eqp. Wt.
6. Mechanical Rooms, typical	150	Assumed Eqp. Wt.

NOTES:
1. Live Load Reduction. Live loads have been reduced on any member supporting more than 150 square feet, including flat slabs, except for floors in places of public assembly and for live loads greater than 100 pounds per square foot in accordance with the following formula:

$$R = r(A-150)$$

The reduction,, R, shall not exceed 40 percent for members supporting one level only, 60 percent for other members, or R as calculated in the following formula:

$$R = 23.1\ (1 + \frac{D}{L})$$

R = Reduction in percent.
r = Rate of reduction equal to .08 percent for floors.
A = Area of floor supported by the member.
D = Total dead load supported by the member.
L = Total, unreduced, live load supported by the member.

2. For storage loads exceeding 100 pounds per square foot, no reduction has been made, except that design live loads on columns have been reduced 20 percent.

D. ELEVATOR LOADS
Machine Beam, Car Buffer, Counterweight Buffer, and Guide Rail Loads. Assumed elevator loads to the supporting structure are shown on the drawings, including machine beam reactions, car buffer reactions, counterweight buffer reactions, and horizontal and vertical guide rail loads. The General Contractor shall submit to the Structural Engineer final elevator shop drawings showing all loads to the structure prior to the installation of the elevators for verification of load carrying capacity.

E. MECHANICAL EQUIPMENT LOADS
The General Contractor shall submit actual weights of equipment to be used in the project to the Structural Engineer for verification of loads used in the design at least three weeks prior to fabrication and construction of the supporting structure.

F. WIND LOADS
1. Wind pressures are based on the American Society of Civil Engineers, Minimum Design Loads for Buildings and Other Structures, ASCE 7–98 with a Wind Speed = 110 MPH (3 sec. gust), Exposure C, Importance Factor 1.15.
2. Wind pressures used in the design of the cladding are shown on these Drawings.

II. FOUNDATION

A. GEOTECHNICAL REPORT
Foundation design is based on the geotechnical investigation report as follows:
1. Reports of Geotechnical exploration, M.E. Rinker Sr. Hall (Revised Location) Near the southeast Corner of Newell Drive and Inner Road, Gainesville, Alachua County, Florida. Law Engineering and Environmental Services, Inc. January 2, 2001.

The geotechnical report is available to the General Contractor upon request to the Owner. The information included therein may be used by the General Contractor for his general information only. The Architect and Engineer will not be responsible for the accuracy or applicability of such data therein.

B. FOUNDATION TYPE
1. Spread Footing.
   a. Design Pressures:
      1. All footings have been designed assuming an allowable bearing pressure of 4000 PSF.
      Allowable pressures are increased 33% for combined gravity and wind loads.

C. SLAB–ON–GRADE
Radon resistant construction guidlines are being followed on this project. The details and specifications for slab–on–grade construction must be adhered to without deviation.
Slab–on–Grade shall be immediately underlain by a 8 mil. vapor barrier. Seams shall be lapped 12 inches and sealed with 2" wide pressure sensitive vinyl tape. All penetrations shall be sealed with tape.

D. CONSTRUCTION DEWATERING
The Contractor shall determine the extent of construction dewatering required for the excavation. The Contractor shall submit to the Geotechnical Engineer for review the proposed plan for construction dewatering, prior to beginning the excavation.

III. REINFORCED CONCRETE

A. CLASSES OF CONCRETE
All concrete shall conform to the requirements as specified in the table below unless noted otherwise on the drawings:

Usage	28 Day Comp. Strength (PSI)	Conc Type	Max Size Agg.	W/C Ratio
1. Elevated Floors	4000	NWT	3/4"	0.48
2. Spread Footings	3000	NWT	1"	0.55
3. Slab–On–Grade	4000	NWT	1"	0.48
4. Fnd. Walls & Plinths	4000	NWT	1"	0.48

All concrete shall be proportioned for a maximum allowable unit shrinkage of 0.03% measured at 28 days after curing in lime water as determined by ASTM C 157 (using air storage).

B. HORIZONTAL CONSTRUCTION JOINTS IN CONCRETE POURS
There shall be no horizontal construction joints in any concrete pours unless shown on the drawings. The Architect/Engineer shall approve all deviations or additional joints in writing.

C. REINFORCING STEEL SPECIFICATION
1. All Reinforcing Steel shall be ASTM A615 Grade 60 unless noted otherwise on the drawings or in these notes.
2. Welded Reinforcing Steel. Provide reinforcing steel conforming to ASTM A706 for all reinforcing steel required to be welded and where noted on the drawings.
3. Galvanized Reinforcing Steel. Provide reinforcing steel galvanized according to ASTM A767 Class II (2.0 oz. zinc PSF where noted on the drawings.
4. Deformed Bar Anchors. ASTM A496 minimum yield strength 70,000 PSI as noted on the drawings. Reinforcing bars shall not be substituted for deformed bar anchors.
5. Welded Wire Fabric. Welded smooth wire fabric, ASTM A 185, yield strength 65,000 PSI where noted on the drawings. Welded deformed wire fabric for, ASTM A 497, yield strength 70,000 PSI where noted on the drawings.

D. PLACEMENT OF WELDED WIRE FABRIC
Wherever welded wire fabric is specified as reinforcement, it shall be continuous across the entire concrete surface and not interrupted by beams or girders and properly lapped one cross wire spacing plus 2".

E. REINFORCEMENT IN TOPPING SLABS
Provide welded smooth wire fabric minimum 6 x 6 W2.9 x W2.9 in all topping slabs unless specified otherwise on the drawings.

F. REINFORCEMENT IN HOUSEKEEPING PADS
Provide welded smooth wire fabric 6 x 6 W2.9 x W2.9 minimum in all housekeeping pads supporting mechanical equipment whether shown on the drawings or not unless heavier reinforcement is called for on the drawings.

G. REINFORCING STEEL COVERAGE
Concrete Cover for reinforcement layer nearest to the surface unless specified otherwise on the drawings.
1. Concrete surfaces cast against and permanently exposed to earth. 3 inches
2. Concrete surfaces exposed to earth or weather or where noted on the drawings 2 inches.
3. Concrete surfaces not exposed to weather or in contact with the ground.
   a. #3 to #11 bars 1 inch

H. SPLICES IN REINFORCING STEEL
1. All unscheduled splices shall be Class A tension splice.

IV. STRUCTURAL STEEL

V. STEEL DECKS

VI. CURTAIN WALL

VII. CONCRETE MASONRY

VIII. MISCELLANEOUS

IX. SUBMITTALS

X. DRAWING INTERPRETATION

Note: Image not to scale.

Figure 26 General notes for structural plans.

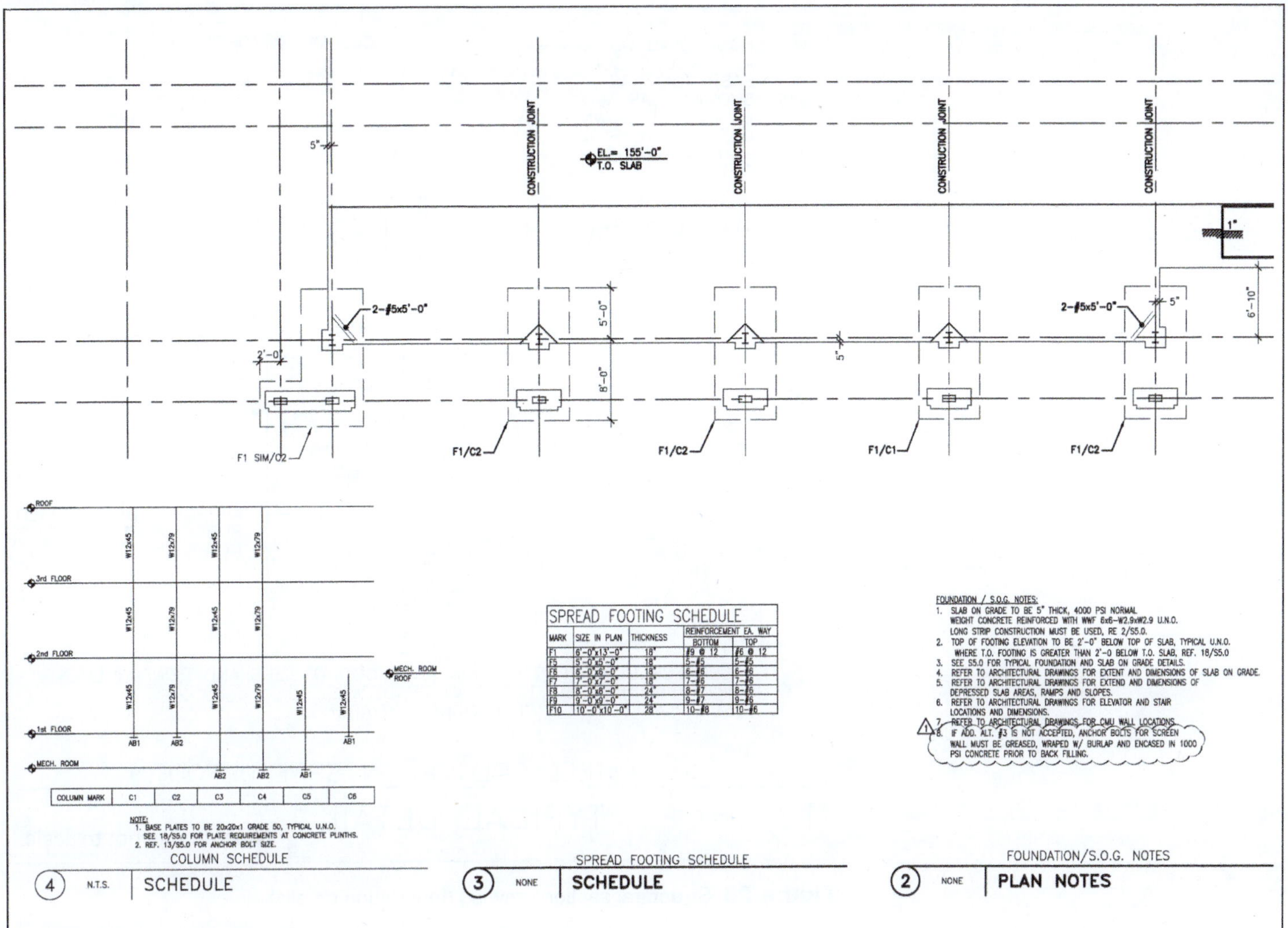

Note: Image not to scale.

Figure 27 Foundation plan (foundation/slab-on-grade plan).

The structural plans include structural section drawings (*Figure 28*), which are like architectural section drawings but show only the structural requirements. Miscellaneous structural details may also be shown in these sections to provide a better understanding of such things as connections and attachments of accessories.

1.5.4 Mechanical Plans

Mechanical plans are engineered plans for motors, pumps, piping systems, and piping equipment. These plans incorporate general notes (*Figure 29*) containing specifications ranging from what the contractor is to provide to how the contractor determines the location of **grilles** and **registers**. A mechanical legend (*Figure 30*) defines the symbols used on the mechanical plans. A list of abbreviations (*Figure 31*) spells out abbreviations found on the plans. Symbols and abbreviations may vary depending on job type and trade. Always refer to drawings to ensure accuracy when translating symbols and abbreviations.

Mechanical plans: Engineered drawings that show the mechanical systems, such as motors and piping.

Grilles: HVAC components that pull air from a room and return it to the heating or cooling system. They are different from registers in that they do not have a damper to control airflow.

Registers: HVAC components that direct air from the heating or cooling system into a room. A damper is used to control the airflow of a register.

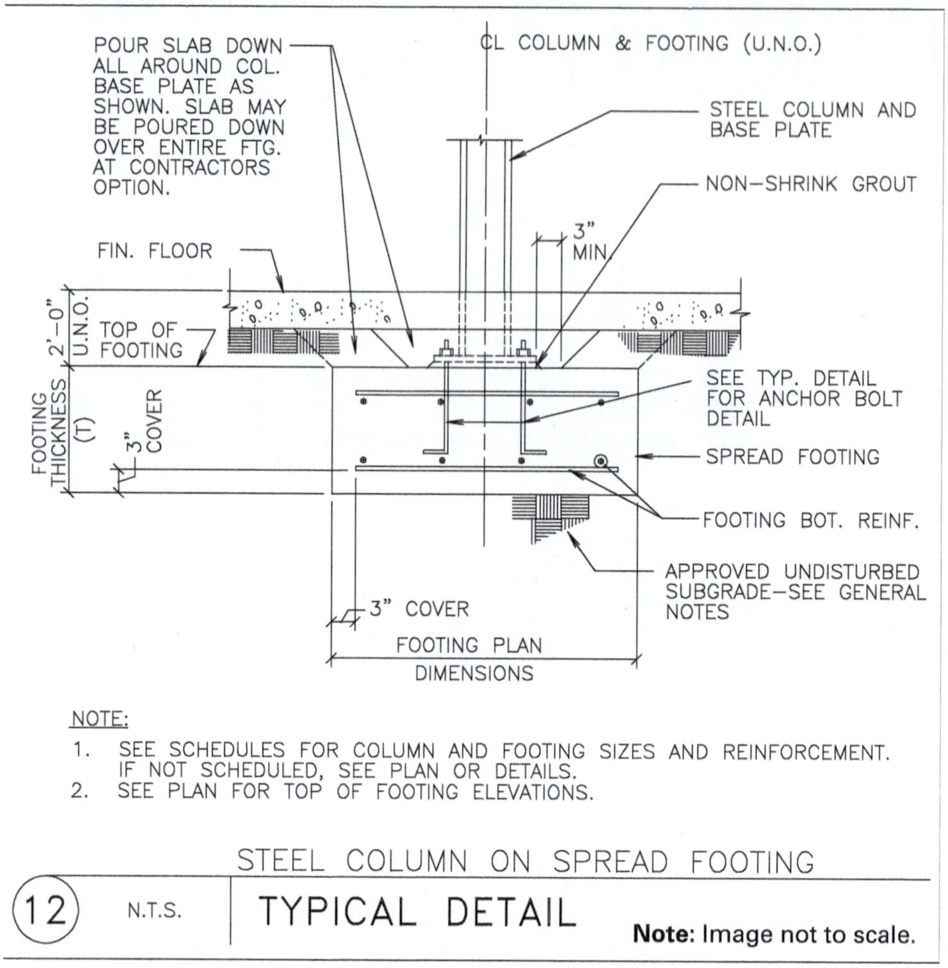

Figure 28 Structural section drawing (foundation details).

Piping and instrumentation drawings (P&IDs): Schematic diagrams of a complete piping system.

Schematic: A one-line drawing showing the flow path for electrical circuitry or the relationship of all parts of a system.

Plumbing plans: Engineered drawings that show the layout for a plumbing system.

Plumbing isometric drawing: A type of 3D drawing that depicts a plumbing system.

Piping and instrumentation drawings (P&IDs) (*Figure 32*) are **schematic** diagrams of a complete piping system that show the process flow. They also show all the equipment, pipelines, valves, instruments, and controls needed to operate the system. P&IDs are not drawn to scale because they are only meant to give a representation, or a general idea, of the work to be done. Additionally, P&IDs do not indicate north, south, east, and west directions.

For more complex jobs, a separate heating, ventilating, and air conditioning (HVAC) plan is added to the set of plans. Piping system plans for gas, oil, or steam heat may be included in the HVAC plan. The mechanical plans include the layout of the HVAC system, showing specific requirements and elements for that system, including a floor, a reflected ceiling, or a roof. HVAC drawings (*Figure 33A* and *Figure 33B*) often include an electrical schematic that shows the electrical circuitry for the HVAC system. HVAC plans can be both mechanical and electrical drawings in one plan. Be aware that a page with a series of mechanical detail drawings may be included in the mechanical plans. These drawings show specific details of certain components within the mechanical system. *Figure 34* is an example of a mechanical detail drawing.

1.5.5 Plumbing/Piping Plans

Plumbing plans (*Figure 35*) are engineered plans showing the layout of a plumbing system that supplies hot and cold water for a sewage disposal system and for the location of plumbing fixtures. For commercial projects, each system may be on a separate plan.

A **plumbing isometric drawing** is a 3D drawing that depicts a plumbing system and serves as part of the plumbing plan. *Figure 36* shows a plumbing isometric drawing for a sanitary riser system.

GENERAL NOTES
(FOR ALL MECHANICAL DRAWINGS)

1. CONTRACTOR IS TO PROVIDE COMPLETE CONNECTIONS TO ALL NEW AND RELOCATED OWNER FURNISHED EQUIPMENT.

2. CONTRACTOR TO COORDINATE THE LOCATION OF ALL DUCTWORK AND DIFFUSERS WITH REFLECTED CEILING PLAN AND STRUCTURE PRIOR TO BEGINNING WORK.

3. DIMENSIONS FOR INSULATED OR NON-INSULATED DUCT ARE OUTSIDE SHEET METAL DIMENSIONS.

4. DRAWINGS ARE NOT TO BE SCALED FOR DIMENSIONS. TAKE ALL DIMENSIONS FROM ARCHITECTURAL DRAWINGS, CERTIFIED EQUIPMENT DRAWINGS AND FROM THE STRUCTURE ITSELF BEFORE FABRICATING ANY WORK. VERIFY ALL SPACE REQUIREMENTS COORDINATING WITH OTHER TRADES, AND INSTALL THE SYSTEMS IN THE SPACE PROVIDED WITHOUT EXTRA CHARGES TO THE OWNER.

5. LOCATION OF ALL GRILLES, REGISTERS, DIFFUSERS AND CEILING DEVICES SHALL BE DETERMINED FROM THE ARCHITECTURAL REFLECTED CEILING PLANS.

6. THE OWNER AND DESIGN ENGINEER ARE NOT RESPONSIBLE FOR THE CONTRACTOR'S SAFETY PRECAUTIONS OR TO MEANS, METHODS, TECHNIQUES, CONSTRUCTION SEQUENCES, OR PROCEDURES REQUIRED TO PERFORM HIS WORK.

7. ALL WORK SHALL BE INSTALLED IN ACCORDANCE WITH PLRC'S SAFETY PLAN AND ALL APPLICABLE STATE AND LOCAL CODES.

8. ALL EXTERIOR WALL AND ROOF PENETRATIONS SHALL BE SEALED WEATHERPROOF. REFERENCE SPECIFICATION SECTION 15050.

9. ALL MECHANICAL WORK UNDER THIS CONTRACT IS TO FIVE (5) FEET OUTSIDE THE BUILDING.

Figure 29 Mechanical plan general notes.

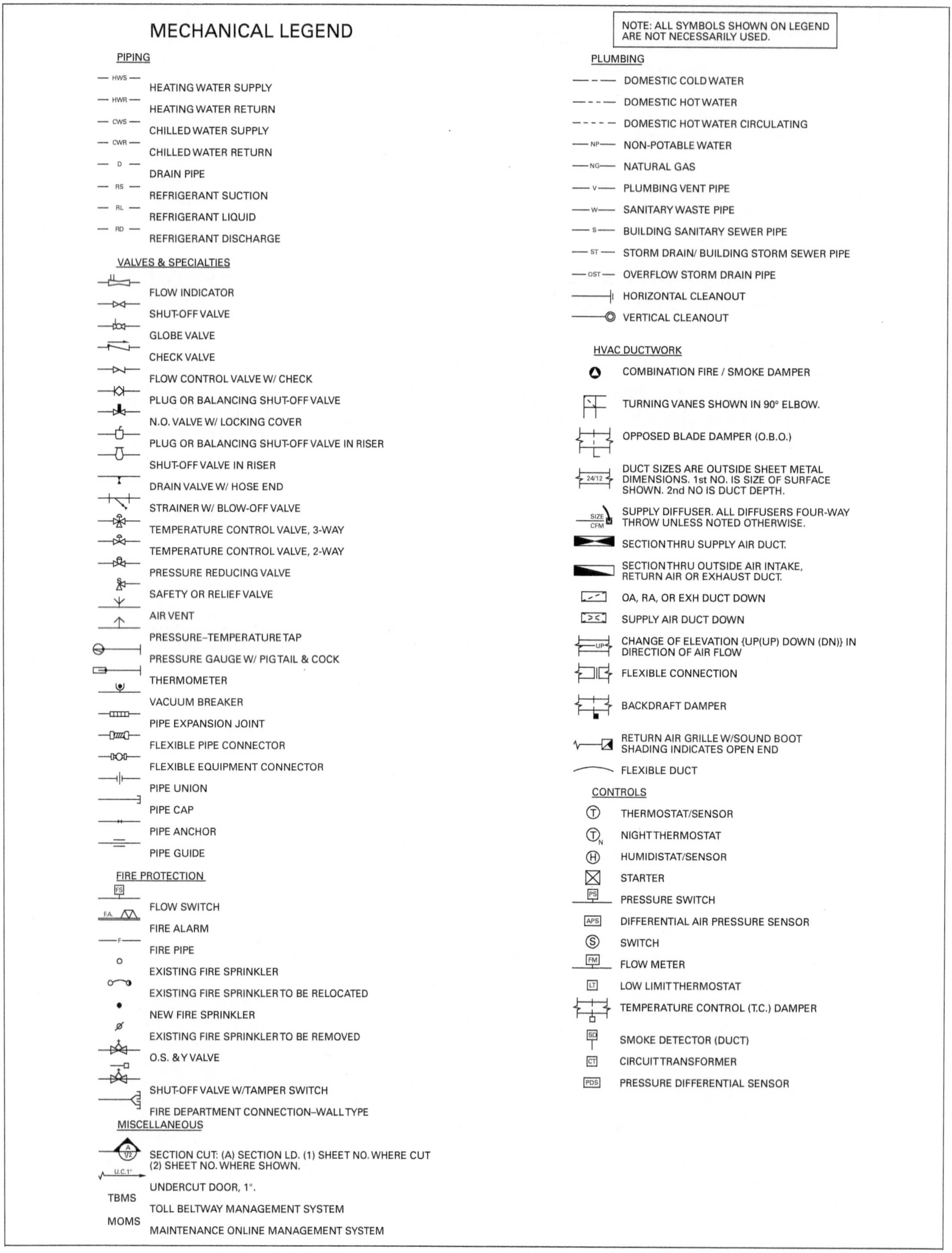

Figure 30 Mechanical plan legend.

ABBREVIATIONS

AFF	ABOVE FINISHED FLOOR
ALT	ALTITUDE
BHP	BRAKE HORSEPOWER
BTU	BRITISH THERMAL UNIT
Cv	COEFFICIENT, VALVE FLOW
CU FT	CUBIC FEET
CU IN	CUBIC INCH
CFM	CUBIC FEET PER MINUTE
SCFM	CFM, STANDARD CONDITIONS
dB	DECIBEL
DCW	DOMESTIC COLD WATER
DEG OR °	DEGREE
DHW	DOMESTIC HOT WATER
DIA	DIAMETER
DB	DRY-BULB
EAT	ENTERING AIR TEMPERATURE
EFF	EFFICIENCY
ELEV or EL	ELEVATION
ESP	EXTERNAL STATIC PRESSURE
EWT	ENTERING WATER TEMPERATURE
EXH	EXHAUST
F	FAHRENHEIT
FLA	FULL LOAD AMPS
FPM	FEET PER MINUTE
FPS	FEET PER SECOND
FT	FOOT OR FEET
FU	FIXTURE UNITS
GA	GAUGE
GAL	GALLONS
GPH	GALLONS PER HOUR
GPM	GALLONS PER MINUTE
HD	HEAD
HG	MERCURY
HGT	HEIGHT
HORZ	HORIZONTAL
HP	HORSEPOWER
HR	HOUR(S)
HWC	HOT WATER CIRCULATING (DOMESTIC)
HZ	HERTZ
ID	INSIDE DIAMETER
IE	INVERT ELEVATION
IN	INCHES
IN W.C.	INCHES WATER COLUMN
KW	KILOWATT
KWH	KILOWATT HOUR
LAT	LEAVING AIR TEMPERATURE
LBS OR #	POUNDS
LF	LINEAR FEET
LRA	LOCKED ROTOR AMPS
LWT	LEAVING WATER TEMPERATURE
MAX	MAXIMUM
MCA	MINIMUM CIRCUIT AMPS
MBH	BTU PER HOUR (THOUSAND)
MIN	MINIMUM
NC	NOISE CRITERIA
N.O.	NORMALLY OPEN
N.C.	NORMALLY CLOSED
N/A	NOT APPLICABLE
NIC	NOT IN CONTRACT
NTS	NOT TO SCALE
NO	NUMBER
OA	OUTSIDE AIR
OD	OUTSIDE DIAMETER
PPM	PARTS PER MILLION
%	PERCENT
PH OR f	PHASE (ELECTRICAL)
PSF	POUNDS PER SQUARE FOOT
PSI	POUNDS PER SQUARE INCH
PSIA	PSI ABSOLUTE
PSIG	PSI GAUGE
PRESS	PRESSURE
RA	RETURN AIR
RECIRC	RECIRCULATE
RH	RELATIVE HUMIDITY
RLA	RUNNING LOAD AMPS
RPM	REVOLUTIONS PER MINUTE
SL	SEA LEVEL
SENS	SENSIBLE
SPEC	SPECIFICATION
SQ	SQUARE
STD	STANDARD
SP	STATIC PRESSURE
SA	SUPPLY AIR
TEMP	TEMPERATURE
TD	TEMPERATURE DIFFERENCE
TSP	TOTAL STATIC PRESSURE
TSTAT	THERMOSTAT
TONS	TONS OF REFRIGERATION
VAV	VARIABLE AIR VOLUME
VEL	VELOCITY
VERT	VERTICAL
V	VOLT
VOL	VOLUME
W	WATT
WT	WEIGHT
WB	WET-BULB

Figure 31 Mechanical plan list of abbreviations.

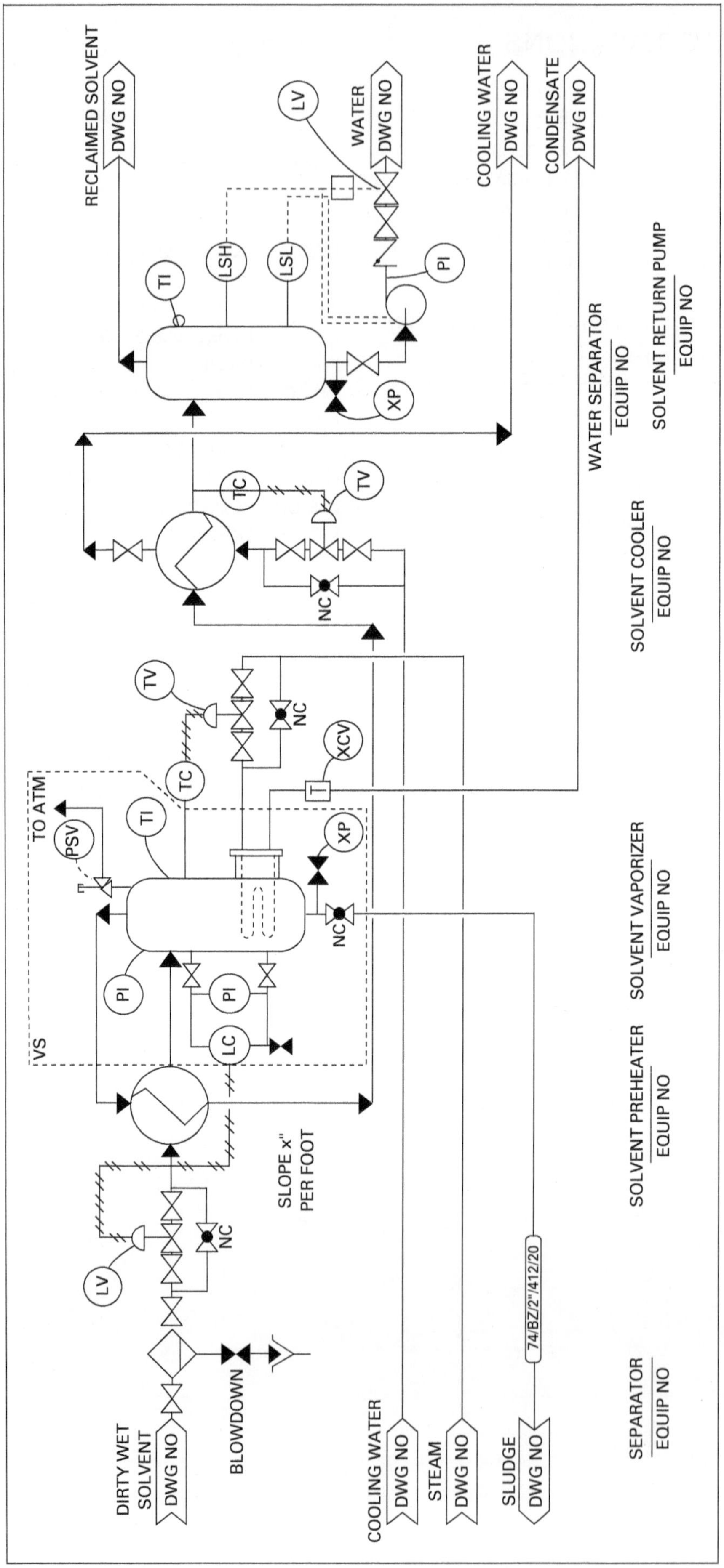

Figure 32 P&ID drawing.

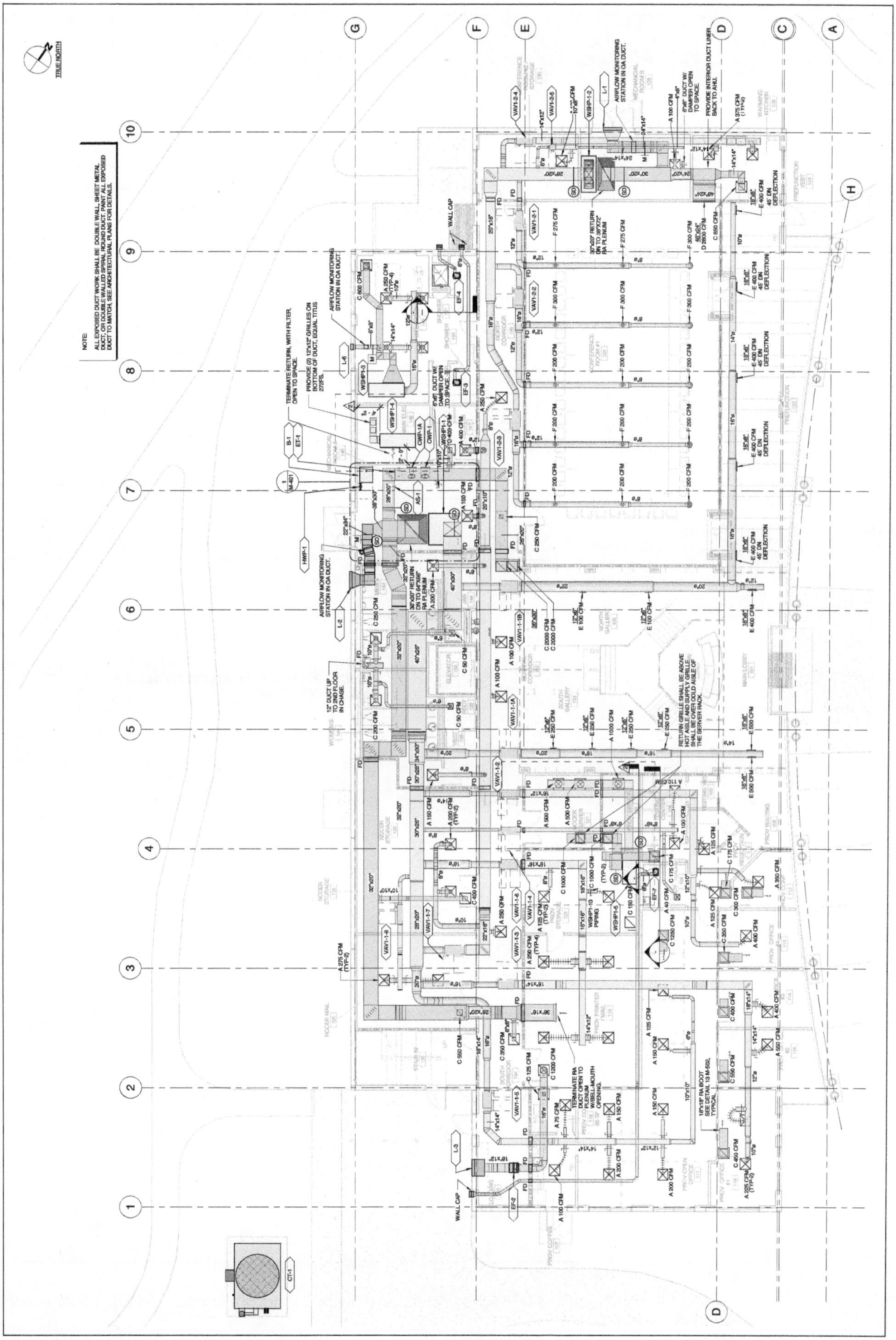

Figure 33A HVAC drawing (1 of 2).

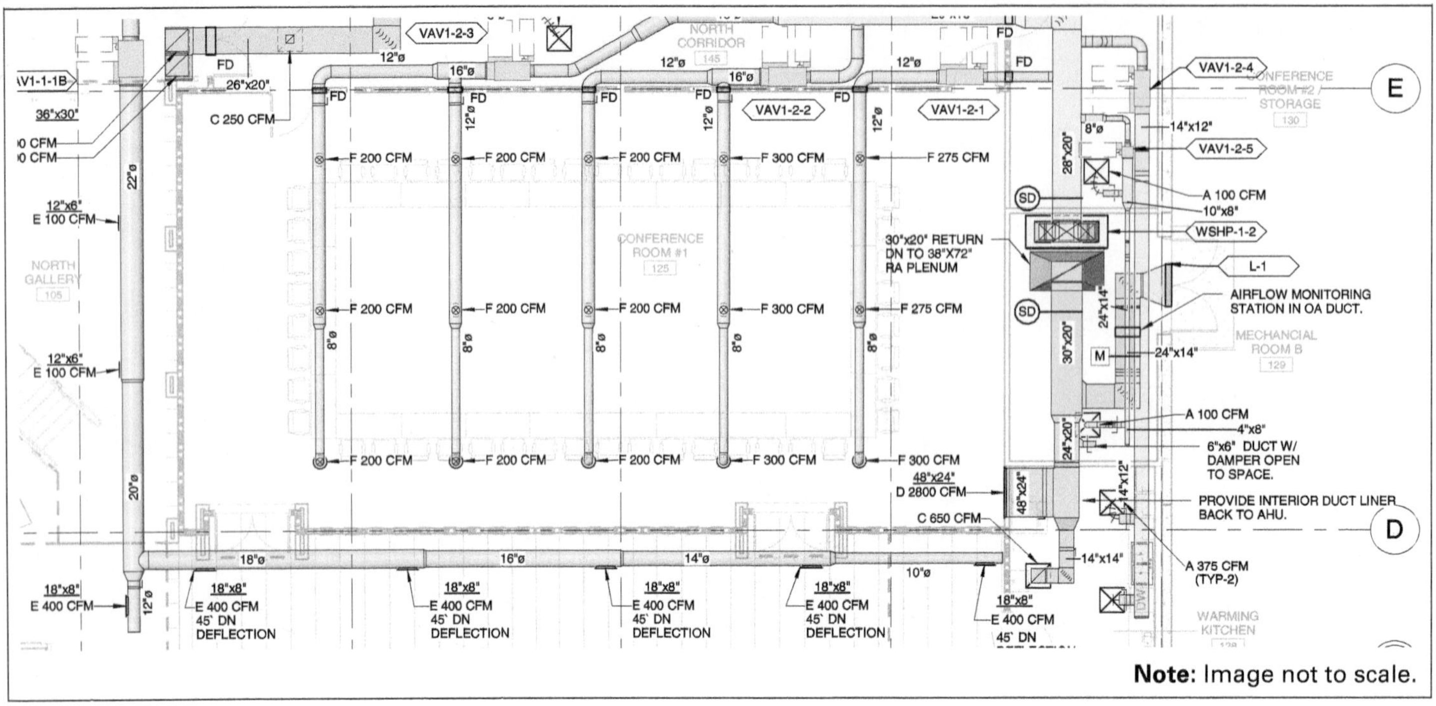

Figure 33B HVAC drawing (2 of 2).

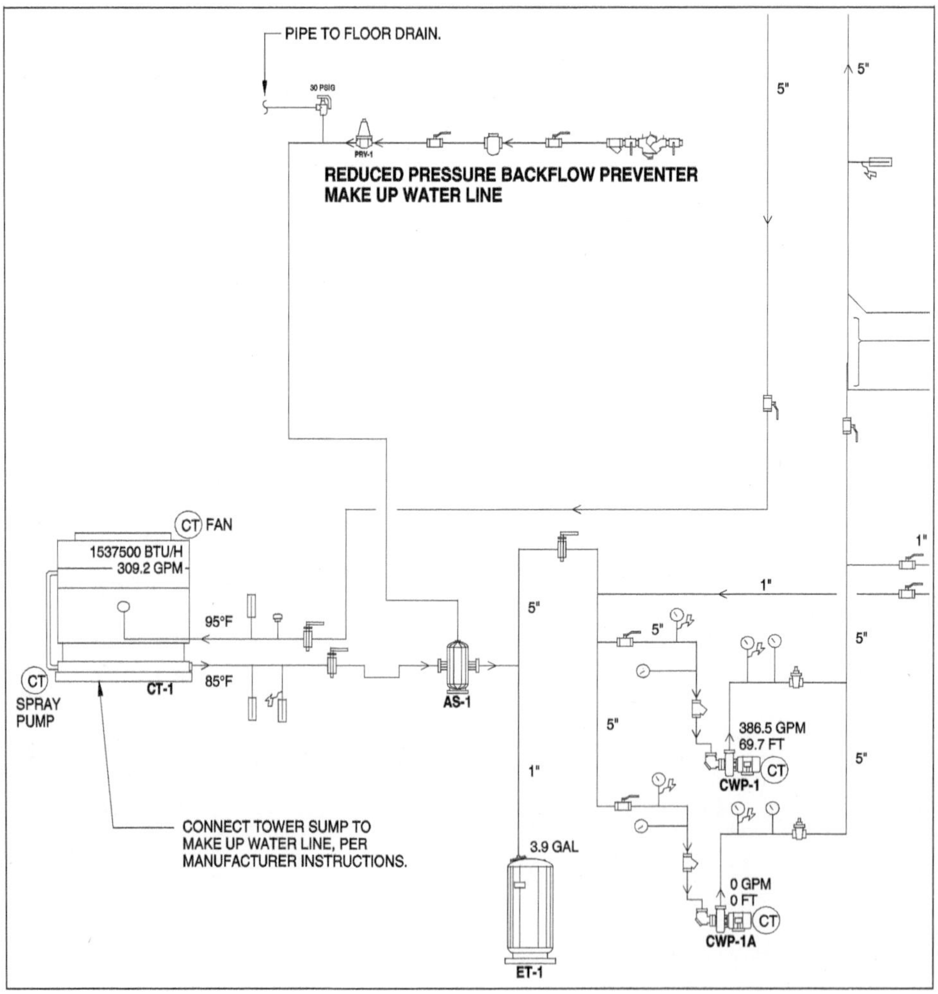

Figure 34 Mechanical detail drawing for hot water riser and drain connections.

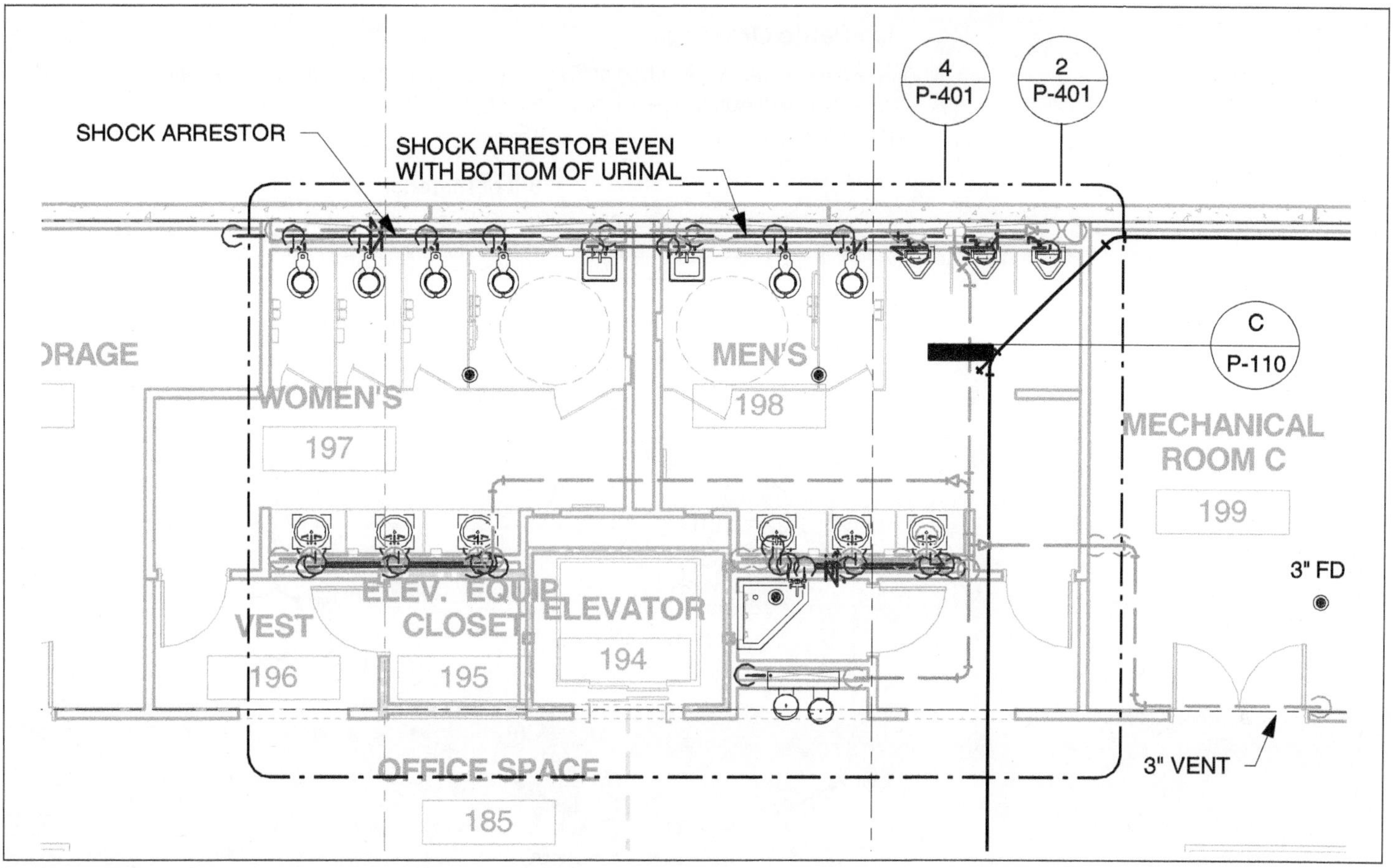

Figure 35 Plumbing plan.

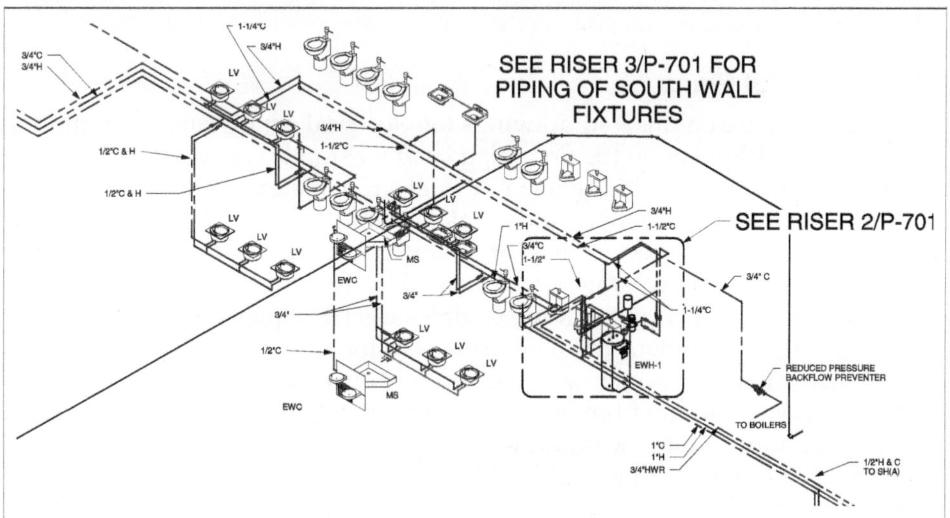

Figure 36 Plumbing isometric drawing (sanitary riser diagram).

Isometric Drawings

An isometric drawing is a type of 3D drawing also known as a *pictorial illustration.* Isometric construction drawings typically depict objects at a 30° angle to provide a 3D perspective rather than a flat, two-dimensional (2D) view.

Bill of Material

P.M.	Req'd	Size	Description
1		1½"	PIPE SCH/40 ASTM-A-120 GR.B
2		¾"	PIPE SCH/40 ASTM-A-120 GR.B
3	5	1½"	90° ELL ASTM-A-197 BW
4	1	1½"	TEE ASTM-197 STD
5	2	¾"	45° ELL ASTM-197 STD
6	1	1½" x ¾"	BELL RED. CONC.
7	2	1½"	GATE VA. BW ASTM-B62
8	1	1½"	CHECK VA. SWING BW 150#

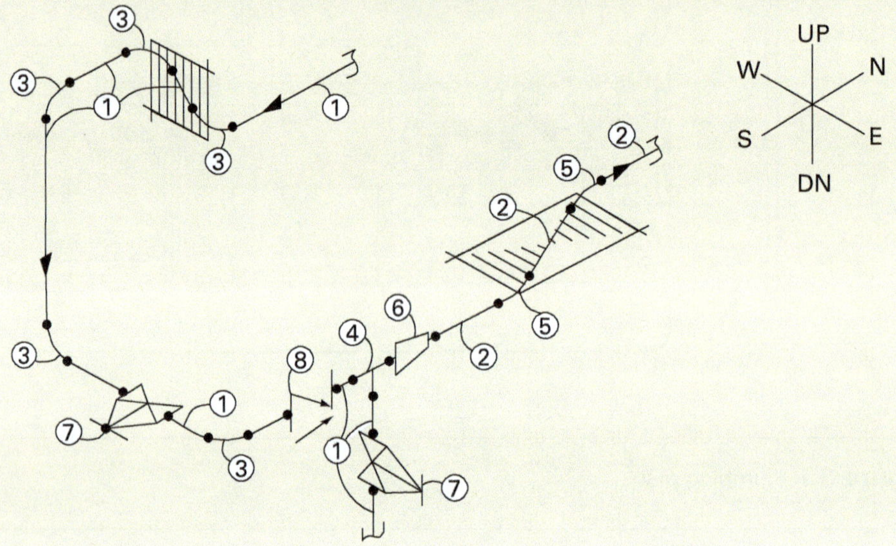

1.5.6 Electrical Plans

Electrical plans: Engineered drawings that show all electrical supply and distribution.

Electrical plans are engineered drawings for electrical supply and distribution. These plans may appear on the floor plan itself for simple construction projects. Electrical plans include locations of the electric meter, distribution panel, switchgear, convenience outlets, and special outlets.

For more complex projects, the information may be on a separate plan added to the set of plans. This separate plan leaves out construction-related details and shows just the electrical layout. More complex electrical plans include locations of switchgear, transformers, main breakers, and motor control centers.

The electrical plans usually start with a set of general notes (*Figure 37*). These notes cover items ranging from main transformers to the coordination of underground penetrations into the building.

The electrical plans can include lighting plans, which show the locations of lights and receptacles, power plans (*Figure 38*), and panel schedules (*Figure 39*). Electrical plans have an electrical legend, which defines the symbols (*Figure 40*) used on the plan, and a key to the abbreviations (*Figure 41*) used on the plan. Depending on the size and scope of the drawings, the legend is often on a separate drawing sheet of its own, rather than on each individual drawing.

GENERAL NOTES (FOR ALL ELECTRICAL SHEETS)

1. COORDINATE LOCATION OF LUMINARIES WITH ARCHITECTURAL REFLECTED CEILING PLANS.
2. COORDINATE LOCATION OF ALL OUTLETS WITH ARCHITECTURAL ELEVATIONS, CASEWORK SHOP DRAWINGS AND EQUIPMENT INSTALLATION DRAWINGS.
3. COORDINATE LOCATION OF MECHANICAL EQUIPMENT WITH MECHANICAL PLANS AND MECHANICAL CONTRACTOR PRIOR TO ROUGH-IN.
4. PROVIDE (1) 3/4"C WITH PULL WIRE FROM EACH TELEPHONE, DATA OR COMMUNICATION OUTLET SHOWN, TO ABOVE ACCESSIBLE CEILING, AND CAP.
5. 3-LAMP FIXTURES SHOWN HALF SHADED HAVE INBOARD SINGLE LAMP CONNECTED TO EMERGENCY BATTERY PACK FOR FULL LUMEN OUTPUT. SEE SPECIFICATIONS.
6. SITE PLAN DOES NOT INDICATE ALL OF THE UG UTILITY LINES, RE: CIVIL DRAWINGS FOR ADDITIONAL INFORMATION. CONTRACTOR TO FIELD VERIFY EXACT LOCATION OF ALL EXISTING UNDERGROUND UTILITY LINES OF ALL TRADES PRIOR TO ANY SITE WORK.
7. THE LOCATIONS OF ALL SMOKE DETECTORS SHOWN ARE CONSIDERED TO BE SCHEMATIC ONLY. THE ACTUAL LOCATIONS (SPACING TO ADJACENT DETECTORS, WALLS, ETC.) ARE REQUIRED TO MEET NFPA 72.
8. ANY ITEMS DAMAGED BY THE CONTRACTOR SHALL BE REPLACED BY THE CONTRACTOR.
9. "CLEAN POWER" AND COMMUNICATION/COMPUTER SYSTEM REQUIREMENTS SHALL BE COORDINATED WITH COMMUNICATION/COMPUTER SYSTEMS CONTRACTOR.
10. REFER TO ARCHITECTURAL PLANS, ELEVATIONS AND DIAGRAMS FOR LOCATIONS OF FLOOR DEVICES AND WALL DEVICES. LOCATION WILL INDICATE VERTICAL AND/OR HORIZONTAL MOUNTING. IF DEVICES ARE NOT NOTED OTHERWISE THEY SHALL BE MOUNTED LONG AXIS HORIZONTAL AT +16" TO CENTER.
11. ALL PLUGMOLD SHOWN SHALL BE WIREMOLD SERIES V2000 (IVORY FINISH) WITH SNAPICOIL #V20GB06 (OUTLETS 6" ON CENTER). PROVIDE ALL NECESSARY MOUNTING HARDWARE, ELBOWS, CORNERS, ENDS, ETC. REQUIRED FOR A COMPLETE SYSTEM.
12. ALL EMERGENCY RECEPTACLE DEVICES SHALL BE RED IN COLOR.
13. ALL BRANCH CIRCUITS SHALL BE 3-WIRE (HOT, NEUTRAL, GROUND).
14. COORDINATE EXACT EQUIPMENT LOCATIONS AND POWER REQUIREMENTS WITH OWNER AND ARCHITECT PRIOR TO ROUGH-INS.
15. ADA COMPLIANCE: ALL ADA HORN/STROBE UNITS SHALL BE MOUNTED +90" AFF OR 6" BELOW FINISHED CEILING, WHICH EVER IS LOWER. ELECTRICAL DEVICES PROJECTING FROM WALLS WITH THEIR LEADING EDGES BETWEEN 27" AND 80" AFF SHALL PROTRUDE NO MORE THAN 4" INTO WALKS OR CORRIDORS. ELECTRICAL AND COMMUNICATIONS SYSTEMS RECEPTACLES ON WALLS SHALL BE 15" MINIMUM AFF TO BOTTOM OF COVERPLATE.
16. COORDINATE ALL UNDERGROUND PENETRATIONS INTO THE BUILDING AND TUNNEL WITH STRUCTURAL ENGINEER, DUE TO EXPANSIVE SOILS.
17. ELECTRONIC STRIKES, MOTION DETECTORS AND ALARM SHUNTS ARE PROVIDED BY OTHERS. PROVIDE ALL NECESSARY ROUGH-INS FOR THESE ITEMS. COORDINATE WORK WITH SECURITY SYSTEM PROVIDER.

Figure 37 Electrical plan general notes.

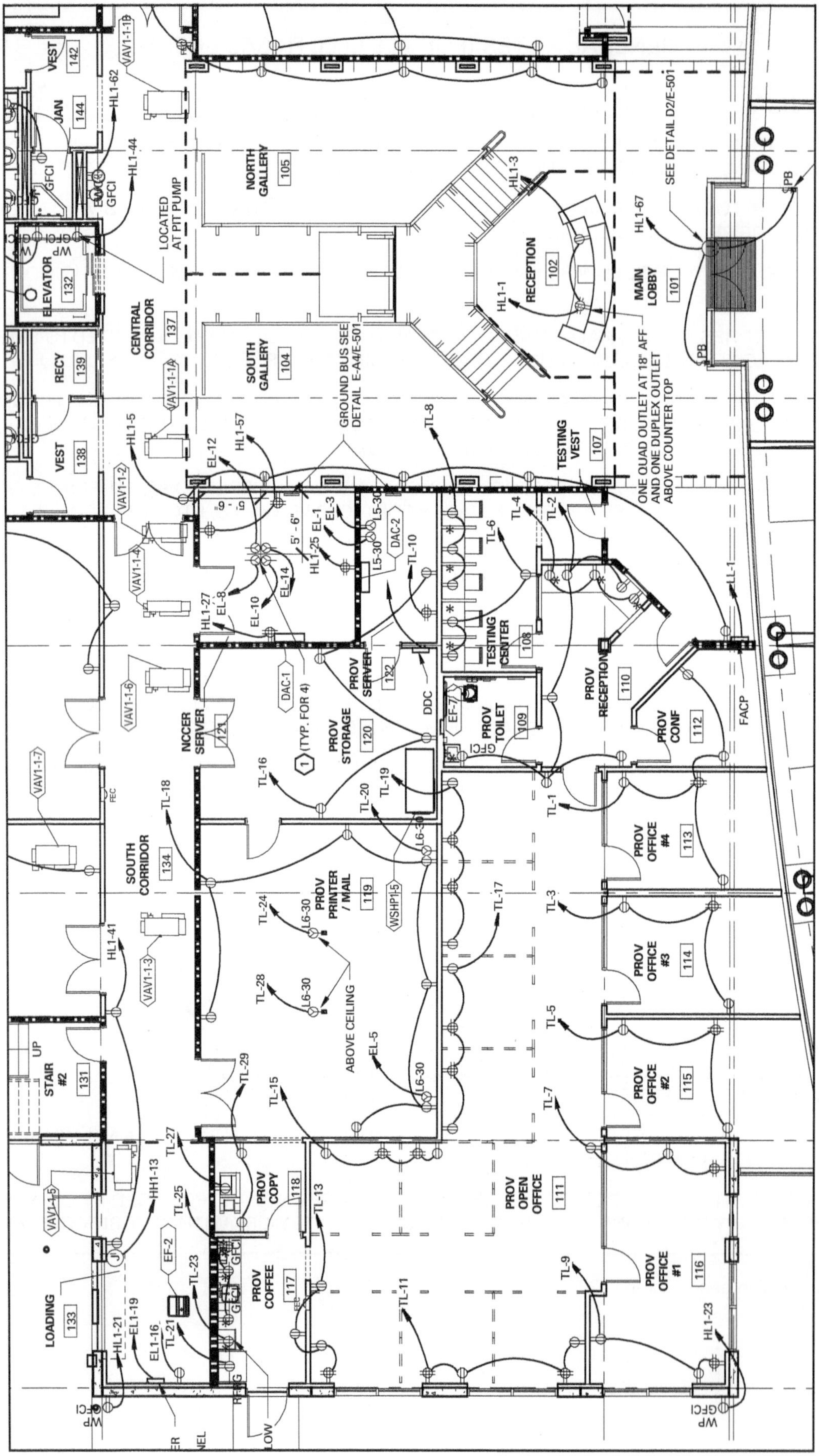

Note: Image not to scale

Figure 38 Portion of a power plan.

PANEL SCHEDULES

VOLTS: 277/480
PHASE: 3
WIRES: 4+G

PANEL DESIGN.	MOUNTING	BUS SIZE AMPERES	MAINS: BREAKERS SYM A.I.C.	MAINS: BREAKERS TRIP AMPS	MAINS: LUGS ONLY	MAINS: SWITCH AMPS	MAINS: FUSE AMPS	MAINS: FUSE TYPE	BRANCH CIRCUITS: POLES	BRANCH CIRCUITS: TRIP AMPS	BRANCH CIRCUITS: SYM A.I.C.	BRANCH CIRCUITS: ACTIVE	BRANCH CIRCUITS: SPARES	BRANCH CIRCUITS: SPACES	BRANCH CIRCUITS: TOTAL	REMARKS
HLP–1–EM	S	100	22K	50					1	20	10K	12			12	
HLP–2–EM	S	100	22K	30					1	20	10K	6			6	
HLP–3–EM	S	100	22K	30					1	20	10K	6			6	

PANEL SCHEDULES

VOLTS: 120/208
PHASE: 3
WIRES: 4+G

PANEL DESIGN.	MOUNTING	BUS SIZE AMPERES	MAINS: BREAKERS SYM A.I.C.	MAINS: BREAKERS TRIP AMPS	MAINS: LUGS ONLY	MAINS: SWITCH AMPS	MAINS: FUSE AMPS	MAINS: FUSE TYPE	BRANCH CIRCUITS: POLES	BRANCH CIRCUITS: TRIP AMPS	BRANCH CIRCUITS: SYM A.I.C.	BRANCH CIRCUITS: ACTIVE	BRANCH CIRCUITS: SPARES	BRANCH CIRCUITS: SPACES	BRANCH CIRCUITS: TOTAL	REMARKS
PP–1A	S	225	22K	100					1	20	10K	40			42	
									2	40	10K	1	–	–		
PP–1B	S	225	22K	100					1	20	10K	42			42	
PP–2A	S	225	22K	100					1	20	10K	42			42	
SECT–I																
PP–2A	S	225	22K	100					1	20	10K	–	12	6	18	
SECT–II																
PP–2B	S	225	22K	100					1	20	10K	42			42	
PP–3A	S	225	22K	100					1	20	10K	40			42	
SECT–I									2	40	10K	1				
PP–3A	S	225	22K	100					1	20	10K		12	6	18	
SECT–II																
PP–3B	S	225	22K	100					1	20	10K	42			42	

Figure 39 Panel schedules.

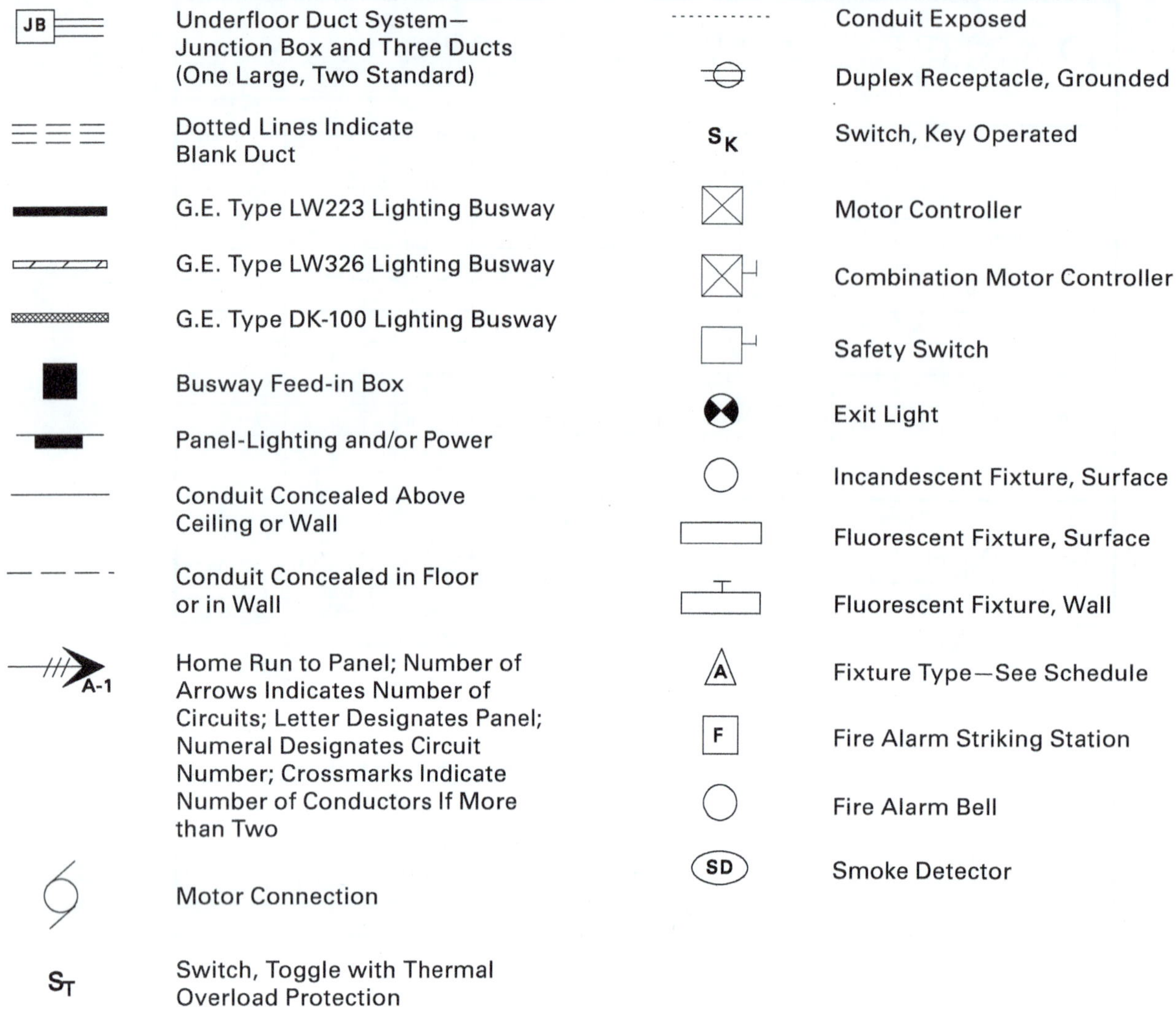

Figure 40 Electrical symbols list.

1.5.7 Other Drawings and Documents

Other types of drawings and documents may be used to supplement the drawing set, depending on each building's type, usage, and installed subsystems.

Fire protection plan: A drawing that shows the details of the building's sprinkler system.

An important drawing that may be included in a set of drawings for public buildings, multifamily residential buildings, and similar structures is a **fire protection plan**. This drawing shows the piping, valves, heads, and switches that make up a building's fire sprinkler system. A fire sprinkler symbols list is usually included on a separate sheet along with the fire sprinkler specifications, details and assembly drawings, and riser diagrams.

Shop drawings provide additional details about the design of specific building components and prefabricated items, such as sheet metal ductwork, piping, plumbing, and electrical components. They include information and step-by-step instructions on how those components are to be fabricated and installed and provide greater detail than what is shown on construction (contract) drawings. Shop drawings help to estimate the type and quantity of required materials, identify dimensions and instructions for fabrication and installation, as well as estimate the time required to fully complete the task. For large commercial jobs, a draftsperson creates shop drawings based on a design drafted by an architect or engineer. On smaller jobs, the draftsperson may work from freehand sketches based on field measurements. Shop drawings, like section and detail drawings, are drawn to a larger scale than the engineer's design.

ELECTRICAL ABBREVIATIONS

A	AMPERE(S)
AC	ALTERNATING CURRENT
ACB	AIR CIRCUIT BREAKER
AFF	ABOVE FINISHED FLOOR
AFG	ABOVE FINISHED GRADE
AL	ALUMINUM
ALT	ALTERNATE
ASYM	ASYMMETRICAL
ATS	AUTOMATIC TRANSFER SWITCH
AWG	AMERICAN WIRE GAUGE
BC	BOTTOM CONDUIT
BD	BUS DUCT
BFG	BELOW FINISHED GRADE
BIL	BASIC IMPULSE LEVEL
BLDG	BUILDING
BX	ARMORED CABLE
C	CONDUIT
CATV	CABLE ANTENNA TELEVISION SYSTEM
CCAB	CONTROL CABINET
CCTV	CLOSED CIRCUIT TELEVISION
CH	CABINET HEATER
CKT	CIRCUIT
CKT BKR/CB	CIRCUIT BREAKER
CL	CLOSET
CLG	CEILING
COND	CONDUCTOR
CO	CONDUIT ONLY
CT	CURRENT TRANSFORMER
CU	COPPER
DB	DUCT BANK
DMB	DIMMER BOARD
DC	DIRECT CURRENT
DH	DUCT HEATER
DIM	DIMMER CONTROL
DISC	DISCONNECT
DM	DAMPER MOTOR
DN	DOWN
DP	DISTRIBUTION POWER PANEL(BOARD)
DT	DOUBLE THROW
DWG	DRAWING
EA	EACH
E.C.	ELECTRICAL CONTRACTOR
E.HTR	ELECTRIC HEATER
ELEV	ELEVATOR
EL	ELECTRIC
EM	EMERGENCY
EMT	ELECTRICAL METALLIC TUBING
ENT	ELECTRICAL NON-METALLIC TUBING
EWC	ELECTRIC WATER COOLER
EX	EXISTING
F	FUSE
FA	FIRE ALARM
FACP	FIRE ALARM CONTROL PANEL
FBO	FURNISHED BY OTHERS
FCC	FLAT CONDUCTOR CABLE
FCU	FAN COIL UNIT
FDR	FEEDER
FL	FLOOR
FLUOR	FLUORESCENT
F.P.C.	FIRE PROTECTION CONTRACTOR
FS	FUSIBLE SWITCH
F.S.C.	FOOD SERVICE CONTRACTOR
FT	FEET OR FOOT
G.C.	GENERAL CONTRACTOR
GEN	GENERATOR
GF	GROUND FAULT
GG	GROUND GRID
GRD	GROUND
HC	HUNG CEILING
H.I.D.	HIGH INTENSITY DISCHARGE
HP	HORSEPOWER
H.P.S.	HIGH PRESSURE SODIUM
HPU	HEAT PUMP UNIT
HT	HEIGHT
HV	HIGH VOLTAGE
HW	HEAVY WALL RIGID CONDUIT
HZ	FREQUENCY IN CYCLES PER SECOND
IC	INTERRUPTING CAPACITY
IG	ISOLATED GROUND
IMC	INTERMEDIATE METALLIC CONDUIT
INC	INCANDESCENT
JB	JUNCTION BOX
K	KEY OPERATED
kVA	KILOVOLT AMPERE(S)
kVAR	KILOVAR(S)
kW	KILOWATT(S)
kWhr	KILOWATT(S) HOUR(S)
LP	LIGHTING PANEL
L.P.S.	LOW PRESSURE SODIUM
LTG	LIGHTING
LV	LOW VOLTAGE
MATV	MASTER ANTENNA TELEVISION SYSTEM
MC	METAL CLAD CABLE
MCC	MOTOR CONTROL CENTER
MCM	THOUSAND CIRCULAR MIL(S)
MCP	MOTOR CONTROL PANEL
M.C.	MECHANICAL CONTRACTOR
MH	MANHOLE
MIC	MICROPHONE
MIN	MINIMUM
MS	MAGNETIC STARTER
MTD	MOUNTED
MTG	MOUNTING
MTR	MOTOR
MTS	MANUAL TRANSFER SWITCH
N	NEUTRAL
NA	NON-AUTOMATIC
NC	NORMALLY CLOSED
NF	NON-FUSE
N.I.C.	NOT IN CONTRACT
NL	NIGHT LIGHT
NO	NORMALLY OPEN
NP	NETWORK PROTECTOR
NTS	NOT TO SCALE
OC	ON CENTER
OL	OVERLOAD ELEMENT
P	POLE
PA	PUBLIC ADDRESS
PB	PULL BOX
P.C.	PLUMBING CONTRACTOR
PF	POWER FACTOR
Ø	PHASE
PL	PILOT (INDICATOR) LIGHT
PNL	PANEL (PANELBOARD)
PP	POWER PANEL
PRI	PRIMARY
PT	POTENTIAL TRANSFORMER
PVC	POLYVINYL CHLORIDE
PWR	POWER
R	RECESSED
RC	REMOTE CONTROL
REC	RECEPTACLE
S	SURFACE
SC	SEPARATE CIRCUIT
SDB	SUB-DISTRIBUTION BOARD
SEC	SECONDARY
SMR	SURFACE METAL RACEWAY
SP	SINGLE POLE
SPK	SPEAKER
ST	SINGLE THROW
SW	SWITCH
SWBD	SWITCHBOARD
SYM	SYMMETRICAL
T	THERMOSTAT
TEL	TELEPHONE
TB	TERMINAL BOX
TC	TOP CONDUIT
TCAB	TELEPHONE CABINET
T.C.C.	TEMPERATURE CONTROL CONTRACTOR
TP	TAMPER PROOF
TV	TELEVISION
TYP	TYPICAL
UG	UNDERGROUND
UH	UNIT HEATER
UNG	UNGROUNDED
UON	UNLESS OTHERWISE NOTED
UPS	UNINTERRUPTED POWER SYSTEM
V	VOLT(S)
VA	VOLTAMP(S)
VAR	VOLT AMPERES REACTIVE
VP	VAPORPROOF
W	WATT(S)
WP	WEATHERPROOF
WT	WATERTIGHT
XFR	TRANSFORMER
XP	EXPLOSION PROOF

Figure 41 Electrical abbreviations.

Visualize Before Building

It is important for a builder to be able to visualize a finished project before starting it. Detail drawings help builders visualize different parts of the structure long before they measure a board or hammer a nail. By visualizing, builders can anticipate and plan for potential problems. This saves time and money on the job. Sometimes a rendering is used to convey visual representations and aesthetics of a project. Although renderings do not have technical details or dimensions, they can aid in the planning process by providing insight into how a property or structure will look after work is complete. Look at the 3D rendering of a milking parlor on a dairy farm. Can you visualize the final structure in operation?

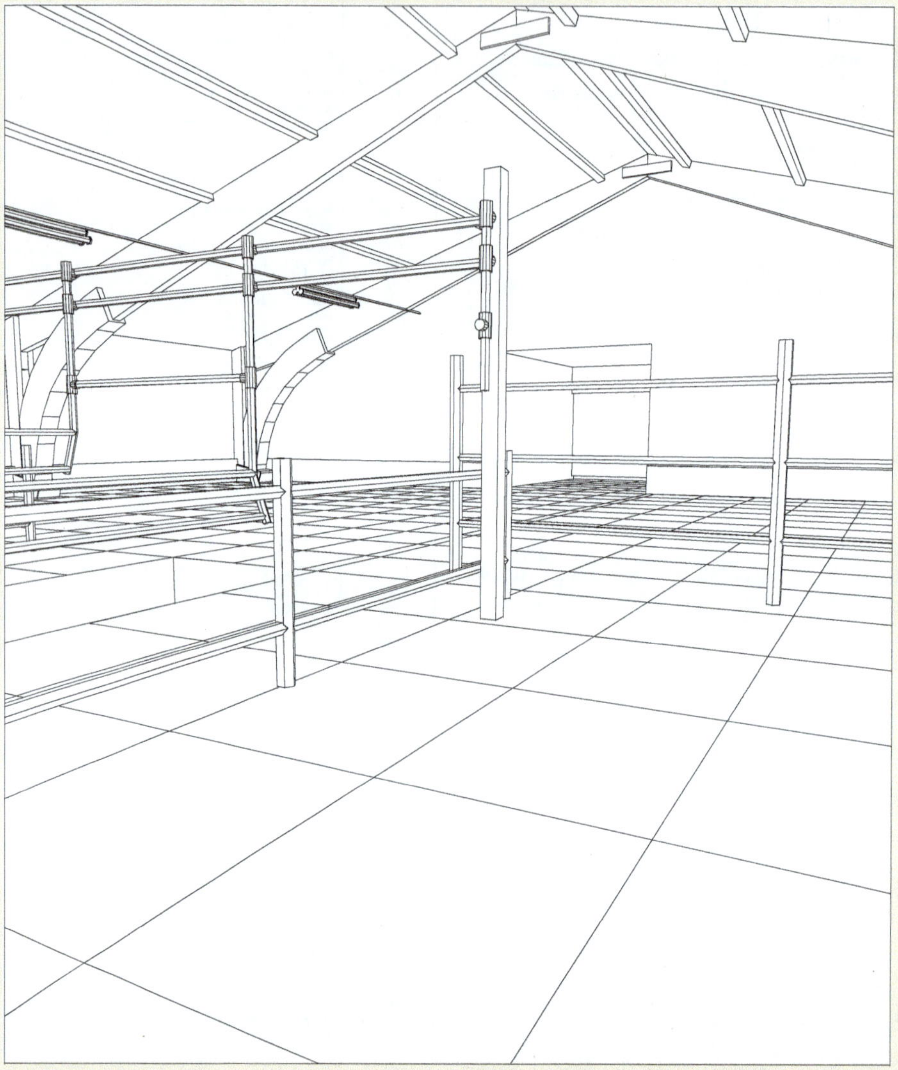

Source: iStock

What Is Reality Capture?

Reality capture is the process of capturing existing site or building conditions with the use of 3D laser scanning and photogrammetry. The scan and photo data can then be brought into the CAD environment, which allows designers to create an accurate 3D model of the existing site and structures to utilize during their design process. This helps to address any site challenges long before the construction process. Benefits from reality capture include:

- Greater design accuracy
- Improved accuracy in the prefabrication process
- Improved scheduling which can help to decrease overall construction time
- Reduced potential for change orders
- Quick and accurate as-built process

As-built drawings reflect the final as-constructed conditions of a project. Construction projects occasionally encounter unexpected obstacles and conditions that mean changes need to be made on-site that conflict with the original design. As-built drawings are usually created by marking any field-built deviations on the contract drawings and then illustrating those on the final drawing set.

Importance of Specifications

Specifications clarify information that cannot be shown on the drawings. Specifications are very important to the architect and owner to ensure compliance to the standards are set. The figure provided shows one page of the specifications for a building's air handling units.

M.E. RINKER SR. HALL
SCHOOL OF BUILDING CONSTRUCTION
BR-191

AIR HANDLING UNITS
SECTION 15760

2.1 AIR HANDLING UNITS

A. Units shall be of the type, size and capacity as set forth in the schedule. The fan outlet velocities and coil and filter face velocities shall be within 5% of the values specified in the schedule. Units shall be double wall McQuay, or approved equal.

B. The units, as assembled, shall be complete with fans, coils, insulated casing, filters, drives and accessories. Each unit, including the fan enclosure, shall have essentially constant cross-sectional dimensions as to width and height. Internal baffles shall be provided as required to prevent bypassing of coils and filters.

C. The casing shall consist of an independent structural steel frame, properly reinforced and braced for maximum rigidity, having individually removable, flush mounted, insulated panels. The casing shall be of sectionalized construction, consisting basically of individual fan section, coil section, access sections, filter section, and drain pan. Sections shall be joined with continuous gasketing to form an air tight closure. Sections shall be so designed that the method of joining can be performed with relative ease and without damage to the insulation and vapor barrier.

The framework shall be constructed of AISC structural rolled shapes having minimum thickness of 1/8" (3 mm) or die formed sheet steel having the minimum gauges set forth in the following schedules:

Maximum Individual Casing Cross Section	Minimum Framework Gauge
Up to 30 ft^2 (2.8 m^2)	14
30.5 ft^2 to 47 ft^2 (2.81 to 4.4 m^2)	12
48 ft^2 (4.45 m^2) and up	10

D. Framework shall be designed with recesses suitable to receive enclosure panels, providing neat appearance, airtight enclosure, and ease of panel removal.

E. Enclosure panels 12 ft^2 in area and larger shall be constructed of not less than 18 gauge die formed sheet steel. Should the sides or top of a casing section exceed 20 ft^2 in area, the panels shall be fabricated of more than one piece, with the individual panels recessed into intermediate structural members.

Protection for the insulation edges shall be provided around the perimeter of each.

GEA 0300-0600

15760 - 2
Revision No. 1
Dec. 19, 2021

1.0.0 Section Review

1. Revisions to the drawing are entered in the revision block and *must* include _____.
   a. the project tag and the date the drawing was approved
   b. engineering approvals and the intended date of completion
   c. the date and the initials of the person who made the revision
   d. dates and signatures for customer approval documentation

2. Codebooks and specifications for a construction project often include the meanings of _____.
   a. drawings
   b. references
   c. glossaries
   d. symbols

3. When a plan is marked NTS, the dimensions as measured on the drawing *cannot* be used to build the project.
   a. True
   b. False

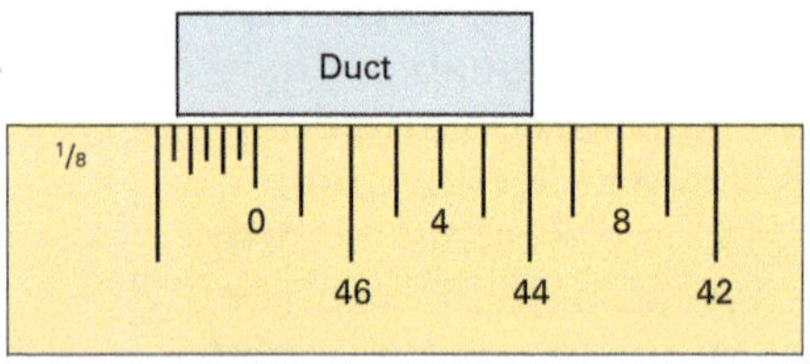

Figure SR01

4. What is the length (in feet and inches) of the section of duct using the architect's scales shown in *Figure SR01*?
   a. 6'-$^3/_4$"
   b. 6'-5"
   c. 44'-$^3/_4$"
   d. 48'-4"

5. A site plan _____.
   a. does not show features such as trees and driveways
   b. shows the location of the building from an aerial view
   c. is drawn after the floor plan is drawn and before the HVAC is designed
   d. includes information about plumbing fixtures and equipment

Module 93106 Review Questions

1. The title block generally contains _____.
   a. revision blocks
   b. special marks used in the drawing
   c. the mechanical plans
   d. the legend

2. The latest revision date on a set of construction drawings can be found _____.
   a. inside the detail
   b. in the schedule
   c. inside the title block
   d. in the legend

3. The Alphabet of Lines consists of _____.
   a. the line types used on a construction drawing
   b. lines that are indicated using letters of the alphabet
   c. lines that match up detail and section drawings
   d. lines that indicate land boundaries on the site plan

4. On a grid line system, a grid divides the area into small parts called _____.
   a. segments
   b. bays
   c. sections
   d. pods

5. When the metric system is used, dimensions are written in _____.
   a. inches, feet, and yards
   b. degrees, radians, and gradians
   c. CADS, RADS, and KRADS
   d. meters, centimeters, and millimeters

6. If the scale on a site plan reads SCALE: 1" = 20'-0", then every _____.
   a. $^1/_{20}$ of an inch on the drawing represents 20 feet, 0 inches
   b. 20 inches on the drawing represents 1 foot, 0 inches
   c. inch on the drawing represents 20 feet, 0 inches
   d. 20 inches on the drawing represents 20 feet, 0 inches

7. What is the length in meters at the point indicated by the arrow on the metric scale in *Figure RQ01*? The unit of length used on the scale is the meter.
   a. 3,500
   b. 350
   c. 35
   d. 3.5

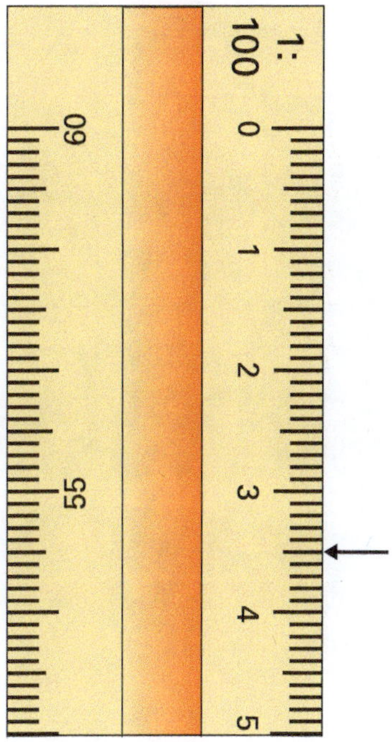

Figure RQ01

8. An engineer's scale is set up in multiples of _____.
   a. 5
   b. 10
   c. 12
   d. 39.37

9. Which type of plan shows the layout of the HVAC system?
   a. Plumbing
   b. Structural
   c. Mechanical
   d. Foundation

10. Electrical plans can include _____.
    a. exterior elevations
    b. section drawings
    c. plumbing isometrics
    d. lighting plans

Answers to odd-numbered questions are found in the Review Question Answer Keys at the back of this book.

Answers to Section Review Questions

Answer	Section Reference	Objective
Section 1.0.0		
1. c	1.1.4	1a
2. d	1.2.2	1b
3. a	1.3.2	1c
4. b	1.4.1	1d
5. b	1.5.1	1e

MODULE 93107

Introduction to Basic Rigging

Source: Slingmax® Rigging Solutions

NOTE

This module is an elective. It is not required for successful completion of *Agricultural Technology*.

Objective

Successful completion of this module prepares you to do the following:

1. Identify and describe various types of rigging slings, hardware, and equipment.
   a. Identify and describe various types of slings.
   b. Describe how to inspect various types of slings.
   c. Identify and describe how to inspect common rigging hardware.
   d. Identify and describe various types of hoists.
   e. Identify and describe basic rigging hitches and the related Emergency Stop hand signal.

Performance Tasks

Under supervision, you should be able to do the following:

1. Demonstrate the proper American Society of Mechanical Engineers (ASME) Emergency Stop hand signal.
2. Demonstrate the ability to report the load capacity of a sling and if the sling is too damaged to use.

Overview

A common activity at nearly every construction site is moving material and equipment from one place to another using various types of lifting gear. The procedures involved in performing this task are known as rigging. Not every worker will participate in rigging operations, but nearly all will be exposed to them at one time or another. This module provides an overview of the various types of rigging equipment, common hitches used during a rigging operation, and the related Emergency Stop hand signal.

NCCER Industry-Recognized Credentials

If you are training through an NCCER-accredited sponsor, you may be eligible for credentials from NCCER. The ID number for this module is 93107. Note that this module may have been used in other NCCER curricula and may apply to other level completions. Contact NCCER at 1.888.622.3720 or go to **www.nccer.org** for more information.

You can also show off your industry-recognized credentials online with NCCER's digital credentials. Transform your knowledge, skills, and achievements into credentials that you can share across social media platforms, send to your network, and add to your resume. For more information, visit **www.nccer.org**.

Digital Resources for Agricultural Technology

SCAN ME

Scan this code using the camera on your phone or mobile device to view the digital resources related to this craft.

1.0.0 Basic Rigging Equipment

Performance Tasks

1. Demonstrate the proper American Society of Mechanical Engineers (ASME) Emergency Stop hand signal.
2. Demonstrate the ability to report the load capacity of a sling and if the sling is too damaged to use.

Objective

Identify and describe various types of rigging slings, hardware, and equipment.

a. Identify and describe various types of slings.
b. Describe how to inspect various types of slings.
c. Identify and describe how to inspect common rigging hardware.
d. Identify and describe various types of hoists.
e. Identify and describe basic rigging hitches and the related Emergency Stop hand signal.

Slings: Wire rope, alloy steel chain, metal mesh fabric, synthetic rope, synthetic webbing, or jacketed synthetic continuous loop fibers made into forms, with or without end fittings, used to handle loads.

Hoists: A device that applies a mechanical force for lifting or lowering a load.

Qualified person: A person who, by possession of a recognized degree, certificate, or professional standing, or by extensive knowledge, training, and experience, has demonstrated the ability to solve or prevent problems relating to a certain subject, work, or project.

Rigging is an essential step in the planned movement of material and equipment from one location to another using **slings**, **hoists**, or other types of equipment. Although rigging and lifting are often used interchangeably, important distinctions exist between these two tasks in the material handling industry. Lifting is raising and moving a load from one place to another using a crane or other hoisting system. Rigging is the process of preparing and securing that load for movement. This includes selecting and assembling the equipment to connect the load to the crane or other hoisting device.

Two common cranes are overhead cranes (*Figure 1*) and mobile cranes (*Figure 2*). As with other types of equipment, cranes are available in many different configurations and capacities.

Figure 1 Overhead crane.

Rigging operations can be extremely complicated and dangerous. Do not experiment with rigging operations, and never attempt a lift without the supervision of an officially recognized, **qualified person**. A lift may appear simple while it is in progress. That is because the people performing the lift know exactly what they are doing; there is no room for guesswork. Whether rigging operations involve simple or complicated equipment, only qualified persons may perform rigging operations without supervision.

WARNING!

Although this module provides instruction and information about common rigging equipment and hitch configurations, it does not provide any level of certification. Any questions about rigging procedures or equipment should be directed to an instructor or supervisor. Always refer to the manufacturer's instructions for any rigging equipment.

(A) (B)

Figure 2 Mobile cranes.
Source: Courtesy of Mammoet

1.1.0 Slings

During a rigging operation, the **load** being lifted or moved must be connected to an apparatus, such as a crane, to provide movement power. The connector—the link between the load and the apparatus—is often a sling made of synthetic, chain, or **wire rope** materials. This section focuses on three types of slings:

- Synthetic slings
- Alloy steel chain slings
- Wire rope slings

Load: The total amount of what is being lifted, including all slings, hitches, and hardware.

Wire rope: A rope made from steel wires that are formed into strands and then laid around a supporting core to form a complete rope; sometimes called *cable*.

1.1.1 Sling Tagging Requirements

All slings are required to have identification tags (*Figure 3*). An identification tag must be securely attached to each sling and clearly marked with the information required for that type of sling. For all three types of slings, that information will include the manufacturer's name or trademark and the rated capacity for at least one **hitch** type. The **working load limit (WLL)** is the maximum load weight that the sling is designed to carry. The WLL of a sling must never be exceeded.

Hitch: The rigging configuration by which a sling connects the load to the hoist hook. The three basic types of hitches are vertical, choker, and basket.

Working load limit (WLL): The maximum load weight a sling or piece of hardware or equipment can hold or lift.

WARNING!

Under no circumstances should the WLL ever be exceeded. Overloading a piece of rigging equipment can result in catastrophic failure.

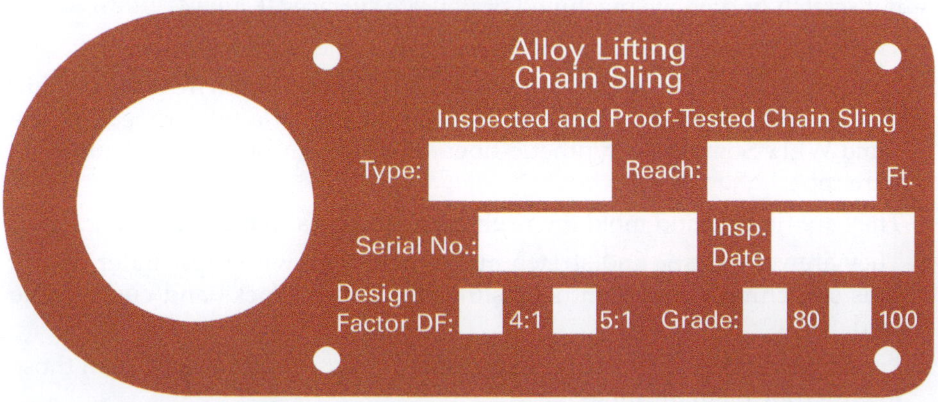

Figure 3 Identification tag.

Sling reach: A measure taken from the master link of the sling, where it bears weight, to either the end fitting of the sling or the lowest point on the basket.

Sling legs: The parts of the sling that reach from the attachment device around the object being lifted.

The following are the tagging requirements for synthetic slings:

- Manufacturer's name or trademark
- Manufacturer's code or stock number (unique for each sling)
- Rated load for the types of hitches used
- Type of synthetic material used in the manufacture of the sling

The following are the tagging requirements for alloy steel chain slings:

- Manufacturer's name or trademark
- Manufactured grade of steel
- Link size (diameter)
- Rated load and the angle on which the rating is based
- **Sling reach**
- Number of **sling legs**

The following are the tagging requirements for wire rope slings:

- Manufacturer's name or trademark
- WLL in a vertical hitch (other hitches optional)
- Diameter of wire rope (optional)
- Manufacturer's code or stock number of sling (optional)

Any sling without an identification tag must be removed from service immediately since its characteristics and rating can no longer be identified. In some cases, a qualified person may be able to install a new tag.

Regulations and Site Procedures

The information in this module is intended as a general guide. The techniques shown here are not the only methods that can be used to perform a lift. Many techniques can be safely used to rig and lift different loads.

Some techniques for certain kinds of rigging and lifting are spelled out in requirements issued by federal government agencies. Some will be provided at the jobsite, where written site procedures addressing any special conditions that affect lifting procedures on that site can be found. Questions about any of these procedures should be directed to the supervisor at the site.

1.1.2 Synthetic Slings

Synthetic slings are widely used to lift loads, especially loads with fragile or finished/exposed materials that may get scratched or marred if you attempt to rig them with wire ropes. This section covers two types of synthetic slings: synthetic web slings and round slings.

Synthetic web slings provide several advantages over other types of slings:

- They are softer and wider than chain or wire rope slings. Therefore, they do not scratch or damage machined or delicate surfaces (*Figure 4*).
- They do not rust or corrode and will not stain the loads they lift.
- They are lightweight, making them easier to handle than wire rope or chain slings. Most synthetic slings weigh less than half of a wire rope with the same WLL. Some new synthetic fiber slings weigh one-tenth as much as wire rope.
- They are flexible and mold themselves to the load's shape (*Figure 5*).
- They are very elastic and stretch much more than wire rope under a load. This stretching allows synthetic slings to absorb shocks and cushion the load.
- Loads suspended in synthetic web slings are less likely to twist than those in wire rope or chain slings.

Figure 4 Web slings provide surface protection.
Source: Slingmax® Rigging Solutions

Figure 5 Synthetic web sling shaping.
Source: Shutterstock

Keep in mind that for all the advantages synthetic web slings provide, there are some concerns related to their use. For example, synthetic web slings should not be exposed to temperatures above 180°F (82°C). They are also susceptible to cuts, abrasions, and other wear-and-tear damage. Riggers use protective pads to prevent damage to synthetic web slings (*Figure 6*).

Most synthetic web slings are manufactured with red-core **warning yarns**. These are used to let the rigger know whether the sling has suffered too much wear or damage to be used. When the red yarns are exposed, the synthetic web sling should not be used (*Figure 7*).

Synthetic web slings are available in several designs. The most common are the following (*Figure 8*):

- Endless web slings, which are also called *grommet slings.*
- Synthetic web eye-and-eye slings, which are made by sewing an end of the sling directly to the sling body. Standard eye-and-eye slings (*Figure 9*) have eyes on the same **plane** as the sling material; twisted eye-and-eye slings have eyes at right angles to the main portion of the sling and are primarily used for choker hitches (which are discussed in *Section 1.5.2*).
- Round slings are made in a continuous circle out of polyester filament yarn. A woven sleeve then covers the yarn.

CAUTION

If the sling does not come with protective pads, use other softeners of sufficient strength or thickness to protect the sling where it contacts the load. Pieces of old sling, fire hose, canvas, or rubber can be used, and manufactured softeners can also be purchased.

Warning yarns: A component of the sling that shows the rigger whether the sling has suffered too much damage to be used.

Plane: A surface in which a straight line joining two points lies wholly within that surface.

Figure 6 Protective pads.
Source: iStock

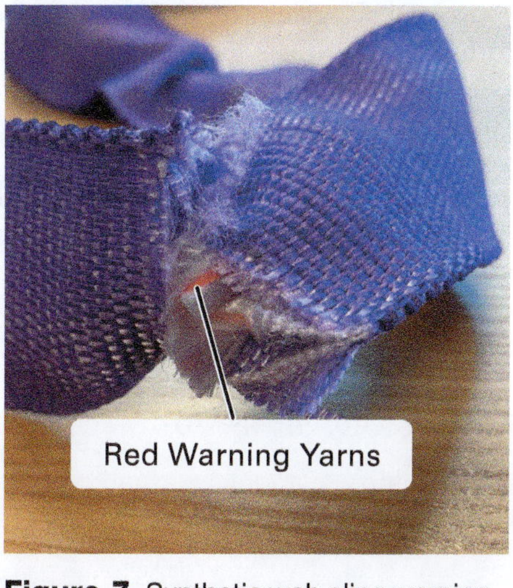

Figure 7 Synthetic web sling warning yarns.

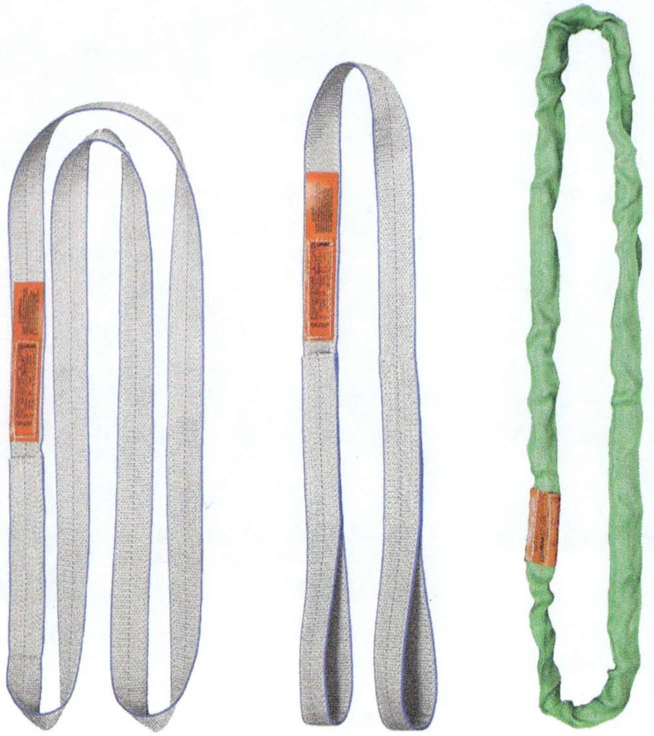

(A) Endless (B) Standard Eye-and-Eye (C) Endless Round

Figure 8 Synthetic web slings.
Source: Lift-All Company, Inc.

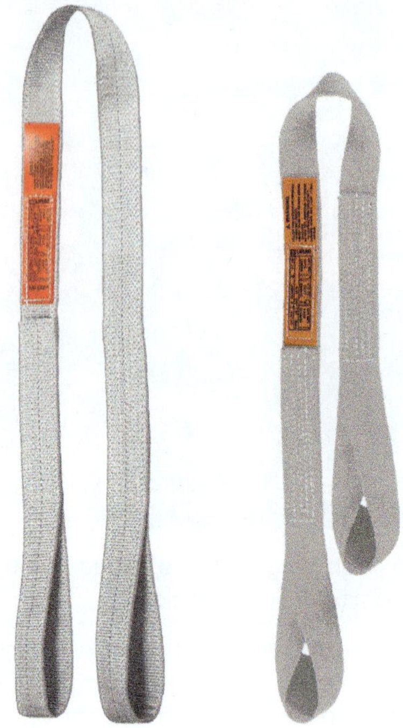

(A) Standard Eye-and-Eye (B) Twisted Eye-and-Eye

Figure 9 Eye-and-eye synthetic web slings.
Source: Lift-All Company, Inc.

Synthetic web eye-and-eye slings are also available with hardware end fittings instead of fabric eyes. The standard end fittings are made of either aluminum or steel. They come in male and female configurations (*Figure 10*).

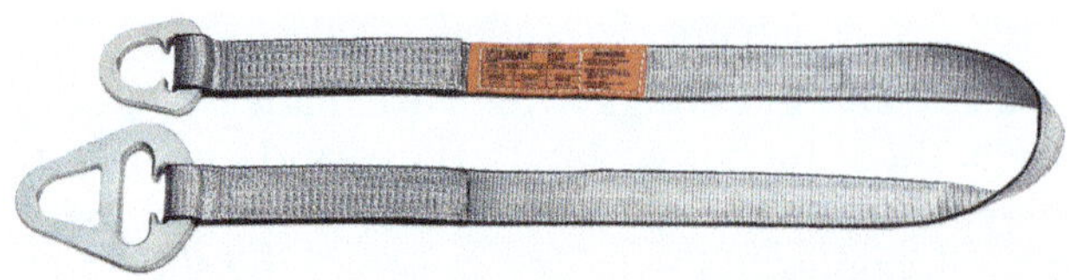

Sling with Triangle and Choker Hardware

Figure 10 Synthetic web sling hardware end fittings.
Source: Lift-All Company, Inc.

Core: Center support member of a wire rope around which the strands are laid.

Strands: A group of wires wound, or laid, around a center wire, or core. Strands are laid around a supporting core to form a rope.

Round slings are made by wrapping synthetic yarn around a set of spindles to form an endless loop. A protective jacket encases the **core** yarn (*Figure 11*). One type of round sling is a Twin-Path® sling. Twin-Path® is a trademarked sling made by Slingmax®, Inc., although the term is sometimes used generically.

Twin-Path® slings are made of synthetic fiber, such as polyester. The material used to make up the **strands** and the number of wraps in a loop determine the sling's WLL. Aramid round slings, for example, have a greater WLL for their size than web-type slings.

Twin-Path® slings are also available with two separately wound loops of strands jacketed together side by side (*Figure 12*). This design greatly increases the sling's lifting capacity.

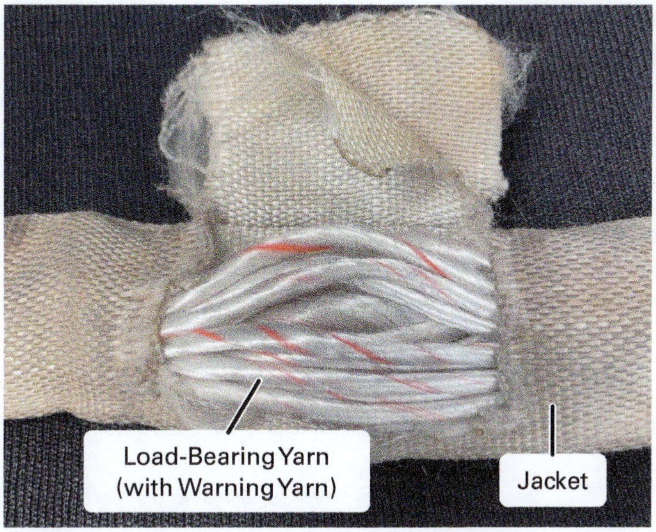

Figure 11 Synthetic endless-strand jacketed sling.

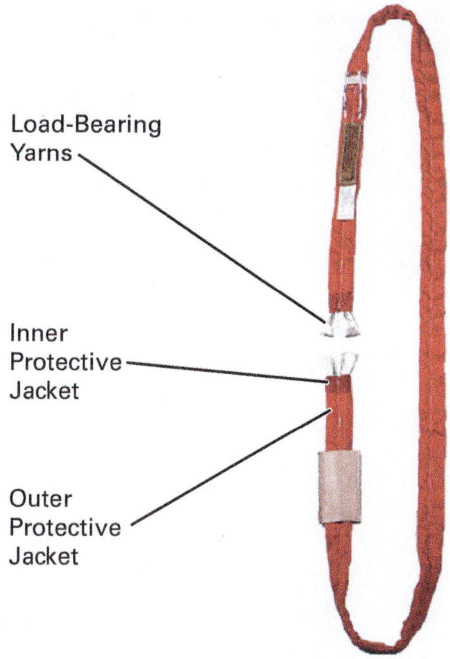

Figure 12 Twin-Path® sling.
Source: Lift-It® Manufacturing

The jackets of these slings are available in several materials for various purposes, including heat-resistant Nomex®, polyester, and bulked nylon (Covermax™), as shown in *Figure 13*. Twin-Path® slings featuring K-Spec® yarn weigh at least 50% less than a polyester round sling of the same size and capacity.

Twin-Path® slings are equipped with **tattle-tail** yarns to help riggers determine whether the sling has become overloaded or stretched beyond a safe limit (*Figure 14*). Note that these devices differ from red-core yarn, which is incorporated directly into the sling's fabric.

Tattle-tail: Cord attached to the strands of an endless loop sling that protrudes from the jacket. A tattle-tail determines whether an endless sling has been stretched or overloaded.

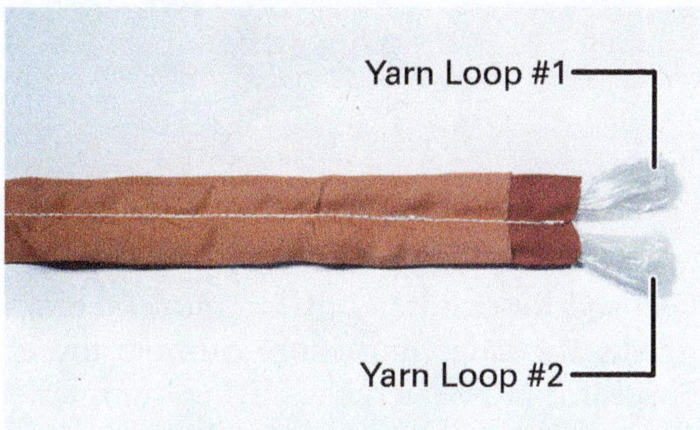

Figure 13 Twin-Path® sling makeup.
Source: Lift-It® Manufacturing

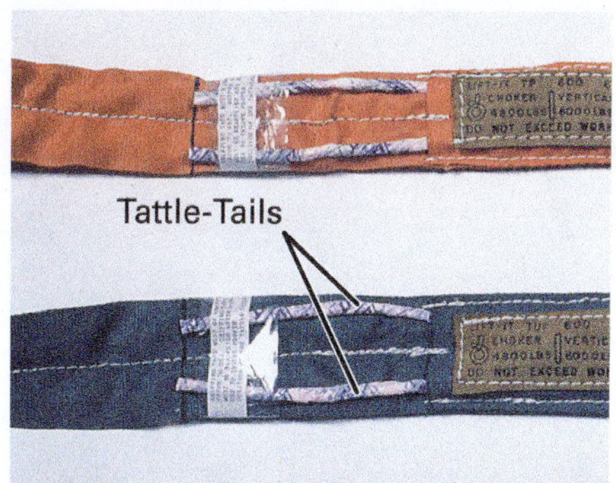

Figure 14 Tattle-tails.
Source: Lift-It® Manufacturing

These slings are also available with a fiber optic inspection cable running through the strand (*Figure 15*). When light is directed at one end of the fiber, the other end will light up to show that the strand has not been broken.

The way the sling will be used determines the material used for the jackets of round slings. Polyester is normally used for light- to medium-duty sling jackets. Covermax™, a high-strength material with much greater resistance to cutting and abrasion, is used to make a sturdier jacket for heavy-duty uses.

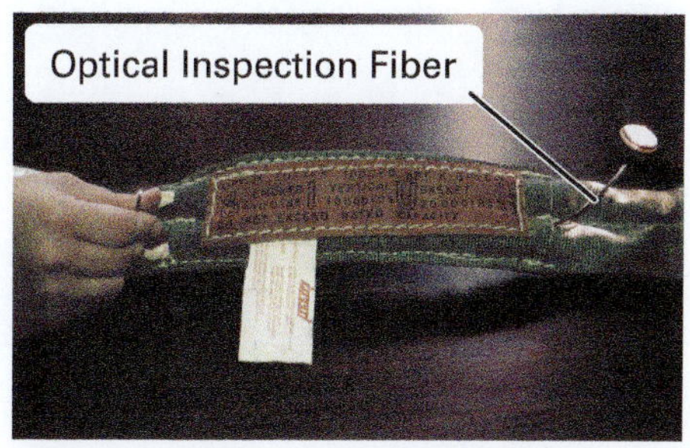

Figure 15 Fiber optic inspection cable.
Source: Lift-It® Manufacturing

1.1.3 Alloy Steel Chain Slings

Alloy steel chain, like wire rope (covered later in *Section 1.1.4*), can be used in many rigging operations. Chain slings are often used for lifts in high heat or rugged conditions. Chain slings can be adjusted over a center of gravity, making them versatile. They are also the most durable. However, a chain sling weighs much more than a wire rope sling and may be harder to inspect. Workers will encounter both types of slings in the field and must decide which type is best for each situation.

Around the World

Heavy Lifters

The world's largest cranes can lift astonishing amounts of weight. Some mega cranes are permanently stationary, some can be assembled and disassembled on-site, and some are mobile. Taisun, a fixed dual-beam gantry crane in China, which can lift over 22,000 tons (almost 20,000 metric tons). Mammoet, a manufacturer of heavy lifting equipment in the Netherlands, has built numerous cranes that are referred to as PTCs (Platform Twinring Containerized). These cranes can be disassembled and packed into shipping containers, unloaded at any dock in the world, and hauled by truck to a jobsite. This type of crane, as shown here, is assembled onto a stationary ring that enables the crane and counterweights to move in a 360° radius. These giant cranes can lift up to 3,200 tons (about 2,900 metric tons).

One of the largest mobile cranes is Liebherr's LR 13000 crawler crane. It has a more conventional design and appearance. Its maximum lifting capacity is over 3,300 tons (about 3,000 metric tons).

Source: Courtesy of Mammoet

Steel chain slings used for overhead lifting must be made of alloy steel. The higher the alloy grade, the safer and more durable the chain is. Alloy steel chains commonly used for most overhead lifting are marked with the number 8, the number 80, the number 800, or the letter A (*Figure 16*).

Steel chain slings have two basic designs with many variations:

Master link: The main connection fitting for chain slings.

Bridle: A configuration using two or more slings to connect a load to a single hoist hook.

- Single- and double-basket slings (*Figure 17*) do not require end-fitting hardware. The chain is attached to the **master link** in a permanent basket hitch or hitches.
- Chain **bridle** slings are available with two to four legs (*Figure 18*).

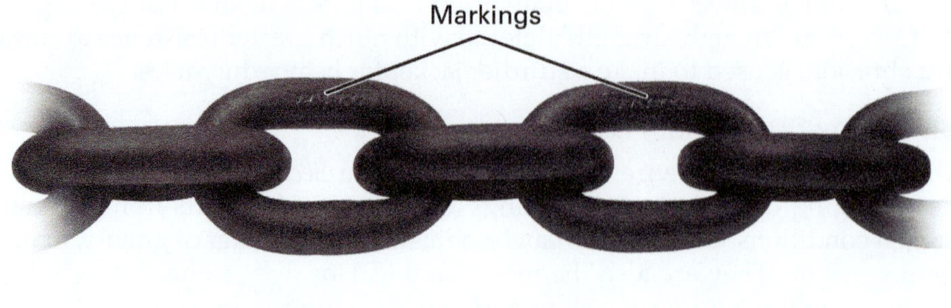

Figure 16 Markings on alloy steel chain slings.
Source: Columbus McKinnon Corporation

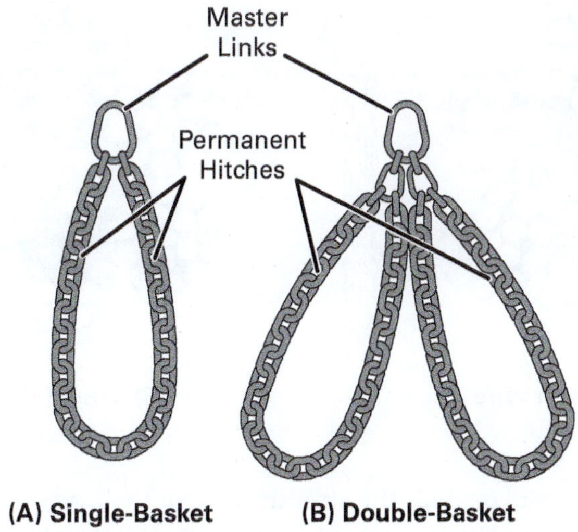

Figure 17 Single- and double-basket chain slings.

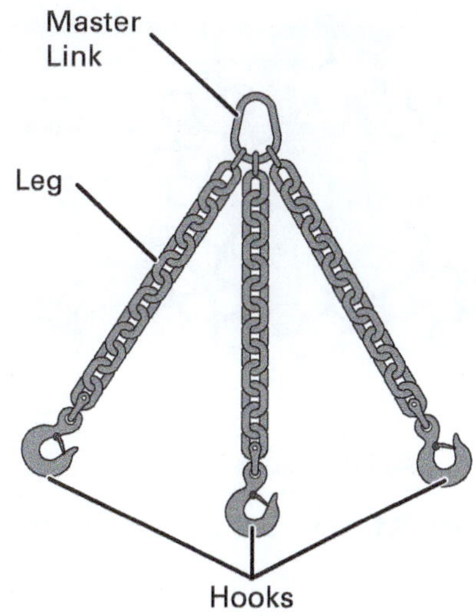

Figure 18 Three-leg chain bridle sling.

Alloy steel chain slings are used for lifts when temperatures are high or when the slings will be subjected to steady and severe abuse. Most alloy steel chain slings can be used in temperatures up to 500°F (260°C) with little loss in WLL.

Even though alloy steel chain slings can withstand extreme temperature ranges and abusive working conditions, these slings can be damaged if loads are dropped on them, if they are wrapped around loads with sharp corners (unless softeners are used), or if they are exposed to intense temperatures.

CAUTION

Never drag alloy steel chain slings across hard surfaces, especially concrete. Friction with abrasive surfaces wears out the chain, weakens the sling, and often erodes identifying markings.

1.1.4 Wire Rope Slings

Wire rope slings (*Figure 19*) are made of high-strength steel wires formed into strands and wrapped around a supporting core. While there are many types of wire rope, they all share this common design. Generally, wire rope slings are lighter than chains, can withstand substantial abuse, and are easier to handle than chain slings. They can also withstand relatively high temperatures.

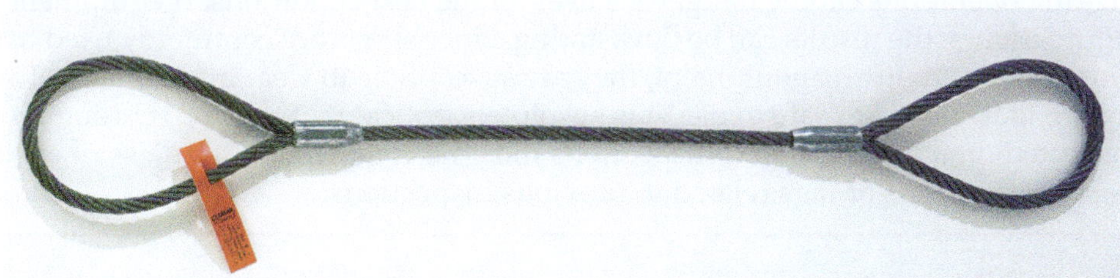

Figure 19 Wire rope sling.
Source: Lift-All Company, Inc.

The wire rope that makes up a wire rope sling is designed so that the strand wires and the supporting core interact by sliding and adjusting. This movement compensates for the ever-changing stresses placed upon the rope. It makes it less likely that the wires and the core will be damaged when the rope is bent around **sheaves** or loads or when placed at an angle during rigging.

Sheaves: Grooved pulley wheels used for changing the direction of a rope's pull; often found on a crane.

The wires must be able to move as designed. A wire rope's WLL depends on whether it is in good condition since its rating is tested in new, pristine condition. Any kinks, corrosion, or other defects will negatively impact the WLL.

Figure 20 shows the three basic components of a wire rope: the supporting core, high-grade steel wires, and multiple center wires.

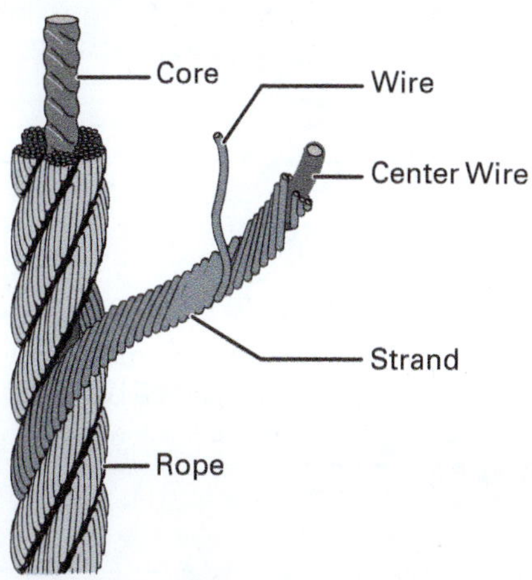

Figure 20 Wire rope components.

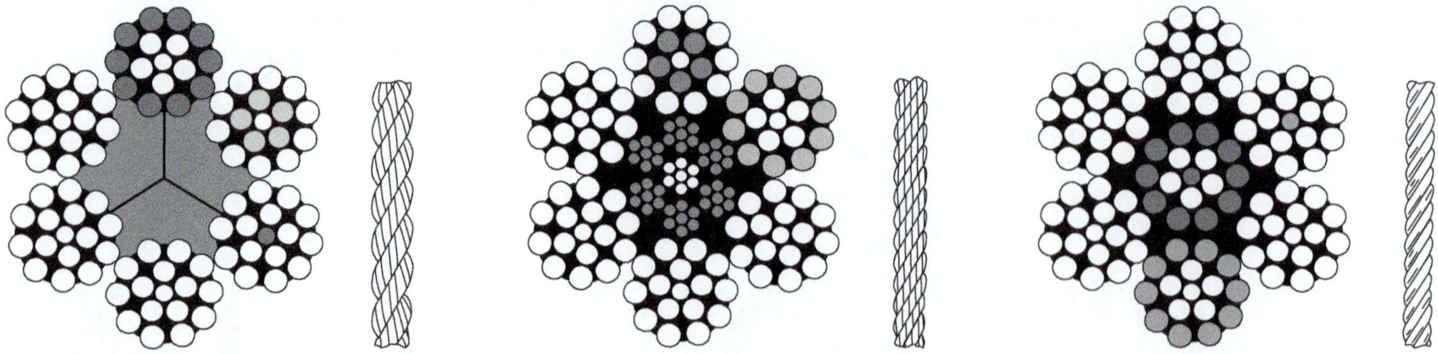

(A) Fiber Core **(B) Independent Wire** **(C) Strand (Steel)**

Figure 21 Wire rope supporting cores.

The supporting core enables the strands to keep their original shapes. There are three basic types of supporting cores for wire rope (*Figure 21*):

- *Fiber cores* — Usually made of synthetic fibers, but can also be made from natural vegetable fibers, such as sisal.
- *Independent wire rope cores* — Made of a separate wire rope with its own core and strands; the core rope wires are much smaller and more delicate than the strand wires in the outer rope.
- *Strand cores* — Made by using one strand of the same size and type as the rest of the strands of rope.

The materials used for the supporting core have both desirable characteristics and drawbacks, depending on how they are to be used. Fiber core ropes, for example, may be damaged by heat at relatively low temperatures (180°F to 200°F [82°C to 93°C]), as well as by exposure to caustic chemicals.

1.2.0 Sling Inspection

Rigging operations can be dangerous even in the best conditions. If equipment failure occurs, the results can be devastating. Since slings are commonly used to connect loads to lifting equipment, they are susceptible to wear and damage. To ensure that slings are safe to use, it is absolutely critical that they be thoroughly inspected on a regular basis. Inspections must be conducted regardless of any existing markings or tags related to previous inspections.

Did You Know?

Sling Cut Protection

Cuts are the primary cause of synthetic sling failure. Often, these cuts come from contacting sharp edges on the load. Besides sling jackets, or *wear pads*, devices called *edge protectors* are often placed on the edges of a load to help reduce the likelihood of cutting the sling.

WARNING!

Although this module provides instruction and information about lifting component inspection, it does not provide any level of certification. Any questions about rigging procedures or equipment should be directed to an instructor or supervisor. Always refer to the manufacturer's instructions for any type of rigging equipment.

1.2.1 Synthetic Sling Inspection

Like all slings, synthetic slings must be inspected before each use to determine whether they are in good condition and can be used. They must be inspected along the sling's entire length visually (looking at them) and manually (feeling them). If any of the **rejection criteria** presented in this section are met, the sling must be removed from service immediately.

Rejection criteria: Standards, rules, or tests on which a decision can be based to remove an object or device from service because it is no longer safe.

WARNING!

A competent person must inspect synthetic slings before they are used every time. The inspection must involve a visual and touch examination. Sometimes, defects and damage can be seen; other times, they can be felt.

In addition, the rigger must exercise sound judgment. Along with looking for any single major problem, the rigger must also watch for combinations of relatively minor defects in the sling. Combinations of minor damage may make the sling unsafe to use, even though the specific defects found may not be listed in the rejection criteria.

Workers who help inspect synthetic slings must alert a qualified person if any defects are suspected, especially if any of the following synthetic sling damage rejection criteria are found (*Figure 22*):

- A missing identification tag or a tag that cannot be read. Any synthetic sling without an identification tag must be removed from service immediately.
- Abrasion that has worn through the outer jacket or has exposed the load-bearing yarn of the sling, or abrasion that has exposed the warning yarn of a web sling.
- A cut that has severed the outer jacket or exposed the load-bearing yarn (single-layer jacket) of a round sling or has exposed the warning yarn of a web sling.
- A tear that has exposed the inner jacket or the load-bearing yarn (single-layer jacket) of a round sling or has exposed the warning yarn of a web sling.
- A puncture.
- Broken or worn stitching in the **splice** or stitching of a web sling.
- A knot that cannot be removed by hand in either a web or round sling.
- A snag in the sling that reveals the warning yarn of a web sling, tears through the outer jacket, or exposes the load-bearing yarns of a round sling.
- Crushing of either a web sling or a round sling. Crushing in a web sling feels like a hollow pocket or depression in the sling. Crushing in a round sling feels like a hard, flat spot underneath the jacket.
- Damage from overload, often called *tensile damage* or *overstretching*, in a web or round sling. Tensile damage in a web sling is evident when the weave pattern of the fabric begins to pull apart. Twin-Path® slings have tattle-tail features. When the tattle-tail has been pulled into the jacket, it indicates that the sling may have been overloaded.
- Chemical damage, including discoloration, burns, and melting of the fabric or jacket.
- Heat damage, ranging from friction burns to melting of the sling material, the load-bearing strands, or the jacket. Friction burns give the webbing material a crusty or slick texture. Heat damage to the jacket of a round sling looks like glazing or charring. In round slings with heat-resistant jackets, you may not be able to see heat damage to the outside of the sling; however, the internal yarns may have been damaged. You can detect that damage by carefully handling the sling and feeling for brittle or fused fibers inside the sling or by flexing and folding the sling and listening for the sound of fused yarn fibers cracking or breaking inside.
- Ultraviolet (UV) damage. Evidence of UV damage is a bleaching of the sling material, which breaks down the synthetic fibers. Web slings with UV damage will release a powder-like substance when flexed and folded. Round slings, especially those made of Cordura® nylon and other specially treated synthetic fibers, have a much greater resistance to UV damage. In round slings, UV damage appears as a roughening of the fabric texture where no other sign of damage, such as abrasion, can be found.
- Loss of flexibility caused by the presence of dirt or other abrasives. A sling that has lost its flexibility becomes stiff. Abrasive particles embedded in the sling material act like tiny blades that cut apart the internal fibers of the sling every time it is stretched, flexed, or wrapped around a load.

Splice: To join together.

(A) Jacket and Web Abrasion
(B) Jacket and Web Sling Abrasion
(C) Minor Abrasion
(D) Outer Jacket Cut
(E) Inner and Outer Jacket Cuts
(F) Cut
(G) Cut with Warning Threads Showing
(H) Broken Splice or Stitching
(I) Snag Showing Red Thread
(J) Tensile Damage
(K) Overload Damage Tattle-Tails Pulled In
(L) Friction Burn from Abrasion and Heat Damage
(M) Severe Heat Damage
(N) One-Ply Penetration
(O) Exposed Yarn

Figure 22 Sling damage rejection criteria.
Source: Ed Gloninger (22A–22M)

Sling Jackets

Under the rejection criteria, a sling can be sent back to the manufacturer for a new jacket when only the outer jacket is damaged. The manufacturer must test and certify the sling before it is used again. It is much less costly to do this than to replace the sling. However, slings removed because of heat- or tension-related jacket damage cannot be returned for repair and must be disposed of by a qualified person.

1.2.2 Alloy Steel Chain Sling Inspection

Alloy steel chain slings must be carefully inspected before each use to determine whether they are safe to use. There are numerous rejection criteria that indicate when an alloy steel chain sling must be removed from service. A chain sling may also need to be removed from service if something shows up that does not exactly match the rejection criteria. Not all riggers are qualified to make decisions about the equipment's condition. A qualified person should be consulted if questions arise about whether a sling is defective or unsafe. Note that the chain links in a chain sling cannot be repaired by replacement.

An alloy steel chain sling must be removed from service for any of the following defects (*Figure 23*):

- Missing or illegible identification tag
- Cracks
- Heat damage
- Stretched links; damage is evident when the link grows long and when the barrels—the long sides of the links—start to close up
- Bent links
- Twisted links
- Evidence of replacement links that have been used to repair the chain
- Excessive rust or corrosion, meaning rust or corrosion that cannot be easily removed with a wire brush
- Cuts, chips, or gouges resulting from impact on the chain
- Damaged end fittings, such as hooks, clamps, and other hardware
- Excessive wear at the link-bearing surfaces
- Scraping or abrasion

1.2.3 Wire Rope Sling Inspection

Like other slings, wire rope slings must be inspected before each use by a **competent person**. A competent person is defined by the Occupational Safety and Health Administration (OSHA). They can identify hazards, such as unsanitary, dangerous, or complex conditions that increase the risk of injuries. Jobsite safety depends on the authorized competent person to take preventative steps when they identify predictable hazards that may not be obvious to others.

Competent person: A worker who can identify hazards and take the corrective action necessary to address them.

If the wire rope is damaged, it must be removed from service. Only a competent person can decide whether to use a wire rope in a rigging operation or discard it if damaged. Any worker who suspects that a wire rope sling may be damaged must bring it to the attention of a competent person, stopping an active lift if necessary to do so. If a wire rope sling is simply missing its tag or the tag has become unreadable due to wear, it can be retagged by a qualified person. Remember, a qualified person is defined by OSHA as a person who has the necessary training or certifications and has demonstrated their knowledge and ability to resolve problems related to the work or project.

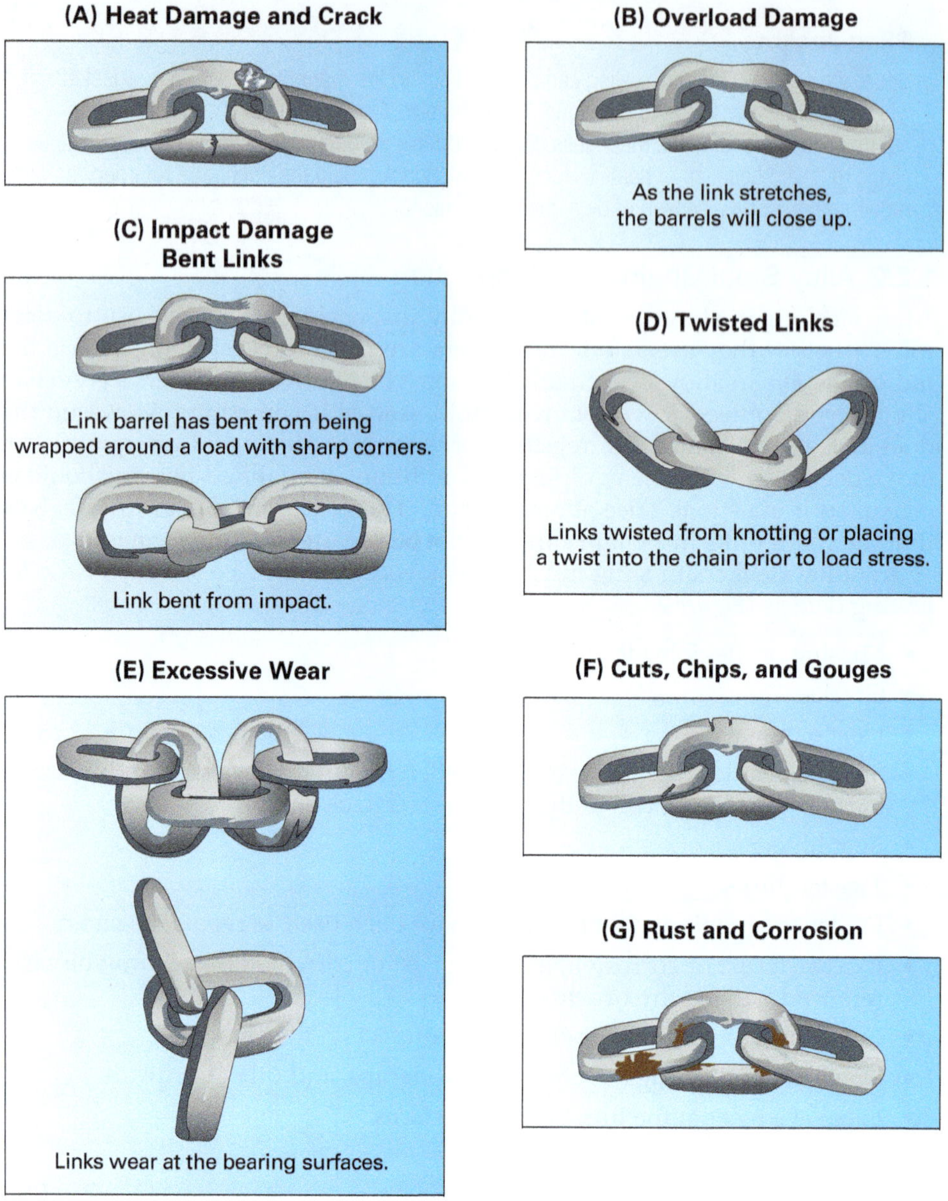

Figure 23 Damage to chains.

Following an inspection, a wire rope sling may be rejected based on several common types of damage (*Figure 24*):

- *Broken wires* — Broken wires in the strands of a wire rope weaken the material strength of the rope and interfere with the interactions among the rope's moving parts. External broken wires usually mean normal fatigue, but internal or severe external breaks should be investigated closely. Internal or severe breaks in a wire rope mean it has been used improperly. Rejection criteria for broken wires consider how many wires are broken in one lay length of rope, or **one-rope lay**. One-rope lay (a single rope lay and not a rope lay made with one rope) is a term that defines the lengthwise distance it takes for one strand of wire to make one complete turn around the core (*Figure 25*). Different wire ropes have different one-rope lays, so a competent person needs to inspect each wire rope closely when looking for broken wires.

One-rope lay: The lengthwise distance it takes for one strand of a wire rope to make one complete turn around the core.

WARNING!

Broken wires in a sling are extremely sharp and can easily cut or puncture the skin. Never use a bare hand to handle wire rope slings or inspect them.

(A) Broken Wires

(B) Kinking

(C) Birdcaging

(D) Crushing

(E) Corrosion

Figure 24 Common types of wire rope damage.
Source: Shutterstock

- *Kinks* — Kinking, or rope distortion, is a very common type of damage. Kinking can result in serious accidents. Sharp kinks restrict or prevent the movement of wires in the strands at the area of the kink. This means the rope is damaged and must not be used. Ropes with kinks in the form of large, gradual loops in a corkscrew configuration must be removed from service.
- *Birdcaging* — This damage occurs when a load is released too quickly, and the strands are pulled or bounced away from the supporting core. The wires in the strands cannot compensate for the change in the stress level by adjusting inside the strands. The built-up stress then finds its own release through the strands. Birdcaging usually occurs in an area where existing damage prevents the wires from moving to compensate for changes in stress, position, and bending of the rope. Any sign of birdcaging is cause to remove the rope from service immediately.
- **Unstranding** — Unstranding describes wire rope strands that have become untwisted. This weakens the rope and makes it easier to break.
- *Signs of core failure* — Failure or breakage within the core is usually spotted by the protrusion of core strands through or into the outer jacket of strands.
- *Crushing* — Crushing results from setting a load down on a sling or from hammering or pounding a sling into place. Crushing of the sling prevents the wires from adjusting to changes in stress, changes in position, and bends. A crushed sling usually crushes or breaks the core wires directly beneath the damaged strands. If crushing occurs, the sling must be removed from service immediately.
- *Corrosion and rust* — Corrosion and rust of wire rope are the result of improper or insufficient lubrication. Corrosion and rust are considered excessive if there is surface scaling, rust that cannot be easily removed with a wire brush, or if they occur inside the rope. The rope must be removed from service if there is excessive corrosion and rust.
- *Heat damage* — Heat damage makes wires in the strands and core in the affected area become brittle. Heat damage appears as discoloration and sometimes the actual melting of the wire rope. A wire rope that has been damaged by heat must be removed from service.
- *Integrity of end connections* — Any end connections applied to the wire rope must be inspected for any signs of damage or failure, such as marks indicating the connection has slipped or moved.

Figure 25 One-rope lay.

Unstranding: Describes wire rope strands that have become untwisted. This weakens the rope and makes it easier to break.

1.3.0 Rigging Hardware

Rigging hardware is as crucial as the crane and slings. If the hardware that connects the slings to either the load or the hoisting device were to fail, the load would drop. Hardware failure related to improper attachment, selection, or inspection contributes to a great number of deaths, serious injuries, and property damage events in rigging accidents. The importance of hardware selection, maintenance, inspection, and proper use cannot be stressed enough. The requirements for rigging hardware are as stringent as those governing cranes and slings.

1.3.1 Shackles

Shackle: A piece of rigging hardware used in an appropriate lifting apparatus to connect the rope to eye fittings, hooks, or other connectors.

A **shackle** is rigging hardware used to attach an item to a load or to couple slings together. For example, it can be used to couple the end of a wire rope to eye fittings or hooks. It consists of a U-shaped body and a removable pin.

Shackles used for overhead lifting should be made from forged steel not cast steel. Quenched and tempered steel is the preferred material because of its increased toughness, but at a minimum, shackles must be made of drop- or hammer-forged steel.

All shackles must have a clearly visible stamp showing the manufacturer's trademark, the size of the shackle (determined by the diameter of the shackle's body, not by the diameter of the pin), and the WLL of the shackle.

Shackles are available in two basic classes, identified by their shapes: anchor shackles and chain shackles. Both anchor and chain shackles have two basic types of pin designs, each one unique, as shown in *Figure 26*.

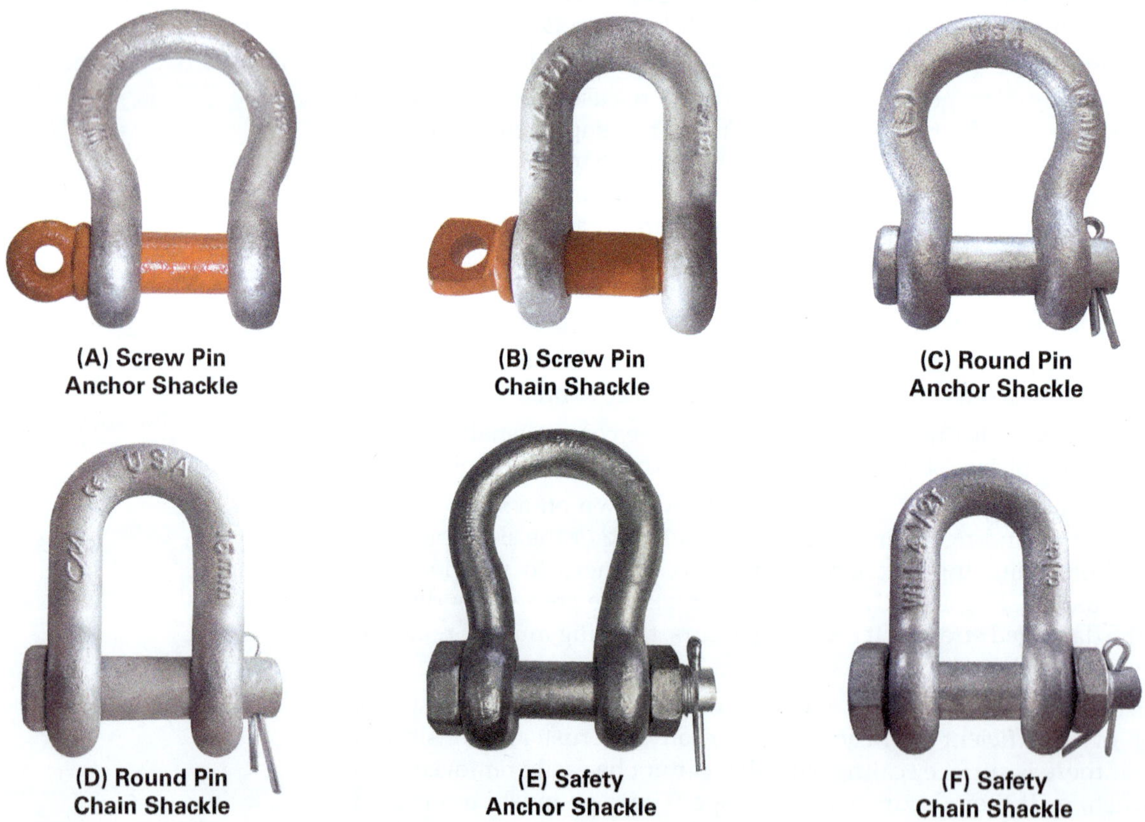

Figure 26 Shackles.
Source: Kito Crosby

The screw pin is threaded and screws directly into the shackle's lugs, stopping at the pin's shoulder. They're easy to use and allow for fast connection or disconnection, but they must be tightened thoroughly before each lift. The bolt-type enclosure, on the other hand, is only threaded at the very tip. It passes through the lugs and receives both a nut and a cotter pin for ultimate security. This shackle is often called a *safety shackle* because the pin is very secure.

Specialty shackles are available for specific applications where a standard shackle would not work well. For example, wide-body shackles (*Figure 27*) are for heavy-lifting applications.

Synthetic web sling shackles (*Figure 28*) are designed with a wide throat opening and a wide bow that is contoured to provide a larger, nonslip surface area to accommodate the wider body of synthetic web slings.

When a screw pin shackle is used, it is important to tighten the pin the proper amount. The pin should be screwed in until it is fully engaged, and the shoulder of the pin is in contact with the shackle body (*Figure 29*). If the pin is left loose, the shackle can stretch under load. If the pin is overtightened, the **load stress** from the sling can torque the pin even more. This can stretch and jam the pin's threads.

When any type of shackle is used in a rigging operation, the load should always be applied to the bow of the shackle and not the pin (*Figure 30*). If the shackle uses a screw pin and the load is positioned on the pin, the shifting of the load could cause the pin to twist and unscrew.

Shackles, like any other type of hardware, must be inspected by the rigger before each use to ensure there are no defects that would make the shackle unsafe. Each lift may cause some degree of damage or further reveal existing damage.

If any of the following conditions exists, a shackle must be removed from use:

- Bends, cracks, or other damage to the shackle body
- Incorrect shackle pin or improperly substituted pin
- Bent, broken, or loose shackle pin
- Damaged threads on threaded shackle pin
- Missing or illegible capacity and size markings

Figure 27 Wide-body shackle.
Source: Kito Crosby

Load stress: The strain or tension applied on the rigging by the weight of the suspended load.

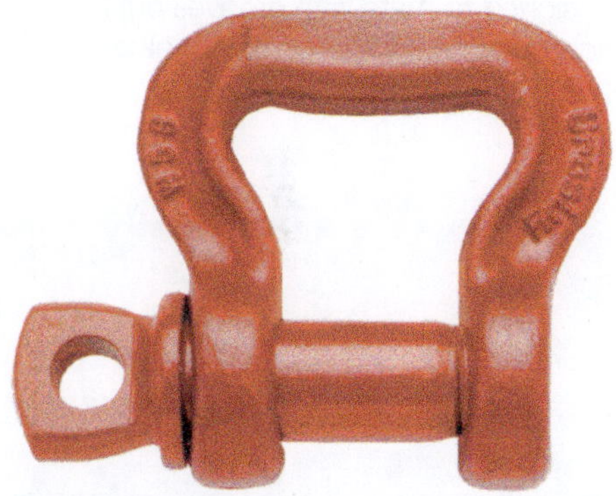

Figure 28 Synthetic web sling shackle.
Source: Kito Crosby

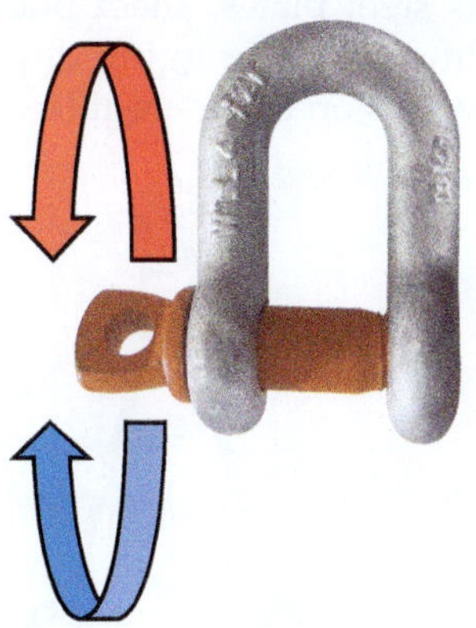

Figure 29 Don't overtighten the shackle pin.
Source: Kito Crosby

Figure 30 Shackle correctly positioned.
Source: Volvo Construction Equipment

1.3.2 Eyebolts

An eyebolt is an item of rigging hardware with a **threaded shank**. The eyebolt's shank end is attached directly to the load, and the eyebolt's eye end is used to attach a sling to the load.

Threaded shank: A connecting end of a fastener, such as a bolt, with a series of spiral grooves cut into it. The grooves are designed to mate with grooves cut into another object to join them together.

Eyebolts for overhead lifting should be made of drop- or hammer-forged steel. They are available in three basic designs with several variations (*Figure 31*).

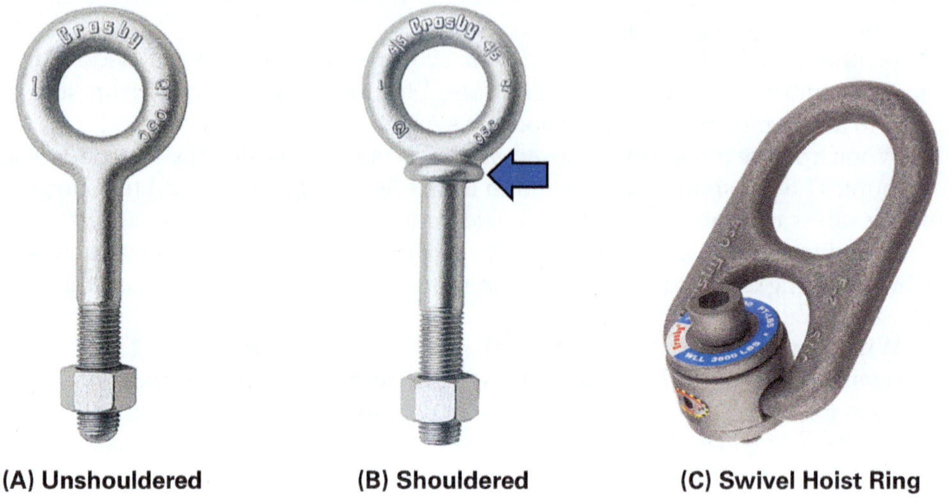

Figure 31 Eyebolt variations.
Source: Kito Crosby

Unshouldered eyebolts are designed for straight vertical pulls only. Shouldered eyebolts have a shoulder used to help support the eyebolt during slightly angular pulls. Swivel eyebolts, also called *swivel hoist rings*, are designed for angular pulls from 0° to 90° from the horizontal plane of the load.

In a few cases, eyebolts must be torqued to a specific value when installed for lifting use.

1.3.3 Lifting Clamps

Lifting clamps are used to move loads, such as steel plates, sheet piles, large pipes, or concrete panels, without the use of slings. These rigging devices are designed to bite down on a load and use the jaw tension to secure it (*Figure 32*).

Lifting clamps: A device used to move loads such as steel plates or concrete panels without the use of slings.

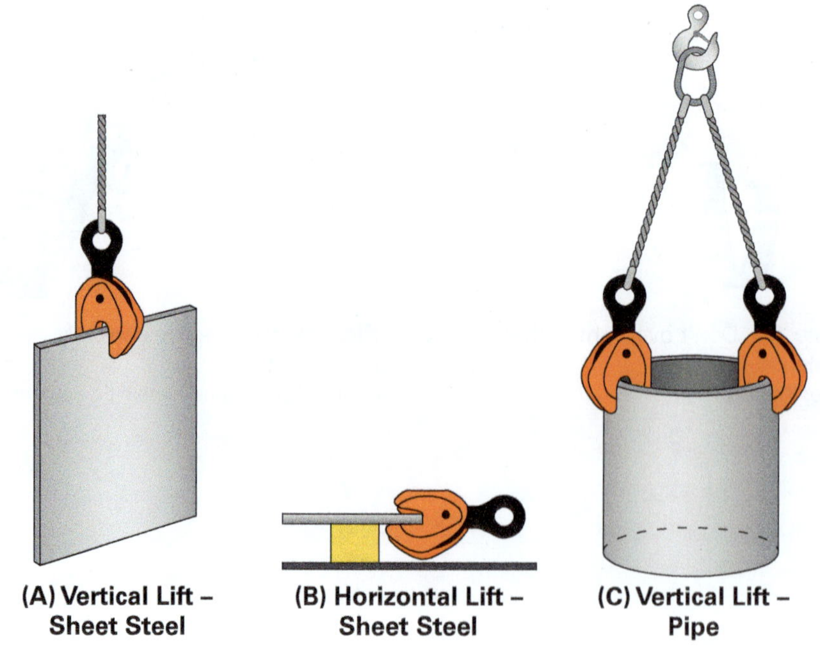

Figure 32 Using lifting clamps.

Did You Know?

Shackles and Pins

Pins should never be swapped in shackles. The thread forms of different brands of shackles and different brands of pins may not engage properly with one another. In addition, there is no easy way to tell how much reserve strength is left in a pin or a shackle. Further, the rated capacities of different shackles and pins may not be compatible. The shackle pin should be placed and secured into a single shackle between uses.

All lifting clamps must be made of forged steel and stamped with their WLL. Some clamps, such as the one shown in *Figure 33*, use the weight of the load to produce and sustain the clamping pressure; the grip tightens as the load increases. Others use an adjustable cam that is set and tightened to maintain a secure grip on the load.

Lifting clamps are designed to carry one item at a time, regardless of the capacity or jaw dimensions of the clamp or the thickness or weight of the item being lifted. For the clamp to hold an item securely, the cam and the jaw must bite or grip both sides of a single item. Placing more than one plate or sheet into the clamp prevents both the cam and the jaw from securing both sides of the plate or sheet. The clamps must be placed to ensure that the load remains balanced. For larger loads, two lifting clamps are typically required.

Lifting clamps are available in various designs, so it is important to match the type of clamp with the intended application. *Figure 34* shows some lifting clamps designed for specific uses.

Lifting clamps, like any other type of rigging hardware, must be inspected by the rigger before every use to ensure there are no defects that would make the clamp unsafe. Each lift may cause some degree of damage or further reveal existing damage.

If any of the following conditions exists, a clamp must be removed from use (*Figure 35*):

- Cracks
- Abrasion, wear, or scraping
- Any deformation or other impact damage to the shape that is detectable during a visual examination
- Excessive rust or corrosion, meaning rust or corrosion that cannot be removed easily with a wire brush
- Excessive wear of the teeth
- Heat damage
- Loose or damaged screws or rivets
- Worn springs

Figure 33 Basic nonlocking clamp.
Source: Photo courtesy of The Caldwell Group, Inc., Rockford, IL.

(A) Locking Clamp **(B) Screw-Adjusted Clamp** **(C) Non-Marring Clamp**

Figure 34 Lifting clamps.
Source: Photo courtesy of The Caldwell Group, Inc., Rockford, IL.

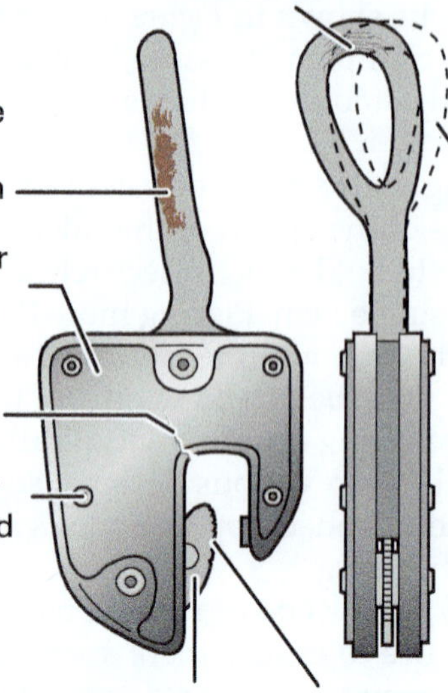

Figure 35 Rejection criteria for lifting clamps.

1.3.4 Rigging Hooks

Rigging hook: An item of rigging hardware used to attach a sling to a load.

A **rigging hook** is rigging hardware used to attach a sling to a load. Most rigging hooks have safety latches or gates to prevent slings or other connectors from accidentally coming out of the hook during use. Hooks that do not have safety latches are usually designed so that a latch can be added.

Many classes of rigging hooks are used, but most rigging hooks fall into a few basic categories (*Figure 36*):

- *Eye hooks* — These hooks are the most common type of end-fitting hook. A rigging eye hook has a large eye that can accommodate large couplers.
- *Reverse eye hooks* — These hooks position the point of the hook perpendicular to the eye.
- *Sliding choker hooks* — These hooks are installed onto the sling when it is made. The hooks, which can be positioned anywhere along the sling body, are used to secure the sling eye in a choker hitch. Sliding choker hooks are available for steel chain slings.
- *Grab hooks* — These hooks are used on steel chain slings. Grab hooks fit securely in the chain link, allowing for choker hitches and chain shortening.
- *Shortening clutches* — Shortening clutches, a more efficient version of grab hooks, provide a secure grab of the shortened sling leg with no reduction in the chain's capacity because the clutch fully supports the links.

Hooks used for rigging must be made of drop- or hammer-forged steel. Although most rigging hooks have safety latches or gates, some hooks used for special applications may not have them. If a safety latch is installed in a rigging hook, the latch must be in good working condition. Damaged safety latches can be easily replaced. Any damage to a safety latch must be reported to an instructor or supervisor.

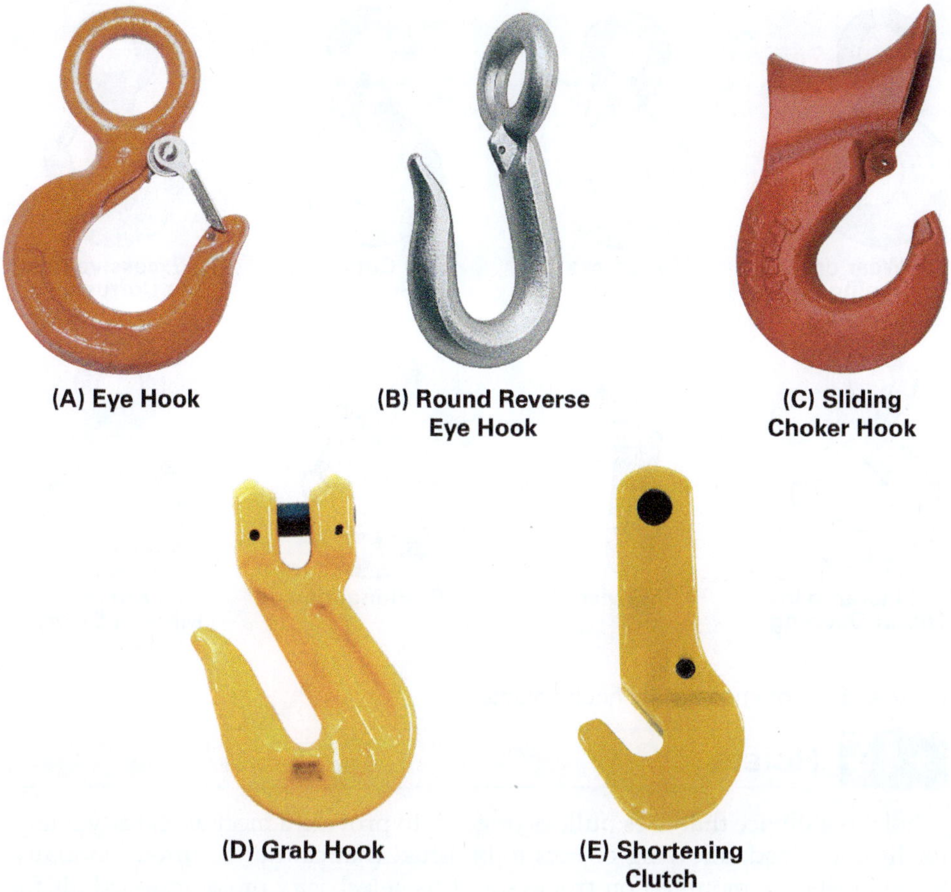

Figure 36 Rigging hooks.
Sources: Columbus McKinnon Corporation (36A); Kito Crosby (36B–36C)

When hooks are installed as end fittings, they must be inspected along with the rest of the sling before each use. Slings with hook-type end fittings need to be removed from service for any of the following defects (*Figure 37*):

- Wear, scraping, or abrasion
- Cracks
- Cuts, gouges, nicks, or chips
- Excessive rust or corrosion, meaning rust or corrosion that cannot be easily removed with a wire brush
- An increase in the throat opening of the hook—easy to detect if the hook is equipped with a safety latch because the latch will no longer bridge the throat opening
- A twist in the hook
- An elongation of the hook
- A broken or missing safety latch

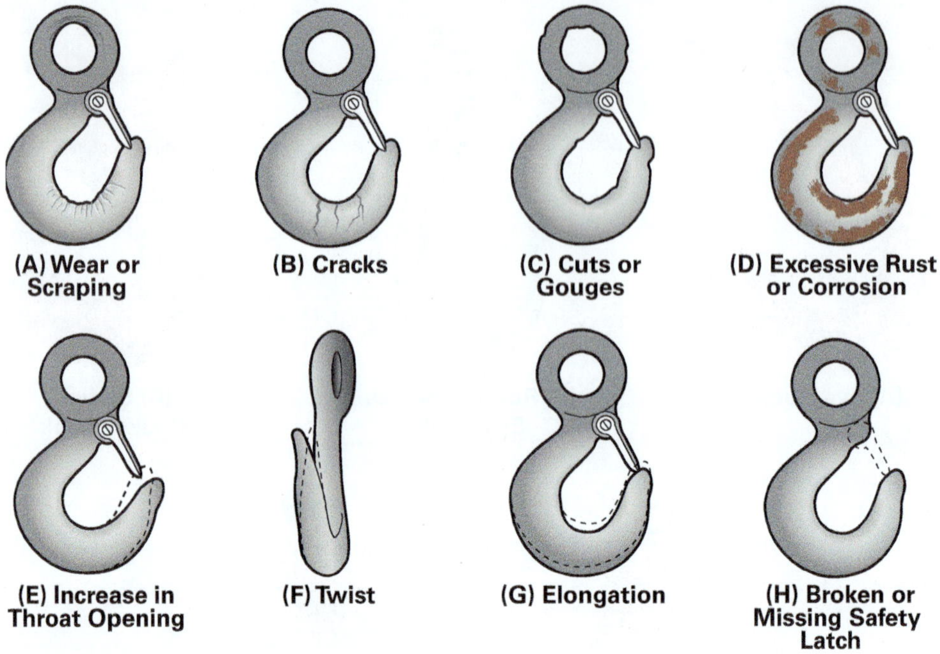

Figure 37 Common rigging hook defects.

1.4.0 Hoists

Block and tackle: A simple rope and pulley system used to lift loads.

A hoist is a device that uses pulleys or gears to provide a mechanical advantage for lifting a load, allowing objects to be lifted that cannot be lifted manually. Some hoists are mounted on trolleys and use electricity or compressed air for power. A simple hoisting mechanism called a **block and tackle** and a more complex hoisting mechanism called a *chain hoist*.

A block and tackle is a simple rope and pulley system used to lift loads. By using fixed pulleys and a wire rope attached to a load, a rigger can raise and lower the load by pulling the rope or by using a winch to lift the load. A block and tackle assembly distributes the weight of an object across the supporting ropes. This distribution reduces the amount of force required to lift an object. In a two pulley system, a worker can lift twice the load compared to what could be lifted without a pulley system (*Figure 38*).

Chain hoists may be operated manually or mechanically. There are three types of chain hoists (*Figure 39*): manual, electric, and pneumatic. Because electric and pneumatic chain hoists use mechanical power, they are known as powered chain hoists. All chain hoists use a gear system to lift heavy loads (*Figure 40*). The gearing is coupled to a sprocket that has a chain with a hook attached to it. The load is hooked onto a chain, and the gearing turns the sprocket, causing the chain to travel over the sprocket and move the load. The hoist can be suspended by a hook connected to an appropriate anchorage point, or it can be suspended from a trolley system (*Figure 41*).

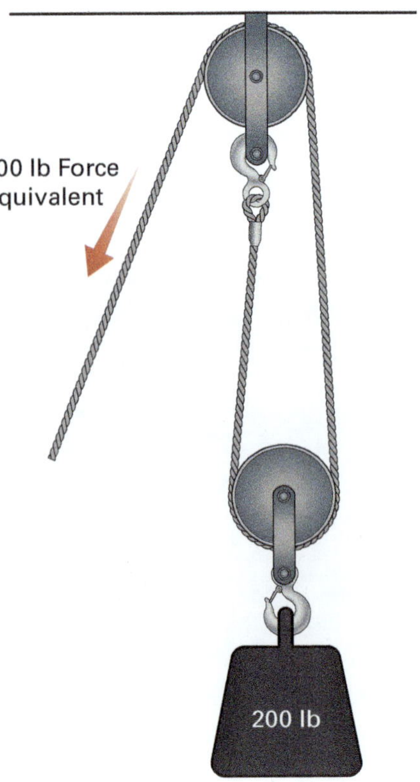

Figure 38 Block and tackle hoist system.

1.4.1 Operation of Chain Hoists

Chain hoists are operated by hand or by electric or pneumatic power. The operating procedures for hand chain hoists and powered chain hoists are as follows:

- *Hand chain hoist* — To use a hand chain hoist, the rigger suspends the hoist above the load to be lifted, using either the suspension hook or the trolley mount. The rigger then attaches the hook to the load and pulls the hand chain drop to raise the load. Depending on which side of the chain drop is pulled, the load will either rise or fall.
- *Powered chain hoist* — To use an electric or a pneumatic chain hoist, the rigger positions the chain hoist on the trolley above the load to be lifted, attaches the hook to the load, and uses the control pad to operate the hoist. Only qualified people may use powered chain hoists.

Figure 39 Types of chain hoists.
Source: Courtesy of JPW Industries, Wilton Tools, JET Tools, Baileigh Industrial

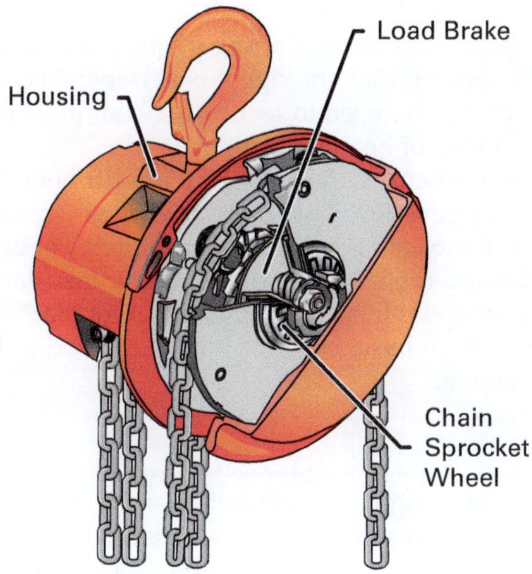

Figure 40 Chain hoist gear system.
Source: Columbus McKinnon Corporation

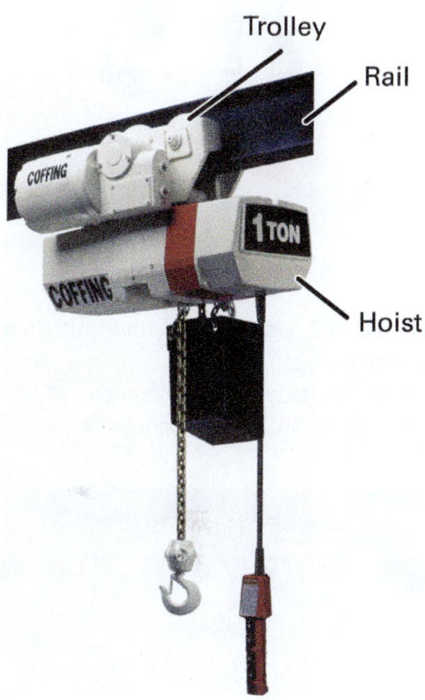

Figure 41 Hoist suspended from a trolley system.
Source: Columbus McKinnon Corporation

1.4.2 Hoist Safety and Maintenance

General safety rules are presented in NCCER Module 93102, *Basic Safety*. In addition to those general safety rules, there are some specific safety rules for working with hoists, including:

- Always wear the appropriate personal protective equipment (PPE) when working with and around any lifting operations. This includes a hard hat, safety glasses, safety shoes, and a high-visibility vest (at a minimum).
- Before attempting a lift, ensure the load is properly balanced and attached to the hoist. Unbalanced loads can slide or shift, causing the hoist to fail.
- Keep gears, chains, and ropes clean. Improper maintenance can shorten their working lives.
- Lubricate gears periodically to keep the wheels from freezing up. A competent person must do all such maintenance work.
- Never perform a lift of any size without proper supervision.

Load control: The safe and efficient practice of load manipulation and using proper communication and handling techniques.

Ratchet Lever Hoist

Figure 42 Use ratchet lever hoists for vertical lifting.
Source: Courtesy of JPW Industries, Wilton Tools, JET Tools, Baileigh Industrial

Figure 43 Single vertical hitch.
Source: Kito Crosby

Tag line: Rope that runs from the load to the ground. Riggers hold on to tag lines to keep a load from swinging or spinning during the lift.

Bull ring: A single ring used to attach multiple slings to a hoist hook.

Never use a common come-along for vertical overhead lifting. Use a come-along only to move loads horizontally over the ground. Be careful not to confuse a come-along with a ratchet lever hoist. Ratchet lever hoists (*Figure 42*) have both a friction-type holding brake and a ratchet-and-pawl **load control** brake. Cable come-alongs only have a spring-load ratchet that holds the pawl in place to secure the cable. If the ratchet and pawl fail, an overhead load falls. Be certain that any device used to lift is designed and specified for that task. Among manufacturers, the terms *come-along* and *ratchet lever hoist* are often used when describing a single item, creating confusion in the marketplace and on the job. In most cases, cable come-alongs are not designed for lifting loads.

1.5.0 Hitches

In a rigging operation, the link between the load and the lifting device is often a sling made of synthetic, alloy steel chain, or wire rope material. The way the sling is arranged to hold the load is called the rigging configuration, or hitch. Hitches can be made using just the sling or by using connecting hardware as well. There are three basic types of hitches:

- Vertical
- Choker
- Basket

One of the most important parts of a rigger's job is making sure that the load is held securely. The type of hitch the rigger uses depends on the type of load to be lifted. Different hitches are used to secure, for example, a load of pipes, a load of concrete slabs, or a load of heavy machinery.

Controlling the movement of the load once the lift is in progress is another important part of a rigger's job. Therefore, the rigger must also consider the intended movement of the load when choosing a hitch. For example, some loads are lifted straight up and then straight down. Other loads are lifted up, turned in midair, and then set down in a completely different place. This section examines how each of the three basic types of hitches is used to both secure a load and control its movement.

WARNING!

All rigging operations are dangerous, and extreme care must always be used. A straight up-and-down vertical lift is as dangerous as a lift that involves rotating a load in midair and moving it to a different place. Only a qualified person may select the hitch for any rigging operation.

1.5.1 Vertical Hitch

The single vertical hitch is used to lift a load straight up. This configuration forms a 90° angle between the hitch and the load. With this hitch, some type of attachment hardware, such as a shackle, is needed to connect the sling to the load (*Figure 43*). The single vertical hitch allows the load to rotate freely. To prevent the load from rotating, some load control method, such as a **tag line**, must be used.

Another classification of hitch is the bridle hitch (*Figure 44*). The bridle hitch consists of two or more vertical hitches attached to the same hook, master link, or **bull ring**. The bridle hitch allows the slings to be connected to the same load without the use of such devices as a spreader beam, which is a stiff bar used when lifting large objects with a crane hook.

The multiple-leg bridle hitch (*Figure 45*) consists of three or four single hitches attached to the same hook, master link, or bull ring. Multiple-leg bridle hitches provide increased stability for the load being lifted. A multiple-leg bridle hitch is always considered to have only two legs supporting much of the load and the rest of the legs balancing it.

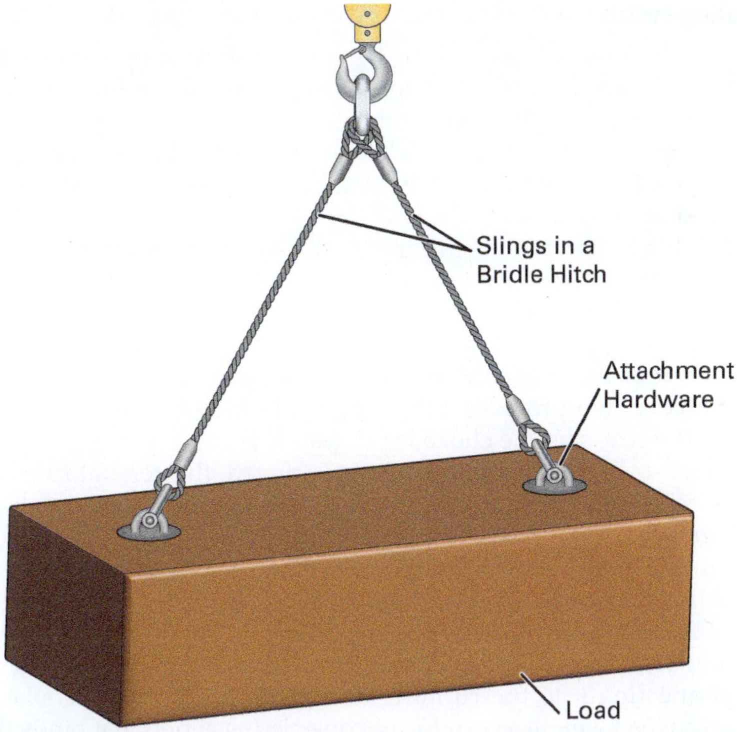

Figure 44 Bridle hitch.

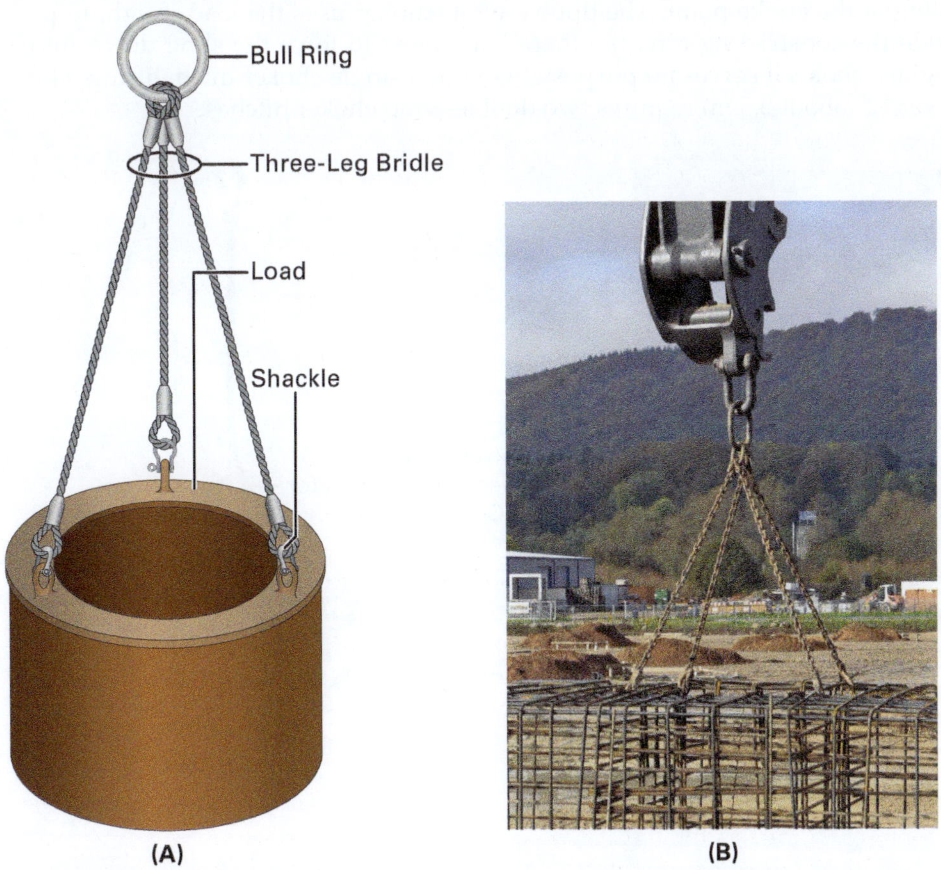

Figure 45 Multiple-leg bridle hitch.
Source: Volvo Construction Equipment (45B)

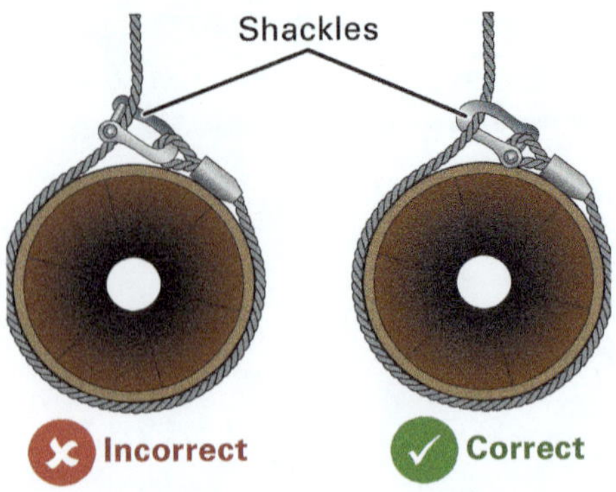

Figure 46 Choker hitches.

1.5.2 Choker Hitch

The choker hitch is used when a load has no attachment points or when the attachment points are not practical for lifting. The choker hitch is made by wrapping the sling around the load and passing it through a shackle to form a constricting loop around it. The shackle used in the choker hitch must be oriented properly, as shown in *Figure 46*. Remember, the pressure of the load should never be placed on the shackle's pin.

The choker hitch affects the sling's capacity, reducing it by a minimum of 25%. This capacity reduction must be considered when choosing the proper sling. Further, note that the choker hitch does not grip the load securely. It is not recommended for loose bundles of materials because they tend to push loose items up and out of the choke. Many beginner riggers mistakenly believe that forcing the choke down provides a tighter grip. In fact, it serves only to drastically increase the stress on the choke leg (*Figure 47*).

When an item over 12' (about 3.5 m) long is rigged, the general rule is to use two choker hitches spaced far apart to provide the stability needed to transport the load (*Figure 48*).

To lift a bundle of loose items, such as pipes and structural steel, or to maintain the load in a certain position during transport, double-wrap choker hitches (*Figure 49*) may be useful. A double-wrap choker hitch is made by wrapping the sling completely around the load and then wrapping the choke end around again and connecting it to the running end like a conventional choker hitch. This enables the load weight to produce a constricting action that binds the load into the middle of the hitch, holding it firmly in place throughout the lift.

Forcing the choke down will drastically increase the stress placed on the sling at the choke point. The double-wrap choker uses the load weight to provide the constricting force, so there is no need to force the sling down into a tighter choke; it serves no purpose. As with a single choker hitch, lifting a load over 12' (about 3.5 m) requires two double-wrap choker hitches.

Figure 47 Choker hitch constriction.

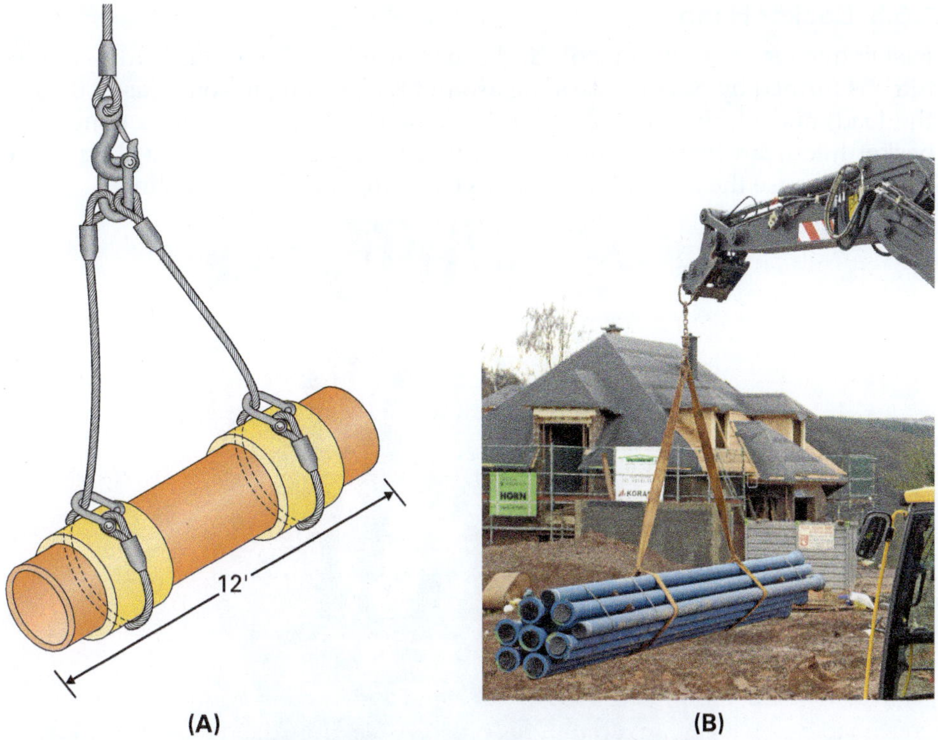

Figure 48 Double choker hitches.
Source: Volvo Construction Equipment (48B)

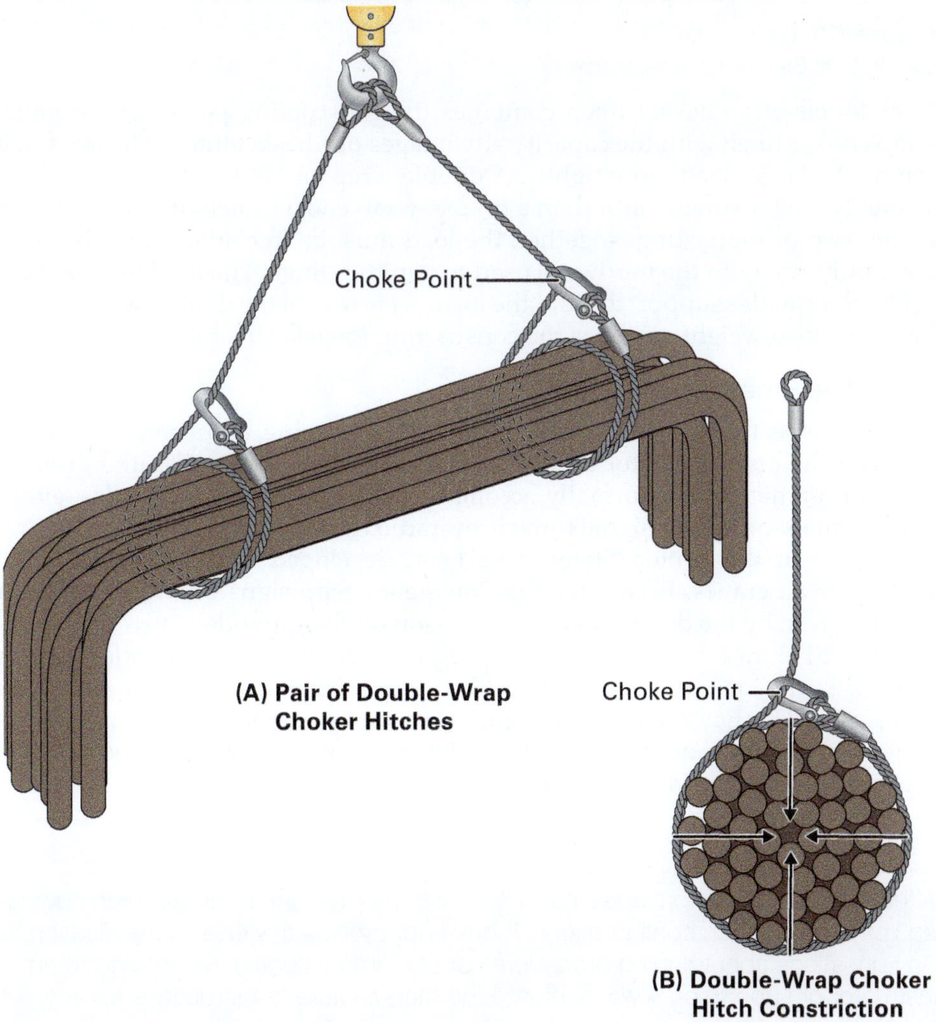

Figure 49 Double-wrap choker hitches.

1.5.3 Basket Hitch

Basket hitches are very versatile and can be used to lift various loads. A basket hitch is formed by passing the sling around the load (or, in some cases, through the load) and placing both eyes in the hook (*Figure 50*). Placing a sling into a basket hitch can increase the sling's capacity (depending on the sling angle). This is because the basket hitch creates two sling legs from one sling.

Figure 50 Basket hitch.
Source: SME Brett Richardson, Starcon

A double-wrap basket hitch combines the constricting power of a double-wrap choker hitch with the capacity advantages of a basket hitch. This means it can hold a larger load more tightly. A double-wrap basket hitch requires a considerably longer sling length than a double-wrap choker hitch. If it is necessary to join two or more slings together, the load must be in contact with the sling body only, not with the hardware used to join the slings. The double-wrap basket hitch provides support around the load. As is true of the double-wrap choker hitch, the load weight provides the constricting force for the hitch.

CAUTION

A basket hitch should not be used to lift loose materials. Loads placed in a basket hitch should be balanced.

1.5.4 The Emergency Stop Signal

A major aspect of rigging safety involves maintaining clear communication between the crane operator and the designated signal person on the ground. This communication is normally accomplished using common signals—either hand signals or verbal signals given by radio. Hand signals used in rigging operations in the United States have been developed and standardized by OSHA for all cranes. Except for the Emergency Stop signal, hand signals can only be given by the designated signal person on the ground.

Figure 51 shows the Emergency Stop signal used in rigging operations. In an emergency, anyone on the ground within sight of the crane operator can give this signal. The Emergency Stop signal is made by extending both arms horizontally with palms down and then quickly moving the arms back and forth by repeatedly extending and retracting them.

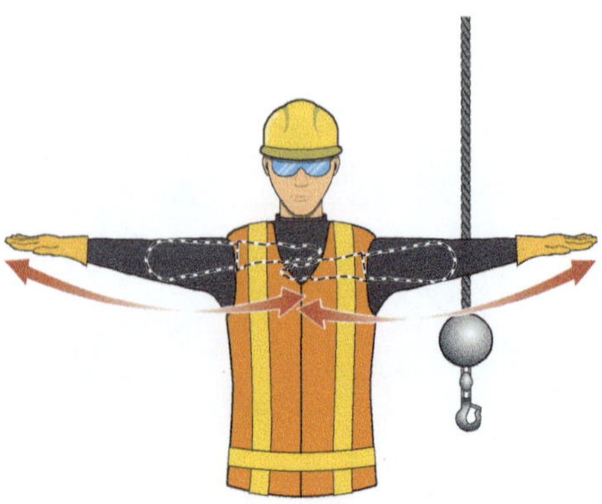

Emergency Stop —
Extend and retract arms rapidly.

Figure 51 Emergency Stop hand signal.

WARNING!

Although this module provides instruction and information about common rigging equipment and hitch configurations, it does not provide any level of certification. Any questions about rigging procedures or equipment should be directed to an instructor or supervisor. Always refer to the manufacturer's instructions for any type of rigging equipment.

1.0.0 Section Review

1. Slings are commonly made of _____.
   a. synthetic fibers, cast iron links, or wire rope
   b. fiberglass strands, exotic metals, or synthetic rope
   c. synthetic material, alloy steel, or wire rope
   d. carbon fiber, stainless steel, or twisted pair cable

2. If the identification tag on a synthetic sling is missing or illegible, the sling *must* be _____.
   a. retagged by a qualified rigger on-site
   b. removed from service immediately
   c. used only for minimal loads
   d. tested on a load to determine its capacity

3. The size of a shackle is determined by the diameter of its _____.
   a. body
   b. shank
   c. eye
   d. pin

4. A type of rigging equipment that uses a pulley system to provide a mechanical advantage for lifting a load is a _____.
   a. fulcrum
   b. hitch
   c. hoist
   d. come-along

5. When a load has no attachment points or when the attachment points are not practical for lifting, it is best to use a _____.
   a. bridle hitch
   b. vertical hitch
   c. multiple-leg hitch
   d. choker hitch

Module 93107 Review Questions

1. Identification tags for slings *must* include the _____.
   a. type of protective pads to use
   b. type of damage sustained during use
   c. color of the tattle-tail
   d. manufacturer's name or trademark

2. The type of wire rope core susceptible to heat damage at relatively low temperatures is the _____.
   a. fiber core
   b. strand core
   c. independent wire rope core
   d. metallic link supporting core

3. Synthetic slings *must* be inspected _____.
   a. once every month
   b. visually at the start of each work week
   c. before every use
   d. once wear or damage becomes apparent

4. An alloy steel chain sling *must* be removed from service if there is evidence that _____.
   a. the sling has been used in different hitch configurations
   b. replacement links have been used to repair the chain
   c. the sling has been used for more than one year
   d. strands in the supporting core have weakened

5. A piece of rigging hardware used to couple the end of a wire rope to eye fittings, hooks, or other connections is a(n) _____.
   a. eyebolt
   b. hitch
   c. shackle
   d. U-bolt

6. A lifting clamp is most likely to be used to move loads such as _____.
   a. steel plates
   b. piping bundles
   c. concrete blocks
   d. plastic tubing

7. Chain hoists are able to lift heavy loads by utilizing a _____.
   a. rope and pulley system
   b. rigger's strength
   c. stationary counterweight
   d. gear system

8. Before attempting to lift a load with a chain hoist, make sure the _____.
   a. hoist is secured to a come-along
   b. load is properly balanced
   c. tag lines are properly anchored
   d. tackle is connected to its power source

9. A hitch configuration that allows slings to be connected to the same load without using a spreader beam is a _____.
   a. double-wrap hitch
   b. choker hitch
   c. bridle hitch
   d. basket hitch

10. To make the Emergency Stop signal that is used by riggers, extend both arms _____.
    a. horizontally with palms down and quickly move both arms back and forth
    b. directly in front and then move both arms up and down repeatedly
    c. vertically above the head and wave both arms back and forth
    d. horizontally with clenched fists and move both arms up and down

Answers to odd-numbered questions are found in the Review Question Answer Keys at the back of this book.

Answers to Section Review Questions

Answer	Section Reference	Objective
Section 1.0.0		
1. c	1.1.0	1a
2. b	1.2.1	1b
3. a	1.3.1	1c
4. c	1.4.0	1d
5. d	1.5.2	1e

User Update

Did you find an error? Submit a correction by visiting **www.nccer.org/olf** or by scanning the QR code using your mobile device.

SCAN ME

MODULE 93108

Basic Communication Skills

Source: iStock

Objectives

Successful completion of this module prepares you to do the following:

1. Describe the communication, listening, and speaking processes and their relationship to job performance.
   a. Describe the communication process and the importance of listening and speaking skills.
   b. Describe the listening process and identify good listening skills.
   c. Describe the speaking process and identify good speaking skills.
2. Describe good reading and writing skills and their relationship to job performance.
   a. Describe the importance of good reading and writing skills.
   b. Describe job-related reading requirements and identify good reading skills.
   c. Describe job-related writing requirements and identify good writing skills.

Performance Tasks

Under supervision, you should be able to do the following:

1. Perform a given task after listening to oral instructions.
2. Fill out a work-related form provided by your instructor.
3. Read and interpret a set of instructions for properly donning a safety harness and then orally instruct another person on how to don the harness.

Overview

Communicating skillfully will help you become a better worker and a more effective leader in construction or agriculture. This module guides active listening and speaking with clarity and purpose. It explains how to use and understand written materials and provides techniques and guidelines that will help you improve your writing skills for emails, business documents, and other important correspondence.

Digital Resources for Agricultural Technology

SCAN ME

Scan this code using the camera on your phone or mobile device to view the digital resources related to this craft.

1.0.0 Communication

Performance Task

1. Perform a given task after listening to oral instructions.

Objective

Describe the communication, listening, and speaking processes and their relationship to job performance.

a. Describe the communication process and the importance of listening and speaking skills.
b. Describe the listening process and identify good listening skills.
c. Describe the speaking process and identify good speaking skills.

Every construction and agricultural professional learns how to use tools. Depending on your trade, your tools could include anything from welding machines and plasma cutters to surveyor's levels and pipe threaders. However, some of the most important tools you will use on the job are not physical. Instead, you will be challenged in the following module to learn and master soft skills. These tools include reading, writing, listening, and speaking abilities.

At first, you may think that these are not really construction tools. However, workplace communication types are very specialized and technical.

Good communication results in a job being done safely. In a way, you are learning another language—a special language that only trained professionals know how to use. Even though you will use professional language that others may not understand, the same communication skills apply to all professions.

The following are some specific examples of why these skills are so important in the construction and agricultural industries:

- *Listening* — Your supervisor tells you where to set up safety barriers, but you missed a spot because you did not listen carefully. As a result, your coworker fell and was injured.
- *Speaking* — You must train two coworkers to do a new task, but you mumble, use jargon they don't understand, and don't answer their questions clearly. Your coworkers do the task incorrectly, and your team must work overtime to fix simple mistakes.
- *Reading* — Your supervisor tells you to read the manufacturer's basic operating and safety instructions for the new drill press before you use it. You don't really understand the instructions, but you are afraid to ask them and seem inexperienced. Doing what you thought was correct, you damage the drill press, resulting in expensive equipment loss and reduced production while your company acquires a replacement.
- *Writing* — Your supervisor asks you to write up a material takeoff (supply list) for a project. You rush through the list and don't check what you've written. As a result, the supplier delivers 250' of PVC piping instead of 25'.

These are all hypothetical scenarios, but as you can imagine, good communication directly correlates to safety, schedules, and budgets. Much like a physical tool in your personal toolbox, the ability to communicate verbally and in writing is something that you can take with you to any job.

Did You Know?

Safety Standards Prevent Accidents

Many accidents are the result of not following safety guidelines. According to a study by the US Occupational Safety and Health Administration (OSHA), 39% of crane operator deaths in the United States resulted from electrocution caused by contact with electrical power lines over a 10-year period. This was the single largest cause of death in the study. How many of those accidents could have been prevented if both the employer and employees had followed OSHA standards and regulations?

1.1.0 The Communication Process

There are two basic steps to clear communication (*Figure 1*). First, a sender transmits a spoken or written message through a communication channel such as a meeting, phone call, two-way radio, or email. When the receiver gets the message, they figure out what it means by listening or reading carefully. If anything is unclear, the receiver gives the sender feedback by requesting more information and clarification.

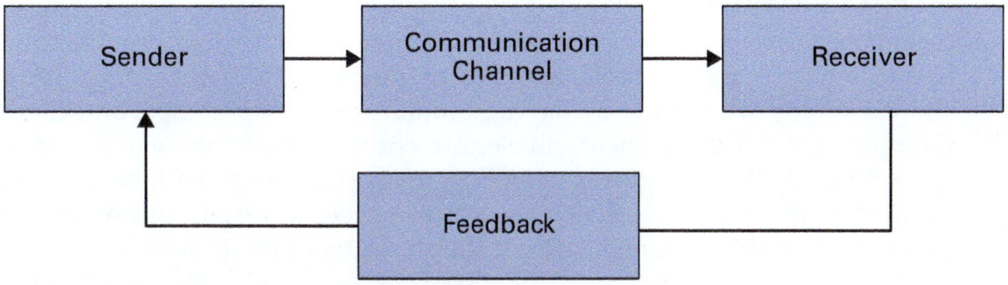

Figure 1 The communication process.

This process is called *two-way communication*, and it is the most effective way to ensure everyone understands what's going on. It sounds simple, so why is good communication so hard to achieve? Physical and mental noise can obstruct our comprehension when we try to communicate. The following includes common examples of communication noise you may experience on a construction or agricultural site:

- The sender uses work-related words, or **jargon**, that the receiver does not understand.
- The sender does not speak clearly.
- The sender's written message is disorganized or contains spelling and grammatical errors.
- The sender is not specific and communicates abstract or vague ideas.
- The sender does not comprehend the "why" of the message.
- The receiver is tired, distracted, or inattentive.
- The receiver has poor listening or reading skills.
- Actual noise from heavy equipment and power tools on the construction site makes it physically challenging to hear a message.
- There is a mechanical problem with the communication equipment, such as static on a radio or poor cell coverage.

Jargon: Specialized terms used in a specific industry.

1.1.1 Nonverbal Communication

People often communicate verbally. If a mechanic asks for a ratchet set, they expect to be given a ratchet set because the words have a shared meaning between the speaker and listener. However, humans are also talented at pattern recognition and often communicate with or understand each other through body language and other forms of **nonverbal communication**.

Nonverbal communication: All communication without words, including appearance, personal environment, use of time, and body language.

Imagine that you are talking to a doctor in a clinic. The whole time you are talking, they are yawning, checking their cell phone messages, and cleaning their fingernails. After you finish, they write out a prescription. Would you trust the diagnosis and the value of the prescription? Their nonverbal communication conveyed that they were bored and inattentive. Perhaps the doctor missed the most important thing you said about your condition, including an allergy to certain medications.

Similarly, you can express feelings and attitudes in various ways without intention. You may show nervousness by fidgeting in your chair or fiddling with your hands. You may show anger by raising your voice, folding your arms, or furrowing your eyebrows. You may show you are happy to see someone by smiling widely and giving them a warm handshake.

There are several basic categories of nonverbal communication:

- *Grooming* — Generally, people who maintain a clean appearance have more successful careers because it communicates a sense of self-discipline and awareness. Shaving or keeping a trimmed beard could make the difference between getting promoted at work and being passed up by less experienced coworkers.
- *Dress* — Dress appropriately and neatly. This doesn't necessarily mean to dress formally—working on a hot rooftop in a suit would be ridiculous. Instead, observe and follow the accepted standards for your workplace.
- *Condition of one's personal environment* — People make their spaces their own by arranging things to their personal preference. An office worker could show a sloppy work ethic by having binders, pens, paper clips, and candy wrappers strewn about their cubicle. A technician could show a lack of responsibility by having gloves, coupons, magazines, and fast food trash on the dashboard of their work vehicle. Whoever sees those environments subconsciously makes judgments about the character of the person.
- *Use of time* — People show respect and care by arriving on time or early to their scheduled events. Workers who show up late may be considered lazy or inconsiderate by those around them. Furthermore, they may hinder the entire organization's productivity by making others wait for them. Not only should you arrive on time, but you should also try to end on schedule when you hold meetings. Doing so shows a humility that others will appreciate.
- *Facial expressions* — One of the main ways humans read and express emotions is through facial expressions. A smile can express interest and excitement, while a frown can express displeasure or pain. An expressionless face expresses an emotion too—perhaps boredom or the desire to get away from the situation. Direct eye contact can express interest, understanding, intelligence, and confidence. On the other hand, a lack of eye contact could show inattention, a lack of confidence, or deceit. This doesn't mean staring rudely, as too much eye contact could give the impression of a challenge or intimidation. Different cultures have different rules, so there is no solid rule for what amount of eye contact is appropriate.
- *Posture and gestures* — Slouching may feel comfortable, but it may also give off an air of laziness or disinterest. Likewise, folding your arms in front of you may make you feel warmer on a cold day, but it will make you seem distant from others and unwilling to talk. A confident and powerful posture shows other people you are confident and powerful. Assume a powerful posture, even if you feel shy. Research shows that people who maintain a powerful posture for a few minutes—such as a wide-legged athletic stance with back straight—experience less stress and behave more boldly when doing a task.
- *Physical distance* — Various cultures have different acceptable personal distances. Also, different social situations call for different personal ranges. People stand closer to their close friends and family members than they do to their business associates. One rule of thumb is to let the other person choose the distance.

Regarding safety and the accuracy of information communicated on the job, what you say plays a greater role than how you say it. However, beyond that, your nonverbal communication habits have a major effect on the quality of what is said and how it is received.

Figure 2 Teaching and learning are often accomplished by speaking and listening.
Source: iStock

A Sense of Humor Goes a Long Way

Maintain a sense of humor! A good sense of humor will get you through many situations. You should always be professional but don't take yourself too seriously. This means speaking well and conveying the proper professional attitude but using work-appropriate humor to diffuse tension and relieve frustration over things that aren't working properly. Never play practical jokes as they can lead to accidents and injuries on the job.

1.1.2 Listening and Speaking Skills

Every day on the job can be a learning experience. The more you learn, the more you can help other team members develop their skills (*Figure 2*). Verbal communication is a very effective method of teaching and learning. The following are some of the ways that verbal teaching and learning take place on the job:

- *Giving and taking instructions* — One worker may read the steps in a calibration process while a second worker accomplishes the task.
- *Offering and listening to presentations* — Equipment manufacturers may visit the jobsite or offices to provide operating instructions for a new piece of equipment.
- *Participating in team discussions* — Safety is often discussed among teams in prework meetings; offer your input and recommendations to improve safety protocols.
- *Talking with your coworkers and supervisor* — Listen carefully as they speak without distracting yourself by thinking of a response too quickly. Taking a slight pause in the conversation to prepare a response after a speaker is finished encourages others to listen more carefully to you.

You will probably do more listening than speaking at this early stage in your career. You may be wondering why it is so important to be a good listener. The answer is simple: experience. People learn by listening, not by speaking. You are only beginning to learn how the construction industry works; there is much to learn! Teachers, supervisors, and experienced workers can guide you to ensure you are learning what you need to know (*Figure 3*).

Figure 3 Your supervisor can help you learn what you need to know.
Source: iStock

1.2.0 Active Listening on the Job

Active listening: A process that involves respecting others, listening to what is being said, and understanding what is being said.

You might think that listening automatically happens when someone says something and someone else hears it. However, real listening—the process of not only hearing but also understanding what is said—is active. You must be involved and pay attention to really listen.

In other words, understanding begins with **active listening**. You must develop good listening skills to be able to listen actively. This section presents some tips and suggestions that you can use to develop good listening skills.

You should understand the possible consequences of not listening. Poor listening skills can cause mistakes that waste time and money (*Figure 4*) and may even result in injury or fatalities. Recognize that other people may have important things to say. Even if what they say is boring, they may mention something vitally important, like what to do when an alarm goes off.

Figure 4 Listen carefully and ask questions to achieve mutual understanding.
Source: iStock

Avoid distractions. If a work-related conversation or presentation occurs, be in the room listening, not darting around from place to place in your mind. If it helps you to stay focused, keep a pad of paper with you and jot down notes. Even scribbling and doodling have been shown to increase comprehension as you listen. When you are sidetracked by thoughts of something you need to do later, jot down some notes about it. Then refocus on the conversation at hand.

One way to stay focused is to show that you are listening with your **body language**, including facial expressions, posture, and gestures (*Figure 5*). While you listen, maintain eye contact and acknowledge what is being said by nodding or making appropriate sounds to show that you are paying attention.

Body language: A person's facial expression, physical posture, gestures, and use of space, all of which communicate feelings and ideas.

While the other person is speaking, do not interrupt or criticize. Instead, wait until they have finished and then ask plenty of questions, especially if they gave incomplete or unclear instructions. Asking questions can minimize misunderstandings and maximize your memory of the content. Of course, your mind will never remember perfectly, so take notes on what is said, especially if the content is technical.

Finally, end a conversation by **paraphrasing** what you heard to the conversation partner. Paraphrasing is the act of repeating what you heard in your own words. Using your own words, you can check the accuracy of your understanding. Therefore, it is important not to repeat exactly what the person said. The original speaker must listen carefully to ensure the paraphrased information is accurate.

Paraphrasing: Expressing something heard or read using different words.

Imagine how many hours of wasted work you can avoid by confirming what you think you hear. For example, your supervisor may tell you, "Recalibrate the machines according to the specifications on this paper." You can respond by saying, "So you want me to recalibrate these three machines in this building according to the chart posted here." Then they can reply with, "Oh, no! I mean, recalibrate the machines like these in the building next door, and follow the chart on the back of this paper since they are slightly different models."

Providing a simple summary in your own words of what was said could save time, money, or even someone's life. Paraphrase instructions whenever you can. It completes listening just as installing a roof completes building construction.

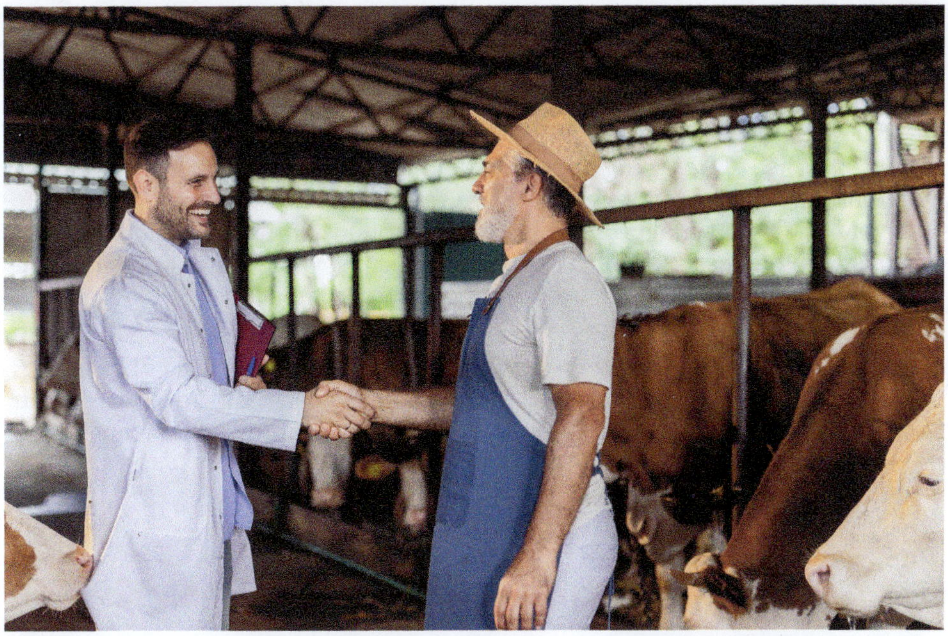

Figure 5 Example of attentive body language.
Source: iStock

Figure 6 Body language can indicate a miscommunication.
Source: iStock

1.2.1 Barriers to Listening

Listening well takes some work on your part. However, even when you have mastered effective listening skills, you will still have to overcome some barriers that keep the message from getting through. Just like the warning signs on a construction site that indicate hazards to be avoided, some signs can visibly indicate problems in listening and understanding (*Figure 6*). Some of those barriers are listed below, along with tips to overcome them:

- *Emotion* — People tend to stop actively listening when angry or upset. Try counting to 10 or asking the speaker to excuse you for a minute. Get a drink of water and take some deep breaths. When you are ready, come back and focus your mind on the task at hand.
- *Boredom* — Maybe the speaker is dull or overbearing. Maybe you think you know it all already. There is no easy tip for overcoming this barrier. You must force yourself to stay focused. Keep in mind that the speaker has important information you need to hear.
- *Distractions* — Anything from too much noise on the site to problems at home can steal your attention. If the problem is noise, ask the speaker to avoid it. If a personal problem keeps you from listening, concentrate harder on staying focused. In some cases, it may help if you explain to your supervisor why you are having trouble concentrating.
- *Your ego* — Do you finish people's sentences for them? Do you interrupt others a lot? Do you think about what you will say next instead of listening? Your ego may be preventing you from effectively listening. Be aware of your ego and try to tone it down a bit so you can get the information you need.

Ten Tips for Dealing with Conflict

Conflict can cause emotions to heat up, ruin relationships, and decrease productivity. Even if you don't care to befriend someone who bothers you at work, you still must get along to accomplish your tasks. Here are some tips for dealing with conflicts at work:

- *Deal with it quickly*— While the problem is still small, it will be easier to solve.
- *Set aside time to discuss the problem*— This will help emotions cool, providing a neutral emotional atmosphere. You will also have more time to discuss the issue fully.
- *Discuss the problem with civility*— Getting angry may make the other person angry, escalating the situation.
- *Ask questions first*— If a coworker has a reason for doing something that makes you angry, you should know what it is. Simply knowing why they behaved that way could solve the problem.
- *Never say "never"*— Avoid saying "always," too. "You never help me" and "You always criticize me" are both statements that are likely untrue and may upset the other person.
- *Be specific; don't generalize*— Notice the difference between "You are a careless person who hurts the whole team" and "On our last two jobs, you forgot to bring the key to the building. We all had to wait while you went back for it. Is there anything we can do to ensure this doesn't happen again?"
- *Do not bring up things from the past*— If it is unrelated to the current situation, bringing up something that a coworker did last year would be unproductive.
- *Put yourself in the other person's shoes*— Understand what the other person needs to come out of the conflict happy. People need to keep their sense of honor intact. If you publicly insult someone or force them with overwhelming use of logic or authority, you will implant a grudge that will inevitably hurt you later. Beyond giving someone the honor they need, see if you can also compromise on some of the issues you disagree on.
- *Praise the person*— This will show that you are not attacking their character and help focus attention on the issue rather than the interpersonal tension.
- *Take responsibility*— Sincerely accepting some blame for the issue can make a huge difference. However, it is important to avoid doing so in a self-righteous way.

1.3.0 Speaking on the Job

Although you can use the skills presented in this module to give a speech if necessary, speaking skills do not refer merely to your ability to give a speech or make a presentation to a group of people. It also refers to your ability to communicate effectively, one-on-one, to others on the job daily.

Effective listening depends on effective speaking. After all, you cannot be expected to understand what has not been made clear to you. Look at the following examples of sentences spoken by one worker to another. Which one is the clearest and most effective? Which one would you like to hear if you were the listener?

- "Hand me that tool there."
- "Hand me the grinder on that bench."
- "Hand me the four-inch angle grinder on the bench behind you."

The third example has enough information to identify the correct tool and location. You do not have to ask the speaker, "Which tool? Where is it?" You will not accidentally give the other worker the wrong tool. As a result, time is not wasted trying to clear up confusion. When someone has to stop what they are doing and re-explain something because it was not clear the first time, the job is not getting done. Time lost this way can add up quickly.

One of the best ways to learn to speak effectively is to listen to someone who speaks well. Think about what makes that person such an effective speaker. Is it the person's choice of words, their body language, or their ability to make something complex sound simple? Keep the following things in mind when you speak:

- Think about what you will say beforehand, but not at the expense of actively listening.
- Introduce your topic with a brief summary, share the detailed information, and end with a paraphrased summary of what you said. Remember the saying, "Tell them what you will tell them, tell them, then tell them what you told them."
- As with writing, take time to organize your ideas logically.
- Choose an appropriate place and time. For example, if you need to give detailed assembly instructions to your team, pick a quiet place and do not hold the meeting just before lunch or the end of the shift.
- Do not talk on the phone, send text messages, or listen to music while communicating with the work crew.
- Encourage your listeners to take notes if necessary.
- Do not overexplain if people are already familiar with the topic.
- Always speak clearly and maintain eye contact with the people you are speaking to.
- When using jargon, ensure everyone knows what the terms mean.
- Give your listeners enough time to ask questions and take the time to answer questions thoroughly.
- When you are finished, ensure everyone understands what you are saying.

Remember these things when you speak; they will make a difference.

1.3.1 Placing Phone Calls

Today, cell phones allow you to make and receive calls from just about anywhere. A cell phone can be a useful tool on the jobsite, but keep the following guidelines in mind:

- Cell phones can distract you from your job, so avoid making or answering personal calls while working.
- Wait until a designated break time to make or receive calls.
- Do not operate cell phones when they pose a safety hazard, such as while operating a piece of machinery, a power tool, or a vehicle.
- Be aware of the regulations regarding cell phone use at your workplace. While some companies allow cell phone use, others do not even allow a phone to be in your possession. Companies make these rules to avoid accidents and wasted time.
- Recognize that cell phone cameras can potentially threaten intellectual property. Workers could photograph secret equipment or data and sell the photographs, causing the company to lose its competitive edge. This is especially worth noting on projects for the government or military. The military protects its secrets for good reason and may ban employees and contractors from using cell phones in certain areas.

When you speak to people face-to-face, you can see them and judge how they react to what you say. When you are on the telephone, you don't have these clues, so effective speaking is even more important. When making a call, keep the following points in mind:

- Start by identifying yourself and ask who you are speaking to.
- Speak clearly and explain the purpose of your call (*Figure 7*).
- Take notes to help you remember the conversation later.

Figure 7 A business call requires speaking clearly.
Source: iStock

If you leave a message for someone, remember the following:

- Keep it brief.
- Prepare your message beforehand so you will know what to say.
- Be sure to leave a number where you can be reached, along with the best time to reach you.

1.3.2 Receiving Phone Calls

How you answer a phone call is just as important as how you place a phone call. Remember to be professional and courteous when answering your phone because you don't know who will be on the other end. When you receive a phone call, remember the following guidelines:

- Don't just say "hello." Identify yourself immediately by giving your name and the company name.
- Don't keep people on hold for long. Instead, ask the caller if you can call back later.
- Transfer calls courteously and introduce the caller to the recipient.
- Keep your calls brief.
- If the call is of a personal nature, continue working and do not answer. There are other opportunities throughout the day when you can return the call without interfering with the job or creating a safety hazard.

1.0.0 Section Review

1. Making meetings last longer than scheduled is a good way to show that you are a diligent person that wants to encourage others to work hard too.
   a. True
   b. False

2. If the content of a speech or conversation is technical in nature, such as including numerical data, be sure to _____.
   a. take notes
   b. ask others to summarize
   c. check to see if the information is true
   d. have the speaker repeat the key information

3. When is the *best* place and time to give assembly instructions to the team?
   a. In a quiet place when they are fresh and not too tired
   b. In the lunchroom while they are eating
   c. In the work area at the end of their shift
   d. In the break room while people are getting snacks and watching TV

4. Which of the following is banned in some workplaces because it can cause wasted time, accidents, and the theft of intellectual property?
   a. Pagers
   b. Watches
   c. Hearing aids
   d. Cell phones

5. When making a phone call, start by _____.
   a. making small talk
   b. explaining the purpose of your call
   c. identifying yourself
   d. asking who you are talking to

2.0.0 Reading and Writing

Performance Tasks

2. Fill out a work-related form provided by your instructors.
3. Read and interpret a set of instructions for properly donning a safety harness and then orally instruct another person on how to don the harness.

Objective

Describe good reading and writing skills and their relationship to job performance.

a. Describe the importance of good reading and writing skills.
b. Describe job-related reading requirements and identify good reading skills.
c. Describe job-related writing requirements and identify good writing skills.

Imagine that you wake up one morning and head to work just like any other day, except for one difference: written language has not yet been invented. You grab a diagram (without notes) for installing a heating system but do not understand one part of the drawing. You head back to the office to ask the engineer what he meant. He starts to explain, but then he cannot clearly remember what he had discussed with the client. The discussion with the client was very long, and there was no formal contract or even an informal list of what the client wanted. Therefore, you and the engineer must go to the client's office to resolve the issue. This would take a tremendous amount of time and effort, resulting in very little work getting done. Fortunately, we can utilize written language, allowing us to be efficient and productive.

Reading and writing are not just for academics. Almost all people read or write dozens of times daily, regardless of whether they are farmers, plumbers, professors, statisticians, or stay-at-home parents.

2.1.0 The Importance of Reading and Writing Skills

Like any other, agricultural and construction industries depend on written materials to carry on business, from routine office paperwork to construction drawings and building codes. Written documents allow workers to follow instructions accurately, help project managers ensure work is on schedule, and enable the company to meet its legal obligations. As a construction professional, you need to be able to read and understand the written documents that apply to your work, whether they are printed in a book, letter, or on a computer (*Figure 8*). As you gain professional experience, you will eventually be responsible for writing documents.

Figure 8 A team reviews instructions prior to starting work.
Source: iStock

Other forms of written construction documents that you are familiar with include textbooks, maintenance manuals, and codebooks. Many of these resources are now online (*Figure 9*). A textbook is an instructional document that contains information that a reader needs to know to carry out a task or a series of tasks. A codebook is a guide that provides the codes and standards for certain construction and agricultural areas, such as electrical and building codes. Instruction manuals for tools and installation manuals from manufacturers are other common examples of documents used on the job.

Workers must be able to read and understand these important documents. When a person's reading skills are poor, it may cause them to misinterpret technical data. This can lead to errors and accidents. This section reviews some techniques that you may find helpful in reading and writing construction documents.

Figure 9 Many instruction manuals and other resources are now online.
Source: iStock

2.2.0 Reading on the Job

When your company issues new safety guidelines or the latest version of the local building code is published, you will need to understand these documents to perform your job safely and effectively. Both documents may contain new standards that affect how you do your job. Reading is an essential skill for any craft professional, farmhand, or ranch hand.

The following are some typical examples of things craft professionals read on the job:

- Safety instructions
- Construction drawings
- Manufacturer's installation instructions
- Materials lists
- Signs and labels
- Work orders and schedules
- **Permits**
- Specifications
- **Change orders**
- Industry magazines and company newsletters
- Emails

Permits: Legal documents that allow a task to be undertaken.

Change orders: A written order by the owner of a project for the contractor to make a change in time, amount, or specifications.

This section reviews some simple tips and techniques to help you read faster and more efficiently on the job. These tips and techniques will also help you become a better writer. With practice, these guidelines will become second nature.

You should always have a purpose in mind when you read. This will help you find the information faster. For example, suppose you are looking for a specific installation procedure in a manual. Do not waste time reading other parts of the manual; go straight to the section that deals with that procedure. Read slowly enough to concentrate on what you are reading. Technical publications are packed with detailed information, so you do not want to miss anything because you rushed.

Most books have features to help you locate information, including the following:

- **Table of contents**
- **Index**
- **Glossary**
- **Appendixes**
- **Tables** and **graphs**

A table of contents is a list of chapters or sections, usually located in the front of a book (*Figure 10*). This is the first place experienced readers look when they want to get a general overview of a book, or to find some specific information.

An index is an alphabetical listing of topics, along with page numbers to show where those topics appear. Indexes also list information in documents other than books, such as construction drawings (*Figure 11*). Suppose you want to look up information about transistors. While the table of contents will show you the chapters that focus on transistors, the index will show you every page mentioning transistors—even if the page is about another topic and only briefly mentions transistors.

A glossary is an alphabetical list of terms used in a book, with definitions for each term. It is usually located at the back of a publication.

An appendix is a source of additional information at the end of a section, chapter, or book. Appendixes are separate because their information is more detailed or specific than the rest of the book. If the information were placed in the main part of the book, it could distract readers or slow them down.

Tables and graphs help illustrate important facts and figures in an easily comprehensible way. Tables usually contain text or numbers, while graphs use images or symbols to convey meaning.

Given the importance of written language to your job, you must pay close attention when reading. This section details some techniques to maximize your comprehension of written material.

The noise and activity on a farm or construction site can easily distract you. While you are reading, try to avoid distractions. Radios, cell phones, televisions, conversations, nearby machinery, and power tools can all affect your concentration.

Take notes or use a highlighter to help you remember and find important text. Your notes on the blank pages at the beginning of a book can be invaluable. If you find the information useful now, you may want to look it up again later, so note the topic and record the page number. Also, highlighting the main sentence of an especially useful section will help you find the section quickly when you flip through the book.

When reading an unfamiliar book, skim or scan the chapter titles and section headings before you start to read. This will help you organize the material in your mind. Look for visual clues that indicate important material. For example, bold **font** or **italics** indicate important words or information. Bold fonts are letters and numbers that are heavier and darker than the surrounding text. Italics are letters and numbers that lean to the right (*like this*) rather than stand straight up.

When reading instructions or a series of steps for performing a task, such as turning on a welding machine, imagine yourself performing the task; you may find the steps easier to remember. Be sure to read all the directions before you begin to follow them. You will then understand how all the steps work together.

Always reread to make sure you understand dense or complex reading material. Also, take a break every now and then. If you are getting tired or frustrated, put the material down and do something else to give your mind a break.

Table of contents: A list of book chapters or sections, usually located at the front of the book.

Index: An alphabetical list of topics and the page numbers where each topic appears.

Glossary: An alphabetical list of terms and definitions.

Appendixes: Source of detailed or specific information at the end of a section, a chapter, or a book.

Tables: A way to present important text and numbers so they can be read and understood quickly.

Graphs: Information shown as a picture or chart. Graphs, including line graphs and bar charts, may be represented in various forms.

Font: The type style used for letters and numbers.

Italics: Letters and numbers that lean to the right rather than stand straight up.

CONTENTS

Module 93101 Build Your Future in Construction

Figure 10 A table of contents lists the chapters or sections in a book.

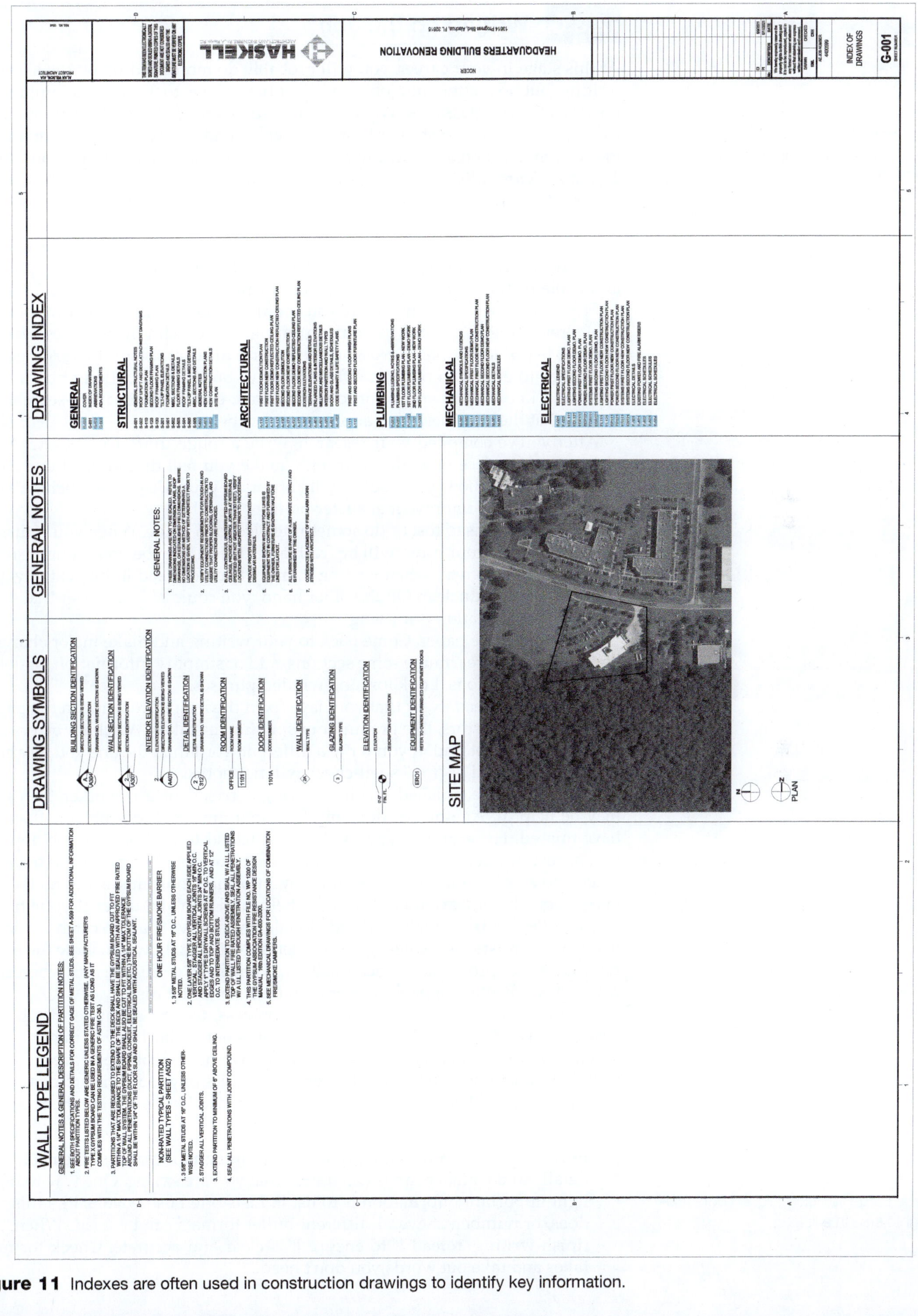

Figure 11 Indexes are often used in construction drawings to identify key information.

2.3.0 Writing on the Job

Punch lists: Written lists that identify deficiencies requiring correction at completion.

Memos: Informal written correspondence. Another term for *memorandum* (plural: *memoranda*).

At this stage in your career, you will probably do more reading on the job than writing. But no matter your job, you'll eventually have to write or edit important documents and messages. Writing skills are essential if you want to succeed. Craft professionals write work orders, health and safety reports, **punch lists**, **memos**, emails, and a whole range of other documents as part of their job (*Figure 12A* and *Figure 12B*).

Writing skills will allow you to perform these tasks as you gain experience on the job. They will help you move into positions of greater responsibility. It is never too early to start brushing up on your writing skills so you will be ready when the opportunity comes. This section reviews the writing process and explains the characteristics of good writing.

First, do some prewriting before you begin to write to organize your thoughts. One prewriting technique is brainstorming, which simply involves writing down a list of ideas. Next, cross out the worst ideas and logically arrange the best ideas to create an outline. After you have built a foundation of basic concepts and know what you want to write about, do the necessary research and ensure that the statements are appropriate for the situation.

After you've compiled your research, write a rough draft. Do not attempt to write a masterpiece now; doing so may make you feel discouraged and overwhelmed. Just follow your outline, incorporate your ideas, and include your research. You can improve it at a later point.

The third step is to rest or do something else for a while. When you return to the project later, your mind will be fresh and critical. Maybe you thought something was easy to read when writing it, but when you read it the next day, you may see that it is unclear. On the other hand, you could also discover that some important information is missing.

Next, revise the paper. Come back to your writing and make major changes. You may need to rearrange some sections, add or simplify information, or delete unnecessary sections. Look for long-winded phrases.

Finally, proofread. Now that you have fixed the major problems in the paper, look for smaller errors, like grammar or spelling mistakes. Watch for errors that spell-check cannot find. If you write a different word than the one you meant to, like "four" instead of "for," spell-check will not notice.

After you have finished writing, asking a friend or a coworker to read it may be helpful. Another reader can often spot mistakes or problems that you have missed. For example, you might ask a friend to review your job application or resume. You can ask a coworker to check over a work permit or material takeoff to ensure you haven't left out any information or made any math mistakes. Don't be embarrassed to ask for help, and don't be upset if your friend or coworker spots a mistake. Professionals are always glad to catch mistakes before those mistakes create bigger problems. Note that how you view the material may affect proofreading accuracy too. Writers often find errors when they read a piece printed on paper rather than on a computer display.

Now that you understand the writing process, we can examine the characteristics of good writing. Being a good writer is not as hard as you might think. You do not need to master English composition or have a college degree to write well. You must remember these five characteristics of good writing: clear, concise, correct, complete, and considerate of the reader. Let's examine them one by one.

Bullets: Large vertically aligned dots that highlight items in a list.

- *Be clear* — Make your main point easy to find and include all the necessary details so the reader can understand what you are saying. One way to highlight important information is to list it. Lists often use **bullets**, like this one does, or numbers. Several different bullet formats can be used. When you finish writing, reread it to ensure it is clear and accurate. Check for mistakes and take out words you don't need.

Burning - Welding - Hot Work Permit

Valid from ____ to ____, ________ Master Card No.______________
(am/pm) (am/pm) DATE

1. Work Description
Equipment Location or Area ______________________________________
Work to be done:

2. Gas Test	☐ None Required			
	☐ Instrument Check	Test Results	Other Tests	Test Results
	☐ Oxygen 20.8% Min			
	☐ Combustible % LFL			

Gas Tester Signature Date Time

3. Special Instructions ☐ None ☐ Check with issuer before beginning work

4. Hazardous Materials ☐ None What did the line / equipment last contain?

5. Personal Protection ☐ Standard Equipment: welder's hood with long sleeves; cutting goggles

☐ Goggles or Face Shield ☐ Respirator ☐ Forced Air Ventilation

☐ Standby Man ☐ Other, specify: ________________

6. Fire Protection ☐ None Required ☐ Portable Fire Extinguisher

☐ Fire Watch ☐ Fire Blanket ☐ Other, specify: ________________

7. Condition of Area and Equipment

Required Yes	No	THESE KEY POINTS MUST BE CHECKED
		a. Lines disconnected & blanked or if disconnecting is not possible, blinds installed?
		b. Lines steamed, purged, or otherwise properly cleared of combustibles?
		c. Area and equipment satisfactorily clean of oil or combustibles?
		d. Trenches, catch basins & sewer connections properly covered or sealed?
		e. Immediate area and/or area under the work barricaded or roped off?
		f. Adjoining equipment & operations checked to have any effect on the job?
		g. Area fire suppression (fire water and sprinkler system) in service?

Comments

Figure 12A A hot work permit is a typical written product in a construction project (1 of 2).

Burning - Welding - Hot Work Permit

8. Approval	Permit Authorization			Permit Acceptance		
	Area Supv.	Date	Time	Maint. Supv./Engineer Contractor Supv.	Date	Time
Issued by						
Endorsed by						
Endorsed by						

9. Individual Review

I have been instructed in the proper Hot Work Procedures

	Signed	Signed
Persons Authorized to Perform Hot Work	__________	__________
	__________	__________
	__________	__________
	__________	__________
	__________	__________
Fire Watch	__________	__________

10. Job Completion

☐ Yes ☐ No Is the work on the equipment completed?

☐ Yes ☐ No Has the work site been cleaned and made safe?

Worker answering above questions ____________________

Issuer's Acceptance ____________________

Forward to Production Superintendent within 7 days of job completion

Figure 12B A hot work permit is a typical written product in a construction project (2 of 2).

- *Be concise* — More words do not equal better writing. Instead of writing, "To prevent unwanted electrical shocks, workers should make sure they disconnect the main power supply before work commences," write, "To avoid being shocked, disconnect the main power supply before starting." Instead of writing "at the present time," write "now." It is also important to consider the intended audience's reading level.
- *Be correct* — Accuracy is vital in the construction industry. For example, if you fill out a form to order supplies from a manufacturer's catalog, use the correct terms and quantities. When writing material takeoffs or other documents that will be used to purchase equipment and assign workers, you need to get it right. Get in the habit of reviewing your writing, whether it's a handwritten note, a supply list, a permit, or an email. Errors cost time and money.
- *Be complete* — When writing something for someone else, even if it is short and simple, always ask yourself the following questions:
  - Have I identified myself to the reader?
  - Have I said why I'm writing this?
  - Will the reader know what to do and, if necessary, how to do it?
  - Will the reader know when to do it?
  - Will the reader know where to do it?
  - Will the reader know whom to call with questions?
- *Be considerate of the reader* — Not everyone who reads what you write will be a professional in the same trade that you are in, and not everyone will understand the same words and concepts that you do. For example, you may need to write a report to a client to describe a technical problem. In that report, you must avoid jargon and use words that uninformed people will understand. Similarly, not every professional in your field will know all the inside information about your company or the history of the problem you are writing about.

2.3.1 Emails

Email has become the standard way to send written information, files, and pictures. Like printed letters and memos, email is used to ask and answer questions and to provide information quickly. However, remember that a direct call may be needed if email does not provide a timely result.

There are advantages and disadvantages to email. Email delivery is much faster than paper-based delivery, and emails can be stored permanently on a computer. They can be replicated quickly and distributed to many people simultaneously.

Business emails should not include private, sensitive, or confidential information. They can easily be sent by accident to people who are not authorized to read them, so information such as this should be sent more securely. Emails may be stored permanently and represent a binding agreement or contract. Negative, aggressive emails can lead to dismissal, just like a verbal attack on someone.

Another advantage to transmitting information electronically is that it reduces the amount of paper, printing materials, and shipping or processing fees associated with written documents. In some cases, however, an email is not considered legally binding. Although handwritten signatures are still used for contracts and other formal agreements, this requirement is changing with the growing popularity of electronic signature software. However, this situation is changing with the growing popularity of electronic signatures.

The Journalist's Questions

Make sure that you cover all the necessary information in your writing by asking yourself the following journalist's questions:

- **Who?** Who told you that? Who wants it done? Who are you?
- **What?** What is the problem? What do you need? What do you want to do?
- **When?** When did it happen? When do you want it finished?
- **Where?** Where is it taking place? Where are you going? Where can you be found?
- **Why?** Why do you want to do that? Why is this problem happening? Why did you decide to write about it now?
- **How?** How will you fix it? How can you be reached? How can they help?

Because email is now used daily in business to communicate with associates, clients, supervisors, and coworkers, it is important to know the proper way to write an email (*Figure 13* and *Figure 14*). When writing an email, the nonverbal part of communication (facial expressions, body language, tone of voice) is lost, so people may misinterpret your message if you are not careful with words. Many rules for writing paper-based documents still apply to emails, but a certain etiquette is involved. Business emailing standards exist so that emails are composed professionally, efficiently, and responsibly.

From: JQSmith@smithcontracting.com Sent: Mon 3/15/2021 1:47 PM

To: WJones@paintersplus.com

Cc:

Subject: Quick Note

Attachments:

View As Web Page

Here are some paint colors and faucets available for your bathroom: Color #1415, Soft Jade; Color #1416, Garden Moss; and Color #1417, Forest Glen. All are available in semi-gloss or eggshell finish. There are also three faucet sets: the Meridian (single handle) $109.88; the Mermaid (dual handles) $83.50; and the Monitor (dual handles) $95.75. All are available in polished brass or polished chrome. I've included paint samples and photos of the faucets. Please tell me your choices by Friday. If you have any questions, call me at 703-555-1212.

Figure 13 Functional but poorly formatted email example.

From: JQSmith@smithcontracting.com Sent: Mon 3/15/2021 1:47 PM

To: WJones@paintersplus.com

Cc:

Subject: Bathroom Paint and Faucet Options

Attachments:

View As Web Page

Dear Mr. Jones,

The paint colors and faucets available for your bathroom are listed below (photos of faucets and paint colors are attached to this email). Please let me know what you decide by 5:00 pm on Friday, March 19th. If you have any questions, please do not hesitate to contact me at 703-555-1212.

Paint Colors (Available in semi-gloss or eggshell finish)

- #1415 – Soft Jade
- #1416 – Garden Moss
- #1417 – Forest Glen

Faucet Sets (Available in polished brass or polished chrome)

Model	Price	Handle Style
• Meridian	$109.88	Single
• Mermaid	$ 83.50	Dual
• Monitor	$ 95.75	Dual

Regards,
John Q. Smith
Smith Contracting

Figure 14 Better example of an email when a sale is at stake.

The following are some general rules for writing a proper business email. Remember that some rules will vary according to the nature of the business and company culture.

- Write a business email the same way you would write a formal business letter or memo.
- Ensure you email the correct individual(s) to maintain confidentiality.
- Always ensure the subject line indicates the purpose of the message.
- Begin the email by addressing the recipient.
- Try to keep the body of the email brief and to the point, so that the recipient does not have to scroll when reading.
- Use a concise format, such as numbered or bulleted points, to clarify what the email is about and what response or action is required of the recipient.
- Write in a positive tone and avoid using negative or accusatory statements.
- Do not type in all capital letters, as it gives the impression of shouting.
- Use italicized, bold, or underlined text to stress a point.
- Avoid sarcasm, as it can be easily misunderstood.
- Do not forward junk mail.
- Do not address private issues and concerns.
- When sending an attachment, state the name of the file and its format.
- Double-check the email for spelling and grammatical mistakes before sending it. Remember, once an email is sent, it cannot be revised.
- Include additional contact information, such as a phone number or a mailing address, and your email address.
- Email is not a substitute for face-to-face interaction. Know when to pick up the phone or schedule a meeting if you cannot express your thoughts or concerns through email. Never send bad news through email, and don't use email to avoid your responsibilities.

2.3.2 Texting

Like email, texting has become nearly universal. Texting is also an immediate form of communication. Although an email may take several hours or more, a text message will likely be seen immediately.

Texting has become an important part of business and industry today. Therefore, the construction professional must be able to text effectively as well. However, how you write a text to your friend is likely quite different than how you may write to your supervisor.

Remember that both texting and email can create a safety hazard. Safety must always be given a higher priority than a text message. Be familiar with and follow company policy on using phones and tablets on the job. If an immediate response is not vital, focus on the job at hand first (*Figure 15*).

You must consider whether sending a text message is the best communication method. Text messages should not be used when the message is long or complex, as entering text into a cell phone can be slower than discussing the same information over the phone or typing it into an email. The average speaking speed is 130 words per minute. Therefore, before you start a long text message conversation, consider whether calling may be more appropriate. Text messages should generally not be used to convey highly emotional messages, apologies, or criticism—emotion cannot be effectively communicated this way.

Brief messages that should be viewed immediately are best communicated by text. Any snippet of information that does not require much interaction can be texted: for example, "I'm leaving now. I'll be at the factory in 10 minutes." Texts are also great for confirming information after a face-to-face or phone conversation; for example, "Looking forward to meeting you at 3:30 Friday afternoon."

Figure 15 Use caution around heavy machinery when texting or speaking on a cell phone.
Source: iStock

When sending a text message, be sure that it is accurate. Texting poses a unique problem in the form of smartphone autocorrection. This feature saves time, but it also frequently causes embarrassment. These mistakes are avoidable if you always take a moment to proofread before sending a text.

Not only should your text message be accurate, it should also be clear. Many people use shorthand in their texts to save time. The use of abbreviations such as LOL (laugh out loud), TTYL (talk to you later), and IDK (I don't know) can be inappropriate. There are a few factors to consider when deciding whether to use shorthand.

If the message is informal, shorthand is usually acceptable. This is especially true if your workgroup has its own abbreviations for certain phrases. If the person you are sending the message to is a friend or close colleague of the same level in the company, you can use shorthand. If the message is to a stranger or someone in upper management, it may be better to send messages with correct spelling and grammar to leave a good impression. If you are sure that the person you are texting knows the shorthand, it is fine to use it. Otherwise, using abbreviations that your friend or coworker does not understand could cause confusion.

The warnings in the previous section about making phone calls while operating machinery also apply to texting. According to research by the Federal Motor Carrier Safety Administration, typing or sending a text takes a driver's eyes off the road for an average of 4.6 seconds. At 55 miles per hour, the car will travel the length of a football field in that time. Never send texts while driving or operating machinery (*Figure 16*).

Figure 16 Maintain focus while driving or operating heavy equipment.
Source: iStock

2.0.0 Section Review

1. When a person's reading skills are poor, it may cause them to _____.
   a. rewrite the material themselves
   b. misinterpret technical information
   c. have another worker read to them
   d. file a complaint

2. Where can a reader look to discover if a book has a chapter on a particular topic?
   a. Table of contents
   b. Index
   c. Glossary
   d. Appendix

3. Bold fonts can be used to indicate that words _____.
   a. have recently been changed
   b. need to be checked for accuracy
   c. are important
   d. are illustrated below

4. After you finish writing a rough draft, you should _____.
   a. ask a colleague to proofread it for you
   b. do some research
   c. check for spelling or grammar errors
   d. have a rest, then review and revise

5. Which of the following is the *best* advice for sending an email?
   a. Treat a business email the same way you would treat a formal business letter.
   b. Type in all capital letters to emphasize the importance of your email and set it apart from others.
   c. Send bad news or emotional information via email so that the recipient can read the message in privacy.
   d. Spelling and grammar do not matter in email communication since it is informal.

Module 93108 Review Questions

1. Good communication on the jobsite _____.
   a. affects safety, schedules, and budgets
   b. will make you unpopular
   c. takes too much time
   d. cannot be learned

2. Nonverbal communication is *best* for communicating _____.
   a. feelings and attitudes
   b. complex ideas
   c. what a person desires
   d. instructions for installing new equipment

3. Which of the following are examples of positive nonverbal communication?
   a. Sitting up straight, keeping a clean workspace, and looking at people who are talking to you
   b. Keeping an expressionless face while looking into the eyes of the speaker the entire time
   c. Speaking with an even tone and looking away most of the time during a conversation
   d. Telling people all the information they need in order to do a task properly and answering questions

4. Which of the following is an example of positive verbal communication on the job?
   a. Talking over or interrupting another person
   b. Giving and taking instructions
   c. Tuning out when someone is speaking
   d. Avoiding team discussions

5. Real listening is _____.
   a. tedious
   b. an active process
   c. unnecessary
   d. an art

6. A coworker asks you to hand her a tool, but does not specify which tool she needs. The most appropriate response would be:
   a. "Get it yourself."
   b. "How am I supposed to figure out which tool you need?"
   c. "I'm sorry; which tool do you need?"
   d. "I am too busy; ask someone else."

7. A supervisor wants an apprentice to insert a plug into a water supply line for a pressure test. The clearest way to instruct the apprentice would be for the supervisor to say:
   a. "Put the plug in."
   b. "Insert that one plug into the pipe."
   c. "Get the water supply line ready for the pressure test."
   d. "Insert the plug into the end of the water supply line so I can do a pressure test."

8. If you are in the middle of a task and you receive a personal phone call, _____.
   a. put the caller on hold until you complete the task
   b. continue working and do not answer
   c. tell the caller you don't have time to talk
   d. hand the phone to a coworker

9. In the construction industry, a codebook provides _____.
   a. the keypad entry codes for locked work areas
   b. codes and standards, such as building codes or electrical codes
   c. passwords for company web services, such as databases and training software
   d. keys for understanding blueprint symbols

10. A reader that wants to find any mention of a certain topic in a book can look in the _____.
    a. table of contents
    b. glossary
    c. appendixes
    d. index

11. Which is a typical example of an item a craft professional might read for work on a regular basis?
    a. Newspaper
    b. Road map
    c. Materials list
    d. Summons

12. When you are looking for a specific installation procedure in a manual, the *best* way to find the information is to _____.
    a. identify the correct section and go straight there
    b. start reading the book from the beginning
    c. study the terms in the glossary
    d. skim the book and take notes on all important information

13. Which of the following is one disadvantage to using email versus paper-based delivery?
    a. Email is an expensive way to communicate.
    b. The spell-checker is unreliable.
    c. Emails cannot be easily replicated.
    d. Emails are not as private as paper-based documents.

14. Many of the rules that apply to writing paper-based documents also apply to _____.
    a. writing emails
    b. leaving voice messages
    c. having a face-to-face conversation
    d. taking orders

15. Text messages are *best* used for _____.
    a. long conversations since a record of what was said will be preserved
    b. telling people how one feels
    c. detailed or complex information
    d. brief messages that should be viewed immediately

Answers to odd-numbered questions are found in the Review Question Answer Keys at the back of this book.

Answers to Section Review Questions

Answer	Section Reference	Objective
Section 1.0.0		
1. b	1.1.1	1a
2. a	1.2.0	1b
3. a	1.3.0	1c
4. d	1.3.1	1c
5. c	1.3.1	1c
Section 2.0.0		
1. b	2.1.0	2a
2. a	2.2.0	2b
3. c	2.2.0	2b
4. d	2.3.0	2c
5. a	2.3.1	2c

MODULE 93109

Basic Employability Skills

Source: iStock

Objectives

Successful completion of this module prepares you to do the following:

1. Describe the opportunities in agricultural and construction businesses and how to enter the workforce.
   a. Describe the agricultural and construction businesses and the opportunities offered by the trades.
   b. Explain how workers can enter the construction workforce.
2. Explain the importance of critical thinking and how to solve problems.
   a. Describe critical thinking and barriers to solving problems.
   b. Describe how to solve problems using critical thinking.
   c. Describe problems related to planning and scheduling.
3. Explain the importance of social skills and identify ways good social skills are applied in the construction and agricultural trades.
   a. Identify good personal and social skills.
   b. Explain how to resolve conflicts with coworkers and supervisors.
   c. Explain how to give and receive constructive criticism.
   d. Identify and describe various social issues of concern in the workplace.
   e. Describe how to work in a team environment and how to be an effective leader.

Performance Tasks

This is a knowledge-based module. There are no Performance Tasks.

Overview

Becoming gainfully employed in the construction and agricultural industries takes more preparation than simply filling out a job application. Understanding how these industries and potential employers operate is essential to finding ways to bring value to an organization in need of your specific skills. Trade skills are extremely important, but employers are also looking for eager people with positive personal characteristics. This module discusses the skills needed to pursue employment successfully.

Digital Resources for Agricultural Technology

SCAN ME

Scan this code using the camera on your phone or mobile device to view the digital resources related to this craft.

NCCER Industry-Recognized Credentials

If you are training through an NCCER-accredited sponsor, you may be eligible for credentials from NCCER. The ID number for this module is 93109. Note that this module may have been used in other NCCER curricula and may apply to other level completions. Contact NCCER at 1.888.622.3720 or go to **www.nccer.org** for more information.

You can also show off your industry-recognized credentials online with NCCER's digital credentials. Transform your knowledge, skills, and achievements into credentials that you can share across social media platforms, send to your network, and add to your resume. For more information, visit **www.nccer.org**.

1.0.0 Opportunities in the Agricultural and Construction Industries

Performance Tasks

There are no Performance Tasks in this section.

Objective

Describe the opportunities in agricultural and construction businesses and how to enter the workforce.

a. Describe the agricultural and construction businesses and the opportunities offered by the trades.

b. Explain how workers can enter the construction workforce.

Whether you want to be a hardworking laborer ensuring the project is completed to a high standard or a project management team member focusing on the data to guarantee the project budget and schedule, the construction and agricultural business models offer abundant opportunities. It is up to everyone to take the necessary steps to advance their career and look for opportunities to learn and grow.

1.1.0 Construction and Agricultural Business Models

The construction and agricultural industries consist of various specialized skills (*Figure 1*). This diverse industry is made up of craft professionals, farmhands, and ranch hands working together toward a common goal. Industry opportunities have a global presence; skilled tradespeople are needed in areas such as war zones and major construction sites worldwide. The trades are not limited solely to building construction. There are also global opportunities for many craftworkers that are not directly related to new construction, such as wind turbine technicians, welders, power line workers, crane operators, and instrumentation technicians.

The construction and agricultural industries consist of commercial and independent companies of all sizes specializing in one or more types of work. For example, a smaller company might install heating, ventilation, and air conditioning (HVAC) systems in residences. In contrast, a larger mechanical contractor might be responsible for the HVAC and piping systems for large-scale farms, dairies, or livestock facilities.

The following are some examples of the workforce needed to build a large-scale project:

- Architects design the structure's shape, color, and spaces.
- Civil engineers analyze the architects' drawings to determine how to make the construction design possible. They are also responsible for determining suitable building materials.
- Project managers plan, coordinate, and control a project's progress.

(A) Wind Turbine Technician

(B) Iron Worker ("Rod Buster")

(C) Welder

(D) HVAC Technician

(E) Surveyor

(F) Supervisor/Foreman

(G) Farmhand

(H) Ranch Hand

Figure 1 The construction and agricultural industries offer many career options.
Source: iStock

- Heavy equipment operators (HEOs) excavate the site and work with mechanical, electrical, and plumbing (MEP) contractors to install underground utilities and connect the site to local infrastructure. On an agricultural site, a heavy equipment operator is essential for maintaining and using specialized vehicles to expedite tasks (*Figure 2*).
- Concrete carpenters and finishers build forms and pour the foundation to a smooth slab for the other trades to build up from.
- Crane operators expedite the movement of building materials around a site.
- Ironworkers cut and fit reinforcement bars and steel beams to provide the structure with the necessary strength to withstand wind, seismic movement, and other external forces.
- Welders secure the framework of buildings and other metal structures.
- Electricians install an array of electrical devices and wiring.
- Plumbers properly install fixtures and other plumbing equipment. This role is especially important in agriculture, where they bring water to and away from livestock facilities and farms (*Figure 3*).
- HVAC technicians install building comfort systems.
- Pipefitters install chilled and hot water systems and process piping.
- Carpenters frame walls, floors, and roofs.
- Instrumentation technicians install measuring and control instruments to ensure the equipment runs efficiently.
- Landscape designers plant flora and install grass and other elements to improve the project's aesthetics.

Mission statement: A statement describing how a company does business.

Many companies have a **mission statement** explaining how a company does business. The company may describe its philosophy in an employee handbook or other materials during a new hire orientation. However a company describes its role, it is important to become familiar with your company's policies and procedures to know how you're expected to conduct yourself. The best time to learn about a potential employer is before applying. The more you know about a company and what is important to them, the better you can plan your responses for an interview.

Figure 2 Specialized agricultural heavy equipment.
Source: iStock

Figure 3 Agricultural plumbing applications.
Source: iStock

1.2.0 Entering the Workforce

Upon completing your training, it will be time to find a job where you can use your newly acquired skills. The first step of a job search is writing a resume to get an employer's attention and highlight your experience.

The following are some guidelines for resume writing:

- Use a readable font, such as Times New Roman or Arial, for the resume's text.
- Write your objective using keywords from the job description to create a relationship between the two. This is especially vital in today's job market, as many employers use AI software to scan your resume before it ends up in the hands of a Human Resources (HR) professional.
- Format your resume chronologically, putting your current or most recent job first.
- Use short, concise sentences and bulleted lists to describe your skills that align with keywords found in the job description.
- Include certifications and training credentials in specific trade areas. Credentials could include things like NCCER training certificates, wallet cards, or automated external defibrillator (AED) certifications.
- Make sure all contact information is current, accurate, and easy to find on the resume.
- Personal information, such as education or volunteer work, can be listed. Employers often appreciate employees with such interests, as they indicate a well-rounded individual. However, such information should be kept brief.
- Check your resume for inaccurate or missing dates. Your career timeline should be easy to follow. Large gaps of undocumented time on a resume may require a reasonable explanation.
- Have someone proofread your resume to ensure it is free of spelling errors, typos, and poor grammar.
- Provide clear personal references or indicate that such a list can be provided upon request.

Look for a job that closely matches your skill level. Typically, a job posting will include a summary of the position, the required qualifications, and an overview of benefits. Applying for positions well above your skill level often leads to disappointment.

Most employers advertise openings on the internet and require online applications. Use a website dedicated to job searching or go directly to a company's website. Libraries and unemployment offices generally provide computer stations for public use if you don't have access to one.

Today, one of the most important and productive methods to identify job opportunities is through networking. Studies indicate that as many as 70% of available positions are not publicly advertised. At the same time, job seekers spend roughly 70% of their time looking through public job postings.

There are many methods to increase your chances and share your interest in finding a specific type of job, including social media. Identify companies you would like to work with, regardless of job postings, and then look for contacts to help you identify opportunities, such as recruiting and temp agencies. Express your interest in finding a job to family and friends. Networking is often the most productive method of finding and landing a job.

Ask your current supervisor, your coworkers, and friends who know you well if they would be **references** for you. A mix of personal and professional references is recommended.

References: People who can confirm to a potential employer that you have the skills, experience, and work habits listed in your resume.

Ensure your resume is up to date, well organized, and easy to read (*Figure 4*). Apply the effective writing techniques you learned in NCCER Module 93108, *Basic Communication Skills,* when writing a resume.

By carefully selecting jobs you are qualified for and submitting an accurate, well-written resume, you improve your chances of being called for interviews (*Figure 5*). The effective communication skills you learned in NCCER Module 93108, *Basic Communication Skills,* will help you present yourself well.

John Q. Smith

Address: 123 Main Street, Anytown, MD 67890
Home phone: (555) 111-2233 **Cell phone:** (555) 333-4455
Email address: johnqsmith@email.com

Employment Objective

- To obtain a position within a construction company where I can use my carpentry and custom woodworking skills.

Skills and Qualifications

- Able to work independently and with a crew.
- Practical knowledge of carpentry hand and power tools, measuring tools, and woodworking tools.
- Trained and certified to use powder-actuated tools.
- Qualified to read residential and commercial construction drawings.
- Experience with energy efficient and conservation building techniques.
- Knowledge of LEED standards.

Work Experience

Journeyman Carpenter
May 2021 – present
LJL Construction, 123 Hammer Heights Road, Fairfax, VA
800-222-2222

- Built custom cabinets and closet systems
- Framed single family homes
- Trained apprentice carpenters

Carpenter
June 2015 – May 2021
Hammer Company, 456 Lathe Lane, Philadelphia, PA
866-555-1234

- Built closets and storage systems for commercial storage units

Carpenter
February 2012 – June 2015
Blue Ridge Carpenters, 789 Mountain Road, Harrisburg, PA
888-123-4567

- Framed condo units
- Built fences
- Raise wood walkways

Construction Education

- Carpentry apprenticeship. MK Builders Inc., Gettysburg, PA. Certificate, 2012.
- Professional carpentry certification program. ABC Training Corp., Philadelphia, PA. Certificate, 2010.
- Powder-actuated tools certification program. All States Community College, Marietta, PA. Certificate, 2010.

Figure 4 Example resume.

Figure 5 Job interviews are an important part of the hiring process.
Source: iStock

Before accepting an offer, ask yourself the following questions to ensure the position will be the right fit:

- Is the salary enough to meet my needs?
- Does the company offer a benefits package?
- Will the work be interesting and challenging?
- Does the company have a good reputation in the industry?
- Do I fit into the company culture?
- Is travel required, and am I able to agree to that?
- Does the company offer training?
- Are advancement opportunities within the organization evident?
- What is the company's safety record?

Before an offer is made, some trades require credentials or qualification testing to assess the applicant's skill level. Many companies require drug testing as part of the preemployment process to ensure workplace safety. Appropriate identification, such as a driver's license, US Social Security card, national identification card, and/or a valid passport must be given to the employer. If a position outside of your country of citizenship is applied for, check the hiring and documentation requirements for that specific country and what may be required of noncitizens.

It is wise to obtain an organizational chart after accepting a position and even before the interview process begins (*Figure 6*). Organizational charts help all employees understand the company's structure and what position and/or individual is responsible for various areas. Understanding the organizational structure and recalling some important details about a prospective employer during an interview can make a strong impression.

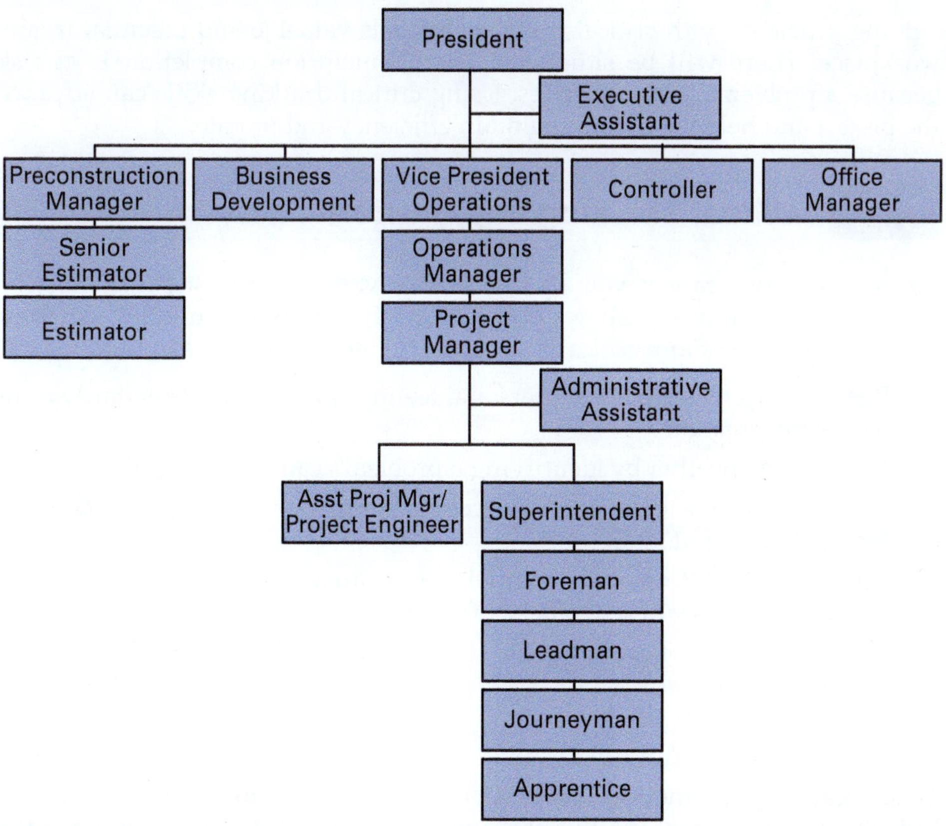

Figure 6 An organization chart example.

Going Green

Green-Collar Jobs

Any job that involves designing, manufacturing, installing, operating, and/or maintaining renewable energy and energy efficient technologies is considered a green-collar job. Many of these jobs have opened new doors or removed barriers to employment. Several opportunities exist in the following areas:

- Installation, construction, maintenance, and repair of energy retrofits, HVAC, solar panels, and whole home performance
- Installation, construction, maintenance, and repair of water conservation and adaptive gray water reuse
- Construction, carpentry, and demolition in green building and recycling

1.0.0 Section Review

1. Who plans, coordinates, and controls a project's progress?
   a. A civil engineer
   b. An architect
   c. A project manager
   d. The funding bank

2. Which of the following is a valid statement about a resume?
   a. Always use a highly creative and unusual font to set your resume apart from all others.
   b. Format your resume chronologically, putting your current or last job first.
   c. Do not allow others to proofread your resume, as they will likely be critical of it.
   d. Resumes should not include certifications; present them only if called for an interview.

2.0.0 Critical Thinking and Problem-Solving

Performance Tasks

There are no Performance Tasks in this section.

Objective

Explain the importance of critical thinking and how to solve problems.

a. Describe critical thinking and barriers to solving problems.
b. Describe how to solve problems using critical thinking.
c. Describe problems related to planning and scheduling.

Solving problems with critical thinking skills is valuable and essential in any workplace. There will be situations where timely job completion is at risk because a problem suddenly arises. Using critical thinking skills can advance the project and help your team maintain efficiency and morale.

2.1.0 Critical Thinking and Barriers

Throughout your career, you will encounter various problems that must be solved. Critical thinking allows you to identify the best solution to complex problems. To think more critically, keep the following strategies in mind:

- Remain objective. Do not let personal feelings prevent you from fairly evaluating information.
- Link events together by identifying a problem's cause and effect.
- Identify people who are good sources of information. Ask experts and mentors for their input.
- Compare new information with what you already know. If it does not fit, question it and try to separate fact from fiction.
- Weigh the merits of each possible solution by analyzing their strengths and weaknesses.

2.1.1 Barriers to Problem-Solving

When searching for the solution to a problem, it is easy to fall into a trap that prevents you from making the best possible decision. The following are the most common barriers to effective problem-solving:

- Closed-mindedness
- Personality conflicts
- General fear of change

To be closed-minded is to distrust any new ideas. Closed-minded people may also resist any suggested changes. Effective problem-solving, however, requires you to be open to new ideas. Sometimes the best solution is one you would have never considered alone. Remember that other people have good ideas too. You must be willing to listen, evaluate their input fairly, and potentially work their recommendations into a compromised plan.

Sometimes you may fail to appreciate the value of information or advice simply because you do not get along with the person offering it. One of the most important skills one can master is separating the message from the messenger. Weigh the value of the information separately from your feelings about the individual. This ability will show people that you are a true professional.

People often fear change, but problems frequently stem from a lack of change and progress. In addition, it is often difficult to clearly see the path to implementing change or to envision the result.

2.2.0 Solving Problems with Critical Thinking

Problems arise when there is a difference between how something is and how it should (or needs to) be. You might feel frustrated or intimidated by a problem or lack the time to solve it. The easiest short-term reaction might be to ignore the problem. Instead, try to see it as an opportunity to demonstrate your skills.

By actively seeking solutions to problems, you will demonstrate to your colleagues and supervisors that you are responsible and capable. Conquering a challenging problem can significantly enhance your reputation as a professional, build confidence in your decision-making skills, and increase your knowledge through valuable experiences.

Use the following five-step process (*Figure 7*) to solve problems that arise on the job, reviewing each step in detail to help you understand how to apply it:

Step 1 *Define the problem.* Before solving a problem, you must know exactly what it is. This step might seem obvious, but people often find themselves trying to solve the wrong problem or one that doesn't even exist.

Step 2 *Analyze and explore alternatives.* Once you have defined the problem, consider different ways to solve it. Collect information from various sources and ask people for their opinions. Perhaps you have a solution in mind, or maybe just one step toward it. It can be very productive to seek opinions on solving small portions of a problem rather than on the big picture. Teamwork is an important part of this step. The more suggestions you get from experts and coworkers, the better your chances are of finding the right answer. Identify and compare the alternatives. Look for solutions that will be the most cost-effective, take the least time, and ensure the highest-quality result. Try to identify the short- and long-term consequences of each possible solution. Eventually, you will have created several plans of attack and given each one some thought.

Step 3 *Choose a solution and plan its implementation.* Move forward by choosing the solution that you think will work best. Next, develop a plan to implement the solution. The plan should include all necessary tools and materials. It should also specify all the tasks involved, who is responsible for them, and estimate how long each task will take.

Step 4 *Implement the solution and monitor the results.* Once the plan is in progress, follow it closely to ensure it is progressing toward the desired results.

Step 5 *Evaluate the result.* If your solution does not work, go back to *Step 3* and choose another solution, as shown in *Figure 7.* Then develop and implement a plan aligned with the new solution.

CAUTION

Remember that not every solution will be successful. Life and the workplace can turn what appears to be a logical solution into a new disaster. This happens to everyone; do not allow such a failure to destroy your confidence. Go back, think through the options once again, and move forward with a new solution and a positive attitude.

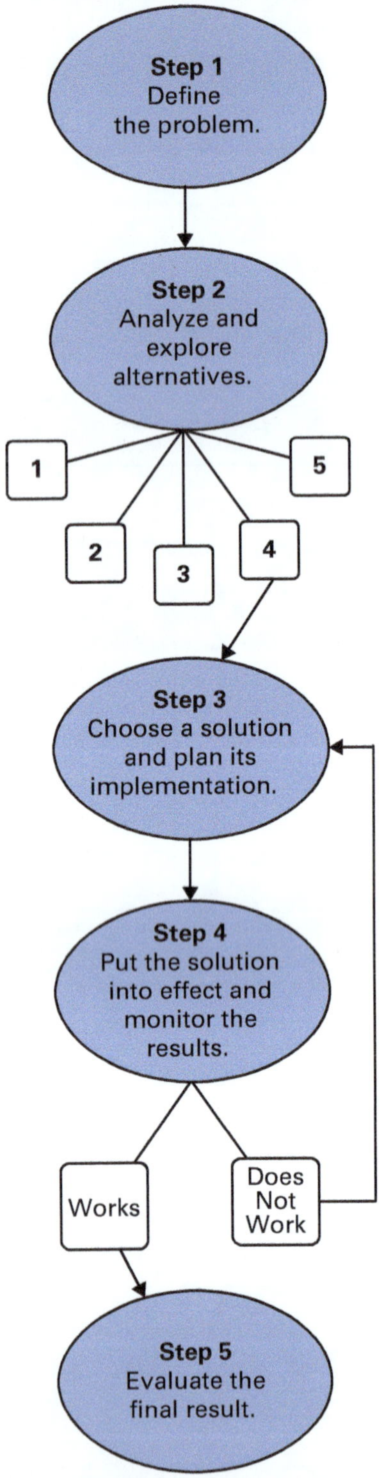

Figure 7 The five-step problem-solving process.

When the problem is finally solved, review your steps and see what lessons can be learned. Perhaps there was something you could have done better. Maybe you could have saved time or money by changing a small part of the plan. When you are satisfied with the solution, remember what worked and what did not. When you face a similar problem in the future, your experience will help you find a solution.

2.2.1 Applying Critical Thinking

Review the following hypothetical case. Imagine that you are responsible for finding a way to prevent the situation from repeating itself.

While on a jobsite, your coworkers ask you to retrieve more welding electrodes. This welding task must be finished by the end of the day, which is cutting it close. You walk to the nearest toolroom—the attendant only has 60 electrodes in stock, but you need several hundred. The attendant checks with the other toolroom, which is fully stocked with the necessary electrodes. It will take some time to get to the other toolroom. What can be done to solve the immediate problem? What can be done to ensure that the shortage doesn't happen again?

Defining the Problem

The proper type of welding electrodes must be readily available to keep the welding job running on schedule. Each shift's welding task must be completed by the end of the shift because the next shift needs to start the next phase of the project. If the welding is incomplete, it will impact the schedule and budget. Currently, there is a shortage of proper welding electrodes at the jobsite.

Analyzing Alternative Solutions

There are several solutions to obtaining the proper type of welding electrodes:

- You could take the 60 electrodes in stock, walk back to the jobsite, and get the job started. Then, you could do a 30-minute bicycle ride to the other toolroom to pick up the remaining electrodes.
- You could take the 30-minute bicycle ride to the other toolroom to pick up the needed electrodes, delaying the start of work until you return.
- You could contact the other toolroom attendant with the extra electrodes and explain the predicament. Then you could ask the attendant to bring the electrodes to you. This would save you (and the welders) the time involved in making the trip. While waiting, you could take the 60 electrodes to your coworkers so work can begin at the start of the next shift.

What are some of the benefits and drawbacks of each alternative?

Choose a Solution and Plan

After weighing each option, you decide to contact the other toolroom attendant, ask for assistance, and then deliver the 60 electrodes to your coworkers. This will keep the job going and will save some of your time by having the electrodes delivered. However, the delivery requires the attendant to close the stockroom while making the trip, which could affect others in need of important materials.

To keep the toolrooms properly stocked with all consumable materials from now on, you suggest to your supervisor that the local stock attendants take an inventory at the end of each shift and restock as appropriate.

Implement and Monitor

At the end of each shift, the attendants now inventory consumable materials to ensure a full stock for the next day. They previously took inventory every three days. After a few days, someone in authority returns to the local stockrooms to see how the change works and how the attendants feel about the new process. Perhaps they will have fresh input and suggest additional improvements to the process.

Evaluate the Result

Since implementation, no toolroom has run low on consumable materials. No one has had to scramble and search for materials, and the attendants do not have to close the toolroom to deliver materials during the middle of the shift.

Remember, the word *critical* means important or indispensable. Critical thinking skills are important and indispensable tools in your personal toolbox (*Figure 8*). As with any other tool, you must learn to use these skills correctly and safely. When you do, you will find that one of the most rewarding experiences you can have as a trade professional is to face a problem and solve it yourself.

Figure 8 Critical thinking is just as important as any tool you carry.
Source: iStock

2.3.0 Planning and Scheduling Problems

Every task has a logical position in the overall project schedule. It must occur before, after, or simultaneously with other tasks. Before beginning each task, ensure you understand what you are expected to do and what is required to accomplish it. You must know where your work begins and ends and which materials to install. Divide the task into individual steps and determine the work's sequence.

No matter how carefully you plan, unexpected problems can suddenly appear. Usually, extra time is built into project schedules so that simple problems will not cause a delay. However, more complex problems can throw the project offtrack.

Project managers and/or site planners (when the position is assigned) are generally responsible for ensuring that a project stays on schedule; therefore, they commonly use project management software (*Figure 9*). Nevertheless, you may be called upon to evaluate and help solve problems as a team member working on a particular job. Schedules must change to accommodate all types of problems that arise. Being familiar with problems that can disrupt a job, including regulatory issues, is important. Typically, problems will fall under one or more of the following categories:

- Materials
- Equipment
- Tools
- Labor

Figure 9 Project management software.
Source: iStock

2.3.1 Materials

The materials required for a job are identified during the estimation stages of a project. Once the contract has been completed, they are ordered from suppliers and delivered according to a prearranged schedule. Many jobsites do not have sufficient room to store all the needed materials for a structure at one time. Material problems include quantity or type errors, delivery delays, unavailability due to back orders, or abrupt business closures. A shortage of materials due to waste, theft, or vandalism is another source of project delays.

Material problems often lead to changes in the material to be used. If the new material was not listed as an option in the contract specifications, a significant amount of time and effort may be required for the change request to go through the approval process.

2.3.2 Equipment

Major construction equipment is usually selected and scheduled well before the need arises. If a site planner is a team member, they may be responsible for ensuring that equipment is on-site at the right time. Otherwise, project managers and other supervisors may be assigned this responsibility. This generally depends on the scope of the project.

Multiple pieces of equipment must often be scheduled simultaneously: backhoes to excavate a foundation and dump trucks (*Figure 10*) to haul away the excavated dirt, for example. Problems with equipment can include unavailability on the scheduled day(s), lack of qualified operators, mechanical breakdowns, and extended maintenance periods.

The lack of major equipment can create significant delays. In many cases, multiple crafts and subcontractors rely on the equipment to complete their jobsite tasks. For example, if a crane required to lift steel is unavailable, ironworkers, welders, and possibly other craftworkers find themselves at a standstill.

2.3.3 Tools

Workers and supervisors are responsible for ensuring the appropriate tools are available. Workers are often expected to have traditional tools that are specific to their tasks when they are employed. Special tools that are expensive or unique are usually provided by the employer. Both parties must do their part to provide and maintain proper and functional tools.

Imagine if all the carpenters arrived for a day's work without hammers—their house project would come to a halt! On the other hand, if pneumatic nail guns are being used extensively but the contractor's supervisor does not ensure an air compressor is on-site, the project will come to a standstill just as quickly. Other common problems include broken, damaged, lost, or stolen tools. A lack of skilled workers also negatively impacts project completion.

Figure 10 Heavy equipment logistics.
Source: iStock

2.3.4 Labor

Labor—the people who work on the jobsite—is the most important component of a project. Labor is often one of the most expensive project components, depending upon the country, project, and type of workforce needed. Project planners estimate the number of people needed for each day and identify the range of skills required of those workers. Foremen ensure the right people are available and working on the right tasks. Common problems related to labor include **tardiness** and **absenteeism**, lack of experience or qualifications to perform a given task, and insufficient available workers. Idle time due to a lack of guidance or laziness can also contribute to delays.

Tardiness: Arriving late for work.

Absenteeism: A consistent failure to show up for work.

2.3.5 Handling Delays

Always be aware of potential delays as you carry out your assigned tasks. Look as far forward as possible into portions of a project you are responsible for and consider the types of delays that could occur. If you see a probable cause of delay, notify your supervisor immediately. Having a plan for delays, especially delays that occur regularly or often, enables a solution to be applied promptly. Always be certain of what your responsibilities are in such situations. Be prepared to solve the problem yourself if that is part of your assignment. Even then, always keep your supervisors updated on delicate situations involving delays.

2.0.0 Section Review

1. In critical thinking, if new information you receive about an issue does not fit with what you already know, _____.
   a. question it and try to separate fact from fiction
   b. change what you previously held as factual in your mind
   c. berate the person and let them know you don't appreciate that kind of input
   d. thank them and forget about the information they offered

2. Why is it important to follow a plan through once it is implemented?
   a. To be sure there are no other options
   b. To be sure no one asks questions
   c. To look for other opinions
   d. To be sure the results are what you are looking for

3. If material *not* specified in contract documents must be substituted for another type, the change will likely need to _____.
   a. pass through an approval process
   b. pass through a testing procedure
   c. be agreeable to all other trades
   d. be considered after the project is done

3.0.0 Relationship and Social Skills

Performance Tasks

There are no Performance Tasks in this section.

Objective

Explain the importance of social skills and identify ways good social skills are applied in the construction and agricultural trades.

a. Identify good personal and social skills.
b. Explain how to resolve conflicts with coworkers and supervisors.
c. Explain how to give and receive constructive criticism.
d. Identify and describe various social issues of concern in the workplace.
e. Describe how to work in a team environment and how to be an effective leader.

A relationship results from interacting with another person or a group of people. Relationships are affected both by real actions and the perceptions of individuals. Every day you interact with coworkers, supervisors, and members of the public who see you working. No one wants to work with, hire, or spend time with an unprofessional person.

Craftworkers need to be aware of the appropriate professional conduct for work situations and always follow it. Your actions reflect your own professional status, that of your colleagues, the company you work for, and the image of your profession as seen by the public.

3.1.0 Personal and Social Skills

Self-presentation: The way a person dresses, speaks, acts, and interacts with others.

Work ethic: Work habits that are the foundation of a person's ability to do their job.

Proper **self-presentation**—how you dress, speak, act, and interact—is vital to any successful work relationship. Self-presentation involves developing good personal and work habits. Personal habits apply to your appearance and general behavior. Work habits, also known as your **work ethic**, apply to how you do your job. Your habits and work ethic make powerful impressions on your colleagues, supervisors, and potential employers. Something as simple as filling out a job application with no errors and good penmanship will give an employer a good first impression.

When you introduce yourself, look the other person in the eyes and stand with confidence. In the United States and several other countries, a firm and sincere handshake has long been considered an indicator of a person's character and personality, especially among men. Recent studies indicate that a firm grip (but not so firm as to imply a competition is at hand) and eye contact at the appropriate moment make a good impression, regardless of gender. Once formed, first impressions are hard to change, so make sure that the first impression you make is a good one.

3.1.1 Personal Habits

Coworkers and supervisors like dependable people. Coworkers know they can trust dependable people to pull their own weight, take their responsibilities seriously, and look after one another's safety. Supervisors know that dependable people will do their best to finish a job correctly and on schedule. When dependable workers say they will do a task, they follow that commitment. It also means showing up for work on time and not stretching out lunch hours and breaks (*Figure 11*).

Organizational skills are also important. Keep your tools and work areas clean and organized. Know which tools you are responsible for and keep them in good working order. This applies to your personal tools and those belonging to your employer. Always plan what you need to do each day before you begin. Although your intended plan is likely to change as the day unfolds, a failure to plan is a plan for failure. Approach your work in an organized fashion and follow the schedule that you have been assigned.

Figure 11 Punctuality is key.
Source: iStock

Ensure that you are technically qualified to perform your duties and specific tasks. This means you can safely use tools, equipment, and machines. Take advantage of opportunities to expand your technical knowledge through classes, books, trade periodicals, and mentoring by experienced colleagues. Being technically qualified also means you will not attempt a task you are unfamiliar with when a safety hazard exists.

Offer to pitch in and help whenever you can. The best workers are willing to take on new tasks and learn new ideas. Supervisors notice employees who are willing but careful not to take on more tasks than they can handle; they will not impress anyone if they cannot fulfill their promise.

Honesty is one of the most important personal habits you can have. Do not abuse your company's system by calling in sick to take a day off or leaving early and asking someone else to punch you off the clock. Never steal from the company or your coworkers; this includes everything from tools and equipment to simple office supplies. If you are struggling with a problem, don't hide it; speak with your supervisor about it truthfully. Be willing to actively look for a solution to problems you are facing.

Professionalism means that you approach your work with integrity. As you learned in NCCER Module 93102, *Basic Safety*, there is no place for horseplay or irresponsible behavior on the construction or agricultural site. Employers want workers who respect the rules and understand that rules exist to keep people safe and projects on schedule. A professional employee always respects company **confidentiality**. In some cases, workers must sign documents indicating that they understand the importance of confidentiality and agree not to discuss certain information shared with them.

Professionalism: Integrity and work-appropriate manners.

Confidentiality: Privacy of information.

Most companies have a dress code, and often they have additional special requirements for specific jobs. Always follow these requirements. Pushing the limits of such policies never impresses anyone. Also, do not forget to attend to the basics of good grooming. People who pay attention to their appearance and develop positive personal habits are more likely to be considered for a job over people who do not have good grooming habits. You reflect your trade, employer, and yourself on the jobsite.

3.1.2 Work Ethic

A strong work ethic and good personal habits are essential for getting hired and promoted. Employers look for people they believe will give them a fair day's work for a fair day's pay. A strong work ethic means you enjoy working and always try your best on each task. When work is important to you, you believe you can positively contribute to any project you are working on (*Figure 12*).

Figure 12 Hardworking ranch hand.
Source: iStock

Construction and agricultural work require people who can work without constant supervision. When there is a problem that you can solve, solve it without waiting for someone to tell you to do it. If you finish your task ahead of time, look for another task to finish in the remaining time or ask coworkers if they need help. This type of positive action is called taking **initiative**. Colleagues and supervisors will respect you more when you demonstrate initiative.

Initiative: The ability to work without constant supervision and solve problems independently.

However, an important part of taking initiative is knowing when and how to implement it. Suppose a supervisor tells you to perform a task. They demonstrate the five steps required to complete the task. Later, as you perform the work, you realize that one of the steps is unnecessary. Leaving out the step could save time—but should you? No, that would not be the right initiative.

Instead, tell your supervisor about your discovery, and ask what you should do. Bringing the options to your supervisor shows initiative. Your supervisor may realize that you are right and allow you to leave out the extra step. After that, whenever anyone performs that task, it will take less time and save the company money; or your supervisor may explain why the step is important. Then you will have learned something more about the work you are doing. The result of taking the initiative in either case is a positive one.

3.1.3 Tardiness and Absenteeism

The two most common problems supervisors face are tardiness and absenteeism. Tardiness is when a worker habitually shows up late to work. Absenteeism is when a worker consistently fails to show up for work at all, with or without excuses. People with a strong work ethic are rarely late or absent.

Construction and agriculture companies operate under tight schedules to make a profit and stay in business. These schedules are primarily built around workers. If you are late or do not show up regularly, expensive adjustments become necessary. Your employer may decide that the best way to save money is to stop wasting it on you (*Figure 13*).

Figure 13 Show up to work on time.
Source: iStock

Consider the following suggestions for improving or maintaining your record of punctuality and attendance:

- Think about what would happen if everyone on the job were late or absent frequently.
- Think about how being late or absent affects your coworkers.
- Know and follow your company's policy for reporting legitimate absences or lateness.
- Keep your supervisor informed if you need to be out for more than a day.
- Allow yourself enough time to get to work.
- Explain a late arrival to your supervisor when you get to work.
- Do not abuse lunch and break time privileges.

3.2.0 Conflict Resolution

Conflict resolution is an important relationship skill because conflict can happen anywhere, anytime. Conflicts between you and your coworkers can arise because of disagreements over work habits, different attitudes about the job or company, differences in personality, appearance, culture, or age, or distractions caused by problems at home.

Conflicts between you and your supervisor can happen because of a disagreement over workload, lateness or absenteeism, or criticism of mistakes and inefficiencies. Significant disagreements with a supervisor should be reported to your employer's HR office.

Most of the time, people are not trying to turn disagreements into conflicts. People are often unaware of the effects of their behavior. Before reacting negatively to a coworker's behavior, remember to be **tactful**. This means considering how others will feel about what you say or do. Do not accuse, embarrass, or threaten the person. This behavior rarely leads to an acceptable solution.

Tactful: Being aware of the effects of your statements and actions on others.

Never let a disagreement or conflict affect job performance, team morale, or—most importantly—jobsite safety. The goal is to keep events from turning into conflicts in the first place. If that is not possible, the next best thing is to address the conflict quickly and resolve it professionally. If a disagreement with a coworker is getting out of hand, try one of the following techniques to cool the situation down:

- Think before you react.
- Walk away, explaining that you need time to think through the situation.
- Try not to take it personally.
- Avoid being drawn into others' disagreements.

Do not let a conflict simmer and then boil over before you act. Not only is it unprofessional, but tempers may also flare unexpectedly and worsen the situation. Being proactive when an issue becomes uncomfortable is usually better than being reactive. This is true unless tempers are already out of bounds. In that case, it may be best to allow some time to pass before attempting to resolve the problem. If the conflict escalates, walk away and notify your supervisor immediately (*Figure 14*).

Figure 14 Don't allow conflicts to escalate.
Source: iStock

However, remember that there are important differences between resolving conflicts with your coworkers and your supervisor. The following sections discuss how to handle these types of conflicts.

3.2.1 Resolving Conflicts with Coworkers

Remember to respect the people you disagree with. After all, they believe they are right too. Always be clear, rational, respectful, and open-minded. Begin by admitting to each other that there is a conflict. Then analyze and discuss the problem. Allow everyone to describe his or her own perception of the conflict. You may realize that the whole problem was simply miscommunication. Begin by asking the following questions:

- How did the conflict start?
- What is keeping the conflict going?
- Is the conflict based on personality issues or a specific event?
- Has this problem been building up for a while or did it start suddenly?

- Did the conflict start because of a difference in expectations?
- Could the problem have been prevented?
- Do both sides have the same perception of what is happening?

Once you have analyzed the situation, discuss all possible solutions. You will probably have to **compromise** to find a solution everyone agrees on. When you agree on the solution, act on it, and see if it works. If it does not, consult your supervisor to help resolve the conflict. Notice that this process is like the problem-solving techniques discussed earlier. If you find that the conflict happened because you were wrong, apologize. An apology is not a sign of weakness but rather a sign of respect.

Compromise: When people are involved in a disagreement, they make concessions to reach a solution that everyone agrees on.

Once an agreement or compromise has been reached, stand by it in every possible way, and do so with a positive attitude.

3.2.2 Resolving Conflicts with Supervisors

You can usually approach coworkers as equals because you work on the same job. However, your supervisor oversees you and your work on the job. You must use a different approach to resolve conflicts with your supervisor.

Before going to your supervisor, take some time to think about the cause of the conflict. Consider writing down your thoughts; organizing the information this way often puts things into perspective. When you approach your supervisor, do so with respect. Wait until your supervisor has a free moment, and then ask if you could arrange a time to talk about something important or leave a note asking your supervisor to meet. Be willing to meet at a time that is convenient for your supervisor. Remember, supervisors have many responsibilities. They do not have much free time during the regular workday, and you may have to meet before or after work.

When meeting with your supervisor, speak calmly and clearly. Do not be emotional, sarcastic, or accusatory. Do not confront your supervisor in a threatening or angry way. State only the facts as you see them; never say anything you cannot prove. Do not mention the names of your coworkers unless they are directly involved. If you want to suggest changes or solutions, explain them clearly and discuss how they will benefit the people involved.

Once you have made your case, allow your supervisor to decide. You should accept and respect your supervisor's final decision. It may not be the one you wanted, but it will be the one that must be followed.

3.3.0 Giving and Receiving Criticism

As a construction professional, farmhand, or ranch hand, you will always learn something new about your job. New technologies, materials, and methods appear all the time. For example, talking to an experienced worker can help you learn a new way to perform a task. As your skills improve with practice, you will be able to use tools and methods you were not able to use before. This is a common way of learning on the job (*Figure 15*).

Another way to learn on the job is through **constructive criticism.** Constructive criticism is advice designed to help you correct a mistake or improve an action. Constructive criticism can improve your job performance and relations with coworkers. You have probably heard the word *criticism* used negatively to indicate fault or blame; that is not the criticism discussed here. Constructive criticism does not mean colleagues and supervisors think little of you; it means the opposite. Colleagues and supervisors who offer constructive criticism do so because they believe it is worth their time to help you improve your skills.

Constructive criticism: A positive offer of advice intended to help someone correct mistakes or improve actions.

As you gain experience on the job, you will be able to give and receive constructive criticism. To make sure that someone does not mistake your constructive criticism for blame, you need to know how to offer it in a positive manner. The following sections offer some general advice on offering and receiving constructive criticism.

Figure 15 A good employee never stops learning on the job.
Source: iStock

3.3.1 Offering Constructive Criticism

You might offer some constructive criticism when training a less experienced person or working with coworkers. How should you offer it? Before you say anything, think about the rules of effective speaking that you learned in NCCER Module 93108, *Basic Communication Skills*. Use positive, supportive words and offer facts, not opinions.

Constructive criticism works best when offered occasionally. Do not constantly comment on another person's work or methods; they may block out your criticism or even become angry. Never criticize people in front of their coworkers or supervisors; they may feel embarrassed and will likely resent you. Constructive criticism should include suggested alternatives. Do not criticize how a coworker does something unless you can suggest another way. Above all, limit your comments to the person's work or methods without targeting them as a person. There is a great deal of difference in how people react to a statement such as "your pipefitting work is unacceptable" compared to "this threaded joint is unacceptable."

Point out improper behavior or incorrect work techniques the first time they occur. If not addressed, the behavior or technique could become a bad habit. Bad habits, in turn, can lead to accidents. Gently and firmly offer constructive criticism for improper behavior as soon as it is noticed (*Figure 16*).

Remember to compliment the person you are criticizing. Compliments positively affect the person receiving them and are easy to offer. You do not have to wait to compliment someone until you are prepared to offer constructive criticism. However, if a compliment is to be offered along with constructive criticism, offer the compliment first.

Try to offer compliments regularly. Your appreciative and respectful approach will help keep team spirits high, and your supervisor will appreciate your professional attitude.

Figure 16 Practice constructive criticism.
Source: iStock

3.3.2 Receiving Constructive Criticism

Constructive criticism can mean the difference between success and failure, but in cases where workplace safety is involved, it can also be the difference between life and death. Think of constructive criticism as a chance to learn and to improve your skills. Take constructive criticism from your supervisor seriously. Treat constructive criticism from experienced coworkers with the same respect; even though they may not be your supervisor, their experience gives them expertise. When someone presents you with constructive criticism in the proper manner, learn from their approach as well.

Always take responsibility for your actions. Never be overly defensive or dispute criticism. If you respond negatively, you might offend someone who is simply trying to help you. As a result, that person will either lose respect for you or be unwilling to help. If that happens, the quality of your work and everybody who works with you will suffer.

When someone offers constructive criticism, demonstrate your positive personal habits and work ethic. If the criticism is vague, ask the person to suggest specific changes you should try to make. If you do not understand the criticism, ask for more details. The deepest respect you can show is to improve your performance based on their advice.

You have a right to disagree with criticism that is unacceptable or incorrect. Someone might honestly misunderstand your situation and offer incorrect advice. Or a coworker might criticize you simply because they think they perform a task better than you. In such cases, clearly and respectfully give your reasons for disagreeing, and explain why you believe you are correct.

3.3.3 Destructive Criticism

Constructive criticism has positive effects. However, you may also encounter destructive criticism. As its name suggests, destructive criticism is designed to hurt, not help, the person receiving it. The destructive criticism could be about something that needs improvement, but because it is offered negatively, there is no desire to listen.

If you receive destructive criticism, the professional thing to do is to stay calm. Do not get into a fight by replying in a negative tone. Find a way to tell the person how the criticism made you feel. If there is legitimate criticism beneath a destructive tone, ask the person to offer positive suggestions for ways you can improve. By taking a positive approach to negative criticism, you can set a good example for others.

3.4.0 Social Issues in the Workplace

The modern construction and agricultural workplace is a cross section of our society (*Figure 17*). A typical project involves men and women from all walks of life, of many ethnic and racial backgrounds, and often from many different countries. Many workers speak more than one language. Workers have grown up and lived in many different income brackets.

Figure 17 Diverse team.
Source: iStock

Construction and agricultural workers face a wide range of mental and physical demands every day on the job. These demands can sometimes feel overwhelming, and people may seek to escape pressure through illegal drugs or alcohol abuse. As a craft professional, farmhand, or ranch hand, you need to be aware of these issues. You also need to know what to do if someone is behaving inappropriately.

You may encounter the following issues in the workplace:

- **Harassment** or **bullying**
- Stress
- Drug and alcohol abuse

Harassment: A type of discrimination that can be based on race, age, disabilities, sex, religion, cultural issues, health, or language barriers.

Bullying: Unwanted, aggressive behavior that involves a real or perceived power imbalance. This form of harassment may include offensive, persistent, insulting, or physically threatening behavior directed at an individual.

3.4.1 Harassment

Harassment is a broad term describing negative social behavior based on race, age, disabilities, gender, religion, cultural issues, health, or language barriers. It often takes the form of ethnic slurs, racial jokes, offensive or derogatory comments, and other verbal or physical conduct. Harassment can create an intimidating, hostile, or offensive working environment and interfere with an individual's work performance.

One type of harassment commonly discussed and reported is **sexual harassment**. While other forms of harassment may not have been talked about as much or as openly in the past, this issue has made headlines for many years.

Sexual harassment: A type of discrimination that results from unwelcome sexual advances, requests for sexual favors, or other verbal or physical behavior with sexual overtones.

When someone makes unwelcome sexual advances or exhibits verbal or physical behavior with sexual overtones, they are guilty of sexual harassment. Sexual harassment can happen between members of the opposite sex or those of the same gender. The harasser may threaten to fire the victim to keep them quiet. Sexual harassment is illegal, and if you experience it, you should report it to your supervisor immediately. Usually, sexual harassment is a pattern of behavior repeated over a period of time, so if you do not report it, you run the risk of experiencing it many times (*Figure 18*).

Figure 18 Always report harassment.
Source: iStock

Bullying may be more common than either sexual harassment or racial discrimination on the job, but it is not reported as often. The term *bullying* is perceived by most people to be related to school-age children. The US government defines bullying as the unwanted, aggressive behavior among school-age children that involves a real or perceived power imbalance. That definition covers a lot of ground. When bullying is discussed among adults, it is typically referred to as hazing, stalking, or simply harassment.

Most adults are uncomfortable using the term bullying when reporting such behavior, as they also perceive it to be related to school-age children. Using the term makes adult victims feel childish to report it as a result. Bullies may target those who appear to be a threat to them and feel the need to control the threat. Sometimes, it occurs simply because the individual feels stronger and wishes to exert control and feel superior.

Workers should understand that dealing with an adult bully can be like trying to work things out with a school-age bully. Adult bullies are often lifetime bullies, having started the practice at a young age. They are not interested in compromise or improvement. Bullying makes them feel important and in control. Adult bullies are hard to change through simple interaction and positive discussion.

Workplace bullying can come from an individual or even a group. Bullying, just like any other form of harassment, can cause serious emotional and physical illnesses such as anxiety, depression, and hypertension among victims.

With the use of today's electronic technology, bullies can even cause harm through social media and other public formats; this is referred to as *cyberbullying*. Once emailed rumors, malicious text messages, embarrassing pictures, videos, or fake profiles are posted on social media, the damage is difficult to undo. While the most common target for cyberbullying is school children, adults are not immune to it.

Once a bullying situation has been reported to your employer, the situation may not change immediately. Your supervisor and/or employer must work within the confines of the law. Careful documentation of events is very important, as it is in all forms of harassment. Clear documentation allows the situation to be handled properly by those in authority. Since there are laws prohibiting such behavior and your coworkers are not schoolchildren, legal action will often occur based on careful documentation.

3.4.2 Drug and Alcohol Abuse

Drinking is a common way to avoid or forget about stress. Alcohol is an accepted, legal part of modern social life. The moderate use of alcohol is socially acceptable and is generally believed to cause little or no harm to most people. Alcohol abuse, or habitually drinking to excess, is not only socially unacceptable but also poses a serious health and safety risk to everyone involved.

Although some countries and US states have legalized or decriminalized the recreational and/or medical use of marijuana—and others are contemplating it—it too can be abused. Marijuana slows reaction times and diminishes decision-making skills, impairing a user's response to critical events. This puts the user and those around them at risk of injury and poor performance. Regardless of whether alcohol and marijuana are considered socially acceptable, there is no room in the workplace for either substance. This is called **zero tolerance** among employers.

Zero tolerance: The policy of applying laws or penalties to even minor code infringements to reinforce its overall importance, typically related to drug and alcohol abuse, when applied to the workplace.

Illegal drug use is never acceptable, nor is the misuse of legal prescription drugs. Many companies require a clean drug test before offering a job or allowing an employee to begin work. Many employers in the construction and agricultural industries are often required to share records related to drug testing as part of their obligations to their clients. Drug testing can also be ordered randomly to test the current workforce and ensure compliance. Further, an accident in the workplace often leads to immediate drug testing to determine whether illegal substances are a factor. Several situations can lead to a drug test beyond a preemployment situation.

WARNING!

Working under the influence of drugs or alcohol can have serious consequences. It can lead to fatal accidents and serious injuries on the jobsite, affecting you and your coworkers, and can cause serious errors that affect the integrity of the work. In addition, causing harm to people or property in the workplace while under the influence of substances can also lead to personal liability for the results of a poor choice.

3.5.0 Teamwork and Leadership

You will interact daily with members of your work crew and your supervisor. Although getting along with people is important, it is not enough. The success of any team depends on its members fulfilling their roles competently. Everyone must contribute to the effort to ensure the team achieves its goals. This cooperation is referred to as *teamwork*.

You will be assigned to different teams from the moment your career begins for different reasons. Teams can be as small as two people and as large as your entire company. Teams often consist of people from different trades who work together to complete a task. For example, a team assigned to burying water and electrical lines might include a backhoe operator to dig the trench, pipefitters to lay and join the pipe, and electricians to run the electrical cable.

Always show respect for the other members of your team. Although you have a specific job, always be willing to help a coworker. Support your coworkers when needed, and they will support you in return.

Good team members are goal-oriented, and those goals are shared among team members. Being goal-oriented means ensuring that all activities benefit the team's final objective. Whatever the result is—fabricating fittings, laying pipe, or framing a house—everyone on the team knows the desired result and concentrates on achieving it.

Use your skills and strengths to the team's advantage (*Figure 19*). At the same time, accept your personal limitations and the limitations of others.

Figure 19 Utilize your team's strengths.
Source: iStock

To be a good team member, strive to do the following:

- Follow your team leader's and/or supervisor's directions.
- Accept that others might be better at some tasks than you.
- Keep a positive attitude when you work with other people.
- Recognize that the work you do is for the benefit of the entire company, not for you personally.
- Learn to work with people who work at different speeds.
- Accept goals that are set by someone else, not by you.
- Trust other team members to perform their tasks.
- Appreciate the work of others as much as you appreciate your own.

Keep in mind that you are not the only person working hard. Everyone on your team should be focusing on the goal too. Offer praise and encouragement to your coworkers, and they will do the same for you. This mutual respect will help you feel confident that your team will reach its goal. Share the credit for good work and be willing to take responsibility for your mistakes and errors. These actions will help you earn the respect of your coworkers.

If goal-oriented teamwork is practiced, the team should be able to meet deadlines and keep projects on schedule. This translates into time and money saved by your company that can be used for pay increases. Sometimes, your coworkers or other trades cannot start their tasks until you have completed yours. Always finish your work promptly—it demonstrates respect for your coworkers and your company's reputation. A team's success reflects on you, just as your personal behavior and characteristics reflect on the team.

An important part of teamwork is training. You will receive guidance and advice from more experienced colleagues as an apprentice. As you gain experience, you will be able to help less experienced colleagues. Training also offers an excellent opportunity to build good relationships.

Being asked to teach someone is an honor. It means someone believes you do your job well enough to teach it to others. Such confidence should inspire you to be the best teacher you can be. Be patient with the person you are teaching and teach by example. Offer encouragement and constructive criticism, as you learned in this module. Teach people in the same manner that you would like to be taught. Keep in mind that your teachers in the workplace are not likely to be trained educators. As a result, their approach may be less than perfect. Discount their limitations and focus on the skills or knowledge they share.

3.5.1 Leadership Skills

As you gain experience and credentials, you will likely assume positions of greater responsibility. These positions are earned through hard work and dedication. Starting as an apprentice, you can work up to team leader, foreman, supervisor, and project manager. With enough hard work, dedication, and ambition, you may even choose to run your own company someday.

Leadership: The ability to set an example for others to follow by exercising authority and responsibility.

To progress steadily through your career, you must develop **leadership** skills and learn how to use them. Leaders set an example for others to follow. Because of their skills, leaders are trusted with the authority to make decisions and are responsible for carrying them out.

You can become a leader at any stage in your career. As an apprentice, you have the authority to perform the task given to you by your supervisor, and your supervisor expects you to be a responsible worker. By carrying out your task quickly, correctly, and independently, you are setting an example for others to follow. In doing so, you are demonstrating leadership skills.

People with the ability to become leaders often exhibit the following characteristics:

- They lead by example.
- They have a high level of drive, determination, and persistence.
- They are effective communicators.
- They can motivate their team to do its best work.
- They are organized planners.
- They have self-confidence.

The functions of a leader vary with the environment, the group of workers being led, and the tasks to be performed. However, certain functions are common to all situations. Some of these functions include the following:

- Organizing, planning, staffing, directing, and controlling
- Empowering team members with authority and responsibility
- Resolving disagreements before they become problems
- Enforcing company policies and procedures
- Accepting responsibility for failures as well as for successes
- Representing the team to different trades, clients, and others

Leadership styles vary (*Figure 20*). If leadership styles are classified according to how a leader makes decisions, there are three broad categories of leadership: autocratic, democratic, and hands-off. An autocratic leader makes all decisions independently, without seeking recommendations or suggestions from the team. A democratic leader involves the team in the decision-making process; such a leader considers team members' recommendations and suggestions before deciding. A hands-off leader leaves all decision-making to the team members themselves.

Figure 20 Choose a leadership style that works for you and your team.
Source: iStock

Select a leadership style appropriate to the situation. Always leading with a single style generally results in real problems. You must consider your authority, experience, expertise, and personality each time a style is chosen. Leaders need to have the respect of people on the team; otherwise, they will be unable to set examples others are willing to follow.

Leaders must make sure that the decisions they make are ethical. Success in the construction and agricultural industries demands the highest standards of ethical conduct. You will encounter three types of ethics on the job: business or legal, professional, and situational. Business or legal ethics involve adhering to all relevant laws and regulations. Professional ethics involve being fair to everybody, and situational ethics involve appropriate responses to a particular event or situation.

Effective leaders motivate or inspire people to do their best. People are motivated by different things at different times. The following are some common ways to motivate people on the job:

- Recognize and praise a job well done.
- Allow people to feel a sense of accomplishment.
- Provide opportunities for advancement.
- Encourage people to feel that their job is important.
- Provide opportunities for change to prevent boredom.
- Reward people for their efforts.

Leaders who motivate people are more likely to have a team with high morale and positive attitudes. Morale and attitude are key components of a successful company with satisfied workers, and such a company is more likely to be successful.

3.0.0 Section Review

1. If you finish a task ahead of time, looking for another task that can be finished in the remaining time is referred to as _____.
   a. stealing someone else's job
   b. being tardy
   c. showing off
   d. taking initiative

2. To find a solution everyone can agree on, you may have to _____.
   a. compromise
   b. discuss it with your supervisor
   c. let someone else solve the problem
   d. come up with more than one solution

3. Constructive criticism works best when offered _____.
   a. often
   b. occasionally
   c. by a member of management
   d. loudly in front of coworkers

4. To put an end to personal bullying, one of the most critical things an employee can do is _____.
   a. confront the person one-on-one
   b. stage an intervention
   c. carefully document the events when they happen
   d. avoid the person in spite of work assignments

5. Before displaying their skills as a leader, all craftworkers must wait until they have a great deal of experience in their chosen craft.
   a. True
   b. False

Module 93109 Review Questions

1. Someone who can vouch for your skills, experience, and work habits is called a(n) _____.
   a. mission statement
   b. entrepreneur
   c. interviewer
   d. reference

2. One of the most important and productive methods to identify job opportunities is by _____.
   a. networking with family and friends
   b. posting a "Seeking Employment" ad on Facebook
   c. trying to call HR offices directly
   d. checking the Help Wanted section of the online local newspaper

3. A common barrier to effective problem-solving includes _____.
   a. overwork
   b. fear of change
   c. inadequate supervision
   d. procrastination

4. After putting a problem's solution into effect, it is important to _____.
   a. stop considering other solutions
   b. insist it become a company procedure
   c. monitor the results
   d. develop a business plan for it

5. Depending on the type of project and region of work, one of the *most* expensive components of a project is often _____.
   a. equipment rental
   b. power tools
   c. hand tools
   d. labor

6. Positive and negative interactions affect _____.
   a. managerial skills
   b. tool use
   c. relationships
   d. work schedules

7. A person who works without constant supervision is showing _____.
   a. initiative
   b. fortitude
   c. respect
   d. self-presentation

8. The process of solving disagreements between coworkers is called _____.
   a. tactfulness
   b. conflict resolution
   c. relationship crafting
   d. addressing conflict

9. Constructive criticism should not be given unless _____.
   a. someone asks your opinion
   b. the supervisor tells you to do so
   c. you feel it is your business
   d. you can also be complimentary

10. A broad term identifying negative social behavior that can be based on race, age, disabilities, gender, religion, cultural issues, health, or language barriers is _____.
    a. absenteeism
    b. racism
    c. cyberbullying
    d. harassment

11. An employer's policy is always "zero tolerance" regarding _____.
    a. being sick
    b. training
    c. alcohol and drug abuse
    d. overtime

12. Marijuana typically causes a person to _____.
    a. hallucinate
    b. accelerate their reaction times
    c. slow their reaction times
    d. be deprived of oxygen

13. When everyone in a group is focused on the final objective, they are _____.
    a. being goal-oriented
    b. being team players
    c. taking initiative
    d. being leaders

14. To progress steadily through your career, you will need to develop _____.
    a. leadership skills
    b. classroom skills
    c. cooking skills
    d. painting skills

15. The three categories of leadership are hands-off, democratic, and _____.
    a. automatic
    b. liberal
    c. autocratic
    d. conservative

Answers to odd-numbered questions are found in the Review Question Answer Keys at the back of this book.

Answers to Section Review Questions

Answer	Section Reference	Objective
Section 1.0.0		
1. c	1.1.0	1a
2. b	1.2.0	1b
Section 2.0.0		
1. a	2.1.0	2a
2. d	2.2.0	2b
3. a	2.3.1	2c
Section 3.0.0		
1. d	3.1.2	3a
2. a	3.2.1	3b
3. b	3.3.1	3c
4. c	3.4.1	3d
5. b	3.5.1	3e

User Update

Did you find an error? Submit a correction by visiting **www.nccer.org/olf** or by scanning the QR code using your mobile device.

SCAN ME

MODULE 93110

Introduction to Material Handling

Source: iStock

Objectives

Successful completion of this module prepares you to do the following:

1. Identify the basic concepts of material handling and common safety precautions.
   a. Describe the basic concepts of material handling and manual lifting.
   b. Identify common material handling safety precautions.
   c. Identify and describe how to tie knots commonly used in material handling.
2. Identify various types of material handling equipment and describe how they are used.
   a. Identify nonmotorized material handling equipment and describe how they are used.
   b. Identify motorized material handling equipment and describe how they are used.

Performance Tasks

Under supervision, you should be able to do the following:

1. Demonstrate safe manual lifting techniques.
2. Demonstrate how to tie two of the following common knots:
   - Square
   - Bowline
   - Half hitch
   - Clove hitch

Overview

Lifting, stacking, transporting, and unloading materials such as bricks, feed, crates, and various supplies are routine tasks on a jobsite. Whether performing these tasks manually or with the aid of specialized equipment, workers must follow basic safety guidelines to keep themselves and their coworkers safe. This module provides guidelines for using the appropriate personal protective equipment (PPE) for the material being handled and using proper procedures and techniques to carry out the job.

Digital Resources for Agricultural Technology

SCAN ME

Scan this code using the camera on your phone or mobile device to view the digital resources related to this craft.

NCCER Industry-Recognized Credentials

If you are training through an NCCER-accredited sponsor, you may be eligible for credentials from NCCER. The ID number for this module is 93110. Note that this module may have been used in other NCCER curricula and may apply to other level completions. Contact NCCER at 1.888.622.3720 or go to **www.nccer.org** for more information.

You can also show off your industry-recognized credentials online with NCCER's digital credentials. Transform your knowledge, skills, and achievements into credentials that you can share across social media platforms, send to your network, and add to your resume. For more information, visit **www.nccer.org**.

1.0.0 Material Handling

Performance Tasks

1. Demonstrate safe manual lifting techniques.
2. Demonstrate how to tie two of the following common knots:
   - Square
   - Bowline
   - Half hitch
   - Clove hitch

Objective

Identify the basic concepts of material handling and common safety precautions.

a. Describe the basic concepts of material handling and manual lifting.
b. Identify common material handling safety precautions.
c. Identify and describe how to tie knots commonly used in material handling.

NOTE

Lower back injuries are common in the agricultural industry. Common activities contributing to the risk include lifting heavy objects, twisting, and falls. A recent survey confirmed that approximately 40% of agricultural workers report back injuries at some point during the year.

Manual material handling is a common task on most construction and agricultural jobsites. Improper or unsafe material handling is one of the leading causes of nonfatal, work-related injuries. According to the Bureau of Labor Statistics, each year more than one million workers suffer from back injuries, and one of every five workplace injuries or illnesses is associated with back injuries.

Most tasks performed during construction activities and daily agricultural operations involve the handling of some type of material or load such as wood, brick, lumber, pipe, or other supplies. A load is simply the quantity of material that can be carried, transported, or relocated at one time by a machine, vehicle, piece of equipment, or person. Because the risk of injury is present whenever materials are being handled, workers should always be conscious of what they are doing and follow basic material handling safety rules.

1.1.0 Material Handling Basics

To reduce the risk of injury when manually handling material, plan the task before doing it, wear the appropriate personal protective equipment (PPE) and follow proper lifting procedures. Also be aware of hazards when working from heights or working near suspended loads.

1.1.1 Pre-Task Planning

When it comes to material handling, it is just as important to be mentally fit as it is to be physically fit. Most material handling accidents occur when workers are new and relatively inexperienced at a job or when very experienced workers think that accidents cannot happen to them. Before handling any material, consciously think about what you are doing.

Before attempting to lift any material, always assess the situation by doing the following:

- Check to make sure the load is not too large, heavy, or hard to grasp.
- Make sure the load does not have protruding nails, wires, or sharp edges.
- Make sure the material is something that you can lift by yourself. If not, ask a coworker for assistance.
- Inspect the path of travel. Look for slip, trip, or fall hazards. If there are hazards in the path of travel, move them, or go around.
- Always read the warning labels or instructions on materials before they are moved and be aware of the potential dangers associated with mishandling a particular product.

1.1.2 Personal Protective Equipment (PPE)

Wear the proper clothing and PPE when moving or handling materials. Remember the following safety guidelines when dressing to perform material handling operations:

- Do not wear loose clothing that can get caught in moving or protruding parts.
- Be sure to button shirt sleeves and tuck in shirt tails.
- Remove all rings and jewelry.
- Tie back and secure long hair underneath your hard hat.
- If wearing a wristwatch, wear one that will easily break away if it gets caught in machinery.
- Wear gloves whenever cuts, splinters, blisters, or other hand injuries are possible. Select gloves that fit properly. Gloves that are too tight may increase hand fatigue. However, the gloves must fit snug and offer reasonable dexterity. Loose gloves reduce grip strength and may get caught on moving objects or machinery.
- Remove gloves when working with rotating machinery and equipment with exposed moving parts.
- Follow the policies of your employer and the jobsite at large.

1.1.3 Proper Lifting and Lowering Procedures

To reduce the risk of back injuries, use proper lifting techniques. Determine the weight of the load and plan the lift prior to lifting. Know where it is to be unloaded and remove any slipping or tripping hazards from the path of travel.

When lifting objects, avoid unnecessary physical stress and strain. Know your limits and what you can physically handle when lifting a load. If the load is too heavy for you to lift on your own, ask an associate or coworker for help.

As you lift an object, make sure you have firm footing. Bend your knees and get a good grip. Lift with your legs, keep your back straight, and keep your head up. Keep the load close to your body (*Figure 1*). Do not reach out with the load. Never turn or twist until you are standing straight, then pivot your feet and body.

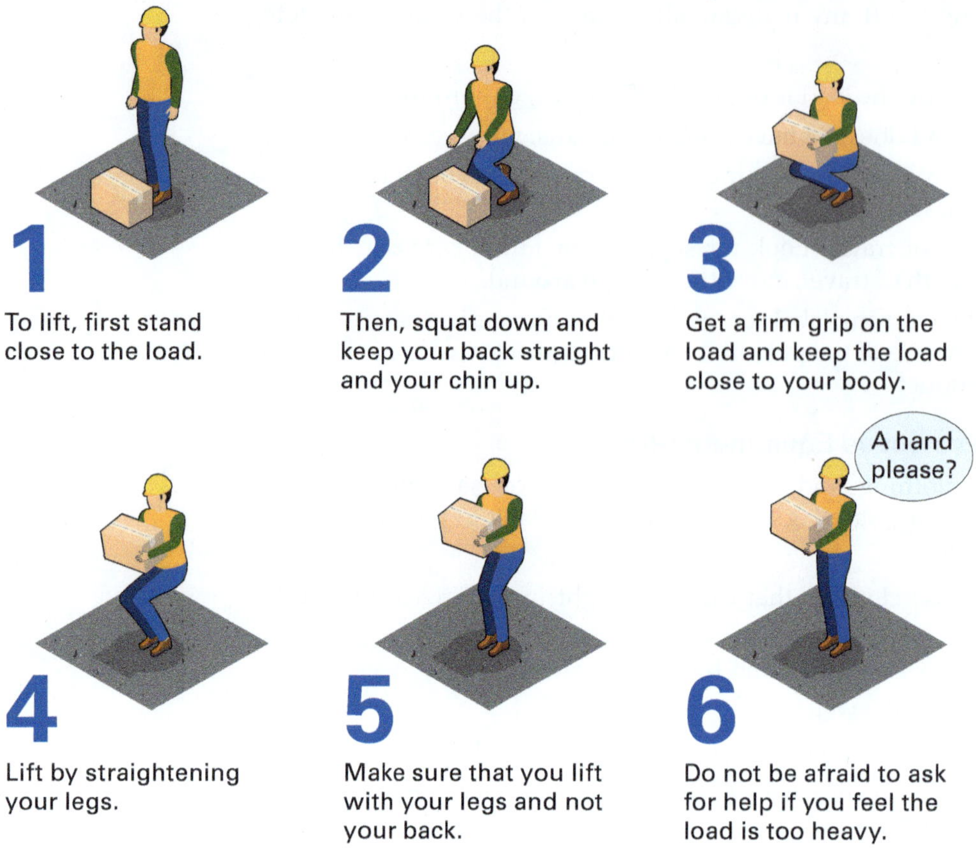

Figure 1 Proper lifting technique.

When unloading materials, use your leg muscles to set the load down. Space your feet far enough apart to maintain good balance and control of the load. When placing the load down, move your fingers out of the way to avoid pinching or crushing them. When possible, attach handles to loads to reduce the chance of injuring your fingers.

Ask for assistance from an associate or coworker in the following situations:

- The load is too heavy (weighing 50 pounds or more) for one person to carry alone.
- The objects to be handled are longer than 10', such as lumber, conduit, pipe, and scaffolding poles.
- The objects to be handled, such as plywood and tarps, can be affected by wind gusts.

Sometimes it may not be the weight of a load that is troublesome, but its configuration. Secure long or cumbersome objects. Floppy rods, pipe, or lumber should be tied together in several places before you try to carry them. Never carry more pieces than you can control.

When handling long objects, such as scaffold poles or planks, be aware of objects and persons that may be struck by them as you turn your body. Never swing around quickly or without checking your surroundings. Never pass any material to another worker unless the person is looking directly at you and expecting to receive it.

Most workers are trained about the dangers of trying to lift heavy loads. Likewise, it is important to talk about lowering heavy loads from overhead (*Figure 2*). Reaching and lowering heavy loads from overhead can be very dangerous. When you are lifting something, you can always stop and put it down if you find that it is too heavy to handle alone. However, if you are lowering something from overhead, it is usually too late by the time you realize it is too heavy. Also, reaching above shoulder height can cause stress to the shoulders and back, and often results in awkward hand and wrist postures.

Figure 2 Plan properly when lifting objects from overhead.
Source: iStock

Think About It

Heavy Objects in Agriculture

Many everyday activities include lifting heavy or awkward objects. On the farm, these objects include heavy hand tools, machinery components, loaded wheelbarrows, and feed containers. Can you think of other objects you lift every day that can cause injury?

Source: iStock

Before attempting to lower overhead loads, consider the following:

- Size up the load. If it looks too heavy to have been lifted by one person to its current location, then it is probably too heavy for one person to take down.
- Ask yourself, "How did it get up there?" Was it put there by a lift truck or by more than one person? The way it got up is probably the best way to get it back down.
- Lower objects down the same way you would lift them up. Keep your knees bent and your back straight. If you must place a load to one side or another, move your feet instead of twisting your body.
- Treat the area that the material is being lowered into as a fall zone. Do not allow people to enter the area while the load is being lowered.

Back injuries can easily occur when lifting and lowering materials. Keep in mind that exercise, good posture, and a healthy weight and diet can help prevent such injuries. You can also ask your doctor about muscle-strengthening techniques for your back.

1.2.0 Material Handling Safety

When working on a jobsite, make sure that all workers, materials, and equipment are safe from unexpected movement, such as falling, slipping, tipping, rolling, blowing over, or any other uncontrolled motion.

1.2.1 Stacking and Storing Materials

To work efficiently and effectively, workers should properly stack and secure materials. Taking care to stack and secure materials correctly can save space and time and helps avoid accidents and injuries. Stacking materials keeps them from interfering with other activities and makes them more organized and readily accessible for use. Stacking materials also helps to create safe storage and allows more room in the work area or on the jobsite.

Always properly store equipment and materials when they are not in use. Learn your company- and site-specific rules, as well as the guidelines of the US Occupational Safety and Health Administration (OSHA) when working in the United States. Failing to store materials properly can lead to accidents, injuries, wasted materials, increased job costs, and project delays. For example, failing to remove nails from reusable lumber prior to storing it can cause puncture wounds when workers take the lumber out again. Stacking materials on the ground when they belong on a pallet can result in having to move them again for safe transport, and can also damage the materials, especially perishable products (*Figure 3*). Leaving tools out when they should be stored can create an untidy worksite, which can lead to accidents, tool damage, and misplaced items. Take the time to learn where and how to store materials properly. If you are unsure about something, ask your supervisor.

Observe the following guidelines when stacking and storing materials:

- Keep aisles and passageways clear in the storage area.
- Keep materials that are stored in cartons away from rain and moisture.
- Never stack cartons higher than the height listed on the carton (*Figure 4*).
- Stack lumber on level, solidly supported surfaces, and remember to remove all nails.
- Stack pipes neatly and chock them so that they cannot roll (*Figure 5*).
- Stack pipes and fittings according to size so workers do not have to dig to find the piece required for the job. Make sure pipes are properly chocked.

Figure 3 Store items for ease of transport and protection from damage.
Source: iStock

- Chock all material and equipment such as pipes, drums, tanks, reels, trailers, and wagons, as necessary to prevent shifting or rolling. Make sure the chock size and shape are appropriate for the material.
- Tie down or band all light, large surface area materials that might be moved by the wind.
- Do not stack bagged material the same width all the way to the top of the pile. Stack bagged material by stepping back the layers and cross-keying the bags at least every 10 bags high (*Figure 6*).
- Stack bricks no higher than 7'. Taper brick stacks back 2" for every foot above 4' (*Figure 7*).
- Taper masonry blocks back one-half block per tier above 6' (*Figure 8*).
- Store flammables, such as gasoline, in a well-ventilated area away from danger of ignition or in an approved flammable storage cabinet.

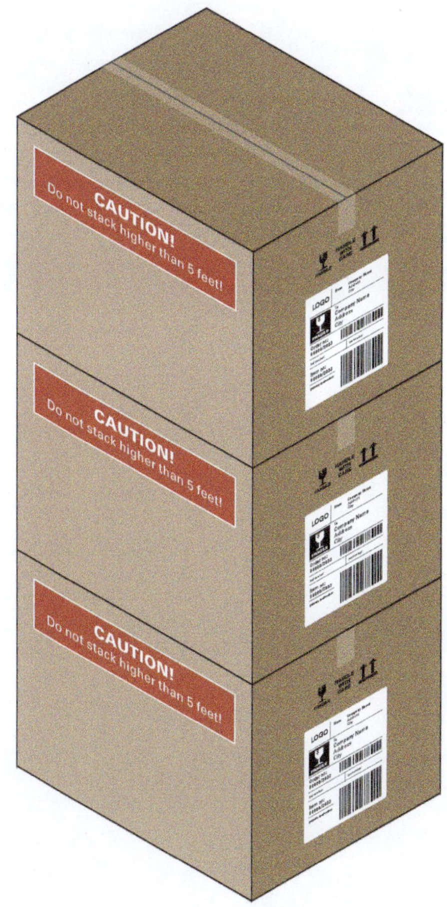

Figure 4 Properly stacked cartons.

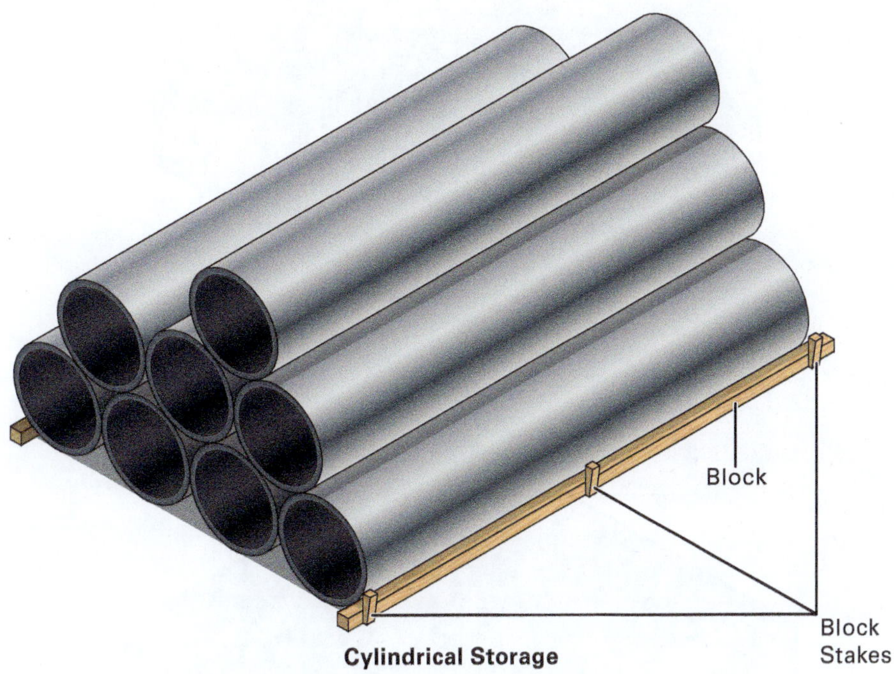

Figure 5 Properly stacked pipes.

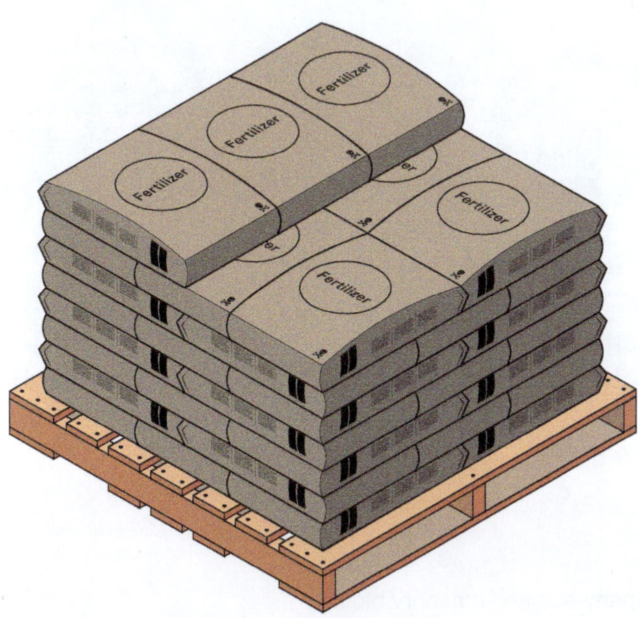

Figure 6 Properly stacked bags.

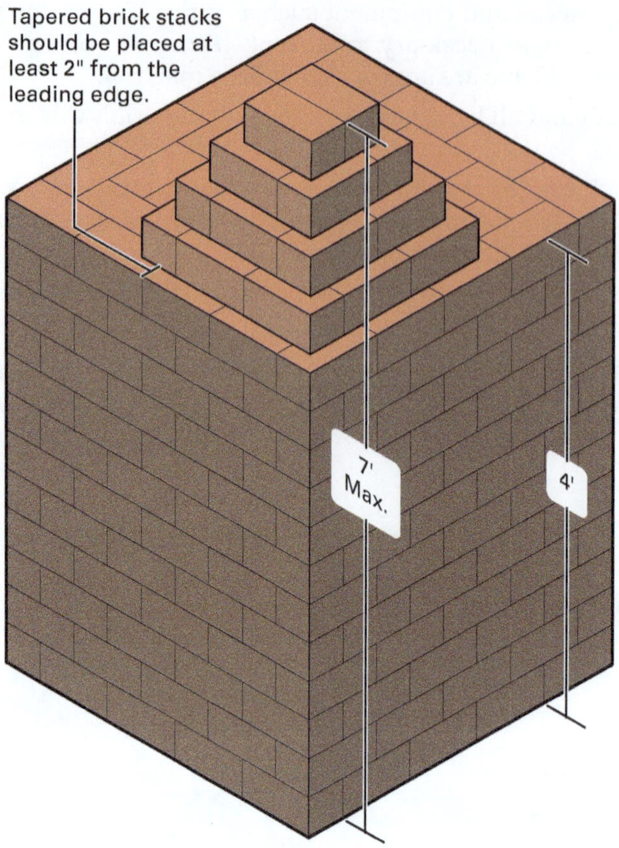

Figure 7 Properly stacked bricks.

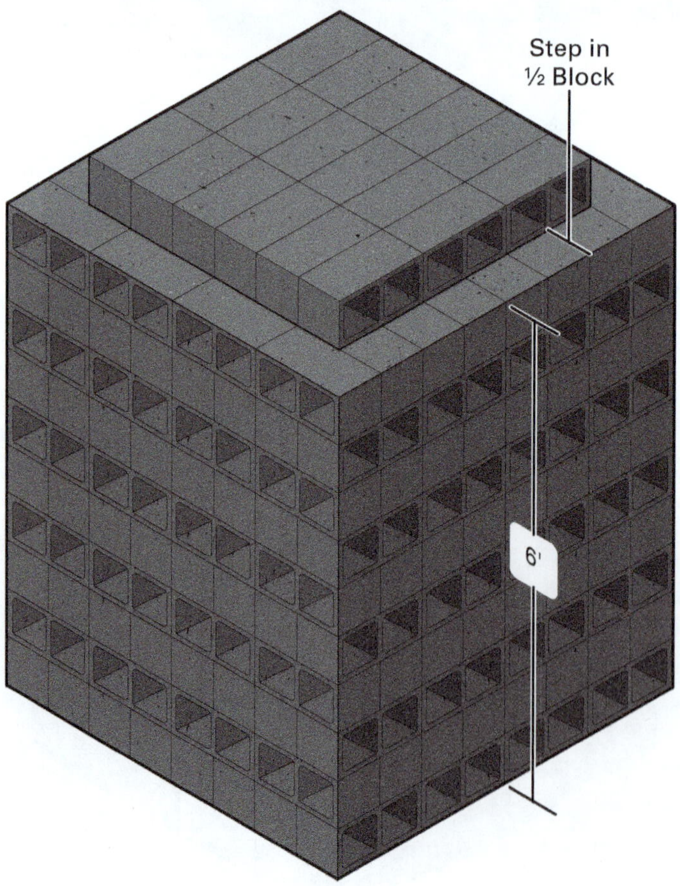

Figure 8 Properly stacked masonry blocks.

1.2.2 Working from Heights

Moving material to a higher location requires good planning and safety awareness. Make sure you use a safety harness with a fall arrest lanyard connected to an appropriate anchor point when working from heights over 6'; a positioning belt with a lanyard can be added if required. Never throw tools or materials up to a worker on a higher level or down to a worker on a lower level. Instead, use a rope to raise or lower tools and materials to and from an elevated work area. Once a material is lifted to an elevated level, tie off the rope and secure the material to keep it from falling. Knots that are commonly used in material handling will be covered later in this module.

When climbing up or down a ladder, check your pockets and tool belt to ensure that no objects will fall out of them or get caught on something along the way. Always keep your hands free when climbing a ladder. Never carry tools or materials by hand up a ladder or throw material up to or down from a work platform. Place objects in a bucket to convey them to and from the platform.

Never stack or store materials on scaffolds or runways. Stacking materials on scaffolds or storing materials in high-traffic areas can increase the chances of falling object hazards.

Going Green

Reduce, Reuse, Recycle!

When cleaning up materials around the jobsite, check to see if they can be reused or recycled. For example, rather than putting plant remains in the trash, look for opportunities like a compost pile for reuse.

Source: iStock

1.3.0 Knots for Material Handling

Knots and hitches are common ways of fastening ropes or lines either to each other or to another object, such as a beam or container. Most knots are meant to be permanent; they must be unwoven to be unfastened. However, some knots, which are generally referred to as hitches, are not permanent. They can be undone by simply pulling the ropes in the reverse direction from which the knots are meant to hold.

Knots and hitches are used for many different material handling tasks, such as securing loads, stabilizing scaffolding, and lifting and lowering tools and supplies. Using the correct type of knot for the task at hand is critical. This section will cover four knots that are commonly used in material handling: the **square knot**, the **bowline**, the **half hitch**, and the **clove hitch**.

First, some knot-tying terminology needs to be reviewed. The **standing end** is the long end of the rope, or the end that is not being knotted. The **working end** is the short end, or the end of the rope that is being used to tie the knot. The **standing part** is the portion of the rope that lies between the standing end and the working end. Finally, to **capsize** means to change the form and rearrange the parts of a knot, usually by pulling on specific ends of the knot. This may be done intentionally to undo a hitch. On the other hand, if certain knots are used inappropriately, they can easily—even spontaneously—capsize, resulting in serious safety hazards.

There are two other basic points to keep in mind when learning how to tie knots. First, many knots can be tied in more than one way. If the chosen technique is followed correctly, the final structure of the knot will be the same, no matter which technique was used to tie it. Second, knots can be tied right-handed or left-handed, depending on the dominant hand of the person who is doing the tying. A knot that is tied left-handed will be the mirror image of the same knot tied right-handed. Note, however, that a few knots may have a right-handed element and a left-handed element. Failing to maintain these elements while tying can produce the wrong knot.

Square knot: A knot made of two reverse half-knots and typically used to join the ends of two ropes of similar diameters; also called a *reef knot*.

Bowline: A knot used to form a loop that neither slips nor jams; sometimes referred to as a *rescue knot* or the *king of knots*.

Half hitch: A knot tied by passing the working end of a rope around an object, across the standing part of the rope, and then through the resulting loop; often used as an element in forming other knots or added to make other knots more secure.

Clove hitch: A knot that consists of two half hitches made in opposite directions; used to temporarily secure a rope to an object.

Standing end: The end of a rope that is not being knotted.

Working end: The end of a rope that is being used to tie a knot.

Standing part: The portion of a rope that is between the standing end and the working end.

Capsize: To change the form and rearrange the parts of a knot, usually by pulling on specific ends of the knot.

1.3.1 Square Knot

A square knot (*Figure 9*), also called a *reef knot*, can be used to tie the ends of a rope around an object, such as a parcel or the neck of a sack. It is commonly used to join two lengths of rope together in low-strain applications. The two ropes should be roughly of the same diameter; otherwise, the knot is more likely to slip.

WARNING!

Square knots can capsize easily and should never be used for critical loads. Additional half-knots are often added to square knots for increased security.

Figure 9 Square knot.

Follow these steps (see *Figure 10*) to tie a square knot:

Step 1 Bring the two working ends together with one end oriented to the left (red rope) and the other to the right (blue rope).

Step 2 Cross them left over right. (The red rope goes left over the blue.)

Step 3 Pass the left-oriented (red) rope under the right-oriented (blue).

Step 4 Bring the left-oriented (red) rope back around the right-oriented (blue) rope so that they are both facing up. This forms a half-knot.

Step 5 Repeat the over-and-under process but go right over left this time. (The red rope goes right over the blue rope.)

Step 6 This forms a second half-knot. Note that this is a key step in tying a square knot. The working end that starts out on the top when you tie the first half-knot must continue to be on top when you switch directions to tie the second half-knot.

Step 7 Pull the ends tight to make a neat, symmetrical square knot.

A very common error when attempting to tie a square knot is to tie the second half-knot in the wrong direction. This produces a highly insecure knot called the granny knot (*Figure 11*). Compare the granny knot to the properly tied square knot in *Figure 9*.

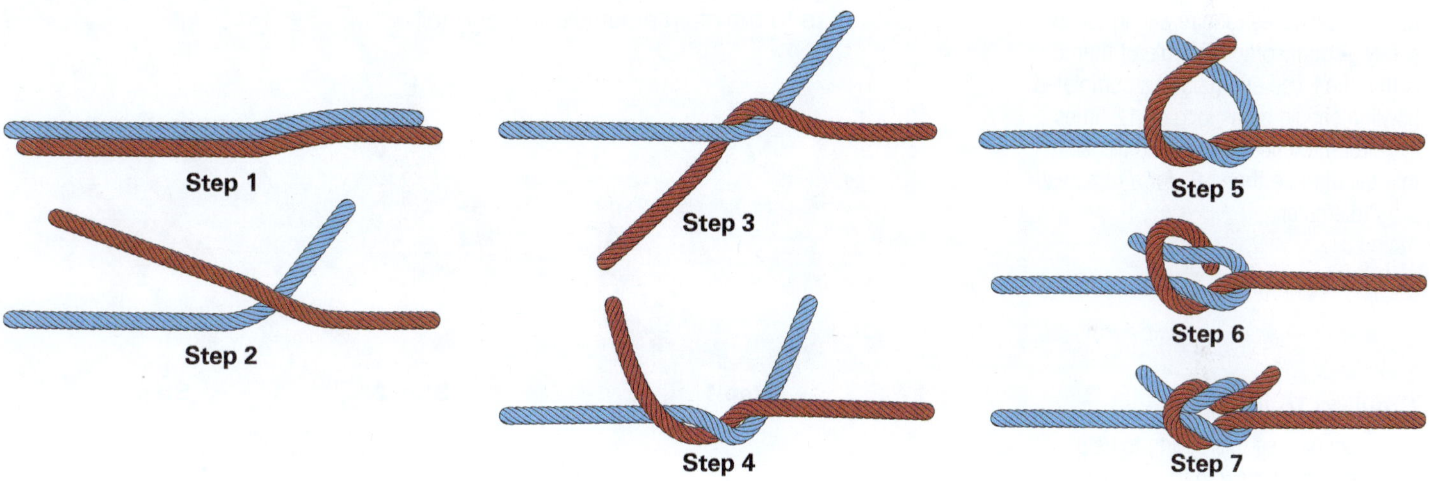

Figure 10 Tying a square knot.

Figure 11 Granny knot.

Figure 12 Bowline.

The Bowline

The bowline was originally used to secure a ship's square sail forward, closer to the wind. This knot was mentioned in *The Seaman's Dictionary* of 1644.

The bowline reduces the strength of the rope that it is tied in by as much as 40%. Bowlines require a long tail for safety. A rarely practiced rule of thumb is that the loose end (tail) of a completed bowline should be as long as 12 times the circumference. This would translate to a tail of more than 18" for a rope with a 1/2" diameter.

Bowline Trivia

Some people use this saying to help them remember how to tie a bowline: "The rabbit comes out of his hole, around a tree, and back into the hole."

1.3.2 Bowline

The bowline (*Figure 12*) is used to form a secure loop in the end of a rope. It is sometimes called a *rescue knot*, or the *king of knots*, because it is reliable enough to be used for rescue work (usually backed up with a stopper knot for extra security or tied twice into a double bowline for extra strength). The bowline is useful for many smaller jobs such as securing a sack of material or a bucket of tools to be hauled up to workers on a platform. Two bowlines can also be linked together to join two ropes when the strain is more than a square knot is capable of bearing.

A bowline does not slip or bind when under load, but it is easy to untie when there is no load. Because a bowline cannot be tied or untied when there is a load on the standing end, avoid using this knot if the rope may have to be released while it is under load.

Follow these steps (*Figure 13*) to tie a bowline:

Step 1 Form a small loop in the rope, leaving a tail long enough to make the desired size of main loop.

Step 2 Pass the tail of the rope through the small loop.

Step 3 Pass the tail under the standing end.

Step 4 Pass the tail back down through the small loop.

Step 5 Adjust the main loop to the required size and tighten the knot, making sure to maintain a sufficiently long tail.

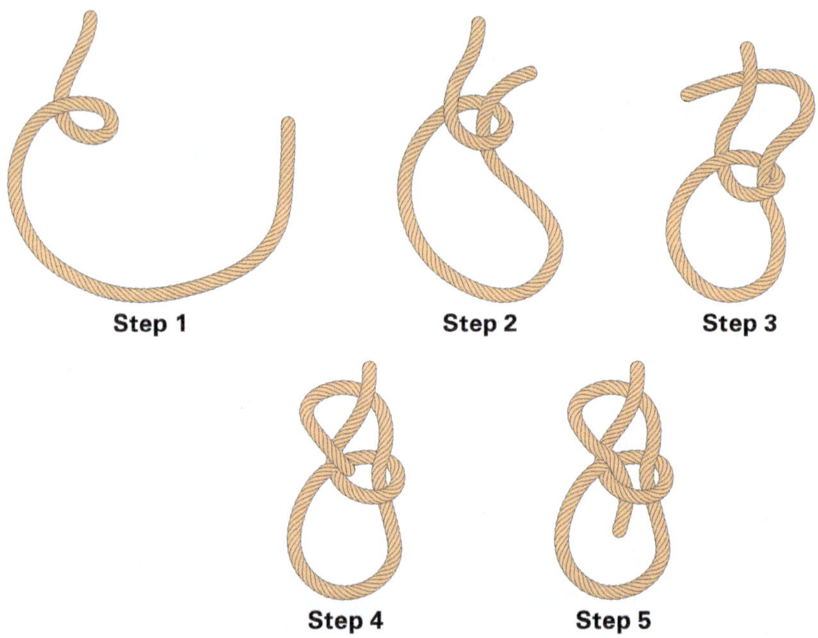

Figure 13 Tying a bowline.

1.3.3 Half Hitch

A half hitch (*Figure 14*) ties a rope around an object such as a rail, bar, post, or ring. It is commonly used for tasks such as suspending items from overhead beams and carrying light loads that must be removed easily. The half hitch is also widely used in making other knots, such as the clove hitch. Because the half hitch is not a very stable knot by itself, it is not suitable for heavy loads or tasks in which safety is paramount. It is often used to back up and secure another knot that has already been tied.

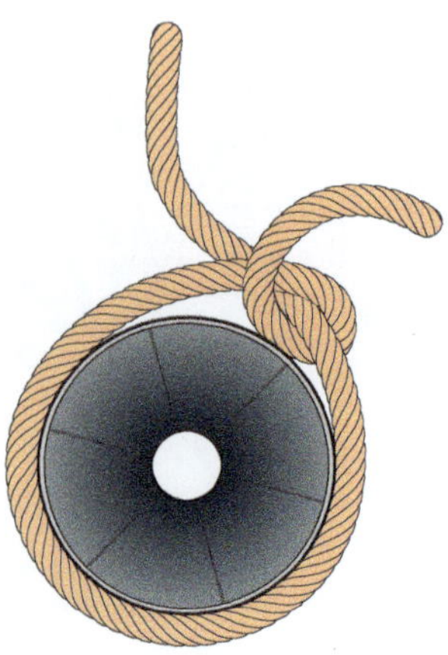

Figure 14 Half hitch.

Follow these steps (*Figure 15*) to tie a half hitch:

Step 1 Form a loop around the object.

Step 2 Pass the tail of the rope around the standing part and through the loop.

Step 3 Tighten the hitch by pulling on the working end and the standing part of the rope simultaneously.

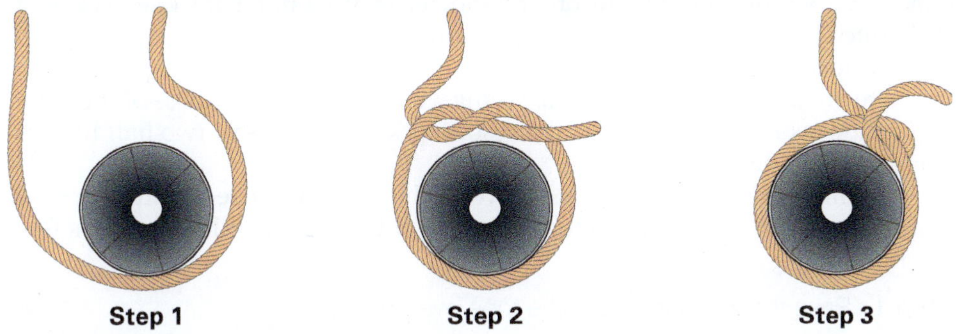

Figure 15 Tying a half hitch.

The Farmer's Loop

The farmer's loop knot is a midline loop made by wrapping a rope around the palm of the hand three times, then pulling the middle turn up from the top edge three times (first right, then left, then right again) until the new middle rope position is turned up to form a loop and tightened for use. Strain can be applied from the loop to either end of the rope, or from between the two ends. This knot is secure and strong, making it a great candidate for hand holds, climbing, scouting, and boating.

Howard W. Riley, a Cornell University professor, referenced this knot in a pamphlet in 1912. The pamphlet is the only known publication of this knot, demonstrated by a farmer at the Genesee County Fair (Bativia, NY) a couple of years earlier in 1910.

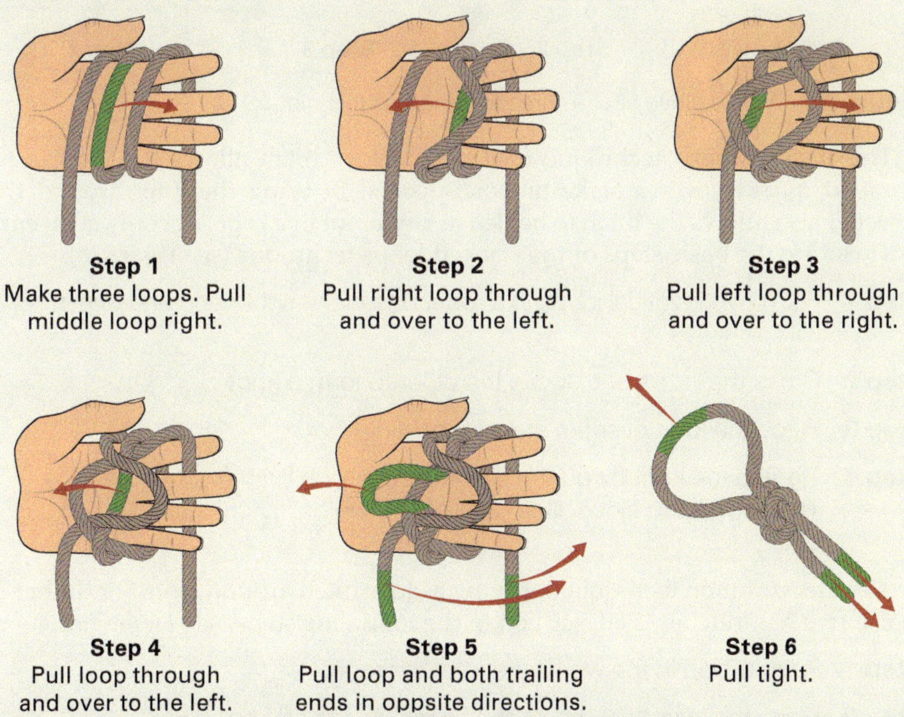

1.3.4 Clove Hitch

NOTE

Records of the clove hitch date back at least 10,000 years, making it an ancient knot.

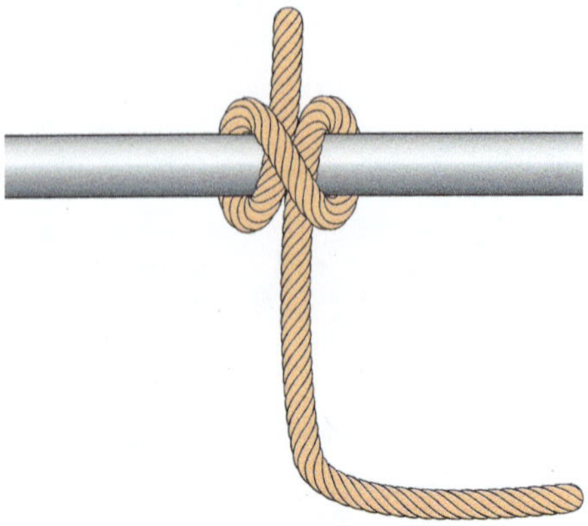

Figure 16 Clove hitch.

A clove hitch (*Figure 16*), sometimes called a *builder's hitch*, is one of the most widely used general hitches. It is typically used to make a quick and secure tension knot on a fixed object that serves as an anchor, such as a post, pole, or beam. A clove hitch can also be used as the first knot when lashing items together.

Because it is a tension knot, a clove hitch comes loose as soon as tension is removed from the rope. On the other hand, it can also bind. For these reasons, a clove hitch should not be used by itself. Additional half hitches can be added to make a clove hitch more secure.

There are alternate techniques for tying a clove hitch. Regardless of the technique, however, the structure of the resulting knot consists of two half hitches made in opposite directions.

A common technique for tying a clove hitch that will be used to attach a rope to a ring or post is to thread the end, as described in the following steps (see *Figure 17*):

Step 1 Pass the working end of the rope over the object.

Step 2 Pass the working end back over the standing part and then over the object. This forms a half hitch.

Step 3 Thread the working end back under itself and up through the loop to form a second half hitch.

Step 4 Pull both the working and standing ends evenly to tighten the hitch.

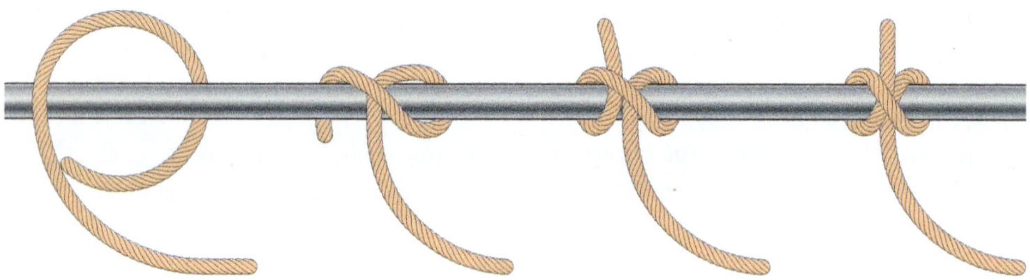

Figure 17 Tying a clove hitch—threading the end technique.

The stacked loops technique for tying a clove hitch allows the rope to be dropped quickly over a stake or post, instead of tying the knot around the object. It also allows the hitch to be tied at any point in a rope, not only at an end.

These are the basic steps of the stacked loops technique (see *Figure 18*):

Step 1 Form two identical loops in a rope, one in the right hand and one in the left.

Step 2 Cross the loops one above the other to form a knot.

Step 3 Place the knot over the stake or post.

Step 4 Tighten the knot by pulling simultaneously on both ends of the rope. The completed clove hitch consists of two half hitches stacked on each other.

Another common technique for tying a clove hitch around a post or timber is to use two half hitches, as described in the following steps (see *Figure 19*):

Step 1 Form a loop in the working end of the rope.

Step 2 Place the loop over the post (this makes the first half hitch).

Step 3 Form a second loop identical to the first.

Step 4 Place this loop over the post as well (making a second half hitch) and tighten to complete the clove hitch.

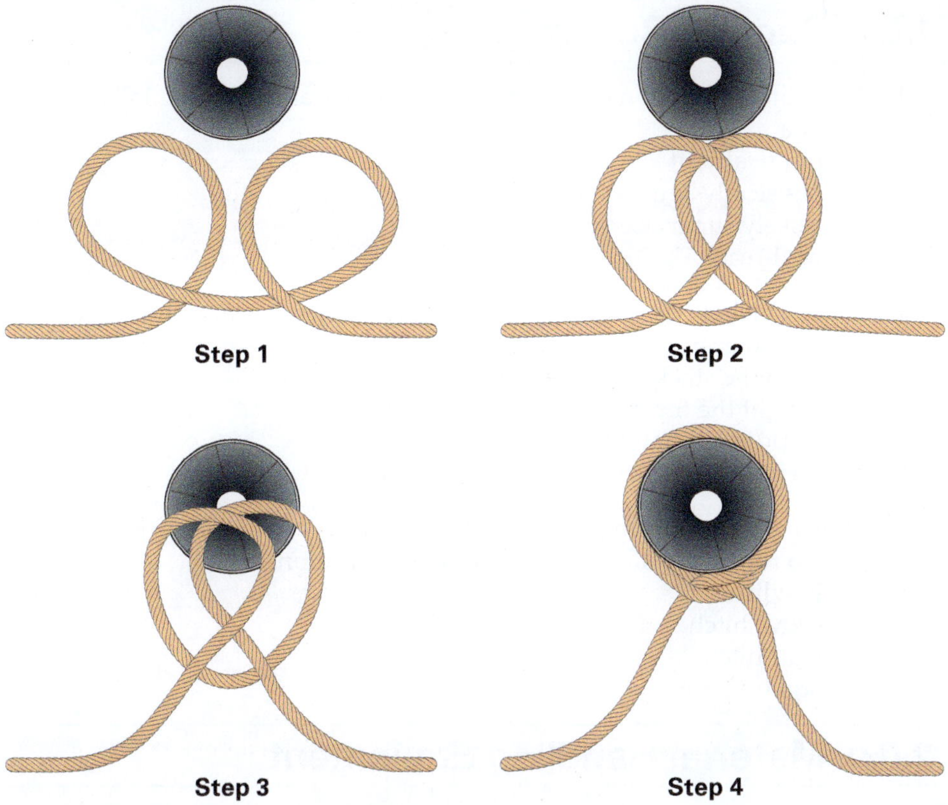

Figure 18 Tying a clove hitch—stacked loops technique.

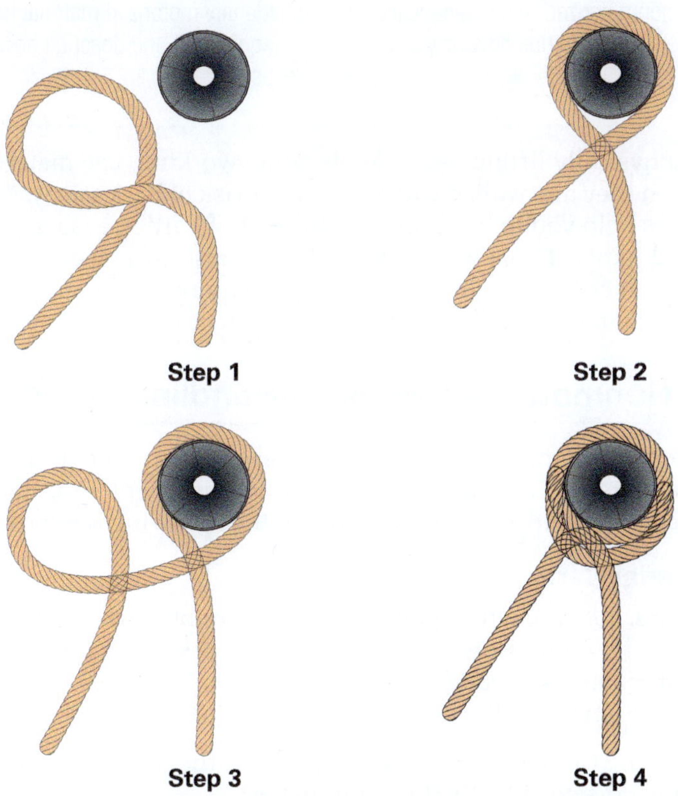

Figure 19 Tying a clove hitch—two half hitches technique.

The Ashley Book of Knots

Although *The Ashley Book of Knots* by Clifford W. Ashley is no longer in print, sailors, climbers, campers, macramé artists—anyone with more than a passing interest in knots—will continue to find this book cited as the definitive work on the subject of knot tying. Originally published by Doubleday & Company in 1944, it contains the histories of and instructions for tying over 3,900 types of knots, accompanied by 7,000 pen-and-ink drawings.

Clifford Ashley was born in 1881, in New Bedford, Massachusetts. By trade he was an author and artist, but while sailing on many types of boats and researching knots, he performed a wide variety of jobs. He spent 11 years writing and drawing the illustrations for *The Ashley Book of Knots*. He died three years after the book's publication. Ashley continues to be regarded as a fine marine painter as well as one of the world's leading authorities on knot tying.

1.0.0 Section Review

1. For material handling tasks, it is just as important to be mentally fit as it is to be _____.
   a. physically fit
   b. physically aggressive
   c. closely supervised
   d. well-read

2. When stacking or storing light materials that have a large surface area and could easily be moved by the wind, _____.
   a. keep the stack height less than 4'
   b. weight the top of the stack with clamps
   c. tie down or band the materials
   d. secure the materials in a flameproof cabinet

3. Which of the following is a type of knot that is often used to join the ends of two ropes in noncritical, low-strain applications?
   a. Bowline
   b. Clove hitch
   c. Half hitch
   d. Square knot

2.0.0 Material Handling Equipment

Performance Tasks

There are no Performance Tasks in this section.

Objective

Identify various types of material handling equipment and describe how they are used.

a. Identify nonmotorized material handling equipment and describe how they are used.

b. Identify motorized material handling equipment and describe how they are used.

Instead of physically lifting heavy loads while working, use material handling devices when they are available to reduce your risk of back injury. Material handling devices help you to work smarter instead of harder, easing stress on your muscles and joints. These work-saving devices also increase productivity and reduce the chance of dropping and damaging equipment or supplies. There are two types of material handling equipment: nonmotorized and motorized.

2.1.0 Nonmotorized Material Handling Equipment

When using nonmotorized, or manual, material handling equipment, you use the force and strength of your body to move the equipment. Examples of manual material handling equipment are covered in the following sections.

2.1.1 Material Carts

Material carts: Four-wheeled devices used to transport materials around a jobsite.

Material carts, also known as platform trucks, are platforms or boards that are laid horizontally on four caster wheels. These types of carts are typically used to transport materials around a jobsite (*Figure 20*). When using a material cart, follow these safety guidelines:

- Before using carts with caster wheels, inspect them to ensure that the casters will roll and swivel freely during transport.
- When using a cart with pneumatic tires, check the tires' air pressure before use. Tires that will not hold pressure or those that have flat spots must be replaced.
- Check to make sure that the surface has adequate traction to avoid slips.

- When moving a cart, keep your hands away from the edges of the handles to avoid pinching, crushing, or cutting your hands.
- Make sure your load is centered and secured so it does not roll or fall off.
- If materials are protruding out in front of the cart, have a **spotter** walk ahead of you to help you avoid contact with objects or other workers.
- Use caution when moving a cart on an inclined or declined surface, and never load a cart past its labeled weight capacity. If the load is too heavy on a decline, it may drag you down with it. If the load is too heavy on an incline, it may back over you. The weight capacity must be marked in plain view on the side of the cart.

Spotter: A person who walks in front of another worker who is carrying or transporting a long load to ensure there is a clear, unobstructed path.

Figure 20 Material cart.

2.1.2 Hand Trucks

Hand trucks, also known as dollies, are two-wheeled carts that are used to transport large, heavy loads, such as appliances or stacks of heavy boxes. Hand trucks are vertical material handling devices that have a metal blade at the bottom that is inserted beneath the load. Many hand trucks are equipped with ratchet straps to secure loads. The entire assembly tilts backward so that it may be easily pulled or pushed (*Figure 21*).

Before using a hand truck, inspect the framework for stress fractures and check the tires. When loading a hand truck, tilt the object to be moved forward and slide the blade of the hand truck under the object. Hold the object against the hand truck's framework as you tilt back on the wheels to load the object onto the hand truck. Use a ratchet strap to secure the load; rubber or bungee-style cords may not hold the load if the weight shifts. Simply buckle the strap around the load and tighten the belt snugly. Then tilt the hand truck back and move the load.

Hand trucks: Two-wheeled carts that are used to transport large, heavy loads; also known as a *dolly*.

Figure 21 Hand truck.

2.1.3 Cylinder Carts

Cylinder carts: Two-wheeled carts that are used to transport cylinders, or bottles, of compressed gases.

High-pressure gas cylinders, or bottles, should be transported on **cylinder carts**, which are designed specifically to transport compressed gases (*Figure* 22). The pressurized gas contained in cylinders can pose explosion risks, even if the gas itself is inert (meaning it is not chemically reactive).

Follow these guidelines when transporting compressed gas cylinders:

- Check the label on the cylinder to identify the gas that it contains and verify that the cylinder is approved for transporting the gas.
- Never transport a gas cylinder without the safety cap screwed all the way down to the body of the cylinder.
- Never lift a gas cylinder by the safety cap.
- Never transport a gas cylinder with the regulator attached.
- Use chains or straps to secure the cylinder to the cart and keep the cylinder upright.
- Many sites require a partition on the cart to separate the two bottles, as shown in *Figure* 22.
- Wear the appropriate PPE whenever handling gas cylinders.

Figure 22 Cylinder cart.
Source: Courtesy of Vestil Manufacturing

WARNING!

If the tank of a compressed gas cylinder ruptures or cracks during transport, the tank can become a deadly projectile or explode into fragments of shrapnel.

2.1.4 Drum Dollies and Carts

Drum dollies: Wheeled circular platforms or caddies with a handle that are used to transport lighter drums/barrels.

Drum carts: Wheeled carts that are used to transport heavier drums/barrels.

Chemicals, lubricants, and other materials are commonly stored in 55-gallon drums or barrels. Drums that are filled with materials can be very heavy and unwieldy to move. Two methods for safely transporting loaded drums around the workplace are **drum dollies** and **drum carts**.

A drum dolly (*Figure* 23) is used to move relatively light drums. The simplest type of drum dolly is a flat wheeled platform that the drum sits on. The dolly may be equipped with a strap to secure it and prevent movement, or a handle to pull and control the load.

A drum cart (*Figure* 24) is typically used to move heavier loads, such as sealed oil drums. One common type is a rotating drum cart. Its main features include prongs, handles, wheels, and a set of heavy-duty casters.

Follow these steps for using a rotating drum cart:

Step 1 With the cart held vertically, slide the two prongs of the cart under the drum.

Step 2 Allow the cart to roll back into the horizontal position and sit on the wheels and casters.

Step 3 With the drum resting horizontally in the well of the cart, use the cart handles to maneuver the load to its destination.

Step 4 Once at the destination, tip the cart into the vertical position to allow the drum contents to be unloaded.

Figure 23 Drum dolly.
Source: Courtesy of Vestil Manufacturing

Figure 24 Drum cart.
Source: Courtesy of Vestil Manufacturing

2.1.5 Roller Skids

Roller skids move materials by pushing them on a table surface that is placed on top of two, three, or four roller skids. These devices are available in different capacities, depending on their intended usage. Roller skids used for moving heavy equipment use steel rollers or steel chain rollers, while those used for moving lighter equipment may use polyurethane or nylon rollers. The table surface of roller skids may differ from manufacturer to manufacturer. Some come equipped with a rotating table surface, some have spikes on the table surface for a better grip, and some have a plain surface like any other table. Many roller skids have a pull-push handle attachment for operators to move them easily (*Figure 25*). As with other material handling equipment, inspect the roller skids before use and use roller skids only for their intended purpose and weight capacity.

Roller skids: Devices that include a surface table and two, three, or four roller skids. Materials that are to be moved are placed on the table surface and then pushed on the skids.

Roller skids must be used on a relatively smooth surface. The rollers can easily be stopped by small rocks or hardware on the floor surface. Ensure that the route ahead is swept, if necessary, prior to the move.

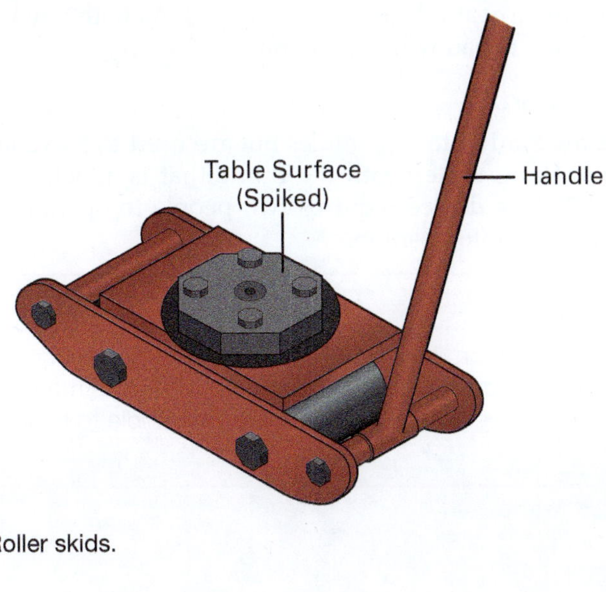

Figure 25 Roller skids.

2.1.6 Wheelbarrows

Wheelbarrows: One- or two-wheeled vehicles with handles at the rear that are used to carry small loads.

Wheelbarrows are one- or two-wheeled vehicles with handles at the rear, used to carry small loads (*Figure 26*). It is recommended that only two-wheeled wheelbarrows be used on jobsites, since they are more stable with heavy loads. Remember to check the air pressure in both tires, and check the body of the wheelbarrow for cracks, breaks, or punctures before using it. Use proper lifting techniques when moving a wheelbarrow. Keep your head up, your back straight, bend your knees, and lift with your legs.

Figure 26 Wheelbarrow.
Source: iStock

2.1.7 Pipe Mules

Pipe mules: Two-wheeled devices used to transport medium-length pieces of pipe, tubing, or scaffolding; sometimes referred to as a *tunnel buggy*.

Pipe mules are two-wheeled devices used to transport medium-length pieces of pipe, tubing, or scaffolding (*Figure 27*). They are sometimes referred to as tunnel buggies. A pipe mule may also be called a V-cart or a grasshopper, which is a brand name. When loading or unloading a pipe mule, use proper lifting techniques. The proper loading method is to have one worker lift one end of the load while another worker rolls the pipe mule under the load. Place the wheels of the pipe mule underneath the center of the load. As with any long load, use a spotter when moving a load with a pipe mule.

2.1.8 Pipe Transports

Pipe transports: Wheeled devices similar to a pipe mule but used to move larger pieces of pipe.

Pipe transports are similar to pipe mules but are used to move larger pieces of pipe. The pipe is slung underneath the frame that is attached to two rubber wheels (*Figure 28*). This device requires two people to operate it. Use caution when moving the pipe into position.

WARNING!

When using a pipe transport, beware of pinch points when securing chains around the pipe. Do not overload the equipment. Follow the manufacturer's recommendations. Always keep both hands on the handle to maintain safe control.

Figure 27 Pipe mule.
Source: Brown & Root Industrial Services

Figure 28 Pipe transport.
Source: This image is provided courtesy of Southwire Company, LLC. All rights are reserved.

2.1.9 Pallet Jacks

Pallet jacks, also known as pallet trucks, are devices that typically use hydraulics to lift and move heavy or stacked pallets (*Figure 29*). Both motorized and nonmotorized versions of this device are available. Motorized pallet jacks are sometimes called walkies (*Figure 30*).

Pallet jacks: Devices used to lift and move heavy or stacked pallets; also known as a *pallet truck*.

Remember the following safety guidelines when using a pallet jack:

- Plan and review the route before using a pallet jack; be very careful about using it on uneven surfaces or ramps.
- Inspect pallet jacks before use, looking carefully for malfunctions or missing parts.
- Check for oil leaks around fittings or hoses of hydraulic pallet jacks and repair any leaks before the jack is put into service.
- Never use a pallet jack to lift or transport a load that exceeds the jack's rated capacity. The load rating should be clearly posted on the body of the jack.
- Inspect the pallet to ensure it is in good condition and can adequately handle the load. Make sure the load is centered between the pallet jack forks (*Figure 31*).

Figure 29 Pallet jack.

Figure 30 Motorized pallet jack or walkie.
Source: iStock

Figure 31 Transport loads placed evenly on the body of the jack.
Source: iStock

2.2.0 Motorized Material Handling Equipment

NOTE

Nameplates must be posted on each material handling device. The nameplate must indicate the capacity of the device, the approximate weight, and any instructional information.

Motorized material handling equipment is powered by gasoline or electric motors. Operators must be trained, certified, and authorized to operate motorized material handling equipment. Training courses must consist of a combination of formal instruction, hands-on training, and evaluation of the operator's performance in the workplace for each type of material handling equipment being used. All operator training and evaluation must be conducted by a person who has knowledge, training, and experience.

WARNING!

All motorized material handling equipment should be operated only by individuals who have the proper training for the specific type of equipment being used. The exact requirements for authorization to operate equipment may differ among the various types. Certification is required to operate many, although not all, types of motorized material handling equipment.

When lifting materials mechanically, carefully secure the materials that are being loaded. Also be aware of any obstructions, such as people or equipment, that are in the path and have the potential to cause accidents.

When handling materials with a machine such as a pallet jack or forklift, always follow these guidelines:

- Know the weight of the object to be handled or moved.
- Know the capacity of the handling device that will be used and never exceed the capacity.
- Ensure that the handling equipment is in good working order and free of damage.

Types of motorized material handling equipment are discussed in the following sections.

2.2.1 Powered Wheelbarrow

A **powered wheelbarrow**, also known as a *power buggy*, is like a manual wheelbarrow, but is powered by an electric or gas motor (*Figure 32*). Before using a powered wheelbarrow, conduct a thorough inspection of the machine. Check the brakes and brake linkages for proper operation and adjustment. Inspect tires and wheels for damage and proper tire pressure. Test the directional, dumping, and speed controls to verify that they are all working properly. Be sure to read safety labels and decals for proper safety instructions. Safety labels should be clearly visible and readable on the side of the machine.

Powered wheelbarrow: A vehicle similar to a manual wheelbarrow but powered by an electric or gas motor; also known as a *power buggy*.

Keep the following safety guidelines in mind when operating a powered wheelbarrow:

- Avoid pinch points on dumping mechanisms.
- Be sure the dump bucket is always down and secure when not dumping.
- Do not move the buggy with the bucket in the dumping position.
- Do not mount or dismount the buggy while it is moving.
- Shut off the engine and lock the parking brake when you are finished using the machine.
- Do not allow passengers to ride in the wheelbarrow.
- Avoid any condition of slope and/or grade which could cause the buggy to tip.
- Do not exceed load limits in weight or height.
- Follow all procedures outlined on the safety labels and decals.

Figure 32 Powered wheelbarrow.
Source: Granite Industries

Tugs

Some companies are reducing their use of forklifts and pallet jacks and are increasing the use of wheeled carts, called *tugs*, to transport loads throughout a facility. Material handling tugs are battery-powered devices that can push and pull loads. One reason the use of tugs has grown is that they can increase productivity by reducing the manpower required to move heavy loads. A single worker operating a tug can safely move loads that weigh hundreds, or even thousands, of pounds. In addition, unlike forklift operators, workers who operate tugs are not required to have special training and certification before they are authorized to operate the device.

Tugs are designed to distribute weight ideally and maximize torque, which minimizes the effort required for operators to pull or push loads. The body of the tug acts as a wedge, slightly lifting the load at an angle and transferring all the weight to the tug's drive wheels. Tugs are equipped with safety features to prevent unintended propulsion, including an emergency reverse switch that automatically stops the tug to prevent it from pinning or crushing the operator when a load is being pulled in reverse.

Source: Courtesy of Vestil Manufacturing

2.2.2 Concrete Mule

Concrete mule: A wheeled device used when a concrete pour is in a location that a concrete delivery truck or pump cannot reach; sometimes referred to as a *Georgia buggy*.

A **concrete mule**, sometimes referred to as a *Georgia buggy*, is a wheeled device used when a concrete pour is in a location that a concrete delivery truck or pump cannot reach. Concrete mules (*Figure 33*) are designed to carry concrete, sand, gravel, and other materials of that nature; it should not be used to transport undesignated materials.

Before using a concrete mule, you must conduct a thorough inspection of the following:

- Tires
- Fluids
- Mechanical hinges and joints
- Throttle cables and steering mechanisms

When refueling a concrete mule, or any other gas-powered material handling machine, always be sure that the engine is turned off and has cooled down before refueling. Only use fuel and oil types that are specified by the manufacturer.

Figure 33 Concrete mule.
Source: Multiquip Inc.

When discharging concrete mix from a concrete mule near an open pit, use a tire stop board or chocks to prevent the mule from rolling forward into the pit during the emptying process.

When operating a concrete mule, make sure there are no obstacles or other workers in the turning radius of the rear platform. Always look behind the mule before operating in reverse, and never allow anyone other than the operator to ride on the machine.

2.2.3 Freight Elevator

On multilevel sites, workers may use a **freight elevator** to transport materials from floor to floor. When freight elevator doors close, they typically travel from the top and bottom simultaneously (*Figure 34*). These doors do not have the safety bumpers that are found on a commercial passenger elevator. Rather, as a person pulls the top door down with a strap or handle, the bottom door rises to meet the top door. When these doors close, there is normally metal-against-metal contact, and this can cause injury. Make sure that all body parts such as arms, hands, and fingers are clear of the door before it closes. Check to make sure that tool belts, body harnesses, and all materials are also clear of the door. Always check the weight capacity shown on the elevator wall before placing material on an elevator.

Freight elevator: An elevator used to transport materials from floor to floor.

Figure 34 Freight elevator.
Source: Brown & Root Industrial Services

Figure 35 Industrial forklift.
Source: iStock

2.2.4 Industrial Forklift

Industrial forklift: A vehicle with a power-operated pronged platform that can be raised and lowered for insertion under a load to be lifted and moved.

Work zone: The area in which the rear end or front forks of a forklift may come in contact with objects or people.

An **industrial forklift** is a vehicle with a power-operated pronged platform that can be raised and lowered for insertion under a load to be lifted and moved (*Figure 35*). Forklifts are typically used to lift, lower, and transport large or heavy loads in areas with smooth terrain, such as warehouses or shops.

When working in the vicinity of a forklift, remember the following safety guidelines:

- All workers must keep a safe distance from the machine. This perimeter is known as the **work zone**; it is the area in which the rear end or front forks of the machine may come in contact with objects or people.
- Workers must also stay clear of the fall zone. This area includes the area directly beneath a load, as well as any portion of the area that suspended materials could possibly fall into.
- Stay in designated walkways and make eye contact with the forklift operator to ensure that they see you. Always avoid the driver's blind spot. Be aware that forklifts pivot on the front wheels and turn with the back wheels, which could cause a potential pinch point.
- Plan a route of escape. Never get caught in a position where you are trapped in a corner or on the side of a truck.

Trained and certified forklift operators must observe certain traffic safety rules and regulations while driving.

Before operating a forklift, put on the proper PPE and fasten the seat belt. If you are not wearing a seat belt and the forklift tips, you can easily be pitched under the forklift or have some of the load land on you. Falling objects are common when stacking and lifting with a forklift, so wear a hard hat. As with most machinery, when working with a forklift, do not wear baggy clothes that can catch on controls, cargo, or stacked objects.

Never play on a forklift or drive recklessly. Accidents often happen because workers forget to respect equipment. A forklift is not a toy and should not be treated like one.

Retrofitting Machinery

The cost of new machinery is a significant expenditure for any business. Because of this expense, many operations look to retrofit and upgrade their current machinery instead. Agricultural machinery is particularly well suited for retrofitting components and extensions, turning heavy machinery into multifunctional tools. As an example, tractors can be retrofitted with a forklift or bucket fork to perform tasks beyond those of a traditional tractor.

Source: iStock

Going Green

Fuel Cell Technology

Source: Plug Power, Inc.

Increasingly, hydrogen fuel cell power units are replacing lead-acid batteries in forklifts, automated guided vehicles (AGVs), and other material handling vehicles. In 2012, more than 4,000 vehicles powered by fuel cells were in operation in the United States at companies such as Walmart, Coca-Cola, BMW, FedEx, Lowe's, Mercedes-Benz, and Stihl Inc. In the same year, approximately 30,000 fuel cell systems were shipped worldwide.

A fuel cell is an electrochemical device that combines hydrogen and oxygen to produce electricity. Since the conversion of fuel to energy takes place using an electrochemical process instead of combustion, the process is clean, quiet, and highly efficient. Fuel cell technology is estimated to be two to three times more efficient than fuel burning. Furthermore, the cost of fuel cells has dropped more than 50% since 2006.

Using material handling equipment powered by fuel cells greatly reduces downtime and increases productivity. The equipment only needs refueling once or twice a day, depending on use, and can be refueled in a few minutes. Traditional battery-powered equipment typically must be taken out of operation about every four to six hours to have the battery replaced or recharged, which usually takes several hours. In addition, unlike lead-acid batteries, hydrogen-powered fuel cells are not adversely affected by low temperatures in the work environment.

For many years, fuel cells were installed in forklifts only as a replacement for lead-acid batteries. Now, however, some lift truck manufacturers have begun producing forklifts that are originally designed to be powered by hydrogen fuel cells, rather than simply being retrofitted with this alternative power source.

Before operating the forklift each day, inspect the following:

- Tires
- Fluid levels
- Battery/fuel level
- Fire extinguisher
- Brakes
- Deadman control
- Warning lights
- Horn
- Backup alarm

Use a checklist if one is available. If you find any defects during the inspection, report them to your supervisor.

Forklift operators with the fewest accidents drive with slow deliberation, take their time, and do not make mistakes. Do not attempt to cut corners to speed up your work. Drive slowly and keep your arms, hands, and legs inside the cage of the forklift. Forklifts are top-heavy and, when driven recklessly, can tip over on a curve, drive off the edge of a dock, or fail to stop in time. Drive at a speed that is appropriate and safe for the location and surface condition and use extra caution on hills, corners, and ramps. Keep the forklift to the right on roadways and in wide aisles and stop at all designated stop signs. Always start and stop smoothly. This helps to preserve the equipment by putting less wear and tear on the forklift. It also keeps loads from shifting, objects from falling, and helps prevent damage to cargo, equipment, and operators.

Remember that pedestrians have the right-of-way. Forklifts, especially electric models, are often so quiet that they are difficult to hear. Pedestrians may not be aware that you are there, so be careful when you are driving near people, especially if they have their back to you. Make sure that the backup alarm is functional. Driving in reverse presents special dangers to pedestrians, especially if you are watching the load in front as you back up.

Workers sometimes attempt to make modifications to forklifts so that they can transport heavier or wider loads than the equipment is designed to lift. This introduces the hazards of broken masts or forks, dropped loads, and tip overs. Never modify a forklift without obtaining consent from the manufacturer or from a mechanical engineer familiar with the equipment.

Going Green

Biodiesel

Cranes and other equipment used in rigging operations consume lots of fuel—just like all the other pieces of equipment at a typical jobsite. Most large trucks and construction equipment run on diesel fuel. These vehicles and machines could go green and use biodiesel instead. Biodiesel is a plant-based oil fuel made from soybeans, canola, and other waste vegetable oils. It is even possible to make biodiesel from recycled frying oil from restaurants. Biodiesel is considered a green fuel since it is made using renewable resources and waste products. Biodiesel can be combined with regular diesel at any ratio or be run completely on its own. This means any combination of biodiesel and regular petroleum diesel can be used or switched back and forth as needed.

Some of the benefits biodiesel has over traditional fuels include:

- It's environmentally friendly. Biodiesel is sustainable and is a much more efficient use of resources than diesel.
- It's nontoxic. Biodiesel reduces health risks such as asthma and water pollution linked with petroleum diesel.
- It produces lower greenhouse gas emissions. Biodiesel is almost carbon-neutral, contributing very little to global warming.
- It can improve engine life. Biodiesel provides excellent lubricity and can significantly reduce wear and tear on your engine.

Think about the environmental impact that would occur if every vehicle and piece of equipment at every jobsite were converted to biodiesel. The use of biodiesel also continues to increase in Europe, where Germany produces most of these fuels. However, even small countries such as Malta and Cyprus have some level of production.

Automation in Material Handling

Automated guided vehicles (AGVs) have long been used in manufacturing. Now, automated forklifts retrofitted with AGV technology are poised to spur big improvements in material handling productivity. Tasks that may be appropriate for forklift automation are similar to the tasks that AGVs handled in manufacturing—tasks involving repetitive horizontal travel along the same route for long distances.

Evolving AGV technology holds promise for material handling in warehouses. One good application for forklift automation is high-volume picking, where the operator is able to remotely advance the forklift. This eliminates the need for the operator to repeatedly climb on and off the vehicle while working in an aisle.

The repetitive routes and well-defined space of farmland make the agricultural industry another promising field for AGV technology. The longevity of farm machinery provides a great opportunity to retrofit new technology to existing machines or to develop new AGVs to more efficiently move materials around the farm.

Source: iStock

Source: iStock

Rough terrain forklift: A vehicle similar to an industrial forklift but designed to be used on rough surfaces. Rough terrain forklifts are characterized by large pneumatic tires, usually with deep treads, that allow the vehicle to grab onto the roughest of roads or ground cover without sliding or slipping.

2.2.5 Rough Terrain Forklift

A **rough terrain forklift** is similar to an industrial forklift but is designed to be used on irregular surfaces. They are typically used for jobs where there are no paved surfaces. Rough terrain forklifts are characterized by large pneumatic tires, usually with deep treads that allow the vehicle to grab onto the roughest of roads or ground cover without sliding or slipping (*Figure 36*). Be sure to apply the same safety guidelines to rough terrain forklifts as you would to industrial forklifts.

Figure 36 Rough terrain forklift.
Source: Courtesy of CASE Construction Equipment

One of the most common accidents involving rough terrain forklifts are tip overs. This is because these types of forklifts are often operated on uneven ground. To prevent tip overs with rough terrain forklifts, observe the following safety guidelines:

- Study load charts carefully and only operate the forklift within its stability limits.
- Before placing a load, apply the brakes, shift to neutral, level the frame, and engage the stabilizers, if so equipped.
- Carry the load low. Avoid driving with an elevated load except for short distances.
- Move carefully on slopes and only operate within specific grade limits.
- Retract the boom and lower the forks before moving a forklift.

2.2.6 Hand Signals

The noise generated by motorized lifting equipment, such as forklifts, and the distance between workers when they are working with the equipment can make verbal communication difficult. Standard hand signals (*Figure 37*) allow both the workers and the forklift operator to go through the lifting, loading, moving, and unloading procedures at a more controlled setting, and help to avoid potential misunderstandings and accidents.

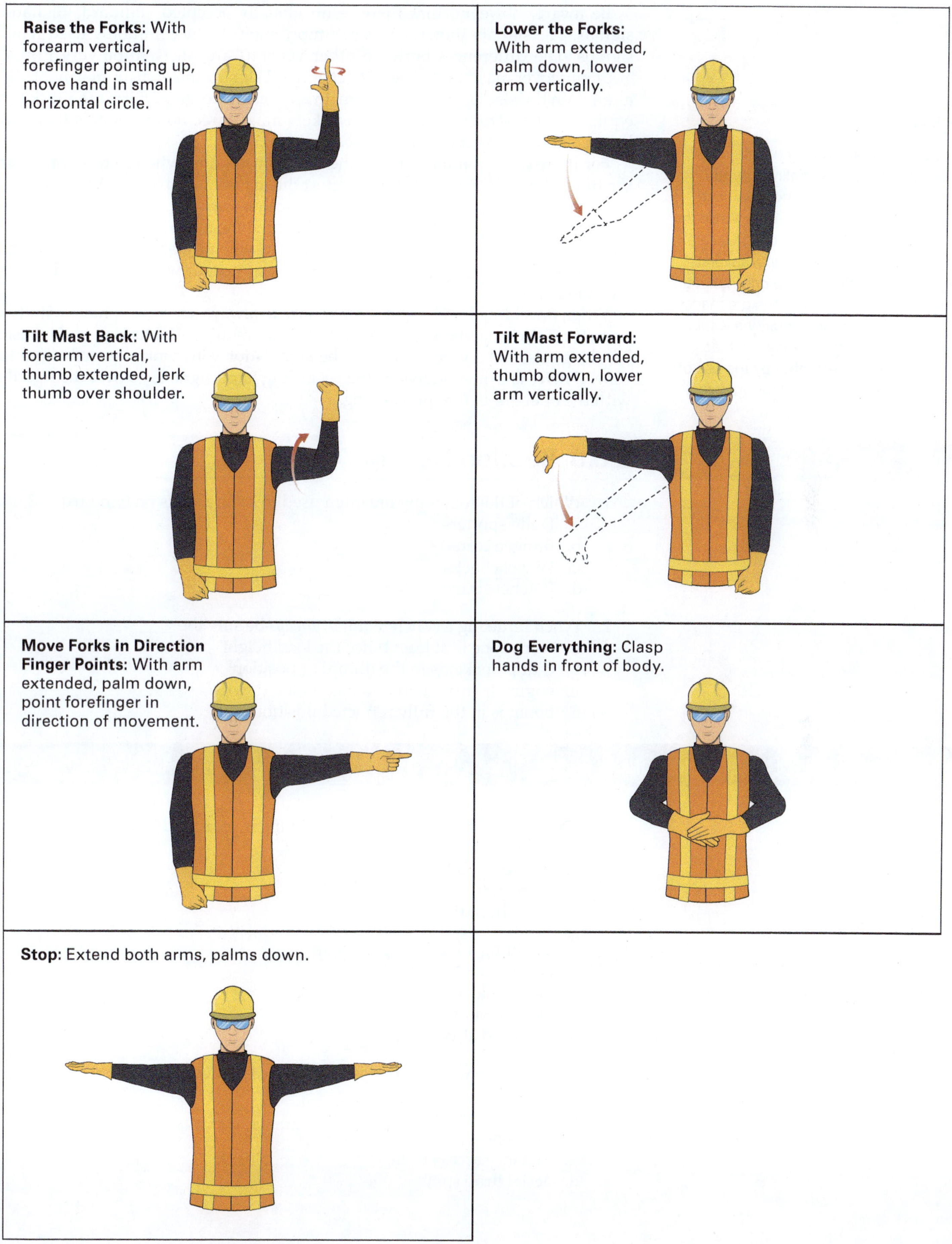

Figure 37 Common forklift hand signals.

CAUTION

The agricultural industry has its own hand signals as defined by the American Society of Agricultural and Biological Engineers (ASABE). While the characteristics of the hand signals are like those in other industries, several signals are modified to be one-handed signals. These modifications are made to address common conditions within the agricultural community, such as having one arm restricted by the use of machinery or being used for grasping.

Be aware, however, that there is no globally accepted standard for hand signals. The signals shown here are simply some of the commonly used hand signals. The differences between other versions are likely to be very minor. Your employer may have a specific set of hand signals that may or may not fully align with the signals presented here. In any case, prior to performing a task the equipment operator and the other workers must agree on the signals that will be used and the exact meaning for each signal.

For instance, it is important for the operator to know when to raise or lower the tines (the L-shaped parts that support the material load) of a forklift to avoid obstructions that may be in their path. Another very important signal for the operator to know is Dog Everything. This signal means to pause. It can be used when potentially risky situations arise, such as when it starts raining, when the load does not fit the space for which it was planned, or when a bystander gets too close.

When providing hand signals, make sure to always keep eye contact with the operator. There may be special operations that require adaptations of the basic hand signals, so be sure to review the signals for your worksite from time to time. If there is a misunderstanding regarding hand signals, go over the signals with the operator before proceeding.

2.0.0 Section Review

1. Which of the following are often used to secure loads on hand trucks?
   a. Dolly spotters
   b. Bungee cords
   c. Wheel chocks
   d. Ratchet straps

2. When refueling a concrete mule, *always* be sure the _____.
   a. work zone is at least twice the load height
   b. dump bucket is in the dumping position
   c. engine is turned off and has cooled down
   d. boom is in the fully retracted position

Module 93110 Review Questions

1. Loose gloves may get caught on moving objects, and they reduce _____.
   a. positive friction
   b. worker mobility
   c. manual torque
   d. grip strength

2. When lifting a load, keep the load _____.
   a. at shoulder height
   b. at arm's length
   c. close to your body
   d. balanced at your hips

3. The area into which a load is being lowered should be treated as a(n) _____.
   a. hot work location
   b. fall zone
   c. primary access route
   d. aerial drop spot

4. How many inches should stacked bricks be tapered for every foot above 4' in the stack?
   a. 2"
   b. 4"
   c. 6"
   d. 12"

5. The proper way to get tools to a worker on a higher level is to _____.
   a. toss them up carefully
   b. carry them by hand up a ladder
   c. climb up with them in your pocket
   d. hoist them up with a rope and bucket

6. A knot that is *not* meant to be permanent is generally referred to as a _____.
   a. hitch
   b. slip
   c. runner
   d. bypass

7. The knot shown in *Figure RQ01* is a _____.
   a. square knot
   b. bowline
   c. half hitch
   d. clove hitch

Figure RQ01

8. To help make sure the path is clear when your view is obstructed while handling materials, use a _____.
   a. mirror
   b. spotter
   c. tow rope
   d. stepladder

9. Before using a hand truck, inspect the framework for _____.
   a. oil residue
   b. popped nails
   c. stress fractures
   d. seal leaks

10. Roller skids for moving heavy equipment use rollers made of _____.
    a. aluminum
    b. brass
    c. polyurethane
    d. steel

11. Which of these devices require two people to operate?
    a. Roller skids
    b. Rotating drum cart
    c. Powered wheelbarrow
    d. Pipe transport

12. Before a worker uses a motorized material handling device to move a load, the worker must consider the _____.
    a. moisture content of the load
    b. fall zone for the load
    c. maximum capacity of the device
    d. universal transport hand signals

13. Which of the following is a motorized material handling device?
    a. Concrete mule
    b. Roller skid
    c. Porta-Power®
    d. Pipe mule

14. An industrial forklift pivots on the front wheels and turns with the back wheels, which could cause a _____.
    a. gravity shift
    b. pinch point
    c. rollover
    d. fall zone

15. When the sound of motorized lifting equipment makes verbal communication difficult, workers should use _____.
    a. written notes
    b. shouted commands
    c. hand signals
    d. a relay person

Answers to odd-numbered questions are found in the Review Question Answer Keys at the back of this book.

Answers to Section Review Questions

Answer	Section Reference	Objective
Section 1.0.0		
1. a	1.1.1	1a
2. c	1.2.1	1b
3. d	1.3.1	1c
Section 2.0.0		
1. d	2.1.2	2a
2. c	2.2.2	2b

MODULE 93111

Introduction to Leadership

Source: iStock

Objectives

Successful completion of this module prepares you to do the following:

1. Explain business and organizational structures in the construction industry.
   a. Understand organizational charts.
   b. Describe organizational roles and relationships, including job descriptions and responsibility, authority, and accountability.
2. Describe frontline leadership and communication skills.
   a. Describe the qualities and role of a leader, including leadership styles and ethical leadership.
   b. Explain the communication process and distinguish between verbal, nonverbal, written, and visual communication types.
   c. Compare problem-solving and decision-making.
3. Identify key skills for team leadership and motivation.
   a. Choose strategies to motivate individuals and a team.
   b. Describe key approaches to building and leading teams.
   c. Recognize how to approach leadership challenges, such as transitioning from peer to leader, addressing performance issues, and resolving conflict.
4. Identify a frontline leader's safety leadership responsibilities.
   a. Assess the impact of incidents.
   b. Explain the purpose of OSHA and describe the role of OSHA in administering worker safety.
   c. Describe the importance of safety and health programs.
   d. Explain the frontline leader's role in incident preparedness and response.
   e. Recognize the importance of supporting your crew's total health.
5. Demonstrate an understanding of construction project phases, project planning and scheduling, field reporting, and cost and resource management.
   a. Identify construction project phases.
   b. Summarize crew planning and scheduling responsibilities, including creating lookahead schedules.
   c. Describe the frontline leader's role in field reporting.
   d. Explain approaches to managing costs and controlling project resources, such as materials, tools, and equipment.

Digital Resources for Agricultural Technology

SCAN ME

Scan this code using the camera on your phone or mobile device to view the digital resources related to this craft.

Performance Tasks

Under supervision, you should be able to do the following:

1. Organize and deliver a safety-related toolbox talk, applying communication skills.
2. Complete a job safety analysis (JSA).
3. Develop and present a lookahead schedule.

Overview

As a frontline leader, you must communicate effectively, provide direction to your crew, plan and schedule work, manage costs, and ensure the jobsite is safe for everyone. Whether you are already a frontline leader or want to become one, this module will help you learn more about the requirements and skills needed to succeed.

NCCER **Industry-Recognized Credentials**

If you are training through an NCCER-accredited sponsor, you may be eligible for credentials from NCCER. The ID number for this module is 93111. Note that this module may have been used in other NCCER curricula and may apply to other level completions. Contact NCCER at 1.888.622.3720 or go to **www.nccer.org** for more information.

You can also show off your industry-recognized credentials online with NCCER's digital credentials. Transform your knowledge, skills, and achievements into credentials that you can share across social media platforms, send to your network, and add to your resume. For more information, visit **www.nccer.org**.

1.0.0 Business and Organizational Structures

Performance Tasks

There are no Performance Tasks in this section.

Objective

Explain business and organizational structures in the construction industry.

a. Understand organizational charts.
b. Describe organizational roles and relationships, including job descriptions and responsibility, authority, and accountability.

When a crew is assembled to complete a job, one person is appointed the leader. Companies give this role different titles, such as **crew leader**, *frontline supervisor, lead person*, or *foreman*. In this module, this role is called the *crew leader* or *frontline leader*. A crew leader or frontline leader is often an experienced craft professional, farmhand, or ranch hand who has demonstrated leadership qualities.

A **superintendent** is an on-site field supervisor responsible for one or more crew leaders or frontline supervisors. A **project manager** may be responsible for managing one or more projects. They typically support the project at an off-site location. This training will concentrate primarily on the leadership role of the frontline leader. These individuals work with others at the jobsite on projects as part of an **organization**.

Crew leader: A person appointed leader of a team or crew, usually an experienced craft professional who has demonstrated leadership qualities. Also known as a *frontline leader*, *supervisor*, *lead person*, or *foreman*.

Superintendent: An on-site supervisor who is responsible for one or more crew leaders or frontline supervisors.

Project manager: A person responsible for managing one or more projects.

Organization: A group of people working together.

1.1.0 Organizations and Organizational Structures

Businesses, clubs, and associations are examples of types of organizations. On a jobsite, the relationships among the people within the company or project create a type of organization.

1.1.1 Organizational Charts

A formal organization will have an **organizational chart** that shows all the positions in the organization and how those positions are related. A detailed organizational chart, or *org chart*, may also show the person's name in each position. *Figure 1* shows an example of an organizational chart for a company. Each position on that organizational chart represents an opportunity for career advancement for a crew leader.

As shown in *Figure 1*, most companies have separate groups or departments that focus on running specific functions. Smaller companies may have one or two individuals who perform multiple functions. Larger companies often create specific roles or departments to handle more complex projects and larger teams.

Companies may have additional roles or teams, such as architectural and engineering design functions. Depending on their size, these groups may become separate departments.

Organizational chart: A diagram that shows how the various management and operational responsibilities relate to each other within an organization. Named positions are ranked top to bottom, from those functional units with the most and broadest authority to those with the least. Lines connect positions and indicate chains of authority and other relationships.

1.2.0 Organizational Roles and Relationships

As a crew leader, you will assign others to perform specific jobs or tasks. Your crew and team members will perform best if the following is true:

- They have a clear job description that explains their job duties.
- They have the required skills and abilities to perform the task.
- They know who has **authority**, **responsibility**, and **accountability**.
- They have a clear understanding of expectations.
- They are trained on policies and procedures.
- They have all the needed tools, equipment, and materials.

Authority: The right to act or make decisions in carrying out an assignment.

Responsibility: The act of accepting a task and completing it to the best of one's ability.

Accountability: Accepting responsibility and being personally answerable for the results.

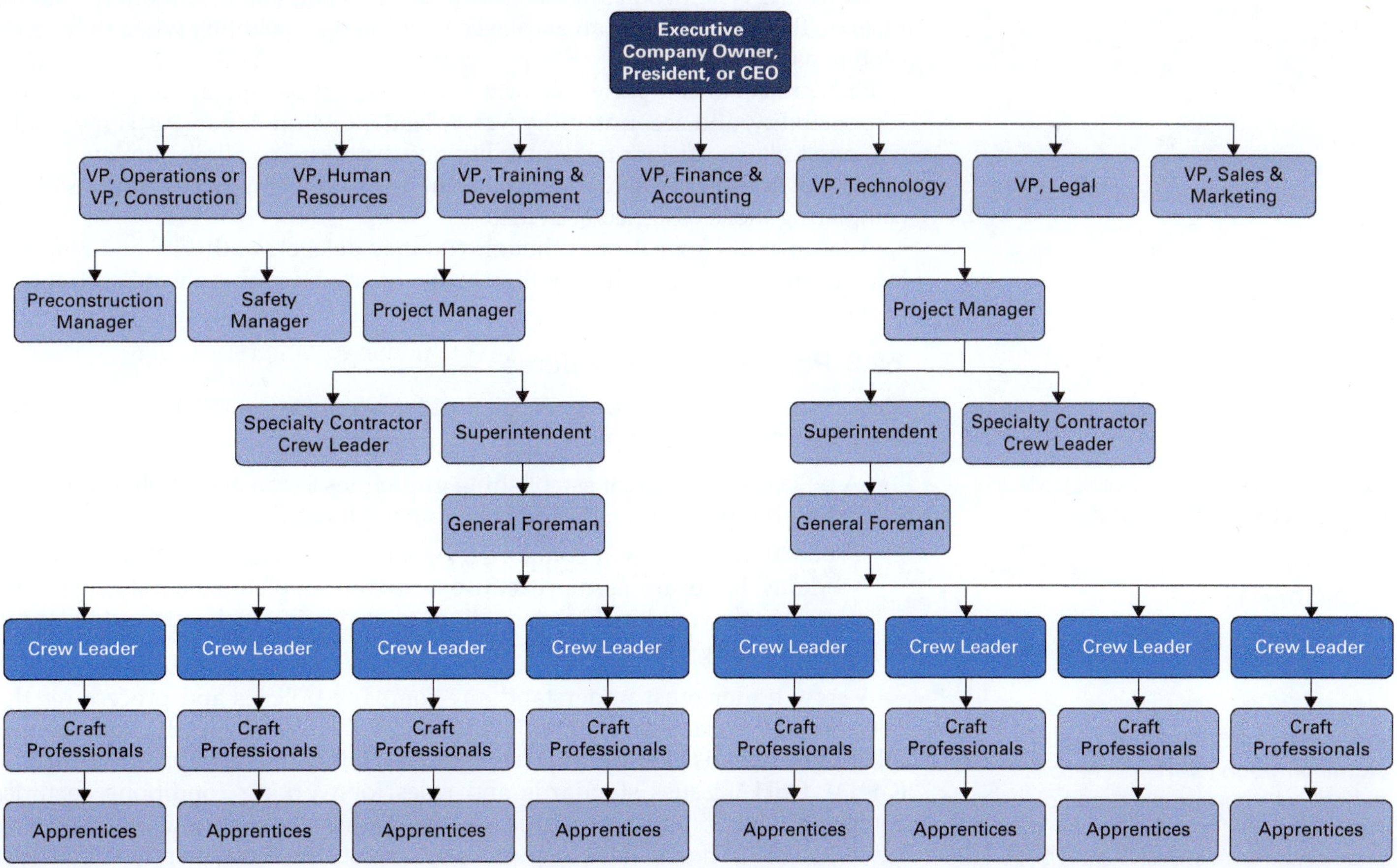

Figure 1 Sample organization chart for a construction company.

1.2.1 Job Descriptions

Job description: A description of the scope and responsibilities of a worker's job so that the individual and others understand what the job entails.

Duty: A work task an employee is expected to perform.

Most companies give each employee a job description as part of the hiring process. A **job description** is a statement that describes the duties of a position with a company. A **duty** is a work task an employee is responsible to perform. Responsibility involves accepting and completing the assigned tasks to the best of one's ability.

A job description should include enough detail so that an employee understands their job. It also should have any information needed to evaluate the employee's job performance. A job description should contain at least the following:

- Job title
- The supervisor to whom the position reports
- A general position description
- Specific duties and responsibilities
- Required and desired qualifications, including certifications and licenses
- Other information, such as working conditions, job type, hours, pay ranges, and more

Job descriptions set a standard for an employee. They outline the tasks that an employee should handle. After reviewing a job description, a new employee should understand all the duties and responsibilities of the job. Having a clear understanding of the job makes it easier for new employees to work well with the rest of the crew while they gain experience. Clear job descriptions help crew leaders judge performance and identify training needs for new employees.

1.2.2 Responsibility, Authority, and Accountability

A good crew leader must delegate work to their team but stay accountable for the outcome. When you assign a crew member work, you give them the authority to do that task. In turn, an employee takes on responsibility when they accept a job or assignment.

Each project or company has differences in the type and amount of authority, responsibility, and accountability a crew leader or worker has. As a crew leader, you must give sufficient responsibility and authority to allow employees to be successful at their job activities. You then must hold employees responsible for completing those assigned activities.

As shown in *Figure 2*, even though you may delegate authority and responsibility to crew members, the frontline leader or crew leader is always accountable for tasks assigned to a crew.

1.2.3 Policies and Procedures

Most companies have formal policies and procedures established to help crew leaders carry out their duties.

Policy: A general statement establishing guidelines for a specific activity.

Procedures: Formal instructions to carry out and meet policies.

- A **policy** is a statement establishing guidelines for an activity like vacations, breaks, workplace safety, and checking out tools.
- **Procedures** are how a company's policies are carried out throughout the workday. For example, a procedure written to implement a policy on workplace safety would include guidelines for reporting incidents and following general safety practices.

Occupational Safety and Health Administration (OSHA): A national public health agency that is part of the US Department of Labor. To promote a safe and healthy work environment, OSHA issues standards and rules for working conditions, facilities, equipment, tools, and work processes.

A crew leader must understand any company policies and procedures that relate to their work, especially safety practices. Company specific policies must follow the rules issued by the **Occupational Safety and Health Administration (OSHA)**. OSHA issues standards and rules for working conditions, facilities, equipment, tools, and work processes to promote a safe and healthy work environment. When OSHA inspectors visit a jobsite, they often question crew leaders and employees about the company's safety policies. During incident investigations, the inspector will verify if the responsible crew leader knew and followed the applicable company policy.

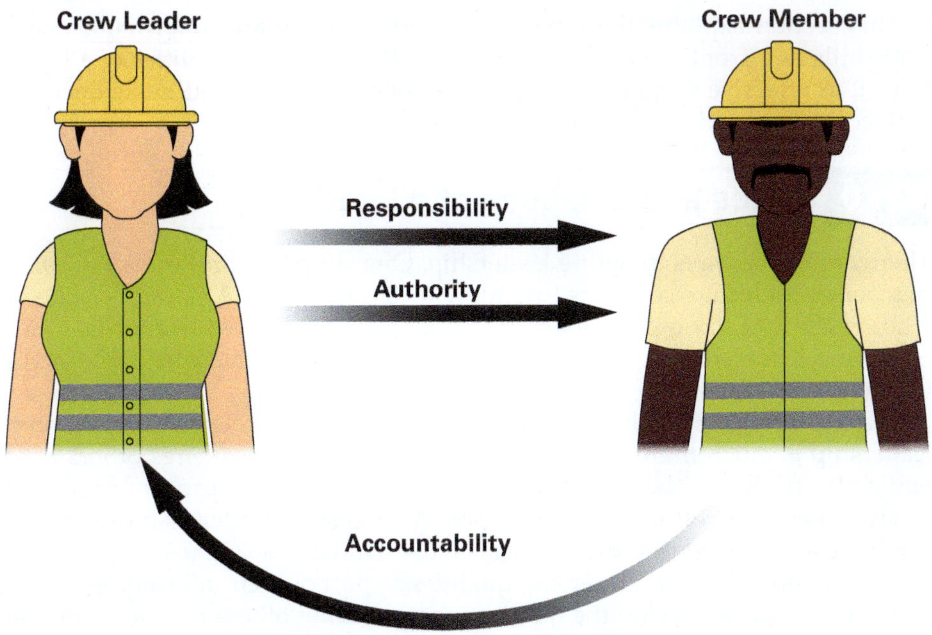

Figure 2 Accountability always rests with the frontline leader.

1.0.0 Section Review

1. Which of the following is shown on an organizational chart?
   a. How to organize materials on a jobsite
   b. How positions in an organization are related
   c. How prevailing wages are determined
   d. How to coordinate work with other crafts

2. Although you may delegate authority and responsibility for tasks to crew members, as a frontline leader, you have the overall _____.
   a. policy
   b. accountability
   c. job description
   d. authority

2.0.0 Leadership and Communication

Objective

Describe frontline leadership and communication skills.

a. Describe the qualities and role of a leader, including leadership styles and ethical leadership.
b. Explain the communication process and distinguish between verbal, nonverbal, written, and visual communication types.
c. Compare problem-solving and decision-making.

Performance Tasks

There are no Performance Tasks in this section.

In addition to being an experienced craft professional, a frontline leader must also demonstrate leadership qualities and strong morals. They must be able to communicate effectively, solve problems, and make good decisions. Frontline leaders are generally promoted from a work crew. A worker's ability to accomplish tasks, get along with others, meet schedules, and stay within the budget significantly influences the selection process. The frontline leader must lead the team safely and provide quality products.

This section explains the importance of developing and improving leadership skills as a frontline leader. It will cover the roles and qualities of a leader, as well as effective ways to communicate, solve problems, and make decisions with coworkers and employees at all levels.

2.1.0 Roles and Qualities of a Leader

There are many ways to define leadership. One simple definition is the ability of a person to influence others to achieve a common goal. A leader is someone who sees how to accomplish goals and rallies people to move toward that vision. Successful leadership requires communicating ideas effectively to inspire action in others.

Leaders are given authority based on their assigned role or title. However, leadership is much more than a title. An effective leader inspires others to act while also directing their actions. A leader must earn the respect of the crew, so crew members follow their directions. A successful leader also uses critical thinking skills to identify the best way to use available resources.

Some people develop leadership qualities as part of their upbringing. Others may work hard to develop the qualities that **motivate** others to follow and perform. You can learn leadership through experience, training, hard work, focus, and practice.

Motivate: To make someone feel determined or enthusiastic or to make someone behave in a particular way.

2.1.1 Functions of a Leader

What is the function of a leader? The answer will vary based on the specific project, the work environment, the team they lead, and the tasks they must perform.

As shown in *Table 1*, a frontline leader has certain functions that apply to almost every work situation.

TABLE 1 Frontline Leadership Functions

Leadership Duties and Responsibilities	Leadership Functions
Planning the work	Organize, plan, staff, direct, and control work.
Safety and compliance	Ensure that the team understands and follows company policies and procedures.
Empathy	Be sensitive to the needs and qualities of a diverse workforce.
Motivating and coaching	Give team members the confidence to make decisions and take responsibility for their work.
Conflict resolution	Maintain a cohesive team by resolving tensions and differences.
Professionalism	Represent the team or group and represent the company or organization.
Accountability	Accept responsibility for the successes and failures of the team's performance.

2.1.2 Leadership Traits and Behaviors

As previously noted, leadership is a skill you can learn and develop through experience and training. As you continue to grow your leadership skills, remember that while each leader has unique approaches, the most effective leaders share the following common **traits** and **behaviors**:

Traits: Distinguishing qualities of a person.

Behaviors: The way a person responds to specific situations.

- Follow company policies and procedures
- Have necessary technical knowledge
- Plan, prioritize, and organize work
- Lead by example

- Are fair and ethical
- Solve problems, make decisions, and assume responsibility
- Communicate effectively
- Promote an idea
- Are enthusiastic and show strong initiative
- Motivate individuals and teams
- Are loyal to their company and crew
- Extend respect and trust to all team members
- Teach others and learn from others

Above all, excellent leaders lead by example. They work and live by the standards that they set for their crew members. A leader will not ask someone to complete a task they are unwilling to do themselves.

Leaders tend to have high drive, determination, and a persistent attitude. When faced with obstacles, effective leaders don't get discouraged. Leaders identify the potential problems and make plans to overcome them. They then work toward achieving the intended goal. Effective leaders learn from their mistakes in the event of failure and apply that knowledge to future situations. They also learn from their successes.

Respect is an important aspect of leadership. Leaders earn respect by being fair to employees and listening to their complaints and suggestions. Using appropriate incentives and **rewards** to motivate a crew also helps leaders earn respect. In addition, frontline leaders with a positive attitude tend to gain the respect of their team and **peers**.

Rewards: Tangible items given to an employee as recognition for good work.

Peers: Team members in positions or levels in an organization equal or similar to yours.

Trust is an equally important part of being a great frontline leader. Trust is a group's belief that their leader is committed, capable, and will do what's best for all involved. To build trust, effective leaders create interactions and conversations with the team. They treat everyone on the team as an individual.

Leaders are good communicators who lay out the goals of a project to their crew members. To do this, you may need to overcome language barriers, bias, or differences in personalities to ensure that each crew member understands the project's established goals.

Effective leaders can motivate their teams to work to their full potential. They will work to develop crew members' skills, encouraging them to improve and learn to contribute more to the team effort. Effective leaders strive for excellence in themselves and their teams, working hard to provide the necessary skills and leadership.

Leaders need to have expert knowledge of the activities they supervise. Crew members need to know they have someone to turn to when they have questions or problems, need guidance, or must make modifications or changes on the job.

In addition, leaders must possess organizational skills. They know what needs to be accomplished and use their resources to make it happen. Because they can't do it alone, leaders require the help of their team members to share in the workload. Effective leaders delegate work to their crew members and implement company policies and procedures to complete the work safely, effectively, and efficiently.

Finally, outstanding leaders have the experience, authority, and self-confidence to make decisions and solve problems. To accomplish their goals, leaders must be able to calculate risks, take in and understand information, set a course of action, make decisions, and assume responsibility for those decisions.

2.1.3 Leadership Styles

Many terms are used to describe different leadership styles. As a frontline leader, you will never lead the exact same project twice. You must adapt to the unique circumstances of each project and apply **situational leadership** to develop the best path to achieving a project's goals and ensuring your company's success.

Situational leadership: A leadership style that involves adjusting to the specific situation or task and the leadership needs of the group.

As shown in *Figure 3*, situational leadership provides a model with four primary leadership styles:

- *Directing* — A leader tells a team what to do and how to do it.
- *Coaching* — A leader coaches individuals to improve their skills and encourages more back-and-forth between leaders and the team.
- *Supporting* — A leader supports the group, participating to help increase motivation. The leader offers less direction and has team members take a more active role in developing ideas and making decisions.
- *Delegating* — A leader allows others to own the task and facilitates completion. Team members make most of the decisions and are responsible for what happens.

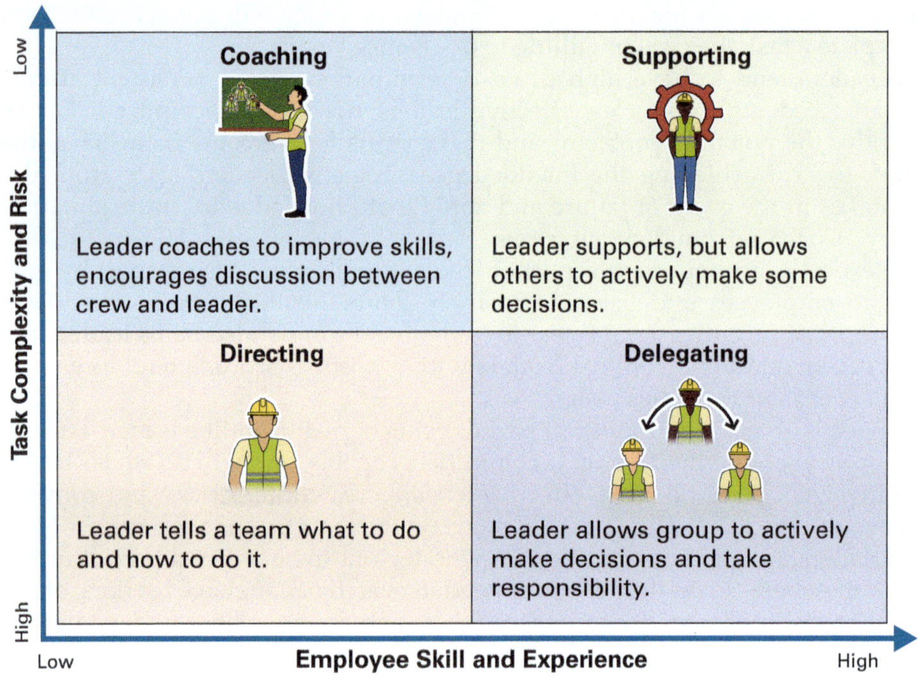

Figure 3 Situational leadership styles.

Knowing which leadership style to apply depends on several key factors:

- *Team relationships* — Consider your team, how well they work together, and how productive they are. An inefficient or productive group might benefit from a directing style. That style emphasizes direction, organization, and clearly defined roles. On the other hand, a highly cohesive group might benefit from a supporting style, which gives the team more input on decisions.
- *Team skills* — Consider the skills and abilities of the individuals completing the task. Giving a task to a team member who lacks the skills to complete it is risky and may fail.
- *Work tasks* — Consider the actual work or tasks at hand. You must have expert knowledge of the tasks and activities to ensure they are done well, completely, and safely. You may need to coach or train workers without experience to help them complete the task.
- *Work situation* — Consider your situation. In some cases, the situation itself will override all other factors.

For example, if the crew lacks experience for the job, you may want to use a *directing* style. A directing style works with people and teams with limited experience or skill in performing a task. It also works when team members are insecure or unmotivated to try a task. The directing style requires close supervision to spot signs of progress, which you can use to motivate ongoing development. If a job task or situation is risky, a directing style can help reduce the risk by providing clear and specific direction.

If your team is working well and showing strong progress on work tasks, you might use a *coaching* style. You still direct how and when specific tasks are completed, but you also discuss why the task matters and where it fits into the overall project or job. With this style, you recognize the team's enthusiasm, interest, and commitment. You also reinforce the importance of learning and gaining task-related experience.

You may want to use a *supporting* style if the team is reasonably skilled and work tasks are not complex, but commitment and attitude are issues. The goal of the supporting style is to create alignment. As a leader, you can discuss a person's commitment by asking open-ended questions. These questions should help you understand what is behind the performance challenges and create possible solutions. The supporting leadership style also works well when work is creative, as it is helpful to brainstorm and exchange ideas with crew members.

The *delegating* style works best with an experienced, strong, committed crew on a well-defined project. The delegating leadership style is also effective when jobs have repetitive tasks that require few decisions. To use the delegating style, your company must give you enough authority as the crew leader to do the job.

Effective leadership takes many forms. The correct style for a particular situation depends on the team and the work you need to accomplish. Knowing the factors to consider can help you decide which type of leadership to apply in which situation. This flexibility allows you to meet each situation with the leadership style that brings out the best in your team and gives them the highest chance of success.

2.1.4 Ethics in Leadership

Ethics are important in business and in life. They are the moral principles that help a person decide between right and wrong. Ethics also guide a person's behavior in different situations. *Figure 4* outlines three basic types of ethics. Frontline leaders must maintain strong ethics and the highest standards of honesty and legality. Every day, a leader makes decisions that have fairness and moral implications. When you make an unethical decision, it reflects poorly on you. It also impacts your team, other project team members, and the company.

As a leader, you will find yourself in situations where you must weigh the ethical results of a choice or decision. For example, imagine a crew member shows symptoms of heat exhaustion. Do you keep that employee working because the superintendent says the project is behind schedule? If you know that your crew did not erect the reinforcing steel properly, do you stop the pour and correct the situation? In both cases, you must act ethically as a leader and make the right choice. If a project manager or superintendent ever asks a frontline leader to make an unethical decision, the leader must inform the next level of authority about the unethical nature of the request.

NOTE

Anyone aware of illegal activity can face legal repercussions for not taking action and reporting the illegal activities. This is true even if one's supervisor requested the activities.

Ethics: The moral principles that guide an individual's or organization's actions when dealing with others.

Business or Legal Ethics	**Professional or Balanced Ethics**	**Situational Ethics**
Adheres to all laws and published regulations related to business relationships or activities.	Carries out all activities in such a manner as to be honest and fair to everyone under one's authority.	Relates to specific activities or events that may initially appear to be a gray area. For example, you may ask yourself, "How will I feel about myself if my actions were going to be published in the news or if I need to justify my actions to my family, friends, and colleagues? Would I still do the same thing?"

Figure 4 Types of ethics.

2.2.0 Communication

Communication: The process of exchanging facts, information, and ideas. Communication comes in many forms, including verbal, written, visual, and nonverbal.

Successful crew leaders learn to communicate well with people at all levels of an organization. **Communication** is the act of accurately and effectively sharing facts, feelings, and opinions with another person or group. It comes in many forms, including verbal, written, visual, and nonverbal.

2.2.1 Why Communication Matters

Effective communication on a jobsite or project is critical for many reasons. First and foremost, clear communication is a key part of safety. Everyone on a jobsite must receive and understand communications about safety policies and procedures. If this does not happen, it creates safety issues for everyone in the group.

Miscommunication is also expensive (*Figure 5*). One study indicated that over half of all rework is due to poor communication and bad information. If communication on a construction or agricultural site fails, it can cause delays and quality issues for the project. It can also result in damage to equipment or property. The good news is that communication is a skill that can be taught, practiced, and improved.

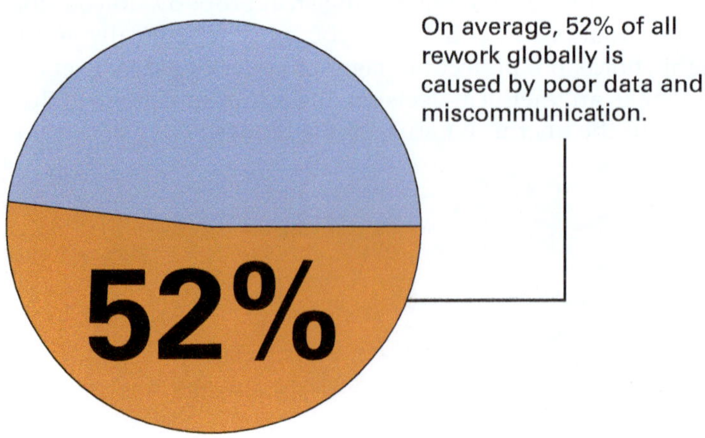

Figure 5 Impacts of miscommunication in the construction industry.

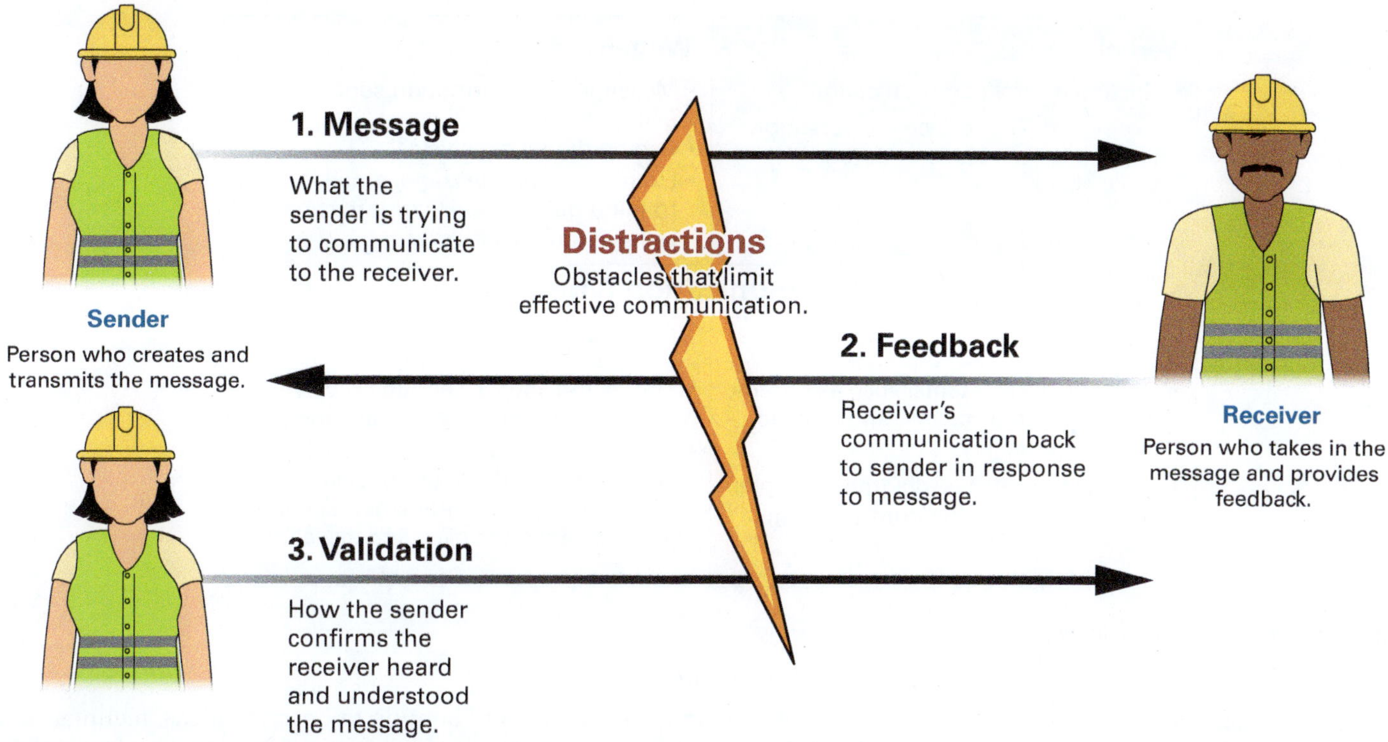

Figure 6 The three-way communication process.

2.2.2 Communication Process

Clear communication has three basic steps, as shown in *Figure 6*.

1. *Message* — A sender sends a message to a receiver, as verbal, written, visual, or nonverbal communication. A message can be a set of directions, an opinion, or a personal message such as praise or a warning. A message is an idea or fact that the sender wants the receiver to know.
2. *Feedback* — The receiver is the person who interprets the message and then repeats it to provide feedback. Feedback is the receiver's communication back to the sender in response to the message. Feedback is a crucial part of the communication process. It shows the sender whether the receiver understood the message as intended. In other words, feedback is a checkpoint to ensure the receiver and sender are on the same page.
3. *Validation* — The sender confirms the receiver heard and understood the message.

This process is called **three-way communication**. Three-way communication is also sometimes called *repeat-back communication*.

Three-way communication is a simple way to ensure clarity. Why is good communication so hard to achieve if this process is so easy? When we try to communicate, many things can get in the way. These obstacles or **distractions** interfere with the meaning, intention, and clarity of a message. Also called *noise*, distractions can include poorly worded messages, emotions, interpretations of a word, or the actual noise from a construction or agricultural site. Distractions can happen at any point in the communication process.

As noted in *Figure 7*, four different types of three-way communication exist. The following sections review these different types of communication and discuss ways to increase the effectiveness of your communication, regardless of type.

Three-way communication: A form of communication with an initial message from a sender, which is then repeated by the receiver to confirm accuracy and then validated by the sender. Also called *repeat-back communication*.

Distractions: Anything that disrupts the communication flow between sender and receiver.

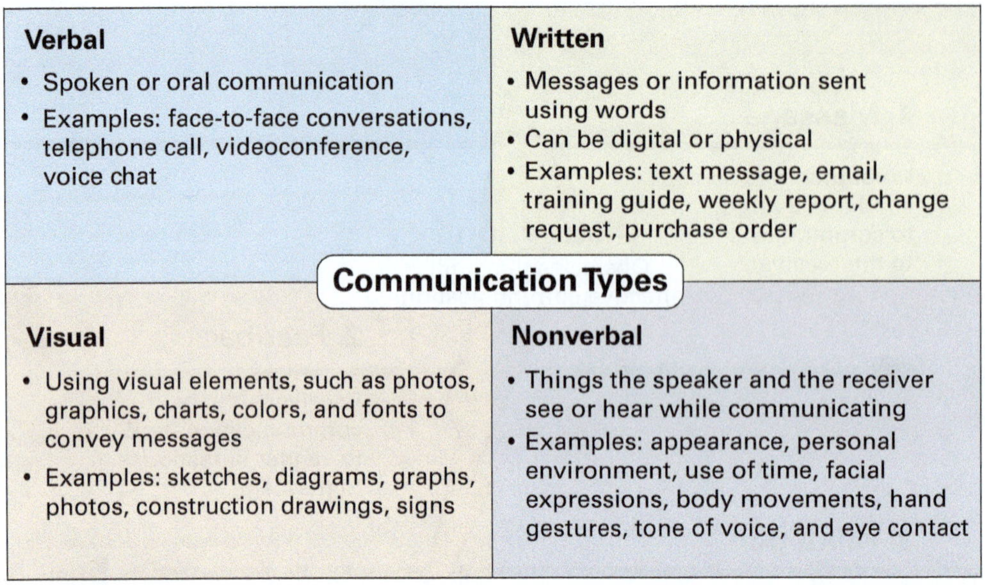

Figure 7 Types of communication.

2.2.3 Verbal Communication

Verbal communication: Messages that are sent using voice or other audible sounds. Also called *spoken communication* or *oral communication*.

Verbal communication uses words and sounds to describe ideas, feelings, and thoughts. It is a powerful way to deliver information and a useful tool to explain complex concepts. Most often, it is referred to as *spoken* or *oral communication*.

With verbal communication, the way you speak is as important as the message itself. As a leader, you should assess your audience before you speak. Even in a daily huddle, using clear verbal communication sets your crew up for a successful day. Consider each person's knowledge of the topic and language, gender, or cultural differences. Do not talk down to anyone in a group. Everyone deserves respect and courtesy, regardless of seniority.

You should also use words and phrases that the receiver can understand. For example, avoid technical language, acronyms, or slang if the receiver does not know these terms. In addition, the speaker should speak at a reasonable pace, which gives the audience time to understand one point at a time.

If you receive a message, listen actively and avoid distractions that interfere with its delivery. Effective listening has many barriers, particularly on a busy jobsite.

As a receiver, you can show that you are actively listening without saying a word. Examples include maintaining eye contact, nodding your head, and taking notes. If you feel unclear about the message, ask questions for additional explanation or information. You can also provide feedback by using your own words or paraphrasing to repeat a message back to the sender. This lets the sender know that you understand the message.

2.2.4 Written or Visual Communication

Written communication: The process of sending messages or information using words, whether digitally or physically.

Much of the information on a job or project is written or visual. **Written communication** includes everything from a text message to an email to a printed training guide. Most project documentation is written, such as weekly reports, requests for changes, purchase orders, and contracts. Many of these items are written to form a permanent record for business and historical purposes.

Visual communication: The process of using visual elements, such as photos, graphics, colors, and fonts, to convey messages.

Visual communication is widely used on the jobsite because it is an effective way to convey information quickly. Complex information is provided through visual communication using documented project plans, drawings, or videos of worksite incidents. Visual communication aids workers who speak different languages or have limited language skills by communicating a message using familiar prompts. Traffic signs, for example, use standard visual symbols because they are clear and well understood.

Written and visual communication used together are even clearer. For example, warning tags and caution signs on a project (*Figure 8*) use both written words and visual elements to help make them clear to all workers on a jobsite.

When writing or creating a written or visual message, you should assess the receiver (the reader). The receiver must be able to read the message and understand the content. If they cannot, the communication process will be unsuccessful.

2.2.5 Nonverbal Communication

Unlike verbal or written communication, **nonverbal communication** does not involve spoken or written words or images. Nonverbal communication includes appearance, personal environment, use of time, and body language. Examples include facial expressions, body movements, hand gestures, tone of voice, and eye contact.

Nonverbal communication: Things the speaker and the receiver see or hear while communicating.

Nonverbal communication can provide an external sign of someone's inner emotions. It often occurs at the same time as verbal communication. You can express feelings in many ways without realizing it. You may show happiness by smiling and giving someone a warm handshake. You may show you are angry by raising your voice.

People observe your physical and nonverbal cues when you communicate. Nonverbal signals can have more impact than verbal communication. Be aware of nonverbal cues to avoid miscommunication based on posture, facial expression, or other signals. For example, if you have your arms crossed in a defensive posture while speaking, the person might assume you are angry, even if you are not.

Figure 8 Communication tags and signs combine written and visual communication.

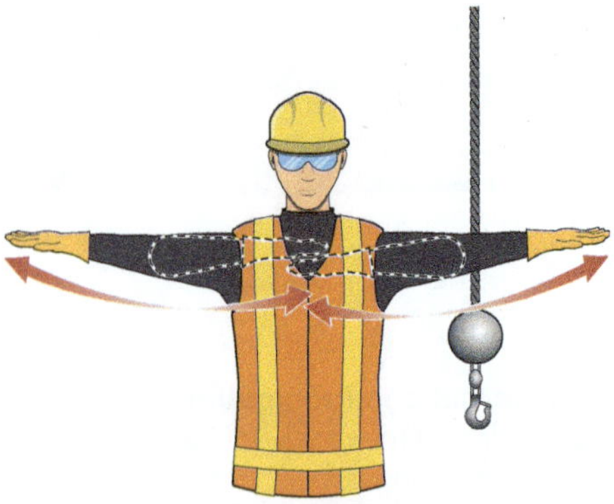

Emergency Stop —
Extend and retract arms rapidly.

Figure 9 The Emergency Stop signal is an example of nonverbal communication.

On a jobsite, nonverbal communication is also a matter of safety. For example, a major aspect of rigging safety is having clear communication between a crane operator and the designated signal person on the ground. This communication relies on common signals, including nonverbal hand signals. The Emergency Stop signal used in rigging operations is a nonverbal communication. Anyone on the ground within sight of the crane operator can use this signal in an emergency (*Figure 9*).

2.2.6 Additional Communication Considerations

As a leader in the field, the way that you communicate has a significant impact on the safety, quality, and productivity of your group. Regardless of your communication method, remember that everyone communicates differently. In some cases, you might need to communicate in multiple ways. You might also need to adjust your level of detail to ensure that everyone understands the intended meaning. For instance, a visual learner may need a map to comprehend directions. Another team member might need a verbal explanation of which turns to take to a destination. *Figure 10* shows an example of ways to tailor the message to your audience.

The construction and agricultural industries are diverse and multicultural. It is common to have a crew with people from different cultural backgrounds working together. Language and cultural differences can make communication even more challenging. Even if employees speak the same language, different understandings of words or symbols can create confusion or offend.

An important part of effective communication is understanding that people are different. Sometimes you must adjust your communication style to meet the needs of the person or people receiving your message. This means field leaders must invest the time to learn how to communicate well with everyone. This requires you to treat everyone as individuals and not just coworkers, which involves getting to know each crew member and recognizing how they learn, deliver feedback, and work with others.

Verbal Instructions (Experienced Crew) "Please drive to the supply shop to pick up our order."	**Verbal Instructions (Inexperienced Crew)** "Please drive to the supply shop. Turn right here and left at Route 1. It's at 75th Street and Route 1. Tell them the company name and that you're there to pick up our order."
Written Instructions 1. Turn right at exit. 2. Drive 2 miles to Route 1. Turn left. 3. Drive 1 mile (past the tire shop) to 75th Street. 4. Look for supply store on right...	**Visual Instructions (Diagram/Map)** 7501 N. Highway 1 (800) 555-7567

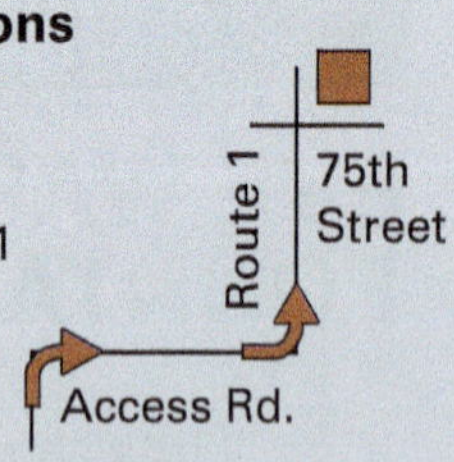

Figure 10 Assess your audience and then tailor your message.

2.3.0 Problem-Solving and Decision-Making

Problem-solving and **decision-making** are a large part of every frontline leader's daily work. They are a part of life for all supervisors, especially in fast-paced, deadline-oriented industries like agriculture and construction. The key difference between problem-solving and decision-making is that solving problems is a process, while making decisions is an action based on experience or information learned during the problem-solving process.

Problem-solving: A process used to identify possible solutions to a situation.

Decision-making: The action or process of making a choice. Decision-making involves starting an action, stopping one, or choosing a different plan or action.

2.3.1 Problem-Solving

The ability to solve problems is an important skill in any workplace. It is especially important for craft professionals, whose workdays are often unpredictable. Your group will look to you as a leader to solve problems regularly.

Figure 11 shows a five-step problem-solving process, which you can apply to any situation:

Step 1 *Define the problem*. Defining a problem is not as easy as it sounds. Thinking through the problem often uncovers additional problems. It will also require setting aside your biases and presumptions toward the situation and individuals involved.

Step 2 *Consider possible solutions*. A problem may have more than one solution, requiring you to consider several solution options. As you do this, get input from the people most affected by the problem, the people who must correct the problem, and the people who will be most affected by any potential solution.

Step 3 *Choose a solution and make an action plan*. Consider each possible solution and pick the best one. Again, you do not have to do this alone; ask for feedback from others who have a stake in solving the problem well and keep an open mind. The best solution might be combining pieces of two different solutions to create a new solution.

Step 4 *Test the solution*. A solution may sound perfect before you try it but turn out to be ineffective. If a solution doesn't work, consider improving it and test your new plan.

Step 5 *Evaluate the process*. After your solution is in place, review the steps you took to discover and implement it. Think about whether you could have done anything better. The next time you face a similar situation, you can use this experience to help solve the problem more effectively.

This problem-solving process can apply to almost any situation. *Figure 12* walks through a scenario to show how problem-solving works in practice. It is important to obtain input from crew members affected by the situation during the process. Their input is valuable when evaluating different solutions, since each solution has pros and cons. For example, workers will probably object to any plan (like the drop-off plan) that leaves their tools vulnerable to theft. Gaining the input of others will help you gain the trust and respect of your team.

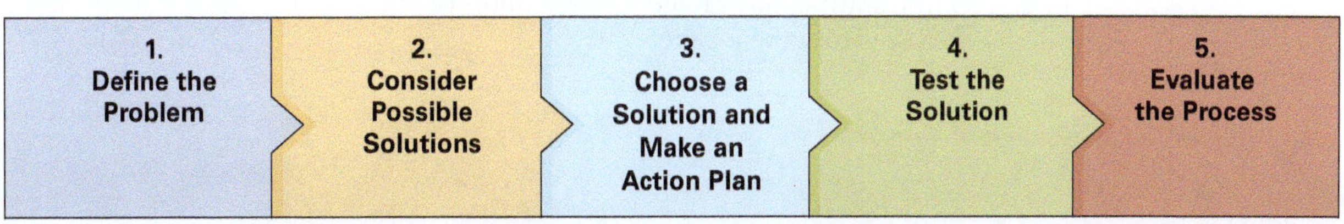

Figure 11 The problem-solving process.

Problem Scenario

You are leading a team assigned to a new shopping mall project. The project will take about 18 months to complete. The only available parking is half a mile from the jobsite. The crew must carry heavy toolboxes and safety equipment from their cars and trucks to the work area at the start of the day, and then carry them back at the end of their shifts.

1. Define the Problem	2. Consider Possible Solutions	3. Choose a Solution and Make an Action Plan	4. Test the Solution	5. Evaluate the Process
1. Problem: Workers are wasting time and energy hauling their tools and equipment to and from the worksite.	**2. Possible solutions:** Think about different ways to solve the problem. Several workers have proposed solutions: • Install lockers for tools and equipment closer to the worksite. • Have workers drive up to the worksite to drop off their tools and equipment before parking. • Bring in another construction trailer.	**3. Solution choice and action plan:** The worksite superintendent doesn't want an additional trailer in your crew's area. You and the team agree the shuttle service is the best option. It solves the time and energy issue and workers can keep their tools with them. The project supervisor plans to arrange a large van and driver to provide the shuttle service. **Note that this is a decision point in the problem-solving process.**	**4. Test the solution:** At first, the solution works, but a new problem arises. Everyone starts and leaves at the same time. The van can only hold half of the team and their equipment. To solve this problem, the supervisor schedules shuttle pickups every 15 minutes and adjusts work schedules so workers board the van at two different times. **Note that this is also a decision point in the problem-solving process.**	**5. Evaluate the process:** This process gave management and workers a chance to express an opinion and discuss the various solutions. Everyone feels pleased with the process and the solution.

Figure 12 Applying the problem-solving process with decision points.

2.3.2 Decision-Making

Figure 12 also shows when decisions are made as part of the problem-solving process. Decision-making is a choice made by using one's judgment. It is based on experience and information gathered during the process. Depending on circumstances, your decision type will fall into one of the three categories shown in *Figure 13*.

Regardless of the type, some decisions are simple or can be made based on past experiences. Other decisions are more difficult. These decisions require more careful thought by using problem-solving techniques to determine how to implement a solution.

All decisions have impacts you will need to consider. For example, if you use a shuttle to transport workers to a jobsite in shifts, as shown in *Figure 12*, the shuttle may create a safety liability while workers are transported back and forth from the jobsite.

Decision Type	Description	Examples
Yes or No Decisions 	Decisions in which the answer is yes or no. Also called *whether* decisions, because you have to decide whether to take an action or not.	• Whether to hire a van to use as a shuttle service • Whether or not to promote a team member • Whether to alert a supervisor to an unsafe condition • Whether to proceed with a concrete pour
Options Decisions 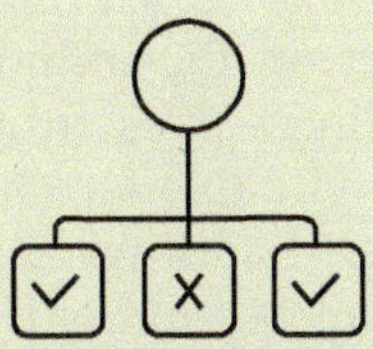	Decisions that require you to choose from a group of options. Also called *which* decisions, because you have to choose which option to take.	• Which of the three options will best address the problem of workers having to haul tools and equipment to and from the worksite? • Which toolbox talk would be most important to deliver to the team today? • Which team member should I have drive to the supply shop to pick up the order?
Contingent Decisions 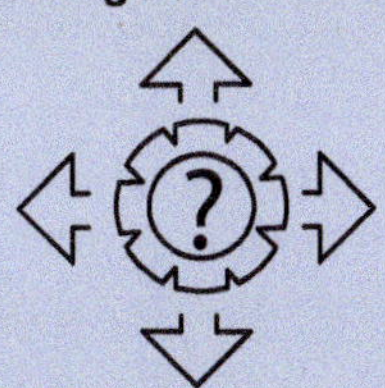	Decisions already made but on hold, until other specific events happen. These decisions can be very powerful because you can "pre-decide" what you will do in the event those events occur.	• If Carlo is late again two more times, I will set up a performance conversation. • If the materials arrive early, we will store them in the back of the jobsite. • If the team completes this task on time, I will find a clear way to recognize and reward the success.

Figure 13 Types of decisions.

2.0.0 Section Review

1. Which of the following is *not* a trait or behavior of effective frontline leaders?
   a. They lead by example.
   b. They are good communicators who lay out project goals to their crew.
   c. They motivate the team to work to their full potential.
   d. They have limited knowledge of the activities they supervise.

2. Which of the following is *not* a good practice to ensure effective communication?
   a. Speaking clearly at a reasonable pace
   b. Using highly technical language, acronyms, and slang
   c. Using visual cues, such as sketches or diagrams, to clarify instructions
   d. Using three-way communication to validate that a message was clear

3. While problem-solving is a process used to identify one or more possible solutions to a problem, the action of making a choice is called _____.
   a. decision-making
   b. a problem statement
   c. a decision statement
   d. action planning

3.0.0 Team Leadership and Motivation

Performance Tasks

There are no Performance Tasks in this section.

Objective

Identify key skills for team leadership and motivation.

a. Choose strategies to motivate individuals and a team.
b. Describe key approaches to building and leading teams.
c. Recognize how to approach leadership challenges, such as transitioning from peer to leader, addressing performance issues, and resolving conflict.

As a frontline leader, you will have an incredible impact on the individuals working on your team. Motivating individuals and building successful teams are important and challenging tasks for any leader. You will also face leadership challenges, including managing performance issues, handling conflict, and leading team members who were once your peers. This section will teach you strategies and approaches to succeed as a crew leader.

3.1.0 Motivating Individuals

Leaders must develop the ability to motivate others to be enthusiastic about a job or behave in a particular way. On a jobsite, a leader uses authority and influence to encourage individuals to use their best effort to accomplish a project's goals. Lack of motivation affects job quality and productivity, reducing profits and limiting growth for companies, businesses, contractors, or organizations that rely on people to complete the work.

As a crew leader, it is important to understand what motivates everyone on your team. There is no single best approach to motivating all your crew members. What motivates one crew member may not motivate another. In addition, what works to motivate a crew member once may not motivate that same person again.

How can you know what motivates the members of your project team? Effective communication skills, specifically active listening, are key to getting to know each person on your team. You can also learn more about what motivates an employee by looking at factors in the employee's performance. Factors that may indicate an individual's motivation include the level or rate of unexcused absenteeism, the number of complaints, and the quality and quantity of work produced.

Generally, the factors that motivate individuals are the same as those that create job satisfaction. These include the following:

- Clear expectations and goals from leaders
- Training to help individuals be successful
- Consistent sense of **accomplishment** and growth
- Recognition and rewards
- A good work environment and culture

Accomplishment: An individual's need to set challenging goals and achieve them.

A crew leader's ability to provide strong support in these areas increases the likelihood of high **morale** within a crew. High morale means that employees will be motivated to work hard and have a positive attitude about coming to work and doing their jobs. A motivated team is generally more productive and safer, which can positively affect project success.

Morale: An individual's emotional outlook toward work and satisfaction gained while performing the assigned jobs.

3.1.1 Clear Expectations and Goals

As a leader, you need to communicate your expectations. A lack of clear expectations can leave your team feeling lost and demotivated. You need to set clear goals and deadlines for completing tasks. Once everyone on the team understands these expectations, they are more likely to be motivated and perform at their best. *Figure 14* includes an example of clear expectations.

Figure 14 Examples of unclear and clear expectations.

Set achievable short-term and long-term goals, including tasks to be done and expected timeframes. Short-term goals allow you to provide regular feedback, positive praise, and rewards. However, because short-term goals can change, setting long-term goals also gives the team additional opportunities for success. Do not set workers up for failure, which can lead to frustration and reduced productivity.

In addition to setting clear goals and expectations, you must also deliver timely feedback. People want and need feedback. They also need positive reinforcement for their contributions and efforts. However, telling someone they did a good job last year or criticizing them for work from a month ago is meaningless. When sharing timely feedback, always include specifics to help the individual know what they did well or where they could improve.

You should always share negative feedback in private. When delivering negative feedback, approach the problem as a coaching moment, working with the individual to understand their challenges and correct the problem. Every failure can be an opportunity for you to help that individual grow if the conversation and feedback are constructive.

Positive feedback can be shared privately or publicly. Praise a person in front of their team, and it may also encourage other team members. Praise can be a valuable tool to reinforce desired behaviors. Bear in mind that not everyone likes to be publicly praised.

3.1.2 Accomplishment and Growth

A sense of accomplishment motivates most people. That is why setting and working toward recognizable goals makes employees more productive. A person with a sense of achievement will likely be better motivated and help motivate others (*Figure 15*).

Figure 15 A sense of accomplishment often translates to growth in a fulfilling career.
Source: iStock

A person who can align a task to a sense of fulfillment is more likely to complete it. Crew leaders can help their team feel fulfilled and accomplished by keeping the work challenging and interesting. For many individuals, overcoming the challenge of new or difficult assignments is much more satisfying than performing repetitive, boring tasks. Workers who complete the same task every day or perform work above or below their skill level can quickly become frustrated and unmotivated.

Remember, however, that too many changes at work can negatively impact morale because people often prefer consistency and predictability. Crew members react to change in many ways and for many different reasons. Change creates anxiety and uncertainty for some people, resulting in a negative attitude toward work. To help manage change with your crew, remember to share why the change is occurring and explain what the employees and company will gain from it.

As a leader, you must ensure your team members feel their skills and abilities are valued and make a difference. Employees who do not feel valued tend to have performance and attendance issues. Leaders should work to ensure the following is true for every crew member:

- They feel like an important part of the team with a skill set that contributes to the project's success.
- They know that every position helps them build skills and experience that will last throughout their careers.
- They can learn new things and develop their skill sets.
- They see opportunities to grow and progress within the company.

Personal growth: Improving one's talents and potential, both in and out of the workplace.

Professional growth: Gaining new skills and work experience that can help one to reach a goal in one's career.

Professional and **personal growth** are also critical factors in motivation. **Professional growth** opens an employee's imagination to new possibilities. As they see this growth, they know that they aren't limited to their current job. When such opportunities occur, let them know that they have a chance to grow with the company and to be promoted as an acknowledgment for excelling in their work.

Show your team that you value their professional growth whenever possible by giving them more advanced responsibilities when they are ready to perform them safely. People who are presented with challenging tasks often are motivated to work harder and improve. This approach is a win-win since the project benefits from a higher level of productivity while the worker experiences professional growth.

Effective leaders encourage each crew member to work to their full potential. They also share information and skills with their employees to help them advance within the organization.

Personal growth is unique to each individual. For example, a team member may find it very rewarding to master a new competency on the job. Experiencing this type of personal growth at work motivates workers to do their best, reduces boredom, and sparks enthusiasm for additional growth.

Personal growth also takes place outside the workplace. Crew leaders should encourage their employees to identify goals for themselves both professionally and personally and then support their efforts to accomplish those goals. Training can improve professional and personal skills, such as time management, communication, financial management, and more.

3.1.3 Training and Upskilling

Another key factor in motivating individuals and ensuring productive and safe projects is providing sufficient training and opportunities for **upskilling**. It provides employees with the tools and knowledge they need to perform their current jobs more effectively and efficiently by using advanced and contemporary skills.

Upskilling: A type of training that provides someone with more advanced skills through additional education.

As people enter the workforce, many will have completed training programs but may not have the specific practical experience needed to complete a task on a project site. When a new employee joins your crew, you will need to take an active role in confirming their skills and helping with on-the-job training (OJT) if there are gaps. A frontline supervisor on a jobsite only has a short time to evaluate their crew members on what tasks they can do or how much experience the crew member has.

In the first week, with any new employee, you should ensure that you train the new employee on any jobsite processes and allow time to do a task verification (*Figure 16*). After establishing the individual's skills, you can work on a plan to leverage those skills while establishing new ones. If you identify gaps, you can create a training plan for that individual to address those areas.

Figure 16 When a new employee joins the crew, be sure to evaluate their skills.
Source: iStock

3.1.4 Recognition and Rewards

Positive feedback is a form of **recognition**. Individuals want to have good work appreciated, applauded, and acknowledged by others. You can accomplish this by thanking employees, offering an employee of the month award, or giving tangible or monetary awards.

Recognition: Acknowledgment of an employee's accomplishments through verbal and nonverbal communications.

The following are tips to remember for giving recognition and praise:

- Be available on the jobsite so that you see good work as it is happening.
- Provide positive feedback publicly or in private.
- Give recognition and praise only when truly deserved. People quickly recognize insincere praise and will lose respect for the person delivering it.
- Encourage improvement by showing confidence in the ability of the crew members to do above average work.
- Use goal setting to build confidence and create opportunities for praise.

Rewards are tangible recognition for good work. Rewards can be monetary, such as opportunities for raises, bonuses, or other incentive pay. Other rewards might include company gear, vouchers for new boots, gift cards, a team outing at the end of the week, or the best spot in the parking area for a month.

3.1.5 Good Work Environment and Culture

A good work environment and culture are critical for creating strong employee motivation and high morale. As a leader, you can create a good work environment and culture for your team in several ways.

Treat employees well. Be considerate, kind, and respectful. Treat people fairly. Mentor your employees. Coaching and supporting employees boosts their self-esteem, self-confidence, and motivation. Remember that small things matter. Having clean lunch areas, good housekeeping, and a comfortable workplace communicates that you respect the group.

Communicate actively. You should be available to ask questions, express your opinions, and share ideas as a leader. Actively engaging and communicating with workers is crucial in motivating your team, as it shows that you appreciate and value their feedback. Involve employees in discussions so that they feel their opinions are valued. This, in turn, leads to a sense of pride and active participation. Making yourself available also prevents team members from feeling isolated from leadership.

Help individuals manage stress. In a job that can be demanding, fast-paced, and tiring at times, managers in the construction and agricultural industries must take measures to ensure that their team is not exposed to undue stress. Work with your supervisors to ensure a safe and reasonable work environment. This means not only physical space but also policies and procedures. For example, allowing workers to take time off to deal with personal matters or emergencies shows that the company prioritizes their well-being. Engage with your supervisors to help address resources and organize shifts in such situations to ensure that project progress and costs are not negatively impacted.

Lead by example. Employees will be far more motivated to follow someone who is willing to do the same things they are required to do. Being a good leader and motivator means leading by example (*Figure 17*). Employees will not be motivated to work if upper management does not display the same level of commitment to the job they are asked to do. For example, if you ask everyone to work late but then leave early to go see a movie, this will be noticed and may create feelings of resentment and decrease team morale. As a leader, you can inspire your team by displaying the behaviors you would like to see from them, including working hard, showing personal and professional integrity, and having a can-do attitude.

NOTE

As much as possible, maintain a sense of humor, especially toward your own failings as a human being. No one is perfect. Your employees will appreciate that and be more motivated to encourage you.

CAUTION

When you communicate, be positive. Approaching your leadership responsibilities by threatening employees with negative consequences can result in higher employee turnover instead of motivation.

Figure 17 Lead by example.
Source: iStock

3.2.0 Leading Teams

As a crew leader, you must lead a team of individuals from various backgrounds and experiences and create a high-performing team. To do so, you must delegate work to that team and implement policies and procedures that support the team. You will also need to address challenges like language barriers and **harassment**.

Harassment: Unwelcome conduct based on race, color, religion, sex, national origin, age, disability, or genetic information. Harassment includes verbal or physical conduct that impacts a person's work performance or creates an intimidating, hostile, or offensive work environment.

3.2.1 Delegating to Your Team

Once the various activities that make up the job have been determined, the crew leader must identify the person or persons who will be responsible for completing each activity. This requires that the crew leader be aware of the crew members' skills and abilities. Then, you must apply this knowledge by matching the crew's skills and abilities to specific tasks needed to complete the job (*Figure 18*).

After matching crew members to specific activities, you must delegate the assignments to the responsible person(s). Generally, when delegating responsibilities, the crew leader verbally communicates directly with the person who will perform or complete the activity.

When delegating work, remember to do the following:

- *Delegate work to a crew member who can do the job properly.* If it becomes evident that the worker doesn't perform to the desired standard, either teach the crew member to do the work correctly or turn it over to someone else who can (without making a public spectacle of the transfer).
- *Ensure crew members understand what to do and the level of responsibility.* Be clear about the desired results, specify the boundaries and deadlines for accomplishing them, and note the available resources.
- *Identify the standards and methods of measuring progress and accomplishment.* Also, identify the consequences of not achieving the desired results. Discuss the task with the crew member and check for understanding by asking questions. Allow the crew member to contribute feedback or suggest how to perform the task safely and well.

Figure 18 Leverage your team's skills to accomplish the task.
Source: iStock

- *Give the crew member the time and freedom to get started.* Don't make them feel pressured by too much supervision. When assigning the work, tell the employee how much time it takes to complete it and confirm that this time is consistent with the job schedule.
- *Check in on progress.* Be sure to periodically check in on the crew member's progress. This will allow you to see how the task is going and allow the crew members to ask questions. Checking in along the way will help you clear up any misunderstandings early and avoid rework.
- *Examine and evaluate the result.* After completing the task, ask the crew members for comments on how the work went. Provide your own feedback. This discussion will help you know what kind of work to assign that crew member in the future. It will also provide a means of measuring your effectiveness in delegating work.

There may be a time when someone else in the company issues written or verbal instructions to your crew or one of its members without going through you as the crew leader. If this happens, work with that individual to understand why they took that approach. Then, help that person understand why it is important for them to work through you for this type of instruction, whenever possible, so that you can manage the work of the entire crew safely and productively.

3.2.2 Implementing Policies and Procedures

Every company establishes policies and procedures that crew leaders are expected to implement, and employees are expected to follow. Company policies and procedures are essentially guidelines for how the organization does business. They can also reflect organizational philosophies, such as prioritizing safety or prioritizing the customer. Examples of policies and procedures include safety guidelines, credit standards, and billing processes.

The following tips can help you effectively implement policies and procedures:

- Learn the purpose of each policy. This will help you follow the policy and apply it appropriately and fairly.
- If unsure how to apply a company policy or procedure, check the company manual or ask your supervisor.
- Always follow company policies and procedures. Remember that they combine what's best for the customer and the company. In addition, they provide directions on handling specific situations and answering questions.

NOTE

Try to obtain a supervisor's policy interpretation in writing or print out their email response so that you can add the decision to your copy of the company manual for future reference.

Crew leaders may need to issue directions to their crew members. A direction is a form of communication that initiates, changes or stops an activity. Directions may be general or specific, written or verbal, formal or informal. It is up to you how you issue directions, but the policies and procedures of the company may govern your choice.

When issuing directions, do the following:

- Make them as specific as possible. Avoid being general or vague unless it is impossible to foresee all the circumstances that could occur in carrying out the directions.
- Recognize that you do not have to write directions for simple tasks unless your company requires this.
- Write directions for more complex tasks, tasks that will take considerable time to complete, or permanent (standing) orders.
- Consider what is being said, the audience to whom it applies, and the situation under which it will be implemented to determine the appropriate level of formality for the direction.

3.2.3 Addressing Language Barriers

Leading a diverse team also means addressing language barriers. Millions of US workers speak languages other than English and jobsites where two or more languages are used are common. Even among speakers of the same language, accents and dialects can present communication barriers (*Figure 19*).

Figure 19 Address communication barriers in a diverse team.
Source: iStock

As you have learned, any communication barrier keeps people from understanding each other clearly. If a person does not fully understand the task a frontline supervisor gives them, the task may be completed incorrectly. This can cause delays, additional expenses, or create an unsafe situation. As a crew leader, you must ensure that each crew member understands the work. The following tips will help when communicating across language barriers:

- Be patient. Give workers time to process the information in a way they can comprehend.
- Humor is easily misunderstood. Keep this in mind if communication barriers exist and always avoid humor at a worker's expense.
- Do not assume an individual is unintelligent simply because they don't understand what you are saying.
- Speak slowly and clearly. Avoid the tendency to raise your voice.
- Use face-to-face communication whenever possible. Other forms of communication can be more difficult when a language barrier is involved.
- Use the plans and drawings or make sketches to ensure people understand clearly.
- For critical written instructions, ensure the individual's literacy level matches the document's difficulty level and language. A person may speak clearly in a second language but struggle with written understanding.
- Communicate in multiple ways or adjust your level of detail or terminology to ensure that everyone understands the meaning as intended.
- Confirm understanding. Ask for feedback with open-ended questions that require more than yes-or-no answers. Use three-way communication and have an individual repeat what they heard to ensure mutual understanding and prevent miscommunication. If a worker is not fluent in English, ask the person to demonstrate their understanding through other means.

Many organizations offer language classes for supervisors and employees to help create more effective communication on jobsites. Whenever possible, take advantage of these training opportunities and encourage your crew to do so.

3.2.4 Preventing Harassment

As a frontline supervisor of a team, you must be aware of harassment as a potential issue. As defined by the US Equal Employment Opportunity Commission (EEOC), harassment is unwelcome conduct that is based on race, color, religion, sex, national origin, age, disability, or genetic information. Harassment includes verbal or physical conduct that impacts a person's work performance or creates an intimidating, hostile, or offensive work environment.

The following are examples of activities that might qualify as harassment:

- Telling offensive jokes
- Calling someone offensive names
- Making verbal or physical advances or assaults
- Making threats or intimidating others
- Displaying offensive objects or pictures

Harassment becomes unlawful when a person must endure offensive conduct to remain employed or the conduct is severe or common enough to create an intimidating, hostile, or abusive work environment. While not all acts of harassing behavior may be serious enough to be considered a violation of the law, all such behavior destroys teamwork and negatively affects jobsite safety and productivity.

Anyone can be a victim of harassment. Historically, many have considered harassment an act perpetrated by someone in power over another. However, a harasser can be a supervisor, a supervisor in another area, a coworker, or any other person in the work environment.

The EEOC enforces sexual harassment laws within industries. When investigating harassment allegations, the EEOC looks at the whole record, including the circumstances and context in which the alleged incidents occurred. Supervision may hold the crew leader responsible if they were aware of sexual harassment and did nothing to stop it. You must serve as a good example for the rest of the crew and take action to stop harassment.

Prevention is the best way to eliminate harassment in the workplace. Your employer will have a harassment policy that communicates zero tolerance for harassment and outlines a complaint or grievance process. This process should include taking immediate and appropriate action when an employee complains. This anti-harassment policy should apply to all employees (including managers, supervisors, and coworkers) and nonemployees (suppliers, owners, vendors, and others). Creating an open and inclusive work environment will discourage harassment.

As a leader, you must understand and uphold your employer's policies and procedures on harassment. If you believe someone has violated the harassment policy, immediately notify your supervisor. If you are uncomfortable reporting the incident to your immediate supervisor, you may report the complaint to any supervisory or human resources (HR) employee. If you are unsure how to proceed, contact your HR partners for guidance.

3.3.0 Leadership Challenges

As a crew leader, you must manage leadership challenges and opportunities every day on the job. While the specific challenges will vary based on your team, the project, and other conditions, you must learn to address these when they arise. Such challenges can include navigating your initial transition from peer to leader, managing performance issues with your crew, or resolving conflicts that arise on the jobsite.

3.3.1 Transitioning from Peer to Leader

Transitioning from crew member to crew leader can be difficult, especially when the new position involves overseeing a group of former peers. When you move from being a peer to a leader, your relationship with your former team members changes in important ways.

When you become a crew leader, you are no longer responsible for your work alone. You are now accountable for the work of a diverse team of individuals with varying skill levels, personalities, work styles, and cultural and educational backgrounds.

Crew leaders must learn to put personal relationships aside and work for the common goals of the entire crew. The crew leader can overcome these problems by working with the crew to set mutual performance goals and by freely communicating with them within permitted limits. Use their knowledge and strengths along with your own so that they feel like they are key players on the team. *Table 2* lists some additional suggestions on how to make the transition from peer to leader easier.

TABLE 2 Making the Transition from Peer to Leader

Strategy	Tactic
Reset boundaries with existing friends	Expectations and boundaries must be set with existing friends. Having a strong, professional working relationship is important, but you may need to reduce time spent with older peers outside of the project. Be clear and explain that, on the job, business comes before personal relationships.
Demonstrate that you have earned this leadership role	Quickly demonstrate good leadership for your team to earn employees' trust and respect. Be fair and consistent in everything you do. The perception of favoritism will tear a crew apart and negatively impact the outcome of any project.
Make the best decisions for the project and your company	Always make decisions that are best for the business. While it can be hard to make a decision that impacts someone who was a peer and is a friend, don't compromise to make a popular decision.

3.3.2 Managing Performance Issues

At the end of the day, if someone is simply not being accountable and completing their work, it might be time for a difficult conversation. In the workplace, a difficult conversation is one in which you have to manage emotions and information sensitively to deal with a workplace issue. A difficult conversation may involve the following:

- Communicating business decisions
- Discussing poor performance or behavior
- Addressing conflict
- Handling complaints
- Ending employment or other changes in job status

As a leader, you must be prepared for difficult conversations at some point in your career. These conversations are never enjoyable, but you can handle them effectively by following some basic steps:

- *Identify the problem to resolve.* Ensure that you have a clear understanding of the problem that you want to resolve. Before preparing for the conversation, ask: Why do you want to have the conversation? What do you want to achieve? Is the outcome you're looking for realistic?
- *Gather the information you need.* Make sure you know the facts of the situation before approaching the employee. Bring any documents you will need to the meeting. For example, if you discuss how an employee is late to work, you will want to have relevant policies and tardiness data. If you are unsure about any internal workplace policies and procedures, you can always seek advice from a supervisor or HR manager.
- *Think about the employee's point of view.* Before you speak, you should think about the matter from your employee's point of view. Engage in the conversation with an open mind and truly seek to understand the situation from the employee's point of view.
- *Get the timing right.* Schedule the conversation as soon as possible, but not if emotions are running high. Choose a time and place appropriate for the discussion, where you will not be rushed or interrupted. Generally, avoid having negative discussions at the end of the week and, if an employee needs time to process the conversation, offer to take a break and meet again later that day or the next.
- *Prepare what you are going to say.* Write down the key points you need to cover so that you can keep your conversation on track and stay in control. While it is important to prepare notes, avoid preparing a rehearsed script, as it may hamper your ability to listen effectively, react accordingly, or find alternative solutions.
- *Give the employee time to prepare.* Make an appointment with the employee and provide some context to give them some time to prepare. Remember that anticipating this conversation may be stressful for the employee, so phrase the approach carefully and don't leave it too long.
- *Seek support if you need it.* If you're comfortable having a one-on-one conversation with the employee, enlist help from your manager or HR person (*Figure 20*).

NOTE

An example of phrasing the appointment request is, "I'd like to talk about a series of quality issues I'm noting and get your point of view. Can you come see me tomorrow?"

Figure 20 Mediation can help build a bridge between two opposing sides.
Source: iStock

After the conversation, be sure to take the following steps:

- *Document the discussion.* Write down any agreements reached and indicate the date, time, and who was present. This will be helpful if you need to revisit the discussion later. Your job diary, which you will learn about later, is useful here.
- *Follow up with the employee.* Confirm what you agreed to in writing and send it to the employee. Be sure to give the employee time to take the steps they agreed to.
- *Take some time to reflect on the matter.* Learn from the experience and consider what you could do differently next time. If you feel you need to discuss the matter, debrief with your supervisor or HR.

Also, remember that you should never discuss this matter with other staff members. You must treat the employee the same way you did before the conversation and remain completely professional.

As previously noted, the best communication strategy is to have an open-door policy or informal one-on-one conversations so that issues can be dealt with as naturally as possible. Be approachable and keep in touch with your team so they will likely come to you with problems before they escalate.

CAUTION

If you need to have this type of difficult conversation, do it as soon as you start to have concerns or see early signals of conflict. The worse a situation becomes, the harder it can be to manage and resolve. In the meantime, your project and the rest of the crew are being affected.

3.3.3 Conflict Resolution

Conflicts will happen during a project. When many groups, including owners, general contractors, and specialty contractors, work together to complete a project, disagreements are likely to happen. Conflicts can also arise within your crew. Each person or group might have a different view or opinion on how something should be done and those differing opinions often lead to conflicts.

As a crew leader, you will have many opportunities to look for positive and proactive means to resolve conflict. Resolving conflicts can lead to a better understanding or a new way of doing things that improve the overall project. If you handle conflict well, it will further earn the respect and trust of your crew. The following are some strategies to use when addressing conflict:

- *Recognize that conflict is normal.* Not all conflict is bad. Conflicts on the site can be good. If the individuals engaged in the conflict are mature, conflict can improve processes and approaches by creating conversation about new work methods.

- *Do not ignore the conflict.* Act early to avoid further complications. Even a small dispute that may appear inconsequential can balloon into a larger problem if it goes unresolved. Even if you allow the parties involved to work it out on their own, you still need to follow up and make sure that there aren't any lingering issues.
- *Understand the real issue behind the conflict.* You must understand the real issues causing the conflict to determine the best solution. Conflicts are often caused by different interpretations of plans and specifications, change orders, scheduling, and poor communication. Work stress can also be an underlying issue. Conflicts, for example, may happen more often later in the day or during a work shift due to fatigue. Use your active listening skills to listen to everyone involved. This shows everyone that their opinions matter and that you are committed to working with them to resolve the issue fairly.
- *Remain neutral and focus on business, not people.* Conflicts can escalate quickly from a simple disagreement to a shouting match. If appropriate, call for a short break before starting the discussion again. Remind everyone that resolving conflicts is about making the best business decision for the project, not personal feelings. Both parties should sit down and stay focused on developing a solution that is in the project's best interest.
- *Compromise and work with the team.* Conflicts rarely have a clear-cut right and wrong. As you resolve the conflict, you must encourage others to compromise and work together to solve the problem (*Figure 21*). The effort here must focus again on the desired business result: "What is the best decision for this situation?" Business is all about cooperation and collaboration. However, if the conflict involves safety rules and regulations or adhering to building codes, any solution must follow the proper policy, procedure, or regulation.
- *Clear communication can help avoid conflict.* Poor communication on the jobsite often leads to conflicts. Make sure that you clearly communicate expectations with everyone on the project. Keeping everyone updated with any changes to the project or schedule can also help avoid conflict.

Figure 21 Working together is always the best strategy to overcome obstacles.
Source: iStock

- *Communicate the resolution and follow up.* Resolving conflicts may not result in a win for everyone. It is important that once a resolution is reached, everyone gets an explanation and is clear on what is expected moving forward. They might not always agree with the decision. Having a clear expectation will help others accept the decision. Whenever an agreement is reached, follow up to determine whether both sides are working as discussed or if issues still exist. You also will remind everyone that, although the conflict might have been difficult, the resolution produced a good outcome for the project.

CAUTION

If you cannot resolve a conflict independently, always seek guidance from your supervisor. Your supervisor or another senior leader can decide and communicate it to the appropriate individuals.

Remember, the goal of any project is to see the job successfully completed on time and within budget. When conflicts arise, they can threaten this goal. As the leader of your crew, you must learn to address and resolve conflicts so that the project work can proceed. Resolving conflicts can be challenging. It all comes down to making the best business decision to keep the project moving forward.

3.0.0 Section Review

1. Which of the following is *not* a way to motivate a team?
   a. Providing training and helping individuals feel a sense of accomplishment and growth
   b. Setting expectations without clear short-term and long-term goals
   c. Offering recognition and rewards when appropriate
   d. Having a good work environment and culture

2. One sign of a high-performing team is _____.
   a. the leader cannot delegate work to the team
   b. harassment is common on the worksite
   c. everyone on the team follows policies and procedures
   d. language barriers prevent effective communication

3. As a frontline leader, steps to successfully resolving conflict may involve any of the following *except* _____.
   a. remaining neutral and focusing on the business, not on people
   b. compromising and working with the team
   c. ignoring the conflict
   d. understanding the real issues causing the conflict

4.0.0 Safety and Safety Leadership

Objective

Identify a frontline leader's safety leadership responsibilities.

a. Assess the impact of incidents.
b. Explain the purpose of OSHA and describe the role of OSHA in administering worker safety.
c. Describe the importance of safety and health programs.
d. Explain the frontline leader's role in incident preparedness and response.
e. Recognize the importance of supporting your crew's total health.

Performance Tasks

1. Organize and deliver a safety-related toolbox talk, applying communication skills.
2. Complete a job safety analysis (JSA).

Craft professionals routinely face **hazards**. These include falling from heights, working on scaffolds, using cranes in the presence of power lines, operating heavy machinery, and working on electrically powered or pressurized equipment. Despite these hazards, applying preventive safety measures can significantly reduce the number of **incidents**.

Hazards: Conditions or activities that, if left uncontrolled, have the potential for harm.

Incidents: Unplanned events that result in property damage, fatalities, injuries, illnesses, or near misses.

Safety matters because people matter. As a crew leader, some of your most important responsibilities are to enforce the company's safety program, ensure that all workers perform their tasks safely, and ensure that everyone on the jobsite remains safe and goes home healthy (*Figure* 22). You must recognize safety risks and work to eliminate or reduce them.

To be successful, the crew leader should do the following:

- Be aware of the human and business impacts of incidents.
- Understand all federal, state, and local governmental safety regulations applicable to your work.
- Be the most visible example of the best safe work practices.
- Be involved in training workers in safe work methods.
- Conduct training sessions.
- Get involved in safety inspections and incident investigations.

Providing employees with a safe working environment by preventing incidents and enforcing safety standards will go a long way toward maintaining the job schedule, ensuring timely job completion within budget, and keeping your team healthy.

Figure 22 Committing to safety is a key part of a crew leader's role.
Source: Getty Images

4.1.0 The Impact of Incidents

Crew leaders are in an important position to ensure that their crew members perform all jobs safely, avoid incidents and prevent *near misses*. A near miss is an unplanned event that does not result in personal injury but may result in property damage or is worthy of recording.

Incidents, including work-related injuries, sickness, and fatalities, cause immense suffering for workers and their families. Project delays and budget overruns can cause huge losses for employers, and worksite incidents damage the overall morale of everyone on a project or jobsite.

Businesses lose billions of dollars every year because of on-the-job incidents. As just one measure of the business impact of incidents, construction companies in the United States pay over one billion dollars per week in **workers' compensation** for nonfatal injuries. Workers' compensation is often called *workers' comp*. It pays for injuries and losses resulting from an incident and it also pays death benefits to families of employees who are killed on the job.

Workers' compensation: Insurance that companies must cover medical expenses, lost wages, and other costs for employees who are injured or become ill while completing work for their job.

Preventing injuries means protecting workers' lives, which also reduces significant costs for their employers. The full impact of workplace injuries and fatalities is estimated to cost companies approximately $151 billion per year. These costs affect the individual employee, the company, and the industry.

Organizations encounter both direct and indirect costs associated with workplace incidents.

- Direct costs are the money companies must pay for workers' compensation claims and sick pay.
- Indirect costs are all the costs a company must account for as the result of a worker's injury or death, such as time off from work.

The direct and indirect costs of incidents are comparable to the visible and hidden portions of an iceberg, as shown in *Figure 23*. The visible part of the iceberg represents direct costs, some of which are covered by insurance. The hidden part of the iceberg represents the indirect costs, which are a greater financial burden than the direct costs.

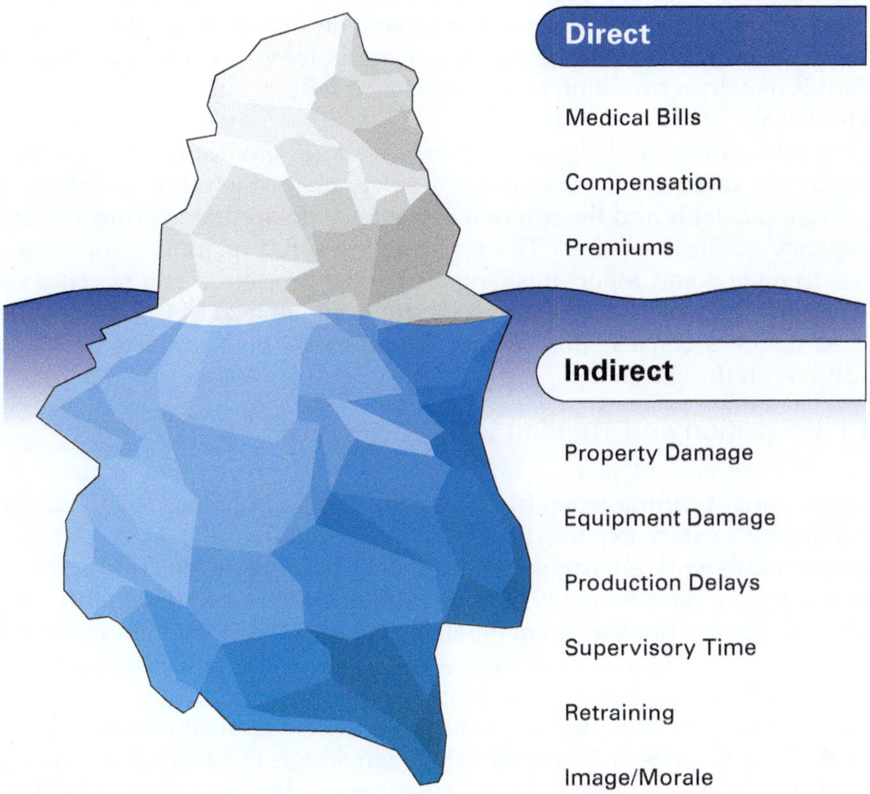

Figure 23 Incidents create both direct and indirect costs.

4.2.0 OSHA

To reduce safety and health risks and the number of injuries and fatalities on the job, the federal government has enacted laws and regulations, including the Occupational Safety and Health Act of 1970 (OSH Act of 1970). This law created the Occupational Safety and Health Administration (OSHA), an agency part of the US Department of Labor. OSHA also provides education and training for employers and workers. Through the administration of OSHA, the US Congress seeks to ensure safe and healthy working conditions for workers by setting and enforcing standards and providing training, outreach, education, and assistance.

OSHA has many functions. It issues standards and rules for working conditions, facilities, equipment, tools, and work processes. It also conducts extensive research into occupational incidents, illnesses, injuries, and deaths to reduce the number of occurrences and adverse effects. In addition, OSHA regulatory agencies conduct workplace inspections to ensure that companies follow the standards and rules. The agencies also have the right to take legal action if companies do not comply with OSHA standards and rules.

OSHA has established significant monetary fines for violating its regulations. In addition to the fines, criminal charges are possible for willful violations resulting in death or serious injury. The employer's attitude and safety history can significantly affect the outcome of a case.

4.3.0 Safety and Health Programs

Under the OSH Act of 1970, employers are responsible for providing a safe workplace. Your employer must set up a program to manage workplace safety and health and to reduce work-related injuries, illnesses, and fatalities. The program must be appropriate for the workplace conditions. It should consider the number of workers employed and the hazards they face. On a project site, a hazard is usually a condition or activity that, if left uncontrolled, has the potential for harm.

One way employers help create a safe working environment is to create a **construction safety plan**. A construction safety plan which should include steps to prevent incidents and the respond if incidents do occur, including calling for emergency medical services. The construction safety plan also includes processes to review and report incidents. For some projects, a **site-specific safety plan (SSSP)** is required to address a jobsite's unique hazards. SSSPs should be created before a construction project is underway, and updated regularly as conditions on the site change.

Construction safety plan: A document that outlines the procedures, rules, and regulations that are or will be put in place to protect workers during a construction project.

Site-specific safety plan (SSSP): A safety plan that includes a focused set of information that accounts for a jobsite's unique hazards.

4.3.1 Education and Training

All employees should receive safety policies and procedures, manual training, and initial training when hired. A company's safety manual should cover the company's safety expectations and the procedures that must be followed to create a safe work environment and meet OSHA requirements. During orientation, appropriate company staff should guide new employees through the safety policies and procedures manual, focusing on the most important sections for their job. An employer must verify that any barriers to worker understanding, such as language barriers, are removed or addressed.

In addition to formal training, safety is a topic you must discuss with your crew daily on the jobsite. You need to help employees be aware of and avoid the hazards to which craft professionals are exposed. As a leader, you should plan to conduct frequent safety training and meetings with your crew, including holding **toolbox talks** (*Figure* 24). These meetings are important to building a strong safety culture and reinforcing your company's commitment to protecting your workers.

Toolbox talks: Short, informal safety meetings held at the start of each day or shift. Sometimes referred to as *tailgate meetings* or *safety briefings*.

Toolbox talks are one way to effectively keep all workers informed of safety issues and guidelines. Toolbox talks are meetings, typically 5 to 10 minutes in length, that review specific health and safety topics at the start of each shift. *Table* 3 provides a checklist for delivering effective safety briefings or toolbox talks.

Always document your toolbox talks or other safety training. Even if certain OSHA standards do not require safety training documentation, keep a record of each safety session signed by all attendees. The record should include the date, a summary of the training session, the trainer, and the attendees.

One-on-one or small group training is also critical to keeping everyone on the project safe. When you assign an inexperienced employee a new task, you must ensure that the employee can safely do the work. You can accomplish this by providing safety information or training for groups or individuals.

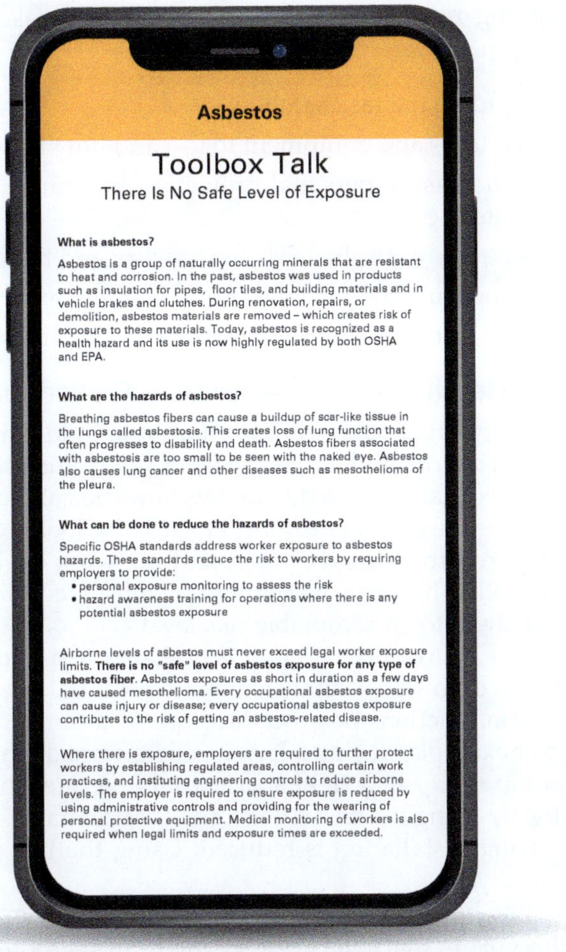

Asbestos

Toolbox Talk

There Is No Safe Level of Exposure

What is asbestos?

Asbestos is a group of naturally occurring minerals that are resistant to heat and corrosion. In the past, asbestos was used in products such as insulation for pipes, floor tiles, and building materials and in vehicle brakes and clutches. During renovation, repairs, or demolition, asbestos materials are removed – which creates risk of exposure to these materials. Today, asbestos is recognized as a health hazard and its use is now highly regulated by both OSHA and EPA.

What are the hazards of asbestos?

Breathing asbestos fibers can cause a buildup of scar-like tissue in the lungs called asbestosis. This creates loss of lung function that often progresses to disability and death. Asbestos fibers associated with asbestosis are too small to be seen with the naked eye. Asbestos also causes lung cancer and other diseases such as mesothelioma of the pleura.

What can be done to reduce the hazards of asbestos?

Specific OSHA standards address worker exposure to asbestos hazards. These standards reduce the risk to workers by requiring employers to provide:

- personal exposure monitoring to assess the risk
- hazard awareness training for operations where there is any potential asbestos exposure

Airborne levels of asbestos must never exceed legal worker exposure limits. **There is no "safe" level of asbestos exposure for any type of asbestos fiber**. Asbestos exposures as short in duration as a few days have caused mesothelioma. Every occupational asbestos exposure can cause injury or disease; every occupational asbestos exposure contributes to the risk of getting an asbestos-related disease.

Where there is exposure, employers are required to further protect workers by establishing regulated areas, controlling certain work practices, and instituting engineering controls to reduce airborne levels. The employer is required to ensure exposure is reduced by using administrative controls and providing for the wearing of personal protective equipment. Medical monitoring of workers is also required when legal limits and exposure times are exceeded.

Figure 24 Safety management software can help companies deliver and track worker training. Some of the solutions help you schedule, track, and deliver toolbox talks.

TABLE 3 Delivering Effective Safety Briefings or Toolbox Talks

Did You...?	Yes/No
Schedule time in advance and create time to allow workers to participate.	
Speak clearly and directly at a natural pace.	
Try to find a spot free of noise and other distractions. If the workers cannot hear you talking, or are distracted by other activities in the area, they won't be focusing on your talk.	
Use props to help make your point and provide visual communication cues. For example, have a safety harness with you for your talk on fall protection.	
Reinforce safety basics, focus on high-risk scenarios, and inform workers about changes to the jobsite and working conditions that may have occurred since their last shift.	
Discuss any incidents or injuries that have occurred and how they could have been prevented.	
Focus the discussion on active work. For example, do not hold a toolbox talk on sitework when electrical work is the only thing being done that day.	
If appropriate, break out a toolbox talk into smaller groups to focus on individual crafts or work activities for certain individuals.	
Focus on the positive too. Highlight safe working practices. Provide positive reinforcement and applaud workers following safety rules and regulations.	
Always give workers an opportunity to ask questions at the end of the toolbox talk.	

When assigning an inexperienced employee a new task, do the following:

- Define the task.
- Explain how to do the task safely.
- Explain what tools and equipment to use and how to use them safely.
- Identify the necessary personal protective equipment (PPE) and train the employee in its use.
- Explain the nature of the hazards in the work and how to recognize them.
- Stress the importance of personal safety and the safety of others.
- Review Safety Data Sheets (SDSs) that may be applicable.

4.3.2 Hazard Identification, Assessment, Prevention, and Control

Job safety analysis (JSA): A technique that focuses on job tasks as a way to identify hazards before they occur. It focuses on the relationship between the worker, the task, the tools, and the work environment. Also known as *job hazard analysis* (*JHA*) or *activity hazard analysis* (*AHA*).

Crew leaders and workers play an important role in identifying and reporting hazards. Hazards are often identified in a **job safety analysis (JSA)**, or *job hazard analysis (JHA)*. This analysis also documents how the hazards will be eliminated or mitigated. It focuses on identifying hazards by looking at the relationship between the worker, the task, the tools, and the work environment. After you identify uncontrolled hazards, you will take steps to control them by eliminating or reducing them to an acceptable risk level.

In a JSA (*Figure 25*), the task at hand is broken down into its individual parts or steps, and then each step is analyzed for its potential hazards. Once a hazard is identified, certain actions or procedures are recommended that will correct that hazard. For example, during a JSA, it is determined that using a chain hoist to install a pump motor in a tight space would be safer than having a worker do it manually. By using the chain hoist, the chance that the worker's hand will be crushed during installation is reduced. Using the JSA process protects the worker from injury.

A JSA can also be used to make certain safety is planned for the job. This approach helps ensure that safety is not compromised for the schedule and productivity goals of the project.

Crew leaders must protect workers from existing or potential hazards in their work areas. They must determine what working conditions are unsafe and inform employees of hazards and their locations. In addition, they should encourage their crew members to tell them about hazardous conditions. To accomplish this, crew leaders must be present and available on the jobsite.

4.3.3 Management, Leadership, and Employee Participation

A culture of commitment to safety and a strong safety and health program requires management, leadership, and employee participation. As a frontline leader, the best way to encourage employee participation in safety and health is to lead by example. Your behavior sets standards for your crew members. If you cut corners on safety, you are telling your crew they may do so as well.

You should always be sure to do the following:

- Stay actively involved in your company's safety and health program.
- Exceed expectations regarding safety. Help others follow the rules and ask your crew for safer ways to do a task.
- Host safety training and meetings. Discuss safety at the start of each shift. Have regular toolbox talks. Discuss safety in every meaning
- Be a visible participant on safety walks.
- Review and provide feedback on hazard analyses.
- Always follow up on safety concerns raised by the workers.

Your employer will use their policies and procedures for incident investigation and documentation. Be sure to follow any of these guidelines when the situation arises.

Job Safety Analysis

Title of Job or Task

Task	Start	End	Hazards	Controls
1.				
2.				
3.				
4.				
5.				
6.				
7.				
8.				

Required Training: Required Personal Protective Equipment (PPE):

Job Name: ______________

Job Number: ______________

Supervisor: ______________

Date: ______________

Weekly Vehicle Checklist: ___ Tire Pressure ___ Oil ___ Air Filter ___ Transmission Fluid ___ Lights ___ Wkly Mileage

Name of Employees:

Print Name	Sign Name	Total Hours

Figure 25 Job safety analysis form.

4.4.0 Incident Preparedness and Response

While the goal is zero incidents on a worksite, you and your team must understand how to respond quickly to a safety incident. If an incident occurs, you should follow the safety policies and procedures in the safety plan. As a front-line leader, you ensure everyone on your team understands the safety plan. You also may have reporting and investigation responsibilities after an incident (*Figure 26*).

4.4.1 Administering First Aid

When someone is hurt or has a medical event, your first goal is to ensure the best outcome: minimizing the health impact on the person. **First aid** provides immediate and temporary medical care to employees involved in incidents and employees experiencing non-work-related health emergencies, such as chest pain or breathing difficulty.

First aid: Help given to a sick or injured worker on the worksite immediately after an injury occurs.

Many companies have very specific policies on administering first aid. If you are unsure of these policies, ask your supervisor and safety team for guidance. You need to know how to handle these situations before an incident occurs. A quick response can be the difference between life and death. The victim of an injury or sudden illness at a jobsite may be at a remote location. The site may be far from a rescue squad, fire department, or hospital, presenting a problem in the rescue and transportation of the victim to a hospital.

Figure 26 Know how to react in an emergency.
Source: iStock

AED (automated external defibrillator): A device used to help those experiencing sudden cardiac arrest.

CPR (cardiopulmonary resuscitation): A lifesaving technique that is used when someone's heartbeat or breathing has stopped.

NOTE

OSHA does not require companies to record when simple first aid addresses the injury or illness.

Your employer must prioritize prompt first aid treatment for any injured employees. To do this, they will provide access to readily available first aid kits. You and everyone on your team should be aware of the location and contents of first aid kits available on the jobsite. Many jobsites also have an **AED (automated external defibrillator)** that is used to help those experiencing sudden cardiac arrest (*Figure 27*). Many lives have been saved due to a prepared team who knew where the AED device was stored and how to use it.

CPR (cardiopulmonary resuscitation) is used to maintain someone's heartbeat during a health emergency when their heart or breathing has stopped. Your company will have an on-site medical professional or at least one member of the workforce who is trained and certified in first aid, including using an AED and CPR. The benefits of having personnel trained in first aid at jobsites include the following:

- Prevents minor injuries from developing into more serious conditions by using immediate and proper treatment of an injury.
- Determines if an injured person requires professional medical attention.
- Stabilizes an injured person before professional medical care arrives.
- Eliminates or reduces medical expenses, lost work time, and sick pay.

If further medical treatment is needed, preparedness and a rapid communication process will mobilize the right people as soon as possible. Emergency numbers should be posted on the jobsite. As a leader, you should be familiar with the safety plan and work with your HR and safety leaders to ensure it stays updated. You need to locate your local off-site medical provider for emergency or nonemergency care before an incident occurs.

4.4.2 Reporting and Investigating Incidents

As a frontline leader, you must know and follow your company's incident reporting processes and ensure your team follows those policies. Your team should know from their training to report any unusual event, incident, or injury immediately. Safety manuals and plans should include the policies and definitions around reporting.

Figure 27 Many jobsites have an AED to help those experiencing sudden cardiac arrest. You should always know where AED devices are stored.
Source: Shutterstock

If your employer is subject to the recordkeeping requirements of the OSH Act of 1970, it must maintain records of all recordable occupational injuries and illnesses using standard OSHA forms. In the event of an incident, the employer is also required to investigate the cause of the incident and determine how to avoid it. According to OSHA regulations, the employer must investigate each work-related death, serious injury or illness, or incident having the potential to cause death or serious physical harm. The employer should document any findings from the investigation, as well as the action plan to prevent future occurrences. The company should complete these actions immediately, with photos or video if possible.

As a crew leader, you may be involved with an incident investigation. If company policy requires it, you must also make a formal investigation and submit a report after an incident. An investigation looks for the causes of the incident by examining the circumstances under which it occurred and talking to the people involved. Investigations are perhaps the most useful tool in the prevention of future incidents. As shown in *Figure 28*, an incident investigation has four major parts.

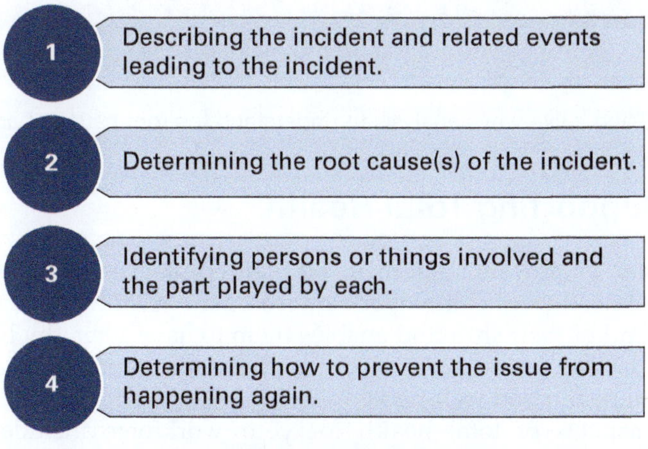

Figure 28 An incident investigation has four major parts.

Root cause: A core issue or fundamental reason for the occurrence of a problem.

Root cause analysis (RCA): A method of problem-solving used for identifying the root causes of faults or problem.

A key step in an investigation is determining the **root cause** of the incident. Understanding the root cause is critical to avoiding similar incidents in the future. In many cases, the root cause was a flaw in the system that failed to identify an unsafe condition or the potential for an unsafe act.

Finding the root cause of a problem can be difficult. A **root cause analysis (RCA)** assesses the impact of different factors involved in an incident. A "five whys analysis" is one type of root cause analysis that is helpful with incident investigations. It is especially useful when the root cause is not that obvious. The method is simple: when a problem occurs, you start with the broadest explanation for an incident and narrow it down to a more specific problem by asking "why" five times. The technique is even more effective when you include experienced team members who were actively part of the situation or have had similar experiences. *Figure 29* shows an example of how a five whys analysis can apply to an incident investigation.

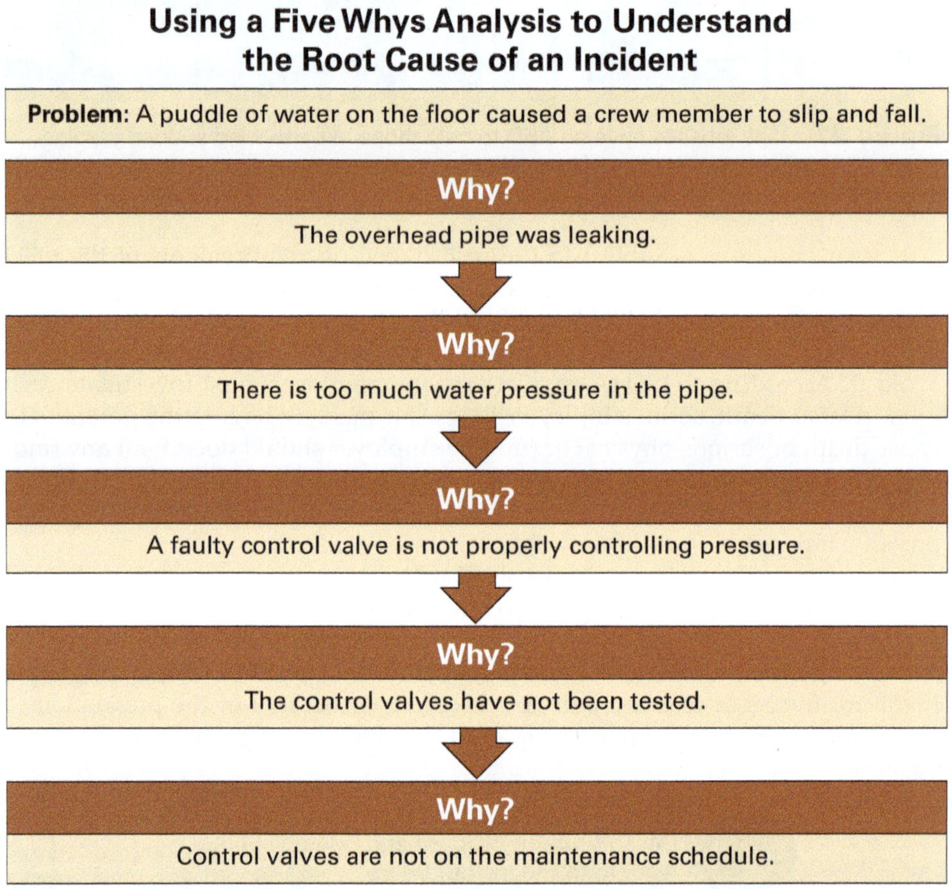

Figure 29 Using a five whys analysis to understand the root cause of an incident.

4.5.0 Supporting Total Health

More than ever, employers must focus on their employees' physical well-being and their total health. A healthy work environment helps workers return home safely at the end of their shift and enables them to have a healthy life outside of work. Many companies are adopting practices that promote workplace safety, as well as overall health and well-being.

Substance abuse: The inappropriate overuse of drugs and chemicals, whether they are legal or illegal.

Two key aspects of total health for your workforce include addressing **substance abuse** and mental health. As a frontline leader, you cannot address the root causes of these issues for your workers. However, you have the important responsibility of enforcing company policies and procedures in these areas and working with your management or HR department to help employees in these situations.

4.5.1 Substance Abuse

Substance abuse is a serious safety issue in the workplace. The overuse of drugs and chemicals results in some form of mental, sensory, or physical **impairment**. Impairment is the point where someone's intake of substances affects their ability to perform appropriately. Alcohol can be abused by consuming to the point of intoxication. Legal prescription drugs can be abused by using the wrong dosage, using other people's medications, or self-medicating. Illegal drugs, such as cocaine or crystal meth, can cause impairment. Even legal recreational drugs inhibit a person's ability to safely and accurately complete tasks.

Impairment: The point where someone's intake of substances affects their ability to perform appropriately.

Crew leaders must enforce company policies and procedures regarding substance abuse. There are legal consequences and safety implications associated with substance abuse. Suppose several crew members go out and smoke marijuana or drink during lunch. Then they return to work to erect scaffolding for a concrete pour in the afternoon. If you can smell marijuana on the worker's clothing or alcohol on their breath, you must step in and act. An impaired worker is more likely to cause an incident that could cause serious injury or death to themselves or others (*Figure 30*).

Figure 30 An impaired worker is a dangerous one.
Source: iStock

If you observe an employee showing impaired behavior, immediately contact your supervisor and/or HR department for assistance. You should work with management to deal with suspected drug and alcohol abuse and should not deal with these situations yourself. Reporting the concern helps you protect the business and the safety of the employees and other workers.

It is often difficult to detect substance abuse because the effects can be subtle. The best way is to look for physical signs or sudden changes in behavior that are not typical of the employee. *Table 4* lists some examples of potential signs and behavioral changes that might indicate substance abuse.

TABLE 4 Potential Signs of Substance Abuse in a Worker

Physical Signs	Behavioral Changes
• Smelling of alcohol or drugs • Slurring of speech or an inability to communicate effectively • Wearing sunglasses indoors or on overcast days to hide dilated or constricted pupils, conditions which impair vision • Significant changes in personal appearance, cleanliness, or health	• Unscheduled absences • Failure to report to work on time • Significant changes in the quality of work • Unusual energy or tiredness • Sudden and irrational temper flare-ups • Sneaky behaviors, such as disappearing to wooded areas, storage areas, or other private locations • Attempting to borrow money from coworkers • Losing tools or company equipment

4.5.2 Mental Health

Mental health is a critical but often overlooked part of overall jobsite safety and health. Construction and agricultural workers have more risk factors than employees in other industries. These risk factors make workers more likely to experience mental health challenges, with a specific set of risks around suicide. Construction workers are six times more likely to die by suicide than they are to die from a fall.

With proper treatment, mental health conditions can be managed and overcome so those experiencing them can have full and productive lives. As with substance abuse, however, these conditions can lead to people being distracted, less productive than normal, and possibly unsafe and unable to perform their normal jobs. Because of this, mental health and suicide prevention must be safety considerations.

As a leader, you cannot solve these challenges for your workers, but you can create a better environment. Use your toolbox talks to discuss mental health as part of total worker safety. In your safety huddles, check in on how your teammates are doing. Encourage your crew to talk to supervisors and colleagues when they feel stressed, anxious, or on the brink of burnout. Help ensure they are aware of any available counseling resources and that they are taking time off when they need it.

Even more so than substance abuse, it can be very difficult to identify a worker struggling with a mental health issue. *Table 5* lists some examples of potential signs and behavioral changes that might indicate someone is struggling with mental health issues.

If you observe an employee showing signs of mental health issues or have concerns for any reason, immediately contact your supervisor and/or HR department for assistance. You should work with management to deal with suspected mental health issues and should not deal with these situations yourself. Reporting the concern helps you protect the business and the employee's and other workers' safety.

TABLE 5 Signs That May Indicate Someone Is Struggling with Mental Health Issues

Behavioral Changes	Performance Changes	Verbal Cues	Life Events
• Becoming withdrawn • Acting anxious or reckless • Misusing drugs and alcohol	• Missing work • Showing up late • Being less productive • Being unable to think clearly or solve problems • Having increased safety incidents	• Talking about feeling trapped • Talking about wanting to die • Saying they feel like a burden • Saying they feel like it would be better if they were gone	• Dealing with a divorce or breakup • Handling a death in the family • Working through a major illness or injury • Having financial issues

4.0.0 Section Review

1. One of a crew leader's most important responsibilities to the employer is to _____.
   a. enforce company safety policies
   b. estimate material costs for a project
   c. make recommendations for setting up a crew
   d. provide input for fixed-price contracts
2. Which of these is *not* a function of OSHA?
   a. Issuing standards and rules for working conditions
   b. Researching occupational incidents and injuries
   c. Conducting workplace inspections
   d. Distributing company safety manuals
3. Which of the following should all employees receive upon being hired?
   a. A copy of the company's safety policies and procedures
   b. A new set of tools
   c. A leadership assessment to help map their career path
   d. First aid and CPR training
4. A problem-solving method that is useful in investigating incidents and identifies the core issue of a problem is _____.
   a. problem-solving
   b. root cause analysis
   c. incident recordkeeping
   d. event playback
5. A focus on overall worker safety, health, and well-being, including mental health, is _____.
   a. home health
   b. complete worker safety
   c. occupational safety
   d. total health

5.0.0 Project Planning, Scheduling, and Cost and Resource Management

Objective

Demonstrate an understanding of construction project phases, project planning and scheduling, field reporting, and cost and resource management.

a. Identify construction project phases.
b. Summarize crew planning and scheduling responsibilities, including creating lookahead schedules.
c. Describe the frontline leader's role in field reporting.
d. Explain approaches to managing costs and controlling project resources, such as materials, tools, and equipment.

Performance Task

3. Develop and present a lookahead schedule.

This section describes the various phases of a project. It examines **estimating**, **planning**, **scheduling**, and **cost control** through managing material, people, tools, and other resources. All workers who participate in a job are responsible at some level for their assigned jobs. For example, the contractor's responsibility begins with obtaining the contract and doesn't end until the client takes ownership of the project. The project manager is generally the person with overall responsibility for coordinating the project. Finally, the superintendent and frontline leader, like a foreman, supervisor, or crew leader, are responsible for coordinating the work of one or more workers, one or more crews of workers within the company, or subcontractors.

Estimating: The process of calculating the cost of a project.

Planning: The process of laying out all details of a construction project before construction begins.

Scheduling: Involves defining which activities happen in what order and assigning dates and times to those tasks.

Cost control: The process that helps keep expenses under control by managing labor, material, and other costs to ensure that the project finishes on budget.

5.1.0 Construction Project Phases

A simple way to think about a construction project is in three phases: the *design and development phase*, the *preconstruction phase*, and the *construction phase*. *Figure 31* shows the flow of a typical project from beginning to end through these phases.

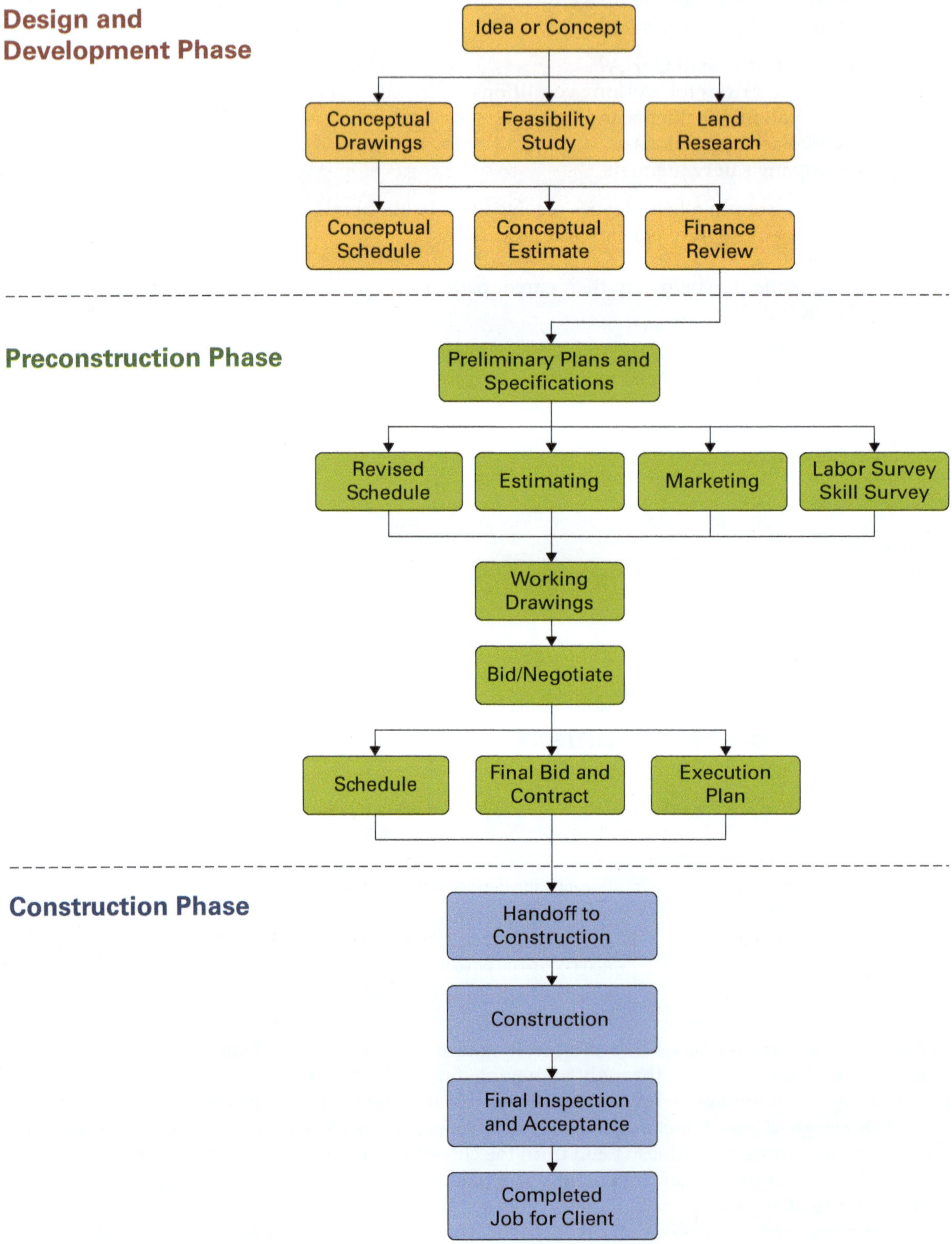

Figure 31 Construction project phases and flow.

5.1.1 Design and Development Phase

The design and development phase is the first stage of a project (*Figure 32*). It starts when an owner decides to build a new facility or add to an existing facility. This process involves land research and a **feasibility study** to determine whether a project can be completed on time and within budget.

Architects or engineers develop conceptual drawings that graphically define the project. They then provide the owner with sketches of room layouts and elevations and suggest what construction materials to use.

During the design and development phase, architects, engineers, and/or the owner develop an estimate for the proposed project and establish a preliminary budget. Once that budget is established, the owner will determine how to pay for or finance the project. The owner will also complete a review to determine whether the project costs will exceed its market value and whether the project will provide a reasonable profit during its existence. If needed, the development team begins preliminary reviews for zoning, building restrictions, landscape requirements, and environmental impact studies.

Feasibility study: An analysis that considers all of a project's relevant factors—including budget, technical elements, legal aspects, and scheduling—to determine the likelihood that the project can be completed successfully.

5.1.2 Preconstruction Phase

If all steps in the design phase are completed successfully, the project moves to the preconstruction phase (*Figure 33*). The preconstruction stage includes planning that occurs before the start of construction. The preconstruction planning process does not always involve you as a crew leader, but you should understand it to do your job well.

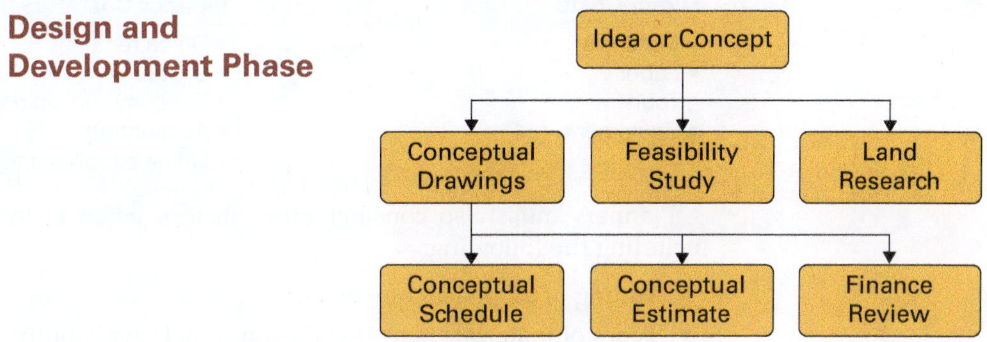

Figure 32 The design and development phase is the first stage of a project.

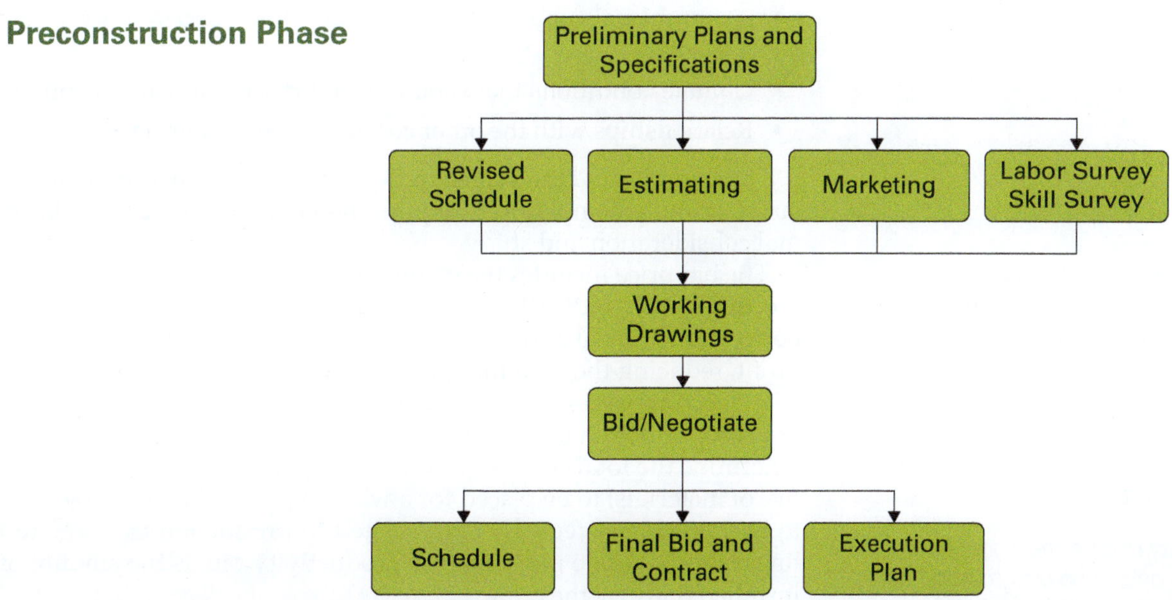

Figure 33 The preconstruction phase is the second stage of a project.

When the architects and engineers develop the project drawings and specifications, they consult with other design professionals such as structural, mechanical, and electrical engineers. They perform the calculations, make a detailed technical analysis, and check project details for accuracy.

The design professionals create preliminary drawings and specifications. They use these drawings and specifications to communicate the necessary information to the contractors, subcontractors, suppliers, and workers contributing to the project.

A key next step is developing the proposal, bid, or negotiated price for the job. This is when the estimator, the project manager, and the field superintendent develop a preliminary plan for completing the work. They apply their experience and knowledge from previous projects to develop the plan. The process involves determining what methods, personnel, tools, and equipment the work will require and what level of productivity they can achieve.

Part of this process is determining how much the project will cost. Before building a project, an estimate must be prepared. Estimating is the process of calculating the cost of a project. Two types of costs are direct and indirect, as shown in *Table 6*. Direct costs, also known as *general conditions*, are those that planners can assign to a budget. Indirect costs are overhead costs shared by all projects. Planners generally calculate these costs as an overhead percentage of labor and material costs.

TABLE 6 Examples of Direct and Indirect Costs

Direct Costs	Indirect Costs
• Materials • Labor • Tools • Equipment	• Office rent • Utilities • Telecommunications • Accounting • Office supplies and signage

Planners must also consider other factors when estimating project costs, including the following:

- Timing of all phases of work
- Types of materials to be installed and their availability
- Equipment and tools required and their availability
- Personnel requirements and availability
- Site and local conditions, such as soil types, accessibility, or available staging areas
- Climate conditions that should be anticipated during the project
- Relationships with the other contractors and their representatives on the job

On a simple job, crews can handle these items almost automatically. However, larger or more complex jobs require the planner to give these factors more formal consideration and study.

The bid price includes the estimated cost of the project as well as the **profit**. If the direct and indirect costs exceed the estimate for the job, the difference between the actual and estimated costs must be deducted from the company's profit, reducing the amount of money the contractor earns.

Many companies employ professional cost estimators to determine the cost of a project. Estimators will focus on **production** and the **productivity rate** to determine the total cost. For example, suppose the production (that is, the quantity of materials) to be placed for a waste pipe installation project is determined to be 1,000 linear feet. The cost related to production is therefore the cost per linear foot of waste pipe, and the productivity rate is the amount of waste pipe installed daily by the crew.

Profit: The amount of money that the contractor will make after paying all the direct and indirect costs.

Production: The amount of construction materials put in place. It is the quantity of materials installed on a job.

Productivity rate: A measure of how much work gets done over a period of time, also called *output per labor hour*.

Production levels are set during the estimating stage. Based on the drawings, plans, and specifications, the estimator determines the production, or total materials to be placed. After the job is complete, the project manager can assess the amount of materials installed and compare the actual production to the estimated production.

During the estimating stage, the estimator uses company productivity records to determine how much time and labor it will take to place a certain quantity of materials. From this information, the estimator calculates the productivity necessary to complete the job on time.

For example, say you need a crew of two people to paint 8,100 ft^2 in nine days. To calculate the required average productivity, divide 8,100 ft^2 by nine days. The result is 900 ft^2/day. As the work progresses, you can compare the daily production of any crew of two painters doing similar work with this average (*Figure 34*).

Production

A quantity or amount of construction materials or work put in place.

Productivity

A rate amount of work done by a worker or crew in a given timeframe.

Production

Paint 8,100 ft^2

Productivity

One crew member can paint 450 ft^2/day.
Two crew members can paint 900 ft^2/day.

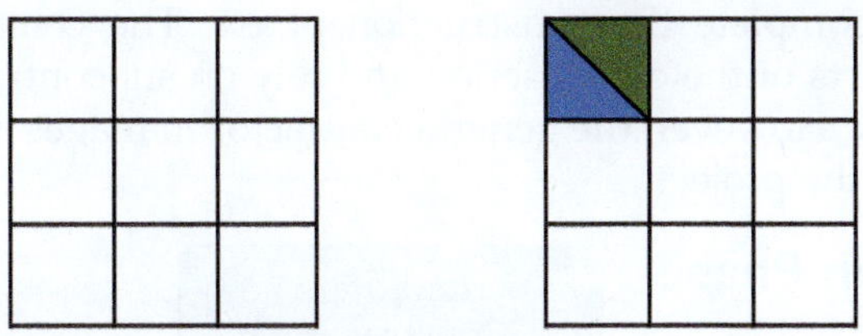

Figure 34 Production and productivity.

During the preconstruction phase, the owners hold meetings to review and refine estimates, adjust plans to conform to regulations, and secure a construction loan. If the project requires a marketing program, such as a condominium, office building, or shopping center, the marketing may start during preconstruction. In such cases, selling the building space often starts before actual construction begins.

Next, the design team produces complete drawings, specifications, and bid documents. After this, the owner will select the method to obtain contractors. The owner may negotiate with several contractors or select one through competitive bidding. Everyone concerned must also consider safety as part of the planning process. A safety crew leader may walk through the site during the pre-bid process.

Contracts for construction projects can take many forms. The most common types of contracts are lump sum and time and materials contracts.

In a lump sum contract, also called a *fixed-price* or *firm-fixed-price contract*, the owner generally provides detailed drawings and specifications, which the contractor uses to calculate the cost of materials and labor. The contractor also adds a percentage to these costs representing company overhead expenses, such as office rent, insurance, and accounting/payroll costs. At the end, the contractor adds a profit factor.

When submitting the bid, the contractor states the conditions and assumptions on which the company based the bid. The **scope**, or *scope of work*, lists the activities that all crew members and suppliers are expected to perform during a project.

Scope: The list of the commitments and activities that all contractors, subcontractors, and suppliers are expected to do during a project. Also called the *scope of work*.

In a time and materials contract, also known as a T&M contract, the owner pays back the contractor for labor, materials, and other costs encountered in the performance of the contract. Typically, the contractor and owner agree in advance on hourly or daily labor rates for different categories of workers. These rates include an amount representing the contractor's overhead expense. The owner also reimburses the contractor for the materials and equipment used on the job. Since the buyer reimburses the contractor for the cost of materials and pays an hourly wage, any unexpected delays, roadblocks, and other work scope changes are covered. T&M contracts thus work best for projects in which the scope of work is not well defined.

On a time and materials contract project, you play an important role by understanding the scope and ensuring that everything is tracked so your company is reimbursed for the work. You must identify changes or problems that increase the scope or amount of planned labor or materials. You also must ensure you and your team follow your company policies and get proper authorization before starting any work.

After the client awards the contract, planners create a final schedule and an execution plan that defines the actual work methods and resources needed to perform the work.

5.1.3 Construction Phase

The construction phase can start after the contract is signed (*Figure 35*). The designated contractor enlists the help of mechanical, electrical, elevator, and other specialty subcontractors to complete the construction phase. The contractor may perform one or more parts of the construction and rely on subcontractors for the remainder of the work. However, the general contractor manages all the trades necessary to complete the project.

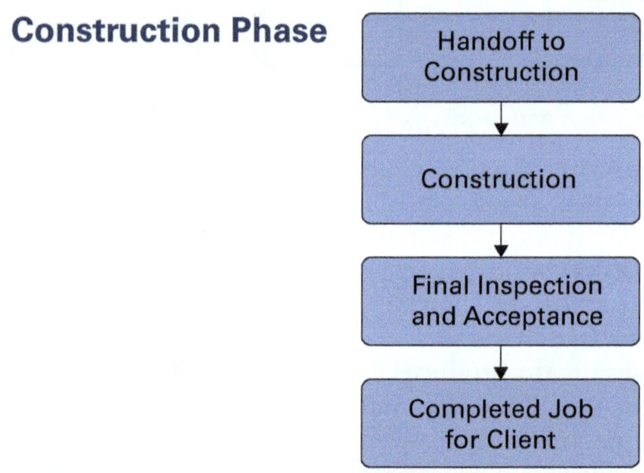

Figure 35 The construction phase is the third stage of a project.

The initial set of drawings for a construction project reflects the completed project conceived by the architect and engineers. During construction, changes may be necessary because of factors unforeseen during the design phase. For example, when electricians must reroute cabling or conduit, or the installed equipment location is different than shown on the original drawing, such changes must be marked on the drawings. Without this record, technicians called to perform maintenance or modify the equipment later will have trouble locating all the cabling and equipment.

Project supervision must document any changes made during construction or installation on the drawings as the changes occur. Before digital drawings, architects would note changes on hard copy drawings using a colored pen or pencil. These marked-up versions are commonly called **redline drawings**. With new digital drawings created using building information modeling tools, architects and engineers can instantly revise and share the latest drawing versions to mobile phones and tablets. The final drawings are called **as-built drawings** after the drawings have been revised to reflect any redline changes. As-built drawings are a revised set of drawings submitted by a contractor upon completion of a project or job. These drawings, sometimes called *record drawings*, reflect all changes made during construction and show the exact specifications of the completed work.

Redline drawings: Marked-up versions of construction drawings documenting any changes made during construction or installation.

As-built drawings: A revised set of drawings that reflect all changes made during the construction process and show the exact specifications of the completed work.

The main part of your role as a frontline leader happens during the construction phase, including daily work planning and scheduling, cost control, and resource management, including material delivery and storage, tool control, and equipment control. All these tasks drive production and productivity, as well as the overall profitability of the project.

NOTE

Scan the QR code in the front of this module to access additional digital resources on construction project delivery methods.

As construction nears completion, the architect/engineer, owner, and government agencies start their final inspections and project acceptance (*Figure 36*). If the general contractor has managed the project, the subcontractors have performed their work, and the architects/engineers have regularly inspected the project to ensure it satisfies the local code, then the inspection process can finish up promptly. This results in a satisfied client and a profitable project for all.

On the other hand, if the inspection reveals faulty workmanship, poor design, incorrect use of materials, or code violations, the inspection and acceptance will take longer. This may result in a dissatisfied client or an unprofitable project.

Figure 36 Final inspections are one of the last steps of project completion.
Source: iStock

5.2.0 Planning and Scheduling

Project delivery methods: Different ways to organize the development, planning, and construction services in order to execute a project.

In the previous sections, you learned about the phases of a construction project, including the design and development phase, the preconstruction phase, and the construction phase. Although there are different construction **project delivery methods**, the main part of your role as a frontline leader happens during the construction phase.

You may be directly involved in daily work planning and scheduling during construction. You also may be involved in cost control and resource management, including material delivery and storage, tool control, and equipment control.

5.2.1 Understanding Overall Schedules

Planning and scheduling refer to determining what will happen during a project, what resources are needed, when key activities will occur, which personnel will be involved, and anything else pertinent to the project.

Construction planning refers to laying out all the details of a construction project beforehand. Scheduling involves defining which activities happen in what order and assigning dates and times to those tasks. Construction schedules are linked to resources such as materials, labor, and equipment to help create the most efficiency in your project. Construction scheduling is deeply connected to planning.

Planning and scheduling are closely related and very important to a successful job. Planning identifies the required activities and how and in what order to complete them. Scheduling involves establishing start and finish times/dates for each activity.

A schedule for a project typically shows the following:

- Operations listed in sequential order
- Units of construction
- Duration, or length of time, of an activity
- Estimated dates to start and complete each activity
- Quantity of materials to be installed

NOTE

Scan the QR code in the front of this module to access additional digital resources on schedule levels.

Figure 37 shows an example of a high-level project schedule. As noted, the development of a schedule for a project starts as early as the design and development phase. As the project moves through each phase and the work is better defined, the schedule becomes more clearly defined.

Level 4 schedule: A schedule that offers additional detail to provide working schedules to the crews doing the work, usually within three to six weeks of completing specific work.

When a crew is close to completing specific work—for example, within three weeks of working on Foundation 1—a **Level 4 schedule** is created. A Level 4 schedule offers additional scheduling details to the work crews (*Figure 38*). It allows a supervisor to create a plan, schedule handoffs from crew to crew, and estimate all the resources that must be in place for the work to happen.

Critical path: In manufacturing, construction, and other types of creative processes, the required sequence of tasks directly controls the project's ultimate completion date.

In reference to *Figure 38*, ironworkers must fabricate and set the rebar, and carpenters must build the forms before the handoff is made to place the concrete. Each of these activities is a critical path activity. A **critical path** is the sequence of dependent tasks with the longest duration. In *Figure 37* and *Figure 38*, the critical path in the schedule is shown in red. In other words, until one activity is complete, workers cannot start the other tasks, and the project will likely be delayed by the amount of time the previous activity was delayed.

Lookahead schedule: For planning and scheduling work that is happening in the short term, usually within one to three weeks; used for anticipating material, labor, tool, and other resource requirements, as well as identifying potential schedule conflicts or other problems. Also called *short-term schedule*.

For work to happen, you will need materials, crew, equipment, completed permits, and more. Developing a **lookahead schedule** helps ensure all resources are available for the project when needed. This is often included in a Level 4 schedule. A lookahead schedule is useful for planning and scheduling work in the short term, usually within one to three weeks (*Figure 39*).

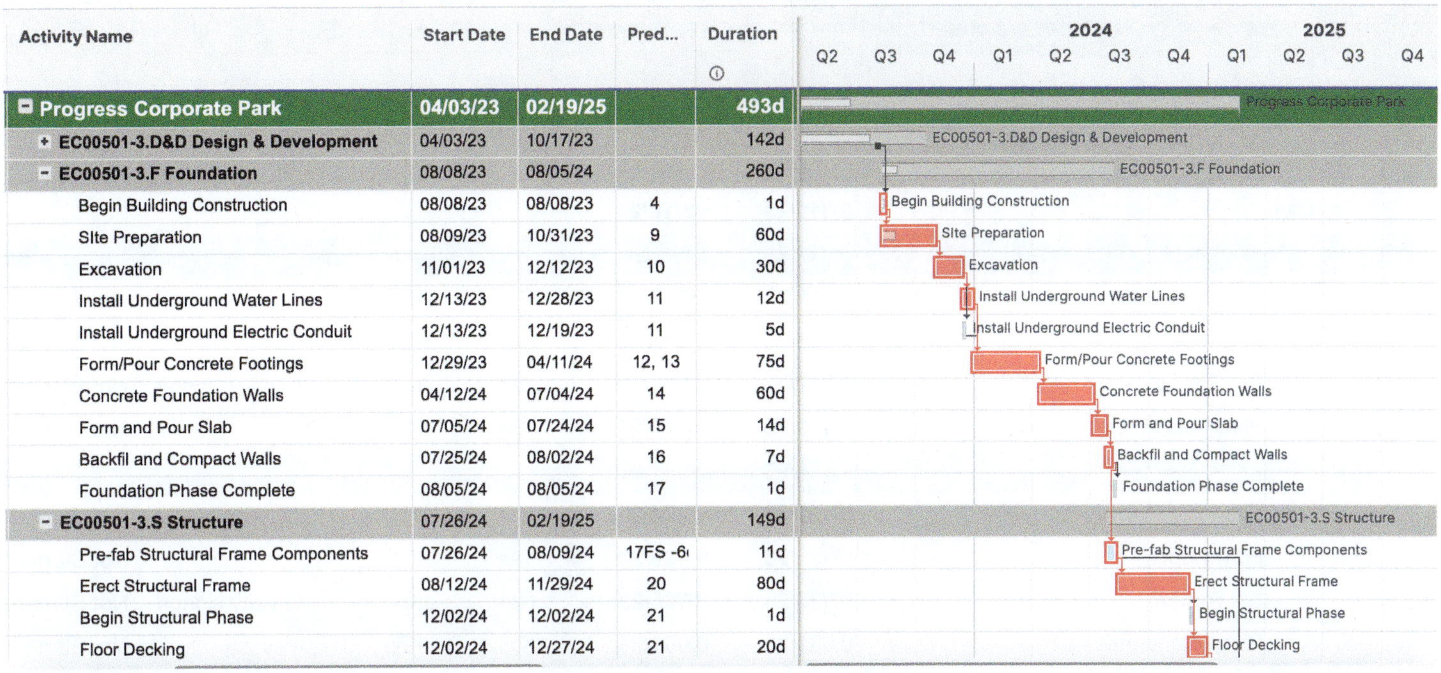

Activity Name	Start Date	End Date	Pred...	Duration
Progress Corporate Park	04/03/23	02/19/25		493d
EC00501-3.D&D Design & Development	04/03/23	10/17/23		142d
EC00501-3.F Foundation	08/08/23	08/05/24		260d
Begin Building Construction	08/08/23	08/08/23	4	1d
SIte Preparation	08/09/23	10/31/23	9	60d
Excavation	11/01/23	12/12/23	10	30d
Install Underground Water Lines	12/13/23	12/28/23	11	12d
Install Underground Electric Conduit	12/13/23	12/19/23	11	5d
Form/Pour Concrete Footings	12/29/23	04/11/24	12, 13	75d
Concrete Foundation Walls	04/12/24	07/04/24	14	60d
Form and Pour Slab	07/05/24	07/24/24	15	14d
Backfil and Compact Walls	07/25/24	08/02/24	16	7d
Foundation Phase Complete	08/05/24	08/05/24	17	1d
EC00501-3.S Structure	07/26/24	02/19/25		149d
Pre-fab Structural Frame Components	07/26/24	08/09/24	17FS -6	11d
Erect Structural Frame	08/12/24	11/29/24	20	80d
Begin Structural Phase	12/02/24	12/02/24	21	1d
Floor Decking	12/02/24	12/27/24	21	20d

Figure 37 Example of a construction project schedule.

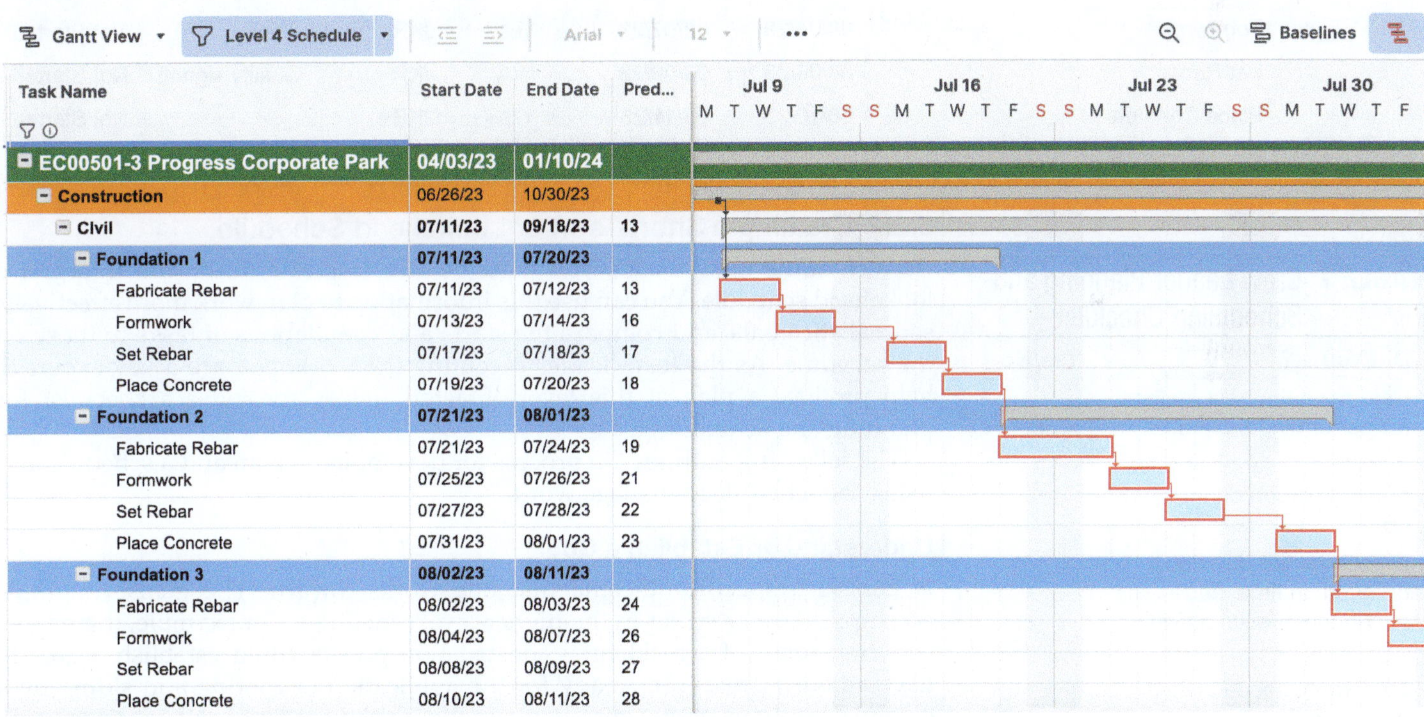

Task Name	Start Date	End Date	Pred...
EC00501-3 Progress Corporate Park	04/03/23	01/10/24	
Construction	06/26/23	10/30/23	
Civil	07/11/23	09/18/23	13
Foundation 1	07/11/23	07/20/23	
Fabricate Rebar	07/11/23	07/12/23	13
Formwork	07/13/23	07/14/23	16
Set Rebar	07/17/23	07/18/23	17
Place Concrete	07/19/23	07/20/23	18
Foundation 2	07/21/23	08/01/23	
Fabricate Rebar	07/21/23	07/24/23	19
Formwork	07/25/23	07/26/23	21
Set Rebar	07/27/23	07/28/23	22
Place Concrete	07/31/23	08/01/23	23
Foundation 3	08/02/23	08/11/23	
Fabricate Rebar	08/02/23	08/03/23	24
Formwork	08/04/23	08/07/23	26
Set Rebar	08/08/23	08/09/23	27
Place Concrete	08/10/23	08/11/23	28

Figure 38 A Level 4 schedule provides enough details for project foremen to forecast handoffs and needed resources.

Grid View | 3 Week Lookahead | Arial | 10

Task Name	Start Date	End Date	Materials Available	Assigned To	Notes	Status
Civil	**07/11/23**	**09/18/23**				
Foundation 1	**07/11/23**	**07/20/23**				
Fabricate Rebar	07/11/23	07/12/23	Yes	JKR		Not Started
Formwork	07/13/23	07/14/23	Yes	JKR		Not Started
Set Rebar	07/17/23	07/18/23	N/A	JKR	Quality signoff	Not Started
Place Concrete	07/19/23	07/20/23	Yes	JKR		Not Started
Foundation 2	**07/21/23**	**08/01/23**				
Fabricate Rebar	07/21/23	07/24/23	Yes	JKR		Not Started
Formwork	07/25/23	07/26/23	Yes	JKR		Not Started
Set Rebar	07/27/23	07/28/23	N/A	JKR	Quality signoff	Not Started
Place Concrete	07/31/23	08/01/23	Yes	JKR		Not Started
Foundation 3	**08/02/23**	**08/11/23**				
Fabricate Rebar	08/02/23	08/03/23	On Order	JKR		Not Started
Formwork	08/04/23	08/07/23	Yes	JKR		Not Started
Set Rebar	08/08/23	08/09/23	N/A	JKR	Quality signoff	Not Started
Place Concrete	08/10/23	08/11/23	Yes	JKR		Not Started

Figure 39 A step-by-step lookahead in a Level 4 schedule provide details for frontline leaders to plan handoffs.

5.2.2 Creating a Short-Term or Lookahead Schedule

As a frontline crew leader, you must be able to create, read, and interpret a lookahead schedule. You can use this information to plan work more effectively, set realistic goals, and compare the starts and completions of tasks to those on the schedule. As the team leader executing the work, you are a key expert to help produce a plan at this level of detail. *Table 7* provides a checklist for planning and scheduling.

The following outlines the process for developing a plan and a schedule and using the schedule to manage work.

TABLE 7 Crew Leader Planning and Scheduling Checklist

Did You...?	Yes/No
Determine the best method for performing the job.	
Identify the responsibilities of each person on the work crew.	
Determine the duration, or length in time, and sequence of each activity.	
Identify the tools and equipment needed to complete the job.	
Ensure that the required materials are at the worksite when needed.	
Make sure that heavy construction and other equipment is available when required.	
Work with other contractors in such a way as to avoid interruptions and delays.	

Understand or Establish a Goal

A goal is a specific outcome that one works toward. The goals you must accomplish are often set in the higher-level schedules. For example, the project superintendent of a residential construction project could establish a goal to have a house dried-in or ready for the application of roofing and siding by a certain date. To meet that goal, you and the superintendent would need to agree to a goal to have the framing completed by a given date. You then establish subgoals for your crew to complete each element of the framing (floors, walls, roof) by a set time.

Identify Required Work Activities and Divide Them into Tasks

To plan effectively, you must be able to break a work activity assignment down into smaller tasks. Large jobs include a greater number of tasks than small ones, but all jobs can be broken down into manageable components.

Make a list of all the activities that the plan requires to complete the job. This includes individual work tasks and special tasks, such as inspections or the delivery of materials. Next, start to divide work activities into tasks. As you do this, remember that almost every work activity has three general parts: preparing for the activity, performing the activity, and cleaning up after the activity.

Figure 40 shows a suitable breakdown for the work activity to install square vinyl floor tiles in a cafeteria. You could create even more detail by breaking down any one of the tasks into subtasks. In this case, however, that much detail is unnecessary and wastes your time and the project's money.

When breaking down an activity into tasks, make each task identifiable and definable. A task is identifiable when you know exactly the types of resources required. A task is definable if it has a specific duration. The activity breakdown does not need to be overly detailed or complex unless the job has never been done before or must be performed efficiently.

At this point, you should just be concerned with generating a list, not with determining how to accomplish the activities, who will perform them, how long they will take, or the necessary sequence to complete them.

Plan Resources Needed for Each Task

Once a job has been divided into activities and tasks, you must identify and account for all the resources needed to complete it. Resource planning includes considerations for safety, materials, site-specific issues, equipment, tools, and labor. *Table 8* discusses each of these in more detail.

Organize Work into a Logical Sequence

The next step in planning is to use the list of activities and resources needed to organize the work activities into a logical sequence. When doing this, remember that certain steps can't happen before the completion of others. For example, concrete placement cannot happen until after the rebar is set. Some construction work activities, such as installing 12" footing forms, are done so often that they require little planning. However, other jobs, such as placing new mechanical equipment, require substantial planning.

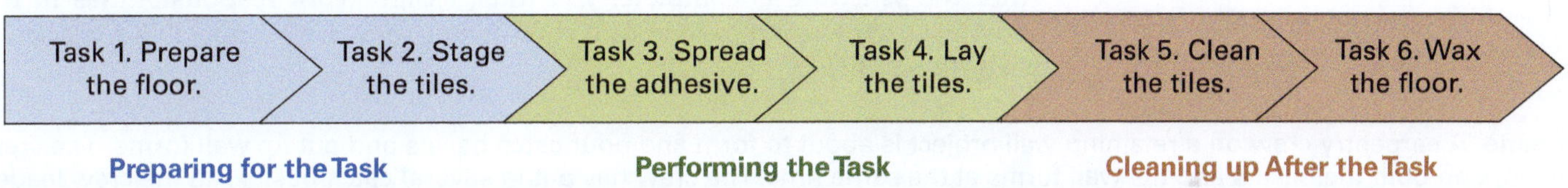

Figure 40 Example task breakdown of the work activity to install square vinyl floor tiles in a cafeteria.

TABLE 8 Key Resource Considerations during Planning

Resource Planning Area	Key Responsibilities
Safety Planning	Using the company safety manual as a guide, you must assess the safety issues associated with the job and take necessary measures to minimize any risk to the crew.
Materials Planning	You may be required to estimate quantities of materials during construction. You also have a key role in the receipt, storage, and control of the materials after they reach the jobsite.
Site Planning	Many planning elements are involved in sitework, including access roads, parking, site security, material and equipment staging and storage, sedimentation control, stormwater runoff, and more.
Equipment Planning	You may need to plan the types of equipment needed, the use of the equipment, and the length of time it will be on the site. You must work with the main office to make certain that the equipment reaches the jobsite on time. You must reserve time for equipment maintenance to prevent equipment failure.
Tool Planning	To plan tool usage, you must determine the tools required, tell your crew who will provide the tools, ensure that the crew is qualified to use the tools safely and effectively, and establish proper tool controls.
Labor Planning	In many companies, the project manager or job superintendent determines the size and makeup of the crew. Then supervision expects you to accomplish the goals and objectives with the crew provided. If you are responsible for planning for labor, you must identify the skills needed to perform the work, determine the number of people having the required skills, and decide who will be on the crew.

Assign Responsibility for Each Task

After identifying the tasks, you must determine who will complete each activity. This requires that you know the experience, skills, and abilities of everyone on your team. You must apply this knowledge by matching the crew's skills and abilities to specific tasks. Your job is to draw from their expertise to get the job done well, safely, and on time.

Assign a Duration to Each Task

Next, assign a duration or length of time that it will take to complete each activity and determine the start date and time for each. Then, place each activity in a scheduled format. This step is important because it helps you compare the task time estimates to the scheduled completion date or time. Do not assign too much work to any person or team.

Make note of any activities that cannot start until previous work is complete. In such cases, do not overlap start times. Instead, assign a start time to new tasks that begins after the other work is complete.

Some information supporting short-term scheduling comes from the estimate or cost breakdown. *Figure 41* shows how you can use estimate or cost breakdown information to set short-term production goals for a team.

In this example, the crew leader used the short-term schedule to translate production into work hours and to schedule work so that the crew could accomplish it within the estimate. A crew can provide input and improve production when they know how much work to accomplish and the time available to complete the work. In addition, setting production targets motivates crews to exceed schedule targets.

Communicate the Schedule and Assigned Responsibilities

After the schedule is complete, you must assign work responsibilities to the crew. Each crew member should know what they are responsible for accomplishing on the job.

Scenario: A carpentry crew on a retaining wall project is about to form and pour catch basins and put up wall forms. The crew will work on both the basins and the wall forms at the same time. The crew has put in several catch basins, so the crew leader is sure that they can perform the work within the estimate. However, the crew leader is concerned about their production of the wall forms and wants to set a clear target for the amount of work to be done each day to hit the schedule.

The scheduling process in this scenario could be handled as follows:

1. Review the estimate or cost breakdown.	2. Consider available workers.	3. Determine the job goal.	4. Analyze how much work the existing workers can complete.	5. Communicate production goal with the crew.
• Production factor for wall forms = 16 worker-hours (w-h) per 100 ft^2 • Work to be done by measurement = 800 ft^2 • Total time: (800 $ft^2 \times$ 16 w-h/100 ft^2) = 128 w-h	The carpentry crew includes: • One carpentry crew leader • Four carpenters • One laborer	The crew leader determines the goal for the job should be set at 128 w-h (from the cost breakdown).	If the crew remains the same (six workers), the work should be completed in about 21 crew-hours (128 w-h ÷ 6 workers/crew = 21.3 crew-hours).	• Crew leader shares the production goal of completing 800 ft^2 in 21 crew-hours with the crew. • Crew leader encourages them to work together to meet the goal of getting the forms erected within the estimated time.

Figure 41 Crew leaders can use lookaheads or short-term scheduling to set production goals.

Use the Schedule to Monitor and Control Work

Schedules are only valuable if they are actively used to monitor and control work. A lookahead or short-term schedule provides you with the information you need to plan for work and track progress. Before starting a job, actively review the schedule to:

- Determine the materials, tools, equipment, and labor needed to complete the job.
- Determine when the various resources are needed.
- Follow up to ensure that the resources are available on the jobsite when needed.

NOTE

Most scheduling apps and programs include this feature that automatically helps you focus on the lookahead period for a job.

You should also verify the availability of required resources three to four working days before the start of the job. This should occur even earlier for larger jobs. Advance preparation will help avoid situations that could delay starting the job or cause it to fall behind schedule. As work continues, use your schedule to monitor whether actual production begins to slip behind estimated production and develop plans to recover the schedule as needed.

Follow Up and Confirm Work Is Completed Well

After you delegate tasks to the appropriate crew members, you must follow up to confirm that the crew completed the tasks correctly and efficiently. To do this effectively, you must be present on the jobsite to make sure all the resources are available to complete the work, to confirm that team members are working on their assigned activities, to answer any questions, and to help to resolve any problems that occur while the work is being done.

5.3.0 Crew Leader Field Reporting Responsibilities

A field reporting system consists of a series of forms completed by you and other leaders on the project to share updates on labor, materials, tools, equipment, and more.

You may not be directly involved in updating overall schedules as a frontline leader. However, the company will use the daily field or progress reports you submit to keep the schedule up to date.

The following sections provide some examples of field reporting. Each company has its own forms and methods for obtaining information. You should always follow your company's policies and procedures for field reporting.

5.3.1 Completing Daily Reports

You will be involved with many activities on a day-to-day basis. You should log those activities in a **job diary** each day. Include entries for job changes, interruptions, and visits. A job diary, also called a *daily report, site diary, construction log,* or *site journal,* is a notebook or digital application in which you record activities or events on the jobsite that may be important later (*Figure 42*).

Job diary: A written record that a supervisor maintains periodically (usually daily) of the events, communications, observations, and decisions made during a project.

For this information to be useful, the entries you and others submit from the field must be accurate. In the event of a legal or contractual conflict with the client over costs, job diaries can be helpful as evidence in legal proceedings or in reaching a settlement. *Table 9* outlines an example scenario where your action of accurate field reporting will help benefit your company.

Today, many companies use mobile applications that allow you to complete daily reporting on your mobile phone or tablet while you are on-site. You can quickly enter information throughout the day and attach photos and other documents to add more information. They also make it easy to send to your supervisor via email or text (*Figure 43*). Your supervisor will be notified when the report is complete.

July 8, 2023

Weather: Hot and Humid

Project: Opal Garage Annex

- The paving contractor crew arrived late (10 am).
- The owner representative inspected the footing foundation at approximately 1 pm.
- The concrete slump test did pass. Two trucks had to be ordered to return to plant, causing a delay.
- David Lopez had an incident on the second floor. I sent him to the on-site medical personnel for medical first aid treatment. The cause of the incident is being investigated.

Figure 42 Sample page from a job diary.

TABLE 9 Accurate Field Reporting Benefits Your Company

Scenario	Field Reporting Action
You are running a crew of five concrete finishers for a specialty contractor. When you and your crew show up to finish a slab, the general contractor (GC) informs you, "We're a day behind on setting the forms, so I need you and your crew to stand down until tomorrow."	• Follow your company's field reporting policy to report the change. For example, you might first call your office to inform them about the delay. Then you would record it in your job diary. • A six-man crew for one day represents 48 worker hours. If your company charges $30 an hour, that's a potential loss of $1,440, which the company would want to recover from the GC. If there is a dispute, your entry in the job diary could result in a favorable decision for your employer.

Whether print or digital, when making entries in a job diary, ensure the information is accurate, factual, complete, consistent, organized, and up to date. Do not include any emotional details; keep the report focused on facts. This is especially true if documenting personnel problems. If the company requires a job diary, follow company policy in determining which events and details you should record. The information needs to be detailed enough so that, if needed, you can determine what happened on any day years later. If you are unsure what to include, having more information than too little is better. Check in with your supervisor to confirm if your daily reporting includes the right information.

5.3.2 Tracking Employee Costs through Time Sheets

Work breakdown structure (WBS): A method of organizing the entire scope of work in a logical manner based on how the work will be executed.

When the project budget is set up, many companies create a work breakdown for each project. A **work breakdown structure (WBS)** organizes each task based on how the work will be performed. Planners assign each major task within the WBS a unique charge number or code (*Figure 44*). Anyone working on that task charges that number or code on their time sheet so project managers can readily track cost performance.

As a frontline leader, you need to ensure everyone completes their time sheets and codes them properly in the WBS. Doing so allows you and your company to track and manage labor costs against specific work.

For example, as shown in *Figure 44*, if all time sheets are correctly coded, you and the company can determine the labor cost to install each of the three types of pipe: carbon steel piping, stainless steel piping, and high-density polyethylene (HDPE) piping. You can also calculate the total cost to install the piping for the project.

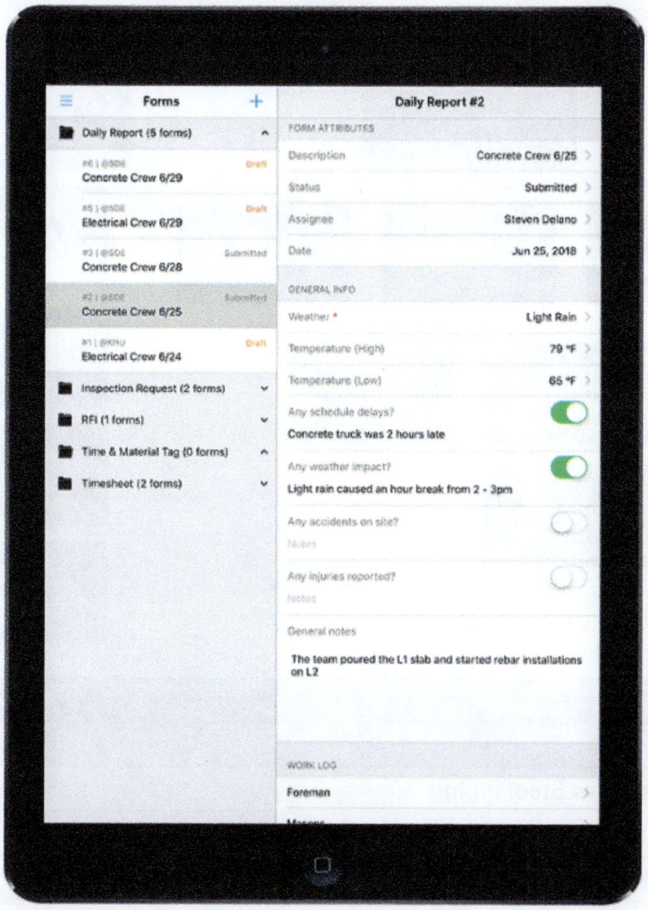

Figure 43 Mobile apps make it easier to complete daily reporting on-site.
Source: Courtesy of Hilti, Inc.

5.4.0 Frontline Leader Cost and Resource Management Responsibilities

Being aware of costs and controlling them is the responsibility of every employee on the job. As a frontline leader, you must ensure that employees uphold this responsibility. Cost control is the process of keeping expenses under control by managing labor, material, and other costs to ensure that the project finishes on budget.

5.4.1 Managing Costs

As you have learned, profit refers to the amount of money that the contractor will make after paying all the direct and indirect costs. If the costs exceed the estimate for the job, the difference between the actual and estimated costs must come from the company's profit. This reduces what the contractor earns on the job.

Early identification of cost changes is important and a key part of your role. As shown in *Figure 45*, an effective cost-control process includes several key steps. Identifying changes early provides enough notice to correct the causes of overruns and manage those costs.

Based on experience, you must learn to anticipate changes in plans and specifications and take measures to prevent them from occurring. If a situation cannot be fully corrected, early notification allows time for communicating and managing expectations. No one likes surprises when it comes to increased project costs.

Table 10 lists some factors that can cause actual costs to exceed estimated costs and some potential actions you can take to avoid these situations.

NOTE

Before taking any action to change costs associated with the project, you should check with company supervisor to ensure that the proposed action is acceptable and within the company's policies and procedures.

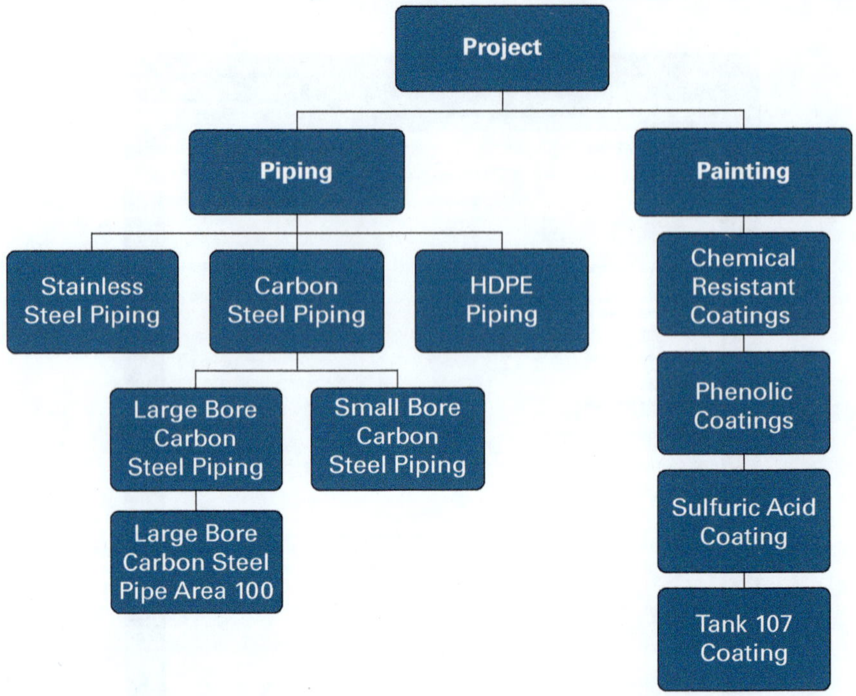

Level	Work	Account Code
Level 0	Piping	B
Level 1	Stainless Steel Piping	B-SS
Level 1	Carbon Steel Piping	B-CS
Level 1	HDPE Piping	B-HP
Level 2	Large Bore Carbon Steel Piping	B-CS-L
Level 2	Small Bore Carbon Steel Piping	B-CS-S
Level 3	Large Bore Carbon Steel Pipe Area 100	B-CS-L-100
Level 0	Painting	L
Level 1	Chemical Resistant Coatings	L-CR
Level 2	Phenolic Coating	L-CR-P
Level 3	Sulfuric Acid Tanks Coating	L-CR-P-ST
Level 4	Tank 107 Coating	L-CR-P-ST-107

Figure 44 An example of a work breakdown structure (WBS) for piping and painting on a project.

A key factor in managing costs is controlling waste. Waste in construction can lead to the loss of critical and costly materials and may result in job delays. You must ensure that every crew member knows how to complete the assigned work and use materials efficiently. One example of waste is a carpenter who saws off a piece of lumber from a full-sized piece, when he could have found the length needed in the lumber scrap pile.

Doing rework is another form of waste (*Figure 46*). For example, less-experienced or less-trained crew members may make mistakes that cause work to be redone. You can help avoid these errors by providing on-the-spot training and checking on newer, less experienced workers more often. Rework can also result from a conflict in craft scheduling that requires the removal and reinstallation of finished work. Rework also increases safety risks on a job. It often involves demolition, time pressure, and frustration from the team, all of which create a higher likelihood of an incident.

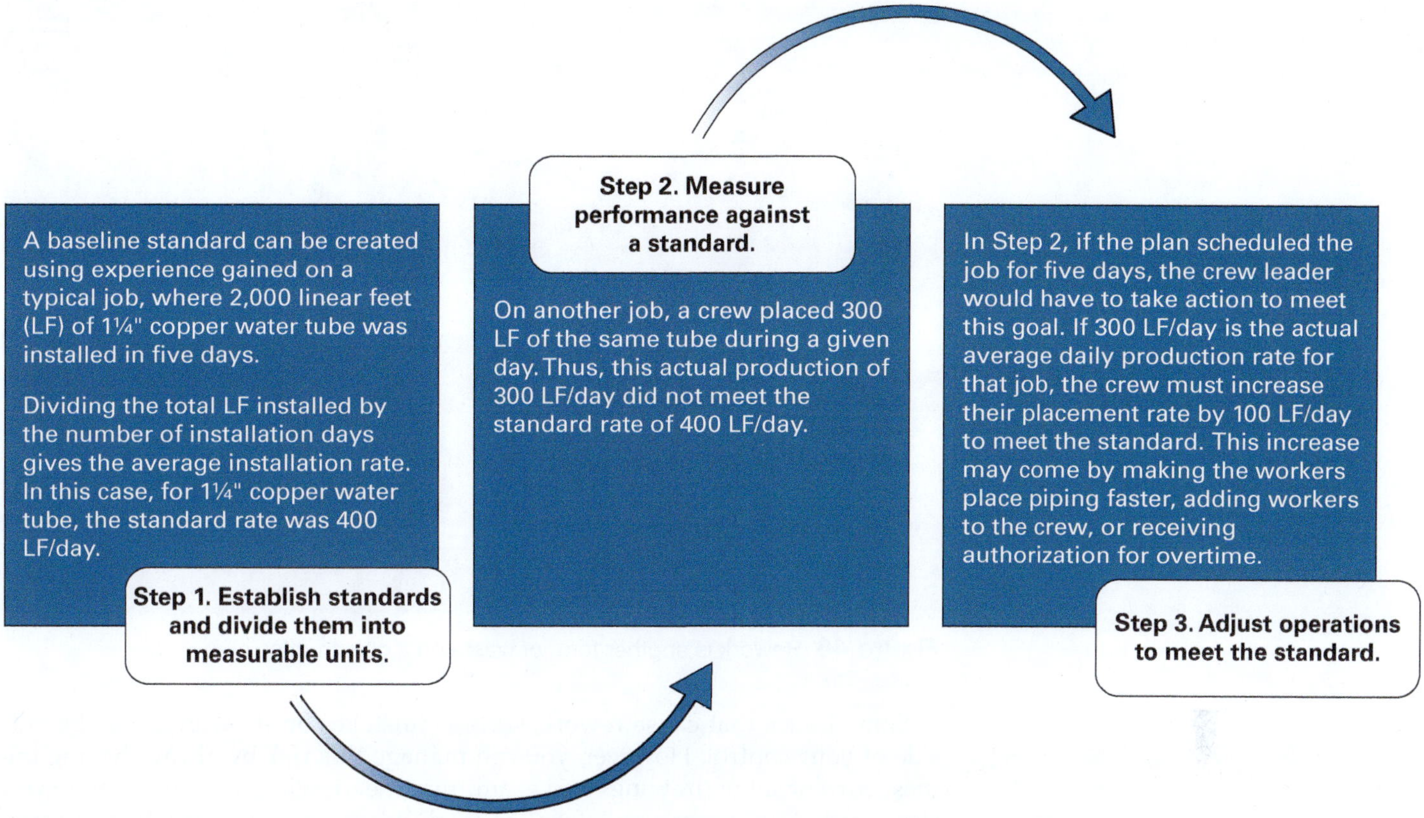

Figure 45 An effective cost control process monitors production and productivity to help manage costs.

TABLE 10 Factors That Drive Cost Changes

Factor	Proactive Action
Client-related changes	Clients may want to change the scope of a project after it has started. You must be able to assess the potential impact of such changes and, if necessary, confer with the employer to determine the course of action. If contractor-related losses occur, you and the superintendent will need to work together to get the costs back in line.
Late delivery of materials, tools, and/or equipment	Review and update your lookahead schedules regularly. Plan to ensure that job resources will be available when needed.
Inclement weather	Work with the superintendent and other field leaders to have alternate plans ready.
Higher-than-expected wages or unexpected overtime due to low manpower	Plan ahead to ensure that labor resources will be available when needed. Ensure you accurately record daily work using your field reporting systems and job diary. Work with your supervisor or the superintendent to identify other staffing approaches.
Contractor or owner delays due to work availability	Ensure you accurately record daily work using your field reporting systems and job diary. Work with the superintendent and other field leaders to have alternate plans ready.
Unmotivated workers	Apply the leadership and team motivation strategies discussed earlier in this course. Remember, different things motivate different people in different ways. This means there is no single best approach to motivating all your crew members.
Incidents	Continue to lead by example, reinforce the critical nature of safety, and follow the guidelines discussed earlier in the course. If needed, work with the superintendent and other field leaders to help evaluate and improve the safety and health program.
Owner-related changes	Owners may want to change the scope of a project after it has started. You must be able to assess the potential impact of such changes and, if necessary, confer with the employer to determine the course of action. If contractor-related losses occur, you and the superintendent will need to work together to get the costs back in line.
Waste due to rework, material usage, and more	Train less-experienced workers in how to complete work. Review and update your lookahead schedules regularly. Ensure crew members know how to use materials efficiently. Ensure you have the latest drawings and reference them daily.

Figure 46 Rework is another form of waste on a construction site.
Source: iStock

Some issues that cause rework, such as mistakes on drawings, may be outside of your control. However, you can manage this risk by always having the most current set of drawings and examining the drawings carefully before you start work. Finding errors before the work begins increases safety and saves time, labor, materials, and more, which saves costs overall.

5.4.2 Ensuring Crew Productivity

People typically represent more than half the cost of a project and thus enormously impact profitability. For that reason, it is essential to manage your crew and their work environment to maximize their productivity.

One way to manage costs and increase productivity is to ensure you have the right skills on your crew to execute the level of tasks that you must complete. Crew mix is a concept that describes the number of experienced or journey-level workers relative to the number of apprentices or less-experienced workers in a crew. For safety, quality, and productivity reasons, you do not want a worker trying to complete a task for which they are not skilled. However, you also do not want higher-wage crew members doing work that can be done by lower-wage crew members. For example, paying a journeyman electrician to pull wire costs more than paying an electrician helper who is skilled enough to do that work under the supervision of a journeyman electrician. As a frontline supervisor, you may not have control of your specific crew mix, but you can work to align their skill levels most appropriate to the specific work. Doing this well can reduce the overall project costs.

Other practices can help to ensure productivity:

- Keeping the crew adequately staffed to avoid job delays
- Ensuring that all workers have the required resources when needed
- Ensuring that everyone knows where to go and what to do after each task is completed
- Making reassignments as needed
- Ensuring that all workers have completed their work properly

The time on the job should be for business, not for handling personal problems. Anything not work-related should be handled after hours, away from the jobsite, if possible. Even the planning of after-work activities, arranging social functions, or running personal errands should occur after work or during breaks. Limited and necessary exceptions to these rules should be permitted (for example, making or meeting medical appointments), which should be spelled out in company or crew policies.

Given the nature of construction work, frequent breaks can be needed to ensure worker safety. However, if crew members often take unscheduled breaks or goofing off, you must confront the behavior and coach these workers. Document any discussions in your job diary. If needed, you should redirect repeated violations to the attention of higher management as guided by company policy.

By carefully tracking the project schedule, you can minimize or avoid delays caused by other crews or contractors. You can anticipate delays that will affect your work and either prevent them or redirect the crew to another task.

The availability and location of materials, equipment, and tools can seriously affect a crew's productivity. In the following sections, you will learn more about managing these resources to minimize such problems.

5.4.3 Handling Material Delivery and Storage

Your specific responsibilities for material delivery and storage will depend on the company's policies and procedures. In general, you will be responsible for ensuring on-time delivery of materials, verifying that the materials were delivered, and storing the materials (*Figure 47*). You also must prevent waste and theft of materials.

You will also be involved in planning materials for tasks such as job-built formwork and scaffolding or acquiring materials if you run out. If you run out of a specific material, consult the appropriate supervisor on how to proceed because most companies have specific purchasing policies and procedures.

Figure 47 The receipt, storage, and control of materials after they reach the jobsite is an important part of a frontline leader's work.
Source: iStock

The materials required for each day's work must be on the jobsite when needed. You should confirm the placement of orders for all materials and that they will arrive on schedule. A week or so before the delivery date, follow up to ensure there will be no delayed deliveries or items on backorder. If other people are responsible for providing the materials for a job, you must follow up to make sure that the materials are available when needed. Otherwise, delays can occur as crew members wait to deliver materials.

You may be responsible for the receipt of materials delivered to the worksite. When this happens, you should require a copy of the shipping invoice or similar document and check each item on the invoice against the actual materials to verify delivery of the correct amounts.

You should also check the condition of the materials to verify that nothing is defective before signing the invoice. This can be difficult and time-consuming because it requires the crew to open cartons and examine their contents. However, this step cannot be overlooked because a signed invoice indicates that the recipient accepted all listed materials and is in an undamaged state. If the crew leader signs for the materials without checking them and then finds damage, no one will be able to prove that the materials came to the site in that condition (*Figure 48*).

Figure 48 When receiving materials, check the condition of the materials to verify that nothing is defective before signing the invoice.
Source: iStock

After checking and signing the shipping invoice, you should give the original or a copy to the superintendent or project manager. The company then will file the invoice for future reference because it serves as the only record to check bills received from the supplier company.

Another essential element of materials control is selecting where to store materials on a jobsite. *Table 11* outlines three factors to consider when selecting an appropriate material storage location.

TABLE 11 Considerations When Selecting a Material Storage Location

Factor	Description
Access	If possible, store materials near where the crew will use them, so the crew can access them easily. Not having to carry the materials long distances will greatly reduce time, effort, and installation costs.
Damage Protection	Store materials in a secure area to avoid damage. Be sure that the storage area suits the materials. For instance, temperature-sensitive materials, such as chemicals or paints, should be stored in climate-controlled areas to prevent waste.
Security	Theft or vandalism of construction materials increases direct and indirect costs. These include replacement materials, lost production time while replacing needed materials, and the costs of additional security precautions. In addition, a contractor's insurance costs may increase due to theft and vandalism.

The best way to avoid theft and vandalism is to have a secure jobsite. At the end of each workday, store unused materials and tools in a secure location, such as a locked construction trailer. If the jobsite is fenced or the building has lockable access, the materials can be stored within. Many sites have security cameras and/or intrusion alarms to help minimize theft and vandalism.

5.4.4 Planning and Controlling Tool Usage

Various company policies govern those who provide hand and power tools to employees. Some companies furnish all the tools, while others furnish only the larger power tools. You should be familiar with and enforce any company policies related to tools.

As a crew leader, you are responsible for planning tool usage for a job. This task includes the following:

- Determining the tools required
- Informing the workers who will provide the tools (company or worker)
- Making sure the workers are qualified to use the tools safely and effectively
- Determining what controls to establish for tools

Tool control involves controlling the issue, use, and maintenance of all company-provided tools. It also involves storing the tools in a location that is easy for your crew to access (*Figure 49*).

Using the proper tools correctly saves time and energy. In addition, proper tool use reduces the chance of damage to the tool during use and injury to the user and nearby workers. This is especially true of edge tools and those with fixed or rotating blades. A dull tool is far more dangerous than a sharp one.

Tools must be adequately maintained and properly stored. Cleaning, drying, and lubricating tools prevent rust and ensure that they are in proper working order. This applies to tools supplied by the company and tools that belong to the workers.

If a tool is damaged, repairing or replacing it promptly is essential. Otherwise, an incident or injury could result. Users should take care of company-issued tools as if they were their own. Workers should not abuse tools simply because they are not their property.

One major source of low productivity on a job is the time spent searching for a tool. Supervision should establish a storage location for company-issued tools and equipment to prevent this. You should ensure crew members return all company-issued tools and equipment to this designated location after use. Similarly, workers should organize their personal toolboxes to readily find the appropriate tools and return them to their toolboxes when they are finished using them. If needed, coach less-experienced team members on how to master this important professional skill.

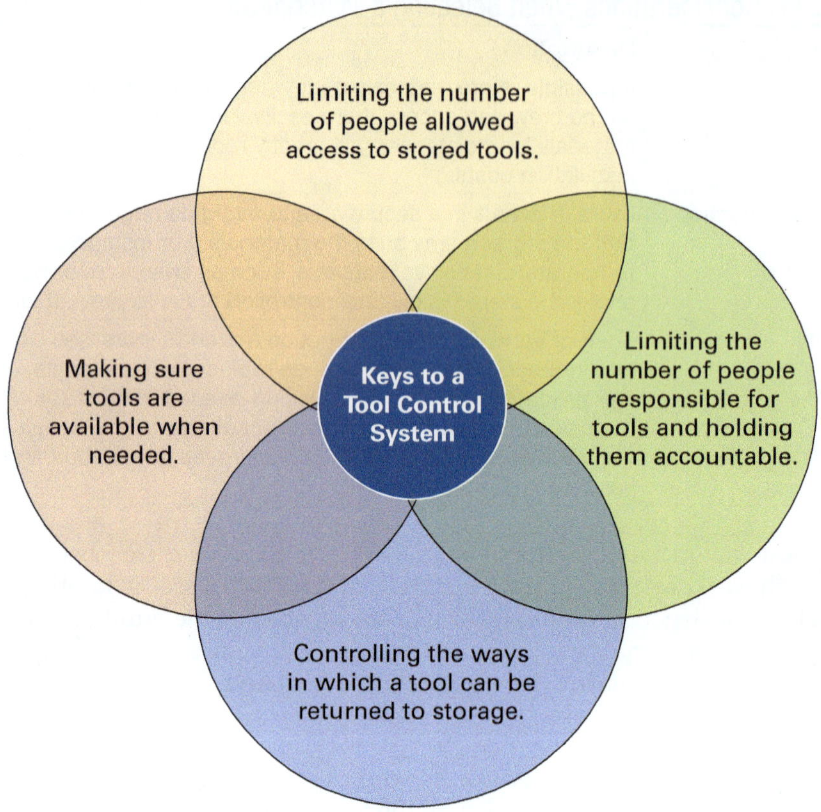

Figure 49 Studies show there are four keys to an effective tool control system.

5.4.5 Ensuring Equipment Control

Equipment planning includes the types of equipment needed, its use, and the length of time it will be on the site. As a frontline crew leader, you may not be responsible for long-term equipment control.

However, the equipment required for a specific job is often your responsibility. The first step is to schedule when a crew member must transport the required equipment from the shop or rental yard. You are responsible for informing the shop of the location of its use and confirming that a worker returns it to the shop when they are finished with it. You must work with the main office to ensure the equipment reaches the jobsite on time.

Coordinating the use of the equipment is also very important. Some equipment operates in combination with other equipment (*Figure 50*). For example, dump trucks are generally required when using loaders and excavators. You should also coordinate equipment with other contractors on the job. Sharing equipment can save time and money and avoid duplication of effort.

It is common for equipment to lie idle at a jobsite because of a lack of proper planning for the job and the equipment arriving early. For example, if wire-pulling equipment arrives at a jobsite before the conduit is in place, this equipment will be out of service while awaiting the installation. In addition to the wasted rental cost, the equipment could be damaged, lost, or stolen while awaiting use.

You need to control equipment use, ensure that the crew operates the equipment according to its instructions, and ensure that they adhere to time and cost guidelines. You must also assign or coordinate responsibility for maintaining and repairing equipment as the applicable preventive maintenance schedule indicates. Delaying maintenance and repairs can lead to costly equipment failures.

You must also ensure that crew equipment operators are properly trained. This includes ensuring that the equipment operators have the necessary credentials to operate the equipment, including applicable licenses.

Figure 50 Coordinating equipment requires knowing which equipment operates in combination with other equipment.
Source: iStock

You are also responsible for properly operating all other equipment resources, including cars and trucks. Reckless or unsafe vehicle operation will likely result in damaged equipment, leading to disciplinary action for the workers, fines and repair costs levied by the vehicle vendors, and a loss of confidence in your ability to control the actions of your crew.

You should also arrange to secure all equipment at the close of each day's work to prevent theft. If continued use of the equipment is necessary for the job, you should lock it in a safe place. If it is not, return it promptly.

5.0.0 Section Review

1. Which of these activities occurs during the design and development phase of a project?
   a. Architect/engineer sketches are prepared and a preliminary budget is developed.
   b. Final inspections are completed.
   c. Detailed project drawings and specifications are prepared.
   d. Contracts for the project are awarded.

2. Which of these activities is *not* part of creating a lookahead schedule?
   a. Identifying the required work activities and dividing them into tasks
   b. Planning resources needed for each task
   c. Organizing the work into a logical sequence
   d. Assigning cost estimates to each task

3. A job diary should typically record any of the following *except* _____.
   a. job interruptions and visits
   b. personal details and emotions
   c. changes needed to project drawings
   d. the actual time each job task took to complete

4. Which of the following is a *correct* statement regarding project cost?
   a. Cost is handled by the accounting department and isn't a concern of the crew leader.
   b. The difference between the estimated cost and the actual cost affects a company's profit.
   c. Wasted material is factored into the estimate and is never a concern.
   d. The contractor's overhead costs aren't included in the cost estimate.

Module 93111 Review Questions

1. Who is generally the on-site field supervisor who is responsible for one or more crew leaders?
   a. Project manager
   b. Lead engineer
   c. Superintendent
   d. Preconstruction manager

2. Accepting responsibility for job completion and being personally answerable for the results is known as _____.
   a. responsibility
   b. delegation
   c. authority
   d. accountability

3. Which of the following is *true* of situational leadership?
   a. It prevents you from making unnecessary adjustments in the way you lead different situations.
   b. It assumes all situations require the same leadership.
   c. It allows you to use a general approach to all situations.
   d. It has four styles, including directing, coaching, supporting, and delegating.

4. When you communicate with your team verbally, be sure to _____.
   a. assess your audience before you speak
   b. use technical language as much as possible
   c. speak quickly to save time
   d. assume everyone has the same basic understanding of the topic

5. The first step in problem-solving should be _____.
   a. choosing a solution to the problem
   b. evaluating the solution to the problem
   c. defining the problem
   d. testing the solution to the problem

6. Which of these is a way to lower the motivation and morale of your employees and team?
   a. Focusing on task completion only, not professional growth
   b. Giving recognition and rewards
   c. Setting clear expectations and goals
   d. Providing opportunities for training and upskilling

7. During an employee's first week of work on your crew, you should be sure to set aside time to _____.
   a. eat lunch together
   b. do a task verification
   c. quiz the employee on company policies
   d. evaluate social engagement

8. Which of these steps is important when delegating work to an individual or a team?
   a. Letting individuals set their own standards of measurement
   b. Checking in only after the work is complete
   c. Ensuring crew members understand what to do and have the skill to do it
   d. Allowing the team to set their own deadlines and get their own resources

9. Which common jobsite documents can help a crew leader overcome a language barrier with a worker?
   a. OSHA incident records
   b. Contracts and schedules
   c. Plans and drawings
   d. Permits

10. As a leader, you can help to manage performance issues by _____.
    a. ignoring the problem that needs resolving
    b. gathering information and preparing well for the employee conversation
    c. discussing the performance issues publicly with the rest of the team
    d. having the conversation when the employee is unprepared

11. The negative impact of incidents may include all of the following *except* _____.
    a. lowering workers' compensation costs
    b. damage to the overall morale of everyone on a project or jobsite
    c. increased project costs to the company
    d. project delays

12. OSHA was established in 1970 to _____.
    a. collect dues for membership
    b. ensure safe and healthful working conditions for workers
    c. manage workers' compensation claims
    d. conduct medical research

13. Which of the following statements regarding safety training sessions is *correct*?
    a. The project manager usually holds them.
    b. They are held only for new employees.
    c. They should be conducted frequently by a frontline leader.
    d. They are required only after an accident has occurred.

14. A key step in an incident investigation is to determine the incident's _____.
    a. root cause
    b. secondary driver
    c. cost impacts
    d. design flaws

15. If you observe an employee showing impaired behavior or signs of mental health issues, you should work with _____.
    a. OSHA
    b. the safety inspector
    c. the owner or client
    d. your supervisor and/or Human Resources

16. What is completed during the design phase to ensure a project can be delivered successfully?
    a. Feasibility study
    b. As-built drawings
    c. Installation work packages
    d. Project schedule

17. A contract in which the client pays the contractor for their labor, materials, equipment, and other costs they incur during the performance of the contract is known as a _____.
    a. fixed-price contract
    b. time-spent contract
    c. time and materials contract
    d. cost-reimbursable contract

18. Which of the following would be tracked in a job diary?
    a. Job changes, interruptions, and visits
    b. The crew's time off
    c. Training needs and career development
    d. Personal information provided by an employee

19. Which of the following is a source of data in the field-reporting system?
    a. The cost estimate for the project
    b. The daily job diary
    c. The Level 1 schedule
    d. The lookahead schedule

20. To prevent job delays due to late delivery of materials, the crew leader should _____.
    a. demand a discount from the supplier to compensate for the delay
    b. tell a crew member to go look for the delivery truck
    c. refuse the late delivery and reorder the materials from another supplier
    d. check with the supplier in advance of the scheduled delivery

Answers to odd-numbered questions are found in the Review Question Answer Keys at the back of this book.

Answers to Section Review Questions

Answer	Section Reference	Objective
Section 1.0.0		
1. b	1.1.1	1a
2. b	1.2.2	1b
Section 2.0.0		
1. d	2.1.2	2a
2. b	2.2.3	2b
3. a	2.3.0	2c
Section 3.0.0		
1. b	3.1.1	3a
2. c	3.2.0	3b
3. c	3.3.3	3c
Section 4.0.0		
1. a	4.0.0	4
2. d	4.2.0	4b
3. a	4.3.1	4c
4. b	4.4.2	4d
5. d	4.5.0	4e
Section 5.0.0		
1. a	5.1.1	5a
2. d	5.2.2	5b
3. b	5.3.1	5c
4. b	5.4.1	5d

APPENDIX 93102A

Basic Safety

OSHA Mission and Standards

The mission of the Occupational Safety and Health Administration (OSHA) is to save lives, prevent injuries, and protect the health of America's workers. To accomplish this, federal and state governments work in partnership with millions of working men and women who are covered by the Occupational Safety and Health Act (OSH Act) of 1970.

Nearly every worker in the United States comes under OSHA's jurisdiction. There are some exceptions, such as miners, transportation workers, many public employees, and the self-employed. These specific groups are not covered by OSHA standards, but most are covered by standards developed by their respective industries.

The Code of Federal Regulations

The *Code of Federal Regulations (CFR) Part 1910* covers the OSHA standards for general industry. *CFR Part 1926* covers the OSHA standards for the construction industry. Either or both may apply to you, depending on where you are working and what you are doing. If a jobsite condition is covered in the *CFR* book, then that standard must be used. However, if a more stringent requirement is listed in *CFR 1910*, it should also be met. Check with your supervisor to find out which standards apply to your job.

29 CFR 1926 is divided into subparts A through Z. As you progress in task-specific training, you will learn about all the subparts applicable to your work. *Subpart C of 29 CFR 1926* applies to all construction and maintenance work. It outlines the general safety and health provisions for the construction industry. It covers the following topics:

- Safety training and education
- Injury reporting and recording
- First aid and medical attention
- Housekeeping
- Illumination
- Sanitation
- PPE
- Standards incorporated by reference
- Definitions
- Access to employee exposure and medical records
- Means of egress
- Employee emergency action plans

For assistance in identifying parts, sections, paragraphs, and subparagraphs of an OSHA standard, refer to *Figure A01*.

All of OSHA's safety requirements in the Code of Federal Regulations apply to residential and commercial construction. In the past, OSHA enforced safety only at commercial sites. The increasing rate of incidents at residential sites led OSHA to enforce safety guidelines for the building of houses and townhomes. Today, however, OSHA still focuses its enforcement efforts on commercial construction.

An OSHA Standard reference may look like this:

29 CFR 1926.501 (a)(1)(i)(A)

and breaks down like this:

29	=	Title (Labor)
CFR	=	Code of Federal Regulations
1926	=	Part (Construction)
.501	=	Section
(a)	=	Paragraph
(1)	=	Subparagraph
(i)	=	Subparagraph
(A)	=	Subparagraph

Figure A01 Reading OSHA standards.

The General Duty Clause

If a standard does not specifically address a hazard, the General Duty Clause must be invoked. Failing to adhere to the General Duty Clause can result in heavy fines for your employer. The General Duty Clause reads as follows:

In practice, OSHA, court precedent, and the review commission have established that if the following elements are present, a General Duty Clause citation may be issued:

- The employers failed to keep the workplace free of a hazard to which employees of that employer were exposed.
- The hazard was recognized. (Examples might include: your safety personnel, employees, organization, trade organization, or industry customs.)
- The hazard was causing or was likely to cause death or serious physical harm.
- There was a feasible and useful method to correct the hazard.

Employee Rights and Responsibilities

While it is the employer's responsibility to keep workers safe by complying with the General Duty Clause and all other OSHA regulations, workers have certain rights and responsibilities on the jobsite as well. First and foremost, workers must follow their employers' safety rules. While workers cannot be cited or fined by OSHA, they can be disciplined for violating their employer's safety rules. Workers must also wear the provided PPE. Workers should also inform their foreman about health and safety concerns on the job.

Section 11(c) of the OSH Act prohibits employers from disciplining or discriminating against any worker for practicing their rights under OSHA, including filing a complaint. You have the right to file a complaint if you do not think that your employer is protecting your health and safety at work. You may submit a written request to OSHA asking for an inspection of your worksite. Workers who file a complaint have the right to have their names withheld from their employers, and OSHA will not reveal this information.

Workers who would like an on-site inspection must submit a written request. You have the following rights when jobsite inspection is conducted:

- You must be informed of imminent dangers. An OSHA inspector must tell you if you are exposed to an imminent danger. An imminent danger is one that could cause death or serious injury now or in the near future. The inspector will also ask your employer to stop any dangerous activity.
- You have the right to accompany the OSHA inspector in the walk-around inspection. Walk-around activities include all opening and closing conferences related to the conduct of the inspection.
- You have the right to be told about citations issued at your workplace. Notices of OSHA citations must be posted in the workplace near the site where the violation occurred and must remain posted for three days or until the hazard is corrected, whichever is longer.

After an inspection has been performed, OSHA will give the employer a date by which any cited hazards must be fixed. Employers can appeal these dates, and appeals must be filed within 15 days of the citation. Workers have the right to meet privately with the OSHA inspector to discuss the results of the inspection.

If you have been discriminated against for asserting your OSHA rights, you have the right to file a complaint with the OSHA area office within 30 days of the incident. Make sure you file your complaint as soon as possible, as the time limit is strictly enforced.

You also have the right to see and copy any medical records about you that the employer has obtained. Your employer is required by *OSHA 29 CFR 1926.33* and *OSHA 29 CFR 1910.1020* to maintain your medical records for 30 years after you leave employment. If you are employed for less than one year, the employer can maintain your records or give them to you when you leave the job.

Inspections

OSHA conducts six types of inspections to determine if employers are complying with standards:

- *Imminent danger inspections* — OSHA's top priority for inspection, conducted when workers face an immediate risk of death or serious physical harm.
- *Catastrophe inspections* — Performed after an incident that requires hospitalization of three or more workers. Employers are required to report fatalities and catastrophes to OSHA within eight hours.
- *Worker complaint and referral inspections* — Conducted due to complaints by workers or a worker representative, or a referral from a recognized professional.
- *Programmed inspection* — Aimed at high-risk areas based on OSHA's targeting and priority methods.
- *Follow-up inspection* — Completed after citations to assure employer has corrected violations.
- *Monitoring inspection* — Used for long-term abatement follow-up or to assure compliance with variances.

Before beginning an inspection, OSHA staff must be able to determine from the complaint that there are reasonable grounds to believe that a violation of an OSHA standard or a safety or health hazard exists. If OSHA has information indicating the employer is aware of the hazard and is correcting it, the agency may not conduct an inspection after obtaining the necessary documentation from the employer.

Complaint inspections are typically limited to the hazards listed in the complaint, although other violations in plain sight may be cited as well. The inspector may decide to expand the inspection based on professional judgment or conversations with workers.

Complaints are not necessarily inspected in first come, first served order. OSHA ranks complaints based on the severity of the alleged hazard and the number of workers exposed. That is why lower priority complaints can often be handled more quickly using the phone/fax method than through on-site inspections.

Inspections are typically performed by conducting a walk-around. During a walk-around inspection, the inspector typically does the following:

- Observes conditions of the jobsite.
- Talks to workers.
- Inspects records.
- Examines posted hazard warnings and signs.
- Points out hazards and suggests ways to reduce or eliminate them.

After the walk-around inspection, there is typically a closing conference held between the inspector and the site contractor or company managers. During this conference, inspectors discuss their findings, citing specific violations and suggested abatement methods. Inspectors may also conduct interviews with the employers, workers, and representatives at this point.

Violations

Employers who violate OSHA regulations can be fined. The fines are not always high, but they can harm a company's reputation for safety. Fines for serious safety violations can cost up to $7,000. Fines for each violation that was done willfully can cost up to $70,000. In 2012, roughly $260,000,000 in fines were issued against employers. In the decade prior to 2012, annual fines averaged approximately $150,000,000. Significant increases and decreases in annual fines tend to be dependent on the current federal administration and OSHA's budget provided by federal lawmakers.

Compliance

Just as employers are responsible to OSHA for compliance, employees must comply with their company's safety policies and rules. Employers are required to identify hazards and potential hazards within the workplace and eliminate them, control them, or provide protection from them. This can only be done through the combined efforts of the employer and employees. Employers must provide written programs and training on hazards, and employees must follow the procedures. You, as the employee, must read and understand the OSHA poster at your jobsite explaining your rights and responsibilities. If you are unsure where the OSHA poster is, ask your supervisor.

To help employers provide a safe workplace, OSHA requires companies to provide a competent person to ensure the safety of the employees. In *OSHA 29 CFR 1926*, OSHA defines a competent person as follows:

Competent person means "one who is capable of identifying existing and predictable hazards in the surroundings or working conditions which are unsanitary, hazardous, or dangerous to employees, and who has authorization to take prompt corrective measures to eliminate them."

In comparison, *OSHA 29 CFR 1926* defines a qualified person as follows:

Someone "who, by possession of a recognized degree, certificate, or professional standing, or who by extensive knowledge, training, and experience, has successfully demonstrated his ability to solve or resolve problems relating to the subject matter, work, or the project."

In other words, a competent person is experienced and knowledgeable about the specific operation and has the authority from the employer to correct the problem or shut down the operation until it is safe. A qualified person has the knowledge and experience to handle problems. A competent person is not necessarily a qualified person.

These terms will be an important part of your career. It is important for you to know who the competent person is on your jobsite. OSHA requires a competent person for many of the tasks you may be assigned to perform, such as confined space entry, ladder use, and trenching. Different individuals may be assigned as a competent person for different tasks, according to their expertise. To ensure safety for you and your coworkers, work closely with your competent person and supervisor.

APPENDIX 93102B

Safety Data Sheet for Regular Bleach

SAMPLE
SAFETY DATA SHEET

Issuing Date January 5, 2023 **Revision Date** June 12, 2023 **Revision Number** 1

1. IDENTIFICATION OF THE SUBSTANCE/PREPARATION AND OF THE COMPANY/UNDERTAKING

Product identifier

Product Name **XXXXX Regular-Bleach$_1$**

Other means of identification

EPA Registration Number 5813-100

Recommended use of the chemical and restrictions on use

Recommended use Household disinfecting, sanitizing, and laundry bleach

Uses advised against No information available

Details of the supplier of the safety data sheet

Supplier Address
The XXXXX Company
1221 Broadway
Oakland, CA 94612

Phone: 1-510-XXX-XXXX

Emergency telephone number

Emergency Phone Numbers For Medical Emergencies, call: 1-800-446-1014
For Transportation Emergencies, call Chemtrec: 1-800-424-9300

Page 1 / 10

Figure B01
Source: Occupational Safety and Health Administration (OSHA)

XXXXX Regular-Bleach₁ **Revision Date** June 12, 2023

2. HAZARDS IDENTIFICATION

Classification

This chemical is considered hazardous by the 2012 OSHA Hazard Communication Standard (29 CFR 1910.1200).

Skin corrosion/irritation	Category 1
Serious eye damage/eye irritation	Category 1

GHS Label elements, including precautionary statements

Emergency Overview

Signal word **Danger**

Hazard Statements
Causes severe skin burns and eye damage
Causes serious eye damage

Appearance Clear, pale yellow **Physical State** Thin liquid **Odor** Bleach

Precautionary Statements - Prevention
Wash face, hands and any exposed skin thoroughly after handling.
Wear protective gloves, protective clothing, face protection, and eye protection such as safety glasses.

Precautionary Statements - Response
Immediately call a poison center or doctor.
If swallowed: Rinse mouth. Do NOT induce vomiting.
If on skin (or hair): Take off immediately all contaminated clothing. Rinse skin with water.
Wash contaminated clothing before reuse.
If inhaled: Remove person to fresh air and keep comfortable for breathing.
Specific treatment (see supplemental first aid instructions on this label).
If in eyes: Rinse cautiously with water for several minutes. Remove contact lenses, if present and easy to do. Continue rinsing.

Precautionary Statements - Storage
Store locked up.

Precautionary Statements - Disposal
Dispose of contents in accordance with all applicable federal, state, and local regulations.

Hazards not otherwise classified (HNOC)
Although not expected, heart conditions or chronic respiratory problems such as asthma, chronic bronchitis, or obstructive lung disease may be aggravated by exposure to high concentrations of vapor or mist.

Product contains a strong oxidizer. Always flush drains before and after use.

Page 2 / 10

Figure B02

Source: Occupational Safety and Health Administration (OSHA)

Unknown Toxicity
Not applicable.

Other information
Very toxic to aquatic life with long lasting effects.

Interactions with Other Chemicals
Reacts with other household chemicals such as toilet bowl cleaners, rust removers, acids, or products containing ammonia to produce hazardous irritating gases, such as chlorine and other chlorinated compounds.

3. COMPOSITION

Chemical Name	CAS-No	Weight %	Trade Secret
Sodium hypochlorite	7681-52-9	5 - 10	*

* The exact percentage (concentration) of composition has been withheld as a trade secret.

4. FIRST AID MEASURES

First aid measures

General Advice Call a poison control center or doctor immediately for treatment advice. Show this safety data sheet to the doctor in attendance.

Eye Contact Hold eye open and rinse slowly and gently with water for 15 - 20 minutes. Remove contact lenses, if present, after the first 5 minutes, then continue rinsing eye. Call a poison control center or doctor for treatment advice.

Skin Contact Take off contaminated clothing. Rinse skin immediately with plenty of water for 15-20 minutes. Call a poison control center or doctor for treatment advice.

Inhalation Move to fresh air. If breathing is affected, call a doctor.

Ingestion Have person sip a glassful of water if able to swallow. Do not induce vomiting unless told to do so by a poison control center or doctor. Do not give anything by mouth to an unconscious person. Call a poison control center or doctor immediately for treatment advice.

Protection of First-aiders Avoid contact with skin, eyes, and clothing. Use personal protective equipment as required. Wear personal protective clothing (see section 8).

Most important symptoms and effects, both acute and delayed

Most Important Symptoms and Effects Burning of eyes and skin.

Indication of any immediate medical attention and special treatment needed

Notes to Physician Treat symptomatically. Probable mucosal damage may contraindicate the use of gastric lavage.

Figure B03
Source: Occupational Safety and Health Administration (OSHA)

XXXXX Regular-Bleach₁ **Revision Date** June 12, 2023

5. FIRE-FIGHTING MEASURES

Suitable Extinguishing Media
Use extinguishing measures that are appropriate to local circumstances and the surrounding environment.

Unsuitable Extinguishing Media
CAUTION: Use of water spray when fighting fire may be inefficient.

Specific Hazards Arising from the Chemical
This product causes burns to eyes, skin, and mucous membranes. Thermal decomposition can release sodium chlorate and irritating gases and vapors.

Explosion Data

Sensitivity to Mechanical Impact	None.
Sensitivity to Static Discharge	None.

Protective equipment and precautions for firefighters
As in any fire, wear self-contained breathing apparatus pressure-demand, MSHA/NIOSH (approved or equivalent) and full protective gear.

6. ACCIDENTAL RELEASE MEASURES

Personal precautions, protective equipment and emergency procedures

Personal Precautions	Avoid contact with eyes, skin, and clothing. Ensure adequate ventilation. Use personal protective equipment as required. For spills of multiple products, responders should evaluate the MSDSs of the products for incompatibility with sodium hypochlorite. Breathing protection should be worn in enclosed and/or poorly-ventilated areas until hazard assessment is complete.
Other Information	Refer to protective measures listed in Sections 7 and 8.

Environmental precautions

Environmental Precautions	This product is toxic to fish, aquatic invertebrates, oysters, and shrimp. Do not allow product to enter storm drains, lakes, or streams. See Section 12 for ecological Information.

Methods and material for containment and cleaning up

Methods for Containment	Prevent further leakage or spillage if safe to do so.
Methods for Cleaning Up	Absorb and containerize. Wash residual down to sanitary sewer. Contact the sanitary treatment facility in advance to assure ability to process washed-down material.

Page 4 / 10

Figure B04
Source: Occupational Safety and Health Administration (OSHA)

7. HANDLING AND STORAGE

Precautions for safe handling

Handling Handle in accordance with good industrial hygiene and safety practice. Avoid contact with skin, eyes, and clothing. Do not eat, drink, or smoke when using this product.

Conditions for safe storage, including any incompatibilities

Storage Store away from children. Reclose cap tightly after each use. Store this product upright in a cool, dry area, away from direct sunlight and heat to avoid deterioration. Do not contaminate food or feed by storage of this product.

Incompatible Products Toilet bowl cleaners, rust removers, acids, and products containing ammonia.

8. EXPOSURE CONTROLS/PERSONAL PROTECTION

Control parameters

Exposure Guidelines

Chemical Name	ACGIH TLV	OSHA PEL	NIOSH IDLH
Sodium hypochlorite 7681-52-9	None	None	None

ACGIH TLV: American Conference of Governmental Industrial Hygienists - Threshold Limit Value. OSHA PEL: Occupational Safety and Health Administration - Permissible Exposure Limits. NIOSH IDLH: Immediately Dangerous to Life or Health.

Appropriate engineering controls

Engineering Measures Showers
Eyewash stations
Ventilation systems

Individual protection measures, such as personal protective equipment

Eye/Face Protection If splashes are likely to occur: Wear safety glasses with side shields (or goggles) or face shield.

Skin and Body Protection Wear rubber or neoprene gloves and protective clothing such as long-sleeved shirt.

Respiratory Protection If irritation is experienced, NIOSH/MSHA approved respiratory protection should be worn. Positive-pressure supplied air respirators may be required for high airborne contaminant concentrations. Respiratory protection must be provided in accordance with current local regulations.

Hygiene Measures Handle in accordance with good industrial hygiene and safety practice. Wash hands after direct contact. Do not wear product-contaminated clothing for prolonged periods. Remove and wash contaminated clothing before re-use. Do not eat, drink, or smoke when using this product.

Figure B05

Source: Occupational Safety and Health Administration (OSHA)

XXXXX Regular-Bleach₁ **Revision Date** June 12, 2023

9. PHYSICAL AND CHEMICAL PROPERTIES

Physical and Chemical Properties

Physical State	Thin liquid		
Appearance	Clear	**Odor**	Bleach
Color	Pale yellow	**Odor Threshold**	No information available

Property	**Values**	**Remarks/ Method**
pH	~12	None known
Melting/freezing point	No data available	None known
Boiling point / boiling range	No data available	None known
Flash Point	Not flammable	None known
Evaporation rate	No data available	None known
Flammability (solid, gas)	No data available	None known
Flammability Limits in Air		
Upper flammability limit	No data available	None known
Lower flammability limit	No data available	None known
Vapor pressure	No data available	None known
Vapor density	No data available	None known
Specific Gravity	~1.1	None known
Water Solubility	Soluble	None known
Solubility in other solvents	No data available	None known
Partition coefficient: n-octanol/water	No data available	None known
Autoignition temperature	No data available	None known
Decomposition temperature	No data available	None known
Kinematic viscosity	No data available	None known
Dynamic viscosity	No data available	None known
Explosive Properties	Not explosive	
Oxidizing Properties	No data available	

Other Information

Softening Point	No data available
VOC Content (%)	No data available
Particle Size	No data available
Particle Size Distribution	No data available

10. STABILITY AND REACTIVITY

Reactivity
Reacts with other household chemicals such as toilet bowl cleaners, rust removers, acids, or products containing ammonia to produce hazardous irritating gases, such as chlorine and other chlorinated compounds.

Chemical stability
Stable under recommended storage conditions.

Possibility of Hazardous Reactions
None under normal processing.

Conditions to avoid
None known based on information supplied.

Incompatible materials
Toilet bowl cleaners, rust removers, acids, and products containing ammonia.

Hazardous Decomposition Products
None known based on information supplied.

Page 6 / 10

Figure B06
Source: Occupational Safety and Health Administration (OSHA)

XXXXX Regular-Bleach₁ **Revision Date** June 12, 2023

11. TOXICOLOGICAL INFORMATION

Information on likely routes of exposure

Product Information .

Inhalation Exposure to vapor or mist may irritate respiratory tract and cause coughing. Inhalation of high concentrations may cause pulmonary edema.

Eye Contact Corrosive. May cause severe damage to eyes.

Skin Contact May cause severe irritation to skin. Prolonged contact may cause burns to skin.

Ingestion Ingestion may cause burns to gastrointestinal tract and respiratory tract, nausea, vomiting, and diarrhea.

Component Information

Chemical Name	LD50 Oral	LD50 Dermal	LC50 Inhalation
Sodium hypochlorite 7681-52-9	8200 mg/kg (Rat)	>10000 mg/kg (Rabbit)	-

Information on toxicological effects

Symptoms May cause redness and tearing of the eyes. May cause burns to eyes. May cause redness or burns to skin. Inhalation may cause coughing.

Delayed and immediate effects as well as chronic effects from short and long-term exposure

Sensitization No information available.

Mutagenic Effects No information available.

Carcinogenicity The table below indicates whether each agency has listed any ingredient as a carcinogen.

Chemical Name	ACGIH	IARC	NTP	OSHA
Sodium hypochlorite 7681-52-9	-	Group 3	-	-

IARC (International Agency for Research on Cancer)
Group 3 - Not Classifiable as to Carcinogenicity in Humans

Reproductive Toxicity No information available.

STOT - single exposure No information available.

STOT - repeated exposure No information available.
Chronic Toxicity Carcinogenic potential is unknown.
Target Organ Effects Respiratory system, eyes, skin, gastrointestinal tract (GI).

Aspiration Hazard No information available.

Figure B07
Source: Occupational Safety and Health Administration (OSHA)

Numerical measures of toxicity - Product Information

The following values are calculated based on chapter 3.1 of the GHS document

ATEmix (oral)
54 g/kg
ATEmix (inhalation-dust/mist)
58 mg/L

12. ECOLOGICAL INFORMATION

Ecotoxicity
Very toxic to aquatic life with long lasting effects.

This product is toxic to fish, aquatic invertebrates, oysters, and shrimp. Do not allow product to enter storm drains, lakes, or streams.

Persistence and Degradability
No information available.

Bioaccumulation
No information available.

Other adverse effects
No information available.

13. DISPOSAL CONSIDERATIONS

Disposal methods
Dispose of in accordance with all applicable federal, state, and local regulations. Do not contaminate food or feed by disposal of this product.

Contaminated Packaging
Do not reuse empty containers. Dispose of in accordance with all applicable federal, state, and local regulations.

14. TRANSPORT INFORMATION

DOT	Not restricted.
TDG	Not restricted for road or rail.
ICAO	Not restricted, as per Special Provision A197, Environmentally Hazardous Substance exception.
IATA	Not restricted, as per Special Provision A197, Environmentally Hazardous Substance exception.
IMDG/IMO	Not restricted, as per IMDG Code 2.10.2.7, Marine Pollutant exception.

Figure B08
Source: Occupational Safety and Health Administration (OSHA)

XXXXX Regular-Bleach[1] **Revision Date** June 12, 2023

15. REGULATORY INFORMATION

Chemical Inventories

TSCA All components of this product are either on the TSCA 8(b) Inventory or otherwise exempt from listing.

DSL/NDSL All components are on the DSL or NDSL.

TSCA - United States Toxic Substances Control Act Section 8(b) Inventory
DSL/NDSL - Canadian Domestic Substances List/Non-Domestic Substances List

U.S. Federal Regulations

SARA 313
Section 313 of Title III of the Superfund Amendments and Reauthorization Act of 1986 (SARA). This product does not contain any chemicals which are subject to the reporting requirements of the Act and Title 40 of the Code of Federal Regulations, Part 372

SARA 311/312 Hazard Categories

Acute Health Hazard	Yes
Chronic Health Hazard	No
Fire Hazard	No
Sudden Release of Pressure Hazard	No
Reactive Hazard	No

Clean Water Act
This product contains the following substances which are regulated pollutants pursuant to the Clean Water Act (40 CFR 122.21 and 40 CFR 122.42)

Chemical Name	CWA - Reportable Quantities	CWA - Toxic Pollutants	CWA - Priority Pollutants	CWA - Hazardous Substances
Sodium hypochlorite 7681-52-9	100 lb			X

CERCLA
This material, as supplied, contains one or more substances regulated as a hazardous substance under the Comprehensive Environmental Response Compensation and Liability Act (CERCLA) (40 CFR 302)

Chemical Name	Hazardous Substances RQs	Extremely Hazardous Substances RQs	RQ
Sodium hypochlorite 7681-52-9	100 lb	-	RQ 100 lb final RQ RQ 45.4 kg final RQ

EPA Statement
This chemical is a pesticide product registered by the Environmental Protection Agency and is subject to certain labeling requirements under federal pesticide law. These requirements differ from the classification criteria and hazard information required for safety data sheets and for workplace labels of non-pesticide chemicals. Following is the hazard information as required on the pesticide label:

> **DANGER: CORROSIVE.** Causes irreversible eye damage and skin burns. Harmful if swallowed. Do not get in eyes, on skin, or on clothing. Wear protective eyewear and rubber gloves when handling this product. Wash thoroughly with soap and water after handling and before eating, drinking, chewing gum, using tobacco, or using the restroom. Avoid breathing vapors and use only in a well-ventilated area.

Page 9 / 10

Figure B09
Source: Occupational Safety and Health Administration (OSHA)

US State Regulations

California Proposition 65
This product does not contain any Proposition 65 chemicals.

U.S. State Right-to-Know Regulations

Chemical Name	New Jersey	Massachusetts	Pennsylvania	Rhode Island	Illinois
Sodium hypochlorite 7681-52-9	X	X	X	X	
Sodium chlorate 7775-09-9	X	X	X		

International Regulations

Canada
WHMIS Hazard Class
E - Corrosive material

16. OTHER INFORMATION

NFPA **Health Hazard** 3 **Flammability** 0 **Instability** 0 **Physical and Chemical Hazards** -

HMIS **Health Hazard** 3 **Flammability** 0 **Physical Hazard** 0 **Personal Protection** B

Prepared By
Product Stewardship
23 British American Blvd.
Latham, NY 12110
1-800-572-6501

Revision Date June 12, 2023

Revision Note Revision Section 14.

Reference 1096036/164964.159

General Disclaimer
The information provided in this Safety Data Sheet is correct to the best of our knowledge, information and belief at the date of its publication. The information given is designed only as a guidance for safe handling, use, processing, storage, transportation, disposal, and release and is not to be considered a warranty or quality specification. The information relates only to the specific material designated and may not be valid for such material used in combination with any other materials or in any process, unless specified in the text.

End of Safety Data Sheet

This material was produced under grant 46F6-HT30 from the Occupational Safety and Health Administration, U.S. Department of Labor. It does not necessarily reflect the views or policies of the U.S. Department of Labor, nor does mention of trade names, commercial products, or organizations imply endorsement by the U.S. Government. Revisions were made to this material under grant number SH-31240-SH7 from the Occupational Safety and Health Administration, U.S. Department of Labor.

Figure B10
Source: Occupational Safety and Health Administration (OSHA)

APPENDIX 93103A

Multiplication Table

Trace across and down from the numbers being multiplied and find the answer at the intersection. In the example, 7 × 7 = 49.

	2	**3**	**4**	**5**	**6**	**7**	**8**	**9**	**10**	**11**	**12**
2	4	6	8	10	12	14	16	18	20	22	24
3	6	9	12	15	18	21	24	27	30	33	36
4	8	12	16	20	24	28	32	36	40	44	48
5	10	15	20	25	30	35	40	45	50	55	60
6	12	18	24	30	36	42	48	54	60	66	72
7	14	21	28	35	42	49	56	63	70	77	84
8	16	24	32	40	48	56	64	72	80	88	96
9	18	27	36	45	54	63	72	81	90	99	108
10	20	30	40	50	60	70	80	90	100	110	120
11	22	33	44	55	66	77	88	99	110	121	132
12	24	36	48	60	72	84	96	108	120	132	144

Figure A01

Conversion Factors and Common Formulas

TABLE A1 Common US Standard Measures

Weight	Length	Volume	Area
1 ton = 2,000 pounds 1 pound = 16 dry ounces	1 yard = 3 feet 1 foot = 12 inches	1 cubic yard = 27 cubic feet 1 cubic foot = 1,728 cubic inches 1 gallon = 4 quarts 1 quart = 2 pints 1 pint = 2 cups 1 cup = 8 fluid ounces	1 square yard = 9 square feet 1 square foot = 144 square inches

TABLE A2 Basic Metric Prefixes

Prefix	Symbol	Number	Multiplication Factor
Giga	G	Billion	$1{,}000{,}000{,}000 = 10^9$
Mega	M	Million	$1{,}000{,}000 = 10^6$
Kilo	k	Thousand	$1{,}000 = 10^3$
Hector	h	Hundred	$100 = 10^2$
Deka	da	Ten	$10 = 10^1$
			Base Units $1 = 10^0$
Deci	d	Tenth	$0.1 = 10^{-1}$
Centi	c	Hundredth	$0.01 = 10^{-2}$
Milli	m	Thousandth	$0.001 = 10^{-3}$
Micro	µ	Millionth	$0.000001 = 10^{-6}$
Nano	N	Billionth	$0.000000001 = 10^{-9}$

TABLE A3 Common Metric Units

Weight Units	Length Units	Volume Units
1 kilogram = 1,000 grams 1 hectogram = 100 grams 1 dekagram = 10 grams 1 gram = 1 gram 1 decigram = 0.1 grams 1 centigram = 0.01 grams 1 milligram = 0.001 grams	1 kilometer = 1,000 meters 1 hectometer = 100 meters 1 dekameter = 10 meters 1 meter = 1 meter 1 decimeter = 0.1 meters 1 centimeter = 0.01 meters 1 millimeter = 0.001 meters	1 kiloliter = 1,000 liters 1 hectoliter = 100 liters 1 dekaliter = 10 liters 1 liter = 1 liter 1 deciliter = 0.1 liters 1 centiliter = 0.01 liters 1 milliliter = 0.001 liters

TABLE A4 Conversions

US Standard to Metric	Metric to US Standard
Weight	
1 ounce = 28.35 grams 1 pound = 453.6 grams or 0.4536 kilograms 1 (short) ton = 907.2 kilograms	1 gram = 0.03527 ounces 1 kilogram = 2.205 pounds 1 metric ton = 2,205 pounds
Length	
1 inch = 2.540 centimeters 1 foot = 30.48 centimeters 1 yard = 91.44 centimeters or 0.9144 meters 1 mile = 1.609 kilometers	1 millimeter = 0.00155 square inches 1 centimeter = 0.3937 inches 1 meter = 3.281 feet or 1.0936 yards 1 kilometer = 0.6214 miles
Area	
1 square inch = 6.452 square centimeters 1 square foot = 929.0 square centimeters or 0.0929 square meters 1 square yard = 0.8361 square meters	1 square millimeter = 0.00155 square inches 1 square centimeter = 0.155 square inches 1 square meter = 10.76 square feet or 1.196 square yards
Volume	
1 cubic inch = 16.39 cubic centimeters 1 cubic foot = 0.02832 cubic meters 1 cubic yard = 0.7646 cubic meters	1 cubic centimeter = 0.06102 cubic inches 1 cubic meter = 35.31 cubic feet or 1.308 cubic yards
Liquid	
1 (fluid) ounce = 0.0296 liter or 28.35 grams 1 pint = 473.2 cubic centimeters 1 quart = 0.9464 liters 1 gallon = 3,785 cubic centimeters or 3.785 liters	1 cubic centimeter = 0.06102 cubic inches 1 liter (1,000 cm^3) = 1.057 quarts, 2.113 pints, or 61.02 cubic inches
Temperature	
To convert degrees Fahrenheit to degrees Celsius, use the following: $°C = \frac{5}{9} \times (°F - 32°)$	To convert degrees Celsius to degrees Fahrenheit, use the following: $°F = (\frac{9}{5} \times °C) + 32°$

TABLE A5 Common Formulas

	Measurement	Formula
Square/Rectangle	Area	length × width
	Perimeter	2 × (length + width)
Triangle	Area	$\frac{1}{2}$ × base × height
	Side Lengths (Pythagorean Theorem)	$a^2 + b^2 = c^2$, or $c = \sqrt{(a^2 + b^2)}$, where c is the length of the hypotenuse, or side opposite of the right angle of the triangle, and a and b are the lengths of the sides that are adjacent to the right angle.
Circle	Area	Area = $\pi \times r^2$, where r is the radius.
	Circumference	Circumference = $\pi \times d$, where d is the diameter.
Cube/Rectangular Prism	Volume	Volume = length × width × height
Pyramid	Volume	Volume = (length × width × height) / 3
Cylinder	Volume	Volume = $\pi \times r^2 \times h$, where r is the radius of the base and h is the height.
Cone	Volume	Volume = $(\pi \times r^2 \times h) / 3$, where r is the radius of the base and h is the height.
Sphere	Volume	Volume = $(4 \times \pi \times r^3) / 3$, where r is the radius.
Pressure	Absolute Pressure	Absolute pressure = gauge pressure – atmospheric pressure The atmospheric pressure at sea level is typically accepted as 14.7 psi.
	Gauge Pressure to Static Pressure	$P = (h \times d) / 144$ P = the pressure in psi, h = the height of the column in feet, and d = the density of the liquid in pound per cubic foot.
Temperature	Convert degrees Celsius to Fahrenheit	$°F = (\frac{9}{5} \times °C) + 32°$
	Convert degrees Fahrenheit to Celsius	$°C = \frac{5}{9} \times (°F - 32°)$

Note: Use the order of operations, which can be remembered using the acronym PEMDAS, whenever evaluating a math problem:

1. Parentheses
2. Exponents
3. Multiplication
4. Division
5. Addition
6. Subtraction

TABLE A6A Inches Converted to Decimals of a Foot (1 of 2)

Inches	Decimals of a Foot	Inches	Decimals of a Foot	Inches	Decimals of a Foot
$\frac{1}{16}$	0.005	$2\frac{1}{16}$	0.172	$4\frac{1}{16}$	0.339
$\frac{1}{8}$	0.010	$2\frac{1}{8}$	0.177	$4\frac{1}{8}$	0.344
$\frac{3}{16}$	0.016	$2\frac{3}{16}$	0.182	$4\frac{3}{16}$	0.349
$\frac{1}{4}$	0.021	$2\frac{1}{4}$	0.188	$4\frac{1}{4}$	0.354
$\frac{5}{16}$	0.026	$2\frac{5}{16}$	0.193	$4\frac{5}{16}$	0.359
$\frac{3}{8}$	0.031	$2\frac{3}{8}$	0.198	$4\frac{3}{8}$	0.365
$\frac{7}{16}$	0.036	$2\frac{7}{16}$	0.203	$4\frac{7}{16}$	0.370
$\frac{1}{2}$	0.042	$2\frac{1}{2}$	0.208	$4\frac{1}{2}$	0.374
$\frac{9}{16}$	0.047	$2\frac{9}{16}$	0.214	$4\frac{9}{16}$	0.380
$\frac{5}{8}$	0.052	$2\frac{5}{8}$	0.219	$4\frac{5}{8}$	0.385
$\frac{11}{16}$	0.057	$2\frac{11}{16}$	0.224	$4\frac{11}{16}$	0.391
$\frac{3}{4}$	0.063	$2\frac{3}{4}$	0.229	$4\frac{3}{4}$	0.396
$\frac{13}{16}$	0.068	$2\frac{13}{16}$	0.234	$4\frac{13}{16}$	0.401
$\frac{7}{8}$	0.073	$2\frac{7}{8}$	0.240	$4\frac{7}{8}$	0.406
$\frac{15}{16}$	0.078	$2\frac{15}{16}$	0.245	$4\frac{15}{16}$	0.411
1	0.083	3	0.250	5	0.417
$1\frac{1}{16}$	0.089	$3\frac{1}{16}$	0.255	$5\frac{1}{16}$	0.422
$1\frac{1}{8}$	0.094	$3\frac{1}{8}$	0.260	$5\frac{1}{8}$	0.427
$1\frac{3}{16}$	0.099	$3\frac{3}{16}$	0.266	$5\frac{3}{16}$	0.432
$1\frac{1}{4}$	0.104	$3\frac{1}{4}$	0.271	$5\frac{1}{4}$	0.438
$1\frac{5}{16}$	0.109	$3\frac{5}{16}$	0.276	$5\frac{5}{16}$	0.443
$1\frac{3}{8}$	0.115	$3\frac{3}{8}$	0.281	$5\frac{3}{8}$	0.448
$1\frac{7}{16}$	0.120	$3\frac{7}{16}$	0.286	$5\frac{7}{16}$	0.453
$1\frac{1}{2}$	0.125	$3\frac{1}{2}$	0.292	$5\frac{1}{2}$	0.458
$1\frac{9}{16}$	0.130	$3\frac{9}{16}$	0.297	$5\frac{9}{16}$	0.464
$1\frac{5}{8}$	0.135	$3\frac{5}{8}$	0.302	$5\frac{5}{8}$	0.469
$1\frac{11}{16}$	0.141	$3\frac{11}{16}$	0.307	$5\frac{11}{16}$	0.474
$1\frac{3}{4}$	0.146	$3\frac{3}{4}$	0.313	$5\frac{3}{4}$	0.479
$1\frac{13}{16}$	0.151	$3\frac{13}{16}$	0.318	$5\frac{13}{16}$	0.484
$1\frac{7}{8}$	0.156	$3\frac{7}{8}$	0.323	$5\frac{7}{8}$	0.490
$1\frac{15}{16}$	0.161	$3\frac{15}{16}$	0.328	$5\frac{15}{16}$	0.495
2	0.167	4	0.333	6	0.500

TABLE A6B Inches Converted to Decimals of a Foot (2 of 2)

Inches	Decimals of a Foot	Inches	Decimals of a Foot	Inches	Decimals of a Foot
$6\frac{1}{16}$	0.505	$8\frac{1}{16}$	0.672	$10\frac{1}{16}$	0.839
$6\frac{1}{8}$	0.510	$8\frac{1}{8}$	0.677	$10\frac{1}{8}$	0.844
$6\frac{3}{16}$	0.516	$8\frac{3}{16}$	0.682	$10\frac{3}{16}$	0.849
$6\frac{1}{4}$	0.521	$8\frac{1}{4}$	0.688	$10\frac{1}{4}$	0.854
$6\frac{5}{16}$	0.526	$8\frac{5}{16}$	0.693	$10\frac{5}{16}$	0.859
$6\frac{3}{8}$	0.531	$8\frac{3}{8}$	0.698	$10\frac{3}{8}$	0.865
$6\frac{7}{16}$	0.536	$8\frac{7}{16}$	0.703	$10\frac{7}{16}$	0.870
$6\frac{1}{2}$	0.542	$8\frac{1}{2}$	0.708	$10\frac{1}{2}$	0.875
$6\frac{9}{16}$	0.547	$8\frac{9}{16}$	0.714	$10\frac{9}{16}$	0.880
$6\frac{5}{8}$	0.552	$8\frac{5}{8}$	0.719	$10\frac{5}{8}$	0.885
$6\frac{11}{16}$	0.557	$8\frac{11}{16}$	0.724	$10\frac{11}{16}$	0.891
$6\frac{3}{4}$	0.563	$8\frac{3}{4}$	0.729	$10\frac{3}{4}$	0.896
$6\frac{13}{16}$	0.568	$8\frac{13}{16}$	0.734	$10\frac{13}{16}$	0.901
$6\frac{7}{8}$	0.573	$8\frac{7}{8}$	0.740	$10\frac{7}{8}$	0.906
$6\frac{15}{16}$	0.578	$8\frac{15}{16}$	0.745	$10\frac{15}{16}$	0.911
7	0.583	9	0.750	11	0.917
$7\frac{1}{16}$	0.589	$9\frac{1}{16}$	0.755	$11\frac{1}{16}$	0.922
$7\frac{1}{8}$	0.594	$9\frac{1}{8}$	0.760	$11\ \frac{1}{8}$	0.927
$7\frac{3}{16}$	0.599	$9\frac{3}{16}$	0.766	$11\ \frac{3}{16}$	0.932
$7\frac{1}{4}$	0.604	$9\frac{1}{4}$	0.771	$11\frac{1}{4}$	0.938
$7\frac{5}{16}$	0.609	$9\frac{5}{16}$	0.776	$11\frac{5}{16}$	0.943
$7\frac{3}{8}$	0.615	$9\frac{3}{8}$	0.781	$11\frac{3}{8}$	0.948
$7\frac{7}{16}$	0.620	$9\frac{7}{16}$	0.786	$11\frac{7}{16}$	0.953
$7\frac{1}{2}$	0.625	$9\frac{1}{2}$	0.792	$11\frac{1}{2}$	0.958
$7\frac{9}{16}$	0.630	$9\frac{9}{16}$	0.797	$11\frac{9}{16}$	0.964
$7\frac{5}{8}$	0.635	$9\frac{5}{8}$	0.802	$11\frac{5}{8}$	0.969
$7\frac{11}{16}$	0.641	$9\frac{11}{16}$	0.807	$11\frac{11}{16}$	0.974
$7\frac{3}{4}$	0.646	$9\frac{3}{4}$	0.813	$11\frac{3}{4}$	0.979
$7\frac{13}{16}$	0.651	$9\frac{13}{16}$	0.818	$11\frac{13}{16}$	0.984
$7\frac{7}{8}$	0.656	$9\frac{7}{8}$	0.823	$11\frac{7}{8}$	0.990
$7\frac{15}{16}$	0.661	$9\frac{15}{16}$	0.828	$11\frac{15}{16}$	0.995
8	0.667	10	0.833	12	1.000

REVIEW QUESTION ANSWER KEYS

MODULE 93101

Answer	Section Reference
1. c	1.1.2
3. a	1.1.3
5. d	1.2.1
7. d	2.2.1
9. d	3.1.0
11. a	3.1.0
13. d	3.2.3
15. b	4.2.0

MODULE 93102

Answer	Section Reference
1. b	1.1.1
3. a	1.2.9
5. a	1.3.0
7. c	1.3.4
9. b	2.3.1
11. d	3.1.1
13. b	3.1.3
15. d	3.2.1
17. d	3.2.1
19. a	4.1.4
21. a	4.2.0
23. c	5.2.3
25. b	6.2.7
27. c	6.3.2
29. a	6.4.4; *Figure 85*

MODULE 93103

Answer	Section Reference
1. b	1.1.0
3. c	1.2.0
5. c	1.3.0
7. a	2.1.3
9. a	2.3.0
11. a	3.2.1
13. b	3.2.2
15. c	3.3.1
17. a	4.1.1
19. c	4.2.1; *Figure 16*
21. a	5.1.2
23. b	5.2.3
25. a	5.4.0
27. b	6.2.1
29. d	6.4.3

MODULE 93103 CALCULATIONS

Question 3

1,478' (total) – 489' (installed) = **989' of cable remaining**

Question 5

400 rolls ÷ 5 sites = **80 rolls per site**

Question 9

$\frac{2}{8} \times \frac{4}{4} = \frac{8}{32}$

$\frac{11}{32} - \frac{8}{32} = \mathbf{\frac{3}{32}}$

Question 11

51.5 nanometers + 89.7 nanometers = **141.2 nanometers**

Question 13

864.5 ft^2 × \$2.37 per ft^2 = \$2,048.865, rounded up to **\$2,048.87**

Question 25

$°C = \frac{5}{9}(15° - 32°) = \frac{5}{9} \times (-17°) =$ **–9.4°C**

Question 29

Volume of cylinder = pi × $radius^2$ × h

$= 3.14 \times (6.25')^2 \times 23'$

$= 3.14 \times 39.06\ ft^2 \times 23'$

= **2,821 ft^3**

MODULE 93104

Answer	Section Reference
1. d	1.1.1
3. a	1.1.5; *Figure 8*
5. b	1.2.1
7. d	1.3.0
9. b	1.4.2; *Figure 19*
11. a	1.5.0
13. a	2.2.1
15. b	2.2.5
17. c	3.2.0
19. d	3.4.1

MODULE 93105

Answer	Section Reference
1. a	1.0.0
3. c	1.1.0
5. d	1.2.1
7. c	1.2.6
9. a	1.3.3
11. b	2.1.2
13. a	2.2.3
15. a	2.3.1
17. a	2.4.2
19. a	3.1.0
21. c	4.2.0

MODULE 93106

Answer	Section Reference
1. a	1.1.1
3. a	1.2.1
5. d	1.3.1
7. d	1.4.2
9. c	1.5.4

MODULE 93107

Answer	Section Reference
1. d	1.1.1
3. c	1.2.1
5. c	1.3.1
7. d	1.4.0
9. c	1.5.1

MODULE 93108

Answer	Section Reference
1. a	1.0.0
3. a	1.1.1
5. b	1.2.0
7. d	1.3.0
9. b	2.1.0
11. c	2.2.0
13. d	2.3.1
15. d	2.3.2

MODULE 93109

Answer	Section Reference
1. d	1.2.0
3. b	2.1.1
5. d	2.3.4
7. a	3.1.2
9. d	3.3.1
11. c	3.4.2
13. a	3.5.0
15. c	3.5.1

MODULE 93110

Answer	Section Reference
1. d	1.1.2
3. b	1.1.3
5. d	1.2.2
7. b	1.3.2; *Figure 12*
9. c	2.1.2
11. d	2.1.8
13. a	2.2.2
15. c	2.2.6

MODULE 93111

Answer	Section Reference
1. c	1.0.0
3. d	2.1.3
5. c	2.3.1
7. b	3.1.4
9. c	3.2.3
11. a	4.1.0
13. c	4.3.1
15. d	4.5.2
17. c	5.1.2
19. b	5.3.1

ADDITIONAL RESOURCES

93101 Build Your Future in Construction

How Are Drones Used in Agriculture? Drone Launch Academy, 2024. **www.dronelaunchacademy.com/resources/how-are-drones-used-in-agriculture/**.

On Average, U.S. Farmers Are Aging, But a Quarter of Them Are Newcomers. Chuck Abbott. Fern's Ag Insider, 2019. **www.thefern.org/ag_insider/on-average-u-s-farmers-are-aging-but-a-quarter-of-them-are-newcomers/**.

Solar Energy + Agriculture. The Nature Conservancy, 2023. **www.nature.org/en-us/about-us/where-we-work/united-states/colorado/stories-in-colorado/colorado-agrivoltaics/**.

93102 Basic Safety

Agricultural Safety and Health Program: Safe Handling of Pesticides on the Farm. Ohio State University Extension. June 2015. **agsafety.osu.edu/newsletter/ag-safety-stat/injury-prevention/safe-handling-pesticides-farm**.

Basic Construction Safety and Health. Fred Fanning. CreateSpace Independent Publishing Platform. 2014.

Construction Project Safety. John Schaufelberger, Ken-Yu Lin. John Wiley & Sons, Inc. 2013.

Farm Safety—Which Mask for Which Task. Fargo, ND: North Dakota State University, 2024. **www.ndsu.edu/agriculture/ag-hub/ag-topics/farm-safety-health/farm-safety/which-mask-which-task**.

Management of Construction Projects: A Constructor's Perspective. John Schaufelberger, Len Holm. 2nd Edition. Routledge, 2017.

Oregon Fatality Assessment and Control Evaluation, Fatality Investigation Report, Oregon Health and Science University, *2003-19*. **www.ohsu.edu/sites/default/files/2019-02/ORFACE-Pub-IR-Ag-FallBeneathCombine.pdf**.

US Occupational Safety and Health Administration. Numerous safety videos are available online at **www.osha.gov/video**.

93104 Introduction to Hand Tools

Essential Guide to the Steel Square: How to Figure Everything Out with One Simple Tool, No Batteries Required. Ken Horner. Mount Joy, PA: Fox Chapel Publishing, 2016.

Hand Tool Basics. Steve Branam. Des Moines, IA: Popular Woodworking Books, 2018.

The Tool Book: A Tool Lover's Guide to More Than 200 Hand Tools. Phil Davy, Luke Edwardes-Evans, Jo Behari, and Matthew Jackson. New York, NY: DK Publishing, 2018.

93106 Introduction to Construction Drawings

How to Start a Farm: Plan Your Operation, US Department of Agriculture. **www.farmers.gov/your-business/beginning-farmers/business-plan**.

93107 Introduction to Basic Rigging

Pulleys. Chris Woodford. Explain That Stuff, 2023. **www.explainthatstuff.com/pulleys.html**.

93109 Basic Employability Skills

The Seven Key Steps of Critical Thinking. Chris Cebollero. Forbes Coaches Council – Council Post, 2021. **www.forbes.com/sites/forbescoachescouncil/2018/04/20/the-seven-key-steps-of-critical-thinking/**.

Safety – The Dreaded Lag Time. Colorado Department of Transportation. State of Colorado, 2024. **www.codot.gov/safety/impaired-driving/druggeddriving/campaign-news/cannabislag**.

93110 Introduction to Material Handling

Agricultural Hand Signals. Penn State Extension. July 2022. **www.extension.psu.edu/agricultural-hand-signals**.

Heavy Equipment Operations. NCCER. Hoboken, NJ: Pearson Education, Inc., 2020.

Knots: The Complete Visual Guide. Des Pawson. New York, NY: DK Publishing, 2012.

Manufacturing Facilities Design & Material Handling. Matthew P. Stevens, Fred E. Meyers. West Lafayette, IN: Purdue University Press, 2013.

Materials Handling Handbook. The American Society of Mechanical Engineers (ASME) and the International Material Management Society (IMMS), Raymond A. Kulwiec, Editor-in-Chief. New York, NY: Wiley-Interscience, 1985.

Simple Solutions: Ergonomics for Construction Workers. US Centers for Disease Control and Prevention, National Institute for Occupational Safety and Health. **www.cdc.gov/niosh/docs/2007-122/**.

Spotlight: Protect Your Back: Ergonomic Strategies for Preventing Back Pain While Farming. Upper Midwest Agricultural Safety and Health Center. October 2022. **umash.umn.edu/spotlight-protect-your-back-ergonomic-strategies-for-preventing-pain-while-farming/**.

GLOSSARY

Abrasive: A substance, such as sandpaper, that is used to wear away material.

Absenteeism: A consistent failure to show up for work.

Accident: According to the US Occupational Safety and Health Administration (OSHA), an unplanned event that results in personal injury and/or property damage.

Accomplishment: An individual's need to set challenging goals and achieve them.

Accountability: Accepting responsibility and being personally answerable for the results.

Active listening: A process that involves respecting others, listening to what is being said, and understanding what is being said.

Acute angle: Any angle greater than 0° and less than 90°.

Adjacent angles: Angles that have the same vertex and one side in common.

AED (automated external defibrillator): A device used to help those experiencing sudden cardiac arrest.

Alternating current (AC): The common power supplied to most all wired devices, where the current reverses its direction many times per second. AC power is the type of power generated and distributed throughout settled areas.

Amp hour (Ah): A rating that describes the maximum amps a battery can provide continuously for up to 60 minutes. Drawing more amps than the battery's rating is possible, but it reduces the time before the battery is drained.

Angle: The shape made by two straight lines coming together at a point. The space between those two lines is measured in degrees.

Appendixes: Source of detailed or specific information at the end of a section, a chapter, or a book.

Apprenticeship/craft training: A program in which you gain valuable skills on the job under the instruction of more experienced professionals, often making money as you obtain training.

Arbor: The end of a circular saw shaft where the blade is mounted.

Arc welding: The joining of metal parts by fusion, in which the necessary heat is produced by means of an electric arc.

Architects: Qualified, licensed people who create and design drawings for a construction project.

Architect's scale: A specialized ruler used in making or measuring reduced scale drawings. The ruler is marked with a range of calibrated ratios for laying out distances, with scales indicating feet, inches, and fractions of inches. Used on drawings other than site plans.

Architectural plans: Drawings that show the design of the project. Also called *architectural drawings*.

Area: The amount of space contained in a two-dimensional object, such as a rectangle, circle, or square.

As-built drawings: A revised set of drawings that reflect all changes made during the construction process and show the exact specifications of the completed work.

Auger bit: A drill bit with a spiral cutting edge for boring holes in wood and other materials.

Authority: The right to act or make decisions in carrying out an assignment.

Base: As it relates to triangles, the base is the line forming the bottom of the triangle.

Beams: Large, horizontal structural members made of concrete, steel, stone, wood, or other structural materials to provide support above a large opening.

Behaviors: The way a person responds to specific situations.

Bevel: The inclination of a surface that intersects another surface at any angle less than 90°; also used as a verb, meaning to cut or trim an edge to an angled profile.

Bisect: To divide into two equal parts. For example, when an angle is bisected, the two resulting angles are equal.

Block and tackle: A simple rope and pulley system used to lift loads.

Blueprints: A traditional method of reproducing technical drawings using a contact print process and light-sensitive paper that resulted in white lines on a blue background. Also, the traditional name used to describe construction drawings.

Body language: A person's facial expression, physical posture, gestures, and use of space, all of which communicate feelings and ideas.

Bowline: A knot used to form a loop that neither slips nor jams; sometimes referred to as a *rescue knot* or the *king of knots*.

Brazing: A process using heat in excess of 800°F (427°C) to melt a filler metal that is drawn into a connection. Brazing is commonly used to join copper pipe.

Bridle: A configuration using two or more slings to connect a load to a single hoist hook.

Building information modeling (BIM): A process in which software reads computer files containing building data to create a virtual picture of a facility before it is built. BIM objects represent many dimensions of construction and allow builders to adjust the plans prior to beginning actual construction.

Bull ring: A single ring used to attach multiple slings to a hoist hook.

Bullets: Large vertically aligned dots that highlight items in a list.

Bullying: Unwanted, aggressive behavior that involves a real or perceived power imbalance. This form of harassment may include offensive, persistent, insulting, or physically threatening behavior directed at an individual.

Capsize: To change the form and rearrange the parts of a knot, usually by pulling on specific ends of the knot.

Carbide: A very hard material made of tungsten carbide and heavy metals such as cobalt. Commonly baked on to the tips of masonry bits and to the edges of saw blades.

Career and technical education (CTE): Programs offered in high school and technical education centers that allow students to explore career paths in construction.

Change orders: A written order by the owner of a project for the contractor to make a change in time, amount, or specifications.

Charge cycles: Describe one complete charge and discharge cycle for a rechargeable battery.

Cheater bar: A length of pipe or rod used to increase the leverage on a wrench or similar tool. The increased leverage provided by a cheater bar often exceeds the tool's capacity and strength, resulting in a damaged tool and workpiece.

Chronic obstructive pulmonary disease (COPD): Lung disease(s) that results in the obstruction of lung airflow and interferes with normal breathing. Chronic bronchitis and emphysema generally fall under the diagnosis of COPD.

Chuck: A clamping device that holds an attachment; for example, the drill's chuck holds the drill bit.

Chuck key: A small, T-shaped steel piece used to open and close the chuck on power drills. Cordless drills typically do not have a chuck key.

Circle: A closed curved line drawn at a constant distance around a central point. A circle measures 360°.

Circumference: The distance around the curved line that forms the circle.

Civil plans: Drawings that show the location of the building on the site from an aerial view, including contours, trees, construction features, and dimensions.

Clean rooms: A room or area that is used for scientific research or specialized processes, such as the manufacture of pharmaceuticals and semiconductors. The cleanliness level of a clean room is generally based on the number of particles measured per cubic meter of area.

Clove hitch: A knot that consists of two half hitches made in opposite directions; used to temporarily secure a rope to an object.

Combustible: Capable of easily igniting and rapidly burning; used to describe a fuel with a flash point at or above 100°F (38°C).

Communication: The process of exchanging facts, information, and ideas. Communication comes in many forms, including verbal, written, visual, and nonverbal.

Competent person: A worker who can identify hazards and take the corrective action necessary to address them.

Compromise: When people are involved in a disagreement, they make concessions to reach a solution that everyone agrees on.

Computer-aided design (CAD): The use of computers to digitally sketch construction designs, allowing other people to see how aspects of a structure would look and work together when built.

Computer-aided drafting (CAD): The making of a set of construction drawings with the aid of a computer.

Concave: Having a shape that curves inward, such as the interior of a circle. *Convex* is an antonym, meaning a shape that curves outward.

Concrete mule: A wheeled device used when a concrete pour is in a location that a concrete delivery truck or pump cannot reach; sometimes referred to as a *Georgia buggy*.

Confidentiality: Privacy of information.

Confined spaces: Work areas large enough for a person to work, but arranged in such a way that an employee must physically enter the space to perform work. Confined spaces have a limited or restricted means of entry and exit. They are not designed for continuous work. Tanks, vessels, silos, pits, vaults, and hoppers are examples of confined spaces. Also see *permit-required confined space*.

Construction safety plan: A document that outlines the procedures, rules, and regulations that are or will be put in place to protect workers during a construction project.

Constructive criticism: A positive offer of advice intended to help someone correct mistakes or improve actions.

Contour lines: Solid or dashed lines showing the elevation of the earth on a civil drawing.

Contractor: Oversees many parts of a construction project, including resource management, budgeting, code adherence, quality assurance, and construction materials.

Core: Center support member of a wire rope around which the strands are laid.

Cost control: The process that helps keep expenses under control by managing labor, material, and other costs to ensure that the project finishes on budget.

Countersink: A bit or drill used to set the head of a screw at or below the surface of the material.

CPR (cardiopulmonary resuscitation): A lifesaving technique that is used when someone's heartbeat or breathing has stopped.

Craft professionals: Trained and skilled individuals who design and build things. Craft professionals conform to the technical and ethical standards of their trade.

Craft/trade: A general term referring to one of the many construction specialties.

Crew leader: A person appointed leader of a team or crew, usually an experienced craft professional who has demonstrated leadership qualities. Also known as a *front-line leader*, *supervisor*, *lead person*, or *foreman*.

Critical path: In manufacturing, construction, and other types of creative processes, the required sequence of tasks directly controls the project's ultimate completion date.

Cross-bracing: Braces (metal or wood) placed diagonally from the bottom of one rail to the top of another rail that add support to a structure.

Cube: A three-dimensional object whose sides are all the same length.

Cylinder carts: Two-wheeled carts that are used to transport cylinders, or bottles, of compressed gases.

Decimal: The part of a number represented by digits to the right of a point, called a *decimal point*. For example, in the number 1.25, .25 is the decimal portion of the number. In this case, it represents the fraction $^{25}/_{100}$.

Decision-making: The action or process of making a choice. Decision-making involves starting an action, stopping one, or choosing a different plan or action.

Degree: A unit of measurement for angles. For example, a right angle measures 90°.

Denominator: The part of a fraction below the dividing line. For example, the 2 in $^{1}/_{2}$ is the denominator. It is equivalent to the divisor in a long division problem.

Detail drawings: Enlarged views of part of a drawing used to show an area more clearly.

Diagonal: Line drawn from one corner of a rectangle or square to the farthest opposite corner.

Diameter: The length of a straight line that crosses from one side of a circle, through the center point, to a point on the opposite side. The diameter is the longest straight line that can be drawn inside a circle.

Difference: The result of subtracting one number from another. For example, in the problem 8 − 3 = 5, 5 is the difference between the two numbers.

Digit: Any of the numerical symbols 0 to 9.

Dimension line: A line on a drawing with a measurement indicating length.

Direct current (DC): An electric power supply where the current flows in one direction only. DC power is supplied by batteries and transformer-rectifiers that change AC power to DC.

Distractions: Anything that disrupts the communication flow between sender and receiver.

Dividend: The number being divided in a division problem. For example, in the problem 16 ÷ 8 = 2, 16 is the dividend.

Divisor: The number that is divided into another number in a division problem. For example, in the problem 16 ÷ 8 = 2, 8 is the divisor.

Dowel: A pin or rod, usually round, that fits into a corresponding hole to fasten or align two workpieces.

Drawing revision control: A system for recording, documenting, and identifying drawing revisions by using a numeric or alphabetic sequence.

Drum carts: Wheeled carts that are used to transport heavier drums/barrels.

Drum dollies: Wheeled circular platforms or caddies with a handle that are used to transport lighter drums/barrels.

Drywall: A large, flat board made of layers of fiberboard and gypsum used primarily for wall construction and finishing. Drywall, also known as *Sheetrock* and *wallboard*, is typically manufactured in 4' × 8' panels and readily accepts paint.

Duty: A work task an employee is expected to perform.

Earn as you learn: An approach to education that pays trainees wages for work completed as part of an on-the-job learning (OJL) program.

Electrical plans: Engineered drawings that show all electrical supply and distribution.

Elevation (EL): Height above sea level, or other defined surface, usually expressed in feet or meters.

Elevation drawing: A side view of a building or object, showing height and width.

Emery cloth: A material made with a thin cloth backing, coated with an abrasive tightly bonded to the backing; typically sold in sheets or narrow rolls.

Engineers: People who apply scientific principles in design and construction.

Engineer's scale: A straightedge measuring device divided uniformly into multiples of 10 divisions per inch so that drawings can be made with decimal values. Used mainly for land measurements on site plans.

Equation: A mathematical statement indicating that the values of two mathematical expressions, such as 2×2 and 1×4, are equal. An equation is written using the equal sign (=) in this manner: $2 \times 2 = 1 \times 4$.

Equilateral triangle: A triangle that has three equal sides and three equal angles.

Equivalent fractions: Fractions having equal values with different numerators and denominators, such as fractions $\frac{1}{2}$ and $\frac{2}{4}$.

Estimating: The process of calculating the cost of a project.

Ethics: The moral principles that guide an individual's or organization's actions when dealing with others.

Excavations: Any man-made cuts, cavities, trenches, or depressions in an earth surface, formed by removing earth. They can be made for anything from basements to highways. Also see *trench*.

Feasibility study: An analysis that considers all of a project's relevant factors—including budget, technical elements, legal aspects, and scheduling—to determine the likelihood that the project can be completed successfully.

Fire protection plan: A drawing that shows the details of the building's sprinkler system.

First aid: Help given to a sick or injured worker on the worksite immediately after an injury occurs.

Flammable: Capable of easily igniting and rapidly burning; used to describe a fuel with a flash point below 100°F (38°C).

Flash burn: The damage that can be done to eyes after even brief exposure to UV light from arc welding. A flash burn requires medical attention.

Flash point: The temperature at which fuel gives off enough gases (vapors) to burn.

Flats: The flat, straight sides of a wrench opening; also, the sides on a nut or bolt head.

Floor plan: A drawing that provides an aerial view of the layout of each room.

Flush: As an adjective, to be even or at the same level with another object. For example, when a nail is fully seated, the top of the head is flush (even) with, or slightly beyond, the surface of the wood.

Font: The type style used for letters and numbers.

Force: A push or pull on a surface. In some contexts, a force is the weight of an object or a fluid.

Formula: A mathematical process used to solve a problem. For example, the formula for finding the area of a rectangle is Side A × Side B = Area, or $A \times B =$ Area.

Forstner bit: A bit designed for use in wood or similar soft material. The design allows it to drill a flat-bottom blind hole in the material.

Foundation plan: A drawing that shows the layout and elevation of the building foundation.

Fraction: A portion of a whole number represented by two numbers. The upper number of a fraction is called the *numerator*, while the bottom number is called the *denominator*.

Freight elevator: An elevator used to transport materials from floor to floor.

Glossary: An alphabetical list of terms and definitions.

Graphs: Information shown as a picture or chart. Graphs, including line graphs and bar charts, may be represented in various forms.

Green construction: Construction practices that involve the creation of structures that adhere to sustainability principles.

Grilles: HVAC components that pull air from a room and return it to the heating or cooling system. They are different from registers in that they do not have a damper to control airflow.

Grit: A granular, sand-like material used to make sandpaper and similar materials abrasive. Grit is graded according to its texture. The grit number indicates the number of abrasive granules in a standard size (inches or centimeters). The higher the grit number, the more particles in a given area, indicating a finer abrasive material.

Ground: The conducting connection between electrical equipment or an electrical circuit and the earth.

Ground fault: Incidental grounding of a conducting electrical wire.

Ground fault circuit interrupter (GFCI): A device designed to protect people from electric shock and equipment from damage by interrupting the flow of electricity if a circuit fault occurs.

Ground fault protection: Protection against short circuits; a safety device cuts power off as soon as it senses any imbalance between incoming and outgoing current.

Guarded: Enclosed, fenced, covered, or otherwise protected by barriers, rails, covers, or platforms to prevent dangerous contact.

Half hitch: A knot tied by passing the working end of a rope around an object, across the standing part of the rope, and then through the resulting loop; often used as an element in forming other knots or added to make other knots more secure.

Hand line: A line attached to a tool or object so a worker can pull it up after climbing a ladder or scaffold.

Hand trucks: Two-wheeled carts that are used to transport large, heavy loads; also known as a *dolly*.

Harassment: Unwelcome conduct based on race, color, religion, sex, national origin, age, disability, or genetic information. Harassment includes verbal or physical conduct that impacts a person's work performance or creates an intimidating, hostile, or offensive work environment.

Hazard Communication Standard (HCS): The Occupational Safety and Health Administration standard that requires contractors to educate employees about hazardous chemicals on the jobsite and how to work with them safely.

Hazards: Conditions or activities that, if left uncontrolled, have the potential for harm.

Hidden line: A dashed line showing an object obstructed from view by another object.

High-efficiency particulate arrestance (HEPA) filters: Special filters designed to capture very small particles. HEPA filters have a minimum efficiency of 99.97 %, capturing particles that are 0.3 microns (0.0000117") or larger in size.

Hitch: The rigging configuration by which a sling connects the load to the hoist hook. The three basic types of hitches are vertical, choker, and basket.

Hoists: A device that applies a mechanical force for lifting or lowering a load.

HVAC: Heating, ventilating, and air conditioning.

Hydraulic: Powered by fluid under pressure.

Impairment: The point where someone's intake of substances affects their ability to perform appropriately.

Improper fraction: A fraction whose numerator is larger than its denominator. For example, $^8/_4$ and $^6/_3$ are improper fractions.

Incident(s): Unplanned events that result in property damage, fatalities, injuries, illnesses, or near misses.

Index: An alphabetical list of topics and the page numbers where each topic appears.

Industrial forklift: A vehicle with a power-operated pronged platform that can be raised and lowered for insertion under a load to be lifted and moved.

Infrastructure: The transportation, water, electrical, and telecommunications systems that allow communities and countries to operate.

Initiative: The ability to work without constant supervision and solve problems independently.

Invert: To reverse the order or position of numbers. In fractions, inverting means reversing the positions of the numerator and denominator, so $^3/_4$ becomes $^4/_3$. When dividing by fractions, one fraction is inverted.

Isosceles triangle: A triangle that has two equal sides and two equal angles.

Italics: Letters and numbers that lean to the right rather than stand straight up.

Jargon: Specialized terms used in a specific industry.

Job description: A description of the scope and responsibilities of a worker's job so that the individual and others understand what the job entails.

Job safety analysis (JSA): A technique that focuses on job tasks as a way to identify hazards before they occur. It focuses on the relationship between the worker, the task, the tools, and the work environment. Also known as *job hazard analysis* (*JHA*) or *activity hazard analysis* (*AHA*).

Joint: The intersection of workpiece components where a physical connection is generally made. Joints are often named by the style of connection, such as butt joints and miter joints.

Joists: Lengths of wood or steel that usually support floors, ceilings, or a roof. Roof joists will be at the same angle as the roof itself, while floor and ceiling joists are usually horizontal.

Kerf: The channel created by a saw blade passing through the material, which is equal to the width of the blade teeth.

Lanyard: A cord, rope, or similar material used to limit the travel of two connected objects.

Lanyards: Short sections of rope or strap, one end of which is attached to a worker's safety harness and the other to a strong anchor point above the work area.

Lath: A thin, flat strip of wood used to form latticework or a foundation for wall plaster.

Leader: In drafting, the line on which an arrowhead is placed and used to identify a component.

Leadership: The ability to set an example for others to follow by exercising authority and responsibility.

Legend: A description of the symbols and abbreviations used in a set of drawings.

Level 4 schedule: A schedule that offers additional detail to provide working schedules to the crews doing the work, usually within three to six weeks of completing specific work.

Level: Used as an adjective in this context, meaning flat or even with another surface or in relation to Earth's horizon.

Lifting clamps: A device used to move loads such as steel plates or concrete panels without the use of slings.

Load: The total amount of what is being lifted, including all slings, hitches, and hardware.

Load control: The safe and efficient practice of load manipulation and using proper communication and handling techniques.

Load stress: The strain or tension applied on the rigging by the weight of the suspended load.

Load-bearing: Carrying a significant amount of weight and/or providing necessary structural support. A load-bearing wall typically carries some portion of the roof's weight and cannot be removed without risking structural failure or collapse.

Lockout/tagout (LOTO): A formal procedure for taking equipment out of service and ensuring that it cannot be operated until an authorized person has removed the lock and/or warning tag.

Lookahead schedule: For planning and scheduling work that is happening in the short term, usually within one to three weeks; used for anticipating material, labor, tool, and other resource requirements, as well as identifying potential schedule conflicts or other problems. Also called *short-term schedule*.

Management system: The organization of a company's management, including reporting procedures, supervisory responsibility, and administration.

Masonry bit: A drill bit with a carbide tip designed to penetrate materials such as stone, brick, or concrete.

Mass: The quantity of matter present.

Master link: The main connection fitting for chain slings.

Material carts: Four-wheeled devices used to transport materials around a jobsite.

Maximum intended load: The total weight of all people, equipment, tools, materials, and loads that a ladder can hold at one time.

Mechanical plans: Engineered drawings that show the mechanical systems, such as motors and piping.

Memos: Informal written correspondence. Another term for *memorandum* (plural: *memoranda*).

Metric scale: A straightedge measuring device divided into centimeters, with each centimeter divided into 10 mm. Usually used for architectural drawings and sometimes referred to as a *metric architect's scale*.

Midrail: Mid-level, horizontal board required on all open sides of scaffolds and platforms that are more than 14" (35 cm) from the face of the structure and more than 10' (3 m) above the ground. It is placed halfway between the toeboard and the top rail.

Mission statement: A statement describing how a company does business.

Miter joints: Joints made by fastening parts together with the ends cut at a similar angle, usually 90°.

Mixed number: A combination of a whole number with a fraction or decimal. Examples of mixed numbers are $3^7/_{16}$, 5.75, and $1^1/_4$.

Morale: An individual's emotional outlook toward work and satisfaction gained while performing the assigned jobs.

Motivate: To make someone feel determined or enthusiastic or to make someone behave in a particular way.

Mushroomed: Refers to the head of a tool that is struck or used for striking when the head becomes wider and distorted, like the head of a mushroom.

Negative numbers: Numbers less than zero. For example, –1, –2, and –3 are negative numbers.

Nonverbal communication: All communication without words, including appearance, personal environment, use of time, and body language.

Not to scale (NTS): Describes drawings that show relative positions and sizes only, without scale.

Numerator: The part of a fraction above the dividing line. For example, the 1 in $\frac{1}{2}$ is the numerator. It is equivalent to the dividend in a long division problem.

Obtuse angle: Any angle greater than 90° and less than 180°.

Occupational Safety and Health Administration (OSHA): An agency of the US Department of Labor with the mission to establish and enforce safe working conditions for the US workforce.

One-rope lay: The lengthwise distance it takes for one strand of a wire rope to make one complete turn around the core.

Opposite angles: Two angles that are formed by two straight lines crossing. They are always equal.

Organization: A group of people working together.

Organizational chart: A diagram that shows how the various management and operational responsibilities relate to each other within an organization. Named positions are ranked top to bottom, from those functional units with the most and broadest authority to those with the least. Lines connect positions and indicate chains of authority and other relationships.

Overstriking: When a hammer misses the head of a nail, with the head going past it and allowing the hammer handle to impact the nail head instead. Overstriking is a common cause of hammer damage.

Pallet jacks: Devices used to lift and move heavy or stacked pallets; also known as a *pallet truck*.

Paraphrasing: Expressing something heard or read using different words.

Peening: The process of bending, shaping, or cutting metal by striking it with a tool.

Peers: Team members in positions or levels in an organization equal or similar to yours.

Perimeter: The distance around the outside of a closed shape, such as a rectangle, circle, square, or any irregular shape.

Permit-required confined space: A confined space that has been evaluated and found to have actual or potential hazards, such as a toxic atmosphere or other serious safety or health hazard. Workers need written authorization to enter a permit-required confined space. Also see confined space.

Permits: Legal documents that allow a task to be undertaken.

Personal growth: Improving one's talents and potential, both in and out of the workplace.

Personal protective equipment (PPE): Equipment or clothing designed to prevent or reduce injuries.

Pi: A mathematical constant value approximately equal to 3.14 (or $\frac{22}{7}$) used to determine the area and circumference of circles. It is sometimes symbolized by the Greek letter π.

Pipe mules: Two-wheeled devices used to transport medium-length pieces of pipe, tubing, or scaffolding; sometimes referred to as a *tunnel buggy*.

Pipe transports: Wheeled devices similar to a pipe mule but used to move larger pieces of pipe.

Piping and instrumentation drawings (P&IDs): Schematic diagrams of a complete piping system.

Place value: The value that a digit represents in a whole number, determined by its place within the whole number or by its position relative to the decimal point. In the number 124, the number 2 represents 20, since it is in the tens position.

Plane geometry: The mathematical study of two-dimensional (flat) shapes.

Plane: A surface in which a straight line joining two points lies wholly within that surface.

Planked: Having pieces of material 2" (5 cm) thick or greater and 6" (15 cm) wide or greater used as flooring, decking, or scaffold decks.

Planning: The process of laying out all details of a construction project before construction begins.

Plumb: Perfectly vertical, meaning a surface is at a right angle (perpendicular) to the horizon or other surface used for comparison.

Plumbing isometric drawing: A type of 3D drawing that depicts a plumbing system.

Plumbing plans: Engineered drawings that show the layout for a plumbing system.

Pneumatic: Powered by air pressure such as a pneumatic tool.

Policy: A general statement establishing guidelines for a specific activity.

Positive numbers: Numbers greater than zero. For example, 1, 2, and 3 are positive numbers. Except for zero, any number without a negative (–) sign in front of it is treated as a positive number.

Powered wheelbarrow: A vehicle similar to a manual wheelbarrow but powered by an electric or gas motor; also known as a *power buggy*.

Problem-solving: A process used to identify possible solutions to a situation.

Procedures: Formal instructions to carry out and meet policies.

Product: The result of a multiplication problem. For example, the product of 6 × 6 is 36.

Production: The amount of construction materials put in place. It is the quantity of materials installed on a job.

Productivity rate: A measure of how much work gets done over a period of time, also called *output per labor hour*.

Professional growth: Gaining new skills and work experience that can help one to reach a goal in one's career.

Professionalism: Integrity and work-appropriate manners.

Profit: The amount of money that the contractor will make after paying all the direct and indirect costs.

Project delivery methods: Different ways to organize the development, planning, and construction services in order to execute a project.

Project manager: Oversees the planning and delivery of construction projects on time and within budget.

Project owner: The initiator of the project who usually finances the building endeavor.

Proximity work: Work done near a hazard but not actually in contact with it.

Punch lists: Written lists that identify deficiencies requiring correction at completion.

Qualified person: A person who, by possession of a recognized degree, certificate, or professional standing, or by extensive knowledge, training, and experience, has demonstrated the ability to solve or prevent problems relating to a certain subject, work, or project.

Quotient: The result of a division problem. For example, when dividing 6 by 2, the quotient is 3.

Radius: The distance from a circle's center point to any point on the curved line, or half the width (diameter) of the circle.

Rafter: One of a series of beams that extend from the perimeter of a building to the high point of a roof, providing structural support for the roof above.

Reciprocating: Moving backward and forward on a straight line.

Recognition: Acknowledgment of an employee's accomplishments through verbal and nonverbal communications.

Rectangle: A four-sided shape with four 90° angles. Opposite sides of a rectangle are always parallel and the same length. Adjacent sides are perpendicular and may or may not be equal in length.

Redline drawings: Marked-up versions of construction drawings documenting any changes made during construction or installation.

References: People who can confirm to a potential employer that you have the skills, experience, and work habits listed in your resume.

Registers: HVAC components that direct air from the heating or cooling system into a room. A damper is used to control the airflow of a register.

Rejection criteria: Standards, rules, or tests on which a decision can be based to remove an object or device from service because it is no longer safe.

Remainder: The amount left over in a division problem. For example, in the problem 34 ÷ 8, 8 goes into 34 four times (8 × 4 = 32) with 2 left over, so the remainder is 2.

Respirators: Devices that provide clean, filtered air for breathing, no matter what is in the surrounding air.

Responsibility: The act of accepting a task and completing it to the best of one's ability.

Revolutions per minute (rpm): The rotational speed of a motor or shaft based on the number of times it rotates each minute.

Rewards: Tangible items given to an employee as recognition for good work.

Rigging hook: An item of rigging hardware used to attach a sling to a load.

Right angle: An angle that measures 90°. The two lines that form a right angle are perpendicular to each other. This is the angle most used in the crafts.

Right triangles: A type of triangle that includes one 90° angle.

Roller skids: Devices that include a surface table and two, three, or four roller skids. Materials that are to be moved are placed on the table surface and then pushed on the skids.

Roof plan: A drawing of the view of the roof from above the building.

Root cause: A core issue or fundamental reason for the occurrence of a problem.

Root cause analysis (RCA): A method of problem-solving used for identifying the root causes of faults or problem.

Rough terrain forklift: A vehicle similar to an industrial forklift but designed to be used on rough surfaces. Rough terrain forklifts are characterized by large pneumatic tires, usually with deep treads, that allow the vehicle to grab onto the roughest of roads or ground cover without sliding or slipping.

Safety culture: The culture created when the whole company sees the value of a safe work environment.

Safety Data Sheet (SDS): A document that must accompany any hazardous substance. The SDS identifies the substance and gives the exposure limits, the physical and chemical characteristics, the kind of hazard it presents, precautions for safe handling and use, and specific control measures of each substance.

Scaffolds: Elevated platforms for workers and materials.

Scale: The ratio between the size of a drawing of an object and the size of the actual object.

Scalene triangle: A triangle with sides of unequal lengths.

Scheduling: Involves defining which activities happen in what order and assigning dates and times to those tasks.

Schematic: A one-line drawing showing the flow path for electrical circuitry or the relationship of all parts of a system.

Scope: The list of the commitments and activities that all contractors, subcontractors, and suppliers are expected to do during a project. Also called the *scope of work*.

Section drawing: A cross-sectional view of a specific location, showing the inside of an object or building.

Sectors: Categories of construction distinguished by specific types of work, materials, equipment, and skills. Sectors of construction include residential, commercial, agricultural, industrial, and heavy civil/infrastructure.

Self-feed bit: A drill bit that uses a lead screw to pull the larger rear portion of the bit into the wood, making it simple to drill larger holes.

Self-presentation: The way a person dresses, speaks, acts, and interacts with others.

Sexual harassment: A type of discrimination that results from unwelcome sexual advances, requests for sexual favors, or other verbal or physical behavior with sexual overtones.

Shackle: A piece of rigging hardware used in an appropriate lifting apparatus to connect the rope to eye fittings, hooks, or other connectors.

Shank: The smooth part of a drill bit that fits into the chuck.

Sheaves: Grooved pulley wheels used for changing the direction of a rope's pull; often found on a crane.

Shielding: A structure used to protect workers in trenches.

Shoring: A support system designed to prevent a trench or excavation cave-in.

Signaler: A person who is responsible for directing a vehicle when the driver's vision is blocked in any way.

Silicosis: A serious lung disease resulting from the inhalation of crystalline silica particles.

Site superintendent: Manages day-to-day activities on a construction site and supervises work performed by the subcontractors.

Site-specific safety plan (SSSP): A safety plan that includes a focused set of information that accounts for a jobsite's unique hazards.

Situational leadership: A leadership style that involves adjusting to the specific situation or task and the leadership needs of the group.

Six-foot rule: A rule stating that platforms or work surfaces with unprotected sides or edges that are 6' (roughly 2 m) higher than the ground or level below it require fall protection.

Sling legs: The parts of the sling that reach from the attachment device around the object being lifted.

Sling reach: A measure taken from the master link of the sling, where it bears weight, to either the end fitting of the sling or the lowest point on the basket.

Slings: Wire rope, alloy steel chain, metal mesh fabric, synthetic rope, synthetic webbing, or jacketed synthetic continuous loop fibers made into forms, with or without end fittings, used to handle loads.

Solid geometry: The mathematical study of three-dimensional shapes.

Specifications: Precise written presentation of the details of a plan.

Splice: To join together.

Spoil: Material such as earth removed while digging a trench or excavation.

Spotter: A person who walks in front of another worker who is carrying or transporting a long load to ensure there is a clear, unobstructed path.

Square knot: A knot made of two reverse half-knots and typically used to join the ends of two ropes of similar diameters; also called a *reef knot*.

Square: A special type of rectangle with four equal sides and four 90° angles; used as an adjective, to have the shape of a square where two components are at precise 90° angles (perpendicular) to each other.

Standing end: The end of a rope that is not being knotted.

Standing part: The portion of a rope that is between the standing end and the working end.

Step drill bit: A drill bit that includes steps that enable the user to cut holes in the metal to a desired diameter.

Straight angle: A 180° angle or flat line.

Strands: A group of wires wound, or laid, around a center wire, or core. Strands are laid around a supporting core to form a rope.

Stripping: Damaging or distorting the head or threads of a fastener.

Structural plans: A set of engineered drawings used to support the architectural design.

Stud: A vertical support inside a wall to which the wall finish material attaches. The base of a stud rests on a horizontal baseplate, and a horizontal cap plate rests on top of a series of studs.

Subcontractor/specialty contractor: Typically employs craft professionals who have specialized skills needed to complete a particular part of the construction project (e.g. HVAC, electrical, plumbing).

Substance abuse: The inappropriate overuse of drugs and chemicals, whether they are legal or illegal.

Sum: The result of an addition problem. For example, in the problem 7 + 8 = 15, 15 is the sum.

Superintendent: An on-site supervisor who is responsible for one or more crew leaders or frontline supervisors.

Sustainability: The practice of minimizing negative impacts to Earth's climate and natural environment.

Symbols: Drawings that represent a material or component on a plan.

Table of contents: A list of book chapters or sections, usually located at the front of the book.

Tables: A way to present important text and numbers so they can be read and understood quickly.

Tactful: Being aware of the effects of your statements and actions on others.

Tag line: Rope that runs from the load to the ground. Riggers hold on to tag lines to keep a load from swinging or spinning during the lift.

Tardiness: Arriving late for work.

Tattle-tail: Cord attached to the strands of an endless loop sling that protrudes from the jacket. A tattle-tail determines whether an endless sling has been stretched or overloaded.

Tempered: Treated with heat to enhance or restore steel hardness.

Threaded shank: A connecting end of a fastener, such as a bolt, with a series of spiral grooves cut into it. The grooves are designed to mate with grooves cut into another object to join them together.

Three-way communication: A form of communication with an initial message from a sender, which is then repeated by the receiver to confirm accuracy and then validated by the sender. Also called *repeat-back communication*.

Title block: A part of a drawing sheet that includes some general information about the project.

Toeboards: Vertical barriers at floor level attached along exposed edges of a platform, runway, or ramp to prevent materials and people from falling.

Toolbox talks: Short, informal safety meetings held at the start of each day or shift. Sometimes referred to as *tailgate meetings* or *safety briefings*.

Top rails: Top-level, horizontal boards required on all open sides of scaffolds and platforms that are more than 14" (36 cm) from the face of the structure and more than 10' (3 m) above the ground.

Torque: The rotating or twisting force applied to an object such as a fastener or the shaft of a motor.

Traits: Distinguishing qualities of a person.

Trench: A narrow excavation made below the surface of the ground that is generally deeper than it is wide, with a maximum width of 15' (4.6 m). Also see excavation.

Triangle: A closed shape that has three sides and three angles.

Trigger lock: A small lever, switch, or part that can be used to activate a locking catch or spring to hold a power tool trigger in the operating mode without finger pressure.

Unit: A standardized quantity used for a particular kind of measurement.

Unstranding: Describes wire rope strands that have become untwisted. This weakens the rope and makes it easier to break.

Upskilling: A type of training that provides someone with more advanced skills through additional education.

Verbal communication: Messages that are sent using voice or other audible sounds. Also called *spoken communication* or *oral communication*.

Vertex: The point at which two or more lines or curves come together.

Virtual reality (VR): Provides a three-dimensional (3D) model of a construction site and allows users to directly immerse themselves into the virtual space. VR enables multiple people to envision the final product, feel what it is like to be in the space, and identify any issues before work begins.

Visual communication: The process of using visual elements, such as photos, graphics, colors, and fonts, to convey messages.

Voltage rating: A measurement of the power a tool can deliver.

Volume: The amount of space contained within a three-dimensional object.

Warning yarns: A component of the sling that shows the rigger whether the sling has suffered too much damage to be used.

Welding: The process of heating two or more metal workpieces to their liquid state and fusing them to create a very strong joint.

Welding curtains: Protective screens set up around a welding operation designed to safeguard workers not directly involved in that operation.

Wheelbarrows: One- or two-wheeled vehicles with handles at the rear that are used to carry small loads.

Whip check: A safety attachment used to prevent whiplashing in hoses when they are inadvertently uncoupled.

Whole numbers: Complete number units without fractions or decimals.

Wind sock: A cloth cone open at both ends mounted in a high place to show which direction the wind is blowing.

Wire rope: A rope made from steel wires that are formed into strands and then laid around a supporting core to form a complete rope; sometimes called *cable*.

Work breakdown structure (WBS): A method of organizing the entire scope of work in a logical manner based on how the work will be executed.

Work ethic: Work habits that are the foundation of a person's ability to do their job.

Work zone: The area in which the rear end or front forks of a forklift may come in contact with objects or people.

Workers' compensation: Insurance that companies must cover medical expenses, lost wages, and other costs for employees who are injured or become ill while completing work for their job.

Working end: The end of a rope that is being used to tie a knot.

Working load limit (WLL): The maximum load weight a sling or piece of hardware or equipment can hold or lift.

Written communication: The process of sending messages or information using words, whether digitally or physically.

Zero tolerance: The policy of applying laws or penalties to even minor code infringements to reinforce its overall importance, typically related to drug and alcohol abuse, when applied to the workplace.